D1483337

The Evolution
of the Genome

The Evolution of the Genome

Edited by

T. Ryan Gregory

ELSEVIER
ACADEMIC
PRESS

AMSTERDAM • BOSTON • HEIDELBERG • LONDON
NEW YORK • OXFORD • PARIS • SAN DIEGO
SAN FRANCISCO • SINGAPORE • SYDNEY • TOKYO

Elsevier Academic Press
525 B Street, Suite 1900, San Diego, California 92101-4495, USA
84 Theobald's Road, London WC1X 8RR, UK

Library of Congress Cataloging-in-Publication Data
Application submitted

British Library Cataloguing in Publication Data
A catalogue record for this book is available from the British Library

ISBN: 0-12-301463-8

For all information on all Academic Press publications
visit our website at www.academicpress.com

Transferred to Digital Printing in 2009

LIST OF CONTRIBUTORS

MICHAEL D. BENNETT, PH.D.
Jodrell Laboratory, Royal Botanic Gardens, Kew, UK.
[Chapter 2]

JUAN PEDRO M. CAMACHO, PH.D.
Departamento de Genética, Universidad de Granada, Granada, Spain.
[Chapter 4]

ROB DeSALLE, PH.D.
Division of Invertebrate Zoology, American Museum of Natural History, New York, USA.
[Chapter 10]

ALAN FILIPSKI, PH.D.
Center for Evolutionary Functional Genomics, Arizona Biodesign Institute, Arizona State University, Tempe, Arizona, USA.
[Chapter 9]

T. RYAN GREGORY, PH.D.
Department of Entomology, The Natural History Museum, London, UK.
[Chapters 1, 8, 10, 11]

MARGARET G. KIDWELL, PH.D.
Department of Ecology and Evolutionary Biology, The University of Arizona, Tuscon, Arizona, USA.
[Chapter 3]

SUDHIR KUMAR, PH.D.
School of Life Sciences, Arizona State University, Tempe, Arizona, USA.
[Chapter 9]

ILIA J. LEITCH, Ph.D.
Jodrell Laboratory, Royal Botanic Gardens, Kew, UK.
[Chapter 2]

BARBARA K. MABLE, Ph.D.
Division of Environmental and Evolutionary Biology, University of Glasgow, Glasgow, UK.
[Chapter 8]

AXEL MEYER, Ph.D.
Department of Biology, University of Konstanz, Konstanz, Germany.
[Chapter 6]

JEROEN RAES, Ph.D.
Department of Plant Systems Biology, Flanders Interuniversity Institute for Biotechnology (VIB), Ghent University, Gent, Belgium.
[Chapter 5]

DOUGLAS E. SOLTIS, Ph.D.
Department of Botany, University of Florida, Gainesville, Florida, USA.
[Chapter 7]

PAMELA S. SOLTIS, Ph.D.
Florida Museum of Natural History, University of Florida, Gainesville, Florida, USA.
[Chapter 7]

JENNIFER A. TATE, Ph.D.
Florida Museum of Natural History, University of Florida, Gainesville, Florida, USA.
[Chapter 7]

JOHN S. TAYLOR, Ph.D.
Department of Biology, University of Victoria, Victoria, British Columbia, Canada.
[Chapter 5]

YVES VAN DE PEER, Ph.D.
Department of Plant Systems Biology, Ghent University, Gent, Belgium.
[Chapter 6]

PREFACE: THE EVOLUTION OF
THE EVOLUTION OF THE GENOME

It has been said that the preface is the most important part of a book, because even the reviewers will read it. That strikes me as a lot of pressure to put on such a small fraction of a book, made no less disconcerting by the fact that the preface can actually be rather challenging to write. How does one encapsulate, in only a few pages, a book that covers (and integrates) as many topics, involved the input of as many people, and has taken as much work to prepare as this one? The only strategy that comes to mind is to follow the old adage that one should stick to writing what one knows best. As such, it seems reasonable (and actually quite appropriate) to discuss this book in biological terms.

A central theme of this volume is that genomes represent a distinct and legitimate level of biological organization, with their own inherent properties and unique evolutionary histories. Books about genomes, too, are organized entities that evolve in sometimes unpredictable ways, and are subject to the same general processes as biological individuals. What is editing, really, but an example of "intellectual selection," whereby the author proposes and the editor disposes (or at least ruthlessly modifies)? As with life at large, the evolution of this book also involved healthy doses of morphological constraint (especially in regard to length), the necessary coordination of parts, hybridization of data from the study of very different taxonomic groups, horizontal transfer and exaptation of concepts among disciplines, and more than a little quirky historical contingency.

The evolution of any entity, living or literary, is a combination of general principles (acting at several levels of organization) and unique historical particulars. The convoluted particulars of this book's genesis are as follows. In 2002, I ventured to Washington, DC to attend a conference. While there, I was kindly hosted by my friend and colleague Grace Wyngaard, whose husband, Roy McDiarmid, is a curator of herpetology at the Smithsonian Institution. Also at the conference was

Chuck Crumly, then a senior editor at Academic Press, who had also trained as a herpetologist and was an old acquaintance of Roy's. Through this connection, Chuck approached me with the notion of preparing a book on genome evolution for Academic Press.

Lacking any comparative data, I can only speculate as to how typical the rest of the book's history is: Rookie editorial hopeful manages to assemble an all-star list of authors and gets a book contract; original publisher is phagocytosed by another company; there ensues a chaotic series of comings and goings on the publishing team involved with the book (including the departure of Chuck Crumly); some authors who planned to contribute are unable to; the chapter list is juggled and reshuffled, and other equally stellar contributors graciously agree to join the revised project; and in the end, a book emerges that bears at least a modest resemblance to the one discussed over a lunchtime meeting more than two years earlier.

Now that the book has come into being, it is worthwhile to consider some aspects of its ecology and morphology—that is to say, the niche it is meant to occupy and the ways in which it is adapted to do so.

Speaking from experience, I can attest that the primary literature of genome biology is, shall we say, less than readily accessible. It can be a significant challenge to follow developments in one's own specialized subdiscipline, and a near impossibility to obtain a broad view of genome evolution at large. The chasms separating researchers working on different genomic questions or with different taxa are as wide as ever, and those dividing the various branches of the life sciences seem to be getting wider. This can be a major barrier to understanding the context and importance of genome evolution, a problem that is particularly acute for newcomers.

The chapters in this book are meant to provide comprehensive summaries of the underlying concepts, histories, current statuses, and future prospects of the major fields of genomic inquiry. Too often, these conceptual and historical underpinnings are overlooked as new discoveries come to dominate discussions of these topics. Overall, this book is aimed at the up-and-coming "genome generation"— advanced graduate students, postdocs, and young faculty members—although it should also serve as a valuable resource for existing experts in these and related fields. Put explicitly, it is written for the people who are, and will be, shaping the future of genome biology. It therefore provides a map to the otherwise bewildering maze of literature on genome evolution, seeks to build much-needed bridges between divergent disciplines, and, perhaps most importantly, stresses the big picture. In some cases, the subject of a given chapter has never been reviewed before, and, as such, is treated as comprehensively as possible. In others, the subject could be (and in some cases *is*) reviewed every few months, though seldom in the broad context emphasized here.

All of this means that this book also is *not* several things. Notably, it is not a comprehensive textbook for undergraduates, meaning that a strong basic background in biology and genetics will be required to get the most out of these chapters. However, neither is it a typical collection of highly technical reviews accessible only to other specialists. By keeping discipline-specific jargon to a minimum (and making all attempts to define it when unavoidable), the chapters provided here are meant to be accessible to anyone with *only* a strong basic biological background. In a phrase, this is the book that I wish had been available when I was first starting out as a student of genome evolution.

Although the various chapters have been written by several different authors and each can stand on its own, the book is designed as a cohesive whole rather than a collection of independent reviews. The volume begins with a discussion of the evolution of genome size (Chapters 1 and 2), which has remained a puzzle in genome biology since before the elucidation of the double helix 50 years ago (and, not coincidentally, is my own primary subject of study). This is followed by a discussion of genomic parasites, first in the form of transposable elements (Chapter 3), and then of B chromosomes (Chapter 4). Next comes a series of chapters dealing with duplications—of individual genes (Chapter 5), of large fragments and entire genomes in the ancient past (Chapter 6), and of whole chromosome sets in more recent times (Chapters 7 and 8). The following two chapters treat the topics of comparative genome sequencing in eukaryotes and prokaryotes (Chapters 9 and 10, respectively), and include information on genome structure and evolution in the two groups. Finally, the new insights derived from genomic analyses in all these areas are placed in the context of evolutionary theory (Chapter 11). Here the policy of integration is taken to the extreme by drawing links between opposite ends of the evolutionary spectrum, from the study of subgenomic elements to the macroevolutionary theory developed by paleontologists.

As Stephen Jay Gould noted in his last tome on evolutionary theory, "[a] book's timescale of production must be labeled as geological compared with a pace of discovery that can only be measured in ecological time." This disconnect in evolutionary rates is especially acute in genomic science, where major discoveries are being made at an ever-accelerating tempo. In fact, the field of genome biology is moving so quickly that even looking over our shoulders as we typed would not guarantee a fully up-to-date review for the reader. Nor can all of the information presently available be treated in complete detail within the constraints of a single book. The net result is that these chapters are, of necessity, both snapshots and sketches of dynamic and wide-ranging subjects. However, even though much new information continues to accrue as we move deeper into the 21st century, the basic historical and conceptual background, the inroads to the existing literature, and the cross-disciplinary bridges provided in these chapters should remain useful for some time. In fact, one of the goals of the book is to help shape the future development of genome biology by highlighting exciting current and potential research and, hopefully, by stimulating young researchers to pursue their own novel avenues of investigation.

A book like this comes to exist only through the encouragement, support, and hard work of many people. So many, in fact, that it is probably prudent to simply apologize in advance to anyone I may fail to list here by name. Believe me, you are appreciated nonetheless.

To start, let me extend sincere thanks to Beth Callaway, David Cella, Bonnie Falk, Sarah Hager, and especially Chuck Crumly and Kelly Sonnack, along with everyone else at Academic Press/Elsevier and Graphic World Publishing Services who helped to guide this project from a rough idea concocted over mediocre Chinese food to a detailed, comprehensive, accessible collection of chapters that makes quite a satisfying thud when dropped on a desk.

Thanks are also owed (and happily given) to Jim Bogart, Thomas D'Souza, Milton Gallardo, Sara Islas, Spencer Johnston, Antonio Lazcano, Steve Le Comber, Serge Morand, Bob Ricklefs, and others for providing access to unpublished information and other assistance. Juan Pedro Camacho thanks J. Cabrero, M.D. López-León, F. Perfectti, and M. Bakkali for fruitful discussions, and Neil Jones and Robert Trivers for helpful comments. John Taylor and Jeroen Raes thank Martine De Cock for help with the retrieval of older papers. Yves Van de Peer and Axel Meyer gratefully acknowledge the numerous graduate students responsible for carrying out much of the analysis described in their chapter. A heartfelt thank you also goes to the various colleagues who read earlier drafts of particular chapters, including Leo Beukeboom, Robert Carroll, Niles Eldredge, and Robert Trivers, with extra-special thanks extended to Jürgen Brosius, Margaret Kidwell, Mike Monaghan, Howard Ochman, Sarah Otto, and Paul Planet, who provided very helpful comments and constructive criticisms.

I wish to personally thank my close colleagues in the genome size world for stimulating conversations, encouragement, and friendship, especially (but not exclusively) Mike Bennett, Dave Hardie, Spencer Johnston, Charley Knight, Ilia Leitch, Ellen Rasch, and Grace Wyngaard. I believe it is safe to say that each of the authors is likewise grateful to his or her various friends, colleagues, and loved ones for support on this project and all others.

A significant thank you must go to the institutions and funding agencies that supported each of the contributors to the book, in particular (but not only) the National Science Foundation (grant number DEB 9815754) (Margaret Kidwell), the Spanish Ministerio de Ciencia y Tecnología and the Plan Andaluz de Investigación (Juan Pedro Camacho), the Natural Sciences and Engineering Research Council of Canada (John Taylor), the Deutsche Forschungsgemeinschaft (Axel Meyer), and the National Institutes of Health, the National Science Foundation, and the Center for Evolutionary Functional Genomics at the Arizona State University (Alan Filipski and Sudhir Kumar). On this note, I wish to extend a very special thank you to the Natural Sciences and Engineering Research Council of Canada (NSERC) for funding my research from the beginning of my graduate training through the end of my postdoc (and hopefully well beyond!), and for giving me the freedom to pursue unorthodox projects such as this one.

I must thank the University of Guelph, the American Museum of Natural History, and the Natural History Museum, London, for hosting me while work was being carried out on this book. I would also like to thank my friends and labmates at each of these fine institutions for putting up with me. A special thank you goes to my former advisor in each lab, Paul Hebert (Guelph), Rob DeSalle (AMNH), and Alfried Vogler (NHM), for always being supportive and for graciously tolerating my rather eccentric work habits (which typically involved prolonged periods of at-home sequestration during the major writing and editing phases).

I owe an enormous debt to the various authors who contributed to this volume. Their hard work, dedication, and expertise have, I believe, made this book something of which we can all be very proud. Thank you!

To my various sets of parents, Bob and Marilyn Gregory, Michele Davis and Frank Brewster, and Steve and Carol Adamowicz, and to my oldest and closest friend (who also happens to be my younger brother), Sean Gregory: thank you for your love, generosity, encouragement, and, well—thank you for everything. And finally, the biggest thank you of all to my wonderful partner and friend, counselor and caretaker, confidant and fellow explorer, Sarah Adamowicz: thank you so much—again, for everything.

I dedicate this book with love to all the members of my family, genetic and otherwise.

T. Ryan Gregory
London, England
Summer 2004

ABOUT THE EDITOR

Dr. T. Ryan Gregory completed his B.Sc. (Honours) in biology at McMaster University in Hamilton, Ontario, Canada in 1997, and his Ph.D. in evolutionary biology and zoology at the University of Guelph in Ontario, Canada in 2002. He has been the recipient of several prestigious scholarships and fellowships and was named the winner of the 2003 Howard Alper Postdoctoral Prize by the Natural Sciences and Engineering Research Council of Canada, one of the nation's  premier research awards. During the preparation of this book, he has been a postdoctoral fellow at the American Museum of Natural History in New York and the Natural History Museum in London, England.

CONTENTS

2 Genome Size Evolution in Plants

Michael D. Bennett and Ilia J. Leitch

PART **II**

The Evolution of Genomic Parasites

3 Transposable Elements

Margaret G. Kidwell

PART **III**

Duplications, Duplications...

5 Small-Scale Gene Duplications

John S. Taylor and Jeroen Raes

6 Large-Scale Gene and Ancient Genome Duplications

Yves Van de Peer and Axel Meyer

PART **IV**

...And More Duplications

7 Polyploidy in Plants

Jennifer A. Tate, Douglas E. Soltis, and Pamela S. Soltis

8 Polyploidy in Animals

T. Ryan Gregory and Barbara K. Mable

PART **V**

Sequence and Structure

9 Comparative Genomics in Eukaryotes

Alan Filipski and Sudhir Kumar

PART **VI**

The Genome in Evolution

xxvi

Contents

PART I

The C-value Enigma

CHAPTER 1

Genome Size Evolution in Animals

T. Ryan Gregory

According to the *Oxford English Dictionary*, the term "genom(e)" was coined by Hans Winkler in 1920 as a portmanteau of *gene* and *chromosome*. This canonical etymology has been challenged by Lederberg and McCray (2001), who suggested that Winkler probably merged *gene* with the generalized suffix *-ome* (referring to "the entire collectivity of units"), and not *-some* ("body") from *chromosome*. In either case, Winkler's intent was to "propose the expression *Genom* for the haploid chromosome set, which, together with the pertinent protoplasm, specifies the material foundations of the species" (translation as in Lederberg and McCray, 2001). Based on this initial formulation, "genome" can accurately be taken to mean either the complete gene complement (in the sense of "genotype"), or the total DNA amount per haploid chromosome set—but not both simultaneously, as will be seen.

As the diversity of topics covered in this volume shows, there are many ways of studying the evolution of the genome. This relates only partially to the multiple meanings of the term, and instead is based primarily on the examination of genome-level phenomena from a variety of different structural and temporal perspectives. The chapters in this book deal with scales ranging from the level of individual sequences through to entire chromosome sets, and from individual replication events to patterns in deep time. The first two chapters begin at the more holistic end of this spectrum, and describe the bulk properties of genomes as reflected by variation in their sizes. On the surface, this may appear to be a relatively simple issue, but in fact it lies at the heart of one of the longest-running puzzles in genome biology.

WHY SHOULD ANYONE CARE ABOUT GENOME SIZE?

Mass is perhaps the most fundamental property of any physical entity, and is usually much more straightforward to characterize than features like composition, structure, or organization. In this sense, even a basic understanding of the nature of a given genome requires information regarding the amount of DNA contained within it. Genome size, which in eukaryotes has traditionally been given as the mass (in picograms, pg[1]) of DNA per haploid nucleus, also sets the context in which analyses of component sequences and organizational characteristics can be interpreted. On these grounds alone, genome size should be considered a crucial aspect of any truly comprehensive program of comparative genomic analysis.

There are also some important practical reasons to be concerned with variation in genome size among eukaryotes. Most obviously, size is a major consideration when choosing targets for complete genome sequencing projects, because DNA amount can be expected to be directly proportional to both the financial and labor costs involved. This applies to all stages of the process, from mapping, to the construction of genomic libraries, to reading the sequence of nucleotides, to reassembling the individual sequenced fragments (see Chapter 9). Indeed, Evans and Gundersen-Rindal (2003) list genome size first among relevant criteria for selecting the next wave of insect species for genome sequencing, along with such parameters as existing gene sequence information, overall biodiversity, and impact on human health and/or agriculture. Targeted genome size measurements of disease vectors and pests are becoming more common as well (e.g., Gregory, 2003a; Panzera *et al.*, 2004).

It is also becoming apparent that smaller-scale genetic studies are influenced by the amount of nuclear DNA. For example, Garner (2002) recently demonstrated a strong negative effect of genome size on the polymerase chain reaction (PCR) amplification potential of microsatellites used in DNA fingerprinting and population genetics studies. Even though the number of microsatellites generally increases with genome size (see later section), this difficulty may arise because more DNA decreases the ratio of target to nontarget DNA and/or dilutes the pool

---

[1] 1 pg = 10^{-12} g. Molecular studies, especially those dealing with smaller genomes, tend to give DNA contents in terms of the number of nucleotide base pairs (bp). In approximate terms, 1 pg ≈ 1 billion bp (i.e., 1000 Mb or 1 Gb). More specifically (Dolezel *et al.*, 2003),

Number of base pairs = Mass in pg × 0.978 × 10^9

and

Mass in pg = Number of base pairs × 1.022 × 10^{-9}.

of available primers by nonspecific binding (Garner, 2002). Likewise, the applicability of fingerprinting by amplified fragment length polymorphisms (AFLPs), which is also commonly used in population genetics studies and is based on PCR amplification, can be strongly influenced by genome size (Fay *et al.*, 2005).

However, from the perspective of this chapter, the most important reasons to study genome size are not so pragmatic in nature, but rather have to do with the major biological and evolutionary significance of this topic (see also Chapters 2 and 11).

GENOME SIZE IN ANIMALS: A HISTORICAL PERSPECTIVE

THE DISCOVERY OF DNA

In the mid- to late 1800s (and to an extent, well into the 20th century), proteins were considered the most significant components of cells. Their very name reflects this fact, being derived from the Greek *proteios,* meaning "of the first importance." In 1869, while developing techniques to isolate nuclei from white blood cells (which he obtained from pus-filled bandages, a plentiful source of cellular material in the days before antiseptic surgical techniques), 25-year-old Swiss biologist Friedrich Miescher stumbled across a phosphorous-rich substance which, he stated, "cannot belong among any of the protein substances known hitherto" (quoted in Portugal and Cohen, 1977). To this substance he gave the name *nuclein,* and published his results in 1871 after confirmation of the remarkable finding by his adviser, Felix Hoppe-Seyler (for reviews, see Mirsky, 1968; Portugal and Cohen, 1977).

Miescher continued his work on nuclein for many years, in part refuting claims that it was merely a mixture of inorganic phosphate salts and proteins. Yet he never departed from the common proteinocentric wisdom, and instead suggested that the nuclein molecule served as little more than a storehouse of cellular phosphorus. In 1879, Walther Flemming coined the term *chromatin* (Gr. "color") in reference to the colored components of cell nuclei observed after treatment with various chemical stains, and in 1888 Wilhelm Waldeyer used the term *chromosome* (Gr. "color body") to describe the threads of stainable material found within the nucleus. For some time, debate existed over whether or not chromatin and nuclein were one and the same. The argument was largely settled when Richard Altman obtained protein-free samples of nuclein in 1889. As part of this work, Altman proposed a more appropriate (and familiar) term for the substance, *nucleic acid.* Over time, the components of the nucleic acid molecules were deduced, and by the 1930s *nuclein* had become *desoxyribose nucleic acid,* and later, *deoxyribonucleic acid.*

The important developments that took place over the ensuing decades are well documented (e.g., Portugal and Cohen, 1977; Judson, 1996), including early hypotheses of DNA's structure (such as Phoebus Levene's failed tetranucleotide hypothesis, or the incorrect helical model of Linus Pauling), Erwin Chargaff's discovery of the constant ratio of the two purines with their respective pyrimidines, Rosalind Franklin's x-ray crystallography of the DNA molecule, and other key developments leading up to Watson and Crick's monumental synthesis in 1953 and the subsequent deciphering of the genetic code.

"A REMARKABLE CONSTANCY" AND THE ORIGIN OF THE "C-VALUE"

During the early stages of investigation into the nature of DNA around the late 1800s and early 1900s, one of the most commonly exploited sources of DNA was cattle tissue. In some cases, as much as 35 kilograms (kg) of cow pancreas would be dissolved in 200 liters (L) of dilute sulfuric acid to yield a workable amount of DNA. Thymus tissue was similarly employed, a legacy that survives in the name *thymine*. By the late 1940s, interest had grown regarding the quantity of DNA located within the cells of these various tissues, in terms of both its abundance relative to RNA and its absolute mass. In 1948, André Boivin and his students Roger and Colette Vendrely began a systematic comparison of DNA contents in different cattle tissues, including pancreas, thymus, liver, and kidney, as well as liver from pig and guinea pig (Boivin *et al.*, 1948; Vendrely and Vendrely, 1948). This was accomplished by computing the DNA concentration of bulk tissue preparations divided by nucleus numbers estimated from suspension counts. These studies revealed nuclei in each of these tissues to contain an approximately equal amount of DNA, that being roughly twice the value contained in sperm from the same species. Thus, Vendrely and Vendrely (1948) reported "a remarkable constancy in the nuclear DNA content of all the cells in all the individuals within a given animal species" (translation by the present author). Shortly thereafter, these findings were extended to several other mammals and to a few fishes and birds (Vendrely and Vendrely, 1949, 1950; Davidson *et al.*, 1950; Mandel *et al.*, 1951).

Strictly speaking, this "DNA constancy hypothesis" was not entirely new. Though Karl Rabl had proposed the general notion as early as 1885, "[Oscar] Hertwig in 1918 was probably the first to state in unequivocal terms the theory that the chromosomal content of all nuclei in an organism is identical" (Swift, 1950a). However, the hypothesis took on a new relevance in the late 1940s and early 1950s when it began to be argued that "from the theoretical point of view the discovery of the constancy of the amount of DNA per nucleus in all tissues of the same animal and the fact that the sperm contains half the DNA content of somatic cells is confirmation of the theory that DNA is an important component of the gene" (Vendrely, 1955). In fact, in the days prior to the famous "blender experiment" of Hershey and Chase (1952), which showed that T1 bacteriophages injected DNA but not proteins into their host cells, and, most important, the elucidation of the double helix model by Watson and Crick (1953), DNA (but not protein) constancy per cell provided an important line of evidence in favor of DNA as the hereditary material.

Although by the late 1940s it had "become a textbook generalization that all cells of an organism have the same chromosomal constitution," some authors remained convinced that "the amount of DNA carried in a given chromosome may vary in different tissues" (Schrader and Leuchtenberger, 1949). As a result, careful studies were conducted using both animal and plant tissues in order to confirm that the amount of DNA per chromosome set was indeed constant within individual organisms and among conspecifics (for reviews of the debate by key participants, see Swift, 1953; Allfrey *et al.*, 1955; Vendrely, 1955; Vendrely and Vendrely, 1956, 1957; Mirsky and Osawa, 1961).

In one of the most influential defenses of DNA constancy, Hewson Swift (1950a) carried out a meticulous analysis of DNA contents in individual nuclei from various tissues of mouse, frog (*Rana pipiens*), and grasshopper (*Dissoteira carolina*), and showed these to conform to the expectations of the hypothesis. In a follow-up study, Swift (1950b) analyzed tissues from maize (*Zea mays*) and three species of *Tradescantia* (relatives of spiderwort) and again found only even multiples of the haploid DNA content in different somatic tissues. In these publications, Swift (1950a,b) referred to different "classes" of DNA; in the first instance with "Class I" being the most common (diploid) value, but in the second case with the "Class 1C value" representing the haploid DNA content. The term "C-value," a derivative of Swift's (1950b) notation, is still used to refer to the haploid nuclear DNA content. When referring to diploid species, "C-value" and "genome size" are interchangeable, but these terms must be used more judiciously when dealing with polypoids (see also Chapter 2).

THE C-VALUE PARADOX

Accepting that genome size constancy within species reflected the role of DNA as the hereditary substance, the obvious implication was that "this constant is probably proportionate to the number of genes" (Vendrely and Vendrely, 1950; translation by the present author). However, this hypothesis began to unravel almost immediately after large comparisons were made across species (Fig. 1.1). In the first broad-scale survey of animal genome sizes, Mirsky and Ris (1951) noted the totally unexpected finding that:

> Comparing the largest and one of the smallest examples among vertebrates, one finds that a cell of amphiuma [an aquatic salamander] contains 70 times as much DNA as is found in a cell of the domestic fowl, a far more highly developed animal. It seems most unlikely that amphiuma contains 70 times as many different genes as does the fowl or that a gene of amphiuma contains 70 times as much DNA as does one in the fowl.

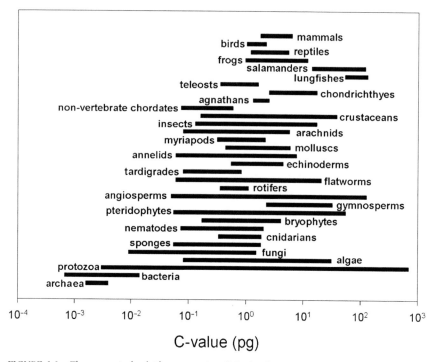

FIGURE 1.1   The ranges in haploid genome sizes ("C-values" in picograms, pg) so far reported in different groups of organisms, showing that genome size is clearly unrelated to intuitive notions of organismal complexity. Note that the very high reported values for protists may be unreliable (see text). From Gregory (2004a), reproduced by permission (© The Paleontological Society).

The repeated confirmation of a total decoupling of DNA content and organismal complexity (taken as a proxy for gene number) only heightened the confusion surrounding this issue over the two decades that followed.[2] In 1971, C.A. Thomas described this vexing problem as the "C-value paradox," which was (and still is) typically described from one of three different perspectives: (1) Some simple organisms have more DNA than complex ones (e.g., Vendrely, 1955), (2) any given genome seems to contain much more DNA than would be needed to account for the predicted gene number (e.g., MacLean, 1973), or (3) some morphologically similar groups exhibit highly divergent DNA contents (e.g., Gall, 1981). These are three ways of saying the same thing: that DNA content does not correlate with the expected number of genes, which is "paradoxical" in the sense

---

[2]Note that more recent claims of a positive correlation between genome size and organismal complexity (e.g., Hughes, 1999; Shabalina and Spiridonov, 2004) are based on highly selective comparisons including bacteria, yeast, and a small number of animals such as nematodes and a few mammals. On the broader scale of eukaryotes at large, there clearly is no such relationship (Fig. 1.1).

that C-values were presumed to be constant precisely because DNA is the stuff of genes. As Commoner (1964) put it 40 years ago, "these observations suggest that the DNA/cell does not correspond to the total gene content of the organism, but they fail to explain why, this being the case, the cellular DNA content of a species is, in fact, a fixed inherited characteristic."

THE MODERN VIEW: FROM PARADOX TO PUZZLE

A paradox, it should be noted, is defined as something that is "self-contradictory," "opposed to common sense," and "seemingly absurd." Under the expectation that genomes consist of the genes, all the genes, and nothing but the genes, and that this is the reason for DNA constancy within chromosome sets, the vast range in C-values across eukaryotes (Fig. 1.1) would indeed qualify as paradoxical. However, within a few years of the coining of the term, the "C-value paradox" was fully resolved by the discovery that the great majority of eukaryotic DNA is noncoding.

Yet, while there is no longer anything "absurd" or "self-contradictory" about the fact that salamanders have much more DNA than chickens, such major disparities in genome size do remain puzzling. Unfortunately, the term "C-value paradox" is still used to refer to this very different problem, having long outlasted its utility by virtue of historical entrenchment. There are several reasons why this is problematic. Most simply, the continued use of the term "paradox" implies a rather embarrassing lack of understanding of even the most basic features of eukaryotic genomes. More important, this phrasing actively befuddles discussions of genome size evolution by turning the issue into one of semantics. Consider, for example, the debate that ensued when Novacek and Norell (1989) stated in a brief review of the "C-value paradox" that "the reason for the remarkable variation in genome size remains one of the mysteries of comparative biology" and that "it is difficult to muster convincing theories for the source of this variation in DNA content." In response, Dover (1990) pointed out that there was no longer any "paradox" to be addressed because several molecular mechanisms can explain the sources of noncoding DNA. Grime and Hendry (1990) also chastised Novacek and Norell (1989) for these statements, but for the opposite reason: that they failed to discuss the many cell- and organism-level phenotypic correlates of genome size. The problem is that these two critiques deal with very different issues, but that both were considered to represent "the" solution to the C-value paradox.

As Wilhelm Johannsen, who coined the terms "gene," "genotype," and "phenotype," noted in 1911, "It is a well-established fact that language is not only our servant...but that it may also be our master, overpowering us by means of the notions attached to the current words." The persistent framing of the larger problem of genome size variation as a "paradox" results in a tendency to treat it as one-dimensional and therefore to seek a single solution to it. This worked well for the legitimate initial "paradox," which was solved by the discovery of noncoding DNA, but it is a counterproductive approach to take with regard to the broader issue.

Genome size evolution is much more accurately outlined as a complex puzzle—a "C-value enigma"—consisting of several independent component questions (Gregory, 2001a, 2002a, 2004a). These include: To what extent, and following what distributional patterns, does genome size vary among taxa? Whence this noncoding DNA? Which types of noncoding sequences predominate in genomes? By what mechanisms are these sequences gained and lost over time? Does any or all of the noncoding DNA have a function? What impacts, if any, does this nongenic majority have on the cellular and organismal phenotype? Why do some groups have so much noncoding DNA, while others have remarkably streamlined chromosomes?

These questions will all be addressed to one extent or another in this and several other chapters in this volume, depending largely on the current availability of pertinent information. The important thing to remember throughout is that even a conclusive answer to one of these components does not solve the entire puzzle. Rather, the resolution of the C-value enigma as a whole will require the application and integration of several very different approaches.

THE STATE OF KNOWLEDGE OF ANIMAL GENOME SIZE

THE ANIMAL GENOME SIZE DATABASE

Most of the questions inherent in the C-value enigma require large amounts of genome size data derived from a broad taxonomic sampling. As discussed in Chapter 2, botanists have been compiling C-value data for plants (especially angiosperms) since the 1970s. These data have been available online since 1997, first as part of the *Angiosperm DNA C-values Database,* and more recently as the expanded *Plant DNA C-values Database* (Bennett and Leitch, 2003). Although Sparrow *et al.* (1972) began with a compilation of C-values that included animals (as well as plants, fungi, and bacteria), only a few taxon-specific summaries have been published in the time since (e.g., Tiersch and Wachtel, 1991). Until very recently, there had been no concerted effort to catalog animal genome sizes, a fact that greatly limited the ability of zoologists to perform large-scale comparisons of the type being conducted in the botanical community.

On January 10, 2001, the *Animal Genome Size Database* was launched online (Gregory, 2001b). At present, the *Animal Genome Size Database* contains almost as many data as the *Plant DNA C-values Database*—nearly 4000 species, including roughly 2500 vertebrates and 1300 invertebrates. Table 1.1 provides more

TABLE 1.1   Summary of the basic patterns of genome size variation in the better-studied groups
of animals. This includes ranges and means, as well as the number of species studied to date,
with the percentage of described species covered given in brackets. Most of the current dataset
comes from vertebrates, where the relative coverage is reasonably good for most groups. Despite
their much higher species diversity, invertebrates remain poorly studied from the perspective of
genome size. See the text for more detailed discussions of patterns within these and a few less
well-known groups, and consult www.genomesize.com/summary.htm for updates to these
summary statistics. Based on the data contained in the *Animal Genome Size Database* (Gregory
2001b) as of early 2004, supplemented with unpublished estimates by T.R. Gregory.

Taxon	C-value range (pg)	Mean C-value (pg)	Species studied
VERTEBRATES			
Jawless fishes	1.3–4.6	2.3	17 (16%)
Cartilaginous fishes	2.7–17.1	5.7	126 (12%)
Lungfishes	50–133	90.4	4 (66%)
Chondrostean fishes	1.5–6.5	3.3	17 (68%)
Teleost fishes	0.4–4.4	1.2	1050 (4%)
Amphibians	0.95–120	16.2	413 (7%)
Reptiles	1.1–5.4	2.3	303 (4%)
Birds	1.0–2.2	1.4	172 (2%)
Mammals	1.7–8.4	3.5	358 (7%)
INVERTEBRATES			
Insects	0.1–16.9	1.6	402 (0.05%)
Crustaceans	0.16–38.0	3.2	230 (0.6%)
Arachnids	0.08–5.7	2.4	118 (0.2%)
Molluscs	0.4–5.9	1.8	174 (0.2%)
Echinoderms	0.5–4.4	1.4	45 (0.9%)
Annelids	0.06–7.6	1.5	121 (1%)
Flatworms	0.06–20.5	2.2	61 (0.3%)
Nematodes	0.03–2.0	0.25	32 (0.3%)

detailed breakdowns of the available data, including the relative coverage of the
major animal groups.

   Although they are maintained separately, the animal and plant databases share
many common goals, such as the free distribution of all available genome size
data, the facilitation of large-scale analyses across many species, the development
of standardized protocols to increase the reliability of future measurements, and
an expansion to include data from previously overlooked taxonomic groups
(Gregory, 2005) (Chapter 2). Both databases are used regularly by researchers

from a wide array of disciplines. As of early 2004, the *Animal Genome Size Database* is receiving 100–150 hits per day on its various pages and has been accessed by visitors from more than 100 countries. A "prokaryote" genome size database that is not limited to species involved in sequencing programs has recently been launched (www.genomesize.com/prokaryotes; see Chapter 10 for a discussion of prokaryote genome size). Unfortunately, equivalent databases have not yet been compiled for fungi or "protists," although this would clearly be a worthy project for experts in those groups to undertake.

PATTERNS OF VARIATION

The following discussion can be considered the first summary of the knowledge acquired to date regarding animal genome sizes, which, unlike similar information from plants, has not been reviewed in detail before. Obviously (and fortunately), the dataset is continuing to expand, such that the ranges and averages provided here will probably be subject to revision in the future. In keeping with this, an updated summary of ranges and other trends in the major animal taxa is provided as part of the *Animal Genome Size Database*.

Animal genome sizes vary extensively, with existing estimates ranging more than 3300-fold, from 0.04 pg in the placozoan *Trichoplax adhaerens* to ~133 pg in the marbled African lungfish *Protopterus aethiopicus*. As with plants (see Chapter 2), the members of most animal groups have relatively small genomes, with only a few taxa exhibiting large and highly variable DNA contents.

The following sections provide more detailed synopses of the patterns within major animal groups, first for vertebrates and then for invertebrates. These summaries are based on the *Animal Genome Size Database* (where the several hundred original source references are listed), supplemented with a few hundred unpublished estimates. Although it is acknowledged that some of these groups (e.g., "reptiles" and "invertebrates") are paraphyletic and therefore not valid from a phylogenetic perspective, they remain convenient on functional grounds and are used here in this capacity.

VERTEBRATES (AND NONVERTEBRATE CHORDATES)

Even though their 50,000 or so species comprise only a relatively small fraction of total metazoan diversity, vertebrates account for ⅔ of the animal genome size dataset. As a result, most vertebrate groups appear reasonably well studied in terms of relative coverage, which ranges from about 2% of described species for birds to 12% in cartilaginous fishes (Table 1.1). Rare groups like jawless fishes, sturgeons, and lungfishes are better covered in relative terms; but in none of these cases is the absolute number of estimates very high (Table 1.1). However, many orders and families within these groups remain understudied. Based on the available data, chordate genome sizes appear to vary a remarkable 1900-fold, from 0.07 pg in the larvacean *Oikopleura dioica* to the aforementioned lungfish. Three other nonvertebrate chordates have been analyzed: the urochordates *Ascidia atra* and *Ciona intestinalis* at roughly 0.16 pg and 0.20 pg, respectively, and the well-known cephalochordate *Branchiostoma lanceolatum* (also called "Amphioxus") at ~0.6 pg. Within the vertebrates alone, the range is about 330-fold, although the variation within any given vertebrate group is considerably lower than this.

Jawless Fishes

Lampreys and hagfishes represent the most primitive group of living vertebrates, although this fact is not reflected in their fairly typical genome sizes. Overall, genome sizes range less than 4-fold in the superclass Agnatha, from ~1.3 pg in the southern brook lamprey, *Ichthyomyzon gagei*, to 4.6 pg in the hagfish *Myxine garmani*. Interestingly, there is apparently no overlap in genome size between lampreys (class Cephalaspidomorphi), with a maximum of ~2.2 pg in the sea lamprey *Petromyzon marinus*, and the hagfishes (class Myxini), with a minimum of ~2.3 pg in *Eptatretus cirrhatus*. Some hagfishes are known to undergo a process of "chromatin diminution" whereby large fragments of the genome (entire chromosomes in this case) that are present in the germline are eliminated from the somatic line (e.g., Kubota *et al.*, 1994, 2001; Nakai *et al.*, 1995). This curious phenomenon is otherwise found only in a few invertebrates (see later section) and certain ciliated protozoa.

Cartilaginous Fishes

All but one of the 126 members of the class Chondrychthyes represented in the present dataset are from the subclass Elasmobranchii (sharks, rays, and skates). The other, the spotted ratfish (*Hydrolagus colliei*), is a member of the subclass Holocephali (~1.5 pg). Elasmobranch genome sizes range from about 2.7 pg in the yellow guitarfish, *Rhinobatos schlegelii*, to more than 17 pg in the angular roughshark, *Oxynotus centrina*. The mean genome size for sharks is 6.7 pg, and for rays and skates it is 5.0 pg. Contrary to previous reports (e.g., Hinegardner, 1976a), this difference between the two groups is statistically significant.

Chondrosteans

Sturgeons, paddlefishes, and bichirs are bony, ray-finned fishes (i.e., members of the class Osteichthyes, subclass Actinopterygii), but are members of the infraclass Chondrostei and are therefore not part of the same group as the more numerous and familiar teleosts. The current dataset includes representatives from two families of sturgeons and paddlefishes (order Acipenseriformes) and one family of bichirs (order Polypteriformes). Sturgeon and paddlefish C-values range from about 1.5 to 6.5 pg (mean of 3.1 pg), although much of this variation is probably due to polyploidy, which is widespread in this group (see Chapter 8). For bichirs, the mean is 5.3 pg, ranging from about 4.7 to 6.0 pg.

Teleost Fishes

Most of the bony fishes (class Osteichthyes) in the current dataset are members of the subclass Actinopterygii and the infraclass Neopterygii, which includes bowfin (order Amiiformes), gar (order Lepisosteiformes), and the members of the highly diverse superorder Teleostei, which makes up about half of all vertebrate species. All three of the nonteleost neopterygian species so far studied have genome sizes of about 1.2 to 1.4 pg.

To date, roughly 1050 teleost genome sizes have been reported. The most striking observation to be made with this dataset is that, notwithstanding their status as the most speciose group of vertebrates, teleosts exhibit very little variation in genome size among species. The mean for teleosts is only 1.2 pg, and the range is a mere 11-fold, from about 0.4 pg in tetraodontid pufferfishes to 4.4 pg in the masked corydoras, *Corydoras metae*. The overwhelming majority of species are near the low end of the overall distribution, and it seems that only groups for which polyploidy has played a significant role commonly exceed a C-value of 2.5 pg.

Lobe-Finned Fishes

The lobe-finned fishes (class Sarcopterygii) are of particular interest to evolutionary biologists for several reasons. The coelacanth, *Latimeria chalumnae* (infraclass Coelacanthimorpha), is perhaps the most famous "living fossil" known, and has apparently remained largely unchanged for at least 200 million years. Lungfishes (infraclass Dipnoi) are of interest because of their phylogenetic proximity to (but *not* placement as) the ancestor of the tetrapod lineage.

The coelacanth C-value is somewhat uncertain, with three very different estimates of 2.8, 3.6, and 6.6 pg reported. Thomson (1991) provided a discussion of these discrepant estimates, and made the case that his own (6.6 pg) is the most accurate based on the methodology and source material used.

Three genera of lungfishes are recognized, each inhabiting a different continent. These include the South American lungfish (*Lepidosiren paradoxa*), the Australian or Queensland lungfish (*Neoceratodus forsteri*), and a few species of African lungfishes (*Protopterus* spp.), all of which possess extraordinarily large genomes. The smallest among them is that of *N. forsteri,* at ~50 pg. The genome size of *L. paradoxa* is ~120 pg, while those of *Protopterus* vary from ~63 pg in *P. annectens* to ~133 pg in *P. aethiopicus*. Vervoort (1980) measured DNA contents in additional *Protopterus* species, but his methods were unreliable and resulted in severe underestimates. For example, Vervoort (1980) reports a C-value of only 40 pg for *Protopterus aethiopicus congicus* and ~82 pg in the polyploid *P. dolloi*. If the C-value is actually the same for both subspecies of *P. aethiopicus*, and assuming that at least the ratios between species reported by Vervoort (1980) are accurate, then this would imply that *P. dolloi* actually possesses a staggering 270 pg of DNA in its nuclei.

Amphibians

Genome sizes vary more extensively among amphibians than in any other group of vertebrates. The overall range in this class is ~130-fold, from 0.95 pg in the ornate burrowing frog *Limnodynastes ornatus* to ~120 pg in the waterdogs (or mudpuppies) *Necturus lewisi* and *N. punctatus*. However, this variation is not distributed evenly among the three living orders of amphibians. In frogs (order Anura or Salientia), genomes range about 12-fold, up to a maximum of 11.5 pg in the European fire-bellied toad, *Bombina bombina*.[3] The lowest genome size among salamanders (order Urodela or Caudata), on the other hand, is ~14 pg in the pygmy salamander *Desmognathus wrighti*, and the mean is nearly 10-fold higher than in frogs (36.7 pg versus 4.6 pg). Caecilians, legless amphibians of the order Gymnophiona (or Apoda), are almost completely unknown in terms of genome size. Only three species have been studied from this perspective, and these range from 3.7 to 14 pg with a mean of 7.4 pg, such that they appear to have somewhat intermediate genome sizes as compared to the other two orders. Obviously, further study on this group would be very useful, although these animals are admittedly rather difficult to come by.

Reptiles

Reported reptilian C-values vary from ~1.1 pg in the skink *Chalcides mionecton* to ~5.4 pg in the Greek tortoise *Testudo graeca*, for an overall range of about 5-fold. The mean reptilian genome size is 2.25 pg, which is almost exactly intermediate between mammals and birds (see next sections). Turtles display the most variable genome sizes of the reptiles, reaching a low of 1.8 pg in the Caspian turtle, *Clemmys caspica*. Crocodilian C-values average the highest at 3.1 pg, ranging from 2.5 pg in the American alligator, *Alligator mississippiensis*, to 3.9 pg in the broad-snouted caiman, *Caiman latirostris*. Snakes and lizards both have mean genome sizes of about 2.1 pg, in the first case ranging from *C. mionecton* at ~1.1 pg to 3.9 pg in girdled lizards of the genus *Cordylus*, and in the second case from ~1.3 pg

---

[3]A value of 19 pg was reported for the round frog, *Arenophryne rotunda*, by Morescalchi (1990), who cites "Mahony, Phil. D. Thesis, 1986" as the source. (Actually, he lists it only as "*Arenophryne*," but this is a monotypic genus according to Frost, 2004.) However, since no further details of Mahony's thesis were given, it has not been possible to confirm this very high frog value.

in the southern American rattlesnake, *Crotalus durissus terrificus,* to 3.7 pg in *Laticauda* sp., a sea snake.

Birds

With roughly 10,000 species, birds are second only to the teleost fishes in terms of diversity, yet they remain one of the least well studied groups of vertebrates with regard to total number of genome size estimates (Table 1.1). Genome sizes are currently available for only about 2% of described bird species, the lowest relative coverage among the major vertebrate taxa. However, despite this undersampling, it is already apparent that avian genome sizes are highly constrained, averaging a mere 1.4 pg. In the entire class, genome sizes vary only 2-fold, from a low of ~1.0 pg in the cut-throat weaver, *Amadina fasciata,* to almost 2.2 pg in the ostrich, *Struthio camelus.*

Mammals

Mammalian C-values range approximately 5-fold, from a low of ~1.7 pg in the bent-winged bat, *Miniopterus schreibersi,* to ~8.4 pg in the red viscacha rat, *Tympanoctomys barrerae.* However, *T. barrerae* has been identified as a polyploid (see Chapter 8), such that the range in genome size (*sensu stricto*) is only ~3.5-fold, with a maximum of ~6.3 pg in some echimyid rodents in the genus *Proechymis.* According to Ohno (1969), "in this respect, evolution of mammals is not very interesting," but in fact the lack of genome size variation despite high levels of species diversification in all three amniote classes is an interesting phenomenon in need of explanation.

The average for mammals is about 3.5 pg, almost exactly equivalent to the genome size of humans. In terms of higher-order divisions, marsupials appear to have mean genome sizes (~4.0 pg) larger than most placentals (~3.4 pg). At the low end of the spectrum are the bats (order Chiroptera), in which the average genome size is only around 2.5 pg. In all 50 or so bats assayed thus far (only one of which, the flying fox *Pteropus giganteus,* is a megachiropteran; all the others have been microchiropterans), genome sizes are well below the mean for the class. Monotremes, primitive egg-laying mammals, have relatively small genomes, with the duck-billed platypus (*Ornithorhynchus anatinus*) at 3.06 pg and the short-nosed echidna (*Tachyglossus aculeatus*) at 2.89 pg. In mammals, birds, and reptiles, genome sizes tend to be relatively normally distributed, unlike the strongly right-skewed distributions observed in amphibians and teleosts.

INVERTEBRATES

Compared with vertebrates, invertebrate genome size data are sorely lacking at present. Indeed, no more than 1% of the known species in any major group is currently covered (Table 1.1), and even this is probably a substantial underestimate given that most invertebrate diversity has yet to be described. In fact, even very common invertebrates, including some of direct economic and medical relevance, have only recently begun to be studied in any serious way. As a result, the following summary of the general patterns in the best (i.e., least poorly) studied invertebrate taxa should be taken as preliminary. However, as interesting trends are already visible, this should also provide an impetus to greatly expand the dataset in the near future.

Insects

No general discussion of metazoan evolution can be complete without reference to the Insecta, for the simple fact that most animals are insects. Put in terms of the C-value enigma, most animal genomes reside within insect cells, so no discussion that ignores this group can be considered comprehensive. Nevertheless, only about 400 insect genome size estimates have been published to date, representing a mere 0.05% coverage of the more than 800,000 species described thus far. Overall, insect genome sizes currently appear to range nearly 170-fold, from ~0.1 pg in the strepsipteran *Caenocholax fenyesi texansis* (J.S. Johnston *et al.*, unpublished data) and several species of parasitic wasps in the family Braconidae (T.R. Gregory, unpublished) to 16.9 pg in the mountain grasshopper, *Podisma pedestris*. Unfortunately, little else can be said about general patterns in insects, given that 90% of published estimates have come from just five of the 32 (or so) insect orders: Coleoptera (beetles), Diptera (true flies), Hemiptera (true bugs; almost all of them aphids), Lepidoptera (moths and butterflies), and Orthoptera (grasshoppers, crickets, and katydids). These groups will be discussed in turn.

Available beetle C-values range from 0.14 pg in the false antlike flower beetle *Pedilus* sp. (family Pedilidae) to 3.7 pg in the leaf beetle *Chrysolina carnifex* (family Chrysomelidae). This range is somewhat misleading, however, because only eight of the more than 220 species studied so far display genomes larger than 1.5 pg, with a mean for the entire beetle dataset of only 0.61 pg. The reported genome size of *C. carnifex* (Petitpierre *et al.*, 1993) itself appears rather anomalous, given that none of the other 76 species of chrysomelids measured in any study exceeds 2 pg. The mean for the entire family is only 0.83 pg, and all six of *C. carnifex*'s congeners studied so far have genome sizes between 0.5 and 1.2 pg. The next largest reported beetle C-value is 2.7 pg in the chafer *Rhizotrogus lepidus* (Bosch *et al.*, 1989). No other beetle species are known to exceed 2 pg, but obviously the relative coverage of this most diverse of all animal groups is particularly poor. The Chrysomelidae (leaf beetles) represent the most variable beetle family so far identified, ranging about 21-fold if *C. carnifex* (3.7 pg) is included and 11-fold if this species is omitted (0.17 to 1.98 pg; $n = 77$). The Coccinellidae (ladybird beetles) range 9-fold in genome size (0.19 to 1.71 pg; $n = 31$), with a mean value of 0.53 pg. In the Tenebrionidae (darkling beetles), genome sizes range about 6-fold (0.16 to 0.87 pg; $n = 66$), with an average of 0.37 pg. The Carabidae (ground beetles) also vary around 6-fold (0.19 to 1.23 pg; $n = 12$), with a mean of 0.51 pg. Amazingly, no other beetle families are represented by more than a dozen species, meaning that even these basic analyses cannot yet be performed.

Although flies are the second best studied insect group behind beetles, the majority of the estimates come from just two families: the Culicidae (mosquitoes, 63%) and the Drosophilidae (fruit flies, 18%). Perhaps this bias is not surprising, given the importance of drosophilids as laboratory models and of mosquitoes as conveyers of the pathogens causing malaria, dengue and yellow fevers, and filialitis. However, this means that the taxonomic coverage of the Diptera is quite limited, and that insights from this order are largely restricted to patterns found within these two families. That said, published dipteran C-value estimates currently range approximately 20-fold, from 0.12 pg in the Hessian fly *Mayetiola destructor* to 1.9 pg in the mosquito *Aedes zoosophus*. The mean genome size for all flies studied to date is 0.71 pg. Genome sizes in the Drosophilidae average about 0.25 pg, and range from 0.15 pg in *Drosophila simulans* to 0.4 pg in *D. nasutoides*. The genome sizes of mosquitoes are generally larger and more variable than those of drosophilids, extending from a low of about 0.25 pg in several species of *Anopheles* to the dipteran maximum of 1.9 pg in *Ae. zoosophus*, with an average of roughly 0.94 pg. The Culicidae is the only dipteran family currently reported to contain members with genome sizes larger than 0.75 pg (roughly ¾ of mosquitoes exceed this level, in fact).

The mean for all hemipterans studied so far is 0.91 pg, but there are important differences between the two suborders, with the Heteroptera having a mean of 1.18 pg, whereas that for the Homoptera is only 0.72 pg. However, this latter value and that of the overall hemipteran mean are heavily biased by the fact that more than half of the species in the present dataset are aphids. The C-values of aphids (families Adelgidae and Aphididae) range 5-fold, from 0.18 pg in *Eoessigia longicauda* to 0.89 pg in *Gypsoaphis oestlundi*, with a mean of 0.51 pg. The four nonaphid homopterans range in genome size from around 1.1 pg in the leafhopper *Athysanus argentarius* to ~5 pg in cicadas (*Tibicen* spp.). The range in heteropterans is roughly equal to that for homopterans, from 0.3 pg in the water strider *Gerris remigis* to 5.4 pg in the milkweed bug *Oncopeltus fasciatus*.

Until recently, very little was known regarding the genome sizes of lepidopterans, despite the long tradition of collecting butterflies and moths and the economic importance of many members of this group (Gregory and Hebert, 2003). The dataset has since been expanded to include nearly 60 species, although all but two of these are (macro)moths. Within this sample, lepidopteran C-values range from 0.29 pg in the monarch butterfly, *Danaus plexippus*, to 1.9 pg in the geometrid moth *Euchlaena irraria*.

By far, the largest insect genomes are found among the Orthoptera. Here, genome sizes average roughly 8.2 pg, and range approximately 11-fold from 1.55 pg in the cave cricket *Hadenoecus subterraneus* to 16.9 pg in *P. pedestris*. As with the Hemiptera, this is heavily biased toward one group, in this case the grasshoppers of the family Acrididae, which make up nearly ¾ of the current dataset. The Acrididae average about 9.5 pg, with a range limited to relatively high C-values (4.5 to 16.9 pg). At the opposite extreme are the Gryllidae (crickets), with a mean C-value of 1.9 pg, ranging only from 1.5 to 2.2 pg. The Gryllacrididae (camel crickets) display genome sizes from 5.2 to 9.6 pg and a mean of 7.3 pg, whereas the Tettigoniidae (katydids) vary from 2.5 to 10.1 pg, with an average of 6.2 pg. The single representative of the family Eumasticidae, the grasshopper *Warramaba virgo*, also falls within this range, with a genome size of 3.9 pg.

The remaining insect orders have been very poorly sampled, or have not been studied at all to date. Based on the existing dataset, it seems that relatively large genome sizes are found in roaches (Blattaria, ~2 to 4 pg), mantids (Mantodea, 3.15 pg), dragonflies (Odonata, 1.5 to 2.2 pg), and stick insects (Phasmida, 2 to 3 pg). Stoneflies (Plecoptera) appear quite variable, already ranging from 0.36 to 2.15 pg. Intermediate C-values have been found in earwigs (Dermaptera, 1.4 pg), webspinners (Embiidina, 1.45 pg), and termites (Isoptera, ~1 pg), whereas smaller genomes appear to predominate in ants, bees, and especially parasitic wasps (Hymenoptera, 0.1 to 0.6 pg), twisted-wing parasites (Strepsiptera, 0.11 to 0.13 pg), fleas (Siphonaptera, 0.27 pg), caddisflies (Trichoptera, 0.66 pg), and mayflies (Ephemeroptera, ~0.75 pg).

Crustaceans

Overall, crustacean C-values range ~250-fold, from 0.16 pg in the water flea *Scapholeberis kingii* to 38 pg in the deep-sea shrimp *Hymenodora* sp. (however, note that the next largest is only 22.2 pg in the Ohio shrimp, *Macrobrachium ohione*). The mean for the entire subphylum is roughly 3.15 pg, with most species displaying genomes smaller than 6 pg. The overall distribution is therefore skewed to the right, in agreement with the pattern found in various other invertebrate groups (Hinegardner, 1976b). Genome sizes in the Crustacea also tend to be relatively constant within classes, families, and genera, but highly variable among families.

Most of the extensive genome size variation in the Crustacea is confined to the class Malacostraca, in particular within the order Decapoda. Interestingly, although decapods are the most variable crustacean group so far analyzed, the majority of species within it have genome sizes between 1 and 5 pg. Only the caridean shrimps (suborder Pleocyemata, infraorder Caridea) display genomes larger than 10 pg; all of the families in this infraorder have members exceeding

7 pg, and none have genomes smaller than 3 pg. Also in the Malacostraca, the Amphipoda range from ~1 to 5.1 pg, and the Isopoda exhibit C-values from 1.7 to 8.8 pg.

The only other highly variable class of crustaceans is the Copepoda, particularly the order Calanoida. In this order, genome sizes range nearly 20-fold, from 0.63 to 12.5 pg, with a mean of roughly 5 pg. In the order Cyclopoida, by contrast, genome sizes are much smaller (mean of 0.9 pg) and vary only from 0.25 to 2 pg. Other copepod orders have not been well studied, but seem to contain at least some genomes smaller than the minimum calanoid value (Harpacticoida: 0.25 pg; Siphonostomatoida: 0.58 pg). Calanoid copepods are of particular interest because they exhibit "quantum shifts" in genome size whereby species vary discontinuously by multiples of the smallest value in the group (McLaren *et al.*, 1988, 1989; Gregory *et al.*, 2000). Many cyclopoids undergo chromatin diminution, in which 35–90% of the germline genome is eliminated during somatic line differentiation, depending on the species (Fig. 1.2) (for reviews, see Wyngaard and Rasch, 2000; Wyngaard and Gregory, 2001).

At the low end of the crustacean distribution are the Branchiopoda, especially the order Cladocera (water fleas). In this order, the mean genome size is only 0.3 pg,

FIGURE 1.2   The process of chromatin diminution in the cyclopoid copepod *Mesocyclops edax*. The presumptive somatic genomes are localized at the poles, with a large fraction of excised heterochromatic DNA (~75% of the germline genome) clustered in the center. Modified from a confocal laser micrograph of Feulgen-stained DNA at 365 nanometers (nm) excitation by Stanley Erlandsen printed in Wyngaard and Gregory (2001), reproduced by permission (© Wiley-Liss Inc.).

ranging from 0.16 to 0.63 pg in the roughly 50 species that have been studied. Members of the order Anostraca (brine and fairy shrimp) have larger genomes, from 0.87 to 2.9 pg. Genome sizes also tend to be small in the class Ostracoda, ranging from 0.46 to 3.1 pg, with a mean of about 1.1 pg. Estimates for barnacles (class Cirripedia) vary from about 0.7 to 2.6 pg.

Arachnids

Many of the highly diverse groups of arachnids, including mites, ticks, scorpions, and harvestmen, have yet to be studied with any significant effort. Until recently, almost no information was available for any arachnid group, but a recent survey has now added about 115 estimates for spiders to the dataset (Gregory and Shorthouse, 2003). Spider genome sizes vary from 0.74 pg in the long-jawed orb-weaver *Tetragnatha elongata* to 5.7 pg in jumping spiders of the genus *Habronattus*, with a mean of 2.4 pg. However, these data are all from the infraorder Araneomorphae; unfortunately, no "tarantulas" (infraorder Mygalomorphae) have been studied thus far. The only other arachnids studied to date include the lone star tick, *Amblyomma americanum*, at ~1.1 pg, and the two-spotted spider mite, *Tetranychus urticae*, which has the smallest known arthropod genome at 0.08 pg.

Molluscs

The Mollusca represent the most speciose aquatic animal phylum, and the second largest group of animals overall. They also make up one of the best-studied invertebrate groups with regard to absolute number of C-value estimates, behind the Insecta and Crustacea. Thus far, about 170 molluscan genome sizes have been assessed, including about 80 species each of bivalves and gastropods, five cephalopods, and seven chitons. Overall, molluscan C-values range nearly 14-fold—with the entire range found among gastropods—from about 0.4 pg in the owl limpet *Lottia gigantea* to 5.9 pg in the Antarctic whelk *Neobuccinum eatoni*. The average C-value for all molluscs is 1.8 pg.

   Following from the first (but very limited) survey of invertebrate genome sizes, Mirsky and Ris (1951) suggested that "among the molluscs the more primitive members have the lowest DNA content." This was based on the notion that "the squid is a far more highly developed animal than are the limpet, snail, and chiton; and the squid has far more DNA per sperm than is found in the lower molluscs." Based on the sizeable dataset now available for molluscs, it is apparent that this pattern no longer holds (of course, neither does the concept of "higher" and "lower" species). Recall that the largest genome is found in a whelk, not a squid, and though it cannot be denied that cephalopods are the most complex of the molluscs, it must be noted that the arrow squid (*Loligo plei*) and the bobtail squid (*Euprymna scolopes*) both have C-values smaller than the members of some supposedly less "highly developed" groups. In accordance with this, Hinegardner (1974a) outlined a tendency for higher DNA contents to be found in generalized groups rather than in the specialized forms. Some recent authors have supported this assertion for the case of bivalves (Rodríguez-Juíz *et al.*, 1996), whereas others have rejected it (González-Tizón *et al.*, 2000; Thiriot-Quievreux, 2002). In molluscs overall, the larger-than-average genomes of cephalopods and terrestrial gastropods would seem to deflate any such trend among major groups. Interestingly, among the pulmonate gastropods, it is apparent that terrestrial snails and slugs (order Stylommatophora) possess genomes roughly twice as large as those of their freshwater relatives (order Basommatophora) (Vinogradov, 2000).

Annelids

At present, the polychaetes (marine worms) remain the best-studied group of annelids. Genome sizes range 120-fold in the Polychaeta, from 0.06 pg in *Dinophilus gyrociliatus* to 7.2 pg in *Nephtys incisa*, with a mean of 1.4 pg. C-values appear to be relatively constant within most polychaete genera, but in some groups there is evidence of quantum shifts in DNA content that are unrelated to polyploidy (Sella *et al.*, 1993). Oligochaetes have so far received only modest attention in terms of genome size variation. In freshwater oligochaetes, C-values vary nearly 10-fold, from 0.8 to 7.6 pg. The genome sizes of terrestrial oligochaetes (earthworms) appear relatively constrained, ranging only 3-fold, from about 0.4 pg in *Lumbricus castaneus* and *L. rubellus* to 1.2 pg in *Dendrobaena octaedra*. As with polychaetes, variation within genera is limited, but when present appears discontinuous. In leeches (class Hirudinea), only seven genome sizes have been measured, and these are all rather small, ranging from about 0.3 pg in the Australian terrestrial leech *Philaemon pungens* to 0.63 pg in the turtle leech *Placobdella parasitica*.

Echinoderms

C-value estimates for echinoderms currently total 44 species, with 39 of these presented 30 years ago by Hinegardner (1974b). Overall, the echinoderm genome sizes currently available vary from 0.5 pg in the sea star *Dermasterias imbricata* to 4.4 pg in the sea cucumber *Thyonella gemmata*, about an 8-fold range. The mean for the phylum is 1.4 pg. Of the major groups studied so far, the asteroidean stelleroids (sea stars) display a mean genome size of about 0.7 pg, whereas the ophiurideans (serpent stars) have much larger genomes averaging 2.6 pg. Echinoids (sea urchins, sand dollars) have a mean genome size of about 1.0 pg, and the Holothuroidea (sea cucumbers) average about 2.0 pg. Sea cucumbers themselves cover almost the entire range observed in the phylum (0.8 to 4.4 pg). Crinoids (feather stars and sea lilies) have not yet been studied in terms of genome size.

Flatworms

To date, genome sizes have been estimated for 56 species in the phylum Platyhelminthes, nearly all of them free-living members of the class Turbellaria. Overall, these vary nearly 350-fold, from a mere 0.06 pg in *Stenostomum brevipharyngium* to 20.5 pg in *Otomesostoma auditivum*; thus, flatworms include some of the smallest and largest genomes found in invertebrates. Relatively large genomes are found in the orders Alloecoela, Rhabdocoela, and Tricladida (means 8.0, 3.6, and 2.1, respectively), whereas the orders Catenulida, Macrostomida, and Proseriata display much smaller and more tightly constrained C-values (0.5, 0.4, and 1.3, respectively). Very few other platyhelminthes have been measured for genome size. Four species of flukes (class Trematoda) have been examined, and range from about 0.9 to 1.3 pg. No tapeworm (class Cestoda) genome size data are currently available.

Miscellaneous Invertebrates

A small number of representatives from a handful of additional animal phyla have been examined with regard to genome size. In the nematodes, for example, reported genome sizes range from 0.03 pg in root-knot nematodes (*Meloidogyne* spp.) to about 2 pg in the horse roundworm, *Parascaris univalens*. This low end will need to be confirmed with more accurate measurements, but with a mean of only 0.25 pg it is clear that the nematodes have some of the smallest genomes among animals. It is therefore not surprising that the first metazoan genome sequenced in its entirety, that of *Caenorhabditis elegans*, belonged to a member of this phylum (see Chapter 9). Nematodes are also of interest because it is in this group that chromatin diminution was first identified 120 years ago (Boveri, 1887). Notably, the large germline genome of *P. univalens* and the much smaller one of *Ascaris lumbricoides* are both reduced to about 0.25 pg in the somatic line, a remarkable 85% reduction in the former case (Moritz and Roth, 1976).

Members of the phylum Tardigrada, commonly known as "water bears," are tiny (0.1 to 1.2 mm) aquatic microinvertebrates whose C-values have been reported to range about 10-fold, from 0.08 to 0.82 pg (mean of 0.38 pg). Species in the phylum Gastrotricha are small (0.5 to 4 mm), colorless, free-living aquatic worms most closely related to nematodes. Gastrotrichs share many features with tardigrades in terms of genome size variation, including a similar number of species studied to date (n = 15), and a small range from 0.08 to 0.63 pg (mean of 0.23 pg).

Rotifers are tiny (most 200 to 500 micrometers [μm]) filter-feeding animals found primarily in freshwater environments and moist soil. The small number of available genome sizes in the phylum Rotifera vary from about 0.35 to 1.1 pg. The first two sponge genome size estimates were provided by Mirsky and Ris (1951), and since then only two more species have been examined (Imsiecke *et al.*, 1995).

Genome size estimates vary about 30-fold in the phylum Porifera, with both species examined by Mirsky and Ris (1951) having reported C-values of 0.06 pg, whereas the two species assessed by Imsiecke *et al.* (1995) had genome sizes around 1.6 to 1.8 pg. Other small phyla for which at least one genome size is known include the Chaetognatha (arrow worms, 0.7 to 1.0 pg), the Ctenophora (comb jellies, 0.3 to 3.2 pg), the Nemertea (ribbon worms, 1.4 pg), the Onychophora (velvet worms, 4.4 pg), the Pogonophora (beard worms, 0.64 pg), the Priapulida (phallus worms, 0.56 pg), and the Sipunculida (peanut worms, 1.3 pg).

INTRASPECIFIC VARIATION IN ANIMALS

One of the more contentious issues in the study of the C-value enigma is the degree to which genome sizes may vary within species. This has been a particularly controversial subject in plants, where genomes previously have been labeled as "fluid," "plastic," or "in constant flux," but in which many former demonstrations of intraspecific variation have since been refuted (see Chapter 2). This is an important issue because large-scale variation in DNA content among conspecifics jeopardizes the assumptions traditionally made in the measurement of genome sizes, and presents a serious challenge to the concept of the C-value itself (see also Chapter 2).

Numerous examples of intraspecific variation have been reported in animals, and in some cases this is to be expected. For example, taxa with chromosomal sex determination systems will necessarily show minor variation between males and females, and indeed this has proved useful for sex identification in several orders of birds (e.g., Rasch and Tiersch, 1992; Tiersch and Mumme, 1993; De Vita *et al.*, 1994; Cavallo *et al.*, 1997; Canon *et al.*, 2000). In some cases, polymorphisms in the X chromosomes can generate a detectable amount of intraspecific variation, as reported for rodent species in the genus *Arvicanthis* from Africa (Garagna *et al.*, 1999). Aneuploidy (the gain or loss of one or a few chromosomes) or B chromosomes (see Chapter 4) can also generate legitimate variation in C-values within species. Examples such as these, with identifiable chromosome-level causes, are known to botanists as "orthodox" cases of intraspecific variation (see Chapter 2).

In some cases, "intraspecific" variation is probably more indicative of a problem with delineating taxonomic boundaries than with the idea of genome size constancy. In pocket gophers (*Thomomys* spp.), for example, Sherwood and Patton (1982) found intraspecific differences that likely correspond to different subspecies. More recently, Ramirez *et al.* (2001) reported significant variation within the South American rodent *Graomys griseoflavus*, which was associated with major karyotypic differences and may relate to the presence of two different species. The same appears to hold in the most prominent case in fishes (Arctic ciscoes,

Coregonus autumnalis), where this almost certainly reflects "a mixture of fish from at least two genetically divergent populations" (Lockwood and Bickham, 1992).

Several other examples are known from teleost fishes (e.g., Gold and Price, 1985; Gold and Amemiya, 1987; Johnson *et al.*, 1987; Ragland and Gold, 1989; Hartley, 1990; Lockwood and Bickham, 1991; Lockwood *et al.*, 1991b; Lockwood and Derr, 1992; Carvalho *et al.*, 1998; Collares-Pereira and Moreira da Costa, 1999), but it should be noted that all of these are from salmonids and cyprinids, which are well known to exhibit more dynamic chromosome-level changes than most teleost groups (including polyploidy) (see Chapter 8). For example, differences in chromosome numbers can be observed among populations of rainbow trout (Thorgaard, 1983). Thus, these families may be atypical in this regard (Johnson *et al.*, 1987; Lockwood and Bickham, 1991), making them of special interest in their own right, but not generally indicative of the situation in fishes (let alone animals or all eukaryotes, as some authors suggest). Perhaps tellingly, a detailed sampling from various wild and domesticated stocks of the channel catfish (*Ictalurus punctatus*, family Ictaluridae) revealed remarkable genome size stability (Tiersch *et al.*, 1990).

More than 30% variation in genome size has been reported within the mosquito *Aedes albopictus*, with island populations in Polynesia having smaller genome sizes than those collected on the Asian mainland (Rao and Rai, 1987). In a follow-up study, Kumar and Rai (1990) examined a total of 37 populations of *Ae. albopictus* sampled from throughout its global range, and found increases in newly colonized areas. These differences in DNA content among populations were also associated with variation in chromosome size (Rao and Rai, 1987), and a subsequent reassociation kinetic analysis showed such differences among *Ae. albopictus* strains to correspond to variation in amounts of highly repetitive DNA (Black and Rai, 1988). The variation in *Ae. albopictus* was correlated positively with both development time and wing size (Ferrari and Rai, 1989). Intriguingly, genome size was also positively associated with the number of years that the strains had been kept in culture, suggesting that "most of the phenotypic variation in genome size and development time was generated in the laboratory" (Ferrari and Rai, 1989). A similar phenomenon may operate in *Drosophila*, where it has been argued that the environmental stresses involved in the invasion of new habitats trigger the expansion of transposable elements and thereby result in intraspecific variation in genome size (Vieira *et al.*, 2002; Vieira and Biémont, 2004). Along these same lines, it has been suggested that the action of transposable elements accounts for the ability of mosquitoes to colonize new environments and evolve resistance to pesticides (Bensaadi-Merchermek *et al.*, 1994).

Intraspecific variation in genome size has also been reported for some tenebrionid beetles, most notably in the genus *Tribolium*, but this was relatively minor and contributed far less to the total variance in the measurements than error within individual samples (Alvarez-Fuster *et al.*, 1991). Within one species, *T. anaphe*, the variation reached 25% (Alvarez-Fuster *et al.*, 1991), but this seems likely due to experimental error because these were all animals from the same laboratory strain, and not from geographically variable populations (moreover, the absolute variation amounted to a mere 0.07 pg). In a more detailed study, Palmer and Petitpierre (1996) examined individuals of the tenebrionid *Phylan semicostatus* from eight island populations in the Mediterranean. They reported an inverse correlation between genome size and body size within this species, which runs counter to that found interspecifically in many invertebrates (see later section). To explain this relationship, Palmer and Petitpierre (1996) invoked a slowing of development by larger DNA contents, which ultimately results in smaller adult sizes. However, the total range in genome sizes was from 0.226 to 0.268 pg in this case, and it is unlikely such a tiny difference in absolute DNA content could affect growth rate and body size to this extent.

Examples of intraspecific variation in genome size have been reported for several other groups of animals, including herpetiles (Lockwood *et al.*, 1991a; MacCulloch *et al.*, 1996), bats (Burton *et al.*, 1989), copepod crustaceans (Escribano *et al.*, 1992), and molluscs (e.g., Rodríguez-Juíz *et al.*, 1996; Vinogradov, 1998a). Interestingly, specimens of fish and amphibians taken from radioactively contaminated areas (specifically, Chernobyl) also show some apparent fluctuations in DNA content (Dallas *et al.*, 1998; Vinogradov and Chubinishvili, 1999). However, most of these cases have not been associated with identifiable chromosomal differences, and therefore qualify as "unorthodox" variation under the botanical terminology (see Chapter 2). This is an important consideration, given that many of the previous reports of unorthodox variation in plants have since been attributed to experimental error, in many cases involving the interference of various phytochemicals with DNA staining. It is possible that similar issues pertain to animals as well, even though such staining inhibitors have not yet been identified in metazoans. As one potential example, Thindwa *et al.* (1994) reported that individuals of the aphid *Schizaphis graminum* raised on sorghum had lower DNA contents than individuals from the same biotypes reared on wheat or johnsongrass. Thindwa *et al.* (1994) proposed the existence of unique growth conditions on sorghum that may select for faster development and lower DNA content, but the absolute differences in C-value between strains were very small (< 0.05 pg). More important, the aphids were simply homogenized prior to measuring DNA content, so it is perhaps more realistic to assume that some botanical compound in their food interfered with the staining.

In summary, there are many reported cases of intraspecific variation in animals. However, as with plants (see Chapter 2), a good number of these may relate to inaccurate taxonomy or experimental error. Nevertheless, some of these examples do qualify as "orthodox" intraspecific variation, with demonstrable chromosome-level underpinnings. In some cases, this is associated with important phenotypic or environmental characters; in others, the variation is probably too small to be of much biological relevance even if there are statistically significant correlations involved.

Overall, the debate surrounding intraspecific variation remains very much open, and is an issue worthy of continued investigation. This should be done with careful consideration of possible sources of measurement error, and with clear guidelines in mind as to what constitutes a biologically significant level of variation.

MECHANISMS OF GENOME SIZE CHANGE

As with all complex evolutionary questions, the C-value enigma involves both proximate and ultimate aspects. One of the most important issues in the former category relates to the mechanisms by which noncoding DNA accumulates in, and/or is deleted from, genomes. Although it has been recognized for some time that many different types of sequences influence genome size, only now that wholescale genome sequence information is available is it becoming possible to examine their relative contributions in detail. In addition to the effect of individual sequences, various chromosomal events play a role in shaping genome size variation. Other evolutionary factors, such as the ultimate consequences of changes in DNA content, are also important in determining the net genome size, but it is best to consider these issues separately.

"SELFISH DNA" AND THE SPREAD
OF TRANSPOSABLE ELEMENTS

The notion that individual fragments of DNA might survive within nuclei as "parasites" of the "host" genome, even if they serve no function or are even somewhat deleterious, dates back to Gunnar Östergren's (1945) discussion of B chromosomes in plants (see Chapter 4). A positive correlation between genome size and the prevalence of B chromosomes was recently identified in flowering plants, but this probably relates to the general tolerance for noncoding DNA or some other indirect association between the two (Trivers *et al.*, 2004). Furthermore, although B chromosomes do contribute to intraspecific variation in DNA content in various plants and animals (see Chapter 4), they are found only in certain species and therefore are not among the more important sources of variation among taxa.

Unlike B chromosomes, mobile genetic elements active within the primary chromosome set have the potential to contribute to genome size variation in eukaryotes at large. As discussed in detail in Chapter 3, these are now known to include several types of transposable elements (TEs) in addition to mobile introns within genes and processed pseudogenes, which are reinserted from RNA after their introns have been spliced out (see Chapters 3 and 5). In 1976, Dawkins suggested that "the simplest way to explain the surplus DNA is to suppose that it is a parasite, or at best a harmless but useless passenger, hitching a ride in the survival machines created by the other DNA" (Dawkins, 1976, p. 47). A few years later, Doolittle and Sapienza (1980) and Orgel and Crick (1980) independently articulated a much more explicit "selfish DNA theory" for the evolution of genome size.

According to the definition of Orgel and Crick (1980), "a piece of selfish DNA, in its purest form, has two distinct properties: (1) It arises when a DNA sequence spreads by forming additional copies of itself within the genome, (2) it makes no [sequence-]specific contribution to the phenotype." Orgel and Crick (1980), and later Sapienza and Doolittle (1981) and Doolittle (1989), suggested that natural selection operating within the genome would tend to favor an increase in DNA amount, which would be counterbalanced by organism-level selection when this became too costly for the host (see Chapter 11). The point at which replication costs became limiting would vary by species according to other biological (e.g., developmental, metabolic) factors (Orgel and Crick, 1980). Under this view, it is possible for noncoding DNA to take on sequence-independent functions, but Orgel and Crick (1980) note that they "prefer to think that the organism has tolerated selfish DNA which has arisen because of the latter's own [intragenomic] selective pressure."

At the time of its inception, the selfish DNA theory was met with much resistance (see Doolittle, 1981, 1982). However, its basic assumptions have since been vindicated, particularly now that the human genome sequence has revealed a very large contribution of "selfish" elements. Specifically, nearly 45% of the human genome, which is very average in size for a mammal and closer to the low end of the overall animal distribution, is composed of transposable elements and (mostly) their inactive remnants (International Human Genome Sequencing Consortium, 2001) (Fig. 1.3). Based on the currently limited data available, it appears that genome size is strongly related to the abundance of TEs across eukaryote species (Kidwell, 2002) (see Chapter 3). Indeed, if any single dominant class of noncoding DNA can be identified within eukaryotic genomes, then transposable elements are likely to be it. However, it bears noting that the specific properties and relative contributions of different TEs may vary considerably among taxa (see Chapters 2 and 3).

In its original formulation, the selfish DNA theory represented a "mutation pressure" theory, meaning that there would be internal pressure for genome size to increase unless mitigated by selection on the organismal level. It is now becoming apparent that decreases in DNA are also important (see later section), and that there are more complex interactions between TEs and their hosts than strict parasitism (see Chapters 3 and 11). Moreover, some early explications of the theory tended to interpret the identification of proximate sources of noncoding DNA as "the" solution to the C-value enigma. As Doolittle and Sapienza (1980) argued, "when a given DNA, or class of DNAs, of unproven phenotypic function can be shown to have evolved a strategy (such as transposition) which ensures its genomic survival, then no other explanation for its existence is necessary."

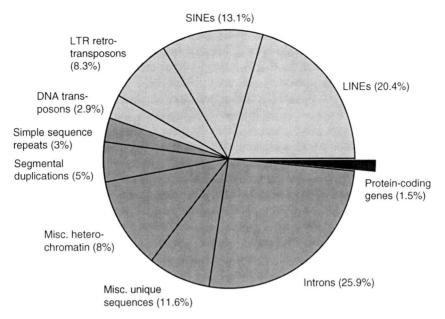

FIGURE 1.3   Summary of the different components of the human genome. Only about 1.5% of the genome consists of strict protein-coding sequences, whereas about 45% of it is composed of transposable elements of various types (light gray sections). Introns make up another 26%, and segmental duplications account for about 5% of the sequence. Data from the International Human Genome Sequencing Consortium (2001).

Although this view was expanded somewhat in the subsequent discussion by Sapienza and Doolittle (1981), the argument has resurfaced since, as for example in Dover's (1990) contribution to the debate outlined earlier in this chapter. As will be seen, there are many other factors to consider regarding the existence and abundance of noncoding DNA, selfish and otherwise.

"Junk DNA" and the Accumulation of Pseudogenes

In this era of front-page genome biology, the term "junk DNA" has become a popular descriptor for noncoding DNA among both scientists and the informed laity. As with "selfish DNA," there were strong objections to the concept of "junk DNA" when first introduced (Ohno and Yomo, 1991). Also as with selfish DNA, the term junk DNA had a specific meaning when coined, although it is now frequently applied to nongenic segments at large (e.g., Pagel and Johnstone, 1992; Moore, 1996).

Specifically, "junk DNA" refers to pseudogenes—defunct copies of protein-coding genes. The "junk DNA theory" for the evolution of genome size was developed as part of Susumu Ohno's early discussions of the role of gene duplication in evolution (see Chapter 5). Although he advocated gene duplication as a crucial requirement for evolutionary diversification, Ohno (1970) noted that most mutations in duplicate genes would not lead to new functions, but rather to inactivation, or a transition to what he originally called "garbage DNA." As he later put it, "the earth is strewn with fossil remains of extinct species; is it any wonder that our genome too is filled with the remains of extinct genes?" (Ohno, 1972a).

In Ohno's usage, as in the vernacular, "garbage" refers to both the *loss* of function and the *lack* of any further utility (i.e., it once was useful, but now is not). Although Brenner (1998) drew a distinction between "junk," which is kept, and "garbage," which is thrown away, Ohno initially used the terms interchangeably, as in his reference to the human genome, in which "at least 90% of our genomic DNA is 'junk' or 'garbage' of various sorts," Interestingly, Ohno used the term "junk DNA" only in the titles of two of his papers (1972a, 1973), and mentioned the phrase only once in passing in a third (1972b). Comings (1972), on the other hand, gave what must be considered the first explicit discussion of the nature of "junk DNA," and was the first to apply the term to *all* noncoding DNA.

As mentioned earlier, processed pseudogenes are more accurately considered as mobile elements, although it appears that they depend on other sequences (e.g., long interspersed nuclear elements, or LINEs) for their movement. Classical pseudogenes and those nuclear pseudogenes of mitochondrial origin ("numts") make up the remainder of "junk DNA," properly defined. As with selfish DNA, the junk DNA theory is a "mutation pressure" theory, which also involves a tendency for DNA to accumulate (this time by chance) unless halted by other selective forces (e.g., Pagel and Johnstone, 1992). However, unlike selfish DNA, true junk DNA (i.e., gene duplicates) appears to constitute a relatively minor fraction of the human genome (International Human Genome Sequencing Consortium, 2001) (Fig. 1.3).

Not only is "junk DNA" an inappropriate moniker for noncoding DNA in general because of the minority status of pseudogenes within genome sequences, but it also has the unfortunate consequence of instilling a strong *a priori* assumption of total nonfunction. As Zuckerkandl and Hennig (1995) pointed out, "given a sufficient lack of comprehension, *anything* (and that includes a quartet of Mozart) can be declared to be junk." Indeed, it is becoming increasingly clear that at least some noncoding sequences play important regulatory or structural roles (e.g., Zuckerkandl, 2002; Nelson *et al.,* 2004; Shabalina and Spiridonov, 2004) (see Chapter 11). Of course, just as there is no reason to expect certain aquatic salamanders to have 70 times more genes than chickens (Mirsky and Ris, 1951), there is no reason to expect them to require 70 times more gene regulation. For this reason, it is unlikely that any one function for noncoding DNA can account for either its sheer mass in some species or its unequal distribution among taxa. However, dismissing it as no more than "junk" in the pejorative sense of "useless" or "wasteful" does little to advance the understanding of genome evolution. For this reason, the far less loaded term "noncoding DNA" is used throughout this chapter and is recommended in preference to "junk DNA" for future treatments of the subject.

Introns

It was recognized in the late 1970s that genes consist of both "expressed regions" (exons) and "intergenic regions" (introns), the latter of which are noncoding. Almost immediately, it was suggested that "the unexpected extra DNA in higher cells, the excess of DNA over that needed to code for the number of products defined genetically, now is ascribed to the introns" (Gilbert, 1978). A similar claim that most noncoding euchromatic (noncondensed) DNA in animal (but not plant) genomes is, in fact, comprised of introns has been advanced recently by Wong et al. (2000). The implication of this proposition is that most of the human genome must be transcribed into RNA despite the fact that less than 2% of this is ultimately translated into proteins (Wong et al., 2001). This view of intron ubiquity has not been widely accepted, but it has been suggested more recently that some 45% of the human genome is transcribed (Shabalina and Spiridonov, 2004). At the least, introns appear to make up about 26% of the human genome sequence (International Human Genome Sequencing Consortium, 2001) (Fig. 1.3). In more general terms, genome size and the lengths of individual introns appear to be positively correlated between species of *Drosophila* (Moriyama et al., 1998), within the class of mammals (Ogata et al., 1996), and across eukaryotes in general (Vinogradov, 1999a), although this latter correlation is relatively weak.

Chromosome-Level Events

There are several mechanisms at the level of individual chromosomes that can lead to both increases and decreases in genome size, provided that they occur in the germline. The duplication or loss of individual chromosomes (aneuploidy) may alter genome size, although this is usually associated with deleterious phenotypic effects. On a slightly smaller scale, segments of chromosomes may break off and can potentially fuse with other chromosomes. Depending on the subsequent segregation of broken/fused chromosomes, this can result in either a net gain or a net loss of DNA. Even the fusion of two existing chromosomes, which need not lead to a change in total DNA content, may be associated with other duplication and rearrangement events. The classic example in this case is the fusion of two ancestral chromosomes into human Chromosome 2, resulting in a difference in karyotype between humans (2n = 46) and other apes (2n = 48) and leading to many smaller-scale secondary changes (Fan *et al.*, 2002). Although transposable element activity appears to be higher in the human genome than in chimpanzees (Liu *et al.*, 2003), the latter displays the larger genome size (1C = 3.75 versus 3.50 pg). This is due to chromosome-level variation such as the possession of terminal repeats not found in humans. Variation in the heterochromatic (highly condensed and repetitive) component of the genome has been suggested to account for differences in C-value among species in some other groups of mammals as well (e.g., Hatch *et al.*, 1976; Gamperl *et al.*, 1982; Pellicciari *et al.*, 1990a,b; Manfredi Romanini *et al.*, 1991; Garagna *et al.*, 1997).

Unequal crossing-over during meiosis and unequal sister chromatid exchange during (germline) mitosis can also create either gains or losses of DNA, depending once again on the way in which the altered chromosomes segregate (see Chapter 5). Similar recombinational processes leading to a loss of DNA have been identified at three additional scales in plants, and may also apply to animals. Specifically, the long terminal repeats of LTR retrotransposons (see Chapter 3) may promote recombination, either between homologous copies of the element or by "illegitimate recombination" between unrelated copies. In the former case, most of the element will be deleted, leaving behind a "solo-LTR." Overall, there will still be a net gain of DNA with each insertion of a new TE, but nevertheless this could help to significantly slow genomic growth (Devos *et al.*, 2002) (see Chapter 2). Illegitimate recombination, on the other hand, may involve the deletion of all intervening sequences between the two LTR elements, leading to a significant net loss of DNA (Devos *et al.*, 2002) (see Chapter 2). Finally, there are indications that differences in recombinational double-strand break repair mechanisms may lead to differential rates of DNA loss from different genomes (e.g., Filkowski *et al.*, 2004) (see Chapter 2).

POLYPLOIDY, C-VALUE, AND GENOME SIZE

It is often stated that polyploidy, the duplication of entire chromosome sets, is the most rapid and substantial means of increasing genome size. However, while polyploidy does indeed increase DNA content and therefore C-value, it does not necessarily change genome size per se, but rather adds a second genome to the nucleus. Strictly speaking, it is more accurate to consider a genome size change to have occurred only after significant rediploidization has taken place, at which time the nucleus again can be considered to contain only one genome. This is a complex issue, and it is not entirely clear exactly how or when rediploidization may occur (Wolfe, 2001). Moreover, the different mechanisms of polyploid formation and variation in the level of rearrangement that occurs after the fact (see Chapters 6, 7, and 8) may have varying implications for the issue of C-value versus genome size. Despite its profound and immediate effects on C-values, polyploidy may in fact be one of the *slowest* mechanisms for changing genome sizes in the strict sense.

That said, and although it is much less common in animals than in plants (see Chapters 7 and 8), polyploidy may play an important role in the long-term evolution of genome size in some metazoan lineages. For example, sirenid salamanders, which have very large genomes, may be ancient polyploids that are now rediploidized (see Chapter 8). On the other hand, various other salamanders and lungfishes with large genomes are not thought to be polyploid (see Chapter 8). In teleost fishes, it is the ancient polyploids (e.g., salmonids) that tend to have the larger genomes. However, in orthopterans, which exhibit the largest genomes among insects, there is only one known case of polyploidy (see Chapter 8).

In discussing the difference between polyploidy and genome size, it seems relevant to point out that the massive variation in eukaryotic genome sizes cited by most authors may be somewhat inflated. At present, the range in published eukaryote genome sizes is between about 0.0023 pg in the parasitic microsporidium *Encephalitozoon intestinalis* (Vivares, 1999) and 1400 pg in another protist, the free-living amoeba *Chaos chaos* (Friz, 1968)—a greater than 600,000-fold difference. The largest genome size is often attributed to *Amoeba dubia* at 700 pg, conveniently 200 times larger than a human's (e.g., Gregory and Hebert, 1999), and sometimes *Encephalitozoon cuniculi* is cited as the smallest at about 0.0029 pg (Biderre *et al.*, 1995), for a more than 200,000-fold range (e.g., Gregory, 2001a). Many previous publications have taken brewers' yeast *Saccharomyces cerevisiae* at 0.008 pg to be the lowest eukaryotic C-value, for a range of about 80,000-fold (e.g., Pagel and Johnstone, 1992; Gregory and Hebert, 1999).

In any event, a few caveats are in order regarding the high end of this range. First, these values for amoebae were based on rough biochemical measurements of total cellular DNA content (Friz, 1968), which probably includes a significant fraction of mitochondrial DNA, food organisms, and endosymbionts. The accuracy of this method is brought into question when one considers that Friz's (1968) value of 300 pg for *Amoeba proteus* is an order of magnitude higher than those reported in subsequent studies (Byers, 1986). Second, some amoebae (e.g., *A. proteus*) contain 500–1000 small chromosomes and are quite possibly highly polyploid (Byers, 1986), while others (e.g., *C. chaos*) are multinucleated (Sparrow *et al.*, 1972), in which case these values would be inappropriate for a comparison of haploid *genome sizes* among eukaryotes. Of course, this has little impact on the need to explain variation in genome size; it may just mean that these impressive (and almost universally cited) examples will require replacement by some based on more reliable estimates.

SENTINEL SEQUENCES AND GLOBAL FORCES

In addition to the types of DNA such as TEs that exert a strong influence on genome size, there are other sequences that tend to be present in proportion to DNA content but which themselves contribute a minor percentage to the total genome. While their presence does not account for much variation in genome size, these do have the potential to act as "sentinel sequences" (Gregory and Hebert, 1999) that provide some insight into the operation of "global forces" (Petrov, 2001) operating on a genome-wide scale.

Satellite DNA

Satellite DNA, also known as tandemly repeated DNA (TR-DNA), represents a class of repetitive elements consisting of clusters of short repeated sequences, and is divided into several categories according to the size of the individual repeats (Li, 1997; Eisen, 1999). *Satellites* are tandemly repeated sequences ranging from two to hundreds of base pairs (bp) per repeat, found in series of thousands of repeats, and localized to the heterochromatic regions of the centromeres and telomeres (and are particularly common in mammalian Y chromosomes). *Minisatellites* vary in repeat size from 9 to 100 bp (usually about 15 bp), and are usually found in clusters of 10 to 100 repeats located in euchromatic subtelomeric regions or dispersed in other areas of the genome. *Short tandem repeats* (STRs) have repeat lengths of 3 to 5 bp, and are repeated 10 to 100 times in various genomic locations. *Microsatellites* consist of very small sequences (1 to 5 bp) repeated 10 to 100 times per cluster, and located in various places in the euchromatic portion of the genome. There is obviously some overlap between these satellite DNA size categories, which can be a cause of confusion. Hancock (1999) recommended that the term "simple sequences" be used "to describe all sequences based on repeats of short motifs," and that "microsatellites" be used in reference to any such motifs less than 6 bp in length.

Satellite DNAs may be generated by some of the chromosome-level processes described in the previous section, namely unequal crossing-over during meiosis and unequal sister chromatid exchange at mitosis. Additionally (and perhaps primarily), short but abundant repeats may result from a process known as "replication slippage" (also called "slipped-strand mispairing") (Levinson and Gutman, 1987). In replication slippage, the nascent DNA strand becomes dislodged from the template strand during replication and reanneals out of phase with the template strand, forming a small loop outside of the double strand. As replication proceeds, this looped segment will be either overlooked or copied twice, such that this slippage mechanism can result in either a gain or loss of small sequences. A (micro)satellite is generated when the same sequence is inserted repeatedly by this mechanism, which is facilitated by the fact that the presence of repetitive

sequences promotes more slippage. It is also possible that microsatellites originate from "proto-microsatellites" in transposable elements, as with the
microsatellite initiating mobile elements ("*mini-me*") of *Drosophila* (Wilder and
Hollocher, 2001).

The sizes, distributions, and sequences of microsatellites as found in the
genomes of the five best-studied model organisms (human, fruit fly, nematode,
mustard weed, and yeast) are discussed by Katti *et al.* (2001). There are some
important differences among these organisms, but it is interesting to note that
microsatellite DNA abundance in general may correlate positively with genome
size (e.g., Hancock, 2002; but see Neff and Gross, 2001). Microsatellites probably make up only a relatively small portion of noncoding DNA in most
cases (Morgante *et al.*, 2002), but it has been suggested that satellite DNA may
account for as much as 30% of the genome in some decapod crustaceans (Lécher
et al., 1995).

Ribosomal DNA

Ribosomal DNA (rDNA) codes for the rRNAs used in the production of ribosomes, the sites of cellular protein synthesis. In most eukaryotes, ribosomal DNA
consists of tandemly repeated arrays of four or five genes located at the nucleolus
organizer region (NOR) of one or more chromosomes. The number of rDNA
copies per genome varies enormously among eukaryotes, with the multiplicity of
these genes generated primarily through unequal crossing-over/sister chromatid
exchange. Although it is believed that no correlation between rDNA multiplicity
and genome size exists in bacteria (Fogel *et al.*, 1999), there has been some
controversy regarding the possibility of a relationship in eukaryotes. For example,
while Birnstiel *et al.* (1971) reported a significant positive association within
a limited eukaryotic sampling, Gall (1981) argued against the existence of a clear-
cut correlation. Other investigations within specific taxonomic groups had failed to
reach consensus (e.g., Ingle *et al.*, 1975; Bobola *et al.*, 1992), but a recent large-scale
sampling across eukaryotes revealed a highly significant positive correlation in
both animals and plants (Prokopowich *et al.*, 2003).

INSERTION–DELETION BIASES

All of the mechanisms of genome size change described thus far either involve only
DNA gain, include both gains and losses, or else are dependent on the properties
of certain DNA elements that are frequently inserted. Although it remains controversial, there is now a suggested mechanism that deals exclusively with the deletion
of DNA, and which therefore helps to fill this theoretical gap. Specifically, it has been
proposed that at the level of small (< 400 bp) insertions and deletions ("indels")

there is a strong bias toward deletion, resulting in a slow but steady negative muta-tion pressure that can remove large amounts of DNA over evolutionary timescales. This mechanism has been widely cited as a cause of genome size variation, even in cases where no relevant data exist, and as such warrants careful consideration.

Such a deletion bias has been recognized to occur within protein-coding regions for many years (e.g., de Jong and Rydén, 1981), and in 1989, Graur and colleagues showed a similar effect in noncoding DNA. In particular, Graur *et al.* (1989) compared the relative rates and sizes of indels in a sample of 22 processed pseudogenes from human, 14 from mouse, and 16 from rat, and reported that not only did deletions outnumber insertions, but that rodent pseudogenes appeared to lose DNA less slowly than those of humans. However, the difference in genome size between murine rodents (~3.2 pg) and humans (3.5 pg) is relatively minor, and it is clear that "[small] deletions and insertions in murid and human genomes do not contribute significantly to genome size" (Ophir and Graur, 1997; see also Mouse Genome Sequencing Consortium, 2002).

The difficulty in conducting a similar analysis with small-genomed model organisms (specifically, *Drosophila*), which have few bona fide pseudogenes, was circumvented in an elegant study by Petrov *et al.* (1996), who examined the indel spectra of deactivated ("dead-on-arrival," or "DOA") copies of the non-LTR retro-transposon *Helena*. Again, small indels were strongly biased toward deletions, and moreover the calculated relative rate of DNA loss (in bp lost per bp substituted) was much higher in *Drosophila* than in mammals, in accordance with its tiny genome (0.18 pg).

In order to test the possible association between genome size and DNA loss rate by small indel bias, additional studies were carried out on another DOA non-LTR transposable element (*Lau1*) in the cricket *Laupala cerasina* at 1.9 pg (Petrov *et al.*, 2000) and nuclear pseudogenes of mitochondrial origin (numts) in the grasshopper *Podisma pedestris* at ~16.9 pg (Bensasson *et al.*, 2001). According to these results, there appeared to be a strong inverse correlation between genome size and relative DNA loss rate across nearly two orders of magnitude.

Based on this, Petrov (2002a) proceeded to develop a "mutational equilibrium model" in which he argued that "differences in genome size may be driven largely by changes in the per nucleotide rate of DNA loss through small indels." Under this view, most of the other mechanisms are relegated to comparatively minor roles. This even holds for rapid increases in genome size, such as the doubling of the maize genome in only a few million years by a surge in transposable element activity (SanMiguel and Bennetzen, 1998) (see Chapter 2), which Petrov (2002a) interprets as little more than "noise around the long-term equilibrium value."

Although the small indel bias approach has made a useful contribution by drawing attention to the importance of genome size reductions, the influence of this particular mechanism has been called into question for several different rea-sons (Gregory, 2003b, 2004b). First, on a statistical level, it has been pointed out that the size distributions of individual deletions are highly skewed, and that the arithmetic means used to calculate average deletion size are inflated. This is especially true of *Drosophila*, in which the use of geometric means, whereby the data are all log(x + 1)-transformed prior to calculating the average, reduces the mean deletion size by a factor of three (Gregory, 2003b). This is a particularly important point, because the entire "rate" correlation is actually produced by differences in average deletion size divided by a measure of time. Notably, the use of geometric means on this skewed dataset negates the difference in DNA loss rate between crickets and grasshoppers (Gregory, 2003b, 2004b).

Second, there are important discrepancies among the types of sequences studied so far. For example, a more general survey of indels in the *Drosophila* genome revealed a 2-fold lower deletion rate, suggesting that "*Helena* may contain regions that are more deletion-prone" than other sequences (Blumenstiel *et al.*, 2002). A similar 2-fold difference was found in pufferfishes depending on whether pseudogenes or DOA transposable elements were assayed (Dasilva *et al.*, 2002; Neafsey and Palumbi, 2003). In *Podisma* numts there is an insertion–deletion hotspot which, if included in the analysis, makes the rate of DNA loss higher in this species than in the cricket with a nearly 10-fold smaller genome.

Third, the value given for "mammals" based on the data of Graur *et al.* (1989) is outdated and should have been replaced with the estimates for rodents and humans made more recently by Ophir and Graur (1997) using a much larger dataset. According to this more extensive sampling, it is apparent that humans delete DNA more slowly than grasshoppers even though the latter has a genome size five times as large (Gregory, 2003b, 2004b) (Fig. 1.4).

Fourth, when the conversion is made from relative to absolute time units, it is clear that in most cases this mechanism is far too slow to exert a significant influence on genome size above a relatively low threshold. For example, whereas the half-life of a new pseudogene is only about 14 million years in *Drosophila* (if one uses an arithmetic mean with the *Helena* element data), the value is 615 million years in *Laupala*, at least 800 million years in "mammals" (using the older dataset), and between 880 and 3500 million years in *Podisma* (Petrov, 2002a). In these latter cases, the time taken to delete *half* a noncoding DNA sequence is longer than the duration of the entire animal lineage to this point. Returning to the notion of an "equilibrium" in maize, by this mechanism it would take about 615 million years, or nearly 10 times the age of the entire grass family (Gaut, 2002), to undo half of the recent doubling of the genome, which itself occurred in only three to six million years.

Even in small-genomed taxa the significance of this mechanism is not yet clear, given that in *Drosophila melanogaster* versus *D. virilis*, which differ 2-fold in genome size, there is, if anything, an inverse relationship with DNA loss rate (Petrov and Hartl, 1998) (Fig. 1.4). The same holds true of smooth (0.4 pg) and spiny (0.8 pg) pufferfishes (Fig. 1.4), and indeed "a decline in the rate of large-scale insertions [rather than increased deletion] is implicated as a probable cause

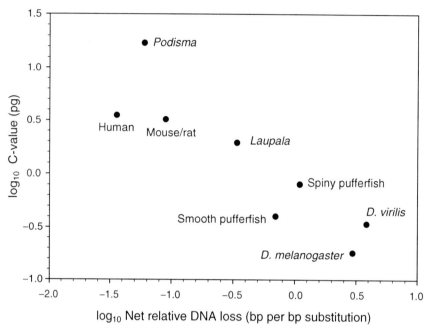

FIGURE 1.4   The relationship between genome size and relative DNA loss rate based on the revised dataset given in Gregory (2004b), which is much less robust than the relationship presented in Petrov (2002a). When the extreme points from *Drosophila* are removed, the relationship is only marginally significant even when log-transformed (Gregory, 2004b). When the highly skewed deletion size data are log-transformed prior to calculating the average (i.e., rather than using simple arithmetic means), the relationship is much weaker (Gregory, 2003b, 2004b). Note that the 2-fold differences in genome size between *D. melanogaster* and *D. virilis* and between smooth (family Tetraodontidae) and spiny (family Diodontidae) pufferfishes do not correspond to differences in DNA loss rate. Note also that the *Drosophila* and pufferfish values may vary up to 2-fold according to which type of sequence is analyzed (Blumenstiel *et al.*, 2002; Neafsey and Palumbi, 2003). Moreover, when the updated mammalian dataset of Ophir and Graur (1997) is used instead of that of Graur *et al.* (1989), it is apparent that humans delete DNA more slowly than grasshoppers (*Podisma*), despite possessing a 5-fold smaller genome.

of the genome size reduction in the tetraodontid (smooth) pufferfish lineage" (Neafsey and Palumbi, 2003).

Again, the reported correlation with DNA loss rate is really a reflection of differences in average deletion size, and in particular the occurrence of relatively large deletions in some species (e.g., *Drosophila*) versus their absence in others. In other words, it would appear that general error-proneness, rather than small neutral deletion bias per se, may be the parameter most closely associated with small genome size. This interpretation is supported by the fact that *Drosophila* has one of the most dynamic genomes so far reported for any eukaryote (Ranz *et al.*, 2001; González *et al.*, 2002). In fact, *Drosophila* is second only to *Caenorhabditis* in terms of genomic malleability (Coghlan and Wolfe, 2002), with the latter having a smaller genome (0.1 pg) and a roughly proportionate DNA loss rate in terms of small indels in *mariner* TEs (although direct comparisons with the other species are problematic because the data were collected in different ways) (Witherspoon and Robertson, 2003). However, it is important to note that *Caenorhabditis* undergoes more frequent duplications, pseudogene formations, and intron gains than *Drosophila* (Semple and Wolfe, 1999; Robertson, 2000; Friedman and Hughes, 2001; Cavalcanti *et al.*, 2003), and also experiences large deletions (up to nearly 800 bp) and even larger insertions (up to almost 4700 bp) (Robertson, 2000).

It has been pointed out several times that selection probably does not act on each small indel in *Drosophila* (Petrov *et al.*, 1996; Petrov and Hartl, 1997, 2000; Moriyama *et al.*, 1998; Lozovskaya *et al.*, 1999; Hartl, 2000; Petrov, 2001, 2002b). However, this does not mean that selection is not operating on the genome size as a whole even if each individual deletion is neutral (Petrov and Hartl, 1997; Hartl, 2000; Petrov, 2001). Moreover, there is evidence that deletions larger than the arbitrary cutoff of 400 bp may indeed be selectively relevant (Blumenstiel *et al.*, 2002). More generally, it is possible that the indel patterns observed in *Drosophila* are a by-product of more deletion-prone genomewide repair mechanisms and other features that could be influenced by selection for small C-value, making the indel bias correlation an effect rather than a cause of genome size dynamics in this species (Gregory, 2004b). Such a shift in interpretation would raise some interesting new questions regarding the reasons for a higher degree of mutational error in species with very small genomes.

GENOME SIZE AND G + C CONTENT

As discussed briefly in Chapter 9, there is variation in the percentage of G + C versus A + T base pairs within different regions of a given animal genome. While variation in the ratio of constituent bases does not contribute directly to differences in DNA content, correlations between C-values and such structural features can provide some important insights into how genome sizes evolve. To this end, a possible correlation between genome size and G + C content has been studied in detail within several groups of vertebrates (e.g., Stingo *et al.*, 1989; Vinogradov and Borkin, 1993; Vinogradov, 1994), and more recently across a sample of 154 species of fishes, amphibians, reptiles, mammals, and birds by Vinogradov (1998b). In this latter study, G + C content was found to be significantly positively correlated with genome size among vertebrates in general, albeit in a nonlinear way. Specifically, G + C content is rather variable at low genome sizes, but stabilizes at about 46% in larger genomes (most of which are in salamanders), making this a triangular relationship (Vinogradov, 1998b). Birds and reptiles tend to have more G+C for a given genome size than amphibians, and within teleost fishes the relationship is actually negative. One possible reason for this association is that larger genomes are more susceptible to mutations (see Chapter 2), suggesting that the greater chemical stability of G+C pairs might be favored at larger C-values (Vinogradov, 1998b).

ASSESSING THE DIRECTIONALITY OF ANIMAL GENOME SIZE EVOLUTION

While the traditional assumption has been that genome size change has been in the direction of increase, the recent identification of mechanisms capable of reducing DNA content has made this issue more complex and interesting. At the broadest scales, it may seem likely that the earliest animal genome sizes were small, as appears to be the case in angiosperm plants (see Chapter 2). This view is supported by the fact that *Trichoplax adhaerens* (0.04 pg), which is the most primitive animal known, has one of the smallest genomes in the kingdom. On the other hand, genomes nearly as small have been found in groups such as tardigrades (0.08 pg), nematodes (possibly as low as 0.03 pg), flatworms (0.06 pg), annelids (0.06 pg), arachnids (0.08 pg), insects (0.1 pg), and even chordates (0.07 pg). Further adding to the confusion, in sponges, another primitive group probably morphologically similar to the early animal ancestor, both very small (0.06 pg) and relatively large (1.8 pg) C-values have been reported. Even in vertebrates the issue is somewhat complex, given that members of primitive groups such as jawless and cartilaginous fishes have larger genomes than derived ones like birds. Moreover, the primitive cephalochordate *Branchiostoma lanceolatum* has a larger genome than some highly specialized teleost fishes.

In angiosperm plants, for which a complete family-level phylogeny is now becoming available, it has been possible to reconstruct some of the probable historical patterns of genome size change. This cannot yet be done for animals at large, although such analyses have been applied within certain families (e.g., Honeycutt *et al.*, 2003). The challenge for this method is that assumptions must be made about the relative probabilities of increases versus decreases in DNA content (e.g., Wendel *et al.*, 2002), and therefore about what the ancestral character states were likely to have been.

Indeed, the only way to reconstruct genome size evolution with some certainty is to consult fossil evidence. Obviously, genome sizes themselves do not fossilize, but they do leave an impression in terms of cell size. Using fossil osteocyte sizes calibrated against living species, it has been possible to determine that genome sizes are secondarily increased in both lungfishes and amphibians (Thomson, 1972;

Thomson and Muraszko, 1978). Thus, although these groups are the most morphologically primitive among the tetrapods, they are actually the most derived in terms of genome size. It would therefore be misleading to anchor a phylogenetic analysis of tetrapod genome size evolution by taking these large values as suggestive of the ancestral condition. It should be borne in mind that the same dangers apply to all such studies, most of which will not be so obvious as this.

Unfortunately, in only one other case were cell sizes used to assess patterns of animal genome size evolution over geological timescales. This is the study of fossil epithelial cells in conodonts (extinct marine animals considered by many to be chordates, or perhaps even early vertebrates) by Conway Morris and Harper (1988). In this case, the sizes of fossilized cell impressions were used to infer a remarkable level of genome size stability for at least the first 50 million years of their history. More variability was apparent through the following 220 million years, but no consistent trends toward increases or decreases could be determined (Conway Morris and Harper, 1988). Moreover, these primitive vertebrates (if such they be) had rather large genome sizes calculated to be between 6 and 17 pg (Conway Morris and Harper, 1988), although this is obviously a very loose estimate. In any case, this intriguing study of fossil C-values provides a necessary caution against assumptions about historical pathways in genome size evolution based only on an examination of a few extant taxa.

These patterns also speak strongly against one common view of genome size evolution, namely that "older" lineages such as lungfishes and salamanders have simply had more time to accumulate noncoding DNA. Notwithstanding that this argument is contradicted by the findings of genome size reduction or stasis in some groups, it is also faulty on logical grounds. That is, it should be noted that unless life arose more than once on the Earth, every extant lineage is of *exactly the same age* and therefore they all have had the same amount of time for genomic growth. One may counter that, say, birds as a group are younger than salamanders, but then the real issue is why avian evolution was associated with a halting (or reversal) of genome growth. In other words, this is actually a question about the genomic and phenotypic properties of the groups in question—a topic discussed in some detail in the following sections.

GENOME SIZE AND CELL SIZE

A strong positive relationship between nucleus size and cell size has been recognized for at least 130 years (Gulliver, 1875) (Figs. 1.5 and 1.6), leading to the concept of the "cytonuclear ratio" (Strasburger, 1893; Trombetta, 1942), or, more commonly, the "karyoplasmic ratio" (e.g., Hertwig, 1903, reviewed in Wilson, 1925). In most early discussions of the karyoplasmic ratio, it was held that a certain cytoplasmic mass was required to support the activities of a nucleus of a given size, and that

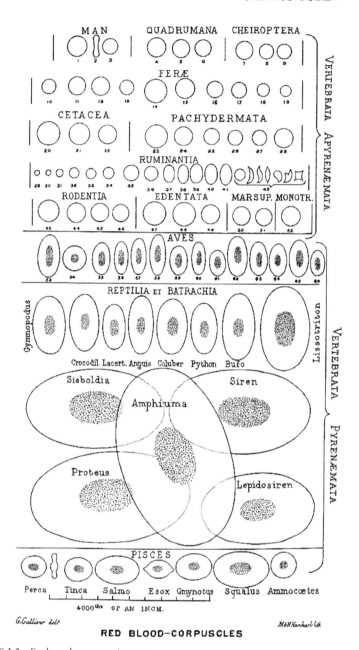

RED BLOOD–CORPUSCLES

FIGURE 1.5   For legend, see opposite page.

cell volumes were adjusted secondarily following alterations in nuclear size (for review see Nurse, 1985). However, because the contents of nuclei were only superficially characterized at this stage, mechanistic explanations for this association were unavailable.

A rough correlation between C-value and cell size has been noted since some of the earliest studies of genome size evolution. As Mirsky and Ris (1951) pointed out, "in the nucleated red cells of vertebrates ... there is an approximately direct relationship between cell mass and DNA content." Importantly, this correlation is not limited to animals, and similar relationships are found in plants (see Chapter 2) and protozoans (e.g., Cavalier-Smith, 1980; Taylor and Shuter, 1981; Wickham and Lynn, 1990). Indeed, a positive correlation with cell size is the most universal implication of genome size variation known.

Among animals, this relationship remains best studied with regard to vertebrate red blood cells (erythrocytes). In 1983, Olmo used compiled cell and genome size data to demonstrate the persistence of the correlation across vertebrates at large, covering a greater than 300-fold range in both parameters. Surprisingly, however, the relationship had not been well examined within several major groups until very recently. Thanks to broad recent comparisons using both existing and original genome and cell size data, it has now been established that the correlation holds within each of the major classes of vertebrates.

In fishes, for example, small initial surveys revealed the general trend (e.g., Pedersen, 1971; Fontana, 1976; Banerjee et al., 1988), and Olmo's (1983) classic study found them to fit well on the overall vertebrate regression, but some more recent comparisons had raised doubts about the existence of the correlation within certain fish taxa (Chang et al., 1995; Lay and Baldwin, 1999). However, these latter studies were deemed methodologically problematic (Gregory, 2001a,c), and a strong positive correlation was subsequently demonstrated in a preliminary way using previously published data from about 50 species of "fishes" (including teleosts, cartilaginous fishes, and lungfishes) (Gregory, 2001c). Far more conclusively, Hardie and Hebert (2003) used original genome size and dry cell area measurements from more than 230 species of teleost and cartilaginous fishes to show that the relationship holds up very well within and across both groups (Fig. 1.7A).

FIGURE 1.5   The relationship between nucleus size and cell size (the "karyoplasmic ratio"), as demonstrated by the figure published by George Gulliver in 1875. Note the immense size range in vertebrate erythrocytes, and the clear association between cell and nucleus size in nonmammals. Note also that mammalian erythrocytes, which do not contain nuclei, are considerably smaller than those of most other vertebrates. Gulliver's suggested classification of vertebrates on the basis of the presence ("Pyrenaemata") and absence ("Apyrenaemata") of nuclei is also illustrated in this figure. The large number of nucleus and cell size measurements provided by Gulliver (1875) were made over a period of 25 years, meaning that he was unaware that nuclei contained DNA, the discovery of which was not published by Miescher until 1871.

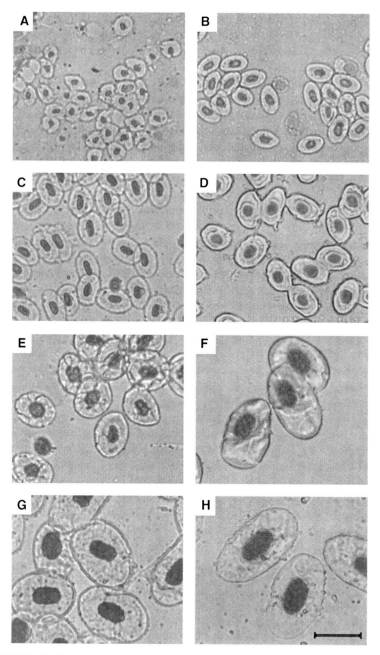

FIGURE 1.6 For legend, see opposite page.

In amphibians, it is clear that genome size is strongly positively correlated with erythrocyte size regardless of whether the latter parameter is measured as dry cell volume (Olmo and Morescalchi, 1975, 1978; Olmo, 1983) (Fig. 1.7B), dry area (De Smet, 1981), or wet volume (Horner and Macgregor, 1983). Neuron size is also correlated positively with genome size in both frogs and salamanders, an association that has some intriguing biological implications (Roth et al., 1994) (see later section). Thus, the sizes of both highly compact (erythrocytes) and expanded (neurons) cells correlate with genome size, although clearly the comparisons must be made within, and not between, these different cell types.

In reptiles, De Smet (1981) found a highly significant positive correlation between genome size and dry erythrocyte area, and in the following year, a similar correlation was reported by Olmo and Odierna (1982) based on dry cell volumes (Fig. 1.7C). De Smet (1981) also compared genome size and dry cell area in a small sample of birds, but did not find a significant correlation, probably because of small sample size ($r = 0.34$, $n = 15$). Other studies have found birds to fit well along the general vertebrate regression line (Commoner, 1964; Olmo, 1983), but the existence of a relationship between genome size and cell size within the class Aves had never been demonstrated convincingly. Olmo's (1983) classic study, for example, contained data for both parameters from only four bird species. In a recent analysis of roughly 50 species, positive relationships were confirmed between genome, nucleus, and cell sizes in this class as well (Gregory, 2002a) (Fig. 1.7D).

Mammals are the only vertebrate class to possess universally enucleated erythrocytes. In other words, mature red blood cells in mammals are genome-free. This ejection of nuclei allows mammalian erythrocytes to reach a level of miniaturization not attainable by any other groups; compared with certain amphibians, mammalian erythrocytes are positively tiny (Figs. 1.5 and 1.8). Nevertheless, a strong positive correlation exists between genome size and dry diameter across a broad array of mammals (Gregory, 2000) (Fig. 1.7E). This also holds when cell size is measured as wet volume (T.R. Gregory, unpublished). Based on a relatively small number of data, there also appears to be a positive correlation with epithelial cell size (Olmo, 1983). No general relationship exists with sperm size (Gage, 1998), although there may be an effect on the smaller taxonomic scale of related species of rodents (Gallardo et al., 1999, 2003), or even between X- and Y-bearing sperm within species (Cui, 1997; van Munster et al., 1999).

FIGURE 1.6   Photomicrographs of Feulgen-stained erythrocytes taken from (A) Siamese fighting fish (Betta splendens; 2C = 1.3 pg), (B) Chicken (Gallus domesticus; 2C = 2.5 pg), (C) Rainbow trout (Oncorhynchus mykiss; 2C = 5.2 pg), (D) African clawed toad (Xenopus laevis; 2C = 6.3 pg). (E) Leopard frog (Rana pipiens; 2C = 13.4 pg), (F) Yellow-spotted salamander (Ambystoma maculatum; 2C ≈ 60 pg), (G) Red-spotted newt (Notophthalmus viridescens; 2C ≈ 80 pg), (H) Australian (Queensland) lungfish (Neoceratodus forsteri; 2C ≈ 105 pg). Photographed at 40x magnification, scale bar = 20 μm. From Gregory (2001a), reproduced by permission (© Cambridge Philosophical Society).

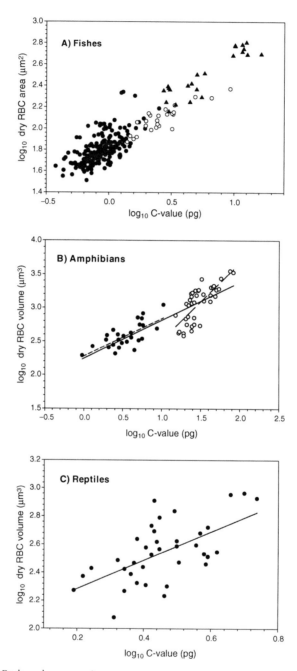

FIGURE 1.7   For legend, see opposite page.

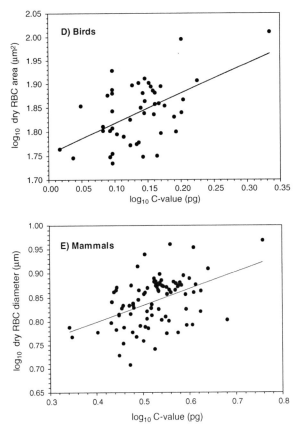

FIGURE 1.7   The strong positive correlation between red blood cell size and genome size in verte-brates. (A) The relationship between haploid nuclear DNA content and dry erythrocyte area (in μm²) in fishes, from the dataset of Hardie and Hebert (2003). This correlation holds across all fishes as well as within diploid (●) and polyploid (○) actinopterygians, and chondrychthyes (▲). (B) The relation-ship with dry cell volume (in μm³) within and across frogs (●) and salamanders (○) using data from Olmo and Morescalchi (1975, 1978), which is also significant using dry cell area and wet volume (see Gregory, 2003c). (C) The relationship with dry erythrocyte volume (in μm³) in reptiles based on the data of Olmo and Odierna (1982). (D) The relationship between genome size and dry erythrocyte area (in μm²) in birds, using cell size data compiled by Gregory (2002a). (E) The relationship between genome size and erythrocyte size in mammals, measured as mean dry diameter (in μm) using the data from Gregory (2000). The correlation also holds using wet cell volume (T.R. Gregory, unpublished), and is of particular interest because mature mammalian erythrocytes do not contain nuclei.

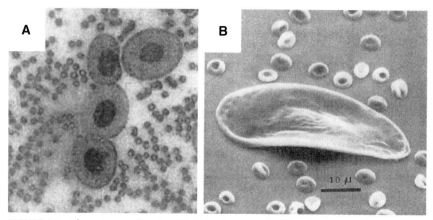

FIGURE 1.8   Photomicrographs using (A) light microscopy by Wintrobe (1933) and (B) scanning electron microscopy by Lewis (1996) showing the vast range in red blood cell sizes among vertebrates. In both figures, the large elliptical cells in the center are those of the aquatic salamander *Amphiuma means* (2C ≈ 165 pg), and the small disks surrounding them are those of humans (2C = 7.0 pg). (B) Reproduced by permission (© Kluwer Academic Publishers).

EXPLAINING THE CORRELATION

Several very different approaches have been taken to explaining the correlation between genome size and cell size. It is important to keep the distinctions between them clear, because they have very different implications for how the large-scale issue of genome size evolution is interpreted. A general overview of the major proposals is given here, with more detailed discussions of these theories available in recent reviews devoted entirely to this topic (Gregory, 2001a,c).

Coincidence: Mutation Pressure Theories

Under the traditional junk DNA and selfish DNA theories, genome size tends to increase by mutation pressure until the costs (e.g., of replication) become too high for the host cell. Cell size, in this case, is seen as being set by the action of genes, and it is assumed that species possessing larger or more slowly dividing cells can simply tolerate the accumulation of more DNA (e.g., Pagel and Johnstone, 1992; Charlesworth *et al.*, 1994). That is to say, the relationship between genome size and cell size is considered purely coincidental.

As Cavalier-Smith (1985) has pointed out, coincidental, tolerance-based explanations like these have considerable difficulty explaining why the relationship between genome and cell sizes should persist across such a wide range of variation. More seriously, no convincing reason has yet been given why tolerance for the accumulation of DNA should scale in a relatively linear way with cell size over even a small range (see also Cavalier-Smith and Beaton, 1999). This approach is also incapable of explaining why the correlation between C-value and cell size should persist following reductions in DNA content. For these and other reasons, the strictly coincidental interpretation has been rejected in more recent discussions of the subject (e.g., Beaton and Cavalier-Smith, 1999; Gregory 2001a,c).

Coevolution: The Nucleoskeletal Theory

Whereas the coincidental interpretation of the genome size–cell size correlation favored under mutation pressure theories might envision nuclei as containers to be filled by the accumulation of noncoding sequences, "optimal DNA" theories postulate a more causative link at the nuclear level. Specifically, Cavalier-Smith (1982) considers a four-step mechanism by which DNA content directly influences nuclear size:

> (a) The C-value determines the basic size range [of the nucleus]; (b) within that range the degree of folding of DNA, probably controlled by proteins, determines the actual nuclear volume of a specific cell type; (c) the protein matrix, fibrous lamina, and pore complexes are assembled in association with, and at a volume determined by, the appropriately folded DNA; (d) the phospholipid-protein membranes of the nuclear envelope attach to the fibrous lamina [see also Collas, 1998; Cavalier-Smith and Beaton, 1999].

Thus, DNA acts as a "nucleoskeleton" around which the nucleus is assembled, with the total amount (including both coding and noncoding sequences) and compaction level exerting a strong effect on the final nucleus size. Such a view is supported by trans-specific DNA injection experiments, which show nucleus-like structures to form around DNA, regardless of its source (e.g., Forbes *et al.*, 1983).

Under the "nucleoskeletal theory," this is as far as causation proceeds. Cell size is not determined by nucleus size in this case, but rather is set adaptively by genes. The correlation between nucleus (and therefore genome) size and cell size arises through a process of coevolution in which nuclear size is adjusted to match alterations in cell size (Cavalier-Smith, 1982, 1985). In its original form, the nucleoskeletal theory was based on the notion that larger nuclei were required for the increased transport of RNA out of the nucleus in response to the greater transcriptional needs of a larger cell. However, this suggestion proved implausible and the theory was subsequently revised. In the most recent formulation, the basis of the nucleoskeletal theory has switched from export to production, with the emphasis now on the needs of larger cells for greater amounts of nuclear space to meet the higher demands for proteins (Beaton and Cavalier-Smith, 1999; Cavalier-Smith and Beaton, 1999). As Cavalier-Smith and Beaton (1999) put it,

> The situation is like that of a car factory aiming for a steady output of cars: engines, wheels, and doors must be made at the same rate; if overall output is to be increased the number of each must be increased by the same proportion. Moreover, if each robot,

machine tool, and operative is already working at the maximal rate, one can increase
output only by increasing the number of assembly lines. As these take up space, the
factory also has to be larger. In a cell the nucleus is the production line for RNA molecules. To produce more per cell cycle one must have more copies of [the] production
machinery... Thus nuclei have to be larger in larger cells.

Several objections have been raised to this approach (Gregory, 2001a,c). First,
although a larger cell may indeed require a heightened level of protein production,
there is no reason to expect this to scale in direct proportion to cell size.
Moreover, there is no reason to expect higher protein production to require a
directly proportional increase in "factory" space, because other strategies like
upregulating gene transcription (that is, speeding up the assembly line rather than
building a second one) can also help to meet this need. Thus, a cell 10 times as
large, for example, might only require seven times as much protein, which can be
produced by increasing the size of the nuclear factory only 4-fold. In such a
case, the strong linear relationship between nucleus size and cell size would tend
to break down even over relatively small ranges. Second, even in cases where
nucleus/genome size and cell size are matched proportionately, there is nevertheless a negative relationship between genome size and cell growth/division rate
(i.e., "balanced growth" is still not maintained). Third, it is not clear why the sizes
of nongrowing differentiated cells such as erythrocytes, especially those without
nuclei, should correlate so strongly with genome size. Finally, this requires that
cell size variation always be considered adaptive, which, as discussed in a later
section, is a very difficult assumption to uphold in several cases.

Overall, the nucleoskeletal theory makes an important contribution by outlining
the mechanism by which DNA content can causally influence nucleus size. Its interpretation of the cell size correlation, on the other hand, seems rather problematic and
has not been widely accepted. Instead, most theorists have tended to extend the idea
of causation to the cell level as well, usually via the intermediate of nucleus size.

Causation: The Nucleotypic Theory

Based on an early comparison of cell and genome sizes in a few vertebrates,
Commoner (1964) suggested that DNA played not one, but two roles in the heredi-
tary process: first, the obvious genic role of coding for proteins, and second, a quantitative one in which bulk DNA exerts an influence on the cell's size and metabolism.
A similar view was provided a short time later by Martin (1966) in reference to plants.
However, it was not until the following decade that this notion of DNA content
causally influencing cellular parameters was developed into an explicit theory.
Referring to plants, Bennett (1971, 1972) coined the term "nucleotype" (cf. "genotype")
to describe "that condition of the nucleus [most notably, DNA content] that affects
the phenotype independently of the informational content of the DNA."

Since then, the "nucleotypic theory," which is also an optimal DNA theory, has
become the most widely implemented approach to understanding the relationship

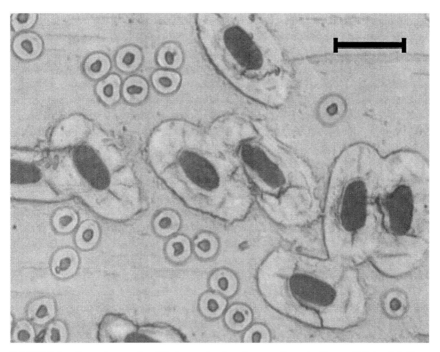

FIGURE 1.9   Photomicrographs of Feulgen-stained erythrocytes from the Siamese fighting fish (*Betta splendens*, 2C = 1.3 pg) and the Australian lungfish (*Neoceratodus forsteri*, 2C ≈ 105 pg), which has a genome roughly 100 times larger. This example provides a rather extreme demonstration of the necessary relationship between genome size and erythrocyte size, given that the nuclei of the lungfish could not physically fit inside the cells of the Siamese fighting fish. Photographed at 40x magnification, scale bar = 20 μm. From Gregory (2001c), reproduced by permission (© Elsevier Inc.).

between genome size and cellular/organismal features.[4] On a certain level, DNA content and cell size must be causally related, due to the physical impossibility of containing very large genomes within small cells (Fig. 1.9). In more general terms, although direct evidence in favor of the theory (e.g., from experimental manipulations of DNA content) is not yet available, several observations do lend considerable support to it. For example, the fact that polyploidization results in an instantaneous and proportionate change in cell size (see Chapters 7 and 8) is difficult to reconcile with the other theories outlined above. Similarly, diploid laboratory hybrids tend to show intermediate genome and cell sizes relative to

---

[4]It bears mentioning that after 1978 Cavalier-Smith rejected the nucleotypic theory in favor of the coevolutionary nucleoskeletal approach. It is therefore inaccurate to cite his later publications as suggesting a *causative* link between genome size and cell size, as unfortunately is often done.

their parental species, again suggesting a causative link between the two parameters. By way of example, this has been found with kidney tubule cells in laboratory hybrids of rodents in the genus *Phyllotis* (Walker *et al.*, 1991) and with erythrocytes in the more familiar case of mules (*Equus caballus* × *E. asinus*) (Vialli and Gerzeli, 1955; Hawkey, 1975). Positive associations between DNA content and cell volume among hybrid fungi have been reported as well (Kuldau *et al.*, 1999).

A few explanatory models have been proposed to account for the nucleotypic effect over the past several decades, but none has proved satisfactory (reviewed in detail in Gregory, 2001a). As Nurse (1985) pointed out, "cell size is determined by an interaction of the function of specific genes with the total DNA content of the cell," and that "such an interactive system can be best understood in terms of cell cycle controls which coordinate progress through the cell cycle with an increase in mass." In keeping with this, an explicit "gene-nucleus interaction model" was developed recently (Gregory, 2001a). This model is based on the strong negative correlation between genome size and cell division rate, which has been best studied in plants (see Chapter 2), but has also been demonstrated in amphibians (Grosset and Odartchenko, 1975a,b; Horner and Macgregor, 1983; Vinogradov, 1999b).

The details of the cell cycle control system are still being worked out, but it is becoming clear that the key regulatory molecules are the proteins known as cyclin-dependent kinases (CDKs), which are activated by cyclin molecules and play various roles in initiating and regulating DNA replication. Whereas CDKs remain intact for long periods of time, cyclin molecules must be synthesized anew in each cell cycle. As such, any mechanisms that slow the production or accumulation of cyclin molecules can delay the onset of cell division and, because growth continues throughout the cell cycle, result in larger daughter cells. Under the "gene-nucleus interaction model," bulk DNA influences the space-filling requirements for cyclins in larger nuclei, and possibly even the influx of regulatory proteins (and therefore cyclin gene expression) owing to effects on nuclear surface area to volume ratios and/or the arrangement of chromatin within the nucleus. In addition, larger amounts of DNA may prolong the DNA synthesis phase (S-phase) and/or the timing of replication initiation (Gregory, 2001a). Importantly, and unlike the coincidental and coevolutionary approaches, such a nucleotypic model is applicable to all cell types, including those in which the nuclei are ejected during final differentiation.

GENOME SIZE AND ORGANISMAL PHENOTYPES

Cell size has long been known to influence key physiological parameters (e.g., Smith, 1925). As cells become larger, the ratio of surface area to volume decreases, and this can have important effects on respiratory gas, ion, and protein exchange rates. Even the transcription of certain genes can be influenced by changes in cell size (e.g., Pritchard and Schubiger, 1996; Lang and Waldegger, 1997; Waldegger and Lang, 1998). It is also a testament to the importance of cell size that, in both mammals and birds, the volume of red blood cells is a such carefully regulated parameter that it can been used as an indicator of health (e.g., Lang and Waldegger, 1997; Bearhop *et al.*, 1999).

At the level of organismal phenotypes, any feature that is impacted by cell size and/or division rate can be affected indirectly by genome size. As Macgregor (1993, p. 11) explained,

> [S]ince genome size is rather well related to cell volume (the bigger the genome the bigger the cells), it is clear that genome size must have considerable developmental significance. The ancient Egyptians built their monuments with blocks of stone the size of a modern motor car. Today we build with blocks of stone the size of a modern motor car. Today we build with blocks of stone the size of a modern motor car. Today we build with 8 x 4 inch bricks. The building techniques are correspondingly different and the buildings themselves reflect the nature of the pieces used to construct them. The fruit fly *Drosophila melanogaster* and the crested newt *Triturus cristatus* provide a close analogy, with genome sizes of [0.18] pg and 20 pg respectively.

The most obvious potential correlates at the organismal level are body size, metabolic rate, and developmental rate, and indeed such relationships are found in animals. However, the nature and existence of the relationships may vary considerably according to the biology of the organisms in question, further emphasizing the complexity of the C-value enigma.

Body Size

In many taxa, variation in body size is primarily a product of differences in cell number rather than cell size. In mammals, for example, the 70 million-fold range in body mass from the tiniest pygmy shrews (e.g., *Microsorex hoyi*, *Suncus etruscus*) and bumblebee bats (*Craseonycteris thonglongyai*) to the monstrous blue whale (*Balaenoptera musculus*) clearly cannot be attributed in any major way to differences in cell size. There is a positive correlation between genome size and body size across mammals at large at the species and genus levels, but this does not hold at higher levels (Gregory, 2002b). Moreover, this relationship does not hold within any orders except rodents (Gregory, 2002b). In birds, there is a more general positive correlation with body size (Gregory, 2002a).

Contrary to most other issues related to genome size, the potential relationship with body size has been best studied in invertebrates. Positive correlations have been reported with wing size in mosquitoes and fruit flies (Ferrari and Rai, 1989; Craddock *et al.*, 2000), and with overall body size in polychaete annelids (Soldi *et al.*, 1994), turbellarian flatworms (Gregory *et al.*, 2000), copepod crustaceans (McLaren *et al.*, 1988; Gregory *et al.*, 2000), and aphids (Finston *et al.*, 1995). Thus, the relationship may persist in both hard- and soft-bodied organisms. In cases where growth is determinate with cell number constancy, a change in DNA content/cell size is expected to exert a particularly notable influence on body size, and indeed this is the case in copepods (McLaren and Marcogliese, 1983). In nematode worms, which also have cell number constancy, it is not genome size variation per se that affects body size, but rather the level of tissue-specific genome duplication ("endopolyploidy") reached in hypodermal nuclei (Flemming et al., 2000).

The correlation between genome size and body size is not ubiquitous among invertebrates, however, and several studies have failed to find any connection between them. For example, although Simonsen and Kristensen (2003) suggested that genome and cell size may influence scale and wing size in moths, there does not appear to be any such link (Gregory and Hebert, 2003). There also does not seem to be any correlation in oligochaete annelids (Gregory and Hebert, 2002). In beetles, no correlations have been found within any of the reasonably well-studied families (i.e., Tenebrionidae, Chrysomelidae, and Coccinellidae) (Juan and Petitpierre, 1991; Petitpierre and Juan, 1994; Gregory et al., 2003). On the other hand, negative correlations have been reported with body size within the genus Pimelia (Palmer et al., 2003) and intraspecifically within Phylan semicostatus (Palmer and Petitpierre, 1996). However, in the former case the relationship is only significant following complex phylogenetic correction when several sub-species are included (i.e., there is no direct correlation), and in the latter the absolute difference in genome size is so small (0.04 pg) that it is unlikely to be of real biological (versus statistical) significance.

METABOLISM

The potential association between genome size and metabolic rate has not been studied in any groups outside of the tetrapod vertebrates, but even here it is evident that the nature of the relationship varies considerably according to organismal biology. At the level of the tissues, the sizes of individual cells can potentially exert a significant influence on aerobic metabolism via effects on both the respiring tissues themselves, in which processes such as ion balance can consume energy in a manner proportional to cell size, and on the surface area to volume ratios, and thus capacity for efficient gas exchange, of the cells responsible for providing the tissues with oxygen, namely erythrocytes (Szarski, 1983; Gregory, 2001a).

Szarski (1970) was perhaps the first to emphasize the general observation that "small cells and a small amount of DNA in the nucleus characterize groups with a high metabolism," which he later developed into the concept of "wasteful" versus "frugal" metabolic strategies (Szarski, 1983). Wasteful metabolisms are those that involve a high level of oxygen consumption and heat production, as found in the endothermic classes (mammals and birds). At the opposite extreme are the very low, and therefore frugal, oxygen consumption rates of aquatic amphibians and lungfishes. It has proved very tempting to attribute much of the variation in genome size among vertebrates to differences in metabolic (and hence, cell size) requirements, but it is becoming increasingly clear that such a one-dimensional explanation is far too simplified.

Mammals and Birds

The high metabolic requirements of endothermy are reflected by the small erythrocytes typical of both mammals and birds. The human red blood cell (RBC), for example, has a volume of about 90 μm^3 and a mass of roughly 95 pg (Albritton, 1952). By comparison, in some amphibians and lungfishes the erythrocyte nuclear DNA content alone is nearly twice as high as this (Fig. 1.8). According to Cavalier-Smith (1985), "strong stabilizing selection for optimal red cell volume is a major selective force that maintains a relatively uniform cell volume in mammals and birds (and secondarily causes the uniformity in C-values)." Moreover, it would seem that mammals were only able to evolve larger genome sizes than other amniotes because the enucleation of their erythrocytes allows small cells to coexist with higher C-values (Cavalier-Smith, 1978; Gregory, 2000).

In 1995, Vinogradov reported a significant inverse correlation between genome size and mass-corrected basal metabolic rate (i.e., oxygen consumption rate) in a variety of mammal species, which also holds within rodents taken separately. Vinogradov (1995) also provided results suggestive, but not demonstrative, of a negative relationship between genome size and mass-corrected resting metabolic rate across a relatively small sample of bird species, and later found such a relationship within a much more restricted set of passerines (Vinogradov, 1997). A more recent comparison across 50 avian species revealed that the correlation applies to birds as well as to mammals (Gregory, 2002a) (Fig. 1.10A). Kozlowski et al. (2003) have since developed a model in which genome size and cell size play a key role in influencing the allometric scaling of metabolic rate and body size.

It is interesting to note that in both mammals and birds, the relationship between genome size and metabolic rate persists even in the absence of the more extreme data that could be included in such an analysis. In mammals, very little is known of the genome sizes of groups with very low metabolisms, such as elephants, sloths, and cetaceans. In birds, representatives of the low end of the metabolic spectrum (e.g., ostrich and emu) are included, but the most metabolically active groups like hummingbirds have yet to be studied from the perspective of genome size.

Given the very high metabolic demands involved, it has been argued that powered flight imposes particularly strong constraints on cell and genome sizes. Notably, some of the smallest genomes among mammals are found in bats, which along with birds are the only truly volant vertebrates. In both bats and birds, there appear to be mechanisms at work to restrict genome growth. For example, repetitive rDNA genes, C-band heterochromatin, and microsatellites are all rarer in bat

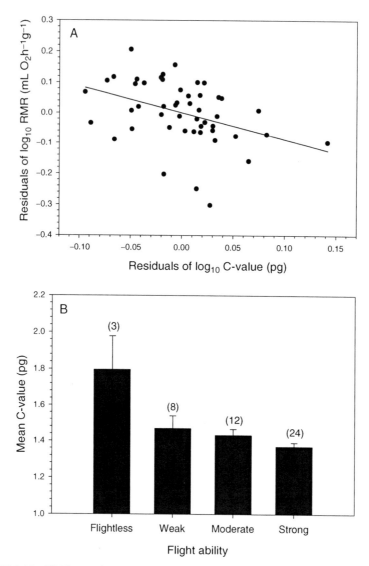

FIGURE 1.10 (A) The significant negative mass-corrected relationship between resting metabolic rate (oxygen consumption in mL $O_2h^{-1}g^{-1}$) and haploid DNA content (C-value, in pg) in birds based on the data from Gregory (2002a). This is a plot of the regression residuals of both parameters versus body mass. (B) Mean C-values for bird families categorized according to flight ability as flightless, weak, moderate, and strong as originally done by Hughes (1999), but with a much larger dataset. Numbers in brackets indicate the number of families included in each category and error bars represent standard error.

genomes than in those of other mammals (Baker *et al.*, 1992; Van Den Bussche *et al.*, 1995), and microsatellites and other repetitive DNA elements are even more sparce in avian genomes (Eden *et al.*, 1978; Epplen *et al.*, 1978; Wagenmann *et al.*, 1981; Venturini *et al.*, 1987; Bloom *et al.*, 1993; Primmer *et al.*, 1997). Neff and Gross (2001) argued that the rarity of microsatellites does not result from flight-related constraints on avian genome sizes because microsatellites may still be common in teleosts with small genomes. However, this does not preclude the possibility that genome size is actually under strong selection in both groups, with a small C-value simply achieved by different routes in the two instances.

As Hughes (1999) demonstrated, strong-flying birds have significantly smaller genome sizes than weak-flying and flightless birds (Fig. 1.10B). Indeed, when viewed in a phylogenetic context, it appears that every time flight ability has been reduced or lost, genome size has tended to increase (Hughes, 1999). There is evidently a link between small genome size and flight in birds, but this leaves open the question as to which of these two came first (Tiersch and Wachtel, 1991; Wachtel and Tiersch, 1993; Gregory, 2002a; Waltari and Edwards, 2002). It has been pointed out many times that introns (but not exons) are significantly smaller in chickens than in humans (Duret *et al.*, 1995; Hughes and Hughes, 1995; Oliver and Marín, 1996; Deutsch and Long, 1999; Hughes, 1999; Waltari and Edwards, 2002). On this basis, it has been proposed that a gradual reduction in genome size had occurred along with or after the evolution of flight (Hughes and Hughes, 1995; Hughes, 1999). By contrast, Vinogradov (1999a) suggested that rodents have even smaller introns on average than chickens, and that humans are therefore not representative of mammals in general (although this may be restricted to GC-rich introns; Oliver and Marín, 1996). As Vinogradov (1999a) pointed out, the more important comparison is between birds and reptiles, rather than mammals. In a direct comparison between chicken and alligator intron sizes, Waltari and Edwards (2002) provided evidence to suggest that a reduction in intron sizes occurred *prior to* the evolution of flight in birds. On the other hand, Holmquist (1989) argued that the paucity of short interspersed nuclear elements (SINEs) (see Chapter 3) in avian but not reptilian genomes implies a secondary loss in birds. Simply put, the currently available data are not sufficient to determine conclusively whether the genomic baggage of the flighted vertebrates has been lost or was never loaded in the first place (Gregory, 2002a).

Amphibians and Reptiles

Individual erythrocytes isolated from different amphibian species have been reported to vary in their cellular metabolic rates according to size (e.g., Goniakowska, 1970, 1973; Monnickendam and Balls, 1973), and in a comparison of seven species (including both frogs and salamanders), Smith (1925) reported an inverse correlation between dry cell area and whole-animal metabolic rate (measured as CO_2 output). Vernberg (1955) suggested that salamanders with relatively active lifestyles possess smaller and more numerous erythrocytes than more lethargic types. Taken together, these findings would seem to suggest that cell (and genome) size are likely to correlate with metabolic rate in amphibians, in keeping with Szarski's (1983) hypothesis that a "frugal" metabolic strategy is achieved via a large genome.

On the other hand, comparisons of related diploid and polyploid taxa have failed to show any differences in metabolic parameters (e.g., Kamel *et al.*, 1985; Licht and Bogart, 1990) (see Chapter 8). In terms of genome size itself, Licht and Lowcock (1991) found a significant inverse correlation with metabolic rate only at certain temperatures in salamanders (15 and 25°C, but not 5 or 20°C). These higher temperatures are above the normal operating range for the salamanders studied, and moreover the correlation at 15°C was dependent on the inclusion of data from *Necturus maculosus* and *Amphiuma means*, two aquatic urodeles with very large genomes. In a subsequent study, Gregory (2003c) used a larger dataset to evaluate the relationship within and between frogs and salamanders, and found that although there is indeed a strong negative correlation across amphibians as a whole, this is based entirely on the difference between the motile frogs with smaller genomes and the more sessile salamanders with large genomes (Fig. 1.11). Within frogs, there were no significant correlations with resting metabolic rate at any temperature in either adults or larvae. In salamanders, temperature- and taxon-dependent correlations similar to those reported by Licht and Lowcock (1991) were noted with regard to resting metabolic rate (i.e., only at 15 and 25°C, again dependent on *Necturus* and *Amphiuma*), whereas none were found with active or larval metabolic rate.

The same basic patterns can be found when comparing cell size and metabolic rate in amphibians, with a strong correlation across the class but not within either major order (Gregory, 2003c). It would therefore appear that neither cell nor genome size plays a significant role in influencing oxygen consumption in this group. This is in accordance with the observation that even the near-total obliteration of erythrocytes in bullfrogs appears to have minimal effects on organismal survivability (Flores and Frieden, 1968), whereas this surely would be fatal for a mammal or a bird. Viewed from the opposite perspective, these results indicate that selection for a frugal metabolism probably has little bearing on the patterns of genome size variation within either frogs or salamanders. This observation from salamanders refutes a major claim of the nucleoskeletal theory, namely that "their exceptionally large genomes are purely the result of selection for exceptionally large cells" (Cavalier-Smith, 1991).

The common suggestion that aquatic salamanders and lungfishes have exorbitant genomes as adaptations for estivation during dry seasons (e.g., Ohno, 1974; Cavalier-Smith, 1991; Gregory and Hebert, 1999) must also be rejected following a more detailed examination of metabolic patterns in the Amphibia. Specifically, some species of estivating frogs with very small genomes are known to reduce their metabolic rates to a low level comparable to that of aquatic salamanders with

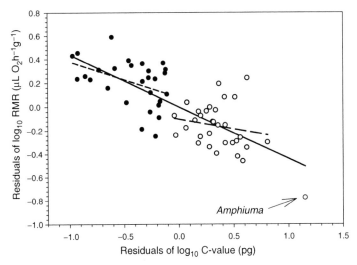

FIGURE 1.11   The relationship between mass-corrected resting metabolic rate (oxygen consumption in mL $O_2h^{-1}g^{-1}$ at 15°C) and haploid DNA content (C-value, in pg). This is a plot of the regression residuals of both parameters versus body mass. Across all amphibians, there is a strong negative correlation (solid line), but this results entirely from the difference between the motile frogs (●) and the sessile salamanders (○). Within frogs (short-dashed line), the correlation is nonsignificant, and in salamanders it is nonsignificant if the single value for *Amphiuma means* is removed (long-dashed line). The relationship is also temperature-dependent. Unlike the case with birds and mammals, metabolic rate does not appear to be very important to genome size evolution within either major group of amphibians. See Gregory (2003c) for sources of data and detailed discussion.

C-values 75 times as large (Pinder *et al.*, 1992). It is therefore extraordinarily unlikely that selection would favor the accumulation of so much noncoding DNA for this purpose (Gregory, 2002c).

The first broad comparison of genome size and metabolic parameters in reptiles was conducted recently by Olmo (2003), who considered oxygen consumption at several temperatures (20, 30, and 35°C), lactic acid production (in milligrams per gram), and field metabolic rate (in kilojoules [kJ] per day). No correlations were found at the species level with any of these features, but significant negative correlations were observed between genome size and O_2 consumption and lactic acid production at the suborder level and with field metabolic rate at the family level. However, the sample sizes were quite small in these analyses ($n = 4$, 3, and 9, respectively), and the suborder-level relationships with oxygen consumption at all three temperatures were clearly anchored by a single datum. Thus, in both amphibians and reptiles, the general pattern is that relationships are only apparent across major taxonomic divisions within the classes, and probably do not play a significant role in affecting species-level patterns in either group.

DEVELOPMENTAL RATE

Organismal growth and the differentiation of tissues are both influenced by cell size and cell division rate, and for this reason it often has been proposed that developmental rate should correlate with genome size in many taxa. Numerous examples are known from plants, in which developmental rate may be correlated either positively or negatively with DNA content, depending on the source of DNA (e.g., polyploidy versus genome size change) and limiting environmental conditions such as temperature and water availability (see Chapter 2). Developmental rate correlations have also been reported in certain animals but appear to be absent from others.

Invertebrates

Genome size has been reported to correlate negatively with overall developmental rate in a small sample of beetles in the genus *Tribolium* (Carreras *et al.*, 1991), and with pupal development in ladybird beetles (Gregory *et al.*, 2003). More generally in leaf beetles, Petitpierre and Juan (1994) noted that species with one generation per year possessed C-values greater than 0.6 pg, whereas those with multiple generations had genome sizes smaller than 0.5 pg. It similarly has been suggested that the rapid life cycles of aphids are linked to their small genome sizes (Ma *et al.*, 1992; Gregory, 2002c). An inverse correlation between genome size and developmental rate has also been found in copepod crustaceans (McLaren *et al.*, 1988; White and McLaren, 2000), which may be inescapable due to their programs of determinate growth. Polychaete annelids inhabiting harsh interstitial environments display smaller C-values than macrobenthic species, which is believed to relate to selection for rapid development and small body size (Soldi *et al.*, 1994; Gambi *et al.*, 1997).

Amphibians

Although clear-cut relationships with metabolic rate have not been found in amphibians, significant negative correlations with developmental rate have been reported several times in both frogs (Goin *et al.*, 1968; Bachmann, 1972; Oeldorf *et al.*, 1978; Camper *et al.*, 1993; Chipman *et al.*, 2001) and salamanders (Pagel and Johnstone, 1992; Jockusch, 1997; Gregory, 2003c). This includes embryonic, larval, and total developmental rate, and also extends to limb regeneration rate in salamanders (Sessions and Larson, 1987).

In their oft-cited analysis, Pagel and Johnstone (1992) reported that when controlled for nucleus and cell size, genome size and hatching time remained significantly correlated, and that when controlled for hatching time, genome size was no longer correlated with cell size. Together, these results were taken to indicate that the most important relationship is between genome size and hatching time (an inverse measure of developmental rate), and not cell size, suggesting that slower-developing animals can tolerate more DNA and do not adjust their genome sizes to compensate for shifts in cell size, or what Pagel and Johnstone (1992) considered "direct support for the view that the nuclear genomes of eukaryotes accumulate junk DNA until the costs to the organism of replicating it become too great." As pointed out a decade later, however, there were several problems with Pagel and Johnstone's (1992) dataset and analyses, and overall the best-supported conclusion is the nucleotypic one under which a larger DNA content causally increases cell size and slows developmental rate in salamanders (Gregory, 2003c).

Fishes

Even though the largest fraction of the available animal genome size data come from fishes, for the most part correlations between genome size and phenotypic traits have not been well studied in this group. In terms of developmental rate, Arkhipchuk (1995) reported a positive correlation between genome size and the shortest length of the reproductive cycle across 45 species. However, this correlation was very weak, and should be reexamined using a much larger dataset. Other parameters such as embryonic development time have yet to be evaluated with regard to genome size variation and may be worth examining in detail.

Amniotes

There has been some disagreement as to whether relationships with development can be found in the amniote classes. For example, John and Miklos (1988) dismissed the possibility of such a correlation in mammals by pointing out that "while genome size shows little variation in mammals, developmental time is extraordinarily variable." Similarly, Hughes (1999) reported that no correlation could be found with fledging time in a sample of 30 bird families. On the other hand, Monaghan and Metcalfe (2000) reported a significant positive relationship between genome size and development time (the inverse of rate) in a similar family-level comparison. A subsequent reanalysis of Monaghan and Metcalfe's (2000) data by Morand and Ricklefs (2001) revealed a significant positive relationship between genome size and development time at the species level.

In the most detailed study conducted thus far, Gregory (2002b) used large datasets to compare genome size with gestation time, lactation time, time to eye opening, "total development time" of young (gestation plus time to eye opening), and "total care period" (gestation plus lactation) in mammals, as well as incubation time, fledging time, "total development time" (incubation plus fledging), and growth rate constants taken from Starck and Ricklefs (1998) in birds. None of these developmental parameters is significantly correlated with C-value at any taxonomic level. The single exception is within the rodents, where developmental rate does appear to be inversely correlated with genome size (Gregory, 2002b).

The same lack of association may apply to reptiles: although it was suggested that the relatively rapid development necessary when eggs are deposited in soil (which maintains a suitable temperature for only a short time) might place constraints on genome size (Olmo, 1991), no correlations have been found between C-value and incubation time in turtles or squamates (Olmo, 1983, 2003). It is interesting to note that endotherms and amphibians are polar opposites in terms of organism-level correlations, with metabolism but not development being important in the former and the reverse being the case in the latter. Reptiles appear to fall somewhere in between these extremes (Olmo, 2003).

DEVELOPMENTAL COMPLEXITY

Developmental rate provides a measure of how quickly morphological differentiation takes place. When comparing this parameter with genome size, there is an implicit assumption that the amount of developing to be done is roughly equivalent among the species under study. However, this is not always the case, and it is therefore worthwhile to consider the flip side of rate on the developmental coin, namely developmental *complexity*. This is the amount of developing to be done, and is the relevant parameter for comparison with genome size when the time available for development is held constant.

There are many reasons that the time available for development may be limited. As discussed in Chapter 2, plants inhabiting climates with short growing seasons tend to have small genome sizes. The same is true of desert-dwelling frogs, which must complete their entire developmental program in the span of a short rainy season. Metamorphosis also represents a strongly time-limited period of intensive morphological differentiation, and appears to impose major constraints on amphibian genome sizes. Thus, these rapidly metamorphosing desert frogs have the smallest genomes in the class, followed by direct-developing and normally metamorphosing (biphasic) frog species. Frog metamorphosis is much more intensive than that of salamanders, and again there is no overlap in genome size between the two groups. Within the salamanders, those with a biphasic life cycle have the smallest genomes, followed by direct-developers. Facultative neotenes, which can eschew metamorphosis under certain conditions, have larger genomes, whereas some of the largest animal genomes known are found in obligate neotenes that never metamorphose (Fig. 1.12). It seems that large genome size and neoteny have coevolved on three separate occasions in this latter group (Gregory, 2002c).

There is some indication that a similar pattern applies to insects. In this case, metamorphosis can be complete (holometabolous development, including distinct

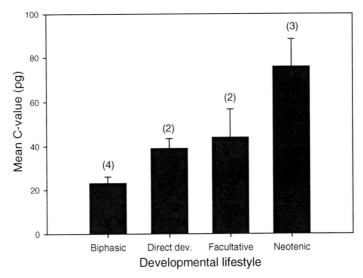

FIGURE 1.12   Mean C-values for salamander families categorized according to developmental lifestyle. Groups with a biphasic developmental program (i.e., with normal metamorphosis) have the smallest genomes, although these are still larger than all frog genomes. Direct-developers have larger average C-values, followed by facultative neotenes that may occasionally metamorphose, and finally by obligate neotenes, which have some of the largest genomes among animals. Numbers in brackets indicate the number of families included in each category and error bars represent standard error.

larval, pupal, and adult stages), incomplete (hemimetabolous, with a series of nymphal molts only), or totally absent (ametabolous, with simple growth from juvenile to adult). Based on the relatively small dataset then available, Gregory (2002c) suggested that species with complete metamorphosis are restricted to genome sizes of 2 pg or less, whereas groups with hemimetabolous or ametabolous development frequently exceed this threshold. With only a few small exceptions, this pattern has held up well with the addition of nearly 250 new insect C-values covering several more orders (Fig. 1.13). Although this hypothesis does not imply that species lacking complete metamorphosis will necessarily have large genomes, it is nevertheless interesting to consider why some of them do not. Aphids (order Hemiptera), for example, are probably under additional constraints related to body size and basic developmental rate (Ma et al., 1992; Finston et al., 1995; Gregory, 2002c). Mayflies (order Ephemeroptera) are particularly interesting because they undergo a very large number of nymphal molts (between 15 and 25 in most species) and are the only hemimetabolous order to undergo a molt (from "subimago" to "imago") as adults (Brittain, 1982). So, while these are not strictly "exceptions," they are interesting counterexamples that lend support to the general rule.

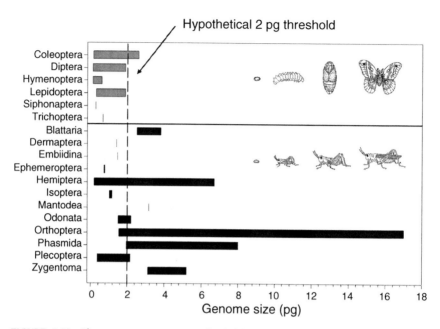

FIGURE 1.13 The apparent genome size threshold associated with complete metamorphosis in insects. The holometabolous orders (those with complete metamorphosis) are shown in the top portion, and those with ametabolous (no metamorphosis) or hemimetabolous (incomplete metamorphosis only) development are shown at the bottom. The 2 pg threshold proposed for orders with complete metamorphosis (Gregory 2002c) seems to hold well, with only one or two beetles (order Coleoptera) out of more than 250 studied reported to exceed it. Data from the *Animal Genome Size Database* and T.R. Gregory (unpublished).

ORGAN COMPLEXITY

In addition to correlations with the complexity of the developmental process, amphibians demonstrate some intriguing relationships with the complexity of developmental products—that is, certain morphological features. This is not to be confused with organismal complexity, which is not related to genome size in eukaryotes, but rather involves the characteristics of specific organs. As with the developmental process, this effect is observed when other parameters such as physical space or available time are limited. Such conditions are most clearly met in brain tissue, which involves substantial cell division and differentiation in a limited space (i.e., the skull). As a result of its causal influence on neuron size and division rate, genome size is inversely correlated with brain complexity in both frogs and salamanders (Roth and Schmidt, 1993; Roth *et al.*, 1994, 1997).

This effect on the brain is especially pronounced in the miniaturized salamanders of the plethodontid tribe Bolitoglossini, in which body (and therefore braincase) sizes have been drastically reduced but genomes are large (these salamanders are direct-developers). In fact, the effect on the visual centers of the brain is so severe that the active predation strategy typical of plethodontids is no longer possible, such that a switch has been made to a lie-in-wait strategy. Along with this ecological shift has come the evolution of highly specialized projectile tongues (Roth and Schmidt, 1993; Roth *et al.*, 1997).

The combination of body miniaturization and genome expansion has other interesting biological consequences. As with their neurons, the erythrocytes of these salamanders are very large but the spaces available to contain them (i.e., blood vessels) are very small. In keeping with this, several species independently have evolved enucleated erythrocytes, a feature otherwise ubiquitous only among mammals (Emmel, 1924; Cohen, 1982; Villolobos *et al.*, 1988; Mueller, 2000). As in mammals, the ejection of nuclei allows the production of smaller cells for a given DNA content (Fig. 1.14). However, whereas in mammals the enucleation of erythrocytes is thought to be an adaptation for efficient gas exchange, cell size does not correlate well with metabolic rate in salamanders. Villolobos *et al.* (1988) hypothesized that, in these salamanders, "enucleation is the result of random cell breakage in circulating blood," but a much more intriguing possibility is that

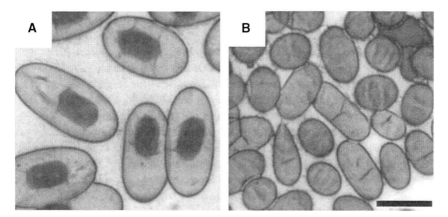

FIGURE 1.14   Photomicrographs of (A) nucleated erythrocytes of the newt *Notophthalmus viridescens* and (B) enucleated erythrocytes from the miniaturized plethodontid salamander *Batrachoseps attenuatus*. Enucleation allows the development of much smaller cells in *Batrachoseps*, even though the two species have similar DNA contents (2C ≈ 40 pg). The same is true of erythrocyte enucleation in mammals, but in that case it probably relates to the physiological constraints of a high metabolism (Gregory, 2000), rather than the possible physical constraint of circulating large cells through small blood vessels (R.L. Mueller, in prep.). Both photographs taken at 63x magnification, scale bar = 20 μm.

enucleation in this case is an adaptive response to *physical*, rather than *physiological*, constraints (R.L. Mueller, in prep.).

LONGEVITY

One of the most controversial potential correlates of genome size discussed to date is longevity. A positive relationship has been reported at the family level in birds (Monaghan and Metcalfe, 2000), within a select sample of fishes (Griffith *et al.*, 2003), and at the suborder level in reptiles (Olmo, 2003). The interactions between longevity and body size, metabolic rate, and development (Ricklefs and Scheuerlein, 2003) may provide a basis for a correlation with cell and genome size. On the other hand, flight involves low natural mortality and therefore presumably favors investment in antisenescence adaptations, which would explain why birds have longer mass-corrected life spans than mammals, why ostrich and emu are much more short-lived in relative terms than other birds, and why bats live longer than rodents of similar size (Partridge and Barton, 1993). In this sense, one might actually expect to find a *negative* correlation between genome size and longevity in mammals and birds.

In birds, Ricklefs and Scheuerlein (2001) found no association between genome size and senescence parameters, with or without correction for variation in development times. Based on a reanalysis of Monaghan and Metcalfe's (2000) data, Morand and Ricklefs (2001) found no correlation with longevity at the species level, and noted that "the significant relationship at the family level represents only a small amount of the total variance (26% for relative life span and 6% for genome size)." Using a much larger dataset, Gregory (2002b) found no correlations with maximum life span at any taxonomic level in either birds or mammals. There also do not appear to be any such correlations within any of the mammalian orders for which sufficient data are available for analysis (Artiodactyla, Carnivora, Primates, Rodentia) (Gregory, 2002b; S. Morand and R.E. Ricklefs, in prep.).

In fishes, the dataset included species from the very distantly related teleosts and chondrosteans, and most of these were from groups in which there is the confounding factor of polyploidy. In fact, in sturgeons (the only group within which a significant correlation was found), the relationship is negative if both body size and chromosome number (i.e., ploidy level) are controlled for, as is the overall correlation within teleosts (Gregory, 2004c). Civetta *et al.* (2004) claim that the correlations across the entire dataset and within sturgeons are positive regardless of ploidy level, but their results can be rejected because they did not log-transform the highly skewed data. In any case, this is based on a very small dataset, and would need to be confirmed with a much broader analysis. Likewise, in reptiles there was no significant relationship across 126 species from various taxa, whereas the significant positive correlation at the suborder level involved only six data points (Olmo, 2003). In conclusion, there do not appear to be any convincing correlations between genome size and longevity in any of the vertebrate groups examined to date.

MEASURING ANIMAL GENOME SIZES

The construction of genome size databases for both animals and plants has allowed broad analyses that would be impossible if all the measurements had to be done by a single set of authors, but it must be borne in mind that both databases contain an appreciable degree of error related to the inclusion of data from hundreds of different studies. In order to improve the consistency and reliability of genome size estimates in the future, both botanists and zoologists have begun developing best practice guidelines. Those for plants are discussed in some detail in Chapter 2, and because most of these also apply to animals, only a brief review of the methodology and some considerations specific to animal studies are provided here.

During the early stages of genome size study, DNA contents were commonly estimated by biochemical means. One such method involved the chemical extraction and quantification of DNA, which was combined with cell counts to give an average DNA amount per nucleus. Another common technique was reassociation kinetics, in which the DNA molecule was denatured and then the time taken for the strands to reanneal used to calculate the amount (and repetitiveness) of DNA present. Finally, some studies analyzed the bulk fluorescence of a sample of cells in suspension, and then again relied on cell counts to get an average per genome. Such methods account for a notable fraction of the existing dataset, even though they have not been in common usage for several decades. However, the great majority of genome size estimates in both animals and plants have come from two main methods: flow cytometry and Feulgen densitometry.

FLOW CYTOMETRY

Initially used for cell counting, and later adapted for identifying the anomalous DNA contents often associated with cancer, flow cytometry has become a mainstay of genome size studies. In basic terms, the method involves staining isolated nuclei with a fluorescent dye, and then measuring the intensity of light emitted when this is stimulated with a laser or other specific light source. This process is automated within a flow cytometer, with several thousand nuclei typically measured for a given sample. To calculate absolute genome size, the peak (i.e., most common) fluorescence level of a species with an unknown genome size is compared against that from a standard species with a known DNA content, which is included in the same measurement run. The primary challenge in this case is obtaining usable (preferably fresh or frozen) material consisting of nuclei of known ploidy. In vertebrates, this usually means working with blood (leukocytes in mammals, erythrocytes in the other classes), which has the advantage that it can be sampled nondestructively and repeatedly from the same individual if necessary. In invertebrates, no such convenient tissue is available. In this case, the preferred tissue will vary from group to group (see, e.g., DeSalle et al., 2005, for arthropods). It is also necessary to choose an appropriate standard within the expected genome size range of the unknown, and to use a fluorochrome that is not base pair–specific (DeSalle et al., 2005).

The major downside to flow cytometry has traditionally been the high cost of the equipment involved, which was often only available in hospitals and large cytogenetics labs. This may soon change, however, as several new compact, relatively low-cost models have recently entered the marketplace.

FEULGEN MICRODENSITOMETRY

In terms of the underlying physics, static densitometry is the opposite of flow cytometry. Most simply, in this method the nuclei are not moving but are instead fixed onto microscope slides. In addition, this involves measuring (or rather, calculating) the amount of light absorbed by stained nuclei rather than the fluorescence emitted. Staining for densitometric purposes has long been performed using the Feulgen reaction, which involves depurinating the DNA molecule by hydrolyzing it with strong acid, and then exposing the resulting free aldehyde groups to Schiff reagent, which turns pink upon contact with them (Feulgen and Rössenbeck, 1924). The staining protocols have since been optimized for use with animal tissue, and best practice guidelines are available in recent reviews (e.g., Hardie et al., 2002; Rasch, 2003).

Because the staining within a nucleus is heterogeneous, and because individual nucleus sizes vary, it is not sufficient to take a single density measurement as might be done with a uniform solution. Instead, a series of individual "point densities" is measured, the sum of which represents the "integrated optical density" (IOD). Mean IODs of unknowns are compared against those of a standard with an established genome size, and the ratio of these used to calculate absolute genome size in the former. These point densities have traditionally been acquired either by slowly passing the nucleus through a static light beam ("scanning stage densitometry") or by moving a light beam point-by-point through a stationary nucleus ("flying spot densitometry"). The disadvantage in such methods is that they are very time-consuming and therefore do not readily permit large samples to be analyzed within or among individual animals. There is also the increasingly relevant issue that the densitometry equipment needed is becoming obsolete (see Chapter 2). So, although such methods have been very important in past genome size studies, their utility is clearly beginning to wane.

FEULGEN IMAGE ANALYSIS DENSITOMETRY

Thanks to advances in computing and imaging technology, it has been possible to not only preserve the time-tested Feulgen staining method, but also to greatly increase the efficiency of densitometric measurements. Specifically, by capturing electronic images of stained nuclei using a camera connected to a computer, and by taking each image pixel as an individual point density, large numbers of nuclei can be analyzed instantly and simultaneously. This new technique of Feulgen image analysis densitometry was first applied to cancer research, and given the life or death consequences of accurate diagnosis, was carefully scrutinized. In this capacity, it has been found that image analysis is as accurate as flow cytometry, while also being vastly more rapid than traditional densitometric methods (see Hardie et al., 2002). Image analysis software packages often have many different uses, which also means that they are likely to be more widely available than the bulky densitometers of the past.

Image analysis densitometry protocols recently have been adapted for use in genome size measurements in both animals (Hardie et al., 2002) and plants (Vilhar et al., 2001; Vilhar and Dermastia, 2002) (see also Chapter 2). Although care must be exercised when choosing material for analysis because DNA compaction levels, and hence stain uptake, vary greatly among different cell types (Hardie et al., 2002), image analysis densitometry has proven to be a versatile method that can be used with a wide variety of tissue types when an appropriate standard is chosen (Fig. 1.15). This approach has already been used in several recent surveys, and indeed, image analysis systems are expected to replace traditional densitometry equipment in the near future. As with the trends toward reduced size and cost in flow cytometers, this should permit new genome size surveys to be carried out on an unprecedented scale.

WHAT ABOUT GENOME SEQUENCING?

The accelerating pace at which the collection of complete genome sequences is growing has begun to raise questions about whether this might become a prominent method for gathering genome size data. Some authors have even suggested that an emphasis on sequencing will render the existing dataset obsolete, as when Civetta et al. (2004) argued that "studies about the evolution of genome size will certainly benefit from more accurate measures of genome size than the currently used literature collection of picogram[s] per [haploid genome]." However, this

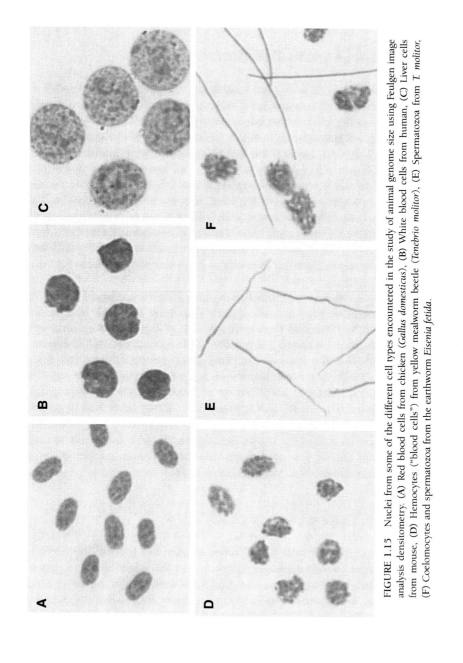

FIGURE 1.15   Nuclei from some of the different cell types encountered in the study of animal genome size using Feulgen image analysis densitometry: (A) Red blood cells from chicken (*Gallus domesticus*), (B) White blood cells from human, (C) Liver cells from mouse, (D) Hemocytes ("blood cells") from yellow mealworm beetle (*Tenebrio molitor*), (E) Spermatozoa from *T. molitor*, (F) Coelomocytes and spermatozoa from the earthworm *Eisenia fetida*.

latter view is based on a misunderstanding of some basic concepts in the study of genome size.

First, it should be obvious that any new genome size data generated by sequencing projects simply will be added to the current databases. Second, and as noted early in the chapter, the influence actually proceeds in the opposite direction, with genome size data collected by other methods playing a major role in directing the course of future sequencing programs (e.g., Evans and Gundersen-Rindal, 2003). Indeed, researchers seeking to propose new genome sequencing projects are frequent visitors to the databases because genome size information is usually a prerequisite for the relevant applications for funding.

Third, complete sequencing is a tremendously inefficient means of obtaining genome sizes, and it is very unlikely that the current (and growing) dataset of about 8000 animal and plant C-values will be supplanted by this method. Moreover, sequencing projects are unlikely to extend to genomes much larger than those of mammals for some time, which excludes many of the most interesting groups from the perspective of genome size evolution.

Finally, the high cost aside, "complete" sequencing is actually a rather inaccurate way of determining the total size of even very small eukaryotic genomes, as illustrated by the case of *Arabidopsis thaliana* (Bennett *et al.*, 2003) (see Chapter 2). So, while genome sizing and sequencing will continue to be intimately related for the reasons outlined at the beginning of this chapter, for the most part they are very likely to remain separate from a methodological standpoint for the foreseeable future.

CONCLUDING REMARKS AND FUTURE PROSPECTS

The view emphasized here is that variation in genome size, which can be generated by a variety of mechanisms, causally influences cell size via effects on nucleus size and cell division rate (Fig. 1.16). These cell-level relationships appear to be ubiquitous, and apply to several different cell types in animals, as well as within plants and protists. Cell size and division rate have the capacity to impact a variety of organismal parameters, but there are no universals at this level. In mammals and birds, effects on metabolic rate are important, probably placing strong constraints on cell and genome size (which are relaxed slightly due to erythrocyte enucleation in mammals). By contrast, developmental parameters appear to be of little relevance in endotherms. The opposite situation obtains in amphibians, in which metabolism has little bearing on genome size evolution but where variation in developmental rate, and especially developmental complexity, are of substantial import. While less well studied from this perspective, there are indications that developmental rate and/or complexity is also relevant in insects, whereas metabolism (but apparently not development) may be significant in reptiles.

These features are far from neutral, and can have cascading effects on the ecology and evolution of the species exhibiting them (Fig. 1.16). For example, small genome size may be either a consequence or a prerequisite of flight, meaning that a flighted lifestyle will be available only to those taxa capable of achieving and/or maintaining small C-values. The evolution of complete metamorphosis in insects has had a profound impact on the global fauna, and may have involved similar genome size-related requirements. The correlation with brain complexity

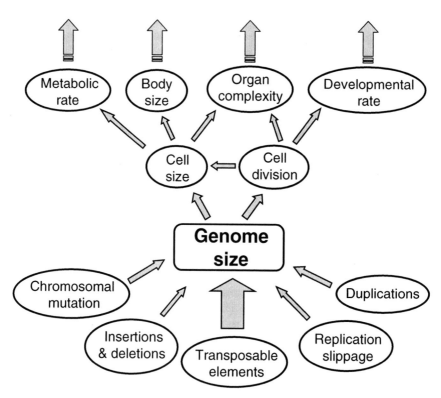

FIGURE 1.16   The cascading consequences of changes in genome size. Many different mechanisms contribute to variation in genome size, but regardless of the source, bulk DNA exerts a strong causative influence on cell size and division rate. As a result, genome size may be linked to one or more features at the organismal level, depending on the biology of the group in question. These, in turn, can have important ecological impacts. The implications of this set of interactions (and the reverse) for evolutionary theory are discussed in detail in Gregory (2004a) and Chapter 11.

in amphibians has clear ecological implications, resulting in a distinct shift in the way these animals interact with other species.

Of course, changes at the organismal level can also influence the evolution of features at the genomic level. The enucleation of mammalian erythrocytes, the loss of metamorphosis in neotenic amphibians, and the abandonment of flight in certain birds all appear to have permitted an increase in genome size. The implication is that influences proceed from both genome to organism (bottom-up) and vice versa (top-down). The net genome size observed in a given species, along with some of its key phenotypic characteristics, will be a product of this hierarchical interaction. Viewing genome size evolution in this way has some important implications for evolutionary theory, as discussed in Chapter 11.

Although major patterns such as these are now coming into focus, there is much work to be done before the C-value enigma can be resolved. Even the basic issue of the extent and patterns of variation has only been addressed in a relatively preliminary way. Many prominent groups of animals remain very poorly studied from the perspective of genome size. This is particularly true of invertebrates, which means that the majority of genome size diversity continues to go uncharacterized. With the methods now available, it should be possible to fill some of the most egregious gaps in the animal genome size dataset in relatively short order, provided that effort is invested in completing this task.

The more complex questions inherent in the C-value enigma are also in need of further study. For example, there is ongoing debate regarding the nature of the processes involved in generating genome size variation, and new mechanisms are still being discovered and proposed. Even the most fundamental phenotypic implications of variation in genome size at the cell level have only recently been addressed in key animal taxa, and remain unexamined in most others. Some important relationships have been discerned at the level of the organism, but the critical importance of differences in other biological traits indicates that this question will need to be approached on a group-by-group basis. In this area, too, invertebrates and even some major groups of vertebrates are in need of much more attention.

It can only be hoped that with a growing appreciation for the practical, as well as cytological, physiological, developmental, and even ecological and evolutionary importance of genome size, a more concerted effort will be devoted to the collection and analysis of data on a broad scale. Without such an undertaking, it is clear that the C-value enigma will continue to defy explanation and thereby to preclude a complete understanding of the evolution of the genome.

REFERENCES

Albritton EC. 1952. *Standard Values in Blood*. Philadelphia: W.B. Saunders.
Allfrey VG, Mirsky AE, Stern H. 1955. The chemistry of the cell nucleus. *Adv Enzymol* 16: 411–500.

Alvarez-Fuster A, Juan C, Petitpierre E. 1991. Genome size in *Tribolium* flour-beetles: inter- and intraspecific variation. *Genet Res* 58: 1–5.

Arkhipchuk VV. 1995. Role of chromosomal and genome mutations in the evolution of bony fishes. *Hydrobiol J* 31: 55–65.

Bachmann K. 1972. Nuclear DNA and developmental rate in frogs. *Quart J Florida Acad Sci* 35: 225–231.

Baker RJ, Maltbie M, Owen JG, *et al.* 1992. Reduced number of ribosomal sites in bats: evidence for a mechanism to contain genome size. *J Mammal* 73: 847–858.

Banerjee SK, Misra KK, Banerjee S, Ray-Chaudhuri SP. 1988. Chromosome numbers, genome sizes, cell volumes and evolution of snake-head fish (family Channidae). *J Fish Biol* 33: 781–789.

Bearhop S, Griffiths R, Orr K, Furness RW. 1999. Mean corpuscular volume (MCV) as a measure of condition in birds. *Ecol Lett* 2: 352–356.

Beaton MJ, Cavalier-Smith T. 1999. Eukaryotic non-coding DNA is functional: evidence from the differential scaling of cryptomonad genomes. *Proc R Soc Lond B* 266: 2053–2059.

Bennett MD. 1971. The duration of meiosis. *Proc R Soc Lond B* 178: 277–299.

Bennett MD. 1972. Nuclear DNA content and minimum generation time in herbaceous plants. *Proc R Soc Lond B* 181: 109–135.

Bennett MD, Leitch IJ. 2003. *Plant DNA C-values Database.* Kew, UK: Royal Botanic Gardens. www.rbgkew.org.uk/cval/homepage.html.

Bennett MD, Leitch IJ, Price HJ, Johnston JS. 2003. Comparisons with *Caenorhabditis* (~100 Mb) and *Drosophila* (~175 Mb) using flow cytometry show genome size in *Arabidopsis* to be ~157 Mb and thus ~25% larger than the Arabidopsis Genome Initiative estimate of ~125 Mb. *Ann Bot* 91: 547–557.

Bensaadi-Merchermek N, Salvado JC, Mouchès C. 1994. Mosquito transposable elements. *Genetica* 93: 139–148.

Bensasson D, Petrov DA, Zhang DX, *et al.* 2001. Genomic gigantism: DNA loss is slow in mountain grasshoppers. *Mol Biol Evol* 18: 246–253.

Biderre C, Pagès M, Méténier G, *et al.* 1995. Evidence for the smallest nuclear genome (2.9 Mb) in the microsporidium *Encephalitozoon cuniculi. Mol Biochem Parasitol* 74: 229–231.

Birnstiel ML, Chipchase M, Speirs J. 1971. The ribosomal RNA cistrons. *Prog Nucl Acids Res Mol Biol* 11: 351–389.

Black WC, Rai KS. 1988. Genome evolution in mosquitoes: intraspecific and interspecific variation in repetitive DNA amounts and organization. *Genet Res* 51: 185–196.

Bloom SE, Delany ME, Muscarella DE. 1993. Constant and variable features of avian chromosomes. In: Etches RJ Gibbons AMV eds. *Manipulation of the Avian Genome.* Boca Raton: CRC Press, 39–59.

Blumenstiel JP, Hartl DL, Lozovsky ER. 2002. Patterns of insertion and deletion in contrasting chromatin domains. *Mol Biol Evol* 19: 2211–2225.

Bobola MS, Smith DE, Klein AS. 1992. Five major nuclear ribosomal repeats represent a large and variable fraction of the genomic DNA of *Picea rubens* and *P. mariana. Mol Biol Evol* 9: 125–137.

Boivin A, Vendrely R, Vendrely C. 1948. L'acide désoxyribonucléique du noyau cellulaire dépositaire des caractères héréditaires; arguments d'ordre analytique. *C R Acad Sci* 226: 1061–1063.

Bosch R, Alvarez-Fuster A, Juan C, Petitpierre E. 1989. Nuclear DNA contents in four species of Scaraboidea (Coleoptera) from the Balearic Islands. *Abstracts Volume, International Congress of Coleopterology, Barcelona, Sept.* 18–23, 95.

Boveri T. 1887. Über Differenzierung der Zellkerne während der Furchung des Eies von *Ascaris megalocephala. Anat Anz* 2: 688–693.

Brenner S. 1998. Refuge of spandrels. *Curr Biol* 8: R669.

Brittain JE. 1982. Biology of mayflies. *Annu Rev Entomol* 27: 119–147.

Burton DW, Bickham JW, Genoways HH. 1989. Flow-cytometric analyses of nuclear DNA content in four families of neotropical bats. *Evolution* 43: 756–765.

Byers TJ. 1986. Molecular biology of DNA in *Acanthamoeba, Amoeba, Entamoeba*, and *Naegleria*. *Int Rev Cytol* 99: 311–341.

Camper JD, Ruedas LA, Bickham JW, Dixon JR. 1993. The relationship of genome size with developmental rates and reproductive strategies in five families of neotropical bufonoid frogs. *Genetics (Life Sci Adv)* 12: 79–87.

Canon NR, Tell LA, Needham ML, Gardner IA. 2000. Flow cytometric analysis of nuclear DNA for sex identification in three psittacine species. *Am J Vet Res* 61: 847–850.

Carreras I, Alvarez-Fuster A, Juan C, Petitpierre E. 1991. Tasa de desarrollo en *Tribolium brevicornis* (Cleoptera: Tenebrionidae) y su relacion con el tamaño del genoma. In *X Bienal de la Sociedad Española de Historia Natural*, Palma de Mallorca, 184.

Carvalho ML, Oliveira C, Foresti F. 1998. Nuclear DNA content of thirty species of Neotropical fishes. *Genet Mol Biol* 21: 47–54.

Cavalcanti ARO, Ferreira R, Gu Z, Li WH. 2003. Patterns of gene duplication in *Saccharomyces cerevisiae* and *Caenorhabditis elegans*. *J Mol Evol* 56: 28–37.

Cavalier-Smith T. 1978. Nuclear volume control by nucleoskeletal DNA, selection for cell volume and cell growth rate, and the solution of the DNA C-value paradox. *J Cell Sci* 34: 247–278.

Cavalier-Smith T. 1980. r- and K-tactics in the evolution of protist developmental systems: cell and genome size, phenotype diversifying selection, and cell cycle patterns. *BioSystems* 12: 43–59.

Cavalier-Smith T. 1982. Skeletal DNA and the evolution of genome size. *Annu Rev Biophys Bioeng* 11: 273–302.

Cavalier-Smith T. 1985. Cell volume and the evolution of eukaryotic genome size. In: Cavalier-Smith T ed. *The Evolution of Genome Size*. Chichester: John Wiley & Sons, 104–184.

Cavalier-Smith T. 1991. Coevolution of vertebrate genome, cell, and nuclear sizes. In: Ghiara G *et al.* eds. *Symposium on the Evolution of Terrestrial Vertebrates*. Modena: Mucchi, 51–86.

Cavalier-Smith T, Beaton MJ. 1999. The skeletal function of non-genic nuclear DNA: new evidence from ancient cell chimaeras. *Genetica* 106: 3–13.

Cavallo D, De Vita R, Eleuteri P, *et al.* 1997. Sex identification in the Egyptian vulture by flow cytometry and cytogenetics. *Condor* 99: 829–832.

Chang, H-Y, Sang T-K, Jan K-Y, Chen C-T. 1995. Cellular DNA contents and cell volumes of batoids. *Copeia* 1995: 571–576.

Charlesworth B, Sniegowski P, Stephan W. 1994. The evolutionary dynamics of repetitive DNA in eukaryotes. *Nature* 371: 215–220.

Chipman AD, Khaner O, Haas A, Tchernov E. 2001. The evolution of genome size: what can be learned from anuran development? *J Exp Zool (Mol Dev Evol)* 291: 365–374.

Civetta A, Griffith OL, Moodie GEE. 2004. Response to Gregory's Letter to the Editor: genome size and its correlation with longevity in fishes. *Exp Gerontol* 39: 861–862.

Coghlan A, Wolfe KH. 2002. Fourfold faster rate of genome rearrangement in nematodes than in *Drosophila*. *Genome Res* 12: 857–867.

Cohen WD. 1982. The cytomorphic system of anucleate non-mammalian erythrocytes. *Protoplasma* 113: 23–32.

Collares-Pereira MJ, Moreira da Costa L. 1999. Intraspecific and interspecific genome size variation in Iberian Cyprinidae and the problem of diploidy and polyploidy, with review of genome sizes within the family. *Folia Zool* 48: 61–76.

Collas P. 1998. Cytoplasmic control of nuclear assembly. *Reprod Fert Dev* 10: 581–592.

Comings DE. 1972. The structure and function of chromatin. *Adv Hum Genet* 3: 237–431.

Commoner B. 1964. Roles of deoxyribonucleic acid in inheritance. *Nature* 202: 960–968.

Conway Morris S, Harper E. 1988. Genome size in conodonts (Chordata): inferred variations during 270 million years. *Science* 241: 1230–1232.

Craddock EM, Kacmarczyk T, Dawley RM. 2000. The roles of genome size and cell size in body size variation among Hawaiian *Drosophila*. *Abstract, 41st Annual Drosophila Research Conference, Pittsburg, PA, 20–26 March*.

Cui K. 1997. Size differences between human X and Y spermatozoa and prefertilization diagnosis. *Mol Hum Reprod* 3: 61–67.

Dallas CE, Lingenfelser SF, Lingenfelser JR, et al. 1998. Flow cytometric analysis of erythrocyte and leukocyte DNA in fish from Chernobyl-contaminated ponds in the Ukraine. *Ecotoxicology* 7: 211–219.

Dasilva C, Hadji H, Ozouf-Costaz C, et al. 2002. Remarkable compartmentalization of transposable elements and pseudogenes in the heterochromatin of the *Tetraodon nigroviridis* genome. *Proc Natl Acad Sci USA* 99: 13636–13641.

Davidson JN, Leslie I, Smellie RMS, Thomson RY. 1950. Chemical changes in the developing chick embryo related to the deoxyribonucleic acid content of the nucleus. *Biochem J* 46: Proceedings, xl.

Dawkins R. 1976. *The Selfish Gene*. Oxford: Oxford University Press.

de Jong WW, Rydén L. 1981. Causes of more frequent deletions than insertions and protein evolution. *Nature* 290: 157–159.

De Smet WHO. 1981. The nuclear Feulgen-DNA content of the vertebrates (especially reptiles), as measured by fluorescence cytophotometry, with notes on the cell and chromosome size. *Acta Zool Pathol Antverp* 76: 119–167.

De Vita R, Cavallo D, Eleuteri P, Dell'Omo G. 1994. Evaluation of interspecific DNA content variations and sex identification in Falconiformes and Strigiformes by flow cytometric analysis. *Cytometry* 16: 346–350.

DeSalle R, Gregory TR, Johnston JS. 2005. Preparation of samples for comparative studies of arthropod chromosomes: visualization, *in situ* hybridization, and genome size estimation. *Methods Enzymol* (in press).

Deutsch M, Long M. 1999. Intron–exon structures of eukaryotic model organisms. *Nucleic Acids Res* 27: 3219–3228.

Devos KM, Brown JKM, Bennetzen JL. 2002. Genome size reduction through illegitimate recombination counteracts genome expansion in *Arabidopsis*. *Genome Res* 12: 1075–1079.

Dolezel J, Bartos J, Voglmayr H, Greilhuber J. 2003. Nuclear DNA content and genome size of trout and human. *Cytometry* 51A: 127–128.

Doolittle WF. 1981. Prejudices and preconceptions about genome evolution. In: Scudder GGE, Reveal JL eds. *Evolution Today: Proceedings of the Second Congress of Systematic and Evolutionary Biology*. Pittsburgh: Hunt Institute for Botanical Documentation, 197–205.

Doolittle WF. 1982. Selfish DNA after fourteen months. In: Dover GA, Flavell RB eds. *Genome Evolution*. New York: Academic Press, 3–28.

Doolittle WF. 1989. Hierarchical approaches to genome evolution. *Can J Philos* 14 (Suppl.): 101–133.

Doolittle WF, Sapienza C. 1980. Selfish genes, the phenotype paradigm and genome evolution. *Nature* 284: 601–603.

Dover G. 1990. Yes, we have no C-value paradox. *Trends Ecol Evol* 5: 62.

Duret L, Mouchiroud D, Gautier C. 1995. Statistical analysis of vertebrate sequences reveals that long genes are scarce in GC-rich isochores. *J Mol Evol* 40: 308–317.

Eden FC, Hendrick JP, Gottlieb SS. 1978. Homology of single copy and repeated sequences in chicken, duck, Japanese quail, and ostrich DNA. *Biochemistry* 17: 5113–5121.

Eisen JA. 1999. Mechanistic basis for microsatellite instability. In: Goldstein DB, Schlötterer C eds. *Microsatellites: Evolution and Applications*. Oxford: Oxford University Press, 34–48.

Emmel VE. 1924. Studies on the non-nucleated elements of the blood. II. The occurrence and genesis of non-nucleated erythrocytes or erythroblastids in vertebrates other than mammals. *Am J Anat* 33: 347–405.

Epplen JT, Leipoldt M, Engel W, Schmidtke J. 1978. DNA sequence organisation in avian genomes. *Chromosoma* 69: 307–321.

Escribano R, McLaren IA, Klein Breteler WCM. 1992. Innate and acquired variation of nuclear DNA contents of marine copepods. *Genome* 35: 602–610.

Evans JD, Gundersen-Rindal D. 2003. Beenomes to Bombyx: future directions in applied insect genomics. *Genome Biol* 4: 107.101–107.104.

Fan Y, Linardopoulou E, Friedman C, *et al.* 2002. Genomic structure and evolution of the ancestral chromosome fusion site in 2q13-2q14.1 and paralogous regions on other human chromosomes. *Genome Res* 12: 1651–1662.

Fay MF, Cowan RS, Leitch IJ. 2005. The effects of DNA amount on the quality and utility of AFLP fingerprints. *Ann Bot* (in press).

Ferrari JA, Rai KS. 1989. Phenotypic correlates of genome size variation in *Aedes albopictus*. *Evolution* 43: 895–899.

Feulgen R, Rössenbeck H. 1924. Mikroskopisch-chemischer Nachweis einer Nucleinsäure vom Typus der Thymonucleinsäure und die darauf beruhende elektive Färbung von Zellkernen in mikroskopischen Präparaten. *Hoppe-Seyler Z Physiol Chem* 135: 203–248.

Filkowski J, Kovalchuk O, Kovalchuk I. 2004. Dissimilar mutation and recombination rates in *Arabidopsis* and tobacco. *Plant Sci* 166: 265–272.

Finston TL, Hebert PDN, Foottit RB. 1995. Genome size variation in aphids. *Insect Biochem Mol Biol* 25: 189–196.

Flemming AJ, Shen Z-Z, Cunha A, *et al.* 2000. Somatic polyploidization and cellular proliferation drive body size evolution in nematodes. *Proc Natl Acad Sci USA* 97: 5285–5290.

Flores G, Frieden E. 1968. Induction and survival of hemoglobin-less and erythrocyte-less tadpoles and young bullfrogs. *Science* 159: 101–103.

Fogel GB, Collins CR, Li J, Brunk CF. 1999. Prokaryotic genome size and SSU rDNA copy number: estimation of microbial relative abundance from a mixed population. *Microb Ecol* 38: 93–113.

Fontana F. 1976. Nuclear DNA content and cytometry of erythrocytes of *Huso huso* L., *Acipenser sturio* L. and *Acipenser naccarii* Bonaparte. *Caryologia* 29: 127–137.

Forbes DJ, Kirschner MW, Newport JW. 1983. Spontaneous formation of nuclear-like structure around bacteriophage DNA microinjected into *Xenopus* eggs. *Cell* 34: 13–23.

Friedman R, Hughes AL. 2001. Gene duplication and the structure of eukaryotic genomes. *Genome Res* 11: 373–381.

Friz CT. 1968. The biochemical composition of the free-living amoebae *Chaos chaos*, *Amoeba dubia*, and *Amoeba proteus*. *Comp Biochem Physiol* 26: 81–90.

Frost DR. 2004. *Amphibian Species of the World: An Online Reference*. Version 3.0. research.amnh.org/herpetology/amphibia.

Gage MJG. 1998. Mammalian sperm morphometry. *Proc R Soc Lond B* 265: 97–103.

Gall JG. 1981. Chromosome structure and the C-value paradox. *J Cell Biol* 91: 3s–14s.

Gallardo MH, Bickham JW, Honeycutt RL, *et al.* 1999. Discovery of tetraploidy in a mammal. *Nature* 401: 341.

Gallardo MH, Bickham JW, Kausel G, *et al.* 2003. Gradual and quantum genome size shifts in the hystricognath rodents. *J Evol Biol* 16: 163–169.

Gambi MC, Ramella L, Sella G, *et al.* 1997. Variation in genome size of benthic polychaetes: systematic and ecological relationships. *J Marine Biol Assoc UK* 77: 1045–1057.

Gamperl R, Ehmann C, Bachmann K. 1982. Genome size and heterochromatin variation in rodents. *Genetica* 58: 199–212.

Garagna S, Civitelli MV, Marziliano N, *et al.* 1999. Genome size variations are related to X-chromosome heterochromatin polymorphism in *Arvicanthis* sp. from Benin (West Africa). *Ital J Zool* 66: 27–32.

Garagna S, Pérez-Zapata A, Zuccotti M, *et al*. 1997. Genome composition in Venezuelan spiny-rats of the genus *Proechimys* (Rodentia, Echimyidae). I. Genome size, C-heterochromatin and repetitive DNAs in situ hybridization patterns. *Cytogenet Cell Genet* 78: 36–43.

Garner TWJ. 2002. Genome size and microsatellites: the effect of nuclear size on amplification potential. *Genome* 45: 212–215.

Gaut BS. 2002. Evolutionary dynamics of grass genomes. *New Phytol* 154: 15–28.

Gilbert W. 1978. Why genes in pieces? *Nature* 271: 501.

Goin OB, Goin CJ, Bachmann K. 1968. DNA and amphibian life history. *Copeia* 1968: 532–540.

Gold JR, Amemiya CT. 1987. Genome size variation in North American minnows (Cyprinidae). II. Variation among 20 species. *Genome* 29: 481–489.

Gold JR, Price HJ. 1985. Genome size variation among North American minnows (Cyprinidae). I. Distribution of the variation in five species. *Heredity* 54: 297–305.

Goniakowska L. 1970. The respiration of erythrocytes of some amphibians *in vitro*. *Bull Acad Pol Sci Ser Sci Biol* 18: 793–797.

Goniakowska L. 1973. Metabolism, resistance to hypotonic solutions, and ultrastructure of erythrocytes of five amphibian species. *Acta Cracov Ser Zool* 16: 113–133.

González J, Ranz JM, Ruiz A. 2002. Chromosomal elements evolve at different rates in the *Drosophila* genome. *Genetics* 161: 1137–1154.

González-Tizón AM, Martínez-Lage A, Rego I, *et al*. 2000. DNA content, karyotypes, and chromosomal location of 18S-5.8S-28S ribosomal loci in some species of bivalve molluscs from the Pacific Canadian coast. *Genome* 43: 1065–1072.

Graur D, Shauli Y, Li WH. 1989. Deletions in processed pseudogenes accumulate faster in rodents than in humans. *J Mol Evol* 28: 279–285.

Gregory TR. 2000. Nucleotypic effects without nuclei: genome size and erythrocyte size in mammals. *Genome* 43: 895–901.

Gregory TR. 2001a. Coincidence, coevolution, or causation? DNA content, cell size, and the C-value enigma. *Biol Rev* 76: 65–101.

Gregory TR. 2001b. *Animal Genome Size Database*. www.genomesize.com.

Gregory TR. 2001c. The bigger the C-value, the larger the cell: genome size and red blood cell size in vertebrates. *Blood Cells Mol Dis* 27: 830–843.

Gregory TR. 2002a. A bird's-eye view of the C-value enigma: genome size, cell size, and metabolic rate in the class Aves. *Evolution* 56: 121–130.

Gregory TR. 2002b. Genome size and developmental parameters in the homeothermic vertebrates. *Genome* 45: 833–838.

Gregory TR. 2002c. Genome size and developmental complexity. *Genetica* 115: 131–146.

Gregory TR. 2003a. Genome size estimates for two important freshwater molluscs, the zebra mussel (*Dreissena polymorpha*) and the schistosomiasis vector snail (*Biomphalaria glabrata*). *Genome* 46: 841–844.

Gregory TR. 2003b. Is small indel bias a determinant of genome size? *Trends Genet* 19: 485–488.

Gregory TR. 2003c. Variation across amphibian species in the size of the nuclear genome supports a pluralistic, hierarchical approach to the C-value enigma. *Biol J Linn Soc* 79: 329–339.

Gregory TR. 2004a. Macroevolution, hierarchy theory, and the C-value enigma. *Paleobiology* 30: 179–202.

Gregory TR. 2004b. Insertion–deletion biases and the evolution of genome size. *Gene* 324: 15–34.

Gregory TR. 2004c. Genome size is not positively correlated with longevity in fishes (or homeotherms). *Exp Gerontol* 39:859–860.

Gregory TR. 2005. The C-value enigma in plants and animals: a review of parallels and an appeal for partnership. *Ann Bot* (in press).

Gregory TR, Hebert PDN. 1999. The modulation of DNA content: proximate causes and ultimate consequences. *Genome Res* 9: 317–324.

Gregory TR, Hebert PDN. 2002. Genome size estimates for some oligochaete annelids. *Can J Zool* 80: 1485–1489.

Gregory TR, Hebert PDN. 2003. Genome size variation in lepidopteran insects. *Can J Zool* 81: 1399–1405.

Gregory TR, Shorthouse DP. 2003. Genome sizes of spiders. *J Hered* 94: 285–290.

Gregory TR, Hebert PDN, Kolasa J. 2000. Evolutionary implications of the relationship between genome size and body size in flatworms and copepods. *Heredity* 84: 201–208.

Gregory TR, Nedved O, Adamowicz SJ. 2003. C-value estimates for 31 species of ladybird beetles (Coleoptera: Coccinellidae). *Hereditas* 139: 121–127.

Griffith OL, Moodie GEE, Civetta A. 2003. Genome size and longevity in fish. *Exp Gerontol* 38: 333–337.

Grime JP, Hendry GAF. 1990. Yes, we have no C-value paradox. *Trends Ecol Evol* 5: 62–63.

Grosset L, Odartchenko N. 1975a. Relationships between cell cycle duration, S-period and nuclear DNA content in erythroblasts of four vertebrate species. *Cell Tiss Kinet* 8: 81–90.

Grosset L, Odartchenko N. 1975b. Duration of mitosis and separate mitotic phases compared to nuclear DNA content in erythroblasts of four vertebrates. *Cell Tiss Kinet* 8: 91–96.

Gulliver G. 1875. Observations on the sizes and shapes of the red corpuscles of the blood of vertebrates, with drawings of them to a uniform scale, and extended and revised tables of measurements. *Proc Zool Soc Lond* 1875: 474–495.

Hancock JM. 1999. Microsatellites and other simple sequences: genomic context and mutational mechanisms. In: Goldstein DB, Schlötterer C eds. *Microsatellites: Evolution and Applications*. Oxford: Oxford University Press, 1–9.

Hancock JM. 2002. Genome size and the accumulation of simple sequence repeats: implications of new data from genome sequencing projects. *Genetica* 115: 93–103.

Hardie DC, Hebert PDN. 2003. The nucleotypic effects of cellular DNA content in cartilaginous and ray-finned fishes. *Genome* 46: 683–706.

Hardie DC, Gregory TR, Hebert PDN. 2002. From pixels to picograms: a beginners' guide to genome quantification by Feulgen image analyses densitometry. *J Histochem Cytochem* 50: 735–749.

Hartl DL. 2000. Molecular melodies in high and low *C. Nat Rev Genet* 1: 145–159.

Hartley SE. 1990. Variation in cellular DNA content in Arctic charr, *Salvelinus alpinus* (L.). *J Fish Biol* 37: 189–190.

Hatch FT, Bodner AJ, Mazrimas JA, Moore DH. 1976. Satellite DNA and cytogenetic evolution: DNA quantity, satellite DNA and karyotypic variations in kangaroo rats (Genus *Dipodomys*). *Chromosoma* 58: 155–168.

Hawkey CM. 1975. *Comparative Mammalian Haematology*. London: William Heinemann Medical Books Ltd.

Hershey AD, Chase M. 1952. Independent functions of viral protein and nucleic acid in growth of bacteriophage. *J Gen Physiol* 36: 39–56.

Hertwig R. 1903. Über Korrelation von Zell- und Kerngrösse und ihre Bedeutung für die geschlechtliche Differenzierung und die Teilung der Zelle. *Biol Centralbl* 23: 49–62.

Hinegardner R. 1974a. Cellular DNA content of the Mollusca. *Comp Biochem Physiol* 47A: 447–460.

Hinegardner R. 1974b. Cellular DNA content of the Echinodermata. *Comp Biochem Physiol* 49B: 219–226.

Hinegardner R. 1976a. The cellular DNA content of sharks, rays and some other fishes. *Comp Biochem Physiol* 55B: 367–370.

Hinegardner R. 1976b. Evolution of genome size. In: Ayala FJ ed. *Molecular Evolution*. Sunderland: Sinauer Associates, Inc., 179–199.

Holmquist GP. 1989. Evolution of chromosome bands: molecular ecology of noncoding DNA. *J Mol Evol* 28: 469–486.

Honeycutt RL, Rowe DL, Gallardo MH. 2003. Molecular systematics of the South American caviomorph rodents: relationships among species and genera in the family Octodontidae. *Mol Phylogenet Evol* 26: 476–489.

Horner HA, Macgregor HC. 1983. C value and cell volume: their significance in the evolution and development of amphibians. *J Cell Sci* 63: 135–146.

Hughes AL. 1999. *Adaptive Evolution of Genes and Genomes*. Oxford: Oxford University Press.

Hughes AL, Hughes MK. 1995. Small genomes for better flyers. *Nature* 377: 391.

Imsiecke G, Custodio M, Borojevic R, *et al.* 1995. Genome size and chromosomes in marine sponges (*Suberites domuncula, Geodia cydonium*). *Cell Biol Int* 19: 995–1000.

Ingle J, Timmis JN, Sinclair J. 1975. The relationship between satellite deoxyribonucleic acid, ribosomal ribonucleic acid gene redundancy, and genome size in plants. *Plant Phys* 55: 496–501.

International Human Genome Sequencing Consortium. 2001. Initial sequencing and analysis of the human genome. *Nature* 409: 860–921.

Jockusch EL. 1997. An evolutionary correlate of genome size change in plethodontid salamanders. *Proc R Soc Lond B* 264: 597–604.

Johannsen W. 1911. The genotype conception of heredity. *Am Nat* 45: 129–159.

John B, Miklos GLG. 1988. *The Eukaryote Genome in Development and Evolution*. London: Allen & Unwin.

Johnson OW, Utter FM, Rabinovitch PS. 1987. Interspecies differences in salmonid cellular DNA identified by flow cytometry. *Copeia* 1987: 1001–1009.

Juan C, Petitpierre E. 1991. Evolution of genome size in darkling beetles (Tenebrionidae, Coleoptera). *Genome* 34: 169–173.

Judson HF. 1996. *The Eighth Day of Creation*. Plainview, NY: Cold Spring Harbor Laboratory Press.

Kamel S, Marsden JE, Pough FH. 1985. Diploid and tetraploid grey treefrogs (*Hyla chrysoscelis* and *Hyla versicolor*) have similar metabolic rates. *Comp Biochem Physiol* 82A: 217–220.

Katti MV, Ranjekar PK, Gupta VS. 2001. Differential distribution of simple sequence repeats in eukaryotic genome sequences. *Mol Biol Evol* 18: 1161–1167.

Kidwell MG. 2002. Transposable elements and the evolution of genome size in eukaryotes. *Genetica* 115: 49–63.

Kozlowski J, Konarzewski M, Gawelczyk AT. 2003. Cell size as a link between noncoding DNA and metabolic rate scaling. *Proc Natl Acad Sci USA* 100: 10480–14085.

Kubota S, Nakai Y, Sato N, *et al.* 1994. Chromosome elimination in Northeast Pacific hagfish, *Eptatretus stoutii* (Cyclostomata, Agnatha). *J Hered* 85: 413–415.

Kubota S, Takano J-I, Tsuneishi R, *et al.* 2001. Highly repetitive DNA families restricted to germ cells in a Japanese hagfish (*Eptatretus burgeri*): a hierarchical and mosaic structure in eliminated chromosomes. *Genetica* 111: 319–328.

Kuldau GA, Tsai H-F, Schardl CL. 1999. Genome sizes of *Epichloë* species and anamorphic hybrids. *Mycologia* 91: 776–782.

Kumar A, Rai KS. 1990. Intraspecific variation in nuclear DNA content among world populations of a mosquito, *Aedes albopictus* (Skuse). *Theor Appl Genet* 79: 748–752.

Lang F, Waldegger S. 1997. Regulating cell volume. *Am Sci* 85: 456–463.

Lay PA, Baldwin J. 1999. What determines the size of teleost erythrocytes? Correlations with oxygen transport and nuclear volume. *Fish Physiol Biochem* 20: 31–35.

Lécher P, DeFaye D, Noel P. 1995. Chromosomes and nuclear DNA of Crustacea. *Invert Reprod Dev* 27: 85–114.

Lederberg J, McCray AT. 2001. 'Ome sweet 'omics—a genealogical treasury of words. *The Scientist*, April 2.

Levinson G, Gutman GA. 1987. Slipped-strand mispairing: a major mechanism for DNA sequence evolution. *Mol Biol Evol* 4: 203–221.

Lewis JH. 1996. *Comparative Hemostasis in Vertebrates*. New York: Plenum Press.

Li W-H. 1997. *Molecular Evolution*. Sunderland: Sinauer Associates, Inc.

Licht LE, Bogart JP. 1990. Comparative rates of oxygen consumption and water loss in diploid and polyploid salamanders (genus *Ambystoma*). *Comp Biochem Physiol* 97A: 569–572.

Licht LE, Lowcock LA. 1991. Genome size and metabolic rate in salamanders. *Comp Biochem Physiol* 100B: 83–92.

Liu G, NISC Comparative Sequencing Program, Zhao S, *et al.* 2003. Analysis of primate genomic variation reveals a repeat-driven expansion of the human genome. *Genome Res* 13: 358–368.

Lockwood SF, Bickham JW. 1991. Genetic stock assessment of spawning Arctic cisco (*Coregonus autumnalis*) populations by flow cytometric determination of DNA content. *Cytometry* 12: 260–267.

Lockwood SF, Bickham JW. 1992. Genome size in Beaufort Sea coastal assemblages of Arctic ciscoes. *Trans Am Fish Soc* 121: 13–20.

Lockwood SF, Derr JN. 1992. Intra- and interspecific genome-size variation in the Salmonidae. *Cytogenet Cell Genet* 59: 303–306.

Lockwood SF, Holland BS, Bickham JW, *et al.* 1991a. Intraspecific genome size variation in a turtle (*Trachemys scripta*) exhibiting temperature-dependent sex determination. *Can J Zool* 69: 2306–2310.

Lockwood SF, Seavey BT, Dillinger RE, Bickham JW. 1991b. Variation in DNA content among age classes of broad whitefish (*Coregonus nasus*) from the Sagavanirktok River delta. *Can J Zool* 69: 1335–1338.

Lozovskaya ER, Nurminsky DI, Petrov DA, Hartl DL. 1999. Genome size as a mutation–selection–drift process. *Genes Genet Syst* 74: 201–207.

Ma RZ, Black WC, Reese JC. 1992. Genome size and organization in an aphid (*Schizaphis graminum*). *J Insect Physiol* 38: 161–165.

MacCulloch RD, Upton DE, Murphy RW. 1996. Trends in nuclear DNA content among amphibians and reptiles. *Comp Biochem Physiol* 113B: 601–605.

Macgregor HC. 1993. *An Introduction to Animal Cytogenetics.* London: Chapman & Hall.

MacLean N. 1973. Suggested mechanism for increase in size of the genome. *Nature New Biol* 246: 205–206.

Mandel P, Métais P, Cuny S. 1951. Les quantités d'acide désoxypentose–nucléique par leucocyte chez diverses espèces de mammifères. *C R Acad Sci* 231: 1172–1174.

Manfredi Romanini MG, Formenti D, Stanyon R, Pellicciari C. 1991. C-heterochromatic-DNA and the problem of genome size variability in Hominoidea. In: Ghiara G *et al.* eds. *Symposium on the Evolution of Terrestrial Vertebrates.* Modena: Mucchi, 387–397.

Martin PG. 1966. Variation in the amounts of nucleic acids in the cells of different species of higher plants. *Exp Cell Res* 44: 84–94.

McLaren IA, Marcogliese SJ. 1983. Similar nucleus numbers among copepods. *Can J Zool* 61: 721–724.

McLaren IA, Sévigny J-M, Corkett CJ. 1988. Body size, development rates, and genome sizes among *Calanus* species. *Hydrobiologia* 167/168: 275–284.

McLaren IA, Sévigny J-M, Frost BW. 1989. Evolutionary and ecological significance of genome sizes in the copepod genus *Pseudocalanus*. *Can J Zool* 67: 565–569.

Miescher F. 1871. Über die chemische Zusammensetzung der Eiterzellen. *Hoppe-Seyler Med-chem Untersuch* 4: 441–460.

Mirsky AE. 1968. The discovery of DNA. *Sci Am* 218 (June): 78–88.

Mirsky AE, Osawa S. 1961. The interphase nucleus. In: Brachet J, Mirsky AE eds. *The Cell: Biochemistry, Physiology, Morphology, Vol. 2.* New York: Academic Press, 677–770.

Mirsky AE, Ris H. 1951. The desoxyribonucleic acid content of animal cells and its evolutionary significance. *J Gen Physiol* 34: 451–462.

Monaghan P, Metcalfe NB. 2000. Genome size and longevity. *Trends Genet* 16: 331–332.

Monaghan P, Metcalfe NB. 2001. Genome size, longevity and development time in birds. *Trends Genet* 17: 568.

Monnickendam MA, Balls M. 1973. The relationship between cell sizes, respiration rates and survival of amphibian tissues in long-term organ cultures. *Comp Biochem Physiol* 44A: 871–880.

Moore MJ. 1996. When the junk isn't junk. *Nature* 379: 402–403.

Morand S, Ricklefs RE. 2001. Genome size, longevity and development time in birds. *Trends Genet* 17: 567–568.

Morescalchi A. 1990. Cytogenetics and the problem of Lissamphibian relationships. In: Olmo E ed. *Cytogenetics of Amphibians and Reptiles*. Basel, Switzerland: Birkhauser Verlag, 1–19.

Morgante M, Hanafey M, Powell W. 2002. Microsatellites are preferentially associated with nonrepetitive DNA in plant genomes. *Nat Genet* 30: 194–200.

Moritz KB, Roth GE. 1976. Complexity of germline and somatic DNA in *Ascaris*. *Nature* 259: 55–57.

Moriyama EN, Petrov DA, Hartl DL. 1998. Genome size and intron size in *Drosophila*. *Mol Biol Evol* 15: 770–773.

Mouse Genome Sequencing Consortium. 2002. Initial sequencing and comparative analysis of the mouse genome. *Nature* 420: 520–562.

Mueller RL. 2000. Who needs a nucleus? Red blood cells in the genus *Batrachoceps*. *Am Zool* 40: 1142A.

Nakai Y, Kubota S, Goto Y, *et al.* 1995. Chromosome elimination in three Baltic, south Pacific and north-east Pacific hagfish species. *Chromosome Res* 3: 321–330.

Neafsey DE, Palumbi SR. 2003. Genome size evolution in pufferfish: a comparative analysis of diodontid and tetraodontid pufferfish genomes. *Genome Res* 13: 821–830.

Neff BD, Gross MR. 2001. Microsatellite evolution in vertebrates: inference from AC dinucleotide repeats. *Evolution* 55: 1717–1733.

Nelson CE, Hersh BM, Carroll SB. 2004. The regulatory content of intergenic DNA shapes genome architecture. *Genome Biol* 5: R25.1–R25.15.

Novacek MJ, Norell MA. 1989. Nuclear DNA content in bats and other organisms: implications and unanswered questions. *Trends Ecol Evol* 4: 285–286.

Novacek MJ, Norell MA. 1990. A paradox by any other name: reply from M.J. Novacek and M.A. Norell. *Trends Ecol Evol* 5: 63.

Nurse P. 1985. The genetic control of cell volume. In: Cavalier–Smith T ed. *The Evolution of Genome Size*. Chichester: John Wiley & Sons, 185–196.

Oeldorf E, Nishioka M, Bachmann K. 1978. Nuclear DNA amounts and developmental rate in holarctic anura. *Z Zool Syst Evol* 16: 216–224.

Ogata H, Fujibuchi W, Kanehisa M. 1996. The size differences among mammalian introns are due to the accumulation of small deletions. *FEBS Lett* 390: 99–103.

Ohno S. 1969. The mammalian genome in evolution and conservation of the original X-linkage group. In: Benirschke K ed. *Comparative Mammalian Cytogenetics*. New York: Springer-Verlag, 18–29.

Ohno S. 1970. *Evolution by Gene Duplication*. New York: Springer-Verlag.

Ohno S. 1972a. So much "junk" DNA in our genome. In: Smith HH ed. *Evolution of Genetic Systems*. New York: Gordon and Breach, 366–370.

Ohno S. 1972b. An argument for the genetic simplicity of man and other mammals. *J Hum Evol* 1: 651–662.

Ohno S. 1973. Evolutional reason for having so much junk DNA. In: Pfeiffer RA ed. *Modern Aspects of Cytogenetics: Constitutive Heterochromatin in Man*. F.K. Stuttgart: Schattauer Verlag, 169–173.

Ohno S. 1974. *Animal Cytogenetics, Vol. 4: Chordata, No. 1: Protochordata, Cyclostomata, and Pisces*. Berlin: Gebrüder Borntraeger.

Ohno S, Yomo T. 1991. The grammatical rule for all DNA: junk and coding sequences. *Electrophoresis* 12: 103–108.

Oliver JL, Marín A. 1996. A relationship between GC content and coding-sequence length. *J Mol Evol* 43: 216–223.

Olmo E. 1983. Nucleotype and cell size in vertebrates: a review. *Bas Appl Histochem* 27: 227–256.

Olmo E. 1991. Genome variations in the transition from amphibians to reptiles. *J Mol Evol* 33: 68–75.

Olmo E. 2003. Reptiles: a group of transition in the evolution of genome size and of the nucleotypic effect. *Cytogenet Genome Res* 101: 166–171.

Olmo E, Morescalchi A. 1975. Evolution of the genome and cell sizes in salamanders. *Experientia* 31: 804–806.

Olmo E, Morescalchi A. 1978. Genome and cell size in frogs: a comparison with salamanders. *Experientia* 34: 44–46.

Olmo E, Odierna G. 1982. Relationships between DNA content and cell morphometric parameters in reptiles. *Bas Appl Histochem* 26: 27–34.

Ophir R, Graur D. 1997. Patterns and rates of indel evolution in processed pseudogenes from humans and murids. *Gene* 205: 191–202.

Orgel LE, Crick FHC. 1980. Selfish DNA: the ultimate parasite. *Nature* 284: 604–607.

Östergren G. 1945. Parasitic nature of extra fragment chromosomes. *Botan Notiser* 2: 157–163.

Pagel M, Johnstone RA. 1992. Variation across species in the size of the nuclear genome supports the junk–DNA explanantion for the C-value paradox. *Proc R Soc Lond B* 249: 119–124.

Palmer M, Petitpierre E. 1996. Relationship of genome size to body size in *Phylan semicostatus* (Coleoptera: Tenebrionidae). *Ann Entom Soc Am* 89: 221–225.

Palmer M, Petitpierre E, Pons J. 2003. Test of the correlation between body size and DNA content in *Pimelia* (Coleoptera: Tenebrionidae) from the Canary Islands. *Eur J Entomol* 100: 123–129.

Panzera F, Dujardin JP, Nicolini P, et al. 2004. Genomic changes of chagas disease vector, South America. *Emerg Infect Dis* 10: 438–446.

Partridge L, Barton NH. 1993. Optimality, mutation and the evolution of ageing. *Nature* 362: 305–311.

Pedersen RA. 1971. DNA content, ribosomal gene multiplicity, and cell size in fish. *J Exp Zool* 177: 65–79.

Pellicciari C, Ronchetti E, Formenti D, et al. 1990a. Genome size and "C-heterochromatic-DNA" in man and the African apes. *Hum Evol* 5: 261–267.

Pellicciari C, Ronchetti E, Tori R, et al. 1990b. Cytochemical evaluation of C-heterochromatic-DNA in metaphase chromosomes. *Bas Appl Histochem* 34: 79–85.

Petitpierre E, Juan C. 1994. Genome size, chromosomes, and egg-chorion ultrastructure in the evolution of Chrysomelidae. In: Jolivet PH Cox ML, Petitpierre E eds. *Novel Aspects of the Biology of Chrysomelidae*. Dordrecht: Kluwer Academic Publishers, 213–225.

Petitpierre E, Segarra C, Juan C. 1993. Genome size and chromosomal evolution in leaf beetles (Coleoptera, Chrysomelidae). *Hereditas* 119: 1–6.

Petrov DA. 2001. Evolution of genome size: new approaches to an old problem. *Trends Genet* 17: 23–28.

Petrov DA. 2002a. Mutational equilibrium model of genome size evolution. *Theor Pop Biol* 61: 533–546.

Petrov DA. 2002b. DNA loss and evolution of genome size in *Drosophila*. *Genetica* 115: 81–91.

Petrov DA, Hartl DL. 1997. Trash DNA is what gets thrown away: high rate of DNA loss in *Drosophila*. *Gene* 205: 279–289.

Petrov DA, Hartl DL. 1998. High rate of DNA loss in the *Drosophila melanogaster* and *Drosophila virilis* species groups. *Mol Biol Evol* 15: 293–302.

Petrov DA, Hartl DL. 2000. Pseudogene evolution and natural selection for a compact genome. *J Hered* 91: 221–227.

Petrov DA, Lozovskaya ER, Hartl DL. 1996. High intrinsic rate of DNA loss in *Drosophila*. *Nature* 384: 346–349.

Petrov DA, Sangster TA, Johnston JS, et al. 2000. Evidence for DNA loss as a determinant of genome size. *Science* 287: 1060–1062.

Pinder AW, Storey KB, Ultsch GR. 1992. Estivation and hibernation. In: Feder ME, Burggren WW eds. *Environmental Physiology of the Amphibians*. Chicago: University of Chicago Press, 250–274.

Portugal FH, Cohen JS. 1977. *A Century of DNA*. Cambridge, MA: MIT Press.

Primmer CR, Raudsepp T, Chowdhary BP, et al. 1997. Low frequency of microsatellites in the avian genome. *Genome Res* 7: 471–482.

Pritchard DK, Schubiger G. 1996. Activation of transcription in *Drosophila* embryos is a gradual process mediated by the nucleocytoplasmic ratio. *Genes Dev* 10: 1131–1142.

Prokopowich CD, Gregory TR, Crease TJ. 2003. The correlation between rDNA copy number and genome size in eukaryotes. *Genome* 46: 48–50.

Ragland CJ, Gold JR. 1989. Genome size variation in the North American sunfish genus *Lepomis* (Pisces: Centrarchidae). *Genet Res* 53: 173–182.

Ramirez PB, Bickham JW, Braun JK, Mares MA. 2001. Geographic variation in genome size of *Graomys griseoflavus* (Rodentia: Muridae). *J Mammal* 82: 102–108.

Ranz JM, Casals F, Ruiz A. 2001. How malleable is the eukaryotic genome? Extreme rate of chromosomal rearrangement in the genus *Drosophila*. *Genome Res* 11: 230–239.

Rao PN, Rai KS. 1987. Inter and intraspecific variation in nuclear DNA content in *Aedes* mosquitoes. *Heredity* 59: 253–258.

Rasch EM. 2003. Feulgen-DNA cytophotometry for estimating C values. In: Henderson DS ed. *Methods in Molecular Biology, vol. 247: Drosophila Cytogenetics Protocols*. Totowa, NJ: Humana Press, 163–201.

Rasch EM, Tiersch TR. 1992. Genome size and sex identification in five species of cranes: comparison of static and flow cytometry. *J Histochem Cytochem* 40: 593.

Ricklefs RE, Scheuerlein A. 2001. Comparison of aging-related mortality among birds and mammals. *Exp Gerontol* 36: 845–857.

Ricklefs RE, Scheuerlein A. 2003. Life span in the light of avian life histories. *Pop Dev Rev* 29 (Suppl.): 71–98.

Robertson HM. 2000. The large *srh* family of chemoreceptor genes in *Caenorhabditis* nematodes reveals processes of genome evolution involving large duplications and deletions and intron gains and losses. *Genome Res* 10: 192–203.

Rodríguez-Juíz AM, Torrado M, Méndez J. 1996. Genome-size variation in bivalve molluscs determined by flow cytometry. *Marine Biol* 126: 489–497.

Roth G, Schmidt A. 1993. The nervous system of plethodontid salamanders: insight into the interplay between genome, organism, behavior, and ecology. *Herpetologica* 49: 185–194.

Roth G, Blanke J, Wake DB. 1994. Cell size predicts morphological complexity in the brains of frogs and salamanders. *Proc Natl Acad Sci USA* 91: 4796–4800.

Roth G, Nishikawa KC, Wake DB. 1997. Genome size, secondary simplification, and the evolution of the brain in salamanders. *Brain Behav Evol* 50: 50–59.

SanMiguel P, Bennetzen JL. 1998. Evidence that a recent increase in maize genome size was caused by the massive amplification of intergene retrotransposons. *Ann Bot* 82 (Suppl. A): 37–44.

Sapienza C, Doolittle WF. 1981. Genes are things you have whether you want them or not. *Cold Spring Harb Symp Quant Biol* 45: 177–182.

Schrader F, Leuchtenberger C. 1949. Variation in the amount of desoxyribose nucleic acid in different tissues of *Tradescantia*. *Proc Natl Acad Sci USA* 35: 464–468.

Sella G, Redi GA, Ramella L, et al. 1993. Genome size and karyotype in some interstitial polychaete species of the genus *Ophryotrocha* (Dorvilleidae). *Genome* 36: 652–657.

Semple C, Wolfe KH. 1999. Gene duplication and gene conversion in the *Caenorhabditis elegans* genome. *J Mol Evol* 48: 555–564.

Sessions SK, Larson A. 1987. Developmental correlates of genome size in plethodontid salamanders and their implications for genome evolution. *Evolution* 41: 1239–1251.

Shabalina SA, Spiridonov NA. 2004. The mammalian transcriptome and the function of non-coding DNA sequences. *Genome Biol* 5: 105.101–105.108.

Sherwood SW, Patton JL. 1982. Genome evolution in pocket gophers (genus *Thomomys*). II. Variation in cellular DNA content. *Chromosoma* 85: 163–179.

Shuter BJ, Thomas JE, Taylor WD, Zimmerman AM. 1983. Phenotypic correlates of genomic DNA content in unicellular eukaryotes and other cells. *Am Nat* 122: 26–44.

Simonsen TJ, Kristensen NP. 2003. Scale length/wing length correlation in Lepidoptera (Insecta). *J Nat Hist* 37: 673–679.

Smith HM. 1925. Cell size and metabolic activity in Amphibia. *Biol Bull* 48: 347–378.

Soldi R, Ramella L, Gambi MC, *et al.* 1994. Genome size in Polychaetes: relationship with body length and life habit. In: Dauvin J-C, Laubier L, Reish DJ eds. *Actes de la 4ième Conférence internationale des Polychètes. Mém Mus natn Hist nat* 162. 129–135.

Sparrow AH, Price HJ, Underbink AG. 1972. A survey of DNA content per cell and per chromosome of prokaryotic and eukaryotic organisms: some evolutionary considerations. In: Smith HH ed. *Evolution of Genetic Systems.* New York: Gordon and Breach, 451–494.

Starck JM, Ricklefs RE. 1998. Avian growth rate data set. In: Starck JM, Ricklefs RE eds. *Avian Growth and Development.* Oxford: Oxford University Press, 381–423.

Stingo V, Capriglione T, Rocco L, *et al.* 1989. Genome size and A-T rich DNA in selachians. *Genetica* 79: 197–205.

Strasburger E. 1893. Über die Wirkungssphäre der Kerne und die Zellgrösse. *Histol Beitr* 5: 97–124.

Swift H. 1950a. The desoxyribose nucleic acid content of animal nuclei. *Physiol Zool* 23: 169–198.

Swift H. 1950b. The constancy of desoxyribose nucleic acid in plant nuclei. *Proc Natl Acad Sci USA* 36: 643–654.

Swift H. 1953. Quantitative aspects of nuclear nucleoproteins. *Int Rev Cytol* 2: 1–76.

Szarski H. 1970. Changes in the amount of DNA in cell nuclei during vertebrate evolution. *Nature* 226: 651–652.

Szarski H. 1983. Cell size and the concept of wasteful and frugal evolutionary strategies. *J Theor Biol* 105: 201–209.

Taylor WD, Shuter BJ. 1981. Body size, genome size, and intrinsic rate of increase in ciliated protozoa. *Am Nat* 118: 160–172.

Thindwa HP, Teetes GL, Johnston JS. 1994. Greenbug DNA content. *Southwest Entom* 19: 371–378.

Thiriot-Quievreux C. 2002. Review of the literature on bivalve cytogenetics in the last ten years. *Cah Biol Marine* 43: 17–26.

Thomas CA. 1971. The genetic organization of chromosomes. *Annu Rev Genet* 5: 237–256.

Thomson KS. 1972. An attempt to reconstruct evolutionary changes in the cellular DNA content of lungfish. *J Exp Zool* 180: 363–372.

Thomson KS. 1991. *Living Fossil.* New York: W.W. Norton & Co.

Thomson KS, Muraszko K. 1978. Estimation of cell size and DNA content in fossil fishes and amphibians. *J Exp Zool* 205: 315–320.

Thorgaard GH. 1983. Chromosomal differences among rainbow trout populations. *Copeia* 1983: 650–662.

Tiersch TR, Mumme RL. 1993. An evaluation of the use of flow cytometry to identify sex in the Florida scrub jay. *J Field Ornithol* 64: 18–26.

Tiersch TR, Wachtel SS. 1991. On the evolution of genome size of birds. *J Hered* 82: 363–368.

Tiersch TR, Simco BA, Davis KB, *et al.* 1990. Stability of genome size among stocks of the channel catfish. *Aquaculture* 87: 15–22.

Trivers R, Burt A, Palestis BG. 2004. B chromosomes and genome size in flowering plants. *Genome* 47: 1–8.

Trombetta VV. 1942. The cytonuclear ratio. *Bot Rev* 8: 317–336.

Van den Bussche RA, Longmire JL, Baker RJ. 1995. How bats achieve a small C-value: frequency of repetitive DNA in *Macrotus. Mamm Genome* 6: 521–525.

van Munster EB, Stap J, Hoebe RA, *et al.* 1999. Difference in volume of X- and Y-chromosome-bearing bovine sperm heads matches difference in DNA content. *Cytometry* 35: 125–128.

Vendrely R. 1955. The deoxyribonucleic acid content of the nucleus. In: Chargaff E, Davidson JN eds. *The Nucleic Acids, Vol. II.* New York: Academic Press, 155–180.

Vendrely R, Vendrely C. 1948. La teneur du noyau cellulaire en acide désoxyribonucléique à travers les organes, les individus et les espèces animales: Techniques et premiers résultats. *Experientia* 4: 434–436.

Vendrely R, Vendrely C. 1949. La teneur du noyau cellulaire en acide désoxyribonucléique à travers les organes, les individus et les espèces animales: Etude particulière des Mammifères. *Experientia* 5: 327–329.

Vendrely R, Vendrely C. 1950. Sur la teneur absolue en acide désoxyribonucléique du noyau cellulaire chez quelques espèces d'oiseaux et de poissons. *C R Acad Sci* 230: 788–790.

Vendrely R, Vendrely C. 1956. The results of cytophotometry in the study of the deoxyribonucleic acid (DNA) content of the nucleus. *Int Rev Cytol* 5: 171–197.

Vendrely R, Vendrely C. 1957. *L' Acide Désoxyribonucléique (D.N.A.): Substance Fondamentale de la Cellule Vivante.* Paris: Amédée LeGrand et Cie.

Venturini G, Capanna E, Fontana B. 1987. Size and structure of the bird genome. II. Repetitive DNA and sequence organization. *Comp Biochem Physiol* 87B: 975–979.

Vernberg FJ. 1955. Hematological studies on salamanders in relation to their ecology. *Herpetologica* 11: 129–133.

Vervoort A. 1980. Tetraploidy in *Protopterus* (Dipnoi). *Experientia* 36: 294–296.

Vialli M, Gerzeli G. 1955. Il tenore di acide desossiribonucleico nei linfociti dell'asino, del cavallo e del mulo. *Rend Inst Lombardo Sci Lett B* 88: 273–281.

Vieira C, Biémont C. 2004. Transposable element dynamics in two sibling species: *Drosophila melanogaster* and *Drosophila simulans*. *Genetica* 120: 115–123.

Vieira C, Nardon C, Arpin C, *et al.* 2002. Evolution of genome size in *Drosophila*. Is the invader's genome being invaded by transposable elements? *Mol Biol Evol* 19: 1154–1161.

Vilhar B, Dermastia M. 2002. Standardisation of instrumentation in plant DNA image cytometry. *Acta Bot Croat* 61: 11–26.

Vilhar B, Greilhuber J, Koce JD, *et al.* 2001. Plant genome size measurement with DNA image cytometry. *Ann Bot* 87: 719–728.

Villolobos M, León P, Sessions SK, Kezer J. 1988. Enucleated erythrocytes in plethodontid salamanders. *Herpetologica* 44: 243–250.

Vinogradov AE. 1994. Measurement by flow cytometry of genomic AT/GC ratio and genome size. *Cytometry* 16: 34–40.

Vinogradov AE. 1995. Nucleotypic effect in homeotherms: body mass–corrected basal metabolic rate of mammals is related to genome size. *Evolution* 49: 1249–1259.

Vinogradov AE. 1997. Nucleotypic effect in homeotherms: body-mass independent metabolic rate of passerine birds is related to genome size. *Evolution* 51: 220–225.

Vinogradov AE. 1998a. Variation in ligand-accessible genome size and its ecomorphological correlates in a pond snail. *Hereditas* 128: 59–65.

Vinogradov AE. 1998b. Genome size and GC-percent in vertebrates as determined by flow cytometry: the triangular relationship. *Cytometry* 31: 100–109.

Vinogradov AE. 1999a. Intron-genome size relationship on a large evolutionary scale. *J Mol Evol* 49: 376–384.

Vinogradov AE. 1999b. Genome *in toto*. *Genome* 42: 361–362.

Vinogradov AE. 2000. Larger genomes for molluskan land pioneers. *Genome* 43: 211–212.

Vinogradov AE, Borkin LJ. 1993. Allometry of base pair–specific DNA contents in Tetrapoda. *Hereditas* 118: 155–163.

Vinogradov AE, Chubinishvili AT. 1999. Genome reduction in a hemiclonal frog *Rana esculenta* from radioactively contaminated areas. *Genetics* 151: 1123–1125.

Vivares CP. 1999. On the genome of Microsporidia. *J Euk Microbiol* 46 (Suppl): 16A.

Wachtel SS, Tiersch TR. 1993. Variations in genome mass. *Comp Biochem Physiol* 104B: 207–213.

Wagenmann M, Epplen JT, Bachmann K, *et al.* 1981. DNA sequence organisation in relation to genome size in birds. *Experientia* 37: 1274–1276.

Waldegger S, Lang F. 1998. Cell volume and gene expression. *J Memb Biol* 162: 95–100.

Walker LI, Spotorno AE, Sans J. 1991. Genome size variation and its phenotypic consequences in *Phyllotis* rodents. *Hereditas* 115: 99–107.

Waltari E, Edwards SV. 2002. Evolutionary dynamics of intron size, genome size, and physiological correlates in archosaurs. *Am Nat* 160: 539–552.

Watson JD, Crick FHC. 1953. A structure for deoxyribose nucleic acid. *Nature* 171: 737–738.

Wendel JF, Cronn RC, Johnston JS, Price HJ. 2002. Feast and famine in plant genomes. *Genetica* 115: 37–47.

White MM, McLaren IA. 2000. Copepod development rates in relation to genome size and 18S rDNA copy number. *Genome* 43: 750–755.

Wickham SA, Lynn DH. 1990. Relations between growth rate, cell size, and DNA content in colpodean ciliates (Ciliophora: Colpodea). *Eur J Prostistol* 25: 345–352.

Wilder J, Hollocher H. 2001. Mobile elements and the genesis of microsatellites in dipterans. *Mol Biol Evol* 18: 384–392.

Wilson EB. 1925. *The Cell in Development and Heredity*. New York: MacMillan.

Winkler H. 1920. *Verbeitung und Ursache der Parthenogenesis im Pflanzen und Tierreiche*. Jena: Verlag Fischer.

Wintrobe MM. 1933. Variations in the size and hemoglobin content of erythrocytes in the blood of various vertebrates. *Folia Haematol* 51: 32–49.

Witherspoon DJ, Robertson HM. 2003. Neutral evolution of ten types of *mariner* transposons in the genomes of *Caenorhabditis elegans* and *Caenorhabditis briggsae*. *J Mol Evol* 56: 751–769.

Wolfe KH. 2001. Yesterday's polyploids and the mystery of diploidization. *Nat Rev Genet* 2: 333–341.

Wong GK-S, Passey DA, Huang Y-Z, *et al.* 2000. Is "junk" DNA mostly intron DNA? *Genome Res* 10: 1672–1678.

Wong GK-S, Passey DA, Yu J. 2001. Most of the human genome is transcribed. *Genome Res* 11: 1975–1977.

Wyngaard GA, Gregory TR. 2001. Temporal control of DNA replication and the adaptive value of chromatin diminution in copepods. *J Exp Zool (Mol Dev Evol)* 291: 310–316.

Wyngaard GA, Rasch EM. 2000. Patterns of genome size in the copepoda. *Hydrobiologia* 417: 43–56.

Zuckerkandl E. 2002. Why so many noncoding nucleotides? The eukaryote genome as an epigenetic machine. *Genetica* 115: 105–129.

Zuckerkandl E, Hennig W. 1995. Tracking heterochromatin. *Chromosoma* 104: 75–83.

CHAPTER 2

Genome Size Evolution in Plants

MICHAEL D. BENNETT AND ILIA J. LEITCH

Every cellular organism possesses a genome, and, because of this, the question of genome size evolution is not limited to any one taxon, but rather is of universal biological interest. Along with animals, plants are the best studied group with regard to variation in DNA content, and have played a critical role since the earliest days of genome size study. This chapter provides an overview of the current state of knowledge concerning genome size evolution in plants. As with the previous chapter on animals, this includes a review of the available data, the patterns of variation both within and among species and higher taxa, the major mechanisms involved in generating disparities among groups, and the impacts of differences in genome size at the nuclear, cellular, tissue, and whole-organism levels. It is evident from this discussion that there are both deep parallels and major divergences between plants and animals in terms of genome size evolution. However, on one point the two kingdoms clearly project the same message: that genome size is a highly relevant biological characteristic whose evolution continues to represent a key puzzle in genomics and evolutionary biology.

A BRIEF HISTORY OF GENOME SIZE
STUDY IN PLANTS

THE FIRST ESTIMATES OF DNA AMOUNTS

As noted in Chapter 1, estimates of nuclear DNA amounts have been made since before the elucidation of the double helix structure in 1953. The earliest approaches were based on analyses of isolated nuclei or cell suspensions, as was first done for several animal species (Boivin et al.,1948; Vendrely and Vendrely, 1948). The constancy of nuclear DNA reported in the early animal studies was examined in greater detail almost immediately for both animals (e.g., Mirsky and Ris, 1949; Swift, 1950a) and plants (e.g., Swift, 1950b). Importantly, such work on Zea mays and Tradescantia paludosa led Swift (1950b) to develop the still widely used term C-value[1] to define the DNA content of an unreplicated haploid nuclear genome. However, these studies dealt only with relative DNA contents in different tissues of a few test species, and did not provide estimates of absolute DNA mass (i.e., in picograms, pg; 1 pg = 10^{-12} g). Probably the first estimate of the absolute amount of DNA in the nuclear genome of a plant was done for Lilium longiflorum cv. Croft by Ogur et al. (1951).

Just 10 years later, published measurements of DNA amount per cell in angiosperms (flowering plants) already ranged more than 50-fold, from 5.5 pg in Lupinus albus to 313 pg in Lilium longiflorum (McLeish and Sunderland, 1961). Soon after, studies reporting 40-fold interspecific variation between 22 diploid species in the family Ranunculaceae (Rothfels et al.,1966), and 5-fold within the genus Vicia (Martin and Shanks, 1966) confirmed that extensive variation occurred within families and even individual genera, independently of ploidy level (e.g., Fig. 2.1).

Possession of nucleated erythrocytes and the absence of a cell wall were important factors in the selection of animal materials for pioneering research on genome size, whereas easy availability as laboratory models with well-studied genetics, amenable cytology, and large nuclei influenced the first plant materials chosen. Over the following decades, plants became increasingly important in studies of genome size, often proceeding several steps ahead of equivalent work on animals (Gregory, 2005).

---

[1]There has been considerable confusion regarding the origin of the term "C-value." Many authors have assumed incorrectly that it refers to "content," "complement," or "characteristic." In coining the term, Swift (1950b) did not provide a clear definition, and made reference only to the 1C (haploid) "class" of DNA, leading to the reasonable conclusion that "C" stood for "class" (e.g., Gregory 2001, 2002). However, in a letter to M.D. Bennett dated June 24, 1975, Hewson Swift stated that: "I am afraid the letter C stood for nothing more glamorous than 'constant,' i.e., the amount of DNA that was characteristic of a particular genotype."

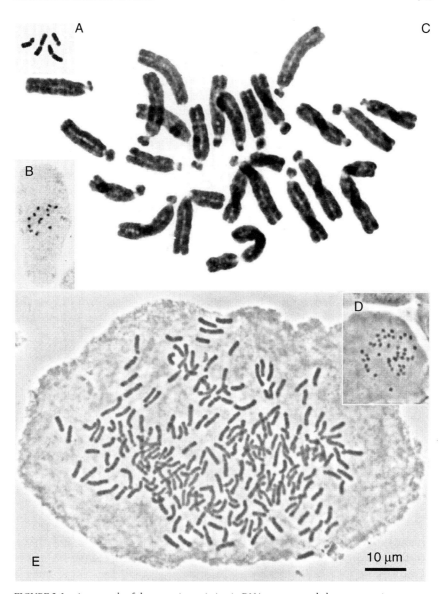

FIGURE 2.1   An example of the extensive variation in DNA amounts and chromosome sizes encountered in plants. (A) *Brachyscome dichromosomatica* 2n = 2x = 4, 1C = 1.1 pg; (B) *Myriophyllum spicatum*
2n = 2x = 14, 1C = 0.3 pg; (C) *Fritillaria* sp. 2n = 2x = 24, 1C ≈ 65 pg; (D) *Selaginella kraussiana* 2n
= 4x = 40, 1C = 0.36 pg; (E) *Equisetum variegatum* 2n ≈ 216, 1C = 30.4 pg. (A) From Kenton
et al. (1993), reproduced by permission (© Elsevier Inc.).

The Main Areas of Focus of Early Genome Size Studies

Once its central role in genetics was clear, research on many aspects of DNA became greatly intensified. Naturally, this included an interest in total genome size, which was soon fuelled by the realization that although this was remarkably constant within individual organisms and species, it varied extensively among different species, often in puzzling ways that were quite independent of an organism's complexity.

Most of the early effort in studying plant genome sizes concentrated on angiosperms, especially crop or model laboratory species. Again, plant studies often led work on animals, which was especially true with regard to early cytological research on chromosomes (from the 1930s to the 1950s). For example, the chromosome number for humans (2n = 46) was still uncertain until 1956 (Ford and Hamerton, 1956; Tjio and Levan, 1956), fully 35 years after the chromosome number and ploidal level of allohexaploid[2] wheat, *Triticum aestivum* (2n = 6x = 42), was firmly established (Sakamura, 1918).

As in animals, much of the initial research was concerned with testing the notion of DNA constancy within species, but it soon diversified to include three main lines of inquiry (which are all still ongoing today):

1. The technical issue of developing methods for estimating genome size, and testing and improving their accuracy.
2. An exploration of the ranges in genome size in different groups and at various taxonomic levels.
3. An investigation of the meaning of genome size variation in terms of the mechanisms responsible, rates of change, and evolutionary significance. In particular, resolving the seeming contradiction between constancy within species (because DNA is the hereditary material) and diversity among species (with the associated lack of correlation with the number of genes), once termed the "C-value paradox" (Thomas, 1971), became a major theme in the study of genome size for several decades.

Although the very broad third issue was well framed in the 1950s and 1960s, progress toward an answer was blocked by a lack of knowledge about the nature of the DNA sequences responsible for variation in genome size, and of suitable techniques to investigate this. It remained so until the molecular revolution, which allowed the copy numbers and karyotypic distributions of specific DNA sequences to be investigated for the first time. In the absence of such information,

---

[2]See Chapter 7 for a discussion of polyploidy in plants, including definitions of allopolyploidy, hexaploidy, and other related terms.

much attention was focused instead on the issue of the phenotypic and other consequences of genome size variation, with particular reference to its practical significance (e.g., in agriculture). Studies of the broader ecological relevance of genome size followed, as discussed in more detail in a later section.

IMPACT OF THE MOLECULAR REVOLUTION ON GENOME SIZE RESEARCH

Although it provided much-needed insights into the structure and content of individual genomes, the molecular revolution also had an inhibitory effect on genome size research. As molecular work on DNA sequences filled the limelight, interest in plant DNA C-values per se began to fade. By the 1980s, the strong emphasis on sequence-based studies made it almost impossible to obtain grant funding to estimate genome sizes in their own right. Consequently, such information was obtained either as a by-product of studies focused primarily on other topics, or by a few laboratories or individuals with basic core funding and equipment. Nevertheless, the process of gathering more data on genome size for plant taxa continued at a low level in an uncoordinated way, although occasionally data for larger samples were gathered in order to test particular hypotheses about the patterns and consequences of DNA content variation.

The careful attention to technical detail characteristic of most early work was also abandoned in many cases, leading to reports of substantial intraspecific variation (i.e., violations of the rule of DNA constancy), including some seemingly related to developmental, environmental, or geographical factors. This was perhaps influenced by molecular studies showing that much of the genome consisted of repetitive DNA sequences that had the potential to change in copy number. In this context, some workers suggested that certain "fluid domains" within the genome were capable of undergoing rapid changes in copy number and hence to alter genome size in response to certain developmental events (e.g., Cavallini and Natali, 1991; Frediani *et al.,*1994). Around this time the concept of the "plastic genome" became popular, in direct contrast to the very notion of the "C-value."

The unquestioning assumption by many researchers that all such variation was real necessitated a second and protracted phase of methodological ground-truthing, the development of best practice techniques, and an emphasis on a more critical approach to claims of substantial variation within species (see, for example, the recommendations arising from the first and second Plant Genome Size Workshops held at the Royal Botanic Gardens, Kew in 1997 and 2003 at www.rbgkew.org.uk/cval/workshopreport.html). This second wave of very careful measurements revealed that much supposed intraspecific variation (though not all) was due to technical artifacts (e.g., Greilhuber 1998, 2005) (see later section).

GENOME SIZE STUDIES IN THE POST-GENOMIC ERA

The advent of large-scale genome sequencing programs in the 1990s focused new interest on the study of genome size, and facilitated detailed studies of the molecular basis of genome evolution in plants. The availability of DNA sequence information for entire genomes (or at least substantial segments thereof) of different taxa has allowed detailed comparisons at several taxonomic levels, including within species (e.g., between the subspecies *indica* and *japonica* of rice, *Oryza sativa*) (Goff *et al.*, 2002), between species within a family (such as the grasses *Oryza*, *Sorghum*, and *Zea*) (Ilic *et al.*, 2003) and between families (such as Poaceae and Brassicaceae) (Bennetzen and Ma, 2003). These provided the first insights into the sorts of intra- and interspecific variation at the DNA sequence level that together result in changes in genome size.

Such work is still in its infancy, and it remains unclear how typical findings based on the study of a small segment of one linkage group really are. Nevertheless, it has already been confirmed that, as expected, changes in genome size mostly involve the gain or loss of families of repeated DNA sequences (especially transposable elements, see Chapter 3) located primarily in intergenic regions. Contrary to what was once thought, it now seems that speciation may have as much to do with changes in such regions as with alterations in the sequences or arrangements of coding regions (Kubis *et al.*, 1998).

Such comparisons will form a main element of future work on plant genome size, and promise to reveal the key molecular mechanisms involved in the gain and/or loss of DNA, and the rate at which such changes can occur either in genomes at large or in different components thereof. This, in turn, will allow a more detailed investigation of the types of sequences involved in generating patterns of variation involving phenotypic and ecological correlates of DNA content. The development and increasingly broad application of new techniques for sequencing, assembling, and comparing genomes make the early 21st century a truly exciting time for all forms of genomics, including the holistic variety concerned with genome size.

THE STATE OF KNOWLEDGE REGARDING PLANT GENOME SIZES

Since 1950, more than 10,000 quantitative estimates of plant C-values have been made, covering roughly 4000 species of plants (Bennett and Leitch, 2003). Such a broad sampling can allow the key questions listed previously to be addressed in a more comprehensive way—but only if the data are accessible. In plants, unlike in animals, there has long been an effort to compile genome size data for the purposes of comparative study. Of course, there is an important feedback process involved in this case, because the compilation of data for one purpose tends to reveal gaps in the dataset and to stimulate the targeted measurement of new values.

In 1972, Bennett collected data for 273 angiosperm species to test a possible relationship between DNA amount and minimum generation time, and incidentally created the largest list of its type available at the time. Although Bennett's (1972) list proved to be a valuable reference resource, it became clear within a few years that the majority of information on plant DNA amounts remained difficult to locate, in part because much of it was widely scattered in a diverse range of journals, and still worse, because a significant proportion of existing estimates were not published anywhere. In recognition of this, Bennett and colleagues began compiling large lists of published and unpublished C-values in plants. To date, seven such lists have been published, providing extensive coverage of the available angiosperm data. Taken together, these lists give DNA amounts for more than 4000 species (1.6% of the global angiosperm flora) derived from 465 original sources (Bennett and Smith, 1976, 1991; Bennett et al., 1982a, 2000a; Bennett and Leitch, 1995, 1997, 2005).

Work in compiling lists of C-values for other plant groups lagged behind that of angiosperms and it was not until 1998 that the first reference list for a non-angiosperm group was published (Murray, 1998). This contained estimates for 117 gymnosperm species (corresponding to 16% of described species), cited from 24 original sources. Since then, C-values for a further 64 gymnosperm taxa have been published from seven new reference sources (e.g., Leitch et al., 2001), but they have not been compiled into a second published list.

For pteridophytes—that is, lycophytes (clubmosses) and monilophytes (including ferns and horsetails)—estimates of DNA amounts were pooled into one list by Bennett and Leitch (2001). The list contained DNA C-values for just 48 species from eight original sources and highlighted the ongoing need for work to increase knowledge in this area. Since then, new pteridophyte C-value data have been published by Obermayer et al. (2002) and Hanson and Leitch (2002).

In bryophytes (mosses, liverworts, and hornworts), some limited data are available but remain scattered, and there is no equivalent compilation of C-values combined from different sources into a pooled list. The largest dataset comes from the work by Voglmayr (2000), who estimated C-values in 138 mosses in a carefully targeted study whose aim was to cover a representative spectrum of taxa. Voglmayr's paper also reviewed C-value estimates made by previous workers and is thus the closest approximation to a single printed reference source for C-values in bryophytes currently available.

C-VALUES IN CYBERSPACE: DEVELOPMENT OF THE PLANT DNA C-VALUES DATABASE

The collected lists of angiosperm DNA amounts were produced to make data more accessible for both reference and analysis purposes. However, as the number of such lists rose, it became more cumbersome to determine whether an estimate for a particular species was listed. By 1997, the problem had become acute, with five collected lists published, containing a total of 2802 species. It was therefore decided to pool available data into a single database and make it available on the Internet. The first version of the *Angiosperm DNA C-values Database* went live in April 1997, and was subsequently updated in 1998 to make it more user-friendly.

Following the publication of a sixth list of angiosperm DNA amounts (Bennett *et al.*, 2000a), which included first C-values for another 691 species, a third release of the database was launched in December 2000. As reference lists for other plant groups were published (i.e., gymnosperms and pteridophytes), the option to make these data more widely available electronically became possible. The C-value data for 48 pteridophytes given in Bennett and Leitch (2001) were also released in December 2000 as the online *Pteridophyte DNA C-values Database*.

These online databases proved to be very useful, and thought was therefore given to constructing counterparts for other plant taxa as data became available. Following one of the key recommendations of the first Plant Genome Size Workshop in 1997, it was decided to assemble one overarching database to cover all land plant groups (i.e., the Embryophyta: angiosperms, gymnosperms, pteridophytes, and bryophytes). The resulting *Plant DNA C-values Database* (www.rbg.kew.org.uk/cval/homepage.html) was launched in September 2001.

Since the first release, work has continued to develop and extend the database. Release 2.0 went live in January 2003 (Bennett and Leitch, 2003) and contains data for 3927 species, comprising 3493 angiosperms, 181 gymnosperms, 82 pteridophytes, and 171 bryophytes (Table 2.1).

Uses and Users of the *Plant DNA C-values* Database

The *Plant DNA C-values Database* is widely used, having received more than 40,000 hits since being launched in September 2001. On average, the database is visited more than 50 times per day, with ~100 C-values commonly taken in a single visit. Not surprisingly, the types of questions for which the database is used are varied. In practical terms, C-value information about individual plant species is important for planning the construction of genomic libraries, for undertaking amplified fragment length polymorphism (AFLP) or microsatellite studies (e.g., Scott *et al.*, 1999; Garner, 2002; Fay *et al.*, 2005), and for deciding which will be the next plant species to have its genome sequenced.

At the other end of the spectrum, the availability of C-value data in one central database has opened up the possibility of carrying out large-scale comparative analyses involving hundreds or even thousands of species. Studies so far reported that have used the database in this way cover diverse fields of biology including ecology, evolution, genomics, and conservation. For example, Knight and Ackerly (2002)

TABLE 2.1 Minimum (min.), maximum (max.), mean, mode, and range (max./min.) of 1C DNA values[a] in major groups of plants, together with the level of species representation of C-value data

	Min. (pg)	Max. (pg)	Mean (pg)	Mode (pg)	Range (max./min.)	No. species with DNA C-values	No. of species recognized[b]	Species representation (%)	No. species in Plant DNA C-values Database[c]	% Representation in the Plant DNA C-values Database[b]
Algae										
Chlorophyta	0.10	19.6	1.75	0.3	196	85	~6500	~1.3	0	0
Rhodophyta	0.10	1.4	0.43	0.2	28	111	~6000	~1.9	0	0
Phaeophyta	0.10	0.9	0.42	0.25	9	44	~1500	~2.9	0	0
Bryophytes	0.17	2.05	0.51	0.45	12.1	171	~18,000	~1.0	171	~1.0
Pteridophytes										
Lycophytes	0.16	11.97	3.81	n/a	74.8	4	~900	~0.4	4	~0.4
Monilophytes	0.77	72.68	13.58	7.8	95.0	63	~9000	~0.7	63	~0.7
Gymnosperms	2.25	32.20	16.99	9.95	14.3	181	~730	~24.8	181	~24.8
Angiosperms	~0.11	127.40	6.30	0.60	~1000	4119	~250,000	~1.6	3493	~1.4
All land plants	~0.11	127.40	6.46	0.60	~1000	4538	~280,000	~1.6	3927	~1.4

[a]C-value data for algae taken from Kapraun (2005), and for bryophytes, lycophytes, monilophytes, gymnosperms and angiosperms from Bennett and Leitch (2003) and Bennett and Leitch (2005).

[b]Numbers of species recognized; taken from Kapraun (2005) for algae, Qiu and Palmer (1999) for bryophytes, lycophytes, and monilophytes, Murray et al. (2001) for gymnosperms, and Bennett and Leitch (1995) for angiosperms.

[c]Plant DNA C-values Database (release 2.0, January 2003) (Bennett and Leitch, 2003).

used the database to extract 401 angiosperm C-values to ask ecological questions such as how DNA amount varied across environmental gradients. This has been aided by the inclusion of information in addition to genome size and taxonomic classification, such as chromosome number, ploidy level, and the method used to analyze DNA content. Specific information relating to the various plant groups is also provided, such as life cycle type (annual, biennial, or perennial) in angiosperms, sperm type (multiflagellate or none) in gymnosperms, and spore type (homosporous or heterosporous), sporangium type (eusporangiate or leptosporangiate), and sperm flagella number (biflagellate or multiflagellate) in pteridophytes.

In the field of genomics, Leitch and Bennett (2004) used data for 3021 angiosperm species to provide insights into the dynamics of C-value evolution in polyploid species. Vinogradov (2003) extracted C-values for 3036 species from the database to reveal a startling negative relationship between the genome size of a species and its current extinction risk status. Most broadly of all, insights into the evolution of DNA amounts across all angiosperms (Soltis *et al.*, 2003), and indeed all land plants (Leitch *et al.*, 2005), have recently been provided by using C-values for more than 140 families included in the database.

PATTERNS IN PLANT GENOME SIZE EVOLUTION

Understanding the evolution of plant genome size involves at least four major components. First, it is necessary to identify the overall distributional patterns of variation observed within and among extant plant taxa. Second, it is important to determine the historical trends that generated the current patterns, and to establish the basic directionality of genome size change through evolutionary time. Third are the questions relating to the mechanisms by which genomes change in size. Fourth are the phenotypic consequences that may influence both the taxonomic and geographical distribution of genome size variation among species. These issues, all of which deal with variation across species, will be treated in order in the following sections, followed later by a discussion of the related issue of variation within species.

THE EXTENT OF VARIATION ACROSS PLANT TAXA

In photosynthetic organisms, reported C-values vary more than 12,000-fold, from ~0.01 pg in the unicellular alga *Ostreococcus tauri* (Prasinophyceae) (Courties *et al.*, 1998) to more than 127 pg in the angiosperm *Fritillaria assyriaca* (Bennett and Smith, 1991). Yet the range, minimum, maximum, mean, and modal C-values vary considerably between the different groups for which data have been compiled (Table 2.1). These differences are further highlighted when data are plotted as histograms (Fig. 2.2). Whereas each group for which data are available contains species with small C-values, the upper limit appears to vary greatly. The smallest range in C-values is found in the Phaeophyta (brown algae, 9-fold), and the largest in the angiosperms (~1000-fold) (Tables 2.1 and 2.2).

However, it should be noted from Table 2.1 that the percent species representation for all but gymnosperms is poor, meaning that the range and distribution of C-values reported may not be entirely representative. (This may be especially true for groups in which fewer than 100 C-values are known.) The exception to this might be the angiosperms, as the ~1000-fold range was first reported more than 20 years ago based on C-values for 993 species (Bennett et al., 1982a) but has not changed even after adding a further 3126 species.

GENOME SIZE IN A PHYLOGENETIC CONTEXT

Given the large range in DNA amounts encountered in plants, any attempt to investigate the directionality of genome size evolution requires that the data are viewed within a rigorous phylogenetic framework. In plants, there are only a few cases where this has been done (Bennetzen and Kellogg, 1997; Cox et al., 1998; Leitch et al., 2001; Obermayer et al., 2002; Wendel et al., 2002). In most cases, this has involved examining variation within individual families of angiosperms, but large-scale analyses across the angiosperm phylogeny have also been conducted, first by Leitch et al. (1998) and then extended using character-state mapping by Soltis et al. (2003). Although an analysis of a continuously varying character such as genome size presents problems when defining character states, it was possible to partially circumvent this difficulty by assigning genome sizes to a series of distinct categories (Soltis et al., 2003): "very small" (C-values ≤1.4 pg), "small" (>1.4 to ≤3.5 pg), "intermediate" (>3.5 to <14.0 pg), "large" (≥14.0 to <35 pg), and "very large" (≥35 pg). With the exception of the "intermediate" category, these size classes were the same as those first defined by Leitch et al. (1998) based on the modal C-value of 0.7 pg for a sample of 2802 species available at the time of analysis. In this sense, the very small and small C-value categories were twice and five times the mode, respectively, while large and very large C-value categories were 20 times and 50 times the mode, respectively.

Using gymnosperms as the outgroup, character-state reconstruction data showed that a very small genome was the ancestral state not only at the root of angiosperms, but also for most major clades within angiosperms (e.g., monocots, magnoliids, all core eudicots, Caryophyllales) (Fig. 2.3) in agreement with the earlier analysis of Leitch et al. (1998). The evolution of very large genomes was shown to be phylogenetically restricted to a few derived families within the monocot and Santalales clades (Fig. 2.3), suggesting that very large genomes have evolved independently more than once during the evolution of angiosperms (Leitch et al., 1998; Soltis et al., 2003).

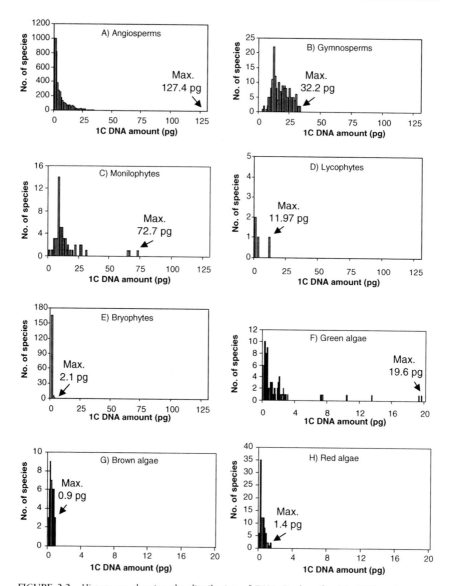

FIGURE 2.2 Histograms showing the distribution of DNA C-values for (A) 4119 angiosperms, (B) 181 gymnosperms, (C) 63 monilophytes, (D) 4 lycophytes, (E) 171 bryophytes, (F) 85 green algae (Chlorophyta), (G) 44 brown algae (Phaeophyta), and (H) 111 red algae (Rhodophyta). The maximum C-value for each group is indicated. (A–E) Redrawn from Leitch *et al.* (2005), reproduced by permission (© Oxford University Press); (F–H) Data from Kapraun (2005).

TABLE 2.2   Some well-known representative species showing the range of 1C DNA amounts in angiosperms

Species	Common name	Chromosome number (2n)	Ploidy level (x)	1C DNA amount	
				pg	Mb[a]
Arabidopsis thaliana	Thale cress	10	2	0.16	157
Oryza sativa	Rice	24	2	0.50	490
Lycopersicon esculentum	Tomato	24	2	1.00	980
Glycine max	Soybean	40	2	1.10	1078
Zea mays	Maize	20	2		
Seneca 60 line				2.50	2450
Zapalote Chico line				3.40	3332
Hordeum vulgare	Barley	14	2	5.55	5400
Secale cereale	Rye	14	2	8.28	8110
Vicia faba	Bean	12	2	13.33	13,060
Allium cepa	Onion	14	2	16.75	16,415
Triticum aestivum	Wheat	42	6	17.32	16,970
Lilium longiflorum	Easter lily	24	2	35.20	34,500
Fritillaria assyriaca	Fritillaria	48	4	127.40	124,850

[a] 1 pg ≈ 980 Mb (see Chapter 1).

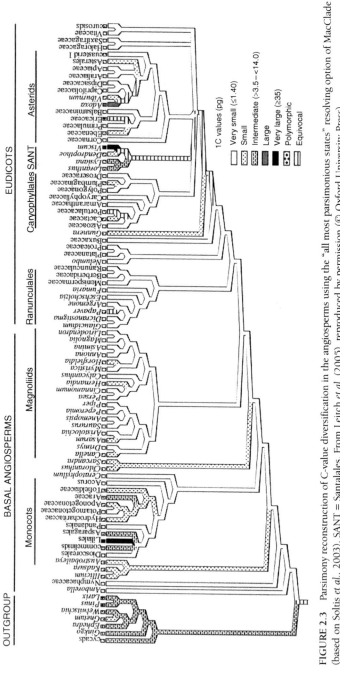

FIGURE 2.3 Parsimony reconstruction of C-value diversification in the angiosperms using the "all most parsimonious states" resolving option of MacClade (based on Soltis et al, 2003). SANT = Santalales. From Leitch et al. (2005), reproduced by permission (© Oxford University Press).

Extending the analysis to include other land plant groups (i.e., gymnosperms, monilophytes, lycophytes, and bryophytes) has provided both insights into the size of ancestral genomes at different parts of the land plant phylogeny and evidence for the bidirectionality of genome size evolution (Leitch *et al.*, 2005) (Fig. 2.4). The main results can be summarized as follows:

1. Different land plant groups are characterized by different ancestral genome sizes. Whereas angiosperms and bryophytes are reconstructed with very small ancestral genomes (i.e., ≤1.4 pg), in gymnosperms and most branches of monilophytes the ancestral genome size is reconstructed as intermediate (i.e., >3.5 to <14.0 pg).

2. Genome size evolution across land plants has been dynamic, with evidence of several independent increases and decreases taking place. Examples of genome size reductions are evident within the monilophytes at the base of the heterosporous water ferns and within the gymnosperms in the branch leading to Gnetaceae (Gnetales). Evidence of large independent increases is seen in the Ophioglossaceae + Psilotaceae clade (monilophytes) and within heterosporous water ferns in Marsileaceae. Thus, observations made within Malvaceae (angiosperms) that both increases and decreases can take place during genome size evolution (Wendel *et al.*, 2002) appear to form a pattern that is repeated across land plants, except perhaps in bryophytes where all species to date have small or very small genomes (Table 2.1).

3. The differences in C-value profiles and patterns of evolution based on reconstruction data strongly suggest that each major group of land plants has been subject to different evolutionary forces. Conversely, it is likely that genome size has itself influenced the shape of the overall plant phylogeny. As will be discussed in a later section, a small genome size correlates with several developmental phenotypic characters (e.g., rapid seedling establishment, short minimum generation time, reduced cost of reproduction, and increased reproductive rate) (Bennett, 1972, 1987; Midgley and Bond, 1991), which together may permit greater evolutionary and ecological flexibility. Thus, the smaller genome sizes of angiosperms may provide one functional explanation as to why they have become so dominant in the global flora relative to the larger-genomed groups like gymnosperms and many monilophytes (Leitch *et al.*, 1998).

HOW DO PLANT GENOME SIZES EVOLVE?

The absence at the time of any known mechanism for decreasing DNA amount led Bennetzen and Kellogg (1997) to speculate that plants may have a "one-way ticket

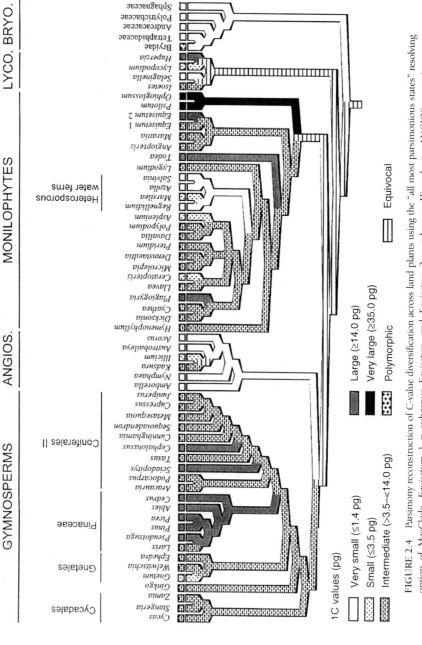

FIGURE 2.4 Parsimony reconstruction of C-value diversification across land plants using the "all most parsimonious states" resolving option of MacClade. *Equisetum* 1 = subgenus *Equisetum*, and *Equisetum* 2 = subgenus *Hippochaete*. ANGIOS. = angiosperms; LYCO. = lycophytes; BRYO. = bryophytes. From Leitch et al. (2005), reproduced by permission (© Oxford University Press).

to genomic obesity" through amplification of retrotransposons and polyploidy. However, as noted previously, there is considerable evidence that both increases and decreases may occur in plant lineages. The revised view is that the net DNA amount of an organism reflects the dynamic balance between the opposing forces of expansion and contraction, in terms of both mechanisms (e.g., Petrov, 2001) and selective consequences (e.g., Gregory, 2001a). Thanks to detailed studies at the DNA sequence level, an understanding of the mechanisms by which genomes change in size is finally taking shape. In many ways, plant studies have led the way on this issue, even though most of the large-scale eukaryotic sequencing projects conducted to date have been from animals and fungi.

SEQUENCES RESPONSIBLE FOR THE RANGE OF GENOME SIZES ENCOUNTERED IN PLANTS

It is generally accepted that different amounts of noncoding, repetitive DNA are primarily responsible for the large range in genome sizes observed in plants (e.g., >70% of some plant genomes are repetitive DNA) (Flavell et al., 1977; Barakat et al., 1997). However, information on the exact nature of the repetitive DNA involved, such as the length of each repeat and their relative contributions, remains elusive in most species (Feschotte et al., 2002).

Based on available data, it is clear that much of the repetitive DNA in plants is composed of transposable elements (TEs). As discussed in detail in Chapter 3, these can be divided into two distinct classes. Class I elements use an RNA-mediated mode of transposition, whereas Class II elements use DNA-mediated transposition mechanisms.

Class I elements are further divided into two subclasses: (1) retrotransposons, which are characterized by direct long terminal repeats (LTRs) (e.g., Ty1-copia and Ty3-gypsy elements) and appear to be ubiquitous in vascular plants (Voytas et al., 1992; Suoniemi et al., 1998), and (2) retroposons, which lack terminal repeats and are referred to as non-LTR retroelements (e.g., long interspersed nuclear elements [LINEs] such as Cin4 in Zea mays, and del2 in Lilium speciosum, and short interspersed nuclear elements [SINEs] such as the S1 element in Brassica). Both subclasses may reach very high copy numbers in plants.

In grasses (Poaceae), LTR-retrotransposons are clearly the most abundant type of transposable elements, and in some species may comprise more than 60% of the nuclear genome (Vicient et al.,1999a; Wicker et al., 2001). In maize, for example, estimates of the size and copy number of retrotransposons in a 240 kilobase (kb) region flanking the adh1 gene suggested that 33–62% of the genome was composed of high copy number retrotransposons, with an additional 16% of the genome containing low to middle copy number retrotransposons (SanMiguel and Bennetzen, 1998). By comparing the structure of the same 240 kb DNA segment in maize (*Zea mays*) and sorghum (*Sorghum bicolor*) it was further shown that the approximately 3-fold larger genome of maize was predominantly due to the presence of retrotransposon sequences that had inserted into the maize genome (SanMiguel and Bennetzen, 1998). Thus, whereas the order of the genes remains largely conserved between the two species, extensive differences in genome size have emerged since their split ~20 million years ago, predominantly reflecting the different extent to which retrotransposons have undergone amplification in the two lineages since that time (Fig. 2.5).

It has been speculated that LTR-retrotransposons play an important role in determining the size of plant genomes in general (Kumar and Bennetzen, 1999). However, although this may be the case in grasses, sequences other than these appear to have a greater influence on genome size differences in other organisms (Wendel and Wessler, 2000), especially those with smaller genomes (Kidwell, 2002).

Class II elements include the *Helitrons, Mu*, and *mutator*-like elements (MULEs), which tend to be large (up to 20 kb), and the smaller miniature inverted repeat transposable elements (MITEs) (0.1–0.5 kb). From available sequence data, Class II elements make up ~ 6% and ~ 12% of the *Arabidopsis thaliana* and *Oryza sativa* genomes, respectively (Feschotte *et al.*, 2002; Jiang *et al.*, 2004).

WHAT TRIGGERS THE SPREAD OF TRANSPOSABLE ELEMENTS?

From the previous section, it is evident that transposable element proliferation plays a major role in increasing plant genome sizes through time. However, the factor(s) responsible for triggering amplification in certain lineages are still not clearly understood in most plant systems. In fact, there remains a puzzling discrepancy between the large number of plant TEs characterized to date and their apparent transcriptional silence observed during normal plant development. The few exceptions to this include the *BARE-1* element from barley (Vicient *et al.*, 1999a), the related *OARE-1* element from oats (Kimura *et al.*, 2001), and *IRRE* elements from some *Iris* species (Kentner *et al.*, 2003), which have been shown to be transcriptionally active under normal growing conditions.

In at least some cases, transcriptional activation can be induced by experimental manipulations of various biotic or abiotic stresses such as wounding, tissue culture, and pathogen attack (e.g., Grandbastien, 1998; Feschotte *et al.*, 2002). Similar effects have also been documented in a natural setting, as with the recent report of retrotransposon activation in natural populations of *Hordeum spontaneum* by Kalender *et al.* (2000). These authors showed that the *BARE-1* LTR-retrotransposon, which comprises ~3% of the *H. spontaneum* genome (Vicient *et al.*, 1999a), had been highly insertionally active in recent times within different plant populations from the single location of "Evolution Canyon" in Israel. They also showed that copy number

107

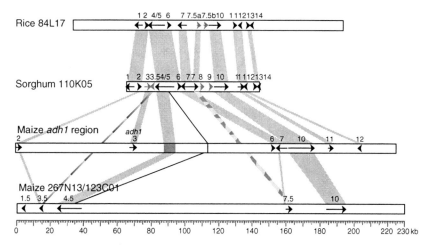

FIGURE 2.5   Comparison between the organization of colinear genomic regions in rice, sorghum, and two homoeologous segments in maize. Shaded areas connecting the regions represent conserved sequences. The length of each segment is drawn to scale and illustrates the large expansion in length, mainly owing to insertion of retrotransposons, that has taken place since *Sorghum* and maize diverged 20 million years ago. From Ilic *et al.* (2003), reproduced by permission (© National Academy of Sciences).

varied 3-fold between different populations and that there was a correlation between copy number and various ecological variables, most notably water availability. Specifically, the population with the highest copy number of *BARE-1* elements occupied the driest site in the canyon, suggesting the possibility that water stress induced the activation of *BARE-1*. Although the question as to the role *BARE-1* might play in the physiological stress response remains to be determined, the data highlight the potential for natural environmental cues to trigger retrotransposon activity. Interestingly, analysis of the draft sequence of the rice genome has also shown that a DNA transposon (*mPing* MITE) has been preferentially amplified in cultivars adapted to environmental extremes (Jiang *et al.*, 2003).

A few studies have suggested that polyploidization and interspecific hybridization may also trigger TE amplification in plants, for example in *Nicotiana* (Fig. 2.6A) (Comai, 2000), *Aegilops-Triticum* allopolyploids (Kashkush *et al.*, 2002), and *Spartina anglica* (Baumel *et al.*, 2002). However, this is clearly not always the case. The doubling of the maize genome as a result of LTR-retrotransposon activity was estimated to have taken place primarily in the last three million years (SanMiguel *et al.*, 1998), roughly eight million years after the polyploidization event that gave rise to the species (Gaut and Doebley, 1997). Similarly, bursts of retrotransposon amplification estimated to have occurred in the last two million years in *Arabidopsis thaliana* (Devos *et al.*, 2002) and the last five million years in *Oryza sativa* (Vitte and Panaud, 2003)

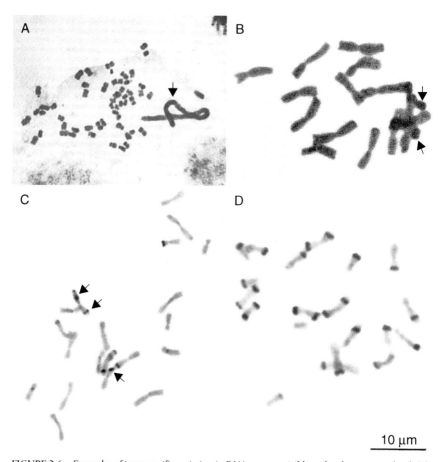

FIGURE 2.6   Examples of intraspecific variation in DNA amount visible at the chromosome level. (A) Megachromosome (arrowed) observed in a corolla metaphase cell from a *Nicotiana tabacum* × *N. otophora* hybrid made by Gerstel and Burns (1966). (B) Loss of heterochromatin blocks of DNA (arrowed) from chromosomes of *Secale kuprijanovii*. (C,D) Intraspecific differences in amount of heterochromatic DNA visible as C-bands (some of which are arrowed) in maize, *Zea mays* ssp. *mays*. (C) Commercial hybrid Seneca 60 (1C = 2.5 pg), a line of maize with few heterochromatic C-bands; (D) Race Zapalote Chico Oaxaca 50 (1C = 3.4 pg), a line of maize with many C-bands. (A) From Reed (1991), reproduced by permission (© Elsevier Inc.), (B) From Gustafson *et al.* (1983), reproduced by permission (© Springer-Verlag), (C,D) From Laurie and Bennett (1985), reproduced by permission (© Nature Publishing Group).

do not coincide with any known hybridization or polyploidization event. Further, where TE amplification has been shown to take place following hybridization, this may be quickly silenced by other genetic and epigenetic events such as methylation (e.g., Liu and Wendel, 2000), so the effects of TE amplification owing to hybridization and polyploidization may be minimal. The recent studies in *Arabidopsis thaliana* showing that TE amplification may be under epigenetic control, together with the development of tools such as transposon display or sequence-specific amplification polymorphism (S-SAP) for detecting TE activity, offer the potential for significant progress to be made in this field in the near future (Feschotte *et al.*, 2002).

In summary, whereas TEs contribute greatly to the large variation in genome size observed among many plant species, especially the grasses, the factors responsible for triggering their amplification still remain unknown in the majority of cases. Moreover, it remains to be determined to what extent TE amplification contributes to changes in total genome size within a species. The study by Kalendar *et al.* (2000) discussed earlier showed that although copy number of the *BARE-1* retrotransposon varied considerably between different ecological sites, no correlation was found between genome size and copy number of the repeat. Such results suggest that although TE amplification clearly has the *potential* to increase genome size, whether or not it actually does so, or instead is compensated for by a reduction in sequence repeats elsewhere in the genome, may depend on other factors within the cell. The possibility that genome size is maintained by internal stabilizing mechanisms will be an important consideration in the discussion of intraspecific variation given later.

Satellite DNA

Tandemly arranged repeats of identical or similar sequences (satellite DNA) can also comprise large fractions of plant genomes. Indeed, some of the first satellite DNAs to be isolated were from rye (*Secale cereale*), with one type comprising ~6% of the genome (Bedbrook *et al.*, 1980). Although satellite sequences are variable in size, the most common monomeric units are 150–180 bp and 320–380 bp. The structure of the repeat may be highly complex, and in some cases may include DNA from other repeat classes. For example, a centromeric satellite sequence in *Brassica campestris* was shown to contain DNA sequences related to both tRNA genes and SINEs (Kubis *et al.*, 1998). Minisatellites (10–40 bp repeats) and microsatellites (2–6 bp, also called simple sequence repeats or SSRs) may also represent an appreciable amount of DNA, such as in telomeric sequences. These are believed to occur ubiquitously in all eukaryotic genomes, but their number and the proportion of the genome they occupy vary significantly among species (Ellegren, 2002; Morgante *et al.*, 2002).

Genome Size Increase by Polyploidy

Polyploidy, resulting from combining three or more basic chromosome sets or genomes in a single nucleus, is a prominent mode of speciation, especially in angiosperms and monilophytes (see Chapter 7), and results in an instant increase in the DNA content of the nucleus (i.e., its C-value). Its prominence in plant evolution has been brought into even sharper focus by recent large-scale genomic analyses that have uncovered the polyploid nature of many plant genomes that were traditionally considered to be diploid, including maize (Moore et al., 1995) and Arabidopsis thaliana (Bowers et al., 2003). The fact that even small-genomed plants such as these appear to be ancient polyploids has led to the suggestion that all angiosperms may have experienced polyploidization at some point in their evolutionary history (Wendel, 2000) (see Chapter 7).

The issue of polyploidy leads to terminological complications when referring to "C-value" versus "genome size." Specifically, for a diploid with just one genome in each gametic nucleus, C-value and genome size are interchangeable. However, in a polyploid, the C-value will represent the total DNA amount of all genomes within the nucleus (e.g., in a tetraploid, each gametic nucleus will have two genomes, whereas a hexaploid will have three). Thus in polyploids, C-value and "basic" genome size sensu stricto are not equivalent: in a polyploid with more than two genomes in the gametic nucleus, basic genome size will always be smaller than the C-value. In general, the basic genome size of a polyploid can be estimated by dividing C-value by the number of genomes in the gametic nucleus (i.e., half the ploidy) but it should be recognized that this only gives an accurate estimate in taxa with equal genome sizes. In taxa with genomes of different sizes (e.g., some allopolyploids) it gives only a mean genome size. The value will be close to the actual genome sizes in most but not all species (e.g., some taxa with bimodal karyotypes). As polyploidy is so prevalent, especially within angiosperms and monilophytes, this has led some authors to suggest that terminology should distinguish between the 1C-value, representing the original definition of C-value (which is independent of ploidy level), and the $1C_x$-value, which indicates the basic genome size (Greilhuber et al., 2005).

In recognizing the difference between basic genome size and C-value it becomes clear that polyploidy will only result in an increase in C-value and not basic genome size. In simple terms, the expectation in new polyploids is that C-value will increase in direct proportion with ploidal level, and that the basic size of the individual genomes included will be unchanged. This expectation is observed in some polyploid series, especially those newly formed (e.g., see Pires et al., 2004). However, there are many examples suggesting that C-values in particular polyploids are less than the expected sum of parental genomes (e.g., Ozkan et al., 2003; Leitch and Bennett, 2004). On a larger scale, a recent analysis of 3008 angiosperms revealed that mean 1C DNA amount did not increase in direct proportion with ploidy (Fig. 2.7A), and thus that mean basic genome size (calculated by dividing 1C value by half the ploidy) tended to decrease with increasing ploidy (Fig. 2.7B) (Leitch and Bennett, 2004). These results suggest that "genome downsizing" following polyploid formation may be a widespread phenomenon of considerable biological significance.

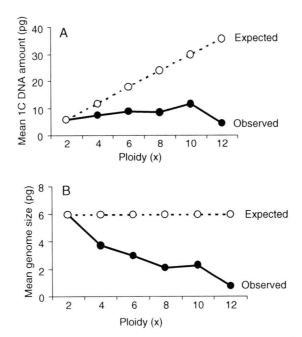

FIGURE 2.7   DNA C-values and basic genome sizes (2C value divided by ploidy) in an "all angiosperms" sample with known C-values and even ploidies between 2x and 12x. (A) Mean 1C DNA values observed (——•——) compared with expectation (- -o- -) assuming C-value and ploidy are directly proportional. (B) Mean basic genome size values observed (——•——) compared with expectation (- -o- -). From Leitch and Bennett (2004), reproduced by permission (© Blackwell Publishing).

In fact, the loss of DNA following polyploidy has been reported both at the gene level (e.g., loss of ribosomal DNA genes, reviewed by Wendel, 2000) and on a genomewide scale (e.g., Shaked *et al.*, 2001; Ilic *et al.*, 2003; Ozkan *et al.*, 2003). For example, a comparison of DNA sequences in orthologous regions of maize, sorghum, and rice suggested that maize may have lost up to 40% of its genes in the analyzed regions since the polyploid event that gave rise to it (Ilic *et al.*, 2003) (see Fig. 2.5). Mechanisms responsible for the loss remain ambiguous, but there are some indications that this involves unequal crossing over.

In some cases, polyploid formation involves the bringing together of genomes that may be similar enough to undergo homoeologous chromosome pairing during meiosis (i.e., between the equivalent chromosomes derived from the different parental species; see Chapter 7). This may lead indirectly to DNA loss as a result of a breakdown in the postreplicative mismatch repair system (Comai, 2000), which corrects any mismatched base pairs or displaced loops in double-stranded DNA following DNA replication and normally also blocks homoeologous recombination by the binding of mismatch repair proteins. In a newly formed polyploid, the excessive number of mismatches caused by homoeologous recombination may lead to the depletion of the mismatch repair proteins, and consequently to a rise in the amount of homoeologous recombination and associated deletion of DNA. A positive feedback loop could be envisaged with increasing homoeologous recombination and associated DNA deletions as the mismatch repair proteins are used up (Comai, 2000). How much DNA is lost from polyploids by this mechanism may be species-dependent, because the extent of homoeologous recombination in different polyploids may vary. It should be noted that this is probably not a ubiquitous mechanism, as there are cases in which polyploidy is associated with only a minimal level of homoeologous recombination (e.g., in *Brassica* polyploids) (Axelsson *et al.*, 2000).

In short, polyploidy will undoubtedly lead to an increase in C-value, but the extent to which this persists over evolutionary time may vary considerably from one case to the next. Also, as discussed in Chapter 1, this will constitute a change in genome size, strictly defined, only following rediploidization.

Mechanisms of Genome Size Decrease

Mechanisms responsible for bringing about a reduction in DNA amount have only recently come to light, and there is still much to learn regarding their significance in genome evolution. However, already the available studies suggest that deletional mechanisms may play a more prominent role in genome size evolution than previously recognized. Several such mechanisms have been identified in plants.

Unequal Intrastrand Homologous Recombination

The process of unequal intrastrand homologous recombination occurs between the long terminal repeats of LTR-retrotransposons, and can lead to deletion of the internal DNA segment and one LTR, leaving behind only one "solo LTR." If this occurs between adjacent LTR-retrotransposons of the same family, several elements may be lost in a single step. That this mechanism could provide a way to counteract retroelement-driven genome expansion was first proposed by Vicient *et al.* (1999b), based on an observed 16-fold excess of solo LTRs to intact *BARE-1* elements in barley.

Subsequent research has shown that the extent to which unequal homologous recombination takes place between individual LTRs, and hence its contribution to genome downsizing, may depend on the particular LTR-retrotransposon in question. Notably, recent studies by Vicient and Schulman (2002) and Vitte and Panaud (2003) found that the proportion of solo LTRs varied considerably among retroelement types in rice. For example, the ratio of solo LTRs to complete retroelements ranged from 6.3 : 1 for the *RIRE1 Ty1-copia*–like element to 0.1 : 1

for the *Retrosat1 Ty3-gypsy*–like element. It was speculated that these differences may be determined by the specific characteristics of the retroelement such as the preferential insertion site (e.g., centromeric versus dispersed along the chromosome).

The length of the LTR element may also be a crucial factor in determining the efficacy of this deletional mechanism. Thus Shirasu *et al.* (2000) suggested that the lower frequency of solo LTRs in maize arose from a lower recombination efficiency between the comparatively short LTRs of maize retroelements (average 450 bp) compared with the longer LTRs of five barley retroelements studied (1.5–4.9 kb). Interestingly, the only two solo LTRs identified in the analysis of a 240 kb segment of the maize genome had longer than average LTRs (1.1 kb) (SanMiguel *et al.*, 1996). Further, Vitte and Panaud (2003) noted that the proportion of solo-LTRs increased with increasing LTR size in three rice retroelements. Overall, the extent to which DNA is lost via unequal homologous recombination may be determined, at least in part, by the characteristics of the LTR-retrotransposons that comprise a species' genome.

Whether unequal homologous recombination occurs continuously through time or is triggered by bursts of retrotransposon amplification (as proposed by Rabinowicz, 2000) remains to be determined, although preliminary data from a study of three *Ty3-gypsy*–like retroelements in rice suggest that solo LTR formation seems to be concomitant with the amplification of active copies of the retro-elements (Vitte and Panaud, 2003).

Illegitimate Recombination

Illegitimate recombination, or recombination that does not require the participation of a recA protein or large (>50 bp) stretches of sequence homology, can be the product of many different mechanisms including slipped strand repair and double strand break repair. Because it does not require such long stretches of homologous sequences to work as unequal homologous recombination, it has the potential to occur within any region of the genome. Indeed, an analysis by Devos *et al.* (2002) of LTR-retrotransposons concluded that illegitimate recombination was the main driving force behind genome size decrease in *Arabidopsis thaliana,* removing at least five times more DNA than unequal homologous recombination because it can act on a larger fraction of the genome. A role for illegitimate recombination as a mechanism to remove DNA has also been suggested by Bennetzen *et al.* (2005) and Ma *et al.* (2004) in rice, and by Wicker *et al.* (2003) in their study of the much larger genomes of *Triticum* species.

Loss of DNA During the Repair of Double Stranded Breaks

The repair of double stranded breaks (DSBs) in plant DNA is often accompa-nied by DNA deletions, although insertions may also occur in some species.

Kirik *et al.* (2000) compared the products of DSBs between *Arabidopsis thaliana* and *Nicotiana tabacum* (tobacco), two species differing more than 20-fold in DNA amount. They found that the size of the deletions differed markedly between the two species: *Arabidopsis* deletions were on average larger than in tobacco, and were not associated with insertions. The apparent negative correlation between the size of the deletions and genome size led to the speculation that species-specific differences in DSB repair pathways may contribute significantly to the evolution of genome size. This has been supported by recent data showing differences in the mechanisms used by *Arabidopsis* and tobacco to repair DSBs (Filkowski *et al.*, 2004). Another component, namely differences in the stability of the free DNA ends resulting from DSBs, was assessed by Orel and Puchta (2003). They established that free DNA ends were more stable in tobacco than *Arabidopsis*, owing to lower DNA exonuclease activity and/or better protection of the DNA break ends. The implication is that if such patterns were observed to occur on a wider range of species, then differences in the degree of exonucleolytic degradation of DNA ends might prove to be an important force in the evolution of genome size (Orel and Puchta, 2003).

KEY CORRELATES OF GENOME SIZE
ACROSS PLANT SPECIES

The recognition that DNA amount in eukaryotes varies over several orders of magnitude, even in related groups of organisms of similar complexity, has provoked considerable interest in the biological effects and other consequences of such differences. The first studies indicating that this is an important line of enquiry were in animals (see Chapter 1), but, as seen in the following section, work on plants has also played a major role in developing this field.

EARLY WORK ON THE PHENOTYPIC CONSEQUENCES
OF GENOME SIZE VARIATION IN PLANTS

Much of the earliest work on the phenotypic effects of genome size variation in plants was motivated by strong practical, and even political, interests. During the early Cold War days of the 1950s and 1960s, much attention was focused on the relationship between genome size and radiosensitivity in plants (e.g., Sparrow and Miksche, 1961; Abrahamson *et al.*, 1973) (Fig. 2.8). That genome size studies were seen as potentially important for national security interests explains the generous funding obtained for such work by a group at the Brookhaven National Laboratory (Upton, NY) led by Arnold Sparrow.

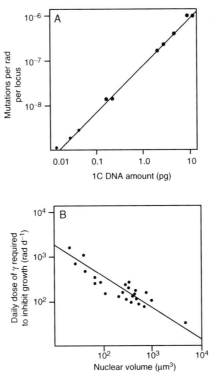

FIGURE 2.8   (A) Relationship between forward mutation rate per locus per rad and the 1C DNA amount; (B) The relationship between nuclear volume and radiosensitivity in 23 species of plants. (A) From a figure redrawn in Bennett (1987) based on a figure originally presented by Abrahamson *et al.* (1973), reproduced by permission (© Blackwell Publishing); (B) Data from Sparrow and Miksche (1961).

Sparrow's analyses included a major emphasis on plant materials to develop a theoretical understanding of the principles involved in radiation sensitivity. This was primarily because plants (1) display a very wide range of DNA content, in terms of both polyploidy and basic genome size; (2) represent staple food sources of national and global strategic significance; (3) are readily available and remain immobile during prolonged exposure to various types of radiation treatment; and (4) are socially more acceptable than animal subjects for work that involves killing large numbers of specimens. Such studies first revealed close correlations between radiosensitivity and nuclear volume (Sparrow and Miksche, 1961), then showed a positive relationship between nuclear volume and DNA amount (Baetcke *et al.*, 1967), and hence revealed an association between DNA content and sensitivity to nuclear radiation. There was considerable discussion as to which was the most meaningful and determining character given the possibly confusing effects of ploidy level (x), heterochromatin, and chromosome number (2n), and it was concluded that interphase chromosome volume (i.e., mean DNA amount per chromosome, equal to nuclear DNA amount ÷ 2n) gave the closest correlation (Baetcke et al., 1967).

From the perspective of genome size evolution, the correlations with sensitivity to intensive radiation were not particularly important. Fortunately, other experimental work undertaken by members of Sparrow's group (e.g., Van't Hof, Price), though still focused on the strategic questions, was of much broader significance and led to seminal discoveries relating to fundamental aspects of cell development (Van't Hof and Sparrow, 1963) and overall patterns of genome size variation (Sparrow and Nauman, 1976).

Meanwhile, other laboratories began parallel studies on the effects of genome size on various additional characters of practical significance. This included a British group of cytogeneticists led by Hugh Rees at University College of Wales, Aberystwyth, whose primary interest was plant breeding. Such work focused on relationships between DNA amount and physical characters (length, volume, mass) at the chromosomal, nuclear, and cellular levels, especially in crops and their close relatives (Rees et al., 1966).

Today, relationships are known between nuclear DNA amount and more than 40 widely different phenotypic characters at the nuclear, cellular, tissue, and organismal levels. Moreover, these extend to all types of plants, and are no longer limited to species of practical interest (e.g., crops or common experimental subjects).

CHROMOSOME SIZE

Early work revealed highly significant positive correlations between nuclear DNA amount and chromosomal characters, such as total mitotic metaphase volume in samples of species in the genera *Lathyrus, Vicia, Lolium* (Rees et al., 1966), and *Allium* (Jones and Rees, 1968). Such findings have since been confirmed in many other comparisons both within and across genera. For example, Figure 2.9A shows the relationship between DNA amount and total mitotic chromosome volume per somatic metaphase cell in 14 angiosperm species estimated from reconstructed cells using quantitative electron microscopy (Bennett et al., 1983), and more recently chromosome area and length were shown to be positively correlated with DNA amount among 12 diploid rice species (Uozo et al., 1997). A positive relationship was also found to apply to specialized chromosome regions; Figure 2.9B shows the tight relationship between DNA C-value and the total volume of centromeres per cell in 11 angiosperm species (Bennett et al., 1981).

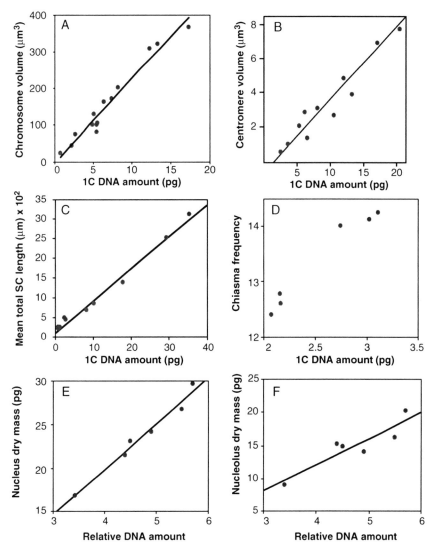

FIGURE 2.9 Relationship between DNA amount and six nuclear characters in angiosperms.
(A) Total somatic chromosome volume (μm³) at mitiotic metaphase in 14 species; (B) Total volume of
centromeres per nucleus in 11 species; (C) Total length of the synaptonemal complex (SC) per cell at
pachytene in 10 species; (D) Mean chiasma frequency in pollen mother cells of six *Lolium* species;
(E) Nuclear dry mass and (F) nucleolar dry mass in six species of *Sorghum*. (A) Redrawn from Bennett
et al. (1983), reproduced by permission (© Company of Biologists Ltd.); (B) From Bennett *et al.*
(1981), reproduced by permission (© Company of Biologists Ltd.); (C) Modified from Anderson *et al.*
(1985), reproduced by permission (© Elsevier Inc.); (D) From Rees and Narayan (1988), reproduced
by permission of the authors; (E, F) Modified from Paroda and Rees (1971), reproduced by permission (© Springer-Verlag).

In addition to volume, total mitotic metaphase chromosome length has been shown to correlate positively with genome size (e.g., Rothfels *et al.*, 1966). However, these relationships are not typically as close as with volume because of the differential condensation of various chromosomal segments (especially euchromatin versus heterochromatin). During meiosis, when chromatin condensation is more relaxed, such differences may be minimized. Thus Anderson *et al.* (1985) showed a tight correlation between genome size and the total length of the haploid chromosome complement at meiosis (measured by tracing the lengths of synaptomental complexes in spread pachytene cells of 10 angiosperms with a roughly 30-fold range of DNA C-values) (see Fig. 2.9C). In addition, total chiasma frequency (i.e., recombination rate) per chromosome and per complete complement are also positively related to genome size (Rees and Narayan, 1988) (see Fig. 2.9D).

NUCLEUS SIZE

Examples of other size-related correlations at the nuclear level include the positive relationship observed between genome size and interphase nuclear volume mentioned in the previous section (e.g., Baetcke *et al.*, 1967). Many of the studies reporting positive relationships between genome size and total chromosome volume also noted similar relationships between genome size and total nuclear dry mass (measured using interference microscopy) in the same plant genera (Rees and Hazarika, 1969; Pegington and Rees, 1970; Paroda and Rees, 1971). For example, Figure 2.9E shows a close relationship between genome size and total nuclear dry mass in six *Sorghum* species. White and Rees (1987) also reported a positive relationship between nuclear dry mass and DNA content in six *Petunia* species. Similar positive relationships have been found with total nuclear histone content (Rasch and Woodard, 1959) and total nucleolar dry mass (Paroda and Rees, 1971) (see Fig. 2.9F). More recently, rRNA gene copy number has been shown to correlate positively with DNA amount in both plants and animals across a wide range of taxa (Prokopowich *et al.*, 2003).

CELL SIZE

As in the well-known example of vertebrate erythrocytes (see Chapter 1), the effects of bulk DNA content extend to characters at the cellular level in plants. For example, Martin (1966) noted a positive correlation between DNA amount and cell mass in root tip cells of 12 angiosperm species, and Holm-Hanson (1969) reported a relationship between DNA content per cell and the total

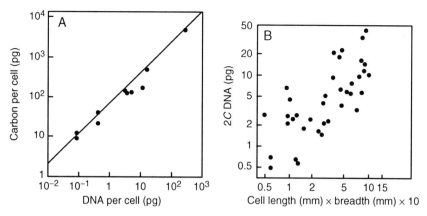

FIGURE 2.10   Relationship between DNA amount and cellular characters. (A) Total inorganic carbon per cell in 10 species of unicellular algae; (B) Mean length × breadth of epidermal cells in mature leaves of a range of herbaceous species. (A) Redrawn from Holm-Hanson (1969), reproduced by permission (© American Association for the Advancement of Science); (B) From a figure presented by Bennett (1987), based on data from Grime (1983), reproduced by permission (© Blackwell Publishing).

weight of carbon per cell in 10 unicellular algae (Fig. 2.10A). Grime (1983) showed a strong positive relationship between nuclear DNA content and the linear dimensions of leaf epidermal cells for 37 British herbaceous angiosperms (Fig. 2.10B).

CELL DIVISION RATE

It has long been appreciated that nucleus size, cell size, and cell division rate are closely linked. As part of the radiosensitivity research program, Van't Hof and Sparrow (1963) noted that "for diploid plants a relationship does exist between the minimum mitotic cycle time, the interphase nuclear volume, and the DNA content per cell. Moreover, the relationship is such that if any one of the three variables is known, an estimate can be made for the remaining two." Figure 2.11A shows their data for a sample of root-tip meristem cells from six angiosperm species, all grown at 23°C, and reveals a strikingly close association between DNA content and mitotic cycle time. Additional experiments showed that a relationship could even be identified between DNA amount and the duration of the DNA synthesis phase (S-phase) in particular (Van't Hof, 1965). These results were first obtained using only small numbers of species, but larger subsequent

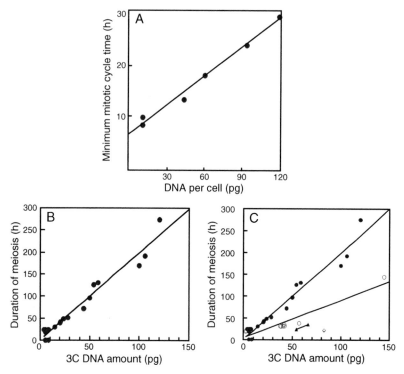

FIGURE 2.11   Relationships between DNA amount and duration of mitosis and meiosis. (A) The relationship between minimum duration of the mitotic cycle in root-tip cells of six angiosperms grown at 23°C; (B) Duration of meiosis in 18 diploid angiosperm species grown at 20°C; (C) Duration of meiosis in 18 diploid ($2x$) (●), ten tetraploid ($4x$) ( ), three hexaploid ($6x$) (▲) and two octoploid ($8x$) (◊) angiosperms. (A) Data from Van't Hof and Sparrow (1963); (B) From Bennett (1977), reproduced by permission (© The Royal Society); (C) Data from Bennett (1977).

comparisons strongly confirmed the original conclusions (Van't Hof, 1974). It has also since been demonstrated that the durations of *all* phases of the cell cycle, not only of S-phase, are positively related to C-value (e.g., Evans and Rees, 1971; Evans *et al.,* 1972).

    Reproductive cell development is both protracted (compared with other stages) and highly canalized, and hence is particularly useful for showing relationships between genome size and other characters. Thus Bennett (1971) found a strong positive correlation between genome size and the duration of meiosis in diploid angiosperms when he examined reproductive cell development in shoots. This was later confirmed for a larger sample of species (Bennett, 1977). By way of example, Figure 2.11B shows the remarkably precise relationship between DNA

C-value and the duration of meiosis in 18 unrelated diploid angiosperm species, all grown at 20°C.

Interestingly, comparisons for plants at different ploidal levels revealed close positive correlations between nuclear DNA C-value and the duration of meiosis, but also showed highly significant differences in the slope of the relationship according to ploidy (see Fig. 2.11C). Surprisingly, the duration of meiosis in plants grown at 20°C was reduced as ploidy increased. In wheat, for example, meiosis was found to take 42 hours in diploids, 30 hours in tetraploids, but only 24 hours in hexaploids, despite the proportionately higher DNA contents of the polyploids (Bennett and Smith, 1972). This shows conclusively that such relationships are not determined by DNA amount alone. Although this is undoubtedly one key factor, its effects can be modified by genetic factors such as genotypic differences and variation in gene dosage owing to polyploidy (Bennett et al., 1974), as well as abiotic factors such as temperature (Bennett et al., 1972). The duration of meiosis in plants can even differ between male and female meiocytes of the same species (Bennett, 1977).

Causation at the Cellular Level: The Nucleotype Concept

It is clear that cell size and division rate are not determined solely by C-value, given that both parameters can vary considerably within and among organisms even with constant nuclear DNA amount. Nevertheless, many relationships discussed in the previous section are strikingly close, a fact that is especially clear when several plots for widely different characters are viewed together, as in Figures 2.9 through 2.11. Indeed, some of these linkages are so strong as to be more reminiscent of physical or chemical relationships than biological ones (Bennett, 1987).

Taken together, these considerations suggest that the relationship between DNA content and chromosomal, nuclear, and cellular parameters is causal in nature, albeit with other factors involved. This is quite evident at the chromosomal level, in which the amount of DNA necessarily impacts upon total chromosome volume and mass, but in which proteins (e.g., histones) also play a role. Figure 2.1, which shows the chromosome complements from species differing ~220-fold in DNA amount all taken at the same magnification, illustrates this point. Similarly, as the nucleus is assembled around a scaffolding of DNA, C-value must causally influence nuclear volume (Cavalier-Smith, 1985).

For biophysical reasons, it is impossible to increase greatly the C-value without also increasing the minimum time needed for cell division. This is because more DNA not only takes longer to replicate (i.e., prolongs S-phase), it also impacts on all stages of the cell cycle (see Gregory, 2001a, and Chapter 1). According to the model developed by Gregory (2001a) this delay in cell division is ultimately responsible for generating the positive correlation between DNA content and cell size (see Chapter 1). At the very least, it is clear that one cannot increase DNA content beyond a certain point without also increasing nucleus and cell size. For example, a comparison between *Fritillaria* sp. (1C ≈ 65 pg) and *Myriophyllum spicatum* (1C = 0.3 pg) shows the impossibility of containing the nucleus, chromosomes, and even the DNA of the former in the small cells of the latter (see Fig. 2.1B,C).

The realization that the relationships discussed in the previous sections were influenced causally by the amount of DNA led to the development of the "nucleotype" concept (Bennett, 1971). Specifically, the nucleotype describes those conditions of the nuclear DNA, most notably its amount, that affect the phenotype independently of its encoded informational content. Whereas the combined set of genes defines the "genotype," the nucleotype consists of all the DNA, both genic and nongenic (Bennett, 1971). The nucleotype can be considered as setting the minimum conditions, or perhaps as exerting a very coarse control, of parameters at the cell level, whereas the genotype is responsible for fine control of these features within these limits (Bennett, 1972; Karp *et al.*, 1982; Gregory, 2001a). For example, while nuclear volume is subject to variation by genetic control during development despite a constant DNA amount (Bennett, 1970), such control can operate only at or above the minimum volume determined by the DNA C-value. The same is true of parameters at the cell level and above.

POLLEN AND SEEDS

There is now considerable evidence that these nucleotypic effects at the cellular level are additive and extend to higher level features of direct relevance to fitness. For example, size relationships are known for reproductive structures such as the male gametophyte (pollen grains) and early sporophyte stages (seeds) in angiosperms. As examples, Figure 2.12A shows the positive relationship between DNA amount and mean pollen grain volume in wind pollinated grasses (Bennett, 1972), and Figure 2.12B shows the correlation between DNA content and minimum seed weight in 24 British legumes (Mowforth, 1985). Similar relationships with pollen size have been shown for species of *Ranunculus*, *Vicia* (Bennett, 1973), and *Petunia* (White and Rees, 1987). Positive relationships between DNA amount and seed weight have likewise been reported for comparisons within populations of the same species (e.g., Caceres *et al.*, 1998), within genera, including *Vicia* and *Allium* (Fig. 2.12C) (Bennett, 1972, 1973; see also Knight and Ackerly, 2002), and across large numbers of species (e.g., Thompson, 1990; Knight *et al.*, 2005). Interestingly, the use of quantile regression analysis by Knight *et al.* (2005) suggests that the relationship between seed weight and genome size may not be linear. Thus it appears that whereas species with small genomes

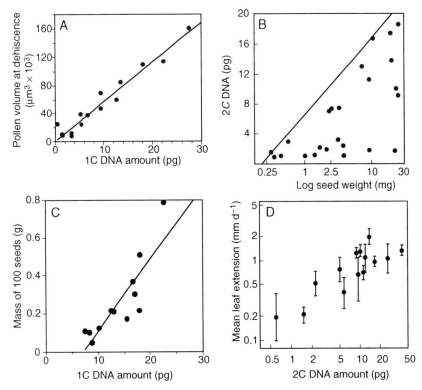

FIGURE 2.12   Relationships between DNA amount and reproductive and growth parameters.
(A) Pollen volume in 16 grass and cereal species; (B) Seed weight in 24 British legume species; (C) Seed
dry mass in 12 *Allium* species; (D) Mean range of leaf extension over the period March 25 to April 5
in 14 grassland species coexisting in the same turf. (A) Redrawn from Bennett (1972), reproduced by
permission (© The Royal Society); (B) From Bennett (1987), originally based on Mowforth (1985),
reproduced by permission (© Blackwell Publishing); (C) Redrawn from Bennett (1972), reproduced
by permission (© The Royal Society); (D) From Bennett (1987), based on a figure first presented by
Grime (1983), reproduced by permission (© Blackwell Publishing).

display a wide range of seed weights, those with large genomes rarely produce
small seeds.

Of course, these relationships are complex, and again it should not be imagined
that genome size alone determines parameters such as seed weight or pollen size.
Nevertheless, genome size may have a major effect in determining the minimum
values possible for these features—an important consideration, given that seed
weight is believed to affect many other ecologically relevant attributes such as
invasiveness and survivability (Westoby *et al.*, 1992; Rejmanek, 1996).

Minimum Generation Time and Developmental Lifestyle

Development at the organism level consists of division and growth at the cell level, suggesting that it can be strongly influenced by variation in genome size. However, because genome size is correlated negatively with division rate, but positively with cell size, the nature of the relationships may vary according to which cellular parameter dominates in a particular developmental process. Thus, although it is generally the case that a large genome size limits the rate at which plants can develop, DNA content may correlate positively with growth rate under certain conditions (Grime and Mowforth, 1982; Knight *et al.*, 2005).

In 1972, as part of the early work on the nucleotype concept, Bennett asked whether DNA content might place a lower limit on the duration of the period from germination until the production of the first mature seed—that is, minimum generation time (MGT). To test this, Bennett (1972) compared the mean and the ranges of nuclear DNA contents for 271 angiosperm species with different life cycle types and different ranges of MGT. For this comparison the species were divided into one of four types of life cycles: (1) ephemerals, which can complete their life cycle in a very short period of time (arbitrarily defined as seven weeks or less), (2) annuals, which by definition complete their life cycle within 52 weeks, (3) facultative perennials, which can potentially set fertile seed within 52 weeks of germination, and (4) obligate perennials, which require more than 52 weeks to produce mature seed. The following intriguing results were obtained:

1. The mean 1C nuclear DNA content for ephemeral species (1.5 pg) was less than for annuals (7.0 pg), which in turn was less than for perennial species (24.6 pg). This was true irrespective of whether the comparison included all species or only diploids.
2. The maximum 1C DNA content was lower for ephemerals (3.4 pg) than for annual species (27.6 pg), which was much lower than the maximum for perennials (127.4 pg).
3. The mean and range of DNA amounts for facultative perennials and annuals (both of which have the same maximum MGT of 52 weeks) were very similar and both were much less than for obligate perennials.
4. Species with very low DNA amounts (i.e., ≤3.4 pg) had life cycle types ranging from ephemeral to long lived perennials.
5. With increasing nuclear DNA content, the MGT increased and the range of life cycle types decreased, such that above 3.4 pg no ephemeral species were found, and above 27.6 pg no annual species or facultative perennials were found. Consequently, all species with 1C values greater than 27.6 pg were obligate perennials.

Taken together, the results clearly suggested a positive relationship between nuclear DNA content and MGT. Based on this, Bennett (1973) developed a model to show the linear relationship between C-value and MGT in which threshold effects play an important role (Fig. 2.13A). As C-value increases, so does MGT, and at various points this results in an inescapable shift in developmental lifestyle. To be an ephemeral, a species would have to have a C-value of less than the limiting value for a MGT of seven weeks. At the opposite extreme, species with C-values greater than the threshold value for a MGT of 52 weeks will be obligate perennials.

Of course, many limiting factors (such as availability of phosphorus, essential for DNA synthesis) can act to slow or delay a plant's generation time. Such effects can occur at many points in the development of multicellular organisms, and therefore can easily confound a relationship between genome size and generation time. However, it is important to note that their effect is only to delay and hence is unidirectional in always increasing generation time above the minimum. Thus the model (see Fig. 2.13A) shows a plot with points distributed either along a linear relationship between genome size and minimum generation time (line A-B), or scattered in the triangle for generation times longer than the minimum (triangle ABC). In this sense, species with a very small genome may include both ephemerals (e.g., *Arabidopsis thaliana,* 1C = 0.16 pg), where the point is expected to fall close to the line A-B, and trees (e.g., birch *Betula populifera,* 1C = 0.2 pg), where the point will fall well to the right of the line A-B. Taxa with very large genomes all have long MGTs, and so are obligate perennials and never ephemerals or annuals. Nevertheless, they still show variation due to delays in development, with some taxa falling on or close to the line A-B (with generation times of just over one year), while others require several or many years to complete a generation and are shifted to the right of the line.

The concept of thresholds is clearly illustrated by the absence of any points in the outer triangle (i.e., to the left of the line A-B; compare this with Fig. 2.12B for seed weight versus DNA amount). However, it should be noted that although it is easy to find generation time data for plant species in general, records of true *minimum* generation times are difficult to obtain. Figure 2.13B shows such "record" minimum generation times for 10 herbaceous angiosperm species (ranging from 31 days in *Arabidopsis thaliana* to at least eight months in *Lilium longiflorum*) plotted against 1C DNA amount. Clearly, the shortest minimum generation time increases with increasing C-value, and the plot suggests that the MGT may be slightly shorter in polyploids (plotted as open circles) than in diploids with the same DNA amount. Once again, the important point is that DNA content sets limits on cellular and organismal development, but does not determine these features by itself. Species with low DNA amounts (which exhibit rapid mitosis and can complete meiosis within one to two days) have the option to express a wide range of life cycle types, from ephemeral to perennial—subject to further genic control. By contrast, species with very high C-values spend so long growing and

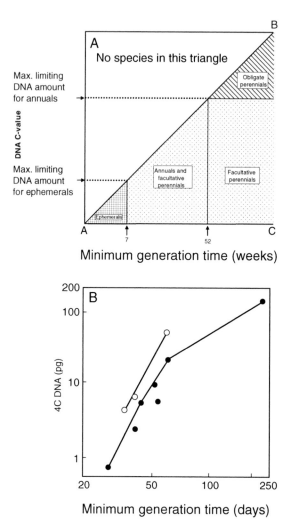

FIGURE 2.13 (A) Diagrammatic illustration of a model for a simple relationship between DNA amount and minimum generation time (MGT) in a temperate environment. Points for the shortest MGT for a given DNA amount lie on the line A-B. All points lie on or to the right of line A-B with none in upper triangle. Species with MGTs of seven or fewer weeks lie in the triangle with cross hatching and hence all have low DNA amounts. Species whose DNA amounts exceed the maximum limiting DNA amount for annuals have MGTs of 52 weeks or more and are therefore nucleotypically deter- mined obligate perennials (striped triangle). (B) Relationship between 4C DNA amount and the shortest MGT known for three polyploid (O) and seven diploid (●) angiosperms. (A) Redrawn based on Bennett (1987), reproduced by permission (© Blackwell Publishing); (B) From Bennett (1987), repro- duced by permission (© Blackwell Publishing).

completing meiosis and sporogenesis that they are simply precluded from expressing an ephemeral (or possibly even an annual) habit, regardless of genotype.

For this reason, there tends to be a change in genome size when plants shift developmental lifestyles from perennial to annual (though not necessarily vice versa). Specifically, such changes from perennial to annual lifestyle appear to occur often when plants move into harsh new environments with short growing seasons, with this derived developmental condition associated with a smaller genome. Such has been the case in species from numerous genera, including *Arachis, Brachyscome, Calotis, Crepis, Haplopappus, Helianthus, Lathyrus, Papaver,* and *Vicia* (Resslar *et al.*, 1981; Sims and Price, 1985; Srivastava and Lavania, 1991; Singh *et al.*, 1996; Naranjo *et al.*, 1998; Watanabe *et al.*, 1999). In one particularly informative example, Watanabe *et al.* (1999) examined the shift from perennial to annual lifestyle in *Brachyscome* using a phylogenetic framework, and were able to provide a clear demonstration that a reduction in genome size was part of the shift in developmental program, and therefore in the adaptation to a new environment.

There is another way of classifying the developmental lifestyles of plants that is of more practical significance to humans—namely, weeds versus nonweeds. Weeds represent a taxonomically eclectic group of plants defined by their annoying habit of growing with great success in places where they are not wanted. Key factors suggested to be important for the success of many weeds include rapid establishment and completion of reproductive development, short minimum generation times, and fast production of many small seeds. All of these factors correlate with low DNA amount, thereby raising the possibility that life as a weed imposes significant constraints on genome size.

Indeed, this is just what is found. In a detailed analysis of the DNA amounts of 156 angiosperm weed species (including 97 recognized as important world weeds) versus 2685 other angiosperms, Bennett *et al.* (1998) provided strong evidence that small genome size is a requirement for "weediness." Specifically, the weed sample had a significantly smaller mean C-value than the nonweed sample, with DNA amounts in weed species restricted to the bottom 20% of the range known for angiosperms. Moreover, a highly significant negative relationship was found between DNA amount and the proportion of species recognized as successful weeds (Fig. 2.14A), demonstrating that the probability of being a recognized weed decreased with increasing DNA amount, reaching zero at a cutoff value of just above 25 pg. In other words, while many angiosperm species have DNA amounts greater than 25 pg, none of them is (or, most likely, could ever become) a weed.

Other unique requirements of developmental lifestyle may impose limits on genome size. An interesting example is provided by the orchid *Erycina pusilla,* which grows as an epiphyte on leaves of tropical trees in South America. In order to survive, it must complete its life cycle before the leaf of the tree falls. This small plant can reach reproductive maturation in only four months, compared with the one to five years commonly taken by other orchid species. Notably, it has

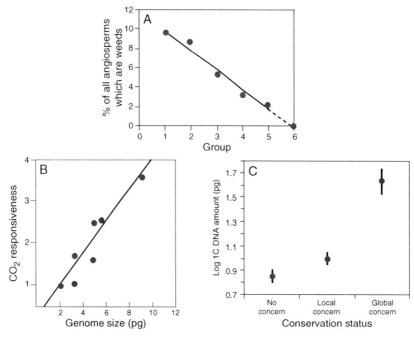

FIGURE 2.14   (A) Relationship between DNA amount and the probability of being a weed. The analysis of the percentage of all angiosperms that are weeds was based on data for 2841 species of known DNA amount (comprising 156 weeds and 2685 other species) ranked in order of increasing size, including 2719 species with 1C values of ≤25.0 pg (the maximum for weeds) divided into five groups with 544 (groups 1–4) and 543 (group 5) and 122 species ≥25.0 pg (group 6). (B) Relationship between the responsiveness (= ratio of average final biomass attained at elevated and ambient CO_2) to elevated CO_2 and 1C nuclear DNA amount among seven annual species of grasses. (C) Relationship between DNA amount and conservation status of 3036 species of angiosperms. Species with known DNA amounts were checked against the United Nations Environment Programme World Conservation and Monitoring Centre Species Database. From the total of 3036 species with known DNA amounts, 305 species were found to be of global concern and 1329 species were of local concern (i.e., threatened only in particular countries). (A) From Bennett *et al.* (1998), reproduced by permission (© Oxford University Press); (B) From Jasienski and Bazzaz (1995), reproduced by permission (© Nature Publishing Group); (C) From Vinogradov (2003), reproduced by permission (© Elsevier Inc.).

a small genome for an orchid (1C = 1.5 pg) compared with the mean for the family (1C = 7.7 pg).

PHYSIOLOGY AND CLIMATE RESPONSE

Physiological correlates of genome size are well known for certain animals, most notably mass-specific oxygen consumption rate in mammals and birds (see Chapter 1). These are based on the effects of genome size on cell size, which also apply to plants. However, other than comparisons between diploids and polyploids, the corresponding implications for plant physiology had remained unaddressed until very recently (see Knight *et al.*, 2005). Importantly, it now appears that mass-specific maximum photosynthetic rate is inversely correlated with genome size (Knight *et al.*, 2005), in parallel with the situation regarding metabolic rate in endothermic animals. This is in keeping with the finding of a significant negative correlation between genome size and specific leaf area (leaf mass ÷ leaf area), which in turn is associated with a range of physiological and ecological traits (Knight *et al.*, 2005).

The more general issue of how plants respond to climate involves both physiological and developmental components, and has therefore been addressed in several different ways. Some of the earliest studies in this area were conducted by Grime and his colleagues. First, Grime and Mowforth (1982) compared genome size with time of shoot expansion and temperature using 24 herbaceous species growing in the Sheffield region of the United Kingdom. They found that the species growing most actively at a given time tended to have smaller and smaller genomes as the seasons progressed from early spring to midsummer. Thus species with large genomes tended to grow early in the spring, and those with smaller genomes started growth later in the year. They suggested that this pattern was related to the fact that cell division is temperature-dependent and cell expansion is not. In this case, species with large genomes grew early in the spring predominantly through the expansion of cells that had divided in the preceding year, whereas small-genomed species growing in the summer did so by normal cell division.

The idea that large DNA contents facilitate growth under moderately cold conditions was reinforced by subsequent studies. In particular, a positive correlation was found between C-value and the mean rate of leaf extension for 14 major plant species in a damp limestone grassland during the cold conditions of an early British spring (Grime, 1983; Grime *et al.*, 1985) (see Fig. 2.12D). More recently, MacGillivray and Grime (1995) also showed larger-genomed plants to be more tolerant to frost.

ECOLOGICAL AND EVOLUTIONARY
IMPLICATIONS OF GENOME SIZE VARIATION

Through its effects on cell size and division rate, genome size can influence a wide range of phenotypic characters, including seed and pollen size, developmental lifestyle (e.g., annual versus perennial, weed versus nonweed), physiology, and climate tolerance—in short, variation in genome size can play a major role in determining when, where, and how plants grow. As discussed in the following sections, this is important for both the natural distribution of plant species and their responses to changes in the environment.

Geographical Distribution and the Large Genome Constraint Hypothesis

The first clues that DNA content might have some relationship with geographical distribution long predated the availability of genome size data. In 1931, Avdulov noted that tropical grasses had uniformly small- to medium-sized chromosomes, whereas most grasses of cool temperate regions had large chromosomes. This is in keeping with the positive correlations between genome size and both chromosome size and capacity for growth under cooler conditions.

The expectation that species with large genome sizes might be concentrated in temperate regions was confirmed in two separate studies carried out more than 45 years after Avdulov's (1931) initial report. Levin and Funderberg (1979) analyzed a broad range of herbaceous angiosperms and showed that the mean 1C-value for temperate species (6.8 pg) was more than double that for tropical species (3.0 pg). However, this difference stemmed from the greater range of DNA amounts and the higher frequency of species with large genomes in temperate versus tropical floras, and not from any exclusion of species with small genomes from temperate regions.

In an analysis of herbaceous cultivated pasture grasses, cereal grain crops, and pulses, Bennett (1976b) showed that the cultivation of species with high DNA amounts per diploid genome tended to be localized at temperate latitudes, or to seasons and regions at lower latitudes where the conditions approximate those normally found in temperate areas. Overall, it seems that a natural positive cline of DNA amount with latitude has been reinforced and exaggerated by the tendency of humans to choose species for cultivation with increasing DNA amounts at successively higher latitudes (Bennett, 1976b, 1987). Importantly, this cline is exhibited by crop species with both C_3 and C_4 photosynthesis and by both annuals and perennials (Bennett, 1976b). Obviously, the choice of species for domestication was done independently of knowledge about DNA content. Rather, the prime factor suiting certain species to human requirements would have been their high yield of seed or leaf per unit area and per unit time. That this indirectly involved choosing species with higher DNA amounts at higher latitudes is an intriguing demonstration of the importance of the nucleotype.

However, as with most of the relationships discussed in this chapter, this pattern involves some important additional complexities. For example, while this DNA–latitude cline for crops is particularly clear in comparisons of the northerly limits of cultivation of cereals in the northern hemisphere in the winter, it may take a very different form in the summer (Fig. 2.15) (Bennett, 1987). Specifically, species with high DNA amounts, which show a pronounced DNA amount–latitude cline in winter, tend to be bunched at the northern end of their range in summer. Indeed, the cline appears to reverse its polarity in summer at the very high latitudes; for example, barley (1C = 5.5 pg) is grown nearer the pole in summer than rye

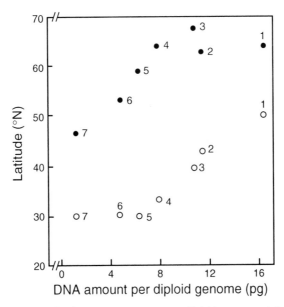

FIGURE 2.15   The relationship between DNA amount per diploid genome and the northern limits of cultivation of several cereal grain species. Key to points: o for a transect from Hudson Bay to Key West in Florida (approx. 82°W) in winter; ● for a transect from near Murmansk by the Arctic Ocean to Odessa by the Black Sea (approx. 32°E) in summer. 1. *Secale cereale*; 2. *Triticum aestivum*; 3. *Hordeum vulgare*; 4. *Avena sativa*; 5. *Zea mays*; 6. *Sorghum* spp.; 7. *Oryza sativa*. From Bennett (1987), reproduced by permission (© Blackwell Publishing).

(1C = 8.3 pg), and, similarly, peas (1C = 4.8 pg) are grown closer to the pole in summer than field beans (13.4 pg)—the reverse of their northern limits during the winter.

This reversal at high latitudes in summer agrees with the general trend for DNA C-value and latitude noted for noncrop species at high latitudes. Thus, for angiosperms from South Georgia and the Antarctic Peninsula, DNA amounts per nucleus (of whatever ploidy) and per diploid genome are within the ranges known for angiosperms at temperate latitudes but, significantly, at their lower end (Bennett *et al.*, 1982b). Bennett *et al.* (1982b) concluded that, above a certain high latitude, maximum DNA content per nucleus and per diploid genome decreases with increasing latitude (and with decreasing temperature). It was therefore suggested that at high latitudes, selection against species with high DNA amounts strongly increases toward the poles as the climate becomes progressively harsher and growing seasons shorter. In particular, this probably relates to limitations in establishment ability among large-genomed plants (Bennett *et al.*, 1982a). Further support for this hypothesis comes from Grime and Mowforth's (1982) survey of British angiosperms, which showed a significantly higher mean genome size for 69 southern species (1C = 6.7 pg) than for 80 widespread species (3.98 pg), with even lower 1C DNA amounts noted in eight northern species (1.13 pg).

Overall, studies of genome size versus latitude, altitude, temperature, and precipitation have tended to show both positive and negative correlations, depending on the particular species and environmental parameters analyzed (reviewed in Knight et al., 2005). In sum, such findings are actually not contradictory, and have recently led to an interesting "large genome constraint hypothesis" (Knight et al., 2005) in which species with low DNA amounts are widespread, whereas those with the highest C-values are progressively excluded from increasingly harsh environments above a midlatitude. Results for both crop and noncrop species fit this view, but the critical latitude above which this exclusion operates is probably higher for crops, owing to the additional input of human selection. Again, within the more intermediate range, large genomes may be favored in the cold where cell expansion is a more efficient means of growth than cell division. Outside this range, however, plants with large genomes appear to be at a disadvantage. This seems to apply to extremes of both temperature and precipitation (Knight and Ackerly, 2002).

GENOME SIZE AND PLANT RESPONSE TO HUMAN ENVIRONMENTAL CHANGE

The power of human activity to alter the global climate and the species inhabiting it should not be underestimated. Anthropogenic effects on the environment pose a major threat to a great many species, including plants. Given the role that genome size plays in setting limits to development, physiology, and geographical distribution, it is important to consider its potential influence in terms of plant responses to continued human activity.

Global Warming

To predict the effects of human production of greenhouse gasses and the resulting increase in global temperature on plants, Grime (1996) looked at how plants responded to elevated temperatures and showed that plants with small genomes exhibited a greater enhancement of growth than those with larger genomes. This led to the suggestion that global warming would stimulate the expansion of species with small genome sizes in certain floras (Grime, 1996), although this could potentially be checked by occasional frosts favoring species with large genomes (MacGillivray and Grime, 1995). On the other hand, Jasienski and Bazzaz's (1995) study of the responses of annual grasses to elevated CO_2 found that growth rates were enhanced in a way *positively* correlated with genome size (see Fig. 2.14B). So, although the most likely net outcome remains somewhat ambiguous, it is clear that genome size variation can be expected to influence the response of plant communities to global warming in some fashion.

Radiation

As noted previously, sensitivity to radiation was one of the first correlates of genome size identified (see Fig. 2.8). Although such experiments obviously employed rather severe doses, the general finding that plants with larger genomes are more susceptible to radiation could have real-world applications. For one thing, nuclear radiation continues to be produced and released by various means (e.g., through the use of nuclear power, and the production of nuclear weapons). Moreover, it could be that large-genomed plants are more sensitive to additional mutagens such as ultraviolet light. If so, then the depletion of the ozone layer might be expected to have especially adverse effects on plants with large C-values.

Chemical Pollution

In a recent study, Vilhar and her colleagues (personal communication and Vidic et al., 2003) investigated the distribution of plants along a gradient of heavy metal pollution produced by a former lead smelter at Zerjav, Slovenia. In this case, there was a negative correlation between the concentration of contaminating metals in the soil and the proportion of species with large genomes. Their results provided the first direct evidence that plants with large genomes are at a selective disadvantage under the extreme environmental conditions caused by pollution.

Generalized Extinction Risk

In an important recent analysis, Vinogradov (2003) showed that large-genomed diploid species are significantly more likely to be listed as rare or in danger of extinction than ones with small genomes (see Fig. 2.14C). This relationship was shown to hold when the analyses were also carried out within families, overcoming to some extent complications arising from phylogenetic issues. Further analysis showed that the effect was independent of ploidy, thereby suggesting that the process of polyploidy itself is not associated with increased risk of extinction. Vinogradov also showed that the maladaptive nature of larger genomes was (at least partially) independent of the duration of the plant life cycle (i.e., whether it was annual or perennial). Vinogradov (2003) and Knight et al. (2005) both reported a negative relationship between genome size and species richness in different plant groups, suggesting that plants with large genomes may indeed be more prone to extinction, and/or that such groups speciate more slowly and would have more difficulty recovering from a loss in biodiversity.

INTRASPECIFIC VARIATION IN GENOME SIZE

Until now, most of the discussion in this chapter has focused on variation in genome size among species and larger clades. Even when dealing with the processes responsible for generating these patterns, the focus has been on mechanisms at the genomic and interspecific levels, without much consideration of variation within species. In part, this is because intraspecific variation remains one of the most controversial topics in the study of plant genome size evolution, and because it is not entirely clear how often such variation even occurs. The following sections provide a review of the controversy in this area, in addition to highlighting some of the known cases in which intraspecific variation has important consequences for phenotypic variation among conspecifics.

Overview of Intraspecific Variation

Again, the earliest work on genome size in both plants and animals established the concept of DNA constancy for a species, given a constant basic chromosome number and type (i.e., with no variation due to aneuploidy, sex chromosomes, or supernumerary chromosomes). For plants, this was supported by several studies in crop species and their wild relatives. For example, whereas Bennett was routinely able to distinguish between lines of wheat that differed by less than 1% in DNA amount (M.D. Bennett, unpublished observation), he failed to find any intraspecific variation in lines of *Vicia faba* with very different ("major" and "minor") seed sizes (Bennett and Smith, 1976) or in land races of *Hordeum vulgare* from widely different geographical regions (Finland, England, Ethiopia, and Iraq) (Bennett and Smith, 1971). Furthermore, several studies reporting intraspecific variation proved to be unrepeatable. For example, although Zakirowa and Vakhtina (1974) reported intravarietal variation of up to 77% in *Allium cepa,* Bennett and Smith (1976) noted clear problems with the techniques used in their study.

During these early stages, Evans *et al.* (1966) did show heritable differences in DNA content of up to 16% induced by environmental factors in flax (*Linum usitatissimum*), and Furuta *et al.* (1975) reported DNA variation of greater than 20% in *Aegilops squarrosa,* but these were considered exceptions to the generally accepted view of DNA constancy in angiosperms.

By the 1980s this view had changed as reports of detectable, and in some cases considerable, intraspecific variation became increasingly common. For example, in 1981 genome size variations of 25% and 27% were reported in two *Microseris* species (Price *et al.,* 1981a,b). This variation, independent of any observable differences in heterochromatin, was reported to be possibly related to environmental factors. In 1985 Bennett reviewed the field, listing 24 angiosperm species in which intraspecific variation was reported, from as low as 4% to 40–50% in *Glycine max*, 80% in *Poa annua*, and a staggering 288% in *Collinsia verna* (Bennett, 1985). In light of such findings, the original concept of the *very* constant genome was replaced by a new view of the *fairly* constant genome.

The 1990s saw the appearance of additional reports (implicit or explicit) of intraspecific variation, including 46% in *Narcissus hedraeanthus* (González-Aguilera *et al.*, 1990), 47% in *Festuca arundinacea* (Ceccarelli *et al.*, 1992), and 287% in *Helianthus annuus* (Johnston *et al.*, 1996; Price and Johnston, 1996). This produced a further shift in perception as the concept of the fairly stable genome was increasingly replaced by the idea of outright genome plasticity (Bassi, 1990). Such studies were often accompanied by molecular data showing that genome size changes could be associated with variability in particular repetitive DNA sequences (e.g., Frediani *et al.*, 1994; Cavallini *et al.*, 1996).

Not everyone was convinced that plant genomes were inherently plastic, however, and some authors began to question whether much of this supposed "intraspecific" variation might be the result of technical artifacts and taxonomic errors. Greilhuber (1998), in particular, sought to distinguish between "orthodox" intraspecific variation owing to genuine chromosomal variation involving duplications and deletions, spontaneous aneuploidy and polyploidy, heterochromatic segments, B chromosomes (see Chapter 4), and in special cases sex chromosomes (see later section) and "unorthodox" intraspecific variation, where genome size variation could not be explained in terms of "orthodox" events but instead was explained by unobserved postulated events such as rapid amplification or deletion of repetitive DNA sequences.

Such concerns stimulated a series of reinvestigations, led by Greilhuber and his colleagues in Vienna, to carefully repeat some of the claims of intraspecific variation using the original material. These studies have resulted in many of the claims for intraspecific variation being rebutted or greatly reduced. For example, several careful studies (Greilhuber and Ebert, 1994; Baranyi and Greilhuber, 1995, 1996) found little or no significant genome size variation in *Pisum sativum*, a species where others (Cavallini and Natali, 1991; Cavallini *et al.*, 1993) had previously reported up to 129% variation. Claims of 15% variation in *Glycine max* related to maturity group (Graham *et al.*, 1994) were unrepeatable according to Greilhuber and Obermayer (1997). A report of 130% variation in *Dactylis glomerata* negatively correlated with altitude (Reeves *et al.*, 1998) was also challenged by Greilhuber and Baranyi (1999), who could not repeat the different hydrolysis curves (see later section) claimed for *Dactylis* compared with the calibration standard *Hordeum vulgare*.

Other reported cases of intraspecific variation were shown to be taxonomic artifacts resulting from insufficient consideration of actual species boundaries and from an incomplete or nonexistent knowledge of the phytogeographical history of the populations in question. For example, *Scilla bifolia* L. *sensu lato* is treated as one species by *Flora Europaea* (McNeill, 1980) but was split into 18 species (of which five are polyploid) by Speta (1980). So, whereas *Scilla bifolia* treated as one species shows a 2-fold genome size variation, its level of "intraspecific" variation diminishes to hardly more than methodological error when split up taxonomically (Greilhuber and Speta, 1985). Further examples disproving claims of intraspecific variation are given in the critical reviews of Greilhuber (1998, 2005). Unfortunately, these do not include a direct reassessment of the largest claim for intraspecific variation of all (288% for *Colinsia verna*) (Greenlee *et al.*, 1984) because no original research material is available. Nevertheless, this case is considered suspect on the grounds of circumstantial evidence (Greilhuber, 1998).

Recent findings in *Helianthus annuus* (sunflower) provide a particularly relevant lesson concerning intraspecific variation. In 1996, Price and Johnston made the surprising claim that the genome size in *H. annuus* varied 2.8-fold depending on light quantity and quality. They concluded that the major factor responsible for inducing a change in genome size was the ratio of red to far-red light, and suggested that phytochromes might be involved in the stability of genome size in sunflowers. However, in a subsequent reevaluation, Price *et al.* (2000) reported that their previous results were not biologically real, but rather a technical artifact caused by staining inhibitors (whose levels were presumably influenced by the quantity and quality of light) and their lack of an internal standard. Similar staining inhibitors caused by compounds in plant tissue have also been noted in other plant species including coffee, yams, roses, oaks (*Quercus* species), and *Allium* species (Ricroch and Brown, 1997; Zoldos *et al.*, 1998; Noirot *et al.*, 2000, 2002, 2003, 2005; Yokoya *et al.*, 2000).

As a result of the careful recent studies, the current view is that many, and perhaps most, previous examples of "unorthodox" intraspecific genome size variation (*sensu* Greilhuber, 1998) must be interpreted with caution until confirmed by independent studies using modern best practice techniques. To be fully accepted, claims of extensive intraspecific variation should also be accompanied by appropriate, wide-ranging molecular studies of the entire repeated sequence profile of the individual materials claimed to show such variation. At present, cases of orthodox variation and confirmed unorthodox variation are relatively few, although some intriguing examples do exist.

GENUINE INTRASPECIFIC VARIATION IN ANGIOSPERMS

Orthodox intraspecific variation associated with observable chromosomal phenomena has been noted in several cases, as with duplications, aneuploidy, and B chromosomes in *Zea mays* (e.g., Poggio *et al.*, 1998). Although much rarer in plants than in animals, the presence of dimorphic sex chromosomes can also generate variation between individual plants, as has been found in *Silene latifolia* (e.g., Costich *et al.*, 1991).

Events visible under the light microscope can also generate detectable gains or losses of repeated sequence DNA in heterochromatin. For example, Gerstel and Burns (1966) noted the occurrence of megachromosomes in F_1 hybrids of *Nicotiana tabacum* × *N. otophora*, where one chromosome in a few cells was enlarged by up to 20 times its normal size (see Fig. 2.6A). Megachromosomes, whose size varied from cell to cell, contained additional DNA in proportion to their size (Collins *et al.*, 1970), and appeared to result from the differential replication of a prominent block of heterochromatin. However, it is unknown whether such pronounced changes in chromosomal DNA content are heritable, and/or whether they might be selected against in later generations.

Loss of telomeric heterochromatin from rye (*Secale cereale*) chromosomes was also seen under the light microscope by Gustafson *et al.* (1983) (see Fig. 2.6B). Selection using cytological screening for reduced telomeric C-bands in hexaploid triticale (× *Triticosecale*) (from 23% to 7% of the rye genome length), was accompanied by a detectable reduction in C-value (0.3–0.7 pg, equivalent to one chromosome), and showed that significant (nondeleterious) reductions in DNA amount can be produced in triticale by artificial selection in just a few years. Interestingly, a line with similarly reduced heterochromatin was the first hexaploid triticale variety to be awarded plant breeders' rights in the United Kingdom (Bennett, 1985).

A few additional examples of genuine intraspecific variation involving differences in heterochromatic segments are known for both crops (e.g., in *Secale cereale*) (Bennett *et al.*, 1977) and noncrops. In this latter category, for example, Greilhuber (1998) cited the subspecies pair *Scilla bithynica* Boiss. ssp. *bithynica*, which has many large C-bands (1C = 29.20 pg) and *S. bithynica* ssp. *radkae*, with few small C-bands (1C = 22.90 pg).

THE SPECIAL CASE OF MAIZE

One of the most widely studied angiosperm species that shows clear and significant intraspecific variation in DNA content is maize, *Zea mays* ssp. *mays*. Pachytene chromosomes of maize can have large heterochromatic knobs, and comparative studies have shown that the distribution of knobs is virtually the same as that of heterochromatin detected as C-bands in mitotic chromosomes. Early studies by Brown (1949) and Bennett (1976a) showed that knob number was negatively correlated with both latitude and altitude, but it remained unclear whether any relationship actually existed between knob number and DNA amount. Independent studies by Rayburn *et al.* (1985), Laurie and Bennett (1985), and Tito *et al.* (1991) have since provided an affirmative answer to this question.

For example, Laurie and Bennett (1985) showed 37% variation in 1C DNA amount in 10 accessions of maize ranging from ~3.35 pg in a Mexican race (Zapalote Chico) down to 2.45 pg in the Seneca 60 race from New York State. Further, this was shown to correlate with the number of C-bands observed (see Fig. 2.6C,D). Results from a study of 21 lines of maize from various locations in North America by Rayburn *et al.* (1985) supported this finding, because they showed significant correlations between DNA content and the numbers of C-bands and heterochromatic knobs. By assigning their 21 lines to five relative maturity zones along a south–north axis, Rayburn *et al.* (1985) were able to demonstrate significant negative correlations between DNA C-value and maturity zone (i.e., latitude of origin), and between maturity zone and both heterochromatin amount and knob number. A similar analysis comparing the zones of origin with DNA C-value for the 11 lines examined by Laurie and Bennett (1985) likewise showed a significant negative correlation. Taken together, these studies have provided convincing evidence that the intraspecific variation in genome size in maize is largely caused by differences in the amount of heterochromatin, and that previously reported correlations between geographical location and knob number are actually related to intraspecific differences in nuclear DNA content.

Rayburn *et al.* (1985) suggested that the pattern of distribution of DNA contents in maize "may relate to selection pressures imposed by man ... which influence the DNA content via its nucleotypic effects." It is generally agreed that maize originated at low latitude and was taken north by humans until environmental barriers (primarily a shorter growing season) prevented normal maturation. Rayburn *et al.* (1985) speculated that the lower DNA C-values of varieties adapted to high latitudes may have resulted from simultaneous selection by humans for earlier maturation and maximum plant size and yield. This may have involved selection for more cells, which could result from the shorter mitotic cycle time that correlates with reduced DNA C-value (see Fig. 2.11A). Significant positive correlations have also been found in maize between DNA content and altitude of cultivation (Rayburn and Auger, 1990) and effective growing season (Bullock and Rayburn, 1991), whereas negative relationships have been reported between genome size and various growth and yield parameters (Biradar *et al.*, 1994). Clearly, where intraspecific variation is real, it can be of considerable adaptive significance—to people as well as plants.

GENUINE INTRASPECIFIC VARIATION IN NONANGIOSPERMS

To date, no studies have reported significant intraspecific variation in pteridophytes or bryophytes. In gymnosperms, on the other hand, cases of intraspecific variation have been claimed in some species. For example, Miksche (1968, 1971) reported intraspecific variation in three gymnosperms, including 58% in *Picea glauca* and

92% in *P. sitchensis*, related to their latitude of growth in North America. However, paralleling the work in angiosperms, subsequent reinvestigations of the material collected from the same range sampled by Miksche failed to repeat these observations (e.g., Teoh and Rees, 1976). Other similar examples are reviewed by Murray (1998). Indeed, a number of more recent studies have failed to find evidence of intraspecific variation even when measurements have been made using different methods in different laboratories (e.g., Ohri and Khoshoo, 1986; Wakamiya *et al.*, 1993) or on highly disjunct populations (Auckland *et al.*, 2001). Intraspecific variation was reported in *Ginkgo biloba* by Marie and Brown (1993), but unfortunately this was done without mention of the sexes of the plants sampled. Because *Ginkgo* is one of the relatively few plants with sex chromosomes, this may represent a case of genuine intraspecific variation. In summary, more detailed studies of a wide range of species are needed to obtain a clearer picture of the extent of intraspecific variation in gymnosperms, following the same principles as outlined for work in angiosperms.

INTRASPECIFIC VARIATION AND SPECIATION

Some closely related species may have very similar C-values (e.g., *Hordeum bulbosum, H. glaucum, H. marinum,* and *H. murinum,* which all have 1C DNA values of 5.5 pg) (Bennett and Smith, 1976), whereas in other species individual lines may have distinct C-values (e.g., different lines of *Zea mays* have C-values that vary by 35%) (Laurie and Bennett, 1985). The implication is that there is no absolute link between the process of speciation and changes in genome size. That is, speciation may occur without any detectable change in C-value, and likewise variation in DNA amount (both gain and/or loss, mostly of repeated sequences) can also precede reproductive isolation and morphological diversification.

Speciation was once thought to depend mainly on changes in informational genes. However, comparative genomics has emphasised striking elements of constancy in this part of the genome. Thus Devos and Gale (1997) noted that "gene content and gene orders are highly conserved between species within the grass family, and that the amount and organization of repetitive sequences has diverged considerably." Such results have led to a rethinking of the role of noncoding repeated DNA sequences in determining diversity, and even to the suggestion that they play a major role in plant speciation (Kubis *et al.*, 1998).

To date, there is no research to indicate in any general way whether species normally diverge before detectable variation in genome size occurs, and/or vice versa. Nor, for species displaying variation in DNA amount before distinct morphological divergence, is there any definitive information on how much intraspecific variation in C-value usually occurs before species diverge, or if there is any limit to the amount of variation that may accrue before species divergence becomes an inevitable consequence. Furthermore, according to Greilhuber (1998), nothing reliable is known about the rate of genome size changes in natural populations. To complicate matters, the answers to these questions will be greatly influenced by the species concept used because amounts of "intraspecific" variation will be much greater for lumpers than for splitters, as illustrated earlier by the example of *Scilla bifolia*.

THE MYSTERY OF DNA CONSTANCY

Part of the current interest in C-values and what determines genome size focuses on a tension between the massive interspecific variation in DNA amounts existing in the angiosperms (Table 2.2), and the surprisingly high degree of genome constancy found in many widely distributed species (e.g., Bennett *et al.*, 2000b). In view of the known molecular mechanisms with the potential to rapidly generate considerable variation in C-value and the clear demonstration in some studies that particular sequences within the genome can fluctuate considerably and often rapidly (e.g., Kalendar *et al.*, 2000), the degree of C-value constancy found in many species is remarkable, and needs explanation. Indeed, it is arguable that such invariance would not be expected without some mechanism(s) to select for constancy (or against changes) in C-value (Bennett *et al.*, 2000b).

Although genome size is widely perceived as free to vary, many results suggest instead that DNA amount may normally be subject to innate controls by "counting mechanisms" that somehow detect, quantify, and regulate genomic size characters within quite tightly defined or preselected limits (Bennett, 1987; Bennett *et al.*, 2000b). For example, several careful studies have shown evidence of "karyotypic orthoselection," whereby large differences in C-value between related plant species involve strictly proportional changes in all chromosomes, which preserves the particular form of a complement (Seal and Rees, 1982; Seal, 1983; Brandham and Doherty, 1998) (Fig. 2.16). This could be a case in which counting mechanisms are at play, acting to produce proportional changes in all linkage groups and/or chromosome arms, presumably driven by some underlying organizational principle.

Clearly, there is still much to be discovered about intraspecific variation in genome size. Fortunately, modern molecular methods can provide powerful new insights into how changes at the DNA sequence level relate to others at the whole genome level, including intraspecific variation in nuclear DNA amount. Comparisons between whole genome sequences for closely related "diploid" subspecies and species with different C-values will soon be possible (see Chapter 9). Meanwhile, complete sequences for homologous chromosome segments in lines of *Zea mays* with different C-values distributed in both knobs and euchromatin should be particularly illuminating. Studies linking DNA information to characters of environmental and ecological interest will provide an important focus for new work

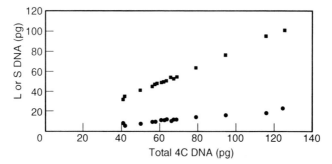

FIGURE 2.16   Karyotypic orthoselection in the genus *Aloe*. The total 4C DNA amounts of 20 *Aloe* species were plotted against the calculated DNA amount for the combined long (L) (■) and combined short (S) (•) chromosomes. For each species the value for the long chromosome is directly above that of the short chromosome. The two highest values are tetraploid, the rest are diploid species. From Brandham and Doherty (1998), reproduced by permission (© Oxford University Press).

on intraspecific variation in plant genome size over the next few years. Such work may also shed light on whether certain types of plants are more predisposed to exhibit gain or loss of repeated sequences, and if this is relevant to understanding patterns of gradualistic versus punctuated changes in genome size over time.

METHODOLOGY FOR ESTIMATING GENOME SIZE IN PLANTS

None of the big questions in plant genome size evolution can be addressed without broad comparative analyses, making the assessment of large numbers of plant genome sizes a crucial first step toward a resolution of the C-value enigma (see also Chapter 1). However, as the previous section showed, it is not the case that any data will do. Rather, these must be collected in such a way as to avoid the numerous technical pitfalls that may otherwise contribute to an erroneous view of genome size variation and evolution. The following sections review the several methods that have been used to estimate genome size in plants, and provide some best practice guidelines for the implementation of the most commonly used techniques. This is meant to facilitate the accurate measurement of plant genome sizes in the future, and to inform critical analyses of estimates reported in the past.

CHEMICAL EXTRACTION AND REASSOCIATION KINETICS

Unlike animals, plants characteristically have cell walls thickened with polymers (e.g., cellulose and lignin) whose presence may complicate the extraction of DNA.

Nevertheless, chemical extraction methods developed for use in animals (Schmidt and Thannhauser, 1945) were later successfully applied to plants (Sunderland and McLeish, 1961; Lyndon, 1963; Rothfels *et al.*, 1966). Under these techniques, the total DNA is extracted from a sample of cells and dissolved in a known volume of solvent. The concentration of DNA is estimated colorimetrically using a modification of the diphenylamine reaction (Burton, 1956, 1968) to give a color reaction whose intensity is proportional to the concentration of deoxyribose sugar, and hence of DNA. The method typically used very large numbers of cells, whose total was estimated using a hemocytometer. Results were usually expressed as mean DNA amount per cell, which consisted mainly of nuclear DNA, but also included small amounts from organelle (mitochondrial and chloroplast) genomes. The results were calibrated in absolute units using either a plot of color intensity against known concentrations of DNA or deoxyribose, or by ultraviolet spectrophotometry with known concentrations of DNA as a standard (Sunderland and McLeish, 1961). Although these methods were usefully reliable, they were complicated and slow, and hence rate-limiting for many larger comparisons; thus these methods have been rarely used since the 1960s.

In the 1970s and 1980s, it became relatively common to employ reassociation kinetics to assess the composition and size of genomes. In this method, DNA is extracted from cells and then denatured by heating. The rates of reassociation of the DNA strands indicate the relative copy number of repeated DNA sequences, and can be calibrated against a standard to give an estimate of absolute DNA content. However, as a method for measuring genome size, reassociation kinetics is very slow (the reassociation of some DNA fragments can last for several days) and not particularly accurate. Not surprisingly, this technique was only rarely used for genome size estimates in plants (Table 2.3). Instead, nearly all modern genome size measurements are made by either Feulgen microdensitometry or flow cytometry.

FEULGEN MICRODENSITOMETRY

The advent and widespread application of photomicrodensitometry represented a major step forward in genome size estimation (Table 2.3). Rather than extracting the DNA from cells, this method involves staining the nuclei and then measuring the amount of light absorbed by the stain. The most commonly employed method of staining is still the Feulgen reaction, first developed by Feulgen and Rössenbeck in 1924. In this case, aldehyde groups are freed by hydrolyzing the DNA with strong acid, followed by staining with Schiff's reagent containing leuco-basic fuchsin, which gives a purple coloration when it complexes with the aldehyde groups. Plant material is typically prepared for Feulgen microdensitometry using various modifications of the method described by McLeish and Sunderland (1961) for plants, itself a variant of the method of Leuchtenberger (1958) for

TABLE 2.3   Summary of the main methods used to estimate genome size in a sample of 5844 angiosperms over the period between 1950 and 2003. Data are grouped into five-year periods (except for 2000–2003). Ch = chemical extraction; Fe = Feulgen microdensitometry; FC = flow cytometry; RK = reassociation kinetics; CGS = complete genome sequencing; Fe/IA = Feulgen image analysis densitometry.

Time period	Estimation method					
	Ch	Fe	FC	RK	CGS	Fe/IA
1950–1954	1	0	0	0	0	0
1955–1959	0	5	0	0	0	0
1960–1964	24	1	0	0	0	0
1965–1969	47	101	0	0	0	0
1970–1974	6	434	0	0	0	0
1975–1979	2	573	0	14	0	0
1980–1984	11	757	26	12	0	0
1985–1989	0	883	36	0	0	0
1990–1994	2	449	479	0	0	0
1995–1999	0	687	550	0	0	0
2000–2003	0	291	438	0	3	12
Total	93	4181	1529	26	3	12

animal tissues. It is generally accepted that Feulgen staining is specific and stoichiometric for DNA, meaning that a measure of nuclear stain concentration (calculated from total optical density), compared against a standard of known DNA amount, provides an accurate estimate of C-value.

Detailed reviews of standardized staining and measurement protocols have been provided recently (Greilhuber and Baranyi, 1999; Greilhuber and Temsch, 2001; Greilhuber, 2005), which the interested reader is urged to consult before performing new estimates of plant genome size. As noted previously, a failure to follow best practice guidelines can lead to substantial errors in results and thence to a false understanding of patterns and mechanisms. Common sources of error include biological, chemical, and physical (optical) factors, some of which are listed here:

1. *Chromatin condensation.* The level of chromatin condensation can vary significantly between different types of plant nuclei (e.g., pollen and egg nuclei), which directly affects the uptake and/or detection of stain, such that DNA amount tends to be underestimated as chromatin condensation increases (Verma and Rees, 1974).
2. *Hydrolysis time.* The acid hydrolysis is perhaps the most sensitive step in the procedure, with both insufficient and excessive hydrolysis resulting in understaining. It is therefore essential that an optimum hydrolysis time is

used to maximize staining of nuclear DNA in all samples being compared (both unknowns and calibration standards). This can be determined by the construction of "hydrolysis curves," which provide measurements of stain densities of nuclei from the same individual hydrolyzed for different times.

3. *Hydrolysis temperature.* For many years "hot hydrolysis" (~12 minutes in 1M HCl at 60°C) was used, giving only a relatively short (typically 3–5 minutes) plateau of maximum staining. Longer treatment (1–2 hours) in 5M HCl at 20°C ("cold hydrolysis") gives greater control of this important step and greatly prolongs the plateau of maximum staining. Moreover, it is important that the hydrolysis step is not done simply at "room temperature," because this may be about 20°C in temperate regions but may be well above 30°C in tropical areas. Failure to recognize the importance of hydrolysis conditions has been a major contributor to false reports of intraspecific variation.

4. *Staining inhibitors.* A quite different, but equally important source of staining error is the effect of cytosolic compounds present in many plant materials that can bind to DNA and greatly reduce its ability to undergo Feulgen staining ("self-tanning") (Greilhuber, 1988).

Provided that these problems are recognized, and adequate steps taken to apply best practice, Feulgen microdensitometry can give quantitative estimates of nuclear DNA amounts of considerable accuracy, with error variation routinely controlled to within 5%, and sometimes to within 1–3%.

FEULGEN IMAGE ANALYSIS DENSITOMETRY

One major factor likely to limit the future applicability of Feulgen methods is the "obsolescence time bomb" of aging microdensitometers, which are no longer manufactured and are becoming increasingly difficult to repair (see www.rbgkew.org.uk/cval/conference.html#outline) (Bennett *et al.,* 2000a). Fortunately, advances in computing and imaging technology have facilitated the development of inexpensive computer-based image analysis densitometry methods that will allow the continued implementation of the time-tested method of Feulgen staining.

In this technique, DNA is Feulgen-stained as usual, but the density of stain in the nucleus is measured using a microscope-mounted video or digital camera to "grab" images and to display them as composites of pixels on a computer screen. The intensity (gray value) of each pixel can be used to calculate an individual point density, allowing the instant and simultaneous measurement of integrated optical density for all nuclei in the field.

The method was originally developed for DNA quantification in cancer diagnosis (Jarvis, 1986), for which accuracy is obviously of extreme importance, and with the result that scientists have reached an international consensus on the methodology and best practice (e.g., Chieco *et al.*, 1994; Bocking *et al.*, 1995; Puech and Giroud, 1999). However, its application to DNA quantification in other organisms has been slow to take off. Probably the first reported use of this method in plants was by Temsch *et al.* (1998), who used it to estimate genome sizes in species of the moss genus *Sphagnum*. This was followed by studies of a variety of plant genera, such as *Crepis* (Dimitrova and Greilhuber, 2000), *Gagea* (Greilhuber *et al.*, 2000), *Hedera* (Obermayer and Greilhuber, 2000), and *Arachis* (Temsch and Greilhuber, 2001).

A recent interlaboratory comparison showed the results of Feulgen image analysis densitometry to be very comparable to those obtained by other methods, over a 100-fold range in plant genome sizes (Vilhar *et al.*, 2001). Studies using animal tissues likewise established the validity of the method (Hardie *et al.*, 2002). Further studies by Vilhar and Dermastia (2002) have led to proposals for standardizing the method to maximize accuracy of the data generated in plants. It is expected that image analysis will become one of the most important sources of new genome size estimates in the near future.

FLOW CYTOMETRY

Perhaps the first description of a flow cytometry apparatus was that of Moldavan (1934) for use in counting the number of red blood cells or yeast in a suspension. With extensive development of the equipment and methodology, the technique of flow cytometry has since been adapted to many different applications, including DNA quantification for cancer detection and, more recently, genome sizing.

In plants, this involves mechanically isolating nuclei, usually from leaf tissue, by chopping. The isolated nuclei are stained with a fluorescent dye that binds quantitatively to DNA and then passed through a flow cytometer, which forces nuclei to pass one at a time past a series of lights, lenses, mirrors, and amplifiers that detect and convert the amount of fluorescent light being emitted by each nucleus into a digital signal. By comparing the intensity of fluorescence with that from a plant of a known DNA amount, the absolute DNA content of the plant can be determined. In physical terms, this is the opposite of densitometric methods.

Although it is now one of the primary methods employed (Table 2.3), the application of this technology to plant genome size studies came relatively slowly, limited largely by the difficulty of isolating nuclei from cells with rigid cell walls. The first successful preparations of intact nuclei suitable for plant flow cytometry were made from root tips of *Vicia faba* by Heller (1973), who used the enzymes pectinase and pepsin to digest the cell wall. Although he noted the potential of this method to analyse cell cycle kinetics, the method was time-consuming and was not adopted by other researchers. Alternative approaches were tried in the early 1980s (e.g., use of intact plant protoplasts) (Puite and Tenbroeke, 1983), but these also suffered from being too laborious and time-consuming. The breakthrough came when Galbraith *et al.* (1983) developed the simple, rapid, and convenient method of chopping to provide isolated nuclei.

Following Galbraith *et al.*'s (1983) paper, various researchers carried out experiments comparing data obtained using flow cytometry with the established method of Feulgen microdensitometry. Generally they found good agreement between the two across a large range of DNA amounts from a broad array of plants (e.g., Hulgenhof *et al.*, 1988; Michaelson *et al.*, 1991; Dickson *et al.*, 1992; Dolezel *et al.*, 1998), so long as best practice techniques were implemented. Two of the most important sources of error turned out to be an inappropriate choice of fluorochrome and the presence of staining inhibitors.

1. *Choice of fluorochromes.* A range of fluorochromes has been used for plant DNA estimations by flow cytometry. These divide into two groups: DNA intercalating dyes (e.g., propidium iodide and ethidium bromide), which bind to DNA independently of the DNA sequence, and base pair–specific dyes, which preferentially bind to AT-rich (e.g., DAPI, Hoechst 33258) or GC-rich (e.g., mithromycin) regions of the DNA. In a comparative study, Dolezel *et al.* (1992) showed that use of base pair–specific dyes could lead to errors approaching 100%, and recommended that only intercalating fluorochromes be used. This recommendation was endorsed at the first Kew Plant Genome Size Workshop in 1997 (see www.rbgkew.org.uk/cval/conference.html#keyrecs).

2. *DNA staining inhibitors and the importance of internal standardization.* Although the use of intercalating fluorochromes overcomes some of the problems encountered in obtaining accurate genome size estimates, in recent years it has become apparent that compounds in the cytoplasm, released during nuclear isolation, can interfere with fluorochrome binding and fluorescence and lead to erroneous genome size data (Noirot *et al.*, 2000, 2002, 2003; Price *et al.*, 2000). Even though the nature of many of these compounds remains elusive, identifying their existence, understanding how their levels may be influenced by environmental and/or genetic factors, and determining how they affect genome size estimations are vital for accurate genome size studies. The importance of testing for the presence of inhibitors was emphasized in the more recent Kew Plant Genome Size Workshop in 2003 (see www.rbgkew.org.uk/cval/workshopreport.html). In practice, this problem can be addressed by chopping, staining, and measuring standards and unknowns together.

If done correctly, flow cytometry has the great advantage of providing rapid and accurate measurements of DNA amount for a large number of nuclei from a small sample of plant tissue. This allows thorough plant population studies to be made *in situ*, enabling the extent and evolutionary significance of intraspecific variation to be more completely assessed and evaluated. For example, flow cytometry has been the method of choice for comparisons of ecotypes of *Medicago* species (Blondon *et al.*, 1994), and of different populations, F_1 hybrids, and inbred lines of maize (Bullock and Rayburn, 1991; Biradar and Rayburn, 1993).

That said, it is important to recognize that flow cytometry is subject to two major constraints. First, running a flow cytometer can be very expensive and requires high levels of technical support and other infrastructure. Second, using flow cytometry does not remove the need for cytological work on the species being studied, because although it may give a highly accurate DNA value for a taxon, this will be of limited value if the chromosome number (2n) of the individual plant (or even tissue) measured is unknown. Also, if chromosomal variations such as aneuploidy, duplications and deletions, and sex and supernumerary chromosomes are not identified along with flow cytometric measurements, then the interpretation of the results could be flawed. So, although flow cytometry is appealingly fast and highly suited to certain types of studies (e.g., population studies, ploidy screening), the importance of cytological analyses, which can be time-consuming, must not be neglected.

COMPLETE GENOME SEQUENCING

Although thousands of DNA amounts have been determined using the previously discussed techniques over the past 50 years, every one of them is but an *estimate,* inevitably subject to technical errors. For this reason, the need for an exact calibration standard whose C-value is not subject to such errors has long been recognized. Since the mid-1990s, a large number of highly accurate determinations of genome size based on complete genome sequences have been published for numerous microbes (see Chapter 10). Because of the high cost and intensive effort currently required, it is unlikely that complete genome sequencing will become a viable and routine method for determining plant genome size in the near future. However, the first complete genome sequence for a plant was eagerly awaited by those in the plant genome size community, because it was expected to provide a highly accurate baseline reference point for calibrating future estimates by Feulgen densitometry and flow cytometry. By 1997 it was clear that the prime candidate for this honor would almost certainly be *Arabidopsis thaliana* (*Arabidopsis* Genome Initiative, 1997).

December 2000 saw the landmark publication giving the genome sequence for *Arabidopsis thaliana* ecotype Columbia (*Arabidopsis* Genome Initiative, 2000), and an estimate for its genome size of 125 megabases (Mb), based on the size of the sequenced regions (115.4 Mb) plus a rough estimate of 10 Mb for the unsequenced centromere and ribosomal DNA regions. Sadly, this was not the long-awaited benchmark standard, because it was not the result of "complete genome sequencing" as these words would be understood by most people. Indeed, later work showed that the amount of DNA in the unsequenced regions had been seriously underestimated and that the genome size of *Arabidopsis thaliana* was in fact about 157 Mb (Hosouchi *et al.*, 2002; Bennett *et al.*, 2003).

SOME COMMENTS ON PLANT GENOME SIZE STANDARDS

The early comparative studies in plants used various plant species as calibration standards, including *Pisum sativum* and *Vicia faba* (McLeish and Sunderland, 1961). By the 1970s, onion (*Allium cepa*), had already emerged as the most frequently used calibration standard. Onion has the advantage that it is widely cultivated and globally available as seed or bulbs, can be readily grown to provide a source of actively growing root-tips over long periods, and is highly amenable to cytological techniques to make root-tip squashes. An analysis of 5871 plant taxa in 2004 showed that ~39% were calibrated either against onion (2259 taxa) or a secondary standard that itself was calibrated against onion (52 taxa). It follows from this that the absolute accuracy of genome size estimates for many plants are directly dependent on the accuracy of the value (1C = 16.75 pg) determined for onion by Van't Hof (1965) using chemical methods. Fortunately, this agrees closely with estimates from four independent studies that used animal species (including *Homo sapiens*) as calibration standards. Importantly, recent work found no significant differences in DNA amount between cultivars from widely different environments, confirming that it has the stable genome size required for a key calibration standard (Bennett *et al.*, 2000b).

For technical reasons, however, it is desirable to use a calibration standard whose genome size is within about 2-fold of the unknown taxon being studied. Thus a need was recognized for other calibration standards with smaller genome sizes than *A. cepa*, spread at useful intervals over the range of genome sizes encountered in plants. The first attempt to provide a range of such standards was made by Bennett and Smith (1976), who listed eight species with 1C-values from 1.5 pg to 17 pg, including *Pisum sativum*, *Hordeum vulgare*, and *Vicia faba*, all calibrated against *A. cepa*. Subsequently, other studies have attempted to refine the values for these taxa, and/or to extend the number and range of standard calibration species to include smaller genomes such as *Vigna radiata*, *Oryza sativa* (Bennett and Leitch, 1995), *Lycopersicon esculentum* (Obermayer *et al.*, 2002), and *Arabidopsis thaliana* (Bennett *et al.*, 2003).

About 10% of all plant C-value estimates have used an animal standard, 89% of these being chicken red blood cells (CRBC). Chicken (*Gallus domesticus*) is used for several reasons. First, like *Allium cepa*, it is readily accessible across the globe. Second, it has long been used in animal studies (Gregory, 2001b) and was

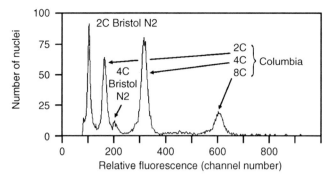

FIGURE 2.17   Flow cytometry histogram showing the relative staining in nuclei of *Arabidopsis thaliana* ecotype Columbia and *Caenorhabditis elegans* variety Bristol N2 (1C = 100 Mb). *Arabidopsis* has approximately 157% of the 2C nuclear DNA fluorescence of *Caenorhabdites,* resulting in a 1C value for *Arabidopsis* of 157 Mb, contrary to the 125 Mb estimate given by the *Arabidopsis* Genome Initiative (2000). From Bennett *et al.* (2003), reproduced by permission (© Oxford University Press).

therefore chosen by Galbraith *et al.* (1983) in their seminal work developing flow cytometry for plant genome size studies. Third, it is an amenable material: one animal can be bled nondestructively at intervals to provide a stable and constant standard for several years. However, in general the use of chicken blood, and animal standards in general, has been discouraged for plant studies because of the different DNA condensation levels and other staining properties of animal versus plant cells. There is one important exception to this, however, namely the potential use of the nematode *Caenorhabditis elegans* (which is the only multicellular organism whose genome has actually been completely sequenced) as the basal calibration standard for genome size comparisons in both plants and animals. Indeed, *C. elegans* was the species used to identify the previous error in the *Arabidopsis thaliana* genome size estimate (Fig. 2.17) (Bennett *et al.,* 2003). Work is already under way to create a ladder of reliable standards, from *A. thalania* on up, using *C. elegans* as the baseline "gold standard."

CONCLUDING REMARKS AND FUTURE PROSPECTS

In an important sense, the field of genome size research in plants may be considered to have reached "the end of the beginning." The first big push for representative and accessible coverage of plant genome size data has been completed by the launch of the *Plant DNA C-values Database* (Bennett and Leitch, 2003). This has allowed at least a basic understanding of the major patterns and consequences of genome size variation, although much work remains to be done in these areas. With a large and growing dataset, the field is poised to tackle some of the key comparative questions relating to the biological, ecological, and evolutionary importance of genome size variation.

EXPANSION OF THE PLANT GENOME SIZE DATASET

Although the *Plant DNA C-values Database* includes representatives from each of the major land plant groups, the percent coverage at the species level remains very poor (generally <2%) for all but the gymnosperms (see Table 2.1). Furthermore, Release 2.0 of the database does not currently contain information for any algae. This latter gap will be filled in the next release of the database because Kapraun (2005) recently compiled three reference lists containing C-value data for 240 algal species, including 85 green algae (Chlorophyta), 111 red algae (Rhodophyta), and 44 brown algae (Phaeophyta). It also bears noting in more optimistic terms that the current state of knowledge of plant genome size is largely due to significant progress made in recent years (e.g., Leitch *et al.*, 2001; Leitch and Hanson, 2002; Hanson *et al.*, 2003), indicating that the rate of data acquisition is accelerating.

One of the key driving forces behind the expansion of the genome size dataset has been the targets set at the Plant Genome Size Workshops held at the Royal Botanic Gardens, Kew in 1997 and 2003. These provided a forum for identifying crucial gaps in plant C-value knowledge, and for setting targets to fill them. In September 2003, the progress achieved after the 1997 meeting was reviewed and new five-year goals were set:

1. *Angiosperms:* To estimate first C-values for the next 1% of species (i.e., an additional 2500 species). Within this, targets of achieving 75% familial (i.e., an additional ~114 families) and 10% generic (i.e., an additional ~400 genera) representation were set.
2. *Pteridophytes:* To estimate first C-values for a further 100 species, with particular emphasis on leptosporangiate ferns (the most diverse group of land plants after the angiosperms).
3. *Bryophytes:* To improve geographical representation by targeting species from the tropical and southern hemisphere floras (no data are currently available for species in these regions). Further, for conservation studies targets were set to estimate C-values in rare taxa in the European flora.
4. *Algae:* Two groups were identified as targets for future C-value research: the Micromonadophyceae, which are considered to hold a place close to the origin of plants, and Charophyceae, now recognized as the sister group to land plants.

MECHANISTIC QUESTIONS

One of the most fundamental questions in the field of genome size research is "How and why do DNA amounts change?" In recent years there have been huge advances in understanding how DNA amounts can increase and decrease, but this probably represents only the tip of the iceberg. Future research should continue to search for novel mechanisms capable of generating changes in DNA amount, and use comparative studies to determine the extent to which such mechanisms are either specific to a particular taxon or universal to all plants. In addition, the natures of the evolutionary forces acting on these mechanisms to promote or restrict their activity are largely unknown, but are becoming increasingly amenable to study with the development of large-scale comparative genomics. All of these issues are essential for shedding light on the molecular bases underlying the C-value enigma.

ECOLOGICAL AND ENVIRONMENTAL QUESTIONS

Another key question for future researchers to address is "What role does genome size play in the response of plants to their environment?" Given that DNA amount is linked to numerous cytological, morphological, and physiological parameters, it will be a critical component of attempts to determine the evolutionary fates of plants subjected to changing ecological and environmental conditions. Already there is evidence that genome size plays an important, yet complex, role in affecting how a plant may respond to anthropogenic changes in the environment such as global warming (Grime, 1996), increased nuclear radiation (Sparrow and Miksche, 1961; Underbrink and Pond, 1976), elevated CO_2 levels (Jasienski and Bazzaz, 1995), and pollution (Vilhar, personal communication). These questions can be approached from various perspectives, including comparative surveys across natural ecological gradients, experimental manipulations in the laboratory, and theoretical modeling designed to examine how organisms will respond to the multifaceted environmental changes induced through human activities. Such work also promises to illuminate the factors that influence the natural distributions of plants with differing genome sizes.

EVOLUTIONARY QUESTIONS

The ability to track changes in C-value over time using characters such as fossil cell size as a proxy for DNA content (Masterson, 1994) will enable the evolutionary patterns of genome size evolution to be viewed over millions of years. Such approaches will provide insights into the role genome size has played during key evolutionary developments in plants, such as the transition onto land, the development of a vascular system for efficient water conduction, the evolution of the seed habit found only in angiosperms and gymnosperms, and the evolution of angiosperms themselves, the most diverse and species-rich group of plants on this planet.

Studies of extant taxa can also allow insights into the origin of long-term evolutionary patterns, for example by combining genome size data with detailed information of plant phylogeny (Leitch *et al.*, 1998, 2005; Soltis *et al.*, 2003) and by further investigating the links between genome size and susceptibility to extinction (Vinogradov, 2003) and propensity for speciation (Knight *et al.*, 2005).

In the roughly 50 years that have elapsed since the first plant genome size estimate was recorded, genome size research has moved from simply documenting C-values and expressing confusion over the "C-value paradox" to identifying the major patterns of variation, illuminating some of the molecular mechanisms and evolutionary forces responsible, and framing a specific series of questions as part of the "C-value enigma." With continued surveys of plant genome size and integration with other fields ranging from molecular biology to ecology, it is possible that the future will finally see the emergence of the "C-value solution(s)." Such an effort will clearly require an appreciation of the biological and evolutionary significance of genome size variation and the development of a more holistic framework for genomics.

REFERENCES

Abrahamson S, Bender MA, Conger AD, Wolff S. 1973. Uniformity of radiation-induced mutation rates among different species. *Nature* 245: 460–462.

Anderson LK, Stack SM, Fox MH, Zhang CS. 1985. The relationship between genome size and synaptonemal complex length in higher plants. *Exp Cell Res* 156: 367–378.

Arabidopsis Genome Initiative. 1997. Objective: The complete sequence of a plant genome. *Plant Cell* 9: 476–478.

Arabidopsis Genome Initiative. 2000. Analysis of the genome sequence of the flowering plant *Arabidopsis thaliana. Nature* 408: 796–815.

Auckland LD, Johnston JS, Price HJ, Bridgwater FE. 2001. Stability of nuclear DNA content among divergent and isolated populations of Fraser fir. *Can J Bot* 79: 1375–1378.

Avdulov NP. 1931. Karyo-systematische Untersuchungen der Familie Gramineen. *Bull Appl Bot Genet Plant Breed (44th Suppl)* 4: 1–428.

Axelsson T, Bowman CM, Sharpe AG, *et al.* 2000. Amphidiploid *Brassica juncea* contains conserved progenitor genomes. *Genome* 43: 679–688.

Baetcke KP, Sparrow AH, Nauman CH, Schwemmer SS. 1967. The relationship of DNA content to nuclear and chromosome volumes and to radiosensitivity (LD_{50}). *Proc Natl Acad Sci USA* 58: 533–540.

Barakat A, Carels N, Bernardi G. 1997. The distribution of genes in the genomes of Gramineae. *Proc Natl Acad Sci USA* 94: 6857–6861.

Baranyi M, Greilhuber J. 1995. Flow cytometric analysis of genome size variation in cultivated and wild *Pisum sativum* (Fabaceae). *Plant Syst Evol* 194: 231–239.

Baranyi M, Greilhuber J. 1996. Flow cytometric and Feulgen densitometric analysis of genome size variation in *Pisum. Theor Appl Genet* 92: 297–307.

Bassi P. 1990. Quantitative variations of nuclear DNA during plant development. A critical approach. *Biol Rev* 65: 185–225.

Baumel A, Ainouche M, Kalendar R, Schulman AH. 2002. Retrotransposons and genomic stability in populations of the young allopolyploid species *Spartina anglica* CE Hubbard (Poaceae). *Mol Biol Evol* 19: 1218–1227.

Bedbrook JR, Jones J, O'Dell M, *et al.* 1980. A molecular description of telomeric heterochromatin in *Secale* species. *Cell* 19: 545–560.

Bennett MD. 1970. Natural variation in nuclear characters of meristems in *Vicia faba*. *Chromosoma* 29: 317–335.

Bennett MD. 1971. The duration of meiosis. *Proc R Soc Lond B* 178: 277–299.

Bennett MD. 1972. Nuclear DNA content and minimum generation time in herbaceous plants. *Proc R Soc Lond B* 181: 109–135.

Bennett MD. 1973. Nuclear characters in plants. *Brookhaven Symp Biol* 25: 344–366.

Bennett MD. 1976a. DNA amount, latitude and crop plant distribution. In: Jones K, Brandham PE eds. *Current Chromosome Research*. Amsterdam: North Holland Publishing Company, 151–158.

Bennett MD. 1976b. DNA amount, latitute and crop plant distribution. *Env Exp Biol* 16: 93–108.

Bennett MD. 1977. The time and duration of meiosis. *Philos Trans R Soc Lond B* 277: 201–277.

Bennett MD. 1985. Intraspecific variation in DNA amount and the nucleotypic dimension in plant genetics. In: Freeling M ed. *Plant Genetics. UCLA Symposia on Molecular and Cellular Biology*. New York: Alan R Liss Inc., 283–302.

Bennett MD. 1987. Variation in genomic form in plants and its ecological implications. *New Phytol* 106: 177–200.

Bennett MD, Leitch IJ. 1995. Nuclear DNA amounts in angiosperms. *Ann Bot* 76: 113–176.

Bennett MD, Leitch IJ. 1997. Nuclear DNA amounts in angiosperms: 583 new estimates. *Ann Bot* 80: 169–196.

Bennett MD, Leitch IJ. 2001. Nuclear DNA amounts in pteridophytes. *Ann Bot* 87: 335–345.

Bennett MD, Leitch IJ. 2003. Plant DNA C-values database (release 2.0, Jan. 2003). www.rbgkew.org.uk/cval/homepage.html.

Bennett MD, Leitch IJ. 2005. Nuclear DNA amounts in angiosperms—progress, problems and prospects. *Ann Bot* (in press).

Bennett MD, Smith JB. 1971. The 4C nuclear DNA content of several *Hordeum* genotypes. *Can J Genet Cytol* 13: 607–611.

Bennett MD, Smith JB. 1972. The effects of polyploidy on meiotic duration and pollen development in cereal anthers. *Proc R Soc Lond B* 181: 81–107.

Bennett MD, Smith JB. 1976. Nuclear DNA amounts in angiosperms. *Philos Trans R Soc Lond B* 274: 227–274.

Bennett MD, Smith JB. 1991. Nuclear DNA amounts in angiosperms. *Philos Trans R Soc Lond B* 334: 309–345.

Bennett MD, Bhandol P, Leitch IJ. 2000a. Nuclear DNA amounts in angiosperms and their modern uses: 807 new estimates. *Ann Bot* 86: 859–909.

Bennett MD, Dover GA, Riley R. 1974. Meiotic duration in wheat genotypes with or without homoeologous meiotic chromosome pairing. *Proc R Soc Lond B* 187: 191–207.

Bennett MD, Gustafson JP, Smith JB. 1977. Variation in nuclear DNA in the genus *Secale*. *Chromosoma* 61: 149–176.

Bennett MD, Heslop-Harrison JS, Smith JB, Ward JP. 1983. DNA density in mitotic and meiotic metaphase chromosomes of plants and animals. *J Cell Sci* 63: 173–179.

Bennett MD, Johnston S, Hodnett GL, Price HJ. 2000b. *Allium cepa* L. cultivars from four continents compared by flow cytometry show nuclear DNA constancy. *Ann Bot* 85: 351–357.

Bennett MD, Leitch IJ, Hanson L. 1998. DNA amounts in two samples of angiosperm weeds. *Ann Bot* 82: 121–134.

Bennett MD, Leitch IJ, Price HJ, Johnston JS. 2003. Comparison with *Caenorhabditis* (~100 Mb) and *Drosophila* (~175 Mb) using flow cytometry show genome size in *Arabidopsis* to be ~157 Mb and thus 25% larger than the *Arabidopsis* genome initiative estimate of ~125 Mb. *Ann Bot* 91: 547–557.

Bennett MD, Smith JB, Heslop-Harrison JS. 1982a. Nuclear DNA amounts in angiosperms. *Proc R Soc Lond B* 216: 179–199.

Bennett MD, Smith JB, Kemble R. 1972. The effect of temperature on meiosis and pollen development in wheat and rye. *Can J Genet Cytol* 14: 615–624.

Bennett MD, Smith JB, Lewis Smith RI. 1982b. DNA amounts of angiosperms from the Antarctic and South Georgia. *Env Exp Biol* 22: 307–318.

Bennett MD, Smith JB, Ward J, Jenkins G. 1981. The relationship between nuclear DNA content and centromere volume in higher plants. *J Cell Sci* 47: 91–115.

Bennetzen JL, Kellogg EA. 1997. Do plants have a one-way ticket to genomic obesity? *Plant Cell* 9: 1509–1514.

Bennetzen JL, Ma JX. 2003. The genetic colinearity of rice and other cereals on the basis of genomic sequence analysis. *Curr Opin Plant Biol* 6: 128–133.

Bennetzen JL, Ma JX, Devos KM. 2005. Mechanisms of recent genome size variation in flowering plants. *Ann Bot* (in press).

Biradar DP, Rayburn AL. 1993. Heterosis and nuclear DNA content in maize. *Heredity* 71: 300–304.

Biradar DP, Bullock D, Rayburn A. 1994. Nuclear DNA amount, growth and yield parameters in maize. *Theor Appl Genet* 88: 557–560.

Blondon F, Marie D, Brown S, Kondorosi A. 1994. Genome size and base composition in *Medicago sativa* and *Medicago truncatula* species. *Genome* 37: 264–270.

Bocking A, Giroud F, Reith A. 1995. Consensus report of the ESCAP task-force on standardization of diagnostic DNA image cytometry. *Analyt Cell Pathol* 8: 67–74.

Boivin A, Vendrely R, Vendrely C. 1948. L'acide desoxyribonucleique du noyau cellulaire dépositaire des caracteres hereditaires: arguments d'ordre analytique. *C R Acad Sci* 226: 1061–1063.

Bowers JE, Chapman BA, Rong JK, Paterson AH. 2003. Unravelling angiosperm genome evolution by phylogenetic analysis of chromosomal duplication events. *Nature* 422: 433–438.

Brandham PE, Doherty MJ. 1998. Genome size variation in the Aloaceae, an angiosperm family displaying karyotypic orthoselection. *Ann Bot* 82: 67–73.

Brown WL. 1949. Numbers and distribution of chromosome knobs in United States maize. *Genetics* 34: 524.

Bullock D, Rayburn A. 1991. Genome size variation in southwestern US indian maize populations may be a function of effective growing season. *Maydica* 36: 247–250.

Burton K. 1956. A study of the conditions and mechanism of the diphenylamine reaction for the colorimetric estimation of deoxyribonucleic acid. *Biochem J* 62: 315–323.

Burton K. 1968. Determination of DNA concentration with diphenylamine. *Methods Enzymol* 12B: 163–166.

Caceres ME, De Pace C, Mugnozza GTS, *et al.* 1998. Genome size variations within *Dasypyrum villosum*: correlations with chromosomal traits, environmental factors and plant phenotypic characteristics and behaviour in reproduction. *Theor Appl Genet* 96: 559–567.

Cavalier-Smith T. 1985. *The Evolution of Genome Size*. Chichester: John Wiley & Sons Ltd.

Cavallini A, Natali L. 1991. Intraspecific variation of nuclear DNA content in plant species. *Caryologia* 44: 93–107.

Cavallini A, Natali L, Cionini PG, Gennai D. 1993. Nuclear DNA variability within *Pisum sativum* (Leguminosae): nucleotypic effects on plant growth. *Heredity* 70: 561–565.

Cavallini A, Natali L, Giordani T, *et al.* 1996. Nuclear DNA changes within *Helianthus annuus* L.: variations in the amount and methylation of repetitive DNA within homozygous progenies. *Theor Appl Genet* 92: 285–291.

Ceccarelli M, Falistocco E, Cionini PG. 1992. Variation in genome size and organization within hexaploid *Festuca arundinacea*. *Theor Appl Genet* 83: 273–278.

Chieco P, Jonker A, Melchiorri C, *et al*. 1994. A user's guide for avoiding errors in absorbency image cytometry: a review with original experimental observations. *Histochem J* 26: 1–19.

Collins GB, Anderson MK, Legg PD. 1970. Cytophotometric determination of DNA content in *Nicotiana* megachromosomes. *Can J Genet Cytol* 12: 377–378.

Comai L. 2000. Genetic and epigenetic interactions in allopolyploid plants. *Plant Mol Biol* 43: 387–399.

Costich DE, Meagher TR, Yurkow EJ. 1991. A rapid means of sex identification in *Silene latifolia* by use of flow cytometry. *Plant Mol Biol Report* 9: 359–370.

Courties C, Perasso R, Chretiennot-Dinet MJ, *et al*. 1998. Phylogenetic analysis and genome size of *Ostreococcus tauri* (Chlorophyta, Prasinophyceae). *J Phycol* 34: 844–849.

Cox AV, Abdelnour GJ, Bennett MD, Leitch IJ. 1998. Genome size and karyotype evolution in the slipper orchids (Cypripedioideae : Orchidaceae). *Am J Bot* 85: 681–687.

Devos KM, Gale MD. 1997. Comparative genetics in the grasses. *Plant Mol Biol* 35: 3–15.

Devos KM, Brown JKM, Bennetzen JL. 2002. Genome size reduction through illegitimate recombination counteracts genome expansion in *Arabidopsis*. *Genome Res* 12: 1075–1079.

Dickson EE, Arumuganathan K, Kresovich S, Doyle JJ. 1992. Nuclear DNA content variation within the Rosaceae. *Am J Bot* 79: 1081–1086.

Dimitrova D, Greilhuber J. 2000. Karyotype and DNA content evolution in ten species of *Crepis* (Asteraceae) distributed in Bulgaria. *Bot J Linn Soc* 132: 281–297.

Dolezel J, Greilhuber J, Lucretti S, *et al*. 1998. Plant genome size estimation by flow cytometry: inter-laboratory comparison. *Ann Bot* 82 (Suppl. A): 17–26.

Dolezel J, Sgorbati S, Lucretti S. 1992. Comparison of three DNA fluorochromes for flow cytometric estimation of nuclear DNA content in plants. *Physiol Plantarum* 85: 625–631.

Ellegren H. 2002. Mismatch repair and mutational bias in microsatellite DNA. *Trends Genet* 18: 552.

Evans GM, Rees H. 1971. Mitotic cycles in dicotyledons and monocotyledons. *Nature* 212: 697–699.

Evans GM, Durrant A, Rees H. 1966. Associated nuclear changes in the induction of flax genotrophs. *Nature* 212: 697–699.

Evans GM, Rees H, Snell CL, Sun S. 1972. The relationship between nuclear DNA amount and the duration of the mitotic cycle. *Chromosomes Today* 3: 24–31.

Fay MF, Cowan RS, Leitch IJ. 2005. The effects of DNA amount on the quality and utility of AFLP fingerprints. *Ann Bot* (in press).

Feschotte C, Jiang N, Wessler SR. 2002. Plant transposable elements: where genetics meets genomics. *Nat Rev Genet* 3: 329–341.

Feulgen R, Rössenbeck H. 1924. Mikroskopisch-chemischer Nachweiss einer Nucleinsäure von Typus der Thymonucleinsäure und die darauf beruhende elective Färbung von Zellkernen in mikroskopischen präparaten. *Hoppe-Seyler Z Physiol Chem* 135: 203–248.

Filkowski J, Kovalchuk O, Kovalchuk I. 2004. Dissimilar mutation and recombination rates in *Arabidopsis* and tobacco. *Plant Sci* 166: 265–272.

Flavell RB, Rimpau J, Smith DB. 1977. Repeated sequence DNA relationships in four cereal genomes. *Chromosoma* 63: 205–222.

Ford CE, Hamerton JL. 1956. Chromosomes of Man. *Nature* 178: 1020–1023.

Frediani M, Colonna N, Cremonini R, *et al*. 1994. Redundancy modulation of nuclear DNA sequences in *Dasypyrum villosum*. *Theor Appl Genet* 88: 167–174.

Furuta Y, Nishikawa K, Makino T. 1975. Intraspecific variation in nuclear DNA content in *Aegilops squarrosa*. *Jap J Genet* 50: 257–263.

Galbraith DW, Harkins KR, Maddox JM, *et al*. 1983. Rapid flow cytometric analysis of the cell cycle in intact plant tissues. *Science* 220: 1049–1051.

Garner TWJ. 2002. Genome size and microsatellites: the effect of nuclear size on amplification potential. *Genome* 45: 212–215.

Gaut BS, Doebley JF. 1997. DNA sequence evidence for the segmental allotetraploid origin of maize. *Proc Natl Acad Sci USA* 94: 6809–6814.

Gerstel DU, Burns JA. 1966. Chromosomes of unusual length in hybrids between two species of *Nicotiana*. *Chromosomes Today* 1: 41–56.

Goff SA, Ricke D, Lan TH, *et al.* 2002. A draft sequence of the rice genome (*Oryza sativa* L. ssp. *japonica*). *Science* 296: 92–100.

González-Aguilera JJ, Ludeña Reyes P, Fernández-Peralta AM. 1990. Intra- and interspecific variation in nuclear parameters of two closely related species of *Narcissus* L. *Genetica* 82: 25–31.

Graham MJ, Nickell CD, Rayburn AL. 1994. Relationship between genome size and maturity group in soybean. *Theor Appl Genet* 88: 429–432.

Grandbastien MA. 1998. Activation of plant retrotransposons under stress conditions. *Trends Plant Sci* 3: 181–187.

Greenlee JK, Rai KS, Floyd AD. 1984. Intraspecific variation in nuclear DNA content in *Collinsia verna* Nutt (Scrophulariaceae). *Heredity* 52: 235–242.

Gregory TR. 2001a. Coincidence, coevolution, or causation? DNA content, cell size, and the C-value enigma. *Biol Rev* 76: 65–101.

Gregory TR. 2001b. *Animal Genome Size Database*. www.genomesize.com.

Gregory TR. 2005. The C-value enigma in plants and animals: a review of parallels and an appeal for partnership. *Ann Bot* (in press).

Greilhuber J. 1988. Self-tanning: a new and important source of stoichiometric error in cytophoto-metric determination of nuclear DNA content in plants. *Plant Syst Evol* 158: 87–96.

Greilhuber J. 1998. Intraspecific variation in genome size: a critical reassessment. *Ann Bot* 82 (Suppl. A): 27–35.

Greilhuber J. 2005. Intraspecific variation in genome size in angiosperms: identifying its existence. *Ann Bot* (in press).

Greilhuber J, Baranyi M. 1999. Feulgen densitometry: importance of a stringent hydrolysis regime. *Plant Biol* 1: 538–540.

Greilhuber J, Ebert I. 1994. Genome size variation in *Pisum sativum*. *Genome* 37: 646–655.

Greilhuber J, Obermayer R. 1997. Genome size and maturity group in *Glycine max* (soybean). *Heredity* 78: 547–551.

Greilhuber J, Speta F. 1985. Geographical variation of genome size at low taxonomic levels in the *Scilla bifolia* alliance (Hyacinthaceae). *Flora* 176: 431–438.

Greilhuber J, Temsch EM. 2001. Feulgen densitometry: some observations relevant to best practice in quantitative nuclear DNA content determination. *Acta Bot Croat* 60: 285–298.

Greilhuber J, Dolezel J, Lysak MA, Bennett MD. 2005. The origin, evolution and proposed stabilization of the terms 'genome size' and 'C-value.' *Ann Bot* (in press).

Greilhuber J, Ebert I, Lorenz A, Vyskot B. 2000. Origin of facultative heterochromatin in the endosperm of *Gagea lutea* (Liliaceae). *Protoplasma* 212: 217–226.

Grime JP. 1983. Prediction of weed and crop response to climate based upon measurements of nuclear DNA content. *Aspects Appl Biol* 4: 87–98.

Grime JP. 1996. Testing predictions of the impacts of global change on terrestrial ecosystems. *Aspects Appl Biol* 45: 3–13.

Grime JP, Mowforth MA. 1982. Variation in genome size: an ecological interpretation. *Nature* 299: 151–153.

Grime JP, Shacklock JML, Band SR. 1985. Nuclear DNA contents, shoot phenology and species co-existence in a limestone grassland community. *New Phytol* 100: 435–445.

Gustafson JP, Lukaszewski AJ, Bennett MD. 1983. Somatic deletion and redistribution of telomeric heterochromatin in the genus *Secale* and in *Triticale*. *Chromosoma* 88: 293–298.

Hanson L, Leitch IJ. 2002. DNA amounts for five pteridophyte species fill phylogenetic gaps in C-value data. *Bot J Linn Soc* 140: 169–173.

Hanson L, Brown RL, Boyd A, *et al.* 2003. First nuclear DNA C-values for 28 angiosperm genera. *Ann Bot* 91: 1–8.

Hardie DC, Gregory TR, Hebert PDN. 2002. From pixels to picograms: a beginners' guide to genome quantification by Feulgen image analysis densitometry. *J Histochem Cytochem* 50: 735–749.

Heller FO. 1973. DNS-Bestimmung an Keimwurzeln von *Vicia faba* L. mit Hilfe der Impulscytophotometrie. *Berichte Deutsch Bot Gesellsch* 86: 437–441.

Holm-Hanson O. 1969. Algae: amounts of DNA and organic carbon in single cells. *Science* 163: 87–88.

Hosouchi T, Kumekawa N, Tsuruoka H, Kotani H. 2002. Physical map-based sizes of the centromeric regions of *Arabidopsis thaliana* chromosomes 1, 2, and 3. *DNA Res* 9: 117–121.

Hulgenhof E, Weidhase RA, Schlegel R, Tewes A. 1988. Flow cytometric determination of DNA content in isolated nuclei of cereals. *Genome* 30: 565–569.

Ilic K, SanMiguel PJ, Bennetzen JL. 2003. A complex history of rearrangement in an orthologous region of the maize, sorghum, and rice genomes. *Proc Natl Acad Sci USA* 100: 12265–12270.

Jarvis LR. 1986. A microcomputer system for video image-analysis and diagnostic microdensitometry. *Analyt Quant Cytol Histol* 8: 201–209.

Jasienski M, Bazzaz FA. 1995. Genome size and high CO_2. *Nature* 376: 559–560.

Jiang N, Bao ZR, Zhang XY, *et al.* 2003. An active DNA transposon family in rice. *Nature* 421: 163–167.

Jiang N, Feschotte C, Zhang X, Wessler SR. 2004. Using rice to understand the origin and amplification of miniature inverted repeat transposable elements (MITEs). *Curr Opin Plant Biol* 7: 115–119.

Johnston JS, Jensen A, Czeschin DG, Price HJ. 1996. Environmentally induced nuclear 2C DNA content instability in *Helianthus annuus* (Asteraceae). *Am J Bot* 83: 1113–1120.

Jones RN, Rees H. 1968. Nuclear DNA variation in *Allium*. *Heredity* 23: 591–605.

Kalendar R, Tanskanen J, Immonen S, *et al.* 2000. Genome evolution of wild barley (*Hordeum spontaneum*) by BARE-1 retrotransposon dynamics in response to sharp microclimatic divergence. *Proc Natl Acad Sci USA* 97: 6603–6607.

Kapraun DF. 2005. Nuclear DNA content estimates in multicellular eukaryotic green, red and brown algae: phylogenetic considerations. *Ann Bot* (in press).

Karp A, Rees H, Jewell AW. 1982. The effects of nucleotype and genotype upon pollen grain development in *Hyacinth* and *Scilla*. *Heredity* 48: 251–261.

Kashkush K, Feldman M, Levy AA. 2002. Gene loss, silencing and activation in a newly synthesized wheat allotetraploid. *Genetics* 160: 1651–1659.

Kentner EK, Arnold ML, Wessler SR. 2003. Characterization of high-copy-number retrotransposons from the large genomes of the Louisiana iris species and their use as molecular markers. *Genetics* 164: 685–697.

Kenton A, Parokonny A, Bennett ST, Bennett MD. 1993. Does genome organization influence speciation? A reappraisal of karyotype studies in evolutionary biology. In: Lees DR, Edwards D eds. *Evolutionary Patterns and Processes*. London: Academic Press, 189–206.

Kidwell MG. 2002. Transposable elements and the evolution of genome size in eukaryotes. *Genetica* 115: 49–63.

Kimura Y, Tosa Y, Shimada S, *et al.* 2001. OARE-1, a *Ty1-copia* retrotransposon in oat activated by abiotic and biotic stresses. *Plant Cell Physiol* 42: 1345–1354.

Kirik A, Salomon S, Puchta H. 2000. Species-specific double-strand break repair and genome evolution in plants. *EMBO J* 19: 5562–5566.

Knight CA, Ackerly DD. 2002. Variation in nuclear DNA content across environmental gradients: a quantile regression analysis. *Ecol Lett* 5: 66–76.

Knight CA, Molinari N, Petrov DA. 2005. The large genome constraint hypothesis: evolution, ecology and phenotype. *Ann Bot* (in press).

Kubis S, Schmidt T, Heslop-Harrison JS. 1998. Repetitive DNA elements as a major component of plant genomes. *Ann Bot* 82: 45–55.

Kumar A, Bennetzen JL. 1999. Plant retrotransposons. *Annu Rev Genet* 33: 479–532.

Laurie DA, Bennett MD. 1985. Nuclear DNA content in the genera *Zea* and *Sorghum*: intergeneric, interspecific and intraspecific variation. *Heredity* 55: 307–313.

Leitch IJ, Bennett MD. 2004. Genome downsizing in polyploid plants. *Biol J Linn Soc* 82:651–663.

Leitch IJ, Hanson L. 2002. DNA C-values in seven families fill phylogenetic gaps in the basal angiosperms. *Bot J Linn Soc* 140: 175–179.

Leitch IJ, Chase MW, Bennett MD. 1998. Phylogenetic analysis of DNA C-values provides evidence for a small ancestral genome size in flowering plants. *Ann Bot* 82: 85–94.

Leitch IJ, Hanson L, Winfield M, *et al.* 2001. Nuclear DNA C-values complete familial representation in gymnosperms. *Ann Bot* 88: 843–849.

Leitch IJ, Soltis DE, Soltis PS, Bennett MD. 2005. Evolution of DNA amounts across land plants (Embryophyta). *Ann Bot* (in press).

Leuchtenberger C. 1958. Quantitative determination of DNA in cells by Feulgen microspectro-photometry. In: Danielli JF ed. *General Cytochemical Methods*. New York: Academic Press, 219–278.

Levin DA, Funderburg SW. 1979. Genome size in angiosperms: temperate versus tropical species. *Am Nat* 114: 784–795.

Liu B, Wendel JF. 2000. Retrotransposon activation followed by rapid repression in introgressed rice plants. *Genome* 43: 874–880.

Lyndon RF. 1963. Changes in the nucleus during cellular development in the pea seedling. *J Exp Bot* 14: 419–430.

Ma J, Devos KM, Bennetzen J. 2004. Analyses of LTR-retrotransposon structures reveal recent and rapid genomic DNA loss in rice. *Genome Res* 14: 860–869.

MacGillivray CW, Grime JP. 1995. Genome size predicts frost resistance in British herbaceous plants: implications for rates of vegetation response to global warming. *Funct Ecol* 9: 320–325.

Marie D, Brown SC. 1993. A cytometric exercise in plant DNA histograms, with 2C values for 70 species. *Biol Cell* 78: 41–51.

Martin PG. 1966. Variation in amounts of nucleic acids in cells of different species of higher plants. *Exp Cell Res* 44: 84–94.

Martin PG, Shanks R. 1966. Does *Vicia faba* have multi-stranded chromosomes? *Nature* 211: 650–651.

Masterson J. 1994. Stomatal size in fossil plants: evidence for polyploidy in a majority of angiosperms. *Science* 264: 421–424.

McLeish J, Sunderland N. 1961. Measurements of deoxyribosenucleic acid (DNA) in higher plants by Feulgen photometry and chemical methods. *Exp Cell Res* 24: 527–540.

McNeill J. 1980. *Scilla* L. In: Tutin TG, Heywood VH, Burges NA, *et al.* eds. *Flora Europaea.* Cambridge: Cambridge University Press, 41–43.

Michaelson MJ, Price HJ, Ellison JR, Johnston JS. 1991. Comparison of plant DNA contents determined by Feulgen microspectrophotometry and laser flow cytometry. *Am J Bot* 78: 183–188.

Midgley JJ, Bond WJ. 1991. Ecological aspects of the rise of angiosperms: a challenge to the reproductive superiority hypotheses. *Biol J Linn Soc* 44: 81–92.

Miksche JP. 1968. Quantitative study of intraspecific variation of DNA per cell in *Picea glauca* and *Pinus banksiana*. *Can J Genet Cytol* 10: 590–600.

Miksche JP. 1971. Intraspecifc variation of DNA per cell between *Picea sitchensis* (Bong.) Carr. provenances. *Chromosoma* 32: 343–352.

Mirsky AE, Ris H. 1949. Variable and constant components of chromosomes. *Nature* 163: 666–667.

Moldavan A. 1934. Photo-electric technique for the counting of microscopical cells. *Science* 80: 188–189.

Moore G, Devos KM, Wang Z, Gale MD. 1995. Cereal genome evolution: grasses, line up and form a circle. *Curr Biol* 5: 737–739.

Morgante M, Hanafey M, Powell W. 2002. Microsatellites are preferentially associated with nonrepetitive DNA in plant genomes. *Nat Genet* 30: 194–200.

Mowforth MAG. 1985. Variation in nuclear DNA amounts in flowering plants: an ecological analysis. Ph.D. thesis, United Kingdom, University of Sheffield.

Murray BG. 1998. Nuclear DNA amounts in gymnosperms. *Ann Bot* 82 (Suppl. A): 3–15.

Murray BG, Leitch IJ, Bennett MD. 2001. *Gymnosperm DNA C-values Database* (Release 1.0). www.rbgkew.org.uk/cval/homepage.html.

Naranjo CA, Ferrari MR, Palermo AM, Poggio L. 1998. Karyotype, DNA content and meiotic behaviour in five South American species of *Vicia* (Fabaceae). *Ann Bot* 82 (Suppl. A): 757–764.

Noirot M, Barre P, Duperray C, et al. 2003. Effects of caffeine and chlorogenic acid on propidium iodide accessibility to DNA: consequences on genome size evaluation in coffee tree. *Ann Bot* 92: 259–264.

Noirot M, Barre P, Duperray C, et al. 2005. Investigation on the causes of stoichiometric error in genome size estimation using heat experiments. Consequences on data interpretation. *Ann Bot* (in press).

Noirot M, Barre P, Louarn J, et al. 2000. Nucleus-cytosol interactions: a source of stoichiometric error in flow cytometric estimation of nuclear DNA content in plants. *Ann Bot* 86: 309–316.

Noirot M, Barre P, Louarn J, et al. 2002. Consequences of stoichiometric error on nuclear DNA content evaluation in *Coffea liberica* var. *dewevrei* using DAPI and propidium iodide. *Ann Bot* 89: 385–389.

Obermayer R, Greilhuber J. 2000. Genome size in *Hedera helix* L.: a clarification. *Caryologia* 53: 1–4.

Obermayer R, Leitch IJ, Hanson L, Bennett MD. 2002. Nuclear DNA C-values in 30 species double the familial representation in pteridophytes. *Ann Bot* 90: 209–217.

Ogur M, Erickson RO, Rosen GU, et al. 1951. Nucleic acids in relation to cell division in *Lilium longiflorum*. *Exp Cell Res* 2: 73–89.

Ohri D, Khoshoo TN. 1986. Genome size in gymnosperms. *Plant Syst Evol* 153: 119–132.

Orel N, Puchta H. 2003. Differences in the processing of DNA ends in *Arabidopsis thaliana* and tobacco: possible implications for genome evolution. *Plant Mol Biol* 51: 523–531.

Ozkan H, Tuna M, Arumuganathan K. 2003. Nonadditive changes in genome size during allopolyploidization in the wheat (*Aegilops-Triticum*) group. *J Hered* 94: 260–264.

Paroda RS, Rees H. 1971. Nuclear DNA variation in eu-sorghums. *Chromosoma* 32: 353–363.

Pegington C, Rees H. 1970. Chromosome weights and measures in the Triticinae. *Heredity* 25: 195–205.

Petrov DA. 2001. Evolution of genome size: new approaches to an old problem. *Trends Genet* 17: 23–28.

Pires JC, Lim KY, Kovarik A, et al. 2004. Molecular cytogenetic analysis of recently evolved *Tragopogon* (Asteraceae) allopolyploids reveals a karyotype that is additive of the diploid progenitors. *Am J Bot* 91: 1022–1035.

Poggio L, Rosato M, Chiavarino AM, Naranjo CA. 1998. Genome size and environmental correlations in maize (*Zea mays* ssp. *mays*, Poaceae). *Ann Bot* 82 (Suppl. A): 107–117.

Price HJ, Johnston JS. 1996. Influence of light on DNA content of *Helianthus annuus* Linnaeus. *Proc Natl Acad Sci USA* 93: 11264–11267.

Price HJ, Chambers KL, Bachmann K. 1981a. Genome size variation in diploid *Microseris bigelovii* (Asteraceae). *Bot Gaz* 142: 156–159.

Price HJ, Chambers KL, Bachmann K. 1981b. Geographic and ecological distribution of genomic DNA content variation in *Microseris douglasii* (Asteraceae). *Bot Gaz* 142: 415–426.

Price HJ, Hodnett G, Johnston JS. 2000. Sunflower (*Helianthus annuus*) leaves contain compounds that reduce nuclear propidium iodide fluorescence. *Ann Bot* 86: 929–934.

Prokopowich CD, Gregory TR, Crease TJ. 2003. The correlation between rDNA copy number and genome size in eukaryotes. *Genome* 46: 48–50.

Puech M, Giroud F. 1999. Standardisation of DNA quantitation by image analysis: quality control of instrumentation. *Cytometry* 36: 11–17.

Puite KJ, Tenbroeke WRR. 1983. DNA staining of fixed and non-fixed plant protoplasts for flow cytometry with Hoechst 33342. *Plant Sci Lett* 32: 79–88.

Qiu YL, Palmer JD. 1999. Phylogeny of early land plants: insights from genes and genomes. *Trends Plant Sci* 4: 26–30.

Rabinowicz PD. 2000. Are obese plant genomes on a diet? *Genome Res* 10: 893–894.

Rasch E, Woodard JW. 1959. Basic proteins of plant nuclei during normal and pathological cell growth. *J Biophys Biochem Cytol* 6: 263–276.

Rayburn AL, Auger JA. 1990. Genome size variation in *Zea mays* ssp. *mays* adapted to different altitudes. *Theor Appl Genet* 79: 470–474.

Rayburn AL, Price HJ, Smith JD, Gold JR. 1985. C-band heterochromatin and DNA content in *Zea mays*. *Am J Bot* 72: 1610–1617.

Reed SM. 1991. Cytogenetic evolution and aneuploidy in *Nicotiana*. In: Tsuchiya T, Gupta PK eds. *Chromosome Engineering in Plants: Genetics, Breeding, Evolution (Part B)*. Amsterdam: Elsevier Inc., 483–505.

Rees H, Hazarika MH. 1969. Chromosome evolution in *Lathyrus*. *Chromosomes Today* 2: 158–165.

Rees H, Narayan RKJ. 1988. Chromosome constraints: chiasma frequency and genome size. In: Brandham PE ed. *Kew Chromosome Conference III*. London: HMSO, 231–239.

Rees H, Cameron FM, Hazarika MH, Jones GH. 1966. Nuclear variation between diploid angiosperms. *Nature* 211: 828–830.

Reeves G, Francis D, Davies MS, *et al*. 1998. Genome size is negatively correlated with altitude in natural populations of *Dactylis glomerata*. *Ann Bot* 82: 99–105.

Rejmanek M. 1996. A theory of seed plant invasiveness: the first sketch. *Biol Conserv* 78: 171–181.

Resslar PM, Stucky JM, Miksche JP. 1981. Cytophotometric determination of the amount of DNA in *Arachis* L. sect. *Arachis* (Leguminosae). *Am J Bot* 68: 149–153.

Ricroch A, Brown SC. 1997. DNA base composition of *Allium* genomes with different chromosome numbers. *Gene* 205: 255–260.

Rothfels K, Sexsmith E, Heimburg M, Krause MO. 1966. Chromosome size and DNA content of species of *Anemone* L. and related genera (Ranunculaceae). *Chromosoma* 20: 54–74.

Sakamura T. 1918. Kurze Mitteilung über die Chromosomenzahlen und die Verwandtschaftsverhältnisse der Triticum-Arten. *Bot Mag* 32: 149–153.

SanMiguel P, Bennetzen JL. 1998. Evidence that a recent increase in maize genome size was caused by the massive amplification of intergene retrotransposons. *Ann Bot* 82 (Suppl. A): 37–44.

SanMiguel P, Gaut BS, Tikhonov A, *et al*. 1998. The paleontology of intergene retrotransposons of maize. *Nat Genet* 20: 43–45.

SanMiguel P, Tikhonov A, Jin YK, *et al*. 1996. Nested retrotransposons in the intergenic regions of the maize genome. *Science* 274: 765–768.

Schmidt G, Thannhauser SJ. 1945. A method for the determination of desoxyribonucleic acid, ribonucleic acid and phosphoproteins in animal tissues. *J Biol Chem* 161: 83–89.

Scott LJ, Cross M, Shepherd M, *et al*. 1999. Increasing the efficiency of microsatellite discovery from poorly enriched libraries in coniferous forest species. *Plant Mol Biol Report* 17: 351–354.

Seal AG. 1983. The distribution and consequences of changes in nuclear DNA content. In: Brandham PE, Bennett MD eds. *Kew Chromosome Conference II*. London: George Allen & Unwin, 225–232.

Seal AG, Rees H. 1982. The distribution of quantitative DNA changes associated with the evolution of diploid Festuceae. *Heredity* 49: 179–190.

Shaked H, Kashkush K, Ozkan H, *et al*. 2001. Sequence elimination and cytosine methylation are rapid and reproducible responses of the genome to wide hybridization and allopolyploidy in wheat. *Plant Cell* 13: 1749–1759.

Shirasu K, Schulman AH, Lahaye T, Schulze-Lefert P. 2000. A contiguous 66-kb barley DNA sequence provides evidence for reversible genome expansion. *Genome Res* 10: 908–915.

Sims LE, Price HJ. 1985. Nuclear DNA content variation in *Helianthus* (Asteraceae). *Am J Bot* 72: 1213–1219.

Singh KP, S.N.R, Singh AK. 1996. Variation in chromosomal DNA associated with the evolution of *Arachis* species. *Genome* 39: 890–897.

Soltis DE, Soltis PS, Bennett MD, Leitch IJ. 2003. Evolution of genome size in the angiosperms. *Am J Bot* 90: 1596–1603.

Sparrow AH, Miksche JP. 1961. Correlation of nuclear volume and DNA content with higher plant tolerance to chronic radiation. *Science* 134: 282–283.

Sparrow AH, Nauman AF. 1976. Evolution of genome size by DNA doublings. *Science* 192: 524–529.

Speta F. 1980. Die frühjahrsblühenden *Scilla*-Arten des östlichen Mittelmeerraumes. *Naturkundl JahrB Stadt Linz* 25 (1979): 19–198.

Srivastava S, Lavania UC. 1991. Evolutionary DNA variation in *Papaver*. *Genome* 34: 763–768.

Sunderland N, McLeish J. 1961. Nucleic acid content and concentration in root cells of higher plants. *Exp Cell Res* 24: 541–554.

Suoniemi A, Tanskanen J, Schulman AH. 1998. *Gypsy*-like retrotransposons are widespread in the plant kingdom. *Plant J* 13: 699–705.

Swift H. 1950a. The desoxyribose nucleic acid content of animal nuclei. *Physiol Zool* 23: 169–198.

Swift H. 1950b. The constancy of desoxyribose nucleic acid in plant nuclei. *Proc Natl Acad Sci USA* 36: 643–654.

Temsch EM, Greilhuber J. 2001. Genome size in *Arachis duranensis*: a critical study. *Genome* 44: 826–830.

Temsch EM, Greilhuber J, Krisai R. 1998. Genome size in *Sphagnum* (peat moss). *Bot Acta* 111: 325–330.

Teoh SB, Rees H. 1976. Nuclear DNA amounts in populations of *Picea* and *Pinus* species. *Heredity* 36: 123–137.

Thomas CA. 1971. The genetic organization of chromosomes. *Annu Rev Genet* 5: 237–256.

Thompson K. 1990. Genome size, seed size and germination temperature in herbaceous angiosperms. *Evol Trends Plants* 4: 113–116.

Tito CM, Poggio L, Naranjo CA. 1991. Cytogenetic studies in the genus *Zea*. 3. DNA content and heterochromatin in species and hybrids. *Theor Appl Genet* 83: 58–64.

Tjio JH, Levan A. 1956. The chromosomes of man. *Hereditas* 42: 1–6.

Underbrink AG, Pond V. 1976. Cytological factors and their predictive role in comparative radiosensitivity. *Curr Topics Radiat Res Quart* 11: 251–306.

Uozo S, Ikehashi H, Ohmido N, *et al.* 1997. Repetitive sequences: causes for variation in genome size and chromosome morphology in the genus *Oryza*. *Plant Mol Biol* 35: 791–799.

Van't Hof J. 1965. Relationships between mitotic cycle duration, S period duration and average rate of DNA synthesis in root meristem cells of several plants. *Exp Cell Res* 39: 48–58.

Van't Hof J. 1974. The duration of chromosomal DNA synthesis, of the mitotic cycle, and of meiosis of higher plants. In: King RC ed. *Handbook of Genetics, vol. 2, Plants, Plant Viruses and Protists.* New York: Plenum Press, 181–200.

Van't Hof J, Sparrow AH. 1963. A relationship between DNA content, nuclear volume, and minimum mitotic cycle time. *Proc Natl Acad Sci USA* 49: 897–902.

Vendrely R, Vendrely C. 1948. La teneur du noyau cellulaire en acide désoxyribonucléique à travers les organes, les individus et les espèces animales: Techniques et premiers résultats. *Experientia* 4: 434–436.

Verma SC, Rees H. 1974. Nuclear DNA and evolution of allotetraploid Brassicae. *Heredity* 33: 61–68.

Vicient CM, Schulman A. 2002. *Copia*-like retrotransposons in the rice genome: few and assorted. *Genome Lett* 1: 35–47.

Vicient CM, Kalendar R, Anamthawat-Jonsson K, *et al.* 1999a. Structure, functionality, and evolution of the BARE-1 retrotransposon of barley. *Genetica* 107: 53–63.

Vicient CM, Suoniemi A, Anamthawat-Jonsson K, *et al.* 1999b. Retrotransposon BARE-1 and its role in genome evolution in the genus *Hordeum*. *Plant Cell* 11: 1769–1784.

Vidic T, Greilhuber J, Vilhar B. 2003. Genome size is associated with differential survival of plant species. *Abstracts of the Second Plant Genome Size Meeting held at The Royal Botanic Gardens, Kew, September 2003.*

Vilhar B, Dermastia M. 2002. Standardization of instrumentation in plant DNA image cytometry. *Acta Bot Croat* 61: 11–25.

Vilhar B, Greilhuber J, Koce JD, *et al.* 2001. Plant genome size measurement with DNA image cytometry. *Ann Bot* 87: 719–728.

Vinogradov AE. 2003. Selfish DNA is maladaptive: evidence from the plant Red List. *Trends Genet* 19: 609–614.

Vitte C, Panaud O. 2003. Formation of solo-LTRs through unequal homologous recombination counterbalances amplifications of LTR retrotransposons in rice *Oryza sativa* L. *Mol Biol Evol* 20: 528–540.

Voglmayr H. 2000. Nuclear DNA amounts in mosses (Musci). *Ann Bot* 85: 531–546.

Voytas DF, Cummings MP, Konieczny A, *et al.* 1992. *Copia*-like retrotransposons are ubiquitous among plants. *Proc Natl Acad Sci USA* 89: 7124–7128.

Wakamiya I, Newton RJ, Johnston JS, Price HJ. 1993. Genome size and environmental factors in the genus *Pinus*. *Am J Bot* 80: 1235–1241.

Watanabe K, Yahara T, Denda T, Kosuge K. 1999. Chromosomal evolution in the genus *Brachyscome* (Asteraceae, Astereae): statistical tests regarding correlation between changes in karyotype and habit using phylogenetic information. *J Plant Res* 112: 145–161.

Wendel JF. 2000. Genome evolution in polyploids. *Plant Mol Biol* 42: 225–249.

Wendel JF, Wessler SR. 2000. Retrotransposon-mediated genome evolution on a local ecological scale. *Proc Natl Acad Sci USA* 97: 6250–6252.

Wendel JF, Cronn RC, Johnston JS, Price HJ. 2002. Feast and famine in plant genomes. *Genetica* 115: 37–47.

Westoby M, Jurado E, Leishman M. 1992. Comparative evolutionary ecology of seed size. *Trends Ecol Evol* 7: 368–372.

White J, Rees H. 1987. Chromosome weights and measures in *Petunia*. *Heredity* 58: 139–143.

Wicker T, Stein N, Albar L, *et al.* 2001. Analysis of a contiguous 211 kb sequence in diploid wheat (*Triticum monococcum* L.) reveals multiple mechanisms of genome evolution. *Plant J* 26: 307–316.

Wicker T, Yahiaoui N, Guyot R, *et al.* 2003. Rapid genome divergence at orthologous low molecular weight glutenin loci of the A and A(m) genomes of wheat. *Plant Cell* 15: 1186–1197.

Yokoya K, Roberts AV, Mottley J, *et al.* 2000. Nuclear DNA amounts in roses. *Ann Bot* 85: 557–562.

Zakirowa RO, Vakhtina LI. 1974. Cytophotometrical and caryological investigation of some species of the genus *Allium* subgenus *Melanocrommyum* (Webb et Berth.) Wendelbo, Sect. *Melanocrommyum*. *Bot J USSR* 12: 1819–1827.

Zoldos V, Papes D, Brown SC, *et al.* 1998. Genome size and base composition of seven *Quercus* species: inter- and intra-population variation. *Genome* 41: 162–168.

PART II

The Evolution of Genomic Parasites

CHAPTER 3

Transposable Elements

Margaret G. Kidwell

Transposable elements (TEs) are discrete DNA sequences that move from one location to another within the genome. They are found in nearly all species that have been studied and constitute a large fraction of some genomes, including that of *Homo sapiens*. TEs are potent broad-spectrum mutator elements that are responsible for generating variation in the host genome and have a role as key players in the ecology of the genome. This chapter presents an overview that includes coverage of TE structures, regulation, distribution, and dynamics. A wealth of examples provides many illustrations of the diversity of TE types and behaviors as well as the rich variety of interactions between TEs and their host genomes. It is evident from this that knowledge of these elements is essential for a full understanding of genome evolution.

A BRIEF HISTORY OF THE STUDY
OF TRANSPOSABLE ELEMENTS

Transposable elements comprise a group of distinct DNA segments with the capacity to move, or transpose, between many nonhomologous (unrelated) sites in the genome. The properties of these elements provide them with the capacity to mutate the DNA of the host organisms in which they reside in many different ways. Their biology is still the subject of active investigation, although their general existence has now been known for more than half a century. The following sections provide a brief history of the discovery and early study of these important genomic components.

THE DISCOVERY OF TRANSPOSABLE ELEMENTS

The pioneering studies of Barbara McClintock in the mid-20th century led to her discovery of TEs in maize (*Zea mays*). In 1949 she showed that the change of unstable recessive alleles to the dominant form in this plant was due to the movement of a short segment of a chromosome. She first named these segments "controlling elements" because of her emphasis on their role in gene expression. Controlling elements were later described by Fincham and Sastry (1974) as

> TEs of apparently sporadic occurrence, which make themselves visible through their abnormal control of the activities of standard genes. Most simply, a controlling element may inhibit activity of a gene through becoming integrated, in, or close to, that gene. From time to time, either in germinal or somatic tissue, it may be excised from this site, and, as a result, the activity of the gene is often more or less restored, while the element may become reintegrated elsewhere in the genome where it may affect the activity of another gene.

Maize is a highly complex eukaryote whose genome is not easily amenable to analysis. It is therefore rather remarkable that this, rather than a simpler organism, was the species in which the discovery of transposition was made. In fact, it was not until a decade after McClintock's initial work that the insertion sequences of the bacterium *Escherichia coli* were discovered and immediate analogies were drawn and acknowledged between the maize and *E. coli* elements. The ease of study of bacteria at the molecular level allowed much faster progress to be made in the study of bacterial TEs than in maize and other eukaryotic species.

Today the familiar concept of transposition (i.e., the ability of mobile genetic elements to replicate themselves, or be replicated, and move around the genome) is often taken for granted. However, as recently as three decades ago, it was difficult for many traditional geneticists to change their view of genes and chromosomes as static entities, often likened to "beads on a string." The discovery of transposition was important in providing a new concept of genomes as fluid and dynamic entities. It represented a true paradigm shift whose importance in the history of science has probably not yet had time to be fully appreciated.

The explosion of knowledge about TEs that has occurred during the last 30 years is reflected in the size and scope of several sequential collections of papers describing the "state of the art" of TE biology. The first two volumes consisted of the publication of the proceedings of meetings held at the Cold Spring Harbor Laboratory in 1976 (Bukhari *et al.*, 1977) and 1980 (Cold Spring Harbor Symposia on Quantitative Biology, vol. 45: "Movable Genetic Elements," 1981). Three later volumes are collections of invited papers updating research on TEs in an increasingly large range of species (Shapiro, 1983; Berg and Howe, 1989; Craig *et al.*, 2002).

EARLY TE STUDIES IN BACTERIA

Probably because of their small genome size, bacteria were some of the earliest organisms to be intensively investigated with regard to TEs. Two general types of elements were originally recognized. The IS (simple insertion) sequences were first defined as elements generally shorter than 2.5 kilobases (kb) and lacking genes unrelated to insertion function (Campbell *et al.*, 1977). These were distinguished from the second type, the Tn elements (referred to at that time as transposons), which were more complex in structure and generally larger than 2.5 kb. It was recognized that Tn elements often contain IS elements and behave like IS elements, but carry additional genes unrelated to insertion function.

Bacterial IS sequences were discovered during early investigations of the molecular genetics of gene expression in *E. coli* and bacteriophages. They were first isolated as highly polar, somewhat unstable mutations in the galactose and lactose operons of *E. coli*, and in the early genes of some bacteriophages. Many of these mutations were shown to be insertions of the same few segments of DNA in different positions and orientations. When it was realized that these segments of DNA were natural residents of the *E. coli* genome, their similarity to the TEs in maize discovered by McClintock (1952) was recognized.

The more complex Tn elements are members of a class of related transposons that are medically important because they confer antibiotic resistance to many pathogenic bacteria. For example, members of the Tn3 family are usually found on plasmids from antibiotic-resistant bacteria, but they may transpose to bacteriophages and to the chromosome of *E. coli* and many other bacteria. Infectious antibiotic resistance (encoded on plasmids) usually results from production of an enzyme that inactivates the antibiotic. Space constraints prevent adequate coverage of the huge body of information on bacterial TEs in the present chapter, but fortunately several in depth reviews are available elsewhere (e.g., references in Berg and Howe, 1989; Craig *et al.*, 2002).

EARLY TE STUDIES IN FUNGI

The first fungal TEs to be studied were the Ty elements in the bakers' yeast *Saccharomyces cerevisiae* (Cameron *et al.*, 1979). Demonstration of the mobility of *Ty* elements in the yeast genome came from the analysis of mutations at a number of genetic loci and the subsequent demonstration of insertions in those loci in some yeast strains, but not in others. Although interest in using TEs for gene tagging resulted in the cloning and analysis of an element called *Tad* in the bread mold *Neurospora crassa* (Kinsey and Helber, 1989), very little information about the TE complement of other fungal genomes was available until the advent of inexpensive DNA sequencing during the last decade. With the recent sequencing of the *N. crassa* genome (Galagan *et al.,* 2003), an interesting property has come to light concerning the TE complement of this species. Specifically, no active TEs were found, only inactive remnants. This observation is in fact consistent with the earlier discovery of a mechanism known as repeat-induced-point mutation (RIP), by which TEs and other relatively large repetitive sequences are inactivated by mutation (see later section for more details).

EARLY TE STUDIES IN PLANTS

Following Barbara McClintock's original discovery, maize dominated the study of plant TEs, even though other suitable models were apparent. For example, variegation in flower color in the snapdragon *Antirrhinum majus* was described by Darwin and others in the 19th century. Although the genetic instability in *A. majus* was known to share many similarities with that in maize, it was not until the 1980s, when molecular techniques were becoming widely available, that the movement of TEs was identified as the cause (Saedler *et al.,* 1984). In particular, the *nivea (niv)* locus was cloned from *A. majus* genomic DNA, which allowed the isolation of an unstable allele that conferred a variegated flower color phenotype. This allele carried a 15 kb TE named *Tam* that turned out to belong to the same family as the *hobo* element discovered in *Drosophila melanogaster* and the Ac element in maize. This family was subsequently named the *hAT* family after these first three elements to be discovered.

EARLY TE STUDIES IN ANIMALS

Prior to 1980, *Drosophila* predominated in the study of animal TEs. More than 20 years ago, a number of different TE families had been described from *D. melanogaster* (Finnegan *et al.,* 1978). Prominent among these were seven different families of *copia*-like elements, a family of *foldback (FB)* elements (Rubin, 1983), and the *P* element family (Bingham *et al.,* 1982; Rubin *et al.,* 1982). The *copia*-like elements were found to be strikingly similar in structure to the *Ty* elements in yeast (Fink *et al.,* 1981) and to the integrated proviruses of RNA tumor viruses. It was not until the early 1980s that the identification of TEs in other animal species such as humans (Singer, 1982) and the nematode worm (*Caenorhabditis elegans*) (Emmons *et al.*, 1983) was first made.

A number of factors contributed to this emphasis on species of *Drosophila*. Notably, the model organism *D. melanogaster* had been among the most intensively studied species at the genetic level throughout the 1970s and provided a number of clues to the activity of TEs. Among these were various unstable mutations that revert to a wild-type phenotype at unusually high rates, and generate deletions and chromosomal rearrangements that had one endpoint at the site of the mutation (reviewed by Green, 1977). Moreover, the giant polytene chromosomes of *D. melanogaster* salivary glands provide the added advantage that the genomic location of any isolated piece of DNA can be determined by the method of *in situ* hybridization. Thus TE insertion sites can be located even though they may not produce any visible phenotypic effect.

The molecular characterization of the *P* element family in 1982 was preceded by more than a decade of work on the phenotypic manifestations and phenomenology of this element family in natural populations of *D. melanogaster*. Following the discovery of an exception to the general rule of absence of recombination in *D. melanogaster* males (Hiraizumi, 1971) and other related observations (e.g., Kidwell and Kidwell, 1975), the phenomenon of hybrid dysgenesis was first described (Kidwell *et al.*, 1977). When male flies from natural populations of *D. melanogaster* were crossed with female flies from laboratory strains of the same species, the F_1 progeny produced high frequencies of mutation, chromosomal aberrations, and other genetic abnormalities. It was later discovered that these abnormalities were caused by the activation of *P* elements in dysgenic crosses between males from strains that carried *P* elements (*P* strains) and females from strains in which *P* strains were absent (*M* strains). The *P* element was not present in natural *D. melanogaster* populations until after the middle of the 20th century, and most common laboratory stocks were derived from these *M* strains. Following horizontal transfer from another species, the *P* element completely invaded natural populations of *D. melanogaster* during the last half-century.

A RECENT EXPLOSION OF NEW INFORMATION FROM DNA SEQUENCING

Studies of the basic biology of TEs have received a major impetus with the development of inexpensive methods of DNA sequencing. This has led to an explosion of comparative data on a multitude of TE families identified in an ever-increasing range of species, including many nonmodel organisms. The quickening pace of genome sequencing, as well as the advent of sophisticated methods of computational analysis of sequenced genomes, is further fueling this trend. However, it is often not fully appreciated that because of the repetitive nature of these sequences, they are frequently among the last to receive attention and the quality of data available is often relatively poor in comparison to that of gene-rich regions of the genome. For example, at the time of this writing—approximately three years after the publication of the first draft of the sequence of the D. *melanogaster* genome—about 80% of the centromeric heterochromatin had yet to be sequenced (Kapitonov and Jurka, 2003a).

WHO CARES ABOUT TRANSPOSABLE ELEMENTS?

It is informative to consider the range of perspectives represented by those researchers who are interested, either positively or negatively, in TEs. Four main groups of investigators have been identified (Holmes, 2002): (1) Geneticists have been particularly interested in the application of TEs as tools useful in research and phylogenetic analysis during the last two decades. These applications include the use of TEs as vectors for interspecies transformation (i.e., the transfer of naked DNA from one species into another) and as efficient markers for gene tagging and phylogenetic studies. Details of these applications are discussed more fully in a later section; (2) Genome annotators are interested in TEs and other repetitive DNA but in a more negative way. For them, these repetitive sequences often represent a major nuisance and increase the cost of genome sequencing. Again, this raises a continuing problem for those interested in the repetitive portions of the genome because these regions tend to be the last to be completely sequenced; (3) Structural molecular biologists have an interest in TEs because of their homologies with virus replication machinery, transcription factors, and binding proteins; (4) TEs have the potential to be of particular interest to evolutionary biologists because of the interactions with their hosts. For the most part, the focus of this chapter emphasizes the interests and approaches of this latter group.

HOW ARE TEs CLASSIFIED?

One of the most fundamental aspects of biological study is classification, and in this regard TEs are no exception. TEs are classified in two ways: according to their degree of functional self-sufficiency and according to their mechanism of transposition.

AUTONOMOUS AND NONAUTONOMOUS ELEMENTS

TEs are described as being autonomous or nonautonomous based on whether or not they encode their own genes for transposition. Autonomous elements, such as long interspersed nuclear elements (LINEs) in humans, are defined as those elements that essentially encode all the sequences that enable them to move. Nonautonomous elements are structurally deficient in one respect or another, and depend to at least some extent on other elements in the genome in order to move. Many nonautonomous elements are derived from autonomous elements through deletions of part of their structures that leave critical cis-acting sequences (i.e., ones on the same molecule) in place. Alternatively, some nonautonomous elements, such as short interspersed nuclear elements (SINEs) in humans, have evolved independently of, but in parallel with, autonomous counterparts. Sometimes the distinction between autonomous and nonautonomous elements is not clear, such as for the human endogenous retroviruses (HERVs). Also, some autonomous elements are not completely independent in that they require the cellular machinery of their hosts in order to transpose.

CLASSIFICATION BASED ON MODE OF TRANSPOSITION

Most TEs can be assigned to one of two broad classes according to their mechanism of transposition (Finnegan, 1989). Class I elements are generally referred to as "retroelements," and Class II elements are often called "DNA transposons" (or just "transposons"). Figure 3.1 shows the basic structural features of the two TE classes along with examples, and Table 3.1 provides some examples of TE families and host organisms.

Class I Transposable Elements

Class I elements are members of a large group of so-called retroelements that utilize reverse transcription of an RNA template to make additional copies of themselves, a process known as retrotransposition (Fig. 3.2). The original element is maintained *in situ*, where it is transcribed. Its RNA transcript is then reverse transcribed into DNA that integrates into a new location in the genome. This class includes the "retrotransposons" that are characterized by flanking long terminal repeats (LTRs) and the "retroposons" (also called "non-LTR retrotransposons") that lack terminal repeats, as well as the SINEs (see Fig. 3.1 for a comparison of the main structural features of these three subclasses). Class I mobile elements also include the endogenous retroviruses that are closely related to the retrotransposons and the mobile introns. These various types of elements are described in more detail in the following sections.

LTR Retrotransposons and Endogenous Retroviruses

The LTR retrotransposons were the first retroelements to be discovered in eukaryotes and are similar in structure and coding capacity to retroviruses.

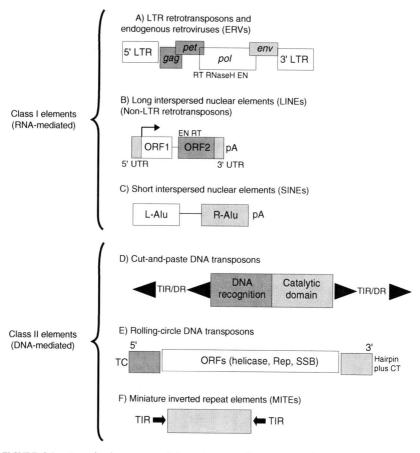

FIGURE 3.1   Generalized structures of the main types of transposable elements. Class I elements are those that transpose via an RNA intermediate, whereas the transposition of Class II elements is DNA-mediated. **(A)** Long terminal repeat (LTR) retrotransposons and endogenous retroviruses (ERVs) consist of partly overlapping coding regions for group-specific antigen (*gag*), protease (*prt*), and polymerase (*pol*) genes. These elements are flanked on both ends by LTRs with promoter capability. The *pol* gene contains domains for reverse transcriptase (RT), RNaseH, and integrase (IN) or endonuclease (EN). The distinction between LTR retrotransposons and ERVs is somewhat blurred by the fact that the envelope (*env*) genes typical of retroviruses have been both acquired by some LTR elements and rendered nonfunctional in some ERVs. This illustration is of the human endogenous retrovirus (HERV), which maintains an intact *env* gene. **(B)** Long interspersed nuclear elements (LINEs), or non-LTR retrotransposons, consist of a 5′ untranslated region (UTR) that has promoter activity, two open reading frames (ORF1 and ORF2) separated by an intergenic spacer, and a 3′ UTR with a poly-A tail (pA). LINEs are capable of autonomous transposition, since their ORF2 contains genes for EN and RT enzymes. ORF1 encodes a protein that binds nucleic acids. **(C)** The *Alu* element, the most common short interspersed nuclear element (SINE) in the human genome, consists of two GC-rich fragments, the left (L-*Alu*) and right (R-*Alu*) monomers which are connected by an A-rich linker and end in a poly-A tail. These elements are transcribed from an internal RNA polymerase III promoter, but do not code for any enzymes and are therefore incapable of transposing autonomously. Rather, SINEs appear to rely on LINEs for their transposition. LINEs and SINEs are also known collectively as "retroposons." **(D)** DNA transposons contain a large ORF that includes a DNA recognition and binding domain and a catalytic domain. These elements are flanked by short terminal inverted repeats (TIR) and may also include direct repeats (DR), and are capable of autonomous transposition.

Continued

TABLE 3.1   Examples of some common TE superfamilies, families, and host genomes

TE class	Superfamily	Family	Host example
I	Gypsy-like	Gypsy	Drosophila virilis
		297	Drosophila melanogaster
	Ty1-copia	Ty1	Saccharomyces cerevisiae
		Copia	Drosophila melanogaster
	Ty3	Ty3	Saccharomyces cerevisiae
		Micropia	Drosophila melanogaster
	LINEs	L1	Homo sapiens
		R1/R2	Bombyx mori
	SINEs	Alu	Homo sapiens
II	Helitrons	Helitrons	Arabidopsis thaliana
	Mariner-Tc1	Mos1	Drosophila mauritiana
		Tc1	Caenorhabditis elegans
	hAT	Hermes	Musca domestica
		Hobo	Drosophila melanogaster
	P	P	Drosophila willistoni
	MuDR	MuDR	Zea mays
	CACTA	Tam1	Antirrhinum majus
		En/Spm	Zea mays
Uncertain	Foldback	Galileo	Drosophila buzzatii

During transposition, a double-stranded DNA intermediate is synthesized by these elements from their RNA template by a mechanism similar to that used by DNA transposons (Craig *et al.*, 2002). The LTRs that are the hallmark of these retroelements play a key role in all aspects of their life cycle. Similar to retroviruses, these elements encode open reading frames (ORFs) called *gag* (specific group antigen) and *pol* (polymerase). The *pol* ORF is subdivided into different domains, including those of reverse transcriptase (RT), integrase (IN) or endonuclease (EN), and RNase H (RH) (Prak and Kazazian 2000; Eickbush and

FIGURE 3.1   *Cont'd.* (E) Rolling-circle DNA transposons replicate by a mechanism similar to the rolling-circle transposition mechanism in prokaryotes. They encode several ORFs, including helicase, replication initiator protein (Rep), and single-stranded DNA binding protein (SSB), and are flanked by a conserved TC dinucleotide on the 5′ end and a conserved hairpin and CT dinucleotide on the 3′ end. (F) Miniature inverted repeat elements (MITEs) have no coding potential and are flanked by TIR. Figure by T.R. Gregory, based on information provided by Prak and Kazazian (2000), reproduced by permission (© Nature Publishing Group), and Z. Tu (personal communication), reproduced by permission of the author.

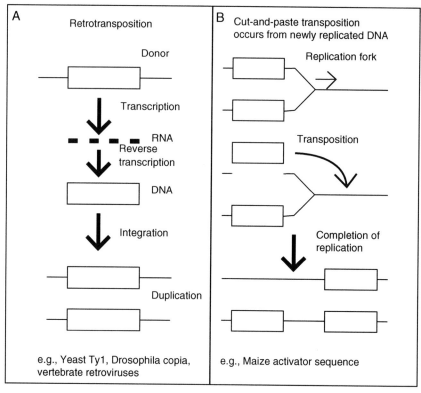

FIGURE 3.2   Two ways in which TEs can increase in copy number. (A) Retrotransposition, whereby a second copy of the TE is reverse transcribed to another location in the genome while the first copy remains in its original location. (B) Cut-and-paste transposition, whereby a DNA transposon is moved from a location on one newly replicated DNA segment into a region of the genome that has yet to be replicated, resulting in one daughter copy of the DNA that contains the TE only in its new location and one copy that includes the TE at both its original and novel locations. Adapted from Brookfield (1995), reproduced by permission (© Oxford University Press).

Malik, 2002) (Fig. 3.1). The arrangement, and even presence, of these different domains varies in different lineages. Recombination between domains is known to occur and thus different domains may exhibit different evolutionary histories. Based on the phylogeny of their reverse transcriptase, retrotransposons include four distinct lineages: *Ty1/copia*, *BEL*, *DIRS*, and *Ty3/gypsy* (Malik *et al.*, 2000). Two of these lineages, *Ty1/copia* and *Ty3/gypsy*, are abundant in animals and plants and have been extensively characterized. The *BEL* and *DIRS* lineages are less abundant and have only recently been described.

Many species also harbor endogenous proviruses in their genomes (for more details, see Bushman, 2002). These proviruses may represent either recent insertions into the genome or ancient molecular fossils. Endogenous retroviruses (ERVs) originate from the retroviral infection of germline cells, followed by fertilization involving the modified gamete. In turn, this gives rise to an offspring that differs from the parent by the presence of a proviral sequence insert. If the modified chromosome becomes fixed (homozygous) in the population then the presence of the provirus will become a permanent genetic property of the population unless changed by subsequent mutation. Human endogenous retroviruses (HERVs) have been particularly well characterized, and the integration of some of these is known to cause changes at the phenotypic level (see later section).

As noted in Figure 3.1, ERVs and LTR retrotransposons are very similar in structure, with the main difference sometimes given as the lack of an envelope (*env*) gene in the latter (Prak and Kazazian, 2000). However, the situation is rendered more complex than this by the fact that at least six lineages of LTR retrotransposons have independently acquired *env*-like ORFs (Eikbush and Malik, 2002). Furthermore, although the exception to the rule, some LTR retrotransposons (e.g., the *Gypsy* element in *D. melanogaster*) have potentially functional *env* genes, whereas many ERVs have nonfunctional *env* genes.

Retroposons (LINEs and SINEs)

Autonomous retroposons, also commonly called LINEs or LINE-like elements, are the second major subclass of TEs that utilizes reverse transcriptase (RT) during transposition. These retroelements are phylogenetically distinct from the retrotransposons. On the basis of RT phylogeny, they are most closely related to the Group II introns (see later section) of mitochondria and bacteria. Retroposons lack LTRs and transpose by simply reverse transcribing a complementary DNA (cDNA) copy of their RNA transcript directly onto the chromosomal target site. Although these elements are abundant and found in all eukaryotic lineages, it was some time before they were first recognized as a distinct self-propagating subclass of autonomous retroelements. Previously, they tended to be grouped with pseudogenes, SINEs, and other nonautonomous elements. Based on the phylogenetic relationship of their RT sequences, combined with the nature and arrangement of their protein domains, retroposons can be divided into five groups, *R2, L1, RTE, I*, and *Jockey*, each named after the first element to be discovered in that group (Eickbush and Malik, 2002).

Short interspersed nuclear elements (SINEs) are nonautonomous retroposons that exploit the enzymatic retrotransposition machinery of LINEs (Kajikawa and Okada, 2002). SINEs have a critical region that is homologous to tRNA, or 7SL RNA (a component of the signal recognition pathway), together with promoter sequences called A and B boxes (Nikaido *et al.*, 2003). SINEs are widely dispersed throughout many eukaryotic genomes and can be present in more than tens of thousands of copies per genome. For example, the SINE family *Alu* constitutes more than 10% of the human genome (International Human Genome Sequencing Consortium, 2001). The enormous number of SINE amplifications per organism makes them important evolutionary agents for shaping the diversity of genomes, and the nature of their mode of insertion makes them useful for diagnosing common ancestry among host taxa (see later section). As such, they represent a powerful new tool for systematic biology that can be strategically integrated with other conventional phylogenetic characters, most notably morphology and DNA sequences.

Mobile Introns

Mobile introns are mobile elements that reside within genes, but the element sequences are spliced out of RNA transcripts after synthesis. Introns are divided into several distinct classes according to their sequence and structure, as well as their splicing mechanism (Belfort *et al.*, 2002). Many Group I introns have been identified in eukaryotes and bacteria, but none have been found in archaea. Group II introns occur in mitochondrial and chloroplast genomes of fungi and plants and in cyanobacteria, proteobacteria, and Gram-positive bacteria, but not in archaea or in the nuclear DNA of eukaryotes. Group II introns are thought to be the likely progenitors of eukaryotic spliceosomal introns. Both Group I and Group II introns have the capacity for "intron homing" (the process by which the intron invades the same site in a cognate intronless allele), but the two groups use a different homing mechanism. Many Group I and Group II introns as well as a few archaeal introns are characterized by the presence of an ORF within the intron that encodes a maturase that promotes mobility of the intron within the genome. Introns can mobilize by two mechanisms, intron homing and intron transposition. Intron homing involves the transposition of the intron between two alleles of the same gene, one of which starts out with a copy of the intron and the other of which does not. By contrast, intron transposition involves the invasion of a new genomic site or locus (Belfort *et al.*, 2002).

Class II Transposable Elements

DNA transposons are subdivided into two subclasses according to their mode of transposition. The majority of Class II elements previously described transpose by means of a classical cut-and-paste mode similar to that of the *Tn10* elements of bacteria. A second group transposes by means of a rolling circle (RC) mechanism reminiscent of the *IS91*, *IS801*, and *IS294* bacterial families of transposons (see Fig. 3.1 for a comparison of the main structural features of these two subclasses). The classification of a third group known as Miniature Inverted Repeat Transposable Elements (MITEs) has previously been unclear, but these elements probably represent nonautonomous Class II elements (see later section and Fig. 3.1).

DNA Transposons That Transpose by a Cut-and-Paste Mechanism

The structural hallmarks of Class II elements that transpose by a cut-and-paste mechanism (Fig. 3.2) include terminal inverted repeats (TIRs) of varying length

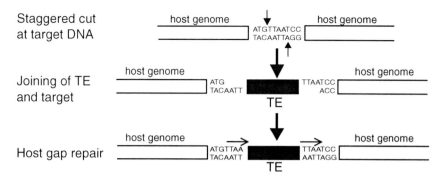

FIGURE 3.3   Generation of target site duplication (TSD) on insertion of a TE in host DNA. TE DNAs are indicated by solid black boxes. TSDs are produced by most TEs, with the exception of *Helitrons* and some long terminal repeat (LTR) retrotransposons. Adapted from Tu (2004), reproduced by permission (© Elsevier, Inc.).

and 2 to 10 base pair (bp) flanking direct repeats generated by target site duplications (TSDs) when the element inserts into a new site in the genome (Fig. 3.3). According to the cut-and-paste transposition model, element-encoded transposases perform both the cleavage and transfer reactions that are required to cut the transposon at both its termini and to insert it into a new position in the genome.

The majority of DNA transposons belong to families that are characterized by transposases of the DDE class named after the highly conserved Asp (D), Asp (D), and Glu (E) amino acid residues belonging to the catalytic core. They form a super-family of Class II transposons that includes the *Tc1* and *mariner* element families that are widely distributed among animals and are prone to horizontal transfer across species. Class II elements that lack DDE signatures are subdivided into a number of superfamilies. Again, the *hAT* superfamily was named after the first members to be described: the *hobo* element in *D. melanogaster,* the *Activator* (*Ac*) element in maize, and the *Tam3* element of snapdragon. Additional superfamilies and families include the *P* elements found mainly in *Drosophila* species, the *piggyBac* family that was first identified in baculoviruses that had been associated with moth cell lines, and the *Mutator* and *En/Spm* families first described in maize.

Miniature Inverted Repeat Transposable Elements (MITEs)

It has been recognized for some time that many nonautonomous members of Class II element families are derived by internal deletion of autonomous elements. Although the position and size of these deletions varies widely within a given family, some internal sequence similarity with the full-sized elements of the same family is commonly observed. A group of TEs collectively referred to as Miniature Inverted Repeat Transposable Elements (MITEs) were first described about ten years ago and were found to possess several properties reminiscent of nonautonomous Class II elements; notably, they are short (~100–500 bp in length) and have conserved terminal repeats. However, they have no coding potential and are frequently present in a copy number much higher than that of typical nonautonomous members of Class II families.

Target site preference is a hallmark of these elements. For example, *Tourist,* one of the first MITE families described, has a target site preference for TAA, and that of another family, *Stowaway,* is TA. MITEs were first discovered in plants, but they have subsequently been identified in many species of animals. They were first mistakenly identified as SINEs, but it is now evident that they are indeed nonautonomous DNA elements that originated from a subset of existing DNA transposons (Feschotte *et al.,* 2002). However, it remains unclear whether all DNA transposon families can give rise to MITEs.

Foldback (FB) Elements

The *FB* elements constitute a heterogeneous family of TEs whose transposition mechanism is still not fully understood. Nevertheless, it does seem clear that their transposition mechanism is quite unrelated to that of retroelements. *FB* elements were first described in *D. melanogaster* and are notable for their large inverted repeat termini that vary in length from several hundred base pairs to several kilobases. The inverted terminal repeats of *FB* elements have an unusual structure in that they are composed primarily of tandem copies of simple sequence DNA. Their central portions are heterogeneous with usually little or no additional information present in these regions. There is good evidence for deletions, inversions, and reciprocal translocations having one or both breakpoints at sites of preexisting *FB* insertions. For example, the origin of two polymorphic chromosomal inversions in *Drosophila buzzatii* can most likely be attributed to an *FB* element named *Galileo* (Caceres *et al.,* 2001; Casals *et al.,* 2003). Copies of this TE have been identified at all four inversion break points of these two inversions, and these breakpoints have become genetically unstable regions and hotspots for the accumulation of TE insertions and other structural changes. Thus *FB* elements appear to represent highly potent agents for generating genome rearrangements.

Rolling Circle Transposons

Previously it was thought that all eukaryotic Class II elements use a DNA-mediated mode of "cut-and-paste" transposition. However, recently a new family, the *Helitrons,* was discovered that is propagated by a mechanism similar to rolling-circle (RC) transposition in prokaryotes (Kapitonov and Jurka, 2001). *Helitrons* tend to be large and, surprisingly, constitute as much as 2% of the genomes of *Arabidopsis thaliana* and *Caenorhabditis elegans* (Kapitonov and Jurka, 2001).

THE RELATIONSHIP BETWEEN
CLASS I AND II ELEMENTS

A bacterial origin, either as transposons or retroelements, is generally assumed for most eukaryotic TEs. The structure and function of eukaryotic DNA elements is in fact very similar to those of bacteria. Eickbush and Malik (2002) have proposed an evolutionary scenario that includes a mosaic origin for LTR retrotransposons (Fig. 3.4). In this scheme, it is proposed that all eukaryotic mobile elements are descended from bacterial elements. LTR retrotransposons are suggested to have evolved from the fusion of bacterial transposons and bacterial retroelements, probably mobile Group II introns (see Fig. 3.4 for more details). This evolutionary scenario provides a rational basis for a new nomenclature

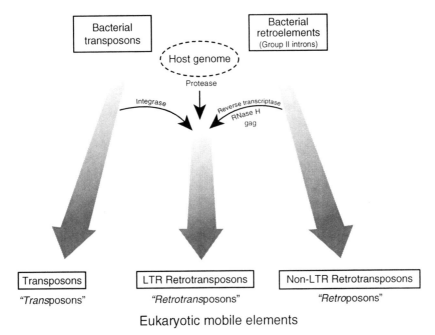

FIGURE 3.4   The probable bacterial origin of eukaryotic TEs. Transposons are DNA-mediated elements (Class II) and are probably derived from bacterial DNA transposons, whereas long terminal repeat (LTR) and non-LTR retrotransposons are RNA-mediated (Class I) (see Fig. 3.1). "Retroposons" refers to long interspersed nuclear elements (LINEs) and short interspersed nuclear elements (SINEs), which may be derived from bacterial Group II mobile introns. LTR retrotransposons are similar to retroviruses, and are believed to have a mosaic origin. Adapted from Eickbush and Malik (2002), reproduced by permission (© American Society for Microbiology Press).

(Eickbush and Malik, 2002) consisting of three main lineages: the "Transposons" (DNA-mediated elements), the "Retroposons" (non-LTR retrotransposons), and the "Retrotransposons" (LTR retrotransposons), the latter group having an origin that is a mosaic of the other two lineages. This nomenclature returns to the original usage of these terms by Howard Temin (1989).

HALLMARKS OF TE SEQUENCES

Most transposable elements share a number of properties that differentiate them from other types of DNA sequences, as discussed in the following sections.

DISPERSED MULTIGENE FAMILIES

There are two major groups of repeats in eukaryotic genomes: tandemly repeated satellites and repeats interspersed with genomic DNA. In contrast to satellite DNA, which tends to be highly repetitive and is usually confined to specific chromosomal regions, TEs belong to dispersed multigene families, which constitute a major component of the middle repetitive DNA that is abundant in many eukaryotic organisms. The degree of repetition (copy number) for any particular element family varies widely from a few to millions of copies, depending on the particular TE family and host species involved.

TARGET SITE DUPLICATIONS

At the DNA level, TEs are almost always recognized by small target site duplications (TSDs), which are induced at the point of insertion (Fig. 3.3). During transposition, the ends of the TE attack the target DNA at staggered positions such that the newly inserted element is flanked by short gaps. Host systems repair these gaps, resulting in the target sequence duplications that are characteristic of transposition. The length of this duplication is characteristic of each element, but typically many different target sites are used in the host DNA and thus different target sequences are duplicated.

TERMINAL REPEATS

The termini of many TEs are characterized by the presence of repeats that are typically homologous among members of the same element family. In DNA-mediated transposition, inverted terminal repeats are binding sites for the transposase that is encoded by complete, autonomous elements and whose role is to fuse the ends of the element with the target DNA. For example, *IS* elements in bacteria carry perfect or nearly perfect inverted repeats of about 10–40 bp. These terminal repeats are believed to serve as recognition sequences for the transposition enzymes (transposases) in their role of fusing the ends of the element with the target DNA. LTRs provide a signature for retrotransposons that is shared with retroviruses, to which they are closely related. The two direct repeat terminal sequences of any single element are identical at the time of insertion into the host DNA. Subsequently the termini of single elements diverge from one another over time and the amount of divergence can be used to estimate the age of the element in the host genome (e.g., SanMiguel *et al.*, 1998). In contrast to the LTR retrotransposons, the retroposons (such as LINEs and SINEs) are characterized by the absence of terminal repeats.

CODING REGIONS AND MOTIFS

In addition to their terminal repeats, autonomous TEs contain various genes. For example, LTR retrotransposons encode genes in an order that is typical of their family (see Fig. 3.5). LINEs include two open reading frames (Fig. 3.1) that can be quite variable in sequence. SINEs are relatively short and are dependent on LINEs for transposition. Both LINEs and SINEs have internal promoters for transcription by RNA polymerase. Cut-and-paste transposons typically encode a transposase gene in one or more ORFs. Sometimes a truncated version of the transposase sequence functions as a repressor of transposition, as seen for the *P* element (Fig. 3.6).

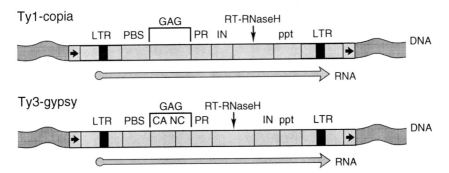

FIGURE 3.5   Structure of the two main families of long terminal repeat (LTR) retrotransposons, *Ty1-copia* and *Ty3-gypsy*. Some members of the *Ty3-gypsy* group also contain an envelope (*env*) gene between the integrase (IN) and polypurine tract (ppt) genes, and therefore replicate as retroviruses (see Fig. 3.1). Adapted from Kumar and Bennetzen (1999), reproduced by permission (© Annual Reviews).

P element

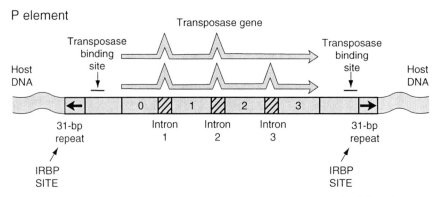

FIGURE 3.6   Structure of the *Drosophila* P element. Two transcripts are produced from the transposase gene, the triply spliced transposase message and the doubly spliced transposase inhibitor. IRBP = inverted repeat binding protein. Adapted from Clark *et al.* (1998), reproduced by permission (© Oxford University Press).

FIXED AND SEGREGATING INSERTION SITES

Chromosomal sites in which TEs have inserted relatively recently will tend to be heterozygous and segregating in natural populations. In contrast, TEs and their derivatives that have been present for longer periods are likely to be homozygous and fixed in natural populations due to chance or selection. This is another way that the relative age of TEs can be assessed.

METHODS USED IN THE IDENTIFICATION AND STUDY OF TEs

GENETIC ANALYSIS OF NATURALLY OCCURRING UNSTABLE MUTATIONS

Before the availability of modern molecular methods for identification and analysis, Barbara McClintock used purely genetic analyses to demonstrate the existence of elements in maize that can transpose to new chromosomal locations, alter the expression of nearby genes, and cause chromosome breakage, all in a developmentally regulated fashion. In *Drosophila*, hints of the presence of TEs were provided by some of their phenotypic manifestations, such as unstable mutations (Green, 1977) and hybrid dysgenesis (Kidwell *et al.,* 1977), long before they could be identified directly by molecular techniques. For example, the molecular nature of the P element responsible for P-M hybrid dysgenesis in D. *melanogaster* was only deducible with the cloning and sequencing of the wild type allele at the *white* locus (Bingham *et al.*, 1981) nearly a decade after the first observations of hybrid dysgenesis at the molecular level. Seven independent mutations at the *white* locus that had been induced by hybrid dysgenesis were examined. It was possible to identify five mutations that were caused by DNA insertions of 0.5, 0.5, 0.6, 1.2, and 1.4 kb in the wild type allele. The DNA insertions in four of these mutations, although heterogeneous in size and pattern of restriction enzyme sites, were homologous in sequence.

Currently, the methods used in the identification and study of TEs are primarily those used in standard molecular genetic analyses. However, phylogenetic methods are also employed in the analysis of the evolutionary history of TEs and with the advent of large-scale genome sequencing, bioinformatics approaches are increasingly being used.

METHODS OF MOLECULAR ANALYSIS

The majority of data on variation in genomic copy number and location of members of a TE family have been obtained using three methods:

1. *Southern blot analysis*. This technique provides a method for detecting TE fragments that are complementary to a specific DNA or RNA sequence by probing Southern blots of restriction-digested genomic DNA. In the case of multigene families, such as TEs, Southern blot analysis can also provide estimates of the copy number of a specific TE family present in a genome. Determining whether the TE of interest is present and in what copy number requires probing the DNA on a nitrocellulose membrane with radio-labeled or biotinylated (biotin-labeled) probes. Probes are chosen with homology to the TE sequences of interest and should not bind to the host genomic DNA or the nitrocellulose in a nonspecific manner. If not fixed within a species, individual TEs are manifested as insertion–deletion polymorphisms. If fixed, they are seen as insertion–deletion variation between related species.
2. *In situ hybridization to polytene chromosomes*. *In situ* hybridization allows TEs and other specific DNA sequences to be localized to particular segments of chromosomes. Because of polytenization (multi-strandedness), *Drosophila* salivary gland chromosomes are relatively large and have a well-defined morphology, which allows the easy localization of individual members of specific families of TEs by *in situ* hybridization.

It has been shown that in several respects Southern blot analysis is inferior to *in situ* hybridization as a method for accurately describing the properties of any specific TE family in the *D. melanogaster* genome (Maside *et al.*, 2001). The Southern blotting technique had serious deficiencies in three ways: underestimating

TE abundance, revealing less than 30% of the new insertions detected by *in situ* hybridization, and spuriously identifying changes in the size of restriction fragments from any source as simultaneous insertion–excision events.

3. *Polymerase Chain Reaction (PCR) amplification using conserved regions.* The PCR technique involves annealing single stranded probe molecules and target DNA to form DNA duplexes. The PCR is a method for amplifying small amounts of DNA or RNA and therefore has obvious useful applications in the study of TEs. It can be used to isolate complete TEs, or fragments of TEs, and also for end labeling, cloning, sequencing, and mutating TE DNAs. Prior to whole genome sequencing, this method, more than any other, has enabled the detailed study of many families of TEs in myriad organisms.

Data from Genome Sequencing Projects

The sequencing of whole genomes should, in theory, be able to provide complete information about an organism's TE content. If the target site duplications that are the hallmarks of TE insertion are still intact, complete genome sequences can allow the identification of new TE families for which probes were not previously available. However, mutation or deletion of TSDs often prevents the identification of members of previously undescribed families, resulting in an underestimation of TE abundance. Furthermore, many early sequence drafts described as "complete" have included only the euchromatic portion of the genome. The heterochromatic fraction that is often particularly rich in TEs is often slow to be sequenced.

The sequencing of the *D. melanogaster* genome provides a good example of the underestimation of TE abundance. The initial draft sequence included only the euchromatic two thirds of the whole genome (~120 Mb of ~180 Mb) (Adams *et al.*, 2000). In this draft, the estimated proportion of TEs was only 3.10% (International Human Genome Sequencing Consortium, 2001). Subsequent study of the abundance and distribution of TEs in a representative part of the euchromatic genome was made by analyzing the sizes and locations of TEs of all known families in the genomic sequences of chromosomes 2R, X, and 4 (Bartolome *et al.*, 2002). This study came up with an even lower estimate of up to 2% TEs. More recently, a systematic computational analysis (Kapitonov and Jurka, 2003a) provided evidence that TEs in the *D. melanogaster* genome are three times more abundant than previously reported by Bartolome *et al.* (2002). Kapitonov and Jurka (2003a) identified about 80 new TE families in addition to the 50 TE families previously reported. TEs were estimated to account for 6% of euchromatin and 60% of heterochromatin, with an overall abundance of 22%. These figures may still be underestimates because even three years after publication of the original draft, the "complete" sequence is still incomplete.

Reconstruction of Ancestral TEs from Incomplete Contemporary Copies

The study of a new TE usually begins with the identification of other homologous family members, followed by sequence alignment, classification into subfamilies, and construction of consensus sequences. Ancestral TEs can be reconstructed from incomplete copies as demonstrated for the *Tc1*-like element *Sleeping Beauty* from fish (Ivics *et al.*, 1997). *Sleeping Beauty* is an active DNA-transposon system from vertebrates that is useful for genetic transformation and insertional mutagenesis. Molecular phylogenetic data were used to construct a synthetic version of *Sleeping Beauty*, which could be identical or equivalent to an ancient element that dispersed in fish genomes in part by horizontal transmission between species. A consensus sequence of a transposase gene of the salmonid subfamily of elements was engineered by eliminating the inactivating mutations.

Databases for Repetitive DNA Sequences

Repbase Update (RU) (www.girinst.org/Repbase_Update.html) is an electronic database assembled to organize the explosively growing number of repetitive sequences from different eukaryotic species. RU is being used in genome sequencing projects worldwide as a reference collection for masking and annotation of repetitive DNA (e.g., by RepeatMasker or CENSOR). It is particularly useful because it includes TE consensus sequences and their biological characterizations that are reported nowhere else.

Phylogenetic Analysis

The phylogenies of TEs based on DNA or amino acid sequences are usually reconstructed to study the evolution of a particular TE lineage over a long period. Frequently it is of interest to compare TE phylogenies with those of host genes from the same species. The main reason for comparing these two types of phylogenies is to determine whether the relationships between elements of a given family detected in different host species agree with the systematic classification of these species. Lack of congruence between the two types of phylogenies may be due to methodological problems, but it can also be due to TE horizontal transfer, which occurs relatively frequently as compared with that of host genes.

Retroelements appear to exhibit modular evolution—that is, different genes in the same element have different evolutionary histories. For example, McClure (1991) compared the phylogenies of the reverse transcriptase (RT) domain with those of the capsid protein and ribonuclease H (RH) among retroposons. The RT and capsid protein trees were congruent with one another, but that from RH was clearly different. These differences could be accounted for by xenologous recombination (replacement of a resident gene by a homologous foreign gene), or by independent assortment (McClure, 1991).

APPLICATIONS OF TEs TO OTHER AREAS OF BIOLOGY

In addition to their basic biological interest, TEs have a number of uses for moving and marking genes in a variety of different genomes. Several examples of these types of applications are described in the following sections.

TRANSFORMATION SYSTEMS BASED ON TRANSPOSABLE ELEMENTS

Genetic transformation involves the transfer and incorporation of foreign DNA into a host genome. In order for this transferred DNA to be transmitted to later generations, transformation of germline or other appropriate cells of the recipient species is essential. The transfer of DNA sequences into eukaryotic species usually requires a "vector" such as a bacterium, virus, or transposable element.

Considerable effort has been expended recently in the development of TE-based transformation systems, particularly in insects, and such systems are now being used in many areas of biology. For example, between 1998 and 2004, four new TE-based vector systems were successfully developed for stable germline transformation of nondrosophilid insects. These systems are based on the *Mos-1* (active *mariner*) element from *Drosophila mauritiana,* the *Hermes* element from *Musca domestica,* the *Minos* element from *Drosophila hydei,* and the *piggyBac* element from *Trichoplusia ni.* In addition to *Drosophila* species, successful transformation of mosquitoes, tephritid fruit flies, and other dipteran species has been achieved. In fact, insect transformation can now be considered routine (Handler, 2001).

TRANSPOSABLE ELEMENT MUTAGENESIS AND GENE TAGGING

TEs are commonly used for mutagenesis and to tag, or mark, host genes that contain an inserted copy. The basic idea is to activate and insert a labeled TE of known sequence, look for a phenotypic effect, and find the marked TE, thereby identifying the gene responsible. Once the gene of interest has been isolated, it is amenable to further genetic analysis. One of the first examples of gene tagging was the use of a retrotransposon to isolate the *white* locus in *D. melanogaster* (Bingham *et al.*, 1981). Shortly following, dysgenesis-induced mutations at the *white* locus were used to identify the molecular basis of *P-M* hybrid dysgenesis (Bingham *et al.*, 1982). Later, the development and use of single genetically marked *P* elements simplified the identification and recovery of induced mutations (Cooley *et al.*, 1988). About the same time, the identification of single *P* elements carrying the transposase gene at a defined chromosomal location (Robertson *et al.*, 1988) made it possible to identify transposition events using genetic crosses rather than by embryo injection. This facilitated the removal of the transposase and subsequent stabilization of the new insertion. Single *P* element mutagenesis has made possible the identification of several thousand lethal *P* element insertions in distinct genes and has played an important role in the *Drosophila* Genome Project (Spradling *et al.*, 1995).

The development of efficient nonviral methodologies for genomewide insertional mutagenesis and gene tagging in mammalian cells is highly desirable for functional genomic analysis. A method of transposon-mediated mutagenesis (TRAMM), using naked DNA vectors, based on the *Drosophila hydei* TE *Minos*, has been developed with this goal in mind (Klinakis *et al.*, 2000). By simple transfections of plasmid *Minos* vectors, a high frequency of cell lines containing one or more stable chromosomal integrations was achieved. The *Minos*-derived vectors insert in different locations in the mammalian genome.

TRANSPOSABLE ELEMENTS AS MARKERS IN EVOLUTIONARY STUDIES

In order to be suitable markers for ecological and evolutionary studies, an element should be represented frequently in a genome, but should not have the potential for excision from its insertion site, nor be subject to frequent horizontal transfer. SINEs are tRNA-derived retroelements that most often have these properties. The irreversible, independent nature of their insertion frequently allows them to be used for diagnosing common ancestry among host taxa with extreme confidence. Also, many SINEs are specific to order, family, genus, and sometimes even species (Nikaido *et al.*, 2003).

It appears that new SINEs were created sporadically in a common ancestor of some lineages during evolution, a fact that makes them particularly useful for phylogenetic reconstruction (Nikaido *et al.*, 2003). For example, a family of SINEs called *AfroSINEs* is distributed exclusively among species of Afrotheria, a taxon that includes elephants, sea cows, and aardvarks. The use of *AfroSINEs* as markers in phylogenetic analysis confirmed the monophyletic relationships of the

Afrotheria that emerged as a result of physical isolation of the African continent from Gondwanaland (Nikaido *et al.*, 2003). In humans, the *Alu* family of SINEs has been used very successfully as markers in population genetic analysis. Again, this success depends on the facts that recently active SINEs can produce a high level of polymorphism and that their transposition does not involve excision.

The Use of Mobile Introns for Targeted Gene Manipulation

Mobile Group II intron RNAs insert directly into DNA target sites and are then reverse-transcribed into genomic DNA by the associated intron-encoded protein. The mechanism of homing employed by these introns has been used to allow reengineered mobile introns to be targeted to new sequences. Thus a highly efficient bacterial genetic assay was developed to determine detailed target site recognition rules and to select introns that insert into desired target sites (Guo *et al.*, 2000). It was shown that Group II introns can be retargeted to insert efficiently into virtually any target DNA and that the retargeted introns retain activity in human cells. The potential applications of this work on targeted Group II introns range from functional genomics to genetic engineering and gene therapy.

THE PREVALENCE OF TEs IN EUKARYOTIC GENOMES

Ancient Origins

It seems likely that some, but not all, eukaryotic TEs (or at least parts thereof) have an ancient bacterial origin. Although the origins of eukaryotic TEs are poorly understood, it appears that they have often evolved by the serial addition of domains. However, the reverse process, namely deletion or loss of domains, also appears to have occurred in some lineages (Capy *et al.*, 1997a). Both the integrase–transposase and the RT domains of Class I elements are likely to have evolved from those of bacteria, and have been reassembled in eukaryotes, leading to retrotransposons and then to retroviruses by the acquisition of an envelope gene (Capy *et al.*, 1997b).

A well-documented example of the ancient origin of TEs is provided by the *R1* and *R2* retroelements in insects. The *R1* and *R2* elements are two distantly related families of non-LTR retrotransposons, which insert at specific sites 74 bp apart in 28S ribosomal RNA genes. These elements have apparently been maintained by vertical transmission at least since the origin of the phylum Arthropoda, approximately 500 million years ago (Burke *et al.*, 1998).

PRESENT-DAY PREVALENCE

TEs have been identified in almost all eukaryotic species that have been examined. They are present in copy numbers ranging from just a few elements to tens or even hundreds of thousands per genome. In contrast to the high proportions found in large genomes, small genomes tend to have a low proportion of TEs. This wide range is illustrated for a sample of species in Table 3.2. In some cases TEs represent a major fraction of the genome, especially in some plants. For example, TEs have been estimated to constitute between 64 and 73% of the maize genome (Meyers *et al.*, 2001). The dispersed repetitive fraction of the human genome is currently estimated as ~46% (International Human Genome Sequencing Consortium, 2001), but may yet prove to be much higher when diverged and degraded TEs are fully included. Of course, when considering these high proportions, it is important to realize that the vast majority of TE-derived sequences in most genomes are inactive.

EXAMPLES OF COMMON TES IN FAMILIAR ORGANISMS

Low Diversity of TEs in Bakers' Yeast

The genome of *Saccharomyces cerevisiae* was the first among eukaryotes to be sequenced, and thereby afforded the earliest glimpse of a full genomic TE complement. It has turned out that this species is exceptional in having only five families of TEs: *Ty1*, *Ty2*, *Ty3*, *Ty4*, and *Ty5*, all of them LTR retrotransposons. A total of 331 insertions were originally identified (Kim *et al.*, 1998) occupying 3.1% of this small genome (~12 Mb). Based on high sequence identities, the *Ty3* and *Ty4* families appear to have been more recent additions to this genome than the other three families. The genomic distribution of *Ty* elements was found to be highly nonrandom, with a high proportion of elements being inserted close to genes transcribed by RNA polymerase III such as tRNA genes.

Retrotransposons in Maize

Although the maize genome is not yet sequenced, analysis of a typical subregion indicated that TEs make up a major component representing numerous families of LTR retrotransposons (SanMiguel *et al.*, 1996). The largest families contain elements such as *Huck* and *Ji* (*copia*-like elements) and *Opie* (a *gypsy*-like element) whose copy numbers are estimated to range between 100,000 to 500,000 per genome (Meyers *et al.*, 2001). These retrotransposons are in general randomly distributed across the maize genome (Meyers *et al.*, 2001). All of the retrotransposons examined appear to have inserted within the last six million years, most in the last

TABLE 3.2   Breakdown of TE content of six species by class

	Yeast	Slime mold	Nematode worm	Mustard weed	Fruit fly	Human
Genome size (Mb)	12	34	100	157	180	3400
Retrotransposons						
No. families	5	6	1	70	22	104
% of genome	3.1	4.4	0.1	6.4		7.9
Retroposons						
No. families	0	7	12	10	5	6
% of genome	0	3.7	0.4	0.7		31.2
DNA elements						
No. families	0	7	12	80	4	63
% of genome	0	1.5	5.3	6.8		2.8
Total						
No. families	5	20	25	180	31	263
% of genome	3.1	9.6	6.5	14	10–22	44.8
References	Kim et al. (1998)	Glockner et al. (2001)	Duret et al. (2000); International Human Genome Sequencing Consortium (2001)	Arabidopsis Genome Initiative (2000); International Human Genome Sequencing Consortium (2001)	Vieira et al. (1999); Kapitonov and Jurka (2003a)	International Human Genome Sequencing Consortium (2001); Li et al. (2001)

three million years (SanMiguel *et al.*, 1998). Amazingly, it seems that retrotransposon insertions have increased the size of the maize genome from approximately 1200 Mb to 2400 Mb in the last three million years (SanMiguel *et al.*, 1998).

P Elements in *Drosophila*

P elements are a family of cut-and-paste TEs that were first described in *D. melanogaster* as the causal agent of *P-M* hybrid dysgenesis (Kidwell, 1994). There is good evidence that P elements were horizontally transferred from a distantly related species, *Drosophila willistoni,* and invaded the cosmopolitan species *D. melanogaster* during the last half of the 20th century. It is important to note that although they represent one of the most intensively studied elements in eukaryotes (such that much is known about their structure, transposition mechanisms, and evolution), P elements are only one of approximately 130 different TE families found in the *D. melanogaster* genome (Kapitonov and Jurka, 2003a).

The canonical P element found in *D. melanogaster* is ~ 3 kb long, and contains four open reading frames (ORFs) that together encode a transposase. In addition, a truncated peptide consisting of only the first three ORFs and part of the third intron encodes a repressor of transposition. Terminal 31-bp perfect inverted repeats and internal 11-bp inverted repeats are required for transposition. The copy number of *D. melanogaster* P elements varies from 0 to about 60 per haploid genome. Usually a minority of these copies are autonomous (transposase-competent) elements and the majority are internally deleted, nonautonomous elements that are generally smaller than autonomous P elements. The induction of internal deletions is associated with active transposition of P elements.

In addition to the canonical P element subfamily, many more diverged sequences have been identified and grouped into 15 subfamilies according to their level of sequence identity. Active P elements have only been found in one other subfamily.

LINEs and SINEs in the Human Genome

Almost half of the ~ 76% fraction of the human genome available for analysis has been estimated to consist of four main types of TEs (International Human Genome Sequencing Consortium, 2001; Li *et al.*, 2001). By total composition, LINEs are the most abundant type and make up about one fifth of the genome (protein-coding genes, by stark contrast, constitute only about 1.5%). A single LINE family, *L1*, accounts for ~16% of the human genome and is present in more than 800,000 copies. However, the majority of these are truncated or rearranged, with a mere 4000 or so being full-length and only 40 to 60 active (Prak and Kazazian, 2000).

Although not yet fully understood, LINE elements are thought to integrate into genomic DNA by a process called target primed reverse transcription (TPRT).

L1 insertions are frequently found within AT-rich DNA, or within other *L1* insertions. This distribution pattern may reflect a successful strategy to avoid elimination from the genome because coding sequences tend to be found most frequently in GC-rich regions. However, some LINE elements are found to be associated with human protein-coding genes (Nekrutenko and Li, 2001) and their associated regulatory regions (Jordan *et al.*, 2003).

*L1*s have shaped the human and other mammalian genomes in several major ways: (1) they have greatly expanded the genome both by their own retrotransposition and by providing the machinery necessary for the retrotransposition of other TEs such as SINEs; (2) they have shuffled host genome sequences by comobilization of flanking sequences, a process often referred to as 5′ transduction (discussed in more detail in a later section); (3) they have affected gene expression by a number of mechanisms. From an applied perspective, *L1* elements are useful as phylogenetic markers, in gene discovery, and in the delivery of therapeutic genes.

Numerically, the most abundant TEs in the human genome are SINEs, which are present in far more copies than LINEs even though they constitute a smaller total percentage (about 13%) of the sequence. The predominant family of SINEs, *Alu,* is present in more than a million copies in the human genome. In light of this staggering abundance, Doolittle (1997) has commented that human genomes "might be ironically viewed as vehicles for the replication of *Alu* sequences." Other types of TEs are less well represented but still contribute a significant portion of the total DNA. Thus LTR retrotransposons make up roughly 8%, whereas DNA transposons contribute only about 3%. In general, although their total contribution is large, human TEs are lacking in the diversity of families seen in some other genomes. Indeed, low diversity among TE families appears to be a feature common to many mammalian genomes, and may reflect competition between these families for replicative dominance, resulting in the survival of single rather than multiple lineages (Furano *et al.*, 2004).

Recent publication of the mouse genome sequence provides additional insights into the evolution of human TEs. Although both contain about 30,000 potential coding genes, the mouse genome (3250 Mb) is about 14% smaller than the human genome (3400 Mb) (Mouse Genome Sequencing Consortium, 2002). A lower proportion of the mouse genome (38.5%) appears to be TE-derived than that of humans (46%). However, this large difference is likely to be something of an artifact owing to the higher nucleotide substitution rate in mice than in humans, which makes it more difficult to recognize ancient repeat sequences in the mouse. Interestingly, marked differences exist in the present-day activity of TEs in the two genomes. The rate of transposition appears to have remained fairly constant in mice, whereas in humans, transposition activity apparently reached a peak about 40 million years ago and has since plummeted to its present low level. This difference in contemporary TE activity is reflected in the large difference in TE-initiated mutations in mice ($\sim 10\%$) versus humans ($<1\%$) (Prak and Kazazian, 2000).

THE DISTRIBUTION OF TEs WITHIN GENOMES

Many aspects of the transposition process are random and, in general, there are few (if any) genomic compartments in which TEs are never found. However, the observed genomic distribution of TEs is often highly nonrandom and distribution patterns appear to differ widely among different groups of organisms. Although reliable data are not yet available for many species, a few general patterns are starting to emerge.

It has previously been postulated that TEs can be loosely divided into two types that occupy two very different niches in the ecology of the genome (Kidwell and Lisch, 1997). One type preferentially inserts into regions distant from host gene sequences, such as heterochromatin or the regions between genes (e.g., the many retrotransposons found inserted between the genes on the third chromosome in maize). This type escapes inactivation (via methylation or heterochromatinization) in regions outside of single copy host genes that are relatively AT-rich and in which recombination is minimal. The second type lives more dangerously by inserting into, or near, single copy sequences that tend to be relatively GC-rich.

Nonrandomness of TE distributions can be accounted for by three main factors: purifying selection acting on inserts that are detrimental to host fitness, a negative correlation between recombination frequency and TE density, and TE target site specificity that may have evolved in response to differential survival owing to selection and recombination. These are discussed in more detail in the following sections.

SELECTION AS A MECHANISM FOR REDUCING TE COPY NUMBER

Population studies of the distribution of TEs on chromosomes have strongly suggested that copy number increase, due to transposition, is balanced by some form of natural selection (Charlesworth *et al.*, 1997). Negative purifying selection is expected to act against the deleterious effects of insertions, particularly those located in gene-coding regions. Selection is also expected to act against the gross chromosomal rearrangements caused by ectopic exchange between TE copies (unequal recombination). Whereas the action of both mechanisms in controlling copy number appears to be indisputable, there is continuing debate as to the relative importance of each (e.g., Biémont *et al.*, 1997; Charlesworth *et al.*, 1997).

TEs are powerful mutagenic agents and, like other mutagens, the changes they produce have a broad range of fitness values at the organismal level, with a high proportion being lethal, causing sterility, or being otherwise deleterious to a greater or lesser extent. As such, mutations in those regions of the genome that are most susceptible to disruption, the coding regions, will most likely be subject to negative purifying selection. Therefore, it is expected and observed that TEs are less dense in coding than in noncoding regions. Nevertheless, some TEs do survive for variable lengths of time in these coding regions, either because they have a neutral impact on host fitness or possibly because they confer some fitness benefit to the host. These latter elements exemplify the second type of TE strategy outlined earlier (see also Kidwell and Lisch, 1997).

It is possible that TEs that tend to insert in or near coding regions have evolved ways to take advantage of relatively accessible chromosomal architecture, a high concentration of transcription factors, host enhancer sequences, and horizontal transfer, to maximize replication advantage (Kidwell and Lisch, 1997). Likely examples are elements such as *Mu* in maize (which target single copy sequences) and *P* elements in *Drosophila*. In the latter example, at least 65% of insertions are located near enhancers (Spradling *et al.*, 1995). It can be argued that these elements trade the disadvantage of an increased risk of negative selection for the advantages of occupying genomic regions that are enriched for factors promoting efficient transcription and replication. However, teasing out the relative importance of the various factors involved in TE survival in any particular case is difficult.

THE ROLE OF RECOMBINATION IN DETERMINING TE DISTRIBUTIONS

Because of their sequence homology, two TE copies inserted in nonhomologous positions in the same or different chromosomes may misalign and, if exchange occurs, this may result in ectopic recombination. The products of this recombination will include deletions and duplications that are likely to be deleterious to the host. Theory suggests that, as a consequence of deleterious ectopic meiotic exchange between TEs, selection can favor genomes with lower TE copy numbers. This predicts that TEs should be less deleterious, and hence more abundant, in chromosomal regions in which recombination is reduced. Indeed, a number of empirical studies have supported this theory. For example, a study of the distribution of nine families of TEs among a sample of autosomes isolated from a natural population of *D. melanogaster* (Charlesworth *et al.*, 1992) provided evidence to support the hypothesis that TE abundance is influenced by the deleterious fitness consequences of meiotic ectopic exchange between elements. However, Blumenstiel *et al.* (2002) concluded that the accumulation of TEs in heterochromatin and in euchromatic regions of low recombination is not a result of biased transposition but of greater probabilities of fixation in these regions relative to regions of normal recombination.

TE Frequencies in Euchromatin and Heterochromatin

In general, the majority of genes are found in the euchromatic portions of genomes that exhibit relatively high rates of recombination. In contrast, the late replicating and highly condensed heterochromatic portions are usually depauperate with respect to genes, and have relatively low frequencies of recombination. It has been observed that for several relatively small genomes the euchromatic and heterochromatic compartments harbor widely differing frequencies of TEs. For example, TEs make up ~6% and 16% of the euchromatic portions of the *D. melanogaster* and *Anopheles gambiae* genomes, respectively, compared with greater than 60% of the heterochromatin in both species (Holt *et al.*, 2002; Kapitonov and Jurka, 2003b).

This concentration of TEs in heterochromatin is also seen in *Arabidopsis thaliana*, in which 95% of repetitive DNA by bulk is found in the left arm of Chromosome 2, which contains the centromeric region (Kapitonov and Jurka, 1999). Also, in the compact *Tetraodon nigroviridis* genome with a 10% complement of TEs, there is a marked compartmentalization of TEs in heterochromatin that is reminiscent of that in the small *Drosophila* and *Arabidopsis* genomes (Dasilva *et al.*, 2002). Comparison of TE densities in heterochromatin and euchromatin of the larger vertebrate genomes of human and mouse (~3300 Mb) with those of the smaller genomes are informative. In excess of 40% of these genomes is made up of TEs and other repeats. With the exception of the sex chromosomes, these are scattered relatively uniformly in both euchromatin and heterochromatin.

Inter- and Intrachromosomal Variation in TE Density

Although detailed distribution patterns are not yet available for very many species, it is possible to make a few tentative generalizations about the distribution of TEs among and within chromosomes. In agreement with theoretical predictions (Sniegowski and Charlesworth, 1994), it appears that in many cases the density of TEs is negatively correlated with recombination rate. TEs tend to accumulate in the pericentromeric and telomeric regions of chromosomes where recombination is reduced. For example, the abundance and distribution of TEs was studied in a representative part of the euchromatic genome of *D. melanogaster*. TEs were not distributed at random in the chromosomes and their abundance was more strongly associated with local recombination rates than with gene density. The results are compatible with the ectopic exchange model, which predicts that selection on the deleterious products of ectopic recombination is a major factor constraining TE copy number (Charlesworth *et al.*, 1997).

Generally speaking, more TEs are found to be associated with sex chromosomes than with autosomes. This pattern is probably related to a greater concentration of heterochromatin in sex chromosomes than in autosomes. However, cause and effect are difficult to disentangle. On one hand, heterochromatin is characterized by low recombination rates. On the other hand, TEs may be agents in the formation and spread of heterochromatin. For example, Steinemann and Steinemann (1998) observed a massive accumulation of DNA insertions in the neo-Y chromosome of *Drosophila miranda*. They claim to present compelling evidence that the first step in Y chromosome degeneration is driven by the accumulation of TEs, especially retrotransposons. The switch from euchromatic to heterochromatic chromatin structure could be enabled by the enrichment of these elements along an evolving Y chromosome.

Overall differences in TE densities among different chromosomes and chromosome arms are a common feature of many sequenced genomes. In addition to differential recombination rates between sex chromosomes and autosomes, these may be related to genomic features specific to individual genomes. For example, TE densities in the different chromosome arms of *A. gambiae* are 59, 37, 46, 47, and 48 TEs per Mb for chromosomes X, 2R, 2L, 3R, and 3L, respectively. The relatively low overall repeat density for the right arm of the second chromosome (2R) may be related to the comparatively large number of paracentric inversions on this chromosome arm (Holt *et al.*, 2002). Significantly, a major effect of inversions is to reduce recombination rate.

TE TARGET SITE SPECIFICITIES

Some TEs appear to have evolved strategies to minimize the potentially devastating effects of their induced mutations on the fitness of their hosts. Such elements provide examples of the first type of TE strategy outlined previously, in which they preferentially insert into regions of the genome distant from host gene sequences. Some of these elements appear to target regions in which recombination is minimal (such as telomeres and centromeres) and where essential genes are scarce. In practice it is important, though often difficult, to determine whether nonrandom TE distributions result from the effects of selection or whether they can be accounted for by element target site specificity. *P* elements in *Drosophila* provide an example of how it is possible to experimentally discriminate between these two possibilities. The distribution of *de novo P* insertion mutations was examined by

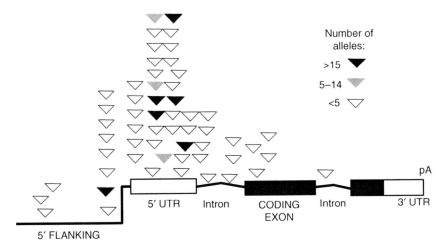

FIGURE 3.7   Illustration of preferential *P* element insertion near the 5′ end of *Drosophila melanogaster* gene transcription units. From Spradling *et al.* (1995), reproduced by permission (© National Academy of Sciences USA).

in situ hybridization before selection had had an opportunity to act on them (Spradling *et al.*, 1995). Whereas the insertion of members of most TE families was essentially random, *P* elements exhibited a preference for a particular subset of genes as well as noncoding regulatory regions within genes (Fig. 3.7). The implications of such target site preferences for the evolution of host gene regulation are discussed later.

THE DYNAMICS OF TE EVOLUTION

LONG-TERM EVOLUTION AND TE LIFE CYCLES

It appears that many TE families have an ancient origin and exhibit common patterns of long-term evolution within their host lineages. Theoretical studies predicted that the ultimate fate of most TEs, over perhaps millions of years, is mutation resulting in loss of transposition activity, rapid divergence, and ultimately the loss of an identity separate from that of the host sequences (Kaplan *et al.*, 1985). In elements subject to horizontal transfer, this scenario is modified to include the possibility of a succession of life cycles in different host lineages with the potential for long-term evolutionary survival.

Mechanisms of Spread and Loss

The main mechanisms responsible for amplification and spread of TEs within and among genomes and populations are transposition, horizontal transfer, and sexual reproduction. The two mechanisms most responsible for loss of TE sequences within a genome are selection against deleterious insertions and selection against the products of ectopic recombination. However, loss can also occur through excision (deletion). There is no general agreement about which mechanism is most important for the removal of TEs. Some argue that ectopic recombination has the greatest influence (e.g., Charlesworth et al., 1997). Others argue that purifying selection on insertions that reduce host fitness is the most important factor (e.g., Biémont et al., 1997).

Transposition

Transposition involves the movement of DNA molecules from one chromosomal location to another in the same cell. This is achieved by means of one of three types of mechanism, depending on the family to which the TE belongs. Transposition may be replicative, nonreplicative (sometimes called conservative), or involve an RNA intermediate. The chemistry of the breakage and reunion reaction is identical for all three types (Plasterk, 1995). Even if the jump is not replicative, it is sufficient for genomic copy number increase that the overall frequency of jumping is higher from replicated to nonreplicated DNA (Plasterk, 1995). Autonomous TEs typically encode the enzymes that are involved in their own transposition, but host factors are often also involved. Transposition, together with out-crossing in sexually reproducing organisms, is responsible for the spread of TEs from one organism to another within a population and species. If unchecked over time, the copy number of a particular TE family may increase to very large numbers, as seen in some plants.

Selection

TEs can be subject to both positive and negative natural selection, which may act at one of two different levels: the molecular (sequence) level and the host organism level. Positive selection at the DNA sequence level results from the ability of TEs to replicate faster than those of the host genome. Such "intragenomic selection" forms the basis of the selfish (or parasitic) DNA hypothesis (Doolittle and Sapienza, 1980). This type of selection is common when TEs invade new genomes, but is probably quite infrequent at other stages of their life cycle. TEs may also be subject to negative selection at the molecular level (because wild type elements coding for a functional transposase tend to transpose more frequently than mutated elements), but several studies have concluded that such selection acts mainly at the time of horizontal transfer (Witherspoon, 1999).

At the host organism level, negative selection commonly results from either TE-induced insertional mutations that are deleterious to hosts or ectopic recombination between homologous TE sequences located in nonhomologous regions of the genome. Obviously, inactive TEs are neutral with regard to this type of selection. Although positive selection of TEs is possible at the organismal level, opinion has been divided about the frequency of the beneficial effects of TEs on host genomes. Some believe that positive selection is so rare that it has virtually no impact on host evolution. A second perspective that is becoming increasingly expressed is that TEs have beneficial effects on their host genomes more frequently than previously acknowledged, and sometimes in unexpected ways (McDonald, 1998; Brosius, 1999a; Kidwell and Lisch, 2001).

As discussed in Chapters 1 and 2, the total amount of DNA contained within a genome can have important fitness-related consequences. Parameters such as cell size, cell division rate, body size, metabolic rate, and development can all be affected by variation in DNA content. Given that a substantial fraction of most eukaryotic genomes is comprised of TEs (see later section), there is good reason to expect selection on genome size to pose limits on TE abundance in certain species. The potentially profound implications of such a process for evolutionary theory are discussed in Chapter 11.

Excision and Deletion

Many, but not all, TEs can excise or remove themselves from their site of insertion in the host genome either precisely or imprecisely. Precise excision leaves a "footprint" in the form of a target site duplication of the original site of insertion in the host genome. Imprecise excision can take many forms, resulting in various mutations in the host genome, including residual partial insertions, deletions, and duplications.

On the scale of small (<400 bp) neutral insertions and deletions, there is a bias toward DNA loss, which some authors have proposed would tend to remove inactive ("dead-on-arrival") TEs over long timescales (Petrov, 2001). Although this is probably true in some organisms with small genomes (e.g., *Drosophila*), the data on which this theory is based are limited, and caution is needed in their interpretation (Gregory, 2003, 2004) (see Chapter 1).

Ectopic Recombination

Multigene families are susceptible to nonhomologous, or ectopic, recombination when homologous regions of different members of the family misalign and genetic exchange takes place either intra- or interchromosomally. Such events can lead to duplications, deficiencies, and new linkage relationships, often with consequent host fitness reduction and constraints on increase in copy number.

A good example of the importance of ectopic recombination is provided by the *BARE-1* retrotransposon in barley (e.g., Kalendar *et al.,* 2000). In addition to full-length copies, the *BARE-1* element is also represented by numerous solo LTRs, which are the relics of intraelement recombination between the LTRs of complete elements, and the consequent loss of their internal domains. The excess of LTRs relative to full-length elements suggests that recombination is additive among elements in the barley genome. Also, the greater the number of solo LTRs relative to full-length *BARE-1* elements in a given part of the genome, the lower the observed density of these elements. This is consistent with the expectation that high recombination rates result in loss of intact elements and an increase in the frequency of solo LTRs.

Horizontal Transfer

In addition to being transmitted vertically from parent to offspring, as with the normal inheritance of host genes, TEs can occasionally be transmitted horizontally (or laterally) from one species to another. A discrepancy between the phylogeny of a mobile element and that of a host gene is one of the most common ways that horizontal transfer is detected. However, it is important to note that

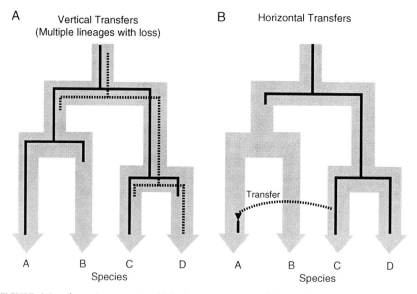

FIGURE 3.8 Alternative ways in which the incongruence of host and TE phylogenies can be explained. (A) Vertical transfer, in which the host gene and TE are lost differentially in the various lineages. (B) Horizontal transfer, in which a host gene or (more likely) a TE is transferred from one unrelated species to another. From Eickbush and Malik (2002), reproduced by permission (© American Society for Microbiology Press).

such discrepancies may have alternative explanations (Fig. 3.8). There is evidence that TEs transfer horizontally more frequently than nonmobile genes (Kidwell, 1993) and that Class II elements are more prone to this behavior than Class I elements. The Class II elements *mariner* and *P* provide good examples (Kidwell, 1994; Lohe *et al.*, 1995), but a major puzzle remains regarding the mechanism by which horizontal transfer is achieved.

REGULATION OF TE ACTIVITY

The unbridled activity of TEs has the potential to cause chaos in the host genome with a consequent reduction of fitness. Not surprisingly, a variety of different mechanisms have evolved to keep transposition in check. These mechanisms can be divided into two groups according to whether they are mediated by the elements themselves, or by the host organism. The regulation of *P* element activity in *Drosophila* is one of the most thoroughly studied in this regard and will serve to provide examples of many of the different kinds of regulation described in the following sections.

ELEMENT-MEDIATED REGULATION

The ability of TEs to regulate themselves is a property not shared with physical and chemical mutagenic agents. Presumably, these self-regulatory mechanisms have evolved to modulate the extent of damage to the host genome upon which the TEs depend.

Regulation by Maternal Inheritance

In the *P* element system in *D. melanogaster,* a *P* element–encoded 66-kDa truncated transposase, formed from mRNA transcripts from which the ORF 2-3 intron is spliced out, acts as one type of repressor of transposition. This was first observed as differences in the frequency of hybrid dysgenesis in reciprocal crosses between *P* (P+) and *M* (P−) strains of this species. *P* strain females produced a maternally inherited repressor that inhibited transposition in F_1 progeny, a condition termed the *P* cytotype. In contrast, *M* females that lacked any *P* elements in their genomes produced no *P* repressor and their progeny were unprotected from the effects of transposition.

Regulation by Multimer Poisoning

Still considering the *P* element system, certain deletion derivatives such as the *KP* element have been shown to encode a product other than the 66-kDa protein that also represses transposition, but is biparentally inherited. These elements may be involved in forming inactive multimers with the transposase, or with a host protein required for transposition.

Regulation by Overproduction Inhibition (OPI)

The *mariner* element is regulated by a mechanism called OPI in which an excess of the transposase reduces the activity of the element as measured by an excision assay. For example, increasing the number of copies of a *Mos1* construct decreased the rate of germline excision by 13% with one copy of the *Mos1* construct and by 37% with two copies of the construct (Lohe and Hartl, 1996).

Regulation by Transposase Titration

Defective TEs that retain their transposase binding sites may play a role in the regulation of transposition through titration of the active transposase. Such titration effects have been suggested to regulate the *P* element in *D. melanogaster* (Simmons and Bucholz, 1985) and the *mariner*-like elements in other *Drosophila* species (Hartl *et al.*, 1997). This mechanism might explain why certain deletions have replaced almost all functional *mariner*-like elements in a number of species (Hartl *et al.*, 1997).

Antisense RNA Regulation

Several TE families are known to produce antisense transcripts that can regulate transposition by means of RNA–RNA interactions. For example, maize contains two antisense transcripts that correspond to the two major transcripts of the *MuDR* TE (Hershberger *et al.*, 1995). Also, evidence for heat shock–induced antisense regulation of the *P* element has been found in *D. melanogaster* (Simmons *et al.*, 1996).

HOST-MEDIATED REGULATION OF TE ACTIVITY

The broad range in copy number of a single TE family over a number of species attests to the importance of host genomic environment in element copy number control. As an example, the copy number of *Ty-copia* group retrotransposons shows extreme variation, particularly among plant species; *A. thaliana* has only 100–200 copies of *Ty-copia* elements, whereas the bean *Vicia faba* has about one million copies (Flavell *et al.*, 1997). The primary mechanisms employed by host genomes to control transposition are outlined in the following sections.

Host Factors

Specific host factors are needed for the excision and transposition of *P* elements. Two good examples are provided by the *P* element Inverted Repeat Binding Protein

(IRBP) and *P* element Somatic Inhibitor (PSI). IRBP is related to the 70-kDa subunit of the human KU autoimmune antigen and is encoded by a mutagen-sensitive gene, *mus309*. IRBP interacts with 16 bp of the terminal 31 bp inverted repeats and may be involved with the repair of double strand DNA breaks.

DNA Methylation

Cytostine methylation is common, but not ubiquitous, among eukaryotes. In mammals and the bread yeast, about 2–3% of cytosines are methylated. Methylation plays an important role in the regulation of repetitive sequences. In both mammals and plants, the vast majority of methylated DNA is within TEs (SanMiguel *et al.*, 1996; Yoder *et al.*, 1997), but whereas most mammalian exons are methylated, plant exons are not. Thus targeting of methylation specifically to transposons appears to be restricted to plants (Rabinowicz *et al.*, 2003). Maize elements, for instance, undergo reversible methylation associated with changes in their activity (Banks *et al.*, 1988). In *Aspergillus*, a highly efficient system operates to methylate repetitive sequences.

The process known as methylation induced premeiotically (MIP) generates heavy methylation of repetitive DNA and is hypothesized to have evolved to silence TEs (Selker, 1997). It is known from studies of several eukaryotes, such as *Arabidopsis* and mice, that otherwise quiescent TEs become transpositionally active in methylation-deficient backgrounds. Following the sequencing of the *N. crassa* genome (Galagan *et al.*, 2003), it was observed that the methylated part of the genome consists almost entirely of relics of TEs (Selker *et al.*, 2003). This observation strongly suggests that, at least in *N. crassa*, DNA methylation has evolved as a partial defense against the invasion of TEs.

REPEAT-INDUCED GENE SILENCING

Repetitive DNA induces diverse silencing mechanisms, including cosuppression in nematodes, repeat-induced-point mutation (RIP) in *Neurospora*, position-effect variegation in *Drosophila*, and RNA-mediated interference (RNAi) in a number of eukaryotic species.

Repeat-Induced-Point Mutation (RIP)

A fascinating mechanism known as "repeat-induced-point mutation," exists in the bread mold *N. crassa*, by which TEs and other relatively large repetitive sequences are inactivated by mutation. The recent sequencing of the genome of this species (Galagan *et al.*, 2003) revealed that the total repeat fraction of the genome is limited to only 10% and, consistent with the hypothesis that RIP has evolved as a defense against the invasion of TEs, not a single intact TE was identified. This situation differs from that of other fungi such as yeast, which have not evolved the RIP mechanism.

RNA-Mediated Interference (RNAi)

RNAi involves the targeting of complementary mRNAs for degradation by double-stranded RNAs (dsRNAs). Proteins of the Dicer family cleave dsRNAs to generate small interfering RNAs, which target the RNA-induced silencing complex (Dernburg and Karpen, 2002). The Dicer proteins play a number of essential roles, including as a host defense mechanism against TEs and viruses. The RNAi mechanism of TE silencing has been particularly well characterized in C. elegans, but has also been found in certain protists, yeasts, insects, and plants (for review, see Vastenhouw and Plasterk, 2004).

DISRUPTION OF TE REGULATION BY ENVIRONMENTAL STRESSES

There is some evidence that various environmental factors affect the transcription and transposition of some TEs, particularly in plants. For example, activation of the Tnt1 family in tobacco, tomato, and Arabidopsis is affected by pathogen attacks, biotic elicitors such as fungal extracts, and abiotic stresses such as wounding. TE-induced mutations have been recorded to occur in transpositional bursts (Gerasimova et al., 1990) whose causes are not well understood, but which are likely related to inbreeding and other forms of genomic or environmental stress, possibly akin to the genomic stress referred to by McClintock (1984) in her Nobel Prize lecture.

Observations of stress-induced high transcription rates of SINEs have been reported in insects. For example, the level of transcripts of Bm1, a SINE found in the silk worm Bombyx mori, increases in response to either heat shock, inhibiting protein synthesis by cycloheximide, or viral infection (Kimura et al., 1999). Post-transcriptional events may partially account for stress-induced increases in Bm1 RNA abundance. In light of these results, it has been proposed that SINE RNAs may serve a role in the cell stress response that predates the divergence of insects and mammals (Kimura et al., 1999), implying that SINEs may have been coopted as a class of cell stress genes.

Temperature is another external factor that may influence TE activation. For example, several cases of disruption of hsp70 regulatory regions by TE insertions have been reported (Lerman et al., 2003), which underlies natural variation in expression of the stress-inducible molecular chaperone Hsp70 in D. melanogaster. It is hypothesized that the distinctive promoter architecture of hsp genes may make them vulnerable to TE insertions. Thus transposition may create quantitative genetic variation in gene expression within populations, on which natural selection can act.

A CONTINUUM OF TE-HOST INTERACTIONS FROM PARASITISM TO MUTUALISM

TEs are often summarily dismissed in the literature as simply being selfish or even junk. It has been argued more recently that a more accurate and enlightened approach is to consider them and their hosts in the coevolutionary terms of host–parasite relationships (Kidwell and Lisch, 2000). Such an approach is considerably more flexible, and envisions a continuum from total parasitism (or selfishness) at one extreme, through a middle ground of neutrality, to "molecular domestication" (or mutualism) at the other extreme. Indeed, the relationship between an element and its host may vary along this continuum over time. For example, when first invading a naïve genome, an element such as the P element in *Drosophila* may multiply rapidly in the host genome and represent the epitome of a selfish element. After perhaps millions of years, in rare cases molecular domestication occurs, representing a mutualistic relationship in which the element and the host genome are interdependent on one another. This very process has been demonstrated for P elements in several *Drosophila* species (Miller *et al.*, 1999). Other examples of important interactions between TEs and their hosts are described in the following section.

TEs AS MUTAGENS AND SOURCES OF GENOMIC VARIATION

CODING SEQUENCES AND THE EVOLUTION OF NOVEL HOST GENES

Because of the deleterious effects of insertion into protein-coding genes, it has not usually been considered likely that TEs play an important role in the evolution of coding regions. Although a number of examples of associations between TEs and coding sequences have been reported, comprehensive data on the genomic frequency of such associations are only just beginning to emerge. For example, Nekrutenko and Li (2001) examined the sequences of 13,799 human genes and found that 533 (~ 4%) of them contained TEs or fragments thereof. Among these 533 genes, ~ 40% were *Alu* elements (SINEs), ~ 27% were *L1* elements (LINEs), ~ 24% were LTR retrotransposons, and 9% were DNA transposons. Extrapolation of this result to the ~ 30,000 genes in the entire human genome suggests that ~ 1200 human genes contain TEs or fragments of TEs (Nekrutenko and Li, 2001).

It is becoming apparent that novel host genes may be mosaic in origin. The stationary *P* element–related gene clusters of *D. guanche, D. madeirensis,* and *D. subobscura* provide an interesting example of mosaic molecular domestication (Miller *et al.,* 1997). Each cluster unit consists of a cis-regulating section composed of different insertion sequences followed by the first three exons of a *P* element that encodes a 66 kDa repressor-like protein. In contrast to this normal repressor function, these stationary *P* element repeats appear to have evolved the function of transcription factors. Remarkably, the *D. guanche* P-protein produces an enhancer-like effect, rather than repressing canonical *P* element activity in transgenic *D. melanogaster* (W. Miller, personal communication). The insertion sequence, which gave rise to the *de novo* A-type promoter of this *P*-gene cluster, has been identified as belonging to a MITE-like TE family (designated *SGM*) that is related to poorly characterized bacterial *IS* elements of other *obscura* group species (Miller *et al.,* 2000).

Of particular interest is the mechanism by which TEs contribute to new coding sequences. As described previously, approximately 4% of coding sequences in the human genome are associated with TE-derived sequences (Nekrutenko and Li, 2001). Two possibilities exist for how TEs are integrated in coding regions: either the TE can insert directly into a protein-coding region or it can first be inserted into a noncoding region (e.g., an intron) and subsequently be recruited as a new exon (Fig. 3.9). Interestingly, the study by Nekrutenko and Li (2001) indicated that only about 10% of TEs have inserted directly into coding regions, whereas almost 90%

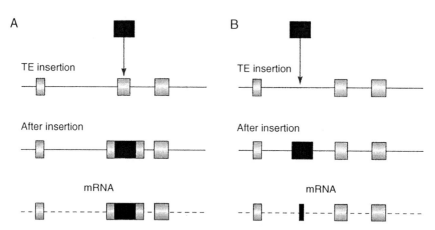

FIGURE 3.9 Two possible ways in which TE insertion into a protein-coding region can alter host genes. (A) A TE (black box) is inserted directly into a protein-coding exon (gray boxes), and therefore would automatically become part of the resulting mRNA if the gene is transcribed. However, note that in the majority of cases, the effects of such direct insertion are probably deleterious because TEs often contain multiple stop codons and would destroy the target exon. (B) A TE (black box) is inserted into an intron (black lines), and later a portion of the TE is recruited as a novel exon. Adapted from Nekrutenko and Li (2001), reproduced by permission (© Elsevier Inc.).

were recruited from nearby noncoding regions. This high rate of recruitment is possible because TEs carry potential splice sites (Nekrutenko and Li, 2001). This result explains the relatively high rate of insertion despite the expectation that the majority of new insertion in coding regions will be selected against. It will be interesting to see whether this pattern is also found in other species.

INTRONS

One way that TEs can increase their probability of survival in the genome is to camouflage themselves as introns in order to mitigate the negative selection pressure that accompanies insertion into host genes. Several lines of evidence suggest that the splicing of TEs may be a generalized mechanism to remove insertions in pre-mRNA. The ability to be spliced appears to be widespread, occurring in various species and both classes of TEs (Purugganan and Wessler, 1992). Several examples have been described in maize, including the *wx-m9* allele. Partial restoration of activity of this allele was shown to be due to the splicing of the *Ds* TE from the gene transcript (Wessler *et al.*, 1987).

ALTERNATIVE SPLICING

Alternative splicing provides an important mechanism for generating the observed proteomic diversity that is derived from a relatively small number of protein-coding genes in vertebrates. Alternative splicing is mediated by alternative exons that are included in only a fraction of mRNAs synthesized from a given gene. This is in contrast to constitutive exons that are included in all mRNAs (Kreahling and Graveley, 2004). For example, at least 5% of human alternative exons are derived from a process called exonization by which *Alu* elements are inserted into mature mRNAs via a splicing-mediated process (Lev-Maor *et al.*, 2003; Kreahling and Graveley, 2004). Although the disease-producing potential of constitutively spliced *Alu* exons is considered important in medical genetics, the evolutionary importance of the more common alternative splicing of these exons should not be overlooked. When alternative splicing occurs, the normal version of the encoded protein may still be synthesized at a level sufficient to maintain its function, enabling an *Alu* exon to diversify and possibly produce a protein with improved functionality (Kreahling and Graveley, 2004) under changing environmental conditions.

GENE REGULATORY SEQUENCES

The potential evolutionary importance of host genome regulatory sequences (as opposed to the protein-coding sequences themselves) has been recognized for some time. TEs can affect gene regulation in several ways. A new insertion can either disrupt an existing host regulatory element, or the TE may contribute its own regulatory sequences to a host gene into which it has inserted at an appropriate location.

Inserted sequences may also provide the potential for future evolution of regulatory sequences. For example, LTR retrotransposons carry a promoter within each terminal repeat that provides the potential for contributing to new patterns of host gene expression. The preference for insertion into genic regions also makes certain elements good candidates for cooption as regulatory elements. MITEs seem to be particularly likely to assume such a role, as shown in examples from the yellow fever mosquito *Aedes aegypti* (Tu, 1997), *Arabidopsis* (Feschotte and Mouches, 2000), and rice (Bureau *et al.*, 1996). Relatively small elements like SINEs are most likely to be coopted, but larger elements may sometimes act in a similar way.

Another important example has recently been provided by analysis of the human genome. Jordan *et al.* (2003) analyzed more than 2000 human promoter sequences, each about 500 bp in length (promoters can be defined as the sequence regions that are located directly 5′ of transcription initiation sites and that regulate their 3′ adjacent genes), and identified TE-derived sequences in almost 25% of these. In fact, TE-derived sequences comprised 8% of the total nucleotides found in all promoters. SINEs were represented in promoters at a higher frequency than in the genome as whole, but LINEs were relatively less frequent.

Many TEs can act as movable carriers of regulatory elements, such as promoters or enhancers, and may integrate into or near genes. They can thus contribute to the functional diversification of genes by supplying cis-regulatory domains and can have the potential to alter tissue-specific expression patterns. A second possibility is that, following their initial insertion into or near genic regions, and after lying immobilized and dormant, perhaps for some time, TEs can mutate to be able to regulate nearby genes. The most common regulatory functions served by coopted TEs include those of transcriptional enhancers, reducers and modulators, and polyadenylation signals (Brosius, 1999b). A number of examples of vertebrate regulatory elements and parts of coding regions that have been generated by retroelements illustrate some of the diversity of possible new host functions (Brosius, 1999b) (see http://exppc01.uni-muenster.de/expath/alltables.htm for a list).

TELOMERES

Although most organisms examined to date use telomerase to prevent the loss of sequences at the ends of chromosomes following DNA replication, insects in the order Diptera use two non-LTR elements, *HetA* and *TART*, for this purpose. These retroelements transpose specifically to the ends of *D. melanogaster* chromosomes and are present in tandem arrays at the ends of these chromosomes (reviewed by

Pardue and DeBaryshe, 2003). It is the serial addition of these elements to the ends of dipteran chromosomes that maintains their length following DNA replication. *HetA* and *TART* are completely dedicated to their role in maintaining the ends of chromosomes. Interestingly, the mechanism that these TEs use for extending *Drosophila* chromosomes is essentially the same as that used by telomerase (Pardue *et al.*, 1997). This suggests that the telomeric TEs may have evolved from telomerase. However, an alternative hypothesis, that *Drosophila* might have lost its telomerase and had the role taken over by a "domesticated" TE, cannot be ruled out.

In addition to the retrotransposons *HetA* and *TART* that transpose specifically to telomeres in flies, several nondipteran retrotransposons are commonly found to be associated with telomeres, although they may also transpose to other sites (Pardue and DeBaryshe, 2003). In the yeast *S. cerevisiae*, Y' retrotransposons are found only in subterminal regions of chromosomes while *Ty5* is found preferentially at telomeres and silent mating loci. The silkworm *Bombyx mori* has at least two non-LTR retroposons that insert specifically in the TTAGG telomerase repeat arrays. The parasitic protozoan *Giardia lamblia* reproduces asexually and has only two active retroelement families, called *GilM* and *GilT*, in its genome. These elements are found in tandem head-to-head arrays in the subtelomeric regions of chromosomes, but do not form the terminal sequences.

CENTROMERES

Centromeres are largely heterochromatic regions of chromosomes devoted to segregation of sister chromatids during cell division. Until very recently the inability of current technology to determine contiguous sequence for highly repetitive regions meant that centromeres largely fell within multimegabase gaps, analogous to black holes from which no information escapes (Henikoff, 2002). Now the previously intractable centromeric boundary is starting to be bridged. For example, analysis of the border of the X chromosome centromere in humans reveals an intriguing gradient (Schueler *et al.*, 2001). Human centromeres consist primarily of alpha satellite DNA (made up of highly repetitive tandemly repeating units). In contrast to the interior centromere sequences, which appear to consist of relatively young alpha satellite repeats that are kept homogeneous by a process such as unequal crossing over, the centromere border reveals repeats that are successively older as the border with euchromatin is approached. Divergence of the repeats increases with age and proximity to the boundary and is associated with an increasing frequency of *L1* insertions (Schueler *et al.*, 2001). Indeed, the frequency of *L1* elements can be used for dating the repeat units in the boundary regions. Recombination among *L1* elements may also be a major mechanism for homogenization of the repeat units. It is argued that this sequence organization is likely to be quite general in complex eukaryotes (Henikoff, 2002), but many more data will be needed to determine the general frequency with which centromeres are formed from or maintained by TEs.

TRANSDUCTION

Transduction is usually defined as the transfer of a small segment of DNA from donor to recipient bacteria via a bacteriophage (see Chapter 10). Recently, however, the term has apparently been used more generally, including the transfer of short DNA sequences from one site to another in the genome by the action of TEs. Specifically, some TEs can provide a vehicle for the mobilization of flanking nucleotide sequences accompanying aberrant transposition events. In addition to nongenic sequences, gene sequences such as exons or promoters can sometimes be transduced and inserted into other existing genes. This may provide a general mechanism for the evolution of new genes. For example, human *L1* retroposons can modify the genome by three separate mechanisms: (1) insertional mutagenesis during retrotransposition, (2) homologous recombination between different *L1* elements, and (3) the initiation of 3′ transduction events by the mobilization of unique DNA sequences downstream of *L1* elements as a result of aberrant transposition events (Fig. 3.10).

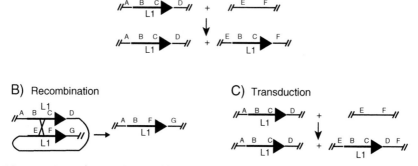

FIGURE 3.10 *L1* elements (a type of long interspersed nuclear element [LINE]) may modify the genome by three means: insertion, recombination, and transduction. Large arrows indicate *L1* elements, letters indicate different regions of the sequence that are affected by *L1* activity. (A) Insertional mutagenesis during retrotransposition. Before insertion, only one of the two genome segments (AD) contains the *L1* element and its genes (BC); the other segment (EF) lacks the element. After insertion, both regions contain the *L1* element. This insertion may disrupt coding regions and act as a powerful mutagen. (B) Intramolecular (as illustrated here) or intermolecular homologous recombination between different *L1* elements can lead to different combinations of genes in host coding regions (e.g., recombination generating the new arrangement ABFG from ABCD and EFG). (C) 3′ transduction occurs when the *L1* element transposes and takes a host sequence (in this case, region D) along with it. This process bears similarities to the transduction of bacterial genes by bacteriophages (see Chapter 10). Redrawn from Goodier *et al.* (2000), reproduced by permission (© Oxford University Press).

GENOME SIZE

Although estimates of genome size are available for several thousand eukaryotes (see Chapters 1 and 2), estimates of the TE fraction are currently available for only a relatively few well-studied species (see Table 3.3 for some examples). At best, these provide only crude estimates of TE proportions in the noneuchromatic regions of genomes, especially because data regarding repetitive fractions of genomes are often slow to be published. In order to explore the extent that TEs contribute directly to genome size variation, the total DNA contributed by TEs was plotted against genome size for 12 species (Kidwell, 2002). Overall, an

TABLE 3.3 Genome sizes and TE proportions for ten species

Species	Genome size (Mb)	No. coding genes	% TE DNA*	References
Saccharomyces cerevisiae	12	6300	3	Kim *et al.* (1998)
Caenorhabditis elegans	100	19,500	6	Waterston and Sulston (1995); International Human Genome Sequencing Consortium (2001)
Arabidopsis thaliana	157	26,000	14	*Arabidopsis* Genome Initiative (2000)
Drosophila melanogaster	180	13,600	10–22	Vieira *et al.* (1999); Kapitonov and Jurka (2003a)
Anopheles gambiae	278	14,000	16 (E), 60 (H)[a]	Holt *et al.* (2002)
Takifugu rubripes	400	31,000	2	Aparicio *et al.* (2002)
Oryza sativa	490	32,000–62,000	16	Yu *et al.* (2002)
Zea mays	2500	50,000	64–75	SanMiguel *et al.* (1996); Meyers *et al.* (2001)
Homo sapiens	3400	30,000	44	International Human Genome Sequencing Consortium (2001)
Mus musculus	3250	30,000	40	Mouse Genome Sequencing Consortium (2002)

[a]E, euchromatin; H, heterochromatin.

approximately linear relationship between total TE DNA and genome size was observed. On the basis of this preliminary analysis, it was suggested that the contribution of TEs to genome size variation is greater, relative to other sources of variation, in larger (>500 Mb) than in smaller (<500 Mb) genomes. Thus, TEs may play a more important role in the increase in size of relatively large plant and animal genomes than in smaller ones. Because increasingly larger genomes provide proportionately more noncoding sites for the insertion of TEs with minimum host damage, the process once started will tend to feed-back on itself and produce larger and larger genomes unless opposed by genomewide selection and/or a mutational proclivity for DNA loss.

Lynch and Conery (2003) have pointed out the potential importance of nonadaptive processes in mediating increases in genome complexity, including increases in the abundance of spliceosomal introns and mobile elements. Under their model, negative selection against the deleterious effects of newly arising introns and insertions is expected to increase in intensity with increasing effective population size. Based on the negative relationship between population size and genome size in their dataset, they argued that TEs have a threshold genome size below which they are unable to become established, an intermediate range in which they are harbored by only a fraction of species, and an upper threshold (~100 Mb) above which all species are infected. On a broader scale, it should be noted that this particular analysis was based on a comparison of sequenced genomes, which were in almost all cases chosen specifically because they are small. It is not at all clear that genome size and population size would correlate strongly across broader samples, and this model therefore is unlikely to explain the extensive variation in genome size found in groups such as animals or plants (see Chapters 1 and 2).

HOST GENOME STRUCTURE

In contrast to mutator mechanisms that act to mediate only small DNA changes such as nucleotide substitutions, TEs can be responsible for both small and large structural changes in the genome, including deletions, inversions, duplications, and translocations. Indeed, TEs in D. melanogaster have been shown to have a major impact on genome structure via such processes (e.g., Montgomery et al., 1991).

Two main mechanisms have been implicated in TE-induced karyotypic changes. The best-known mechanism is ectopic recombination, in which homologous recombination occurs between multiple copies of a TE present in a genome. The second mechanism for inducing genomic rearrangements is alternative transposition of Class II elements in bacteria, plants, and animals (Gray, 2000). Ectopic recombination is a meiotic process that is a significant source of the molecular rearrangements causing human disease (Reiter et al., 1999).

Further, a study of mobilization rates of nine TEs in D. *melanogaster* indicated that most changes in restriction patterns were consistent with rearrangements, rather than with true transposition (Dominguez and Albornoz, 1996). A survey of ectopic recombination in the region flanking the *white* locus of D. *melanogaster* indicated that inter- and intrachromosomal recombinants were generated in about equal numbers.

Several species of *Drosophila* provide good examples of TE-induced genomic structural changes. During the evolution of this genus, the molecular organization of the major chromosomal elements has been repeatedly rearranged via the fixation of inversions, and the rate of chromosomal reshuffling appears to be higher than that of any other animal or plant taxon that has been similarly studied (Ranz *et al.*, 2001), with the notable exception of *C. elegans* (Coghlan and Wolfe, 2002). For example, detailed analysis of the breakpoints of large inversions in natural populations of *Drosophila buzzatii* (Caceres *et al.*, 2001; Casals *et al.*, 2003) demonstrated an unprecedented frequency and complexity of molecular rearrangements in the relatively short chromosomal regions surrounding the inversion breakpoints and the remarkable rapidity of these changes on an evolutionary timescale. The *foldback*-like element *Galileo* was found to be present at inversion breakpoints and is the most likely inducer of the inversions (Caceres *et al.*, 2001; Casals *et al.*, 2003).

The human genome appears to be relatively rich in segmental duplications that are produced by duplicative transposition between nonhomologous chromosomes. Some of these play an important role in human evolution and disease, and are prone to promote further rearrangements because of misalignment (Eichler, 2001). Examination of the junctions of a large number of human segmental duplications revealed that 27% terminated within an *Alu* sequence (Bailey *et al.*, 2003). Although several mechanisms are likely to be responsible for generating these duplications, the role of *Alu–Alu* recombination appears to be among the most important (Bailey *et al.*, 2003).

CONCLUDING REMARKS AND FUTURE PROSPECTS

Until quite recently, the majority of information on TEs was obtained from a restricted number of model organisms. Now that genome sequencing data are becoming increasingly available, it is of considerable interest to determine which TE characteristics commonly vary between species and which have properties in common across a broad range of taxa. Some of the questions to be addressed in this context are: What are the relative frequencies of different families and classes of TEs in different organisms? Why are different element classes differentially abundant in eukaryotic genomes? (See Table 3.2.)

A number of outstanding questions also relate to the relationship between TEs and their hosts. Among the most interesting are: How frequently are TEs found in coding and regulatory sequences of host genes? How do patterns of TE distribution vary within and between different TE families and different host species? What is the relative importance of TEs in determining genome size? Why are some genomes relatively streamlined, whereas others have expanded so greatly as a result of retroelement proliferation?

The propensity of TEs to undergo horizontal transfer is a fascinating aspect of their evolution, and a number of outstanding questions remain in this area, including: How frequent is TE horizontal transfer in nature? Why does TE horizontal transfer appear to be rarer in plants than in animals? Can TEs act as vehicles for ferrying host genes between species? And, most fundamentally, how are TEs transferred (vectored) between species? This last question is one for which very few answers are currently available. Possible vectors currently considered are viruses, bacteria, mites, parasitic wasps, and other parasites. It is possible that multiple vectors might be involved in any particular case, making this an especially challenging puzzle to unravel in the future.

Finally, there are questions of both theoretical and practical significance still to be addressed in the study of TEs. For example, it will be important from the standpoint of evolutionary theory to determine the relative roles of selection and chance in the evolution of TE families. From a pragmatic perspective, there is still a great deal to be learned about the relevance of TEs for genetic engineering in agriculture and medicine, and in creating risks for xenotransplantation (Bromham, 2002).

Although a number of aspects of the relationship between TEs and their hosts are still quite controversial, there does appear to be general agreement that these elements play an important role as mutagenic agents that provide new sources of genomic variation on which natural selection may act. An important aspect of their role as mutators in evolution is the broad spectrum of mutations produced by their activity. TE-induced genetic changes range from substitutions, deletions, and insertions of single nucleotides to modifications in the size and arrangement of entire genomes. TEs can produce small, silent changes that are detectable only at the DNA sequence level or may exert major effects on phenotypic traits. Indeed, the spectrum of TE-induced mutations is broader than that produced by any other mutator mechanism.

TEs produce their mutagenic effects not simply on initial insertion into host DNA, but also when they excise imprecisely, leaving either no identifying sequence or only small "footprints" of their previous presence. Of special evolutionary significance to their hosts may be TE-induced mutations that affect the regulatory sequences of the genome (Britten, 1997). Unfortunately, the identifiable properties of TE sequences present in the genomes of contemporary species of animals and plants may inadequately reflect the full range of elements that have been present in the past because of the rapid divergence of inactive elements.

Simple DNA base substitutions are well suited for the generation, diversification, and optimization of local protein space (Maeshiro and Kimura, 1998), but a hierarchy of natural mutational events is required for the rapid generation of protein diversity (Bogarad and Deem, 1999). Sequence shuffling has the potential to improve protein function significantly better than does point mutation alone. Because they are uniquely competent to reshuffle DNA sequences, TEs and viruses are important generators of the more complex types of mutations in the mutational hierarchy. The relative silence of plant TEs during normal development and their activation by stresses including wounding, pathogen attack, and cell culture provide additional interesting aspects of their mutational roles.

In summary, although many more studies are needed to flesh out the details, it is becoming increasingly accepted that TEs have the potential to exert a major influence on the evolution of their hosts. The characteristics that allowed TEs to be labeled "selfish DNA" might have allowed them to furnish genomes with the plasticity to evolve new mechanisms for generating genetic diversity. Thus it is probable that a balance between fidelity and exploration (Kidwell and Lisch, 2000) has evolved through the operation of natural selection and chance on the products of both recent and ancient interactions between TEs and their hosts. In a very real sense, TEs have played a prominent role in shaping the evolution of the genome.

REFERENCES

Adams MD, Celniker SE, Holt RA, et al. 2000. The genome sequence of Drosophila melanogaster. Science 287: 2185–2195.

Aparicio S, Chapman J, Stupka E, et al. 2002. Whole-genome shotgun assembly and analysis of the genome of Fugu rubripes. Science 297: 1301–1310.

Arabidopsis Genome Initiative. 2000. Analysis of the genome sequence of the flowering plant Arabidopsis thaliana. Nature 408: 796–815.

Bailey JA, Liu G, Eichler EE. 2003. An Alu transposition model for the origin and expansion of human segmental duplications. Am J Hum Genet 73: 823–834.

Banks JA, Masson P, Fedoroff N. 1988. Molecular mechanisms in the developmental regulation of the maize Suppressor-mutator transposable element. Genes Dev 2: 1364–1380.

Bartolome C, Maside X, Charlesworth B. 2002. On the abundance and distribution of transposable elements in the genome of Drosophila melanogaster. Mol Biol Evol 19: 926–937.

Belfort M, Derbyshire V, Parker MM, et al. 2002. Mobile introns: pathways and proteins. In: Craig NL, Craigie R, Gellert M, Lambowitz AM eds, Mobile DNA II. Washington, DC: ASM Press, 761–783.

Berg DE, Howe MM. 1989. Mobile DNA. Washington, DC: American Society for Microbiology.

Biémont C, Tsitrone A, Vieira C, Hoogland C. 1997. Transposable element distribution in Drosophila. Genetics 147: 1997–1999.

Bingham PM, Kidwell MG, Rubin GM. 1982. The molecular basis of P-M hybrid dysgenesis: the role of the P element, a P-strain-specific transposon family. Cell 29: 995–1004.

Bingham PM, Levis R, Rubin GM. 1981. Cloning of DNA sequence from the white locus of Drosophila melanogaster by a novel and general method. Cell 25: 693–704.

Blumenstiel JP, Hartl DL, Lozovsky ER. 2002. Patterns of insertion and deletion in contrasting chromatin domains. *Mol Biol Evol* 19: 2211–2225.

Bogarad LD, Deem MW. 1999. A hierarchical approach to protein molecular evolution. *Proc Natl Acad Sci USA* 96: 2591–2595.

Britten RJ. 1997. Mobile elements inserted in the distant past have taken on important functions. *Gene* 205: 177–182.

Bromham L. 2002. The human zoo: endogenous retroviruses in the human genome. *Trends Ecol Evol* 17: 91–97.

Brookfield JFY. 1995. Transposable elements as selfish DNA. In: Sherratt DJ ed. *Mobile Genetic Elements*. Oxford: IRL Press, 130–153.

Brosius J. 1999a. Genomes were forged by massive bombardments with retroelements and retrosequences. *Genetica* 107: 209–238.

Brosius J. 1999b. RNAs from all categories generate retrosequences that may be exapted as novel genes or regulatory elements. *Gene* 238: 115–134.

Bukhari AI, Shapiro JA, Adhya SL. 1977. *DNA Insertion Elements, Plasmids, and Episomes*. Cold Spring Harbor, NY: Cold Spring Harbor Laboratory.

Bureau TE, Ronald PC, Wessler SR. 1996. A computer-based systematic survey reveals the predominance of small inverted-repeat elements in wild-type rice genes. *Proc Natl Acad Sci USA* 93: 8524–8529.

Burke WD, Malik HS, Lathe WC, Eickbush TH. 1998. Are retrotransposons long-term hitchhikers? *Nature* 392: 141–142.

Bushman F. 2002. *Lateral DNA Transfer: Mechanisms and Consequences*. Cold Spring Harbor, NY: Cold Spring Harbor Laboratory Press.

Caceres M, Puig M, Ruiz A. 2001. Molecular characterization of two natural hotspots in the *Drosophila buzzatii* genome induced by transposon insertions. *Genome Res* 11: 1353–1364.

Cameron JR, Loh EY, Davis RW. 1979. Evidence for transposition of dispersed repetitive DNA families in yeast. *Cell* 16: 739–751.

Campbell A, Berg DE, Botstein D, *et al.* 1977. Nomenclature of transposable elements in prokaryotes. In: Bukhari AI, Shapiro JA, Adhya SL eds. *DNA Insertion Elements, Plasmids, and Episomes*. Cold Spring Harbor, NY: Cold Spring Harbor Laboratory.

Capy P, Bazin C, Higuet D, Langin T. 1997a. *Dynamics and Evolution of Transposable Elements*. Austin, TX: Landes Bioscience.

Capy P, Langin T, Higuet D, *et al.* 1997b. Do the integrases of LTR-retrotransposons and class II element transposases have a common ancestor? *Genetica* 100: 63–72.

Casals F, Caceres M, Ruiz A. 2003. The foldback-like transposon *galileo* is involved in the generation of two different natural chromosomal inversions of *Drosophila buzzatii*. *Mol Biol Evol* 20: 674–685.

Charlesworth B, Langley CH, Sniegowski PD. 1997. Transposable element distributions in *Drosophila*. *Genetics* 147: 1993–1995.

Charlesworth B, Lapid A, Canada D. 1992. The distribution of transposable elements within and between chromosomes in a population of *Drosophila melanogaster*. II. Inferences on the nature of selection against elements. *Genet Res* 60: 115–130.

Clark JB, Kim P, Kidwell MG. 1998. Molecular evolution of *P* transposable elements in the genus *Drosophila*. III. The *melanogaster* species group. *Mol Biol Evol* 15: 746–755.

Coghlan A, Wolfe KH. 2002. Fourfold faster rate of genome rearrangement in nematodes than in *Drosophila*. *Genome Res* 12: 857–867.

Cooley L, Kelley R, Spradling A. 1988. Insertional mutagenesis of the *Drosophila* genome with single P elements. *Science* 239: 1121–1128.

Craig NL, Craigie R, Gellert M, Lambowitz AM. 2002. *Mobile DNA II*. Washington, DC: ASM Press.

Dasilva C, Hadji H, Ozouf-Costaz C, *et al.* 2002. Remarkable compartmentalization of transposable elements and pseudogenes in the heterochromatin of the *Tetraodon nigroviridis* genome. *Proc Natl Acad Sci USA* 99: 13636–13641.

Dernburg AF, Karpen GH. 2002. A chromosome RNAissance. *Cell* 111: 159–162.

Dominguez A, Albornoz J. 1996. Rates of movement of transposable elements in *Drosophila melanogaster*. *Mol Gen Genet* 251: 130–138.

Doolittle WF. 1997. Why we still need basic research. *Ann R Coll Phys Surg Can* 30: 76–80.

Doolittle WF, Sapienza C. 1980. Selfish genes, the phenotype paradigm and genome evolution. *Nature* 284: 601–603.

Duret L, Marais G, Biemont C. 2000. Transposons but not retrotransposons are located preferentially in regions of high recombination rate in *Caenorhabditis elegans*. *Genetics* 156: 1661–1669.

Eichler EE. 2001. Recent duplication, domain accretion and the dynamic mutation of the human genome. *Trends Genet* 17: 661–669.

Eickbush TH, Malik HS. 2002. Origin and evolution of retrotransposons. In: Craig NL, Cragie R, Gellert M, Lambowitz AM eds. *Mobile DNA II*. Washington, DC: ASM Press, 1111–1144.

Emmons SW, Yesner L, Ruan KS, Katzenberg D. 1983. Evidence for a transposon in *Caenorhabditis elegans*. *Cell* 32: 55–65.

Feschotte C, Mouches C. 2000. Evidence that a family of miniature inverted-repeat transposable elements (MITEs) from the *Arabidopsis thaliana* genome has arisen from a pogo-like DNA transposon. *Mol Biol Evol* 17: 730–737.

Feschotte C, Zhang X, Wessler SR. 2002. Miniature inverted repeat transposable elements and their relationship to established DNA transposons. In: Craig NL, Cragie R, Gellert M, Lambowitz AM eds. *Mobile DNA II*. Washington, DC: ASM Press, 1147–1158.

Fincham JRS, Sastry GRK. 1974. Controlling elements in maize. *Annu Rev Genet* 8: 15–50.

Fink G, Farabaugh P, Roeder G, Chaleff D. 1981. Transposable elements (Ty) in yeast. *Cold Spring Harb Symp Quant Biol* 45 Pt 2: 575–580.

Finnegan DJ. 1989. Eukaryotic transposable elements and genome evolution. *Trends Genet* 5: 103–107.

Finnegan DJ, Rubin GM, Young MW, Hogness DS. 1978. Repeated gene families in *Drosophila melanogaster*. *Cold Spring Harb Symp Quant Biol* 42: 1053–1063.

Flavell AJ, Pearce SR, Heslop-Harrison P, Kumar A. 1997. The evolution of Ty1-copia group retrotransposons in eukaryote genomes. *Genetica* 100: 185–195.

Furano AV, Duvernell DD, Boissinot S. 2004. L1 (LINE-1) retrotransposon diversity differs dramatically between mammals and fish. *Trends Genet* 20: 9–14.

Galagan JE, Calvo SE, Borkovich KA, *et al.* 2003. The genome sequence of the filamentous fungus *Neurospora crassa*. *Nature* 422: 859–868.

Gerasimova TI, Ladvishenko A, Mogila VA, *et al.* 1990. Transpositional bursts and chromosome rearrangements in unstable lines of *Drosophila*. *Genetika* 26: 399–411.

Glockner G, Szafranski K, Winckler T, *et al.* 2001. The complex repeats of *Dictyostelium discoideum*. *Genome Res* 11: 585–594.

Goodier JL, Ostertag EM, Kazazian HH. 2000. Transduction of 3′-flanking sequences is common in L1 retrotransposition. *Hum Mol Genet* 9: 653–657.

Gray YH. 2000. It takes two transposons to tango: transposable-element-mediated chromosomal rearrangements. *Trends Genet* 16: 461–468.

Green MM. 1977. The case for DNA insertion mutations in *Drosophila*. In: Bukhari AI, Shapiro JA, Adhya SL eds. *DNA Insertions, Plasmids and Episomes*. Cold Spring Harbor, NY: Cold Spring Harbor Laboratory, 437–445.

Gregory TR. 2003. Is small indel bias a determinant of genome size? *Trends Genet* 19: 485–488.

Gregory TR. 2004. Insertion-deletion bias and the evolution of genome size. *Gene* 324: 15–34.

Guo H, Karberg M, Long M, *et al.* 2000. Group II introns designed to insert into therapeutically relevant DNA target sites in human cells. *Science* 289: 452–457.

Handler AM. 2001. A current perspective on insect gene transformation. *Insect Biochem Mol Biol* 31: 111–128.

Hartl DL, Lohe AR, Lozovskaya ER. 1997. Regulation of the transposable element *mariner*. *Genetica* 100: 177–184.

Henikoff S. 2002. Near the edge of a chromosome's black hole. *Trends Genet* 18: 165–167.

Hershberger RJ, Benito MI, Hardeman KJ, et al. 1995. Characterization of the major transcripts encoded by the regulatory *MuDR* transposable element of maize. *Genetics* 140: 1087–1098.

Hiraizumi Y. 1971. Spontaneous recombination in *Drosophila melanogaster* males. *Proc Natl Acad Sci USA* 68: 268–270.

Holmes I. 2002. Transcendent elements: whole-genome transposon screens and open evolutionary questions. *Genome Res* 12: 1152–1155.

Holt RA, Subramanian GM, Halpern A, et al. 2002. The genome sequence of the malaria mosquito *Anopheles gambiae*. *Science* 298: 129–149.

International Human Genome Sequencing Consortium. 2001. Initial sequencing and analysis of the human genome. *Nature* 409: 860–921.

Ivics Z, Hackett PB, Plasterk RH, Izsvak, Z. 1997. Molecular reconstruction of *Sleeping Beauty*, a *Tc1*-like transposon from fish, and its transposition in human cells. *Cell* 91: 501–510.

Jordan IK, Rogozin IB, Glazko GV, Koonin EV. 2003. Origin of a substantial fraction of human regulatory sequences from transposable elements. *Trends Genet* 19: 68–72.

Kajikawa M, Okada N. 2002. LINEs mobilize SINEs in the eel through a shared 3′ sequence. *Cell* 111: 433–444.

Kalendar R, Tanskanen J, Immonen S, et al. 2000. Genome evolution of wild barley (*Hordeum spontaneum*) by BARE-1 retrotransposon dynamics in response to sharp microclimatic divergence. *Proc Natl Acad Sci USA* 97: 6603–6607.

Kapitonov VV, Jurka J. 1999. Molecular paleontology of transposable elements from *Arabidopsis thaliana*. *Genetica* 107: 27–37.

Kapitonov VV, Jurka J. 2001. Rolling-circle transposons in eukaryotes. *Proc Natl Acad Sci USA* 98: 8714–8719.

Kapitonov VV, Jurka J. 2003a. Molecular paleontology of transposable elements in the *Drosophila melanogaster* genome. *Proc Natl Acad Sci USA* 100: 6569–6574.

Kapitonov VV, Jurka J. 2003b. A novel class of SINE elements derived from 5S rRNA. *Mol Biol Evol* 20: 694–702.

Kaplan N, Darden T, Langley CH. 1985. Evolution and extinction of transposable elements in Mendelian populations. *Genetics* 109: 459–480.

Kidwell MG. 1993. Lateral transfer in natural populations of eukaryotes. *Annu Rev Genet* 27: 235–256.

Kidwell MG. 1994. The evolutionary history of the P family of transposable elements. *J Heredity* 85: 339–346.

Kidwell MG. 2002. Transposable elements and the evolution of genome size in eukaryotes. *Genetica* 115: 49–63.

Kidwell MG, Kidwell JF. 1975. Cytoplasm-chromosome interactions in *Drosophila melanogaster*. *Nature* 253: 755–756.

Kidwell MG, Lisch D. 1997. Transposable elements as sources of variation in animals and plants. *Proc Natl Acad Sci USA* 94: 7704–7711.

Kidwell MG, Lisch DR. 2000. Transposable elements and host genome evolution. *Trends Ecol Evol* 15: 95–99.

Kidwell MG, Lisch DR. 2001. Perspective: transposable elements, parasitic DNA, and genome evolution. *Evolution* 55: 1–24.

Kidwell MG, Lisch DR. 2002. Transposable elements as sources of genomic variation. In: Craig NL, Cragie R, Gellert M, Lambowitz AM eds. *Mobile DNA II*. Washington, DC: ASM Press, 59–90.

Kidwell MG, Kidwell JF, Sved JA. 1977. Hybrid dysgenesis in *Drosophila melanogaster*: a syndrome of aberrant traits including mutation, sterility and male recombination. *Genetics* 36: 813–833.

Kim JM, Vanguri S, Boeke JD, et al. 1998. Transposable elements and genome organization: a comprehensive survey of retrotransposons revealed by the complete *Saccharomyces cerevisiae* genome sequence. *Genome Res* 8: 464–478.

Kimura RH, Choudary PV, Schmid CW. 1999. Silk worm *Bm1* SINE RNA increases following cellular insults. *Nucleic Acids Res* 27: 3380–3387.

Kinsey JA, Helber J. 1989. Isolation of a transposable element from *Neurospora crassa*. *Proc Natl Acad Sci USA* 86: 1929–1933.

Klinakis AG, Zagoraiou L, Vassilatis DK, Savakis C. 2000. Genome-wide insertional mutagenesis in human cells by the *Drosophila* mobile element *Minos*. *EMBO Rep* 1: 416–421.

Kreahling J, Graveley BR. 2004. The origins and implications of *Alu*ternative splicing. *Trends Genet* 20: 1–4.

Kumar A, Bennetzen JL. 1999. Plant retrotransposons. *Annu Rev Genet* 33: 479–532.

Lerman DN, Michalak P, Helin AB, et al. 2003. Modification of heat-shock gene expression in *Drosophila melanogaster* populations via transposable elements. *Mol Biol Evol* 20: 135–144.

Lev-Maor G, Sorek R, Shomron N, Ast G. 2003. The birth of an alternatively spliced exon: 3′ splice-site selection in *Alu* exons. *Science* 300: 1288–1291.

Li WH, Gu Z, Wang H, Nekrutenko A. 2001. Evolutionary analyses of the human genome. *Nature* 409: 847–849.

Lohe AR, Hartl DL. 1996. Autoregulation of mariner transposase activity by overproduction and dominant-negative complementation. *Mol Biol Evol* 13: 549–555.

Lohe AR, Moriyama EN, Lidholm DA, Hartl DL. 1995. Horizontal transmission, vertical inactivation, and stochastic loss of *mariner*-like transposable elements. *Mol Biol Evol* 12: 62–72.

Lynch M, Conery JS. 2003. The origins of genome complexity. *Science* 302: 1401–1404.

Maeshiro T, Kimura M. 1998. The role of robustness and changeability on the origin and evolution of genetic codes. *Proc Natl Acad Sci USA* 95: 5088–5093.

Malik HS, Henikoff S, Eickbush TH. 2000. Poised for contagion: evolutionary origins of the infectious abilities of invertebrate retroviruses. *Genome Res* 10: 1307–1318.

Maside X, Bartolome C, Assimacopoulos S, Charlesworth B. 2001. Rates of movement and distribution of transposable elements in *Drosophila melanogaster*: in situ hybridization vs Southern blotting data. *Genet Res* 78: 121–136.

McClintock B. 1952. Chromosome organization and gene expression. *Cold Spring Harb Symp Quant Biol* 16: 13–47.

McClintock B. 1984. The significance of responses of the genome to challenge. *Science* 226: 792–801.

McClure MA. 1991. Evolution of retroposons by acquisition or deletion of retrovirus-like genes. *Mol Biol Evol* 8: 835–856.

McDonald JF. 1998. Transposable elements, gene silencing and macroevolution. *Trends Ecol Evol* 13: 94–95.

Meyers BC, Tingey SV, Morgante M. 2001. Abundance, distribution, and transcriptional activity of repetitive elements in the maize genome. *Genome Res* 11: 1660–1676.

Miller WJ, McDonald JF, Nouaud D, Anxolabehere D. 1999. Molecular domestication—more than a sporadic episode in evolution. *Genetica* 107: 197–207.

Miller WJ, McDonald JF, Pinsker W. 1997. Molecular domestication of mobile elements. *Genetica* 100: 261–270.

Miller WJ, Nagel A, Bachmann J, Bachmann L. 2000. Evolutionary dynamics of the SGM transposon family in the *Drosophila obscura* species group. *Mol Biol Evol* 17: 1597–1609.

Montgomery EA, Huang SM, Langley CH, Judd BH. 1991. Chromosome rearrangement by ectopic recombination in *Drosophila melanogaster*: genome structure and evolution. *Genetics* 129: 1085–1098.

Mouse Genome Sequencing Consortium. 2002. Initial sequencing and comparative analysis of the mouse genome. *Nature* 420: 520–562.

Nekrutenko A, Li WH. 2001. Transposable elements are found in a large number of human protein-coding genes. *Trends Genet* 17: 619–621.

Nikaido M, Nishihara H, Hukumoto Y, Okada N. 2003. Ancient SINEs from African endemic mammals. *Mol Biol Evol* 20: 522–527.

Pardue ML, Danilevskaya ON, Traverse KL, Lowenhaupt K. 1997. Evolutionary links between telomeres and transposable elements. *Genetica* 100: 73–84.

Pardue ML, DeBaryshe PG. 2003. Retrotransposons provide an evolutionarily robust non-telomerase mechanism to maintain telomeres. *Annu Rev Genet* 37: 485–511.

Petrov DA. 2001. Evolution of genome size: new approaches to an old problem. *Trends Genet* 17: 23–28.

Plasterk RA. 1995. Mechanisms of DNA transposition. In: Sherratt DJ ed. *Mobile Genetic Elements*. Oxford: IRL Press, 18–37.

Prak ET, Kazazian HH. 2000. Mobile elements and the human genome. *Nat Rev Genet* 1: 134–144.

Purugganan M, Wessler S. 1992. The splicing of transposable elements and its role in intron evolution. *Genetica* 86: 295–303.

Rabinowicz PD, Palmer LE, May BP, et al. 2003. Genes and transposons are differentially methylated in plants, but not in mammals. *Genome Res* 13: 2658–2664.

Ranz JM, Casals F, Ruiz A. 2001. How malleable is the eukaryotic genome? Extreme rate of chromosomal rearrangement in the genus *Drosophila*. *Genome Res* 11: 230–239.

Reiter LT, Liehr T, Rautenstrauss B, et al. 1999. Localization of *mariner* DNA transposons in the human genome by PRINS. *Genome Res* 9: 839–843.

Robertson HM, Preston CR, Phillis RW, et al. 1988. A stable genomic source of P element transposase in *Drosophila melanogaster*. *Genetics* 118: 461–470.

Rubin GM. 1983. Dispersed repetitive DNAs in *Drosophila*. In: Shapiro JA ed. *Mobile Genetic Elements*. New York: Academic Press, 329–361.

Rubin GM, Kidwell MG, Bingham PM. 1982. The molecular basis of P-M hybrid dysgenesis: the nature of induced mutations. *Cell* 29: 987–994.

Saedler H, Bonas U, Gierl A, et al. 1984. Transposable elements in *Antirrhinum majus* and *Zea mays*. *Cold Spring Harb Symp Quant Biol* 49: 355–361.

SanMiguel P, Gaut BS, Tikhonov A, et al. 1998. The paleontology of intergene retrotransposons of maize. *Nat Genet* 20: 43–45.

SanMiguel P, Tikhonov A, Jin YK, et al. 1996. Nested retrotransposons in the intergenic regions of the maize genome. *Science* 274: 765–768.

Schueler MG, Higgins AW, Rudd MK, et al. 2001. Genomic and genetic definition of a functional human centromere. *Science* 294: 109–115.

Selker EU. 1997. Epigenetic phenomena in filamentous fungi: useful paradigms or repeat-induced confusion? *Trends Genet* 13: 296–301.

Selker EU, Tountas NA, Cross SH, et al. 2003. The methylated component of the *Neurospora crassa* genome. *Nature* 422: 893–897.

Shapiro JA. 1983. *Mobile Genetic Elements*. New York: Academic Press.

Simmons MJ, Bucholz LM. 1985. Transposase titration in *Drosophila melanogaster*: a model of cytotype in the P-M system of hybrid dysgenesis. *Proc Natl Acad Sci USA* 82: 8119–8123.

Simmons MJ, Raymond JD, Grimes CD, et al. 1996. Repression of hybrid dysgenesis in *Drosophila melanogaster* by heat-shock-inducible sense and antisense P-element constructs. *Genetics* 144: 1529–1544.

Singer MF. 1982. SINEs and LINEs: highly repeated short and long interspersed sequences in mammalian genomes. *Cell* 28: 433–434.

Sniegowski PD, Charlesworth B. 1994. Transposable element numbers in cosmopolitan inversions from a natural population of *Drosophila melanogaster*. *Genetics* 137: 815–827.

Spradling AC, Stern DM, Kiss I, et al. 1995. Gene disruptions using P transposable elements: an integral component of the *Drosophila* genome project. *Proc Natl Acad Sci USA* 92: 10824–10830.

Steinemann M, Steinemann S. 1998. Enigma of Y chromosome degeneration: neo-Y and neo-X chromosomes of *Drosophila miranda* a model for sex chromosome evolution. *Genetica* 103: 409–420.

Temin HM. 1989. Reverse transcriptases. Retrons in bacteria. *Nature* 339: 254–255.

Tu Z. 1997. Three novel families of miniature inverted-repeat transposable elements are associated with genes of the yellow fever mosquito, *Aedes aegypti*. *Proc Natl Acad Sci USA* 94: 7475–7480.

Tu Z. 2004. Insect transposable elements. In: Gilbert L, Latrous K, Gill S eds. *Comprehensive Molecular Insect Science*. Oxford: Elsevier.

Vastenhouw NL, Plasterk RHA. 2004. RNAi protects the *Caenorhabditus elegans* germline against transposition. *Trends Genet* 20: 314–319.

Vieira C, Lepetit D, Dumont S, Biemont C. 1999. Wake up of transposable elements following *Drosophila simulans* worldwide colonization. *Mol Biol Evol* 16: 1251–1255.

Waterston R, Sulston J. 1995. The genome of *Caenorhabditis elegans*. *Proc Natl Acad Sci USA* 92: 10836–10840.

Wessler S, Baran G, Varagona M. 1987. The maize transposable element *Ds* is spliced from RNA. *Science* 237: 916–918.

Witherspoon DJ. 1999. Selective constraints on P-element evolution. *Mol Biol Evol* 16: 472–478.

Yoder JA, Walsh CP, Bestor TH. 1997. Cytosine methylation and the ecology of intragenomic parasites. *Trends Genet* 13: 335–340.

Yu J, Hu S, Wang J, *et al*. 2002. A draft sequence of the rice genome (*Oryza sativa* L. ssp. *indica*). *Science* 296: 79–92.

CHAPTER 4

B Chromosomes

Juan Pedro M. Camacho

Almost every biological system and level of organization is prone to attack by parasites. Parasites take many forms and have representatives in most major groups of organisms—from animals to plants to fungi to bacteria. In some cases, as with viruses, the parasites may be little more than independent strands of nucleotides that usurp the replication machinery of their host cells. In other cases, parasitic genetic elements are not independent at all and reside within the very genomes of their hosts, as illustrated by the transposable elements discussed in detail in Chapter 3.

Chromosomes are vehicles that facilitate the organized and fair transmission of cooperative genes during cell division. In parallel with this cooperation, some rules have evolved to rid this "parliament of genes" (Leigh, 1977) of exploitation by outlaw genes. The Mendelian segregation law, for example, assures that different versions of the same gene (i.e., alleles) in a heterozygote share the same 50% likelihood of transmission to the next generation. This is based on an ordered cell division process (meiosis) in which the two copies of each gene meet when the homologous chromosomes pair and segregate to opposite poles, in the process reducing chromosome number to haploidy. But some genetic elements escape this rule by affecting the viability of those meiotic products that do not carry them, thus increasing their transmission likelihood. This is the case, for example, for

The Evolution of the Genome, edited by TR Gregory

Segregation Distorter in *Drosophila* (see Kusano *et al.*, 2003) and the *t*-alleles in mice (see Harrison *et al.*, 1998), which are transmitted to about 90% of the spermatozoa because of the spermiogenic failure of the spermatids lacking these elements.

In many organisms, meiosis is functionally asymmetrical because some of the products are always unviable. The most widespread example is female meiosis, where one of the products (the polar body) in each of both meiotic divisions is regularly unviable. But there are also cases of male meiosis, such as the lecanoid system in Homoptera, where the paternal set of chromosomes is condensed into heterochromatic form and genetically inactivated (Brown and Nelson-Rees, 1961). These natural asymmetries may provide the escape route used by many parasitic genetic elements that prosper by disobeying the Mendelian segregation law and in so doing assure an unfair transmission share. This is typically the means by which an entire chromosome can become a parasitic element—that is, a B chromosome.

A BRIEF HISTORY OF THE STUDY
OF B CHROMOSOMES

THE NAME GAME

B chromosomes were discovered very early in the history of cytogenetics. Only four years after the development of the Sutton–Boveri chromosome theory of inheritance, which established the parallelism between chromosome behavior and Mendelian inheritance, Wilson (1906, 1907) observed the presence of "extra" chromosomes in the hemipteran insect *Metapodius*. Noting that these did not appear to behave as part of the standard chromosome set, he dubbed them "supernumerary chromosomes." Similar cases were soon discovered in the beetle *Diabrotica* (Stevens, 1908) and later in maize (Kuwada, 1915). Different names have since been used to describe these supernumerary chromosomes, including "extra fragment chromosomes" (Müntzing, 1944; Östergren, 1945), "surplus" or "superfluous" chromosomes (Håkansson, 1945), and "accessory chromosomes" (Håkansson, 1948; Müntzing, 1948). As an interesting aside, it is worth pointing out that the term "accessory chromosome" was originally coined by McClung (1901, 1902) to describe the additional chromosome present only in the females of some insect species, which he (correctly) believed was causally related to sex determination. This same discussion of sex chromosomes in insects also led to the term "nucleolus," which was an early name given to the insect sex chromosome. In one of his figures, Henking (1891) happened to label the "accessory chromosome" of *Pyrrhocoris apterus* with the letter "X," an act that ultimately led to the familiar notion of the "X chromosome." Links between B chromosomes, sex chromosomes, and nucleolar organizer regions (NORs) will resurface in a far less superficial way later in this chapter.

Randolph (1928) was the first to employ the term "B chromosome" in order to distinguish them from the standard "A chromosomes." Although some other names have occasionally been used in the recent literature, such as "paternal sex ratio (PSR) chromosome" in a wasp (Werren, 1991) and "conditionally dispensable chromosome" in fungi (Covert, 1998), the term "B chromosome" maintains prominence today after 75 years in use. Despite the fact that the existence of B chromosomes has been known for nearly a century, a great deal remains to be learned regarding their biological properties and evolutionary dynamics.

WHAT IS A "B CHROMOSOME"?

The term "B chromosome" includes a variety of extra chromosomes that display conspicuous heterogeneity in their natures, behaviors, and evolutionary dynamics. For this reason, it is not easy to define them properly. In looking for a consensus definition, J.P.M. Camacho and J.S. Parker proposed, during the First B Chromosome Conference (see Beukeboom, 1994), that "B chromosomes are additional dispensable chromosomes that are present in some individuals from some populations in some species, which have probably arisen from the A chromosomes but follow their own evolutionary pathway." This definition highlights some of the most universal properties of B chromosomes: (1) their dispensability (that is, they are not necessary for the host to complete a normal life cycle); (2) their origin from A chromosomes (either from within the same species or from another species); and (3) their remarkable differentiation relative to A chromosomes, with which they do not recombine.

There is an additional property of B chromosomes that should be included in the definition, namely that they lack regular meiotic behavior. That is to say, they do not always pair up and segregate during meiosis, a fact that both facilitates their accumulation ("drive") in the germline and impedes their stable integration into the A genome. Thus Jones (1995) accurately defined B chromosomes as "dispensable supernumeraries which do not recombine with any members of the basic A chromosome set and which have irregular and non-Mendelian modes of inheritance." Notably, this definition omits the specification that B chromosomes are derived from A chromosomes—but given the potentially surprising findings of modern molecular analyses, this may yet turn out to be a correct interpretation.

B CHROMOSOMES AS GENOMIC PARASITES

The biological relevance of B chromosomes was poorly understood until Östergren (1945) interpreted them as nothing less than parasites of the A genome. Speaking of B chromosomes in plants, Östergren (1945) opined, "I think reasonable support may be given to the view that in many cases these chromosomes have no useful function at all to the species carrying them, but that they often lead an exclusively parasitic existence." As he put it, "they need only be 'useful' to themselves." Certainly this was the first clear expression of the "selfish DNA hypothesis" (see Chapter 1), predating the application of the concept to transposable elements (Doolittle and Sapienza, 1980; Orgel and Crick, 1980) (see Chapter 3) or "selfish genes" (Dawkins, 1976) by more than 30 years.

The harmful effects that many B chromosomes exert on the host and the variety of mechanisms acting to facilitate their accumulation in the germline (see later section) are strongly reminiscent of the behavior of other types of parasites. Again with incredible foresight, Östergren (1945) argued that "a similar antagonism in the evolutionary tendencies as that between a parasite and its host should be expected between parasitic fragment chromosomes and the plants carrying them." And thus, he continued,

> [the] tendency to the evolution of an 'eliminative system' in the normal complement [of chromosomes] would be counteracted by a tendency to the evolution of an 'accumulative system' in the fragment itself. Selection would favour fragments with a more efficient spreading mechanism ... It should also be realized that the selection pressure for the former change is much weaker than that for the latter. Selection for an eliminative system is at work only in that small fraction of the population which has the fragments, and this selection is rather weak, as only a small difference in viability and fertility is at stake. Selection for accumulative tendencies, on the other hand, is working on a hundred per cent of the fragments, and what is at stake for them is their very existence.

In 1984, M.W. Shaw published the most complete analysis on the population genetics of B chromosomes. In that paper he predicted the existence of modifier genes capable of suppressing B chromosome drive. A year later Shaw and Hewitt (1985) and Nur and Brett (1985) found the first experimental evidence for these genes in the grasshopper *Myrmeleotettix maculatus* and the mealybug *Pseudococcus affinis,* respectively. The existence of these genes implied an "arms race" between B chromosomes and their hosts, in keeping with Östergren's (1945) early predictions and as expected under the Red Queen hypothesis of competitive evolutionary dynamics (Van Valen, 1973).

Not all authors have been convinced of Östergren's hypothesis, however. For some time, the parasitic interpretation coexisted with the alternative view that B chromosomes are maintained because they provide some benefit to the organisms carrying them. This "heterotic" (that is, adaptive) model was anticipated by Darlington (1958) and taken up later by White (1973), and assumes that B chromosomes are beneficial at low numbers but harmful at high numbers. This evolutionary view of B chromosomes was mainly based on an adaptive interpretation of increases in chiasma frequency (see later section), a kind of group selection argument (see Bell and Burt, 1990). It has recently found some support in studies of the chive *Allium schoenoprassum* (Plowman and Bougourd, 1994; Bougourd *et al.*, 1995), in which B chromosomes do not appear to exhibit drive and may confer some benefit on seed germination under dry conditions (see later section).

A few interesting examples such as this notwithstanding, the preponderance of the evidence favors the parasitic view, as discussed in much more detail later in the chapter. As such, B chromosomes will be described using the concepts and terminology of parasitism throughout the course of this chapter. This approach has been very fruitful in elucidating the evolutionary interactions between B chromosomes and their hosts (Bell and Burt, 1990; Camacho *et al.,* 1997, 2002; Frank, 2000), and provides an important conceptual framework for understanding their biology. However, as with most parasites, the interactions with the host are complex and dynamic, and may change considerably through time. It is therefore important to keep an open mind when discussing the nature of B chromosomes.

THE FREQUENCY OF B CHROMOSOME INFECTION

HOW WIDELY DISTRIBUTED ARE B CHROMOSOMES?

B chromosomes are widely distributed among eukaryotes. Table 4.1 summarizes the number of species in which B chromosomes have been found to date in different taxa. It shows that B chromosomes have been reported in 10 species of fungi, nearly 1300 plants (more than 1400 when different ploidy levels of the same species are considered separately), and well over 500 animals. Although the greatest prevalence of B chromosomes seems to be restricted to certain groups (e.g., the Compositae in the order Asterales among dicotyledonous plants, Gramineae within the order Poales and Liliaceae within the order Liliales among monocotyledonous plants, and the order Orthoptera among insects), this is probably a by-product of the higher intensity of study in these groups, itself determined in part by issues of technical ease. In those groups that have been reasonably well studied, B chromosome presence can be substantial. For example, J. Cabrero has recently compiled a database including all available cytogenetic studies in orthopteran insects, showing that B chromosomes have been found in 210 out of 1506 species analyzed cytologically, a total of about 14%. This figure is remarkably similar to the 10–15% B chromosome presence estimated by Jones (1995) for flowering plants.

More detailed study of other groups is almost certain to expand the known distribution of B chromosomes substantially. For example, until the 1990s it was thought that B chromosomes were entirely absent from more primitive eukaryotes, but the introduction of the pulse-field gel electrophoresis technique has since prompted their discovery in several species of fungi (see Covert, 1998). Interestingly, this might provide a link between the simplest form of extra

TABLE 4.1 Number of species with B chromosomes in several types of eukaryote organisms. This score was performed on a B chromosome database kindly supplied by R.N. Jones, which contained all literature up to 1994. It was then updated to 2003 by means of searches in the ISI Science Citation Index. Species were taxonomically grouped with the help of the taxonomy browser provided by the NCBI Entrez Taxonomy Homepage (www.ncbi.nlm.nih.gov/Taxonomy/Browser)

Type of organism	Order	Presence of Bs			
		Number of species with Bs	Including different ploidy levels within species	Maximum number of Bs	Species showing this maximum
Fungi		11			
Plants:		1284	1415		
Ferns		6	6	7	
Conifers	Coniferales	12	12	7	*Podocarpus macrophyllus*
Dicotyledons		687	739		
	Apiales	9	9	9	*Heracleum sphondylium*
	Asterales	271	301	22	*Centaurea scabiosa*
	Brassicales	33	37	8	
	Caryophyllales	17	20	15	*Silene maritma*
	Cornales	3	3	2	
	Cucurbitales	21	21	4	
	Dipsacales	6	6	2	
	Ericales	26	27	13	*Phlox subulata*
	Fabales	56	57	7	
	Gentianales	1	2	3	
	Geraniales	3	3	2	
	Lamiales	84	87	10	*Pedicularis sudentica*
	Laurales	3	3	5	
	Magnoliales	2	2	3	
	Malpighiales	16	16	10	*Viola montana*
	Malvales	4	5	4	
	Myrtales	25	25	17	*Clarkia williamsonii*
	Ranunculales	28	30	15	*Hepatica nobilis*
	Rosales	13	13	5	
	Santalales	13	14	8	
	Sapindales	2	2	2	
	Saxifragales	37	42	50	*Pachyphytum fittkaui*
	Solanales	14	14	3	

(Continues)

TABLE 4.1 (*Continued*)

Type of organism	Order	Presence of Bs		Maximum number of Bs	Species showing this maximum
		Number of species with Bs	Including different ploidy levels within species		
Monocotyledons		579	658		
	Poales	240	286	34	*Zea mays*
	Alismatales	28	29	14	*Schismatoglottis lancifolia*
	Asparagales	103	109	11	*Crocus flavus & Eria microchilos*
	Commelinales	27	30	31	*Gibasis karwinskyana*
	Liliales	175	198	26	*Fritillaria japonica*
	Zingiberales	6	6	4	
Animals:		540			
Platyhelminthes		8		4	
Mollusca		2		6	*Helix pomatia*
Arthropoda					
Arachnida		4		19	*Metagagrella tenuipes*
Crustacea		20		9	*Echinogammarus berilloni*
Insecta	Aphaniptera	1		7	*Nosophyllus fasciatus*
	Coleoptera	69		17	*Alagoasa oblecta*
	Diptera	62		20	*Xylota nemorum*
	Hemiptera	37		14	*Cimex lectularius*
	Hymenoptera	9		12	*Leptothorax spinosior*
	Lepidoptera	23		7	*Euphydryas colon*
	Neuroptera	3		2	
	Orthoptera	210		16	*Gonista bicolor*
Chordata					
Pisces		29		16	*Callichthys callichthys*
Amphibia		27		15	*Leiopelma hochstetteri*
Reptilia		15		6	*Takydromus sexlineatus*
Aves		1		1	*Taeniopygia guttata*
Mammalia		55		24	*Apodemus peninsulae*

chromosomes found in unicellular fungi (e.g., the 2-μm circular plasmid in yeast, with 6318 base pairs) and the more than 1 megabase (Mb) B chromosomes in multicellular fungi, and could therefore be of considerable significance in deciphering the evolutionary origin of B chromosomes.

The wide presence of B chromosomes in fungi suggests that they are probably ubiquitous among eukaryotes. To be sure, their apparent absence from a taxonomical group or even from a given species does not mean that they cannot be found in the future following more intensive study. The polymorphic nature of B chromosomes implies that their detection requires the analysis of relatively large population samples, especially when they are at low frequency. To date, such broad sampling efforts have simply not been undertaken for most taxa.

THE LIKELIHOOD OF INFECTION

Aside from issues of technical ease and intensity of study, the likelihood that a particular eukaryote will be found to harbor B chromosomes depends on the combined probabilities of the origin and maintenance of Bs. Since the *origin* of B chromosomes has much to do with aneuploidy (see later section), the more rigid the control on cell divisions in a kind of organism the more unlikely will be the presence of B chromosomes. Moreover, only a minority of the extra chromosomes that arise through aneuploidy will become true B chromosomes, because their *maintenance* requires that they show either a transmission advantage (drive) or that they confer some benefit to the host. In other words, a newly formed B chromosome can conceivably be maintained by either intragenomic selection, which favors the B at the expense of the A chromosomes, or by phenotypic selection, which depends in part on the tolerance of A chromosomes for the presence of Bs. The action of these two kinds of selection may engender a genetic conflict that can be solved in different ways, as discussed in more detail later.

The key point for the time being is that these two forces are the most important in determining the rise and fate of a B chromosome. Driving Bs can maintain themselves despite being harmful for the host, but nondriving Bs are almost certainly condemned to disappear unless they are beneficial for the host. The two possibilities are not equally likely, however, given that aneuploidy generally produces deleterious effects. Again, this is a good reason to suppose that most B chromosomes have arisen because they showed drive—that is, because they behaved as parasites.

VARIATION IN THE INTENSITY OF INFECTION

Table 4.1 shows that B chromosomes seem to be well tolerated in some organisms, reaching remarkably high maximum numbers of Bs. For example, the highest number of B chromosomes found in a plant is 50 in *Pachyphytum fittkaui* (Crassulaceae); other members of this genus are similarly tolerant, with other species having up to 20 B chromosomes (Uhl and Moran, 1973). Some maize plants have been found carrying 34 B chromosomes (see Jones and Rees, 1982). Other impressive botanical examples include 31 Bs in *Gibasis karwinskyana* (Kenton, 1991), 26 in *Fritillaria japonica* (Noda, 1975), and 22 in *Centaurea scabiosa* (Fröst, 1957).

High numbers of B chromosomes have also been found among animals; for example, up to 24 in the wood mouse *Apodemus peninsulae* (Volobujev and Timina, 1980), 20 in the fly *Xylota nemorum* (Boyes and Van Brink, 1967), and 16 in both the orthopteran *Gonista bicolor* (Sannomiya, 1974) and the frog *Leiopelma hochstetteri* (Green, 1988). Because it is widely agreed that B chromosomes are harmful at high numbers, these extreme values are probably indicative of especially high drive or tolerance for Bs in these species. In most cases, the number of B chromosomes per individual is not nearly so high, and it is rather rare to find individuals of either plants or animals with more than three or four Bs (Fig. 4.1).

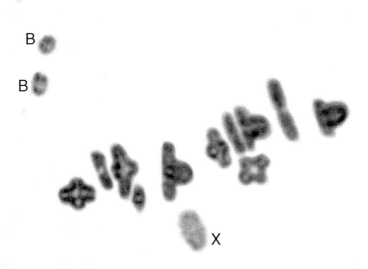

FIGURE 4.1   Primary spermatocyte at metaphase I in the grasshopper *Eyprepocnemis plorans* showing 11 autosomal bivalents at the equatorial plate, the X chromosome univalent, and two B chromosome univalents.

In most species harboring B chromosomes, population studies have shown a variety of situations, ranging from populations lacking B chromosomes to others carrying high B numbers. In some cases, B chromosome–free populations can be difficult to find, even after intensive sampling efforts. For instance, Matsuda and coworkers have shown that polymorphisms for B chromosomes in *Aster* plants are widespread in Japan, and are present in all ploidy levels from diploid to hexaploid (see references in Jones, 1995). More extremely, B chromosomes in the migratory locust, *Locusta migratoria,* are present in most analyzed populations from Asia (Itoh, 1934; Hsiang, 1958; Nur, 1969; Kayano, 1971), Africa (Rees and Jamieson, 1954; Lespinasse, 1973, 1977; Dearn, 1974), Australia (King and John, 1980), and Europe (Cabrero *et al.,* 1984; Viseras *et al.,* 1990). In the grasshopper *Eyprepocnemis plorans,* almost all populations in the Iberian Peninsula, including the whole Mediterranean coast from Tarragona to Huelva, harbor B chromosomes (Henriques-Gil *et al.,* 1984); the only exceptions found thus far are several populations situated at the head of the Segura River basin (Cabrero *et al.,* 1997).

Unlike with the standard A chromosome set, or even genome size (see Chapters 1 and 2), it is not uncommon to find substantial variation in B chromosome number among populations within a species. In addition to drive and phenotypic selection, other factors such as genetic drift, the number of generations since infection, and the ease of transmission among populations may generate intraspecific variation in B chromosome numbers (Camacho *et al.,* 2000). The number of B chromosomes in the different individuals within a population usually follows a binomial distribution. The parameters most commonly employed to measure B chromosome frequencies within populations are the average number of B chromosomes per individual ("mean") and the proportion of B-carrying individuals ("prevalence"). In some cases, it is useful to add a third parameter, namely the mean number of Bs found among B-carrying individuals ("B-load"). The relationship among the three parameters is quite simple: mean equals prevalence times B-load.

A strict drive-selection balance may produce B frequency stability over the years within a population, but B chromosome frequencies sometimes do show temporal, as well as spatial, variation. When these changes oscillate and show no definite tendency, they are most likely a result of genetic drift or may simply reflect a low sample size. But, in a few cases, a sharp increase in B frequency has been found, thereby indicating the recent invasion of a population by a B chromosome. For instance, in the Torrox (Málaga, Spain) population of the grasshopper *E. plorans,* the parasitic B_{24} chromosome showed a change in the mean number of Bs per individual from 0.34 in 1984 to 0.98 in 1992 and 1.53 in 1994. Moreover, significant meiotic drive in females by B_{24} led it to replace the "neutralized" B_2, which was still present in 1984 but absent from the later samples (Zurita *et al.,* 1998). In the fish *Prochilodus lineatus,* B frequency increased from 1.443 Bs per individual to 2.766 in the span of only 10 years (Cavallaro *et al.,* 2000). Interestingly, this increase in frequency was paralleled by a decrease in B mitotic instability, which is presumed to be the drive mechanism for this B chromosome. In the wasp *Trypoxylon albitarse*, B chromosome frequency increased from 0.133 to 0.962 in the Nova Ilha population (Porto Firme, Brazil) between 1997 and 1999 (that is, in only four generations), suggesting a remarkably fast B invasion (Araújo et al., 2002). Finally, in *E. plorans* from Mallorca island (Spain), the mean number of Bs in a northern population at Pollença increased from 0.053 to 0.692 in only 10 years (Riera et al., 2004). Because such B invasions seem to be very rapid, the likelihood of witnessing such an event must be rather low, which probably explains the paucity of cases reported in the literature.

THE BIOLOGY OF B CHROMOSOMES

Although B chromosomes represent a highly heterogeneous class of chromosomes with many particularities that cannot be extrapolated from one species to another, they do share some biological characteristics at different levels, from basic structure to population dynamics. As such, it is useful to examine some of the general features of B chromosomes before moving on to a discussion of their origin and evolutionary histories.

SIZE

To reiterate, B chromosomes constitute a rather diverse collection of extra chromosomes, covering a wide range of sizes and morphologies. In some species, such as the harvest mouse *Reithrodontomys megalotis*, B chromosomes are conspicuously smaller than the smallest A chromosomes (Peppers et al., 1997). The fly *Megaselia scalaris* contains a B chromosome that is little more than an independent centromere, making it just about the smallest conceivable element that could still qualify as a chromosome (Wolf et al., 1991). The Australian rodent *Uromys caudimaculatus* (Baverstock et al., 1982) and the South American characid fish *Astyanax scabripinnis* (Mestriner et al., 2000) exist at the other end of the spectrum by having B chromosomes that may be as large as the largest A chromosome. On occasion, B chromosomes may actually be conspicuously *larger* than the A chromosomes, as in the cyprinid fish *Alburnus alburnus* (Ziegler et al., 2003), although these seem to be isochromosomes. In general, most B chromosomes are intermediate in size, although some size variation can sometimes be found within a single species (e.g., *E. plorans*) (Henriques-Gil et al., 1984; López-León et al., 1993).

As first noted by Hewitt (1979) and later observed in many additional instances, large B chromosomes tend to be mitotically stable, such that they show the same number in all the cells of a given individual. Small Bs, on the other hand, exhibit a tendency to be mitotically unstable, and thus may vary in number from cell to cell within the same individual.

STRUCTURE

Centromere position can be quite heterogeneous, with B chromosomes being either metacentric (with chromosome arms of roughly equal length) or acrocentric (with the centromere near one end of the chromosome), even within single species. Interestingly, the morphology of the predominant B chromosomes in a species seems to reflect that of the A chromosomes. For instance, most B chromosomes in the grasshopper *E. plorans* are acrocentric, the same as all the A chromosomes (Henriques-Gil *et al.*, 1984; López-León *et al.*, 1993). Most B chromosomes found in gomphocerine grasshoppers carrying 2n = 16 + X0/XX chromosomes, with three metacentric A pairs, are also biarmed. This is the case, for instance, in *Myrmeleotettix maculatus* (John and Hewitt, 1965), *Omocestus bolivari* (Camacho *et al.*, 1981), *Chorthippus vagans* (Cabrero and Camacho, 1987), and *Omocestus burri* (Santos *et al.*, 1993). More generally, a recent comparative analysis of 1166 species in mammals has shown this same association between A and B chromosome morphology (Palestis *et al.*, 2004).

Most B chromosomes are heterochromatic—that is, they consist of chromatin characterized by a high level of condensation throughout most of the cell cycle. This compact nature results from the high content of repetitive DNA of various types, especially satellite (satDNA), ribosomal (rDNA), and mobile element DNA (mobDNA). Table 4.2 summarizes the known DNA compositions of B chromosomes as of this writing, although it must be borne in mind that more B chromosomes are continually being analyzed from a molecular perspective.

COMPOSITION

The pioneering work on the analysis of B chromosome DNA was done by means of gradient density ultracentrifugation, and suggested an association between the presence of satellite DNA and B chromosomes in the grasshopper *M. maculatus* (Gibson and Hewitt, 1972). However, this result could not be reproduced by Dover and Henderson (1975) in this same species, nor by Chilton and MacCarthy (1973) in maize or Klein and Eckhardt (1976) in the mealybug. The next significant technical change was introduced by Amos and Dover (1981), who employed restriction endonuclease digestion of satellite DNAs previously isolated by gradient density ultracentrifugation, Southern filter hybridization, and *in situ* hybridization (see Table 4.2). These authors proposed, for the first time, the massive amplification of satellite DNAs in the B chromosomes.

Endonuclease digestion of total genomic DNA from B-carrying individuals followed by gel electrophoresis to look for ladder patterns is a technique that has been used extensively in the study of B chromosomes over the past few decades, largely because of its relative simplicity. A technical approach that has been

TABLE 4.2 Studies on DNA composition of B chromosomes. Technical approaches: DNA hybrid = DNA–DNA hybridization, DGC = density gradient centrifugation, RED = restriction endonuclease digestion, ISH = *in situ* hybridization, FISH = fluorescence *in situ* hybridization, PCR = polymerase chain reaction, PFGE = pulse-field gel electrophoresis, RAPDs = random amplified polymorphic DNAs.

Kind of organism	Species	Technical approach	Name of the DNA sequence	Known homology	B-specific	Authors	Year
Grasshopper	*Myrmeleotettix maculatus*	DGC				Gibson and Hewitt	1972
Maize	*Zea mays*	DGC				Chilton and McCarthy	1973
Grasshopper	*Myrmeleotettix maculatus*	DGC				Dover and Henderson	1975
Triticinae	*Aegilops and Triticum*	DNA hybrid., DGC				Dover	1975
Rye	*Secale cereale*	DNA hybrid.				Rimpau and Flavell	1975
Mealybug	*Pseudococcus affinis*	DGC				Klein and Eckhardt	1976
Tsetse fly	*Glossina austeni*	DGC, RED, Southern, ISH	1682		no	Amos and Dover	1981
Tsetse fly	*Glossina morsitans morsitans*	DGC, RED, Southern, ISH	1685, 1677L and 1677H		no	Amos and Dover	1981
Tsetse fly	*Glossina pallidipes*	DGC, RED, Southern, ISH	1676 and 1686		no	Amos and Dover	1981
Wasp	*Nasonia vitripennis*	Cloning and others	psr2, psr18, psr22		yes	Nur et al.	1988
Raccoon dog	*Nyctereutes procyonides*	ISH		telomeric DNA	no	Wurster-Hill et al.	1988
Smooth Hawkbeard	*Crepis capillaris*	FISH		rDNA	no	Maluszynska and Schweizer	1989
Rye	*Secale cereale*	RED, Southern, cloning	D1100		yes	Sandery et al.	1990
Wasp	*Nasonia vitripennis*	Cloning and others	psr79		no	Eickbush et al.	1992

(Continues)

TABLE 4.2 (Continued)

Kind of organism	Species	Technical approach	Name of the DNA sequence	Known homology	B-specific	Authors	Year
Maize	Zea mays	RED, Southern, cloning	pZmBs	centromere, transposon	yes	Alfenito and Birchler	1993
Rye	Secale cereale	RED, Southern, cloning	E3900		yes	Blunden et al.	1993
Grasshopper	Eyprepocnemis plorans	FISH		rDNA	no	López-León et al.	1994
Grasshopper	Eyprepocnemis plorans	RED, Southern, cloning	pEpD15		no	López-León et al.	1995
Greater glider	Petauroides volans	Microdissection		centromere	no	McQuade et al.	1994
Greater glider	Petauroides volans	Microdissection			yes	McQuade et al.	1994
Smooth Hawkbeard	Crepis capillaris	Microdissection	B134	rDNA	no	Jamilena et al.	1994
Australian daisy	Brachycome dichromosomatica	RED, Southern, FISH		rDNA	no	Donald et al.	1995
Australian daisy	Brachycome dichromosomatica	Microdissection	Bd49		no	Leach et al.	1995
Rye	Secale cereale	FISH			some	Wilkes et al.	1995
Fruit fly	Drosophila subsilvestris	RED, Southern, cloning	pssP216		no	Gutknecht et al.	1995
Maize	Zea mays	RAPDs		Prem-1 retroelements		Stark et al.	1996
Rye	Secale cereale	Microdissection			no	Houben et al.	1996
Australian daisy	Brachycome dichromosomatica	Microdissection	Bdm29		yes	Houben et al.	1997
Maize	Zea mays	RAPDs			yes	Lin and Chou	1997

Wasp	*Nasonia vitripennis*	PCR, Southern	NATE	retrotransposon	yes	McAllister and Werren	1997
Fungus	*Nectria haematococca*	Cloning and others	Nht1	transposon	no	Enkerli et al.	1997
Frog	*Leiopelma hochstetteri*	Microdissection		W chromosome	no	Sharbel et al.	1998
Water rat	*Nectomys squamipes*	FISH		telomeric DNA	no	Silva and Yonenaga-Yassuda	1998
Fungus	*Nectria haematococca*	Cloning and others		transposons		Enkerli et al.	2000
Rat	*Rattus rattus*	FISH, Southern		rDNA	no	Stitou et al.	2000
Salamander	*Dicamptodon tenebrosus*	Microdissection, cloning, Southern			no	Brinkman et al.	2000
Rye	*Secale cereale*	PCR and FISH	E3900	Ty3-gypsy retrotransposon	yes	Langdon et al.	2000
Fish	*Astyanax scabripinnis*	RED, Southern, cloning	As51	transposon	no	Mestriner et al.	2000
Australian daisy	*Brachycome dichromosomatica*	Microdissection	Bdm29		no	Houben et al.	2001
Australian daisy	*Brachycome dichromosomatica*	Microdissection	Bdm54		no	Houben et al.	2001
Maize	*Zea mays*	AFLPs	M8-2D	telomeric repeat	yes	Qi et al.	2002
Raccoon dog	*Nyctereutes procyonoides*	Microdissection and FISH			yes	Trifonov et al.	2002
Wood mouse	*Apodemus peninsulae*	Microdissection and FISH			no	Trifonov et al.	2002
Wood mouse	*Apodemus peninsulae*	Microdissection			no	Karamysheva et al.	2002

(Continues)

TABLE 4.2 (*Continued*)

Kind of organism	Species	Technical approach	Name of the DNA sequence	Known homology	B-specific	Authors	Year
Herb	*Plantago Lagopus*	RED, Southern, FISH		rDNA	no	Dhar et al.	2002
Fungus	*Nectria haematococca*	Cloning and others	Nh2	copia LTR retrotransposon		Shiflett et al.	2002
Fish	*Prochilodus lineatus*	RED, Southern, cloning, FISH	SATH1		no	de Jesus et al.	2003
Fish	*Alburnus alburnus*	AFLPs, Southern, PCR, FISH		retrotransposon	yes	Ziegler et al.	2003
Raccoon dog	*Nyctereutes procyonides*	FISH		rDNA	no	Szczerbal and Switonski	2003
Fungus	*Magnaporthe oryzae*	PFGE, Southern		transposons	no	Chuma et al.	2003
Maize	*Zea mays*	Microdissection, FISH and others	pBPC51	Maize B centromere	yes	Cheng and Lin	2003
			18 DNA sequences	repeats in non-coding regions	no	Cheng and Lin	2003

employed in far fewer cases (because it is very time-consuming, especially in organisms with large genomes), is the construction of genomic libraries of total DNA from B-carrying individuals, followed by a screening of the thousands of clones by double hybridization with B⁺ DNA and B⁻ DNA. This method was first employed by Nur et al. (1988) to isolate several repetitive DNAs specific to the paternal sex ratio (PSR) B chromosome in the wasp *Nasonia vitripennis,* after screening 6000 clones. A similar approach has been followed in the case of the fungus *Nectria haematococca,* although in this case B chromosomes can be isolated by pulse field electrophoresis which, in addition to the small size of the B (1.6 Mb), has permitted the building of the first physical map of a B chromosome by ordering a series of cosmids in a DNA library (Enkerli et al., 2000). Indeed, this B chromosome may soon become the first B to be sequenced in its entirety.

In the 1990s, a revolutionary technique emerged involving the microdissection of a specific chromosome and the extraction and Polymerase Chain Reaction (PCR) amplification of its DNA, which may then either be cloned or marked and used to paint chromosomes by fluorescent *in situ* hybridization (FISH). This technique was first used for isolating B chromosome DNA by McQuade et al. (1994) in the greater glider marsupial *Petauroides volans.* The same technique has been employed in several other animal and plant species (see Table 4.2). The analysis of B chromosomes in a variety of plant and animal species has revealed that B chromosomes are replete with repetitive DNAs, and in particular mobile elements. Cheng and Lin (2003) have recently isolated 19 repetitive DNA sequences from a microdissected maize B chromosome, 18 of which were also present in the A chromosomes and showed homology to noncoding regions of several genes (e.g., *Adh1, Bz1, Gag,* and *Zein*). The remaining repetitive element was B-specific and homologous to the maize B centromere sequence previously isolated by Alfenito and Birchler (1993) (see Table 4.2). It is conceivable that islands of genes may exist among these oceans of repetitive elements, although these have proven difficult to find, with the notable exception of ribosomal RNA genes, whose amplification makes them much more conspicuous.

In fact, many B chromosomes are known to harbor rRNA genes (Green, 1990). On A chromosomes, rRNA genes exist in tandem arrays localized at the nucleolar organizer regions (NORs). The presence of these genes on B chromosomes has been mainly demonstrated by silver impregnation, which detects active NORs, and fluorescent *in situ* hybridization, which detects both active and inactive rRNA genes. In 1990 Green published a list of 21 species with NOR-carrying B chromosomes, 12 of which were plants, and only five years later Jones (1995) reported a list of 23 plant species carrying NORs. New cases are being reported regularly (e.g., Dhar et al., 2002). The list is also increasing for animals such as the rodents *Akodon* aff. *arviculoides* (Yonenaga-Yasuda et al., 1992), *Apodemus peninsulae* (Boeskorov et al., 1995), and *Rattus rattus* (Stitou et al., 2000). In most of these cases, only active rRNA genes have been detected (by silver impregnation), but it is likely that with the increased use of FISH, more inactive genes will also be identified. Some possible reasons for the presence of rDNA on B chromosomes will be discussed later in the chapter.

Interstitial blocks of telomeric DNA have been observed in both large and small B chromosomes of the raccoon dog *Nyctereutes procyonides* (Wurster-Hill *et al.*, 1988), and in a submetacentric B chromosome in the neotropical water rat *Nectomys squamipes* (Silva and Yonenaga-Yassuda, 1998), suggesting that some chromosomal rearrangements involving chromosome tips might occur in the origin of this B chromosome. Likewise, in maize, a B-specific DNA sequence (M8-2D) has recently been found to contain the telomeric repeat unit AGGGTTT conserved in plant telomeres (Qi *et al.*, 2002).

Single-copy genes seem to be scarce in B chromosomes, although this could reflect the difficulty of finding them amid the stretches of repetitive DNA present in B chromosomes. The subject has been reviewed by Green (1990) and Jones (1995) who, in addition to listing a number of species with Bs harboring rRNA genes, mentioned a few cases where B chromosomes are candidates to contain single-copy genes. In most cases, the evidence for these comes from an association between B chromosome presence and an observable phenotypic change. These include genes controlling B chromosome transmission in rye and maize, achene color in *Haplopappus gracilis*, meiotic pairing in *Aegilops*, leaf striping in maize, crown rust resistance in *Avena sativa*, sex determination in the frog *Leiopelma hochstetteri*, pathogen virulence in the fungus *Nectria haematococca*, and male sterility in *Plantago coronopus*. It bears mentioning that this last example, first reported by Paliwal and Hyde (1959), could not be repeated in a later study by Raghuvanshi and Kumar (1983). By contrast, in the case of *N. haematococca*, several pea pathogenicity genes have since been identified, isolated, and sequenced (Han *et al.*, 2001; Funnell *et al.*, 2002). One gene, *PDA1*, encodes a specific cytochrome P450 enzyme that confers resistance to pisatin, an antibiotic produced by the pea plant hosts. Other genes showed similarity to fungal transposases and, as Han *et al.* (2001) pointed out, this set of genes shares some significant structural and molecular characteristics with pathogenicity islands in pathogenic bacteria of plants and animals (see Chapter 10). Finally, it is also worth noting here that there is a recurrent mistake in Green's and Jones's reviews regarding the supposed presence of an esterase gene in the B chromosome of the plant *Scilla autumnallis*, whereas the original publication actually reported a *regulatory effect* of B chromosomes on an A chromosome esterase gene (Ruíz-Rejón *et al.*, 1980; Oliver *et al.*, 1982).

MEIOTIC BEHAVIOR

During meiosis, each A chromosome actively pairs with its homolog to form a bivalent which then segregates (i.e., one homolog moves to each pole) in the first division, thus ensuring a fair share of transmission. This behavior, along with a similar chance of fertilization for the gametes receiving each allele (or homologous A chromosome) in the heterozygote, is the basis for the Mendelian segregation law. In contrast, B chromosomes do not always form pairs and, in fact, their attack and maintenance is mainly based on their individual behavior as univalents during meiosis. Instead of being forced to segregate in a Mendelian fashion (as usual when a chromosome has a counterpart), univalents are free to segregate preferentially to a specific meiotic pole, thereby resulting in transmission shares higher than the 50% established by the Mendelian law.

In many respects, B chromosomes resemble sex chromosomes. Their similar condensation–decondensation cycle during meiosis, meiotic univalency (as is the case with X0 males), and other shared similarities (see Camacho *et al.*, 2000) suggest a close relationship between these two kinds of chromosomes, so that B chromosomes in some species seem to be derived from sex chromosomes. This appears to be the case in the fly *Glossina* (Amos and Dover, 1981), the frog *Leiopelma hochstetteri* (Green, 1988; Sharbel *et al.*, 1998), and the grasshopper *E. plorans* (López-León *et al.*, 1994). It has even been suggested that some sex chromosomes are actually descended from B chromosomes (see later section).

Meiotic behavior is variable among different B chromosomes, either within a single species or among different species, in the same way that X chromosomes may behave differently among species. Rebollo *et al.* (1998), after *in vivo* analysis of male meiotic behavior of several kinds of univalents (X, B, and As), suggested the existence of "chromosomal strategies" for adaptation to univalency, as a result of selection acting on the chromosomes to increase transmission effectiveness. These authors noted that several different strategies may serve the same purpose of B chromosome maintenance. They observed that B chromosomes may vary in dynamism during meiotic metaphase I, some of them being as static as the X, others being slightly more dynamic, and still others being very dynamic univalents. One of the most remarkable behaviors of B univalents during metaphase I is the reorientation from pole to pole, resembling X chromosome behavior in the grasshopper *Melanoplus differentialis* (Nicklas, 1961). This suggests that both poles are able to capture the same univalent, which is migrating from pole to pole up to anaphase I. In *Melanoplus sanguinipes*, however, the X chromosome stably remains in a pole without performing such movements, suggesting the existence of mechanisms of anchorage to the spindle (Ault, 1986). Rebollo *et al.* (1998) found that most Bs analyzed (in several species) were able to reorient from pole to pole, including some Bs showing meiotic drive in females (e.g., B_{24} in *E. plorans*). But, remarkably, the B in the grasshopper *Heteracris littoralis*, which also shows meiotic drive in females, did not reorient, indicating that it shows a static behavior similar to the X in *Melanoplus sanguinipes*, and thus possible anchorage mechanisms could operate for B chromosomes as well.

242

Camacho

MECHANISMS OF DRIVE

The unfair advantage in transmission gained by B chromosomes, by a variety of processes including segregation distortion, is usually known as an "accumulation mechanism," or simply "drive." Drive is the property that qualifies B chromosomes as selfish elements, and is the main weapon that they use to increase in frequency and invade new populations. B chromosome drive has been the subject of intensive research in many species. Drive might occur at several stages during the life cycle, and can be premeiotic, meiotic, or postmeiotic (for reviews, see Jones, 1991 and Camacho et al., 2000).

Premeiotic drive (Fig. 4.2A) implies an increase, during development, of the number of B chromosomes in cells of the germline, so that when they enter meiosis to generate gametes their mean number of B chromosomes is significantly higher than the number of Bs originally contained in the zygote. This mechanism was first described by Nur (1963) in the grasshopper *Calliptamus palaestinensis*, where he found variation of B number among spermatocytes of single individuals, suggesting that mitotic nondisjunction occurs in the germline. Later on, Nur (1969) compared the number of B chromosomes in spermatocytes and gastric caeca in the grasshopper *Camnula pellucida* and found a 37% higher B frequency in spermatocytes, suggesting B accumulation during the development of the testis. This kind of premeiotic accumulation has also been shown in the locust *Locusta migratoria* (Kayano, 1971; Viseras et al., 1990), where the mitotic nondisjunction of B chromosomes during embryogenesis was first visualized (Pardo et al., 1995). Similar mechanisms also operate in plants such as *Crepis capillaris* (Rutishauser and Röthlisberger, 1966).

Meiotic drive (Fig. 4.2B) depends on the functional asymmetry of meiotic products. As noted previously, one of the two cells resulting from each of both meiotic divisions is inherently unviable (the polar bodies) in female meiosis. Cooperative alleles or chromosomes have a 50% chance of being eliminated into a polar body, but B chromosomes can manage to escape this elimination by

FIGURE 4.2 Diagrams showing different accumulation mechanisms (drive) for B chromosomes. (A) In premeiotic accumulation, the B chromosome undergoes mitotic nondisjunction so that its two chromatids go to the same pole. If the chosen pole is destined to be germ cells more frequently than chance, B accumulation results, with a mean number of Bs in germ cells being higher than the original number in the zygote. In the represented example, the zygote had 1B but the three germ cells had 1.67 Bs on average. (B) Meiotic drive results when a B univalent at metaphase I of a functionally asymmetric meiosis (typically in females) preferently migrates to the viable pole (the secondary oocyte in the example). (C) Postmeiotic drive is very frequent in plants during gametophyte maturation. This diagram shows the mitotic nondisjunction of the two chromatids of a B chromosome replicated for passing through the first mitosis for pollen grain maturation, and their preferential migration to the generative nucleus.

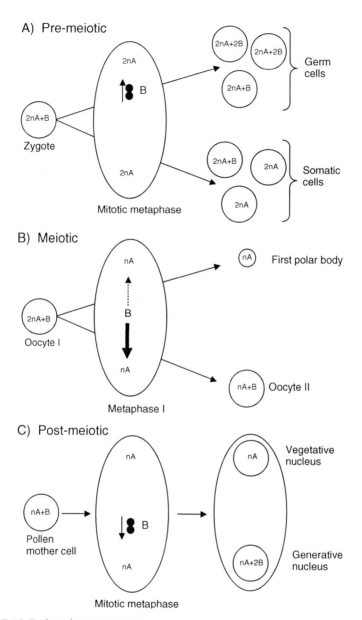

A) Pre-meiotic

2nA+B
Zygote

2nA
B
2nA

Mitotic metaphase

2nA+2B
2nA+2B
2nA+B
Germ cells

2nA+B
2nA
2nA
Somatic cells

B) Meiotic

2nA+B
Oocyte I

nA
B
nA

Metaphase I

nA   First polar body

nA+B   Oocyte II

C) Post-meiotic

nA+B
Pollen mother cell

nA
B
nA

Mitotic metaphase

nA   Vegetative nucleus

nA+2B   Generative nucleus

FIGURE 4.2 For legend, see opposite page.

preferentially migrating to the viable meiotic pole (the oocyte II in the first division or the ootid in the second).

Meiotic drive has been shown, for instance, in the plant *Lillium callosum* (Kayano, 1957) and the grasshoppers *Melanoplus femur-rubrum* (Lucov and Nur, 1973; Nur, 1977), *Myrmeleotettix maculatus* (Hewitt, 1973a, 1976), *Heteracris litoralis* (Cano and Santos, 1989), *Omocestus burri* (Santos *et al.*, 1993), and *Eyprepocnemis plorans* (Zurita *et al.*, 1998; Bakkali *et al.*, 2002). Interestingly, Hewitt (1976) observed that the spindle is conspicuously asymmetrical in the female first meiotic metaphase of *M. maculatus*, with the egg pole being longer than the polar body pole. He proposed a higher chance for a B to orient and segregate at first anaphase toward the egg nucleus. This claim was based on the continuous movement from pole to pole performed by the X univalent observed *in vivo* in meiocytes by Nicklas (1961). Such a movement has also been observed for B chromosome univalents (Rebollo and Arana, 1995; Rebollo *et al.*, 1998). Under continuous movement from pole to pole, the B would spend more time in the longer pole (the oocyte one) and thus would have a higher chance of remaining in it at the end of anaphase. Meiotic drive can also occur in males, e.g., in the mealybug *Pseudococcus affinis* (formerly *obscurus*), where B chromosomes escape the elimination of the paternal A chromosome set that characterizes spermatogenesis in this insect (Nur, 1962).

In plants, B chromosomes have found a way to drive by nondisjunction in the postmeiotic mitoses (Fig. 4.2C), leading to gametophyte maturation with the two B sister chromatids preferentially going to the gamete nucleus. This mechanism was first shown in rye, where the B chromosome preferentially moves with its two nondisjunct chromatids toward the generative nucleus in the first pollen grain mitosis (Hasegawa, 1934). B chromosome drive in rye appears to be exceptionally potent in this regard, because a similar process also takes place for Bs at the first mitosis of the megaspore (Jones, 1995; Puertas, 2002). Other plants, such as maize, show this kind of drive through the male side only (Carlson and Roseman, 1992), but in this case maize Bs are also overtransmitted because of preferential fertilization of B-carrying sperm (Roman, 1948).

In the haplodiploid wasp *Nasonia vitripennis*, the B (specifically, the "paternal sex ratio," PSR) chromosome transforms diploid zygotes (destined to be females) into haploid males by destroying the whole paternal chromosome set in the first mitotic division of the diploid zygote (Werren, 1991). Paradoxically, this system involves drive in "female" zygotes, even though the B is present in adult males only. As a result of this mechanism, the transmission rate of the PSR chromosome is about 100% of fertilized eggs, but this is merely the same as it would be in males for a B chromosome lacking this property, and is actually lower overall than it would be if the B could be transmitted through both sexes (R. Trivers, personal communication). Therefore, although there is no net drive for the PSR chromosome, there is a catastrophic cost to paternal chromosomes as long as the B persists in changing the host genome in each generation.

CENTROMERIC DRIVE

It has recently been suggested that there may exist a bias regarding the number of centromeres during female meiosis, so that when more centromeres are favored over fewer, the karyotype becomes predominately acrocentric, and vice versa (Pardo-Manuel de Villena and Sapienza, 2001). This association would be based on a differential ability for capturing centromeres by the two meiotic poles, so that a meiotic structure in metaphase I with an odd number of centromeres (e.g., a trivalent resulting from heterozygosity for a centric fusion) would not segregate randomly but in a biased way, with the two acrocentrics going to the stronger pole and the single metacentric going to the weaker pole (Fig. 4.3A). If the stronger pole were the oocyte, then the resulting karyotype would be predominantly asymmetrical (i.e., composed of uniarmed chromosomes). This would be a case

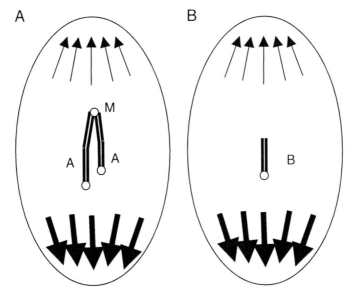

FIGURE 4.3   The theory of centromeric drive. The two poles of anaphase I cells differ in the ability to capture centromeres so that when an odd number of centromeres are segregating, such as in a heterozygote for a centric fusion with one metacentric (M) and two acrocentric (A) chromosomes (A), the fitter pole (e.g., the oocyte II) captures the two As and the M goes to the polar body pole. This would imply that the As would be easily retained, leading to asymmetrical karyotypes mainly composed of uniarmed chromosomes. For the same reason, a univalent B chromosome should be prone to drive in species where the oocyte pole captures more centromeres and hence possesses more asymmetrical karyotypes (B).

in which more centromeres were favored over fewer. If, however, the polar body is the more adept at capturing centromeres, then the resulting karyotype would be predominantly symmetrical (i.e., composed of biarmed chromosomes).

This theory of centromeric drive was based on the finding of mammals heterozygous for several chromosome mutations, in which an odd number of centromeres was observed during meiosis. It has very recently obtained additional support from the work by Palestis *et al.* (2004). The key insight presented by these authors was that B chromosomes could be used to test this theory, because they are predominantly univalents in meiosis and would therefore be greatly affected by any differential ability in capturing centromeres by the poles (Fig. 4.3B). In the presence of such a bias, B chromosomes should be more frequent in species where karyotypes suggest that more centromeres are favored over fewer—that is, species with asymmetrical karyotypes. To test this hypothesis, Palestis *et al.* (2004) performed a comparative analysis in 1166 species of mammals, 50 of which had B chromosomes. Remarkably, they found that 61% of species with Bs have their A chromosome set constituted predominantly by acrocentric chromosomes, a figure significantly higher than the 35% observed for species lacking Bs. These results are consistent with the expectations of the theory of centromeric drive, and indicate that similar tests are warranted in other taxonomic groups.

Very recently Bidau and Martí (2004) provided the first intraspecific evidence for centromeric drive. In 53 Argentinean populations of the grasshopper *Dichroplus pratensis* showing parallel polymorphisms for B chromosomes and Robertsonian fusions for several A chromosomes, B chromosomes were more frequently found in populations with a higher proportion of acrocentric chromosomes. In addition, the population frequency of B-carrying individuals was significantly negatively correlated with that of Robertsonian fusions per individual.

If centromeric drive were found to be widespread, this could imply that B chromosome meiotic drive might be a rather passive consequence of oocyte characteristics. But the existence of different B variants showing different degrees of transmission within a single species, from drive to absence of drive (e.g., in *E. plorans*, see later section), and the control that some B chromosomes seem to exert on their own transmission (e.g., in rye and maize, see Puertas, 2002), suggest that B chromosomes themselves may have much to say with respect to their accumulation. However, it would seem that only a few of the extra chromosomes that have the possibility of becoming B chromosomes actually possess the ability to take advantage of oocyte shortcomings, because otherwise B chromosomes would presumably be even more frequent than they already are.

THE ORIGIN(S) OF B CHROMOSOMES

So far this chapter has included a review of the distribution and basic characteristics of B chromosomes, but the question of where Bs come from has not yet been addressed. As it turns out, B chromosomes can arise by a variety of different mechanisms, making this an interesting but challenging question to answer (Camacho *et al.*, 2000).

DERIVATION FROM A CHROMOSOMES

The most logical source of B chromosomes is the A genome itself, as a simple by-product of karyotypic evolution. Some of the extra A chromosomes or centric fragments that eventually appear in the evolution of many species may become a B chromosome if they manage to drive (referred to here as "neo-Bs"). The most difficult aspect of the puzzle of B chromosome origins is to uncover the mechanism(s) by which a neo-B becomes a B chromosome.

Aneuploidy, the existence of incomplete multiples of the A chromosome set (in particular, the presence of extra copies of certain chromosomes), is one of the main sources of potential B chromosomes. Notably, extra ("polysomic") A chromosomes are usually rapidly heterochromatinized and acquire an appearance and meiotic behavior resembling those of many B chromosomes (Hewitt, 1973a; Peters, 1981; Talavera *et al.*, 1990). Neo-Bs could also be a by-product of some chromosome mutations like Robertsonian translocations, arising as a small centric fragment during the fusion of two acrocentric A chromosomes into a metacentric one. In other cases, the origin of a B chromosome may be traced to a specific A chromosome region. For instance, Keyl and Hägele (1971) found that the polytene banding pattern in a B chromosome in the midge *Chironomus plumosus* was similar to that in the vicinity of the centromere in Chromosome 4. Likewise, in the grasshopper *E. plorans,* the relative order of two repetitive DNAs (rDNA and a 180 bp tandem repeat) with respect to the centromere was coincident only in the B and the X chromosome, suggesting a B origin from the paracentromeric region of the X chromosome (López-León *et al.*, 1994). Aneuploid neo-Bs may also derive from incompletely expelled A chromosomes from the sperm in pseudogamous parthenogens, as seen in the flatworm *Polycelis nigra* (Sharbel *et al.*, 1997), or by escaping from sperm chromosome destruction through cytoplasmic incompatibility, following hybridization between the wasp *Nasonia vitripennis* and a species of *Trichomalopsis* (McAllister and Werren, 1997).

Regardless of the specific mode of origin of the neo-B chromosome, it is expected that the neo-B undergoes a rapid structural modification that renders it incapable of meiotic pairing with the ancestral A chromosome (Camacho *et al.*, 2000). The accumulation of repeat DNA sequences seems to be a very common pathway in B chromosome differentiation, and the presence of rDNA in B chromosomes of many species (see Table 4.2) might be indicative of a possible role of this kind of DNA in B origins. The *de novo* evolution of satellite DNA on the rye B chromosome from euchromatic sequences (Langdon *et al.*, 2000) might also illuminate possible evolutionary pathways by which B chromosomes differentiate molecularly from the ancestral A chromosome. These are important issues still awaiting experimental resolution.

FROM WHICH SPECIES?

Accepting that B chromosomes derive primarily from A chromosomes, the question still remains as to *whose* chromosomes they are derived from. In other words, it is still necessary to determine whether a given B chromosome derives from the A genome of its current host (i.e., intraspecifically) or is of alien origin (i.e., interspecifically). Comparative analyses of DNA sequences on the B and A chromosomes can shed some light on this issue. In rye, B chromosomes are mostly composed of DNA sequences shared with A chromosomes (Wilkes *et al.*, 1995; Houben *et al.*, 1996), suggesting their intraspecific origin (Puertas, 2002). It bears reiterating, however, that Langdon *et al.* (2000) have recently reported the *de novo* formation of satellite DNA repeats in the B that were derived from complex euchromatic sequences. In maize, many DNA sequences in the B chromosomes are highly repetitive and shared with the A chromosomes (Alfenito and Birchler, 1993), which also suggests an intraspecific origin for these Bs (Stark *et al.*, 1996). Intraspecifically-derived Bs have also been found in the plants *Crepis capillaris* and *Brachycome dichromosomatica*, as well as the animals *Chironomus plumosus, Drosophila subsilvestris, Petauroides volans, Reithrodontomys megalotis, Eyprepocnemis plorans*, several *Glossina* species, and *Leiopelma hochstetteri* (see Camacho *et al.*, 2000).

The second possibility, an interspecific origin, can be illustrated by the presence of B-specific DNA sequences that are not present in the host A chromosomes but are found in those of related species. The best example is the PSR chromosome in the wasp *Nasonia vitripennis*, in which a molecular phylogeny of the retrotransposon *NATE* indicated that the PSR has recently been transferred into *N. vitripennis* from a species in the genus *Trichomalopsis* (McAllister and Werren, 1997). Experimental crossing in *Nasonia* has shown that B chromosomes can form *de novo* during interspecific hybridization (Perfectti and Werren, 2001). Some cases are considerably more ambiguous than this, however. For instance, in the daisy *Brachycome dichromosomatica*, the analysis of a B-specific DNA repeat sequence (Bd49) seemed to indicate a possible interspecific origin for this B chromosome (John *et al.*, 1991), but additional analyses showed the presence of this repeat, at low copy number, in the A genome of this and other *Brachycome* species (Leach *et al.*, 1995). More recent analyses of other B chromosome types and additional DNA repeat families have suggested that these B chromosomes most likely came from the standard genome, indicating that the reconstruction of B chromosome formation is a very difficult task (Houben *et al.*, 2001). Other examples of interspecific B chromosome origin have been reported in *Coix* plants (Sapre and Deshpande, 1987) and the gynogenetic fish *Poecilia formosa* (Schartl *et al.*, 1995).

It is clear that both modes of B chromosome origin are occurring, but the frequency with which B chromosomes form intra- or interspecifically is variable among organisms and dependent on such features as karyotype dynamics (especially the frequency of appearance of extra A chromosomes or centric fragments) and the likelihood of interspecific hybridization.

FROM WHICH POPULATION?

Another important issue in the study of B chromosome origins is the elucidation of the place of origin. For this purpose, the geographical distribution of B chromosomes is very illustrative. For example, in the grasshopper *M. maculatus*, B chromosomes were present in southern Britain but absent from continental Europe, which suggested their British origin (Hewitt, 1973a). In this case, the geographical distribution was also informative about the timing of B chromosome emergence, suggesting that it arose after the last physical contact between the British Isles and the European continent around the time of the most recent period of glaciation (8000 years ago) (Hewitt, 1973a).

In the grasshopper *E. plorans*, the B chromosomes in Moroccan populations seem to derive from Spanish Bs based on the observations that (1) they are essentially made of the same DNA sequences (Cabrero *et al.*, 1999), and (2) they still exhibit drive in the most southern population analyzed in Morocco but not in the most northern ones (Bakkali *et al.*, 2002) nor in Spanish populations (López-León *et al.*, 1992), suggesting that Bs are younger in Morocco. Clearly, it is only through the detailed study of B chromosome biology that such inferences can be made.

FROM WHICH CHROMOSOME?

Having established that a given B chromosome derives from the A genome of a particular population of a certain species, it becomes prudent to ask which specific chromosome generated the B. Once again, detailed detective analyses of B and A chromosome DNA may supply some helpful clues, although this remains a difficult question to answer in many cases. In the plant *Crepis capillaris*, for instance, all B chromosome DNA sequences isolated by microdissection were also present in the A chromosomes, but it was impossible to deduce the ancestral A chromosome for the B chromosome (Jamilena *et al.*, 1994). In maize, on the other hand, a DNA repeat has been isolated from the B centromere (Alfenito and Birchler, 1993) which is related to a centric repeat in Chromosome 4, suggesting a possible common origin for both chromosomes (Page *et al.*, 2001). In *E. plorans*, the coincident order of two repetitive DNAs with respect to the centromere suggested that B_2 had most likely originated from the X chromosome

(López-León *et al.*, 1994). Additional analyses have confirmed that B chromosomes in Spain and Morocco are most likely derived from the X chromosome, but those from the Caucasus have probably arisen independently because they seem to derive from the smallest autosome (Cabrero *et al.*, 2003a).

Recently the origin of a heterochromatic B chromosome from Chromosome 2 has been witnessed in the plant *Plantago lagopus* (Dhar *et al.*, 2002). Starting with plants carrying three Chromosome 2s, these authors performed a cross with euploid plants and obtained one progeny plant carrying an extra ring chromosome. In the progeny of this plant, an individual carried an extra heterochromatic isochromosome showing many of the characteristics of a B chromosome. The sequence of processes involved in the formation of this B chromosome, in only two generations, included aneuploidy, the formation of a ring chromosome, chromosome fragmentation, massive amplification of 5S rDNA, and centromere misdivision with chromatid nondisjunction to generate the isochromosome. Dhar *et al.* (2002) claimed that their data do suggest preferential transmission because the transmission rate of the isochromosome (41% and 42% through male and female gametes, respectively) is higher than the 25% usually shown by trisomics. This aside, theoretical work predicts that nondriving Bs (i.e., those with a transmission rate ≤50%) cannot invade a population (Camacho *et al.*, 1997). It therefore seems unlikely that the neo-B in *P. lagopus* could have invaded the population, unless it has rapidly increased its transmission rate in the time since. Notably, the neo-B was only observed because of the artificial selection performed. Nevertheless, this study illuminates some possible pathways for B chromosome origin that might take place in many natural populations.

ACCUMULATION OF TRANSPOSABLE ELEMENTS

It is important to consider the possibility that extra centromeres (like those in *Megaselia scalaris*) may grow by capturing a variety of repetitive DNA sequences from the A chromosomes, most notably mobile elements, and thereby build large B chromosomes (see also Jones and Houben, 2003). Neo-B transformation into true B chromosomes, and the degeneration of the latter, might largely proceed in this way, and indeed the rapid heterochromatinization of extra chromosomes might be a symptom of the speed of this process. The heterochromatic nature of most B chromosomes may make them safe havens for transposable elements (TEs), for two reasons. First, TEs are best tolerated in genomic regions of low gene density because TE insertions into genes are generally deleterious (see Chapter 3). Second, some TEs may show a tendency to target heterochromatin directly, as evidenced by the *I* element in the proximal heterochromatin in *Drosophila melanogaster* Chromosome 2 (Dimitri *et al.*, 1997). Whether these processes apply to B chromosomes remains to be seen, but this is an important possibility.

TE accumulation is typical of chromosome degeneration, as demonstrated by the enriched occurrence of the *TRAM* element in the evolving neo-Y chromosome of *Drosophila miranda* (Steinemann and Steinemann, 1997). Similarly, a variety of DNA sequences related to TEs have been found in B chromosomes (see Table 4.2). The absence of recombination between A and B chromosomes is perhaps the main characteristic favoring TE accumulation in B chromosomes, because this prevents the TEs from inserting into A chromosomes and thereby avoids the consequences of selection on gene function. However, in species with large amounts of heterochromatin in the A chromosomes, the possibility still remains that TEs might move between A and B chromosomes by means of ectopic recombination, which seems to be frequent for some TEs (Montgomery *et al.*, 1991). This is an interesting issue to investigate by intraspecifically comparing TE composition in A and B chromosomes.

Many TEs can be considered as molecular parasites (Kidwell and Lisch, 2001) (see Chapter 3), suggesting that those found in B chromosomes may be parasites of parasites. In fact, if the elements in question are short interspersed nuclear elements (SINEs), which parasitize long interspersed nuclear elements (LINEs) for their transposition machinery, then these would be parasites of parasites of parasites. To add yet another link to this chain of exploitation, consider that some B chromosomes are themselves found within the bodies of parasitoid animals, such as the wasp *N. vitripennis*, which lays its eggs into dipteran pupae!

SIMILARITIES TO SEX CHROMOSOMES: ANALOGY OR HOMOLOGY?

The evolutionary relationship between Bs and sex chromosomes seems to be more than incidental. They may be similar in terms of meiotic behavior, size, morphology, and heteropycnocity (see Camacho *et al.*, 2000), which raises the question of whether these coincidences are the product of a common origin (homology) or merely reflect convergence on a similar "phenotype" (analogy). In cases where the ancestry of the B chromosome can be traced to a sex chromosome, as in the fly *Glossina,* the frog *Leiopelma hochstetteri*, and the grasshopper *E. plorans* mentioned previously, homology is the appropriate interpretation. But Bs and sex chromosomes are also subject to similar molecular evolutionary processes (e.g., chromosome degeneration) and functional constrains (e.g., chromosome inactivation), such that analogy is also a strong possibility in many cases (Camacho *et al.*, 2000).

The fact that sex chromosomes are better tolerated than autosomes in an aneuploid condition (Hewitt, 1973a) supports the possibility of a common origin for Bs and sex chromosomes. Small autosomes are also better tolerated in a polysomic state than large ones, and have likewise been identified as possible sources of

B chromosomes (e.g., in Caucasian specimens of *E. plorans,* as noted earlier). Again, there is also the interesting possibility that this evolutionary relationship between Bs and sex chromosomes might be reversible, with B chromosomes sometimes playing a role as an intermediate stage in the origin of some sex chromosomes.

DRIVE TO SURVIVE!

It is a reasonable expectation that neo-Bs are continually emerging in many eukaryote species, given that they are by-products of normal processes of karyotypic evolution (Camacho *et al.,* 2000). Again, only those neo-Bs showing significant drive from the beginning have a chance to become true B chromosomes because otherwise they will be fated to disappear. It could be argued that they may also prosper if they conferred an advantage to the host genome, but this is quite unlikely for neo-Bs that arise by aneuploidy because all their genes are already present in the host chromosomes, meaning that they contribute nothing new to the genome. Obviously, this is especially true of the tiny Bs in *Megaselia scalaris,* which are essentially just bare centromeres. On the other hand, it is much more easily conceivable that Bs that arise interspecifically may be advantageous to the host, because the alien DNA in the B might include some novel beneficial genes. For instance, the conditionally dispensable (CD) chromosome in the filamentous ascomycete *Nectria haematococca* contains specific DNA repeats that are absent from the A chromosomes (Enkerli *et al.,* 1997), and analysis by FISH has shown that the CD chromosome is not a duplicate of other chromosomes in the genome (Taga *et al.,* 1999). Again, this B chromosome, which has presumably arisen interspecifically, contains genes conferring resistance to the natural antibiotics produced by the pea plants of which *N. haematococca* is a parasite (Miao *et al.,* 1991a,b). Here, the B chromosome may not be parasitic—but only because it increases the fitness of its parasitic host.

INTERACTIONS WITH THE HOST GENOME

As with any parasite, there are important interactions between B chromosomes and their hosts. These interactions take place at several levels of biological organization, including genes, genomes, cells, organisms, populations, and perhaps even species. Most simply, it might be imagined that B chromosomes exert their most important impacts at the levels of organization closest to the chromosomal level, where they would be the most conspicuous to natural selection. This is the level to be discussed first in the following sections, followed by a treatment of their impacts at higher levels of organization.

EFFECTS ON GENE EXPRESSION

One of the most interesting potential impacts of B chromosomes is on gene expression in the A chromosome set. For example, the presence of B chromosomes in the plants *Scilla autumnalis* (Ruiz-Rejón *et al.*, 1980; Oliver *et al.*, 1982) and *Allium schoenoprasum* (Plowman and Bougourd, 1994) have been shown to influence the expression of A genes for an esterase and endosperm protein, respectively. Similarly, B presence can affect the expression of NORs on the A chromosomes in the grasshopper *E. plorans* (Cabrero *et al.*, 1987). Unfortunately, the effects of B chromosomes on the expression of A chromosome genes is a subject that has not received sufficient attention, but is one that should now be amenable to much more intensive study by modern molecular techniques.

RECOMBINATION IN A CHROMOSOMES

In many plant and animal species, the presence of B chromosomes is associated with a change in the level of recombination in the A chromosomes, as deduced from chiasma frequency during meiosis (for reviews, see Jones and Rees, 1982; Bell and Burt 1990; Camacho *et al.*, 2002). In most cases where an effect has been reported, it was an increase, but there are also examples of chiasma frequency decrease (or simply no effect at all). Interestingly, such contradictory effects have been found within a single species, which has led to a chaotic proliferation of at least four different hypotheses to explain such diverse effects.

The first of these is the "adaptation hypothesis," which proposes that B chromosome–induced chiasma effects are adaptive because, when chiasma frequency is increased, the resulting higher level of genetic variation enables rapid population evolution (Darlington, 1958; John and Hewitt, 1965; Hewitt and John, 1967), whereas when it is decreased, adaptive gene combinations are maintained (Fontana and Vickery, 1973). This interpretation has been strongly criticized by Bell and Burt (1990) because it implies that parasites are selected on the grounds of the favorable effects they have on hosts. The second possibility, the "passive effects hypothesis," suggests that chiasma effects represent some necessary mechanical or physiological effect of additional or exotic DNA (Bell and Burt, 1990). However, this hypothesis has been unable to explain contradictory intraspecific results in rye (Jones and Rees, 1967; Zečević and Paunović, 1969) and the grasshoppers *M. maculatus* (John and Hewitt, 1965), *Phaulacridium vittatum* (John and Freeman, 1975; Westerman and Dempsey, 1977), and *E. plorans* (Camacho *et al.*, 1980; Henriques-Gil *et al.*, 1982). The third suggestion is that increased chiasma frequency is favored in the presence of parasitic Bs, because this generates more recombinant progeny, some of which may be resistant to the accumulation of Bs in the germline (Bell and Burt, 1990). Under this hypothesis, parasitized individuals should show greater rates of recombination than unparasitized individuals in the same population, as a result of selection for genes that increase the rate of recombination only when some stimuli associated with parasite activity are detected. Such a response has been called "inducible recombination," and it is a special case of the Red Queen hypothesis, which predicts increased recombination when conditions are poor. Finally, Carlson (1994) has suggested that B effects on chiasma frequency in different species may serve the selfish purposes of each specific B. Therefore, the increase in A chromosome chiasma frequency might be a side effect, in some cases, of increased pairing between Bs to reduce B meiotic loss; in others, it would result from the blocking of B-bivalent formation; and, in still other cases, chiasma effects would result from restricting A-B pairing to facilitate B evolution from an A chromosome.

New results from the study of the grasshopper *E. plorans* have made it possible to discriminate between these different hypotheses (Camacho *et al.*, 2002). In brief, this study showed that the intensity of chiasma effects may change with the evolutionary status of the B chromosome polymorphism. In this case, chiasma frequencies were analyzed in four kinds of population—one lacking Bs, and three others harboring B chromosomes at different evolutionary stages: parasitic (B_{24} in the Torrox population), neutralized (B_2 in Salobreña), and partly neutralized (B_1 in Morocco). In each of these three populations, chiasma frequency increased with B number, but whereas a single parasitic B was enough to produce the effect, at least two neutralized Bs were necessary to produce a significant effect, and an intermediate situation was observed for the partly neutralized Bs. This demonstrated a close relationship between the intensity of drive and genomic effects in this species. A comparison among the four populations showed that chiasma frequency was highest in the population harboring the driving B_{24}, and it was progressively lower in the partly driving B_1 in Morocco, the nondriving B_2 at Salobreña, and the B-lacking population (Fig. 4.4). As discussed by Camacho *et al.* (2002), these results clearly support inducible recombination, the mechanism proposed by Bell and Burt (1990) in the light of the Red Queen hypothesis to explain chiasma effects as an adaptive response of the host to the presence of parasitic Bs. This is hypothesized to be favored by natural selection because the increase in recombination rate may yield new genotypes among the offspring, some of which could be more resistant to Bs. If changes in recombination are evoked by the parasite, then one should expect the response to be dependent on the degree of parasitism (in terms of the intensity of drive and virulence). This was precisely the pattern observed in *E. plorans*, in which drive and virulence seem to be closely associated (see also Camacho *et al.*, 2003).

Based on these results, it appears that the effect of parasitic Bs on recombination in the host genome is dependent on the intensity of B chromosome attack. The contradictory results found previously within some species could thus be explained by differences in the intensity of B parasitism among populations.

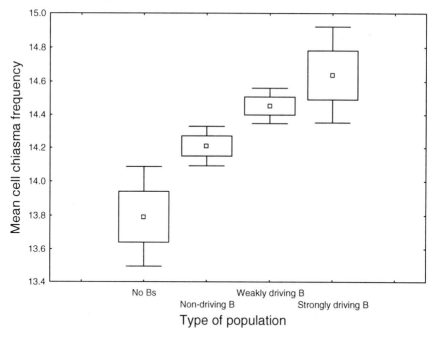

FIGURE 4.4 A-chromosome recombination (measured by chiasma frequency) in three population types harboring B chromosomes at different evolutionary stages (determined by the degree of B drive) and a population lacking them. Box = Mean ± SE, Bars = Mean ± 1.96 SE.

Of course, it is unlikely that this explanation will apply to all species, but it does provide a preliminary means to disentangle this mysterious effect. Many questions remain to be answered, such as how to explain the cases in which chiasma frequency is found to decrease in the presence of B chromosomes. Perhaps this would be expected for heterotic Bs, because a decrease in A chromosome recombination might favor the maintenance of the beneficial Bs. A possible example of such a phenomenon is provided by the ryegrass *Lolium peremne,* in which chiasma frequency decreases with increasing B numbers (Williams, 1970) and B-carrying plants show higher survival than B-lacking ones at high-density experiments (Rees and Hutchinson, 1973). However, more research is needed in this area before such an explanation can be accepted conclusively. Another aspect to be solved in the future is the mechanism by which chiasma frequency increases in the host genome—that is, the kinds of molecular signals that determine this host response to the presence of B chromosomes. Possible candidates in this regard are molecules involved in the response to other kinds of stress. It is known that recombination rate increases as the environment becomes more stressful (Hoffman and Parson, 1991). Notably, Kovalchuk *et al.* (2003) have recently shown evidence for a systemic signal increasing the frequency of homologous recombination in tobacco plants infected with either of two different viruses. The authors suggested that this signal might be an adaptive response to biotic stress, because the increase in recombination may provide new specificities in pathogen resistance genes. The dependence of the degree of chiasma frequency increase on the strength of B drive shown by Camacho *et al.* (2002) gives strong support to the "inducible recombination" hypothesis, and now Kovalchuk *et al.* (2003) have illuminated how such an adaptive response might work.

THE ODD–EVEN EFFECT

One of the most intriguing properties of the B chromosome–host interaction is the dependence of the effects on whether the Bs are present in odd or even numbers. This puzzling effect was discovered by Darlington and Upcott (1941), who found that maize plants with odd numbers of Bs have more chiasmata per B than do plants with even B numbers, although they did not describe the observation in these terms (see Jones, 1995). Jones and Rees (1967) were the first to explicitly recognize the odd–even effect by finding, in rye, that the between-cell variance in the number of chiasmata was significantly higher in plants with odd versus even numbers of B chromosomes. Remarkably, a similar odd–even effect has since been detected for properties such as dry nuclear mass, protein and RNA amounts, external morphological characters, and fitness-related traits (e.g., fertility) in a number of plants and animals (Jones and Rees, 1982). As pointed out by Jones (1995), odd numbers are, in general, more detrimental than even numbers.

In an interesting recent study, Gorlov and Tsurusaki (2000) analyzed a natural Japanese population of the harvestman *Metagagrella tenuipes* by repeated sampling for two consecutive years. They observed that average B chromosome frequency appeared to be stable, with a mean number equal to six in both years. But when individuals with odd and even numbers of B chromosomes were considered separately, seasonal changes appeared, with a sharp decrease in the frequency of individuals bearing an even number of Bs over the collection period (June to November), which was not apparent for individuals with an odd B number. The odd–even effect in this case was associated with differential susceptibility to protozoan parasites. Based on these results, the authors concluded that this B is most likely beneficial—however, they did not measure transmission rate, a parameter necessary to reaching such a conclusion. They argued against the existence of drive for this B on the basis of cytological behavior because the Bs seemed to show regular disjunction in male mitosis and the number of Bs per cell in spermatogonial mitoses was the same as in mitotic cells of the gut epithelium. Unfortunately, the data upon which these claims were based were not presented in the paper, and apparently have not been published since. However, in a previous paper by one of the same authors, it was clearly stated that "the number of B's varied considerably from one cell to the other within a single individual, possibly due to nondisjunction at anaphase during mitosis" (Tsurusaki, 1993). This discrepancy suggests that either there is something wrong with one of these datasets, or else that the B chromosome behaves differently among populations, because the one analyzed in Gorlov and Tsurusaki (2000) was different from the nine analyzed in Tsurusaki (1993). Because of this mitotic instability, Tsurusaki (1993) employed the modal number of Bs as a representative of the number of Bs in each individual. In the case of Gorlov and Tsurusaki (2000), however, the Bs seemed to be mitotically stable. The difference is very important because, for a mitotically unstable B, the odd-even effect would only make sense for effects that might be analyzed at the cytological level (e.g., chiasma frequency, chromosome behavior, etc.). Other traits, such as most of those analyzed by the authors (e.g., body size or parasitism status), can only be measured at the individual level and, for a mitotically unstable B, an individual is actually a mosaic of cells with different numbers of Bs.

Clearly this case is in need of further clarification. Still, it is worth noting that the excess of individuals with even numbers of Bs at the beginning of the season (>70%) might suggest the existence of a drive mechanism that promotes the production of individuals with even numbers of Bs. This is because an undirected binomial distribution of B frequency should involve a similar ratio of individuals with odd and even B numbers. By contrast, the high frequency of mitotic nondisjunction displayed by B chromosomes during postmeiotic gamete maturation in plants like rye and maize (see Jones, 1991) clearly produces an extreme predominance of progenies with even B numbers in 0B × 2B crosses (Chiavarino et al., 1998; Puertas et al., 1998).

New cases of the odd–even effect are continually being reported, such as the recent demonstration of an effect on chiasma frequency in the plant *Acanthophyllum laxiusculum* (Ghaffari and Bidmeshkipoor, 2002). The nature of the odd–even effect remains a mystery, although it might be related to the mechanism of B transmission or with somatic association of Bs leading to differential effects of paired and unpaired Bs (Jones and Rees, 1969). The second possibility cannot be ruled out because multiple Bs do not seem to be randomly dispersed throughout the nucleus in root-tip cells (Morais-Cecílio et al., 1997), but the possible association with B chromosome transmission is certainly an attractive option because the few available data suggest that odd–even effects are usually present in the case of B chromosomes whose drive mechanism has something to do with mitotic nondisjunction. For instance, in grasshoppers, a higher frequency of macrospermatids (a presumptively negative effect for host fertility) is often associated with odd numbers of mitotically unstable B chromosomes (see Camacho et al., 2004). This is contrary to the situation in *M. tenuipes* but is consistent with the usual pattern observed in plants.

It should be borne in mind that the odd–even effect is not always apparent. For instance, it was not found for fecundity in the plant *Aegilops speltoides* (see Cebriá *et al.*, 1994) despite the fact that B chromosomes in this species drive by directed nondisjunction at anaphase of the first pollen mitosis (Mendelson and Zohari, 1972). Likewise, a recent review in grasshoppers (Camacho *et al.*, 2004) has shown that the odd–even effect appears irregularly among traits and species.

HOST RESISTANCE

The general theme of this chapter is that B chromosomes are best viewed as parasitic elements that infect eukaryote genomes by virtue of their unfair transmission (drive) and with some damage to the host (virulence). In response, the host genome (i.e., the A chromosomes) may evolve resistance, defined in this case as the capacity to diminish the harmful effects of B chromosome infection. The primary means of resistance by the host genome is the suppression of drive, which is the only evolutionary response capable of leading to parasite clearance. However, host resistance does not necessarily involve the immediate eradication of the Bs, and varying degrees of tolerance may also develop.

Some form of resistance to B chromosomes has evolved in several species. Indirect evidence for such resistance comes from variable transmission rates for B chromosomes among individuals within the same population, which suggests a possible influence of A chromosome background on the expression of B drive. The existence of such intraspecific variability in B chromosome transmission rates has long been known in species such as rye (Müntzing, 1954), *Festuca pratensis* (Bosemark, 1954), maize (Carlson, 1969), *Myrmeleotettix maculatus* (Hewitt, 1973b), and *Hypochoeris maculata* (Parker *et al.*, 1982).

The first theoretical analysis of B chromosome drive suppression was performed by Shaw (1984) as part of his seminal discussion of the population genetics of a B chromosome polymorphism. In it, he explored the dynamics of both dominant single gene suppression and polygenic modification, and concluded that the drive suppression process may last thousands of generations. Bearing in mind that B-variants with increased drive might be produced frequently, thereby provoking the replacement of suppressed B variants, he suggested that the B chromosome (in the case of *M. maculatus*) will remain in the population indefinitely. Shaw and Hewitt (1985) also demonstrated the existence of an A chromosome gene capable of modifying B transmission rates in *M. maculatus*. These authors concluded that the modifier gene is present as a polymorphism in all *M. maculatus* populations, with the presence of the B simply increasing the equilibrium frequency at which the modifier is maintained. Simultaneously, in a survey of all available data on transmission rates in this species, Shaw *et al.* (1985) showed evidence for a genetic polymorphism defining high and low transmission types, and suggested that "the co-existence of the B with its host genome must be visualized as a dynamic one."

The genetic dissection of B chromosome drive suppression advanced another giant step in the same year with the publication of the first of a series of three papers dealing with the mealybug, *Pseudococcus affinis*. By means of artificial selection for increased and reduced numbers of B chromosomes in a series of isofemale lines, Nur and Brett (1985) demonstrated the existence of genotypes able to reduce the rate of transmission of the B chromosome. In the second study, Nur and Brett (1987) showed that the high and low transmission lines differed at two unlinked loci with additive effects on B transmission rate. The mechanism of drive suppression was then elucidated in the third paper (Nur and Brett, 1988). In male mealybugs, the set of chromosomes of paternal origin becomes heterochromatic and genetically inactive in early embryogenesis. During spermatogenesis, the two sets segregate and only the meiotic products with the euchromatic set (of maternal origin) form sperm. B chromosome drive in *P. affinis* occurs because the B mimics the euchromatic set by decondensing and thus segregating with it to the sperm. The drive suppressor genes affect the decondensation of the B so that its preferential segregation with the euchromatic set is abolished. The possibility that suppression takes place through some epigenetic mechanism, such as DNA methylation, was suggested by Nur and Brett (1988). It bears noting here that, because of the inactivation of the paternal set of chromosomes, males are genetically haploid (Brown and Nelson-Rees, 1961) and transmit only chromosomes of maternal origin, except for the Bs that are transmitted regardless of origin (Nur, 1962).

A certain parallelism exists between the mechanism of transmission for the B chromosomes in *P. affinis* and *N. vitripennis* because, in both cases, it results from escaping the elimination of the paternal set of chromosomes. But these cases differ markedly in terms of the potential response of the host genome. In *P. affinis*, B chromosomes are present in both sexes and resistance genes can evolve in the nondriving sex. In *N. vitripennis*, by contrast, B chromosomes are restricted to males, which makes the evolution of resistance unlikely because every A chromosome gene that comes into contact with the B is eliminated. This explains why A chromosome repressors of PSR action have not been found (see Werren and Stouthamer, 2003). The extreme burden levied on the host (i.e., fitness = 0), which instigates very strong selection against PSR by local extinction of all-male colonies, explains why this B does not reach high frequency in natural populations (see also Werren and Stouthamer, 2003).

In species where B chromosomes do not appear to show drive, the possibility remains that drive has been suppressed by the A genome. A way to test this possibility would be to break the genetic coadaptation by means of interpopulation crosses. Such an approach was used to show that the apparent absence of drive for the main B chromosome variants in the grasshopper *E. plorans* (López-León *et al.*, 1992) was likely owing to the existence of drive repressors (Herrera *et al.*, 1996).

In some cases, it seems that B chromosomes may indirectly affect their own transmission. By means of artificial selection experiments, Romera *et al.* (1991) and Jiménez *et al.* (1995) showed the existence of genotypes in rye displaying high or low B chromosome transmission rates. Puertas *et al.* (1998) showed that the "genes" controlling B transmission rate are actually located on the Bs, and hypothesized that such genes are in fact sites for chiasma formation in the Bs, which is logical because B chromosome drive (by postmeiotic mitotic nondisjunction) depends in this instance on regular meiotic behavior (Jiménez et al., 2000).

In maize, the genetic control of B chromosome nondisjunction (causing one of its two postmeiotic drive mechanisms) is also associated with a region in the B chromosome itself (for a recent review, see González-Sánchez et al., 2003). However, the genetic control of preferential fertilization (the other postmeiotic drive mechanism), when the Bs are transmitted on the male side, is owned by A chromosomes, as was suggested by the observation of an inbred line with the ability, as the female parent, to block the preferential fertilization of the egg by pollen carrying the Bs (Carlson, 1969). The demonstration of this fact came from artificial selection experiments and the generation of high and low transmission rate lines by Rosato et al. (1996) and the finding that B chromosome transmission, in crosses where the B is transmitted through the male side only, depends on the high or low transmission rate status of the 0B female parent (Chiavarino et al., 1998). Subsequent experiments showed that a single major gene (mBt) is involved in the control of preferential fertilization in maize (Chiavarino et al., 2001; González-Sánchez et al., 2003). In addition, a dominant gene in the A chromosomes (fBt^l) seems to favor meiotic elimination of B univalents in females (González-Sánchez et al., 2003). Additional discussions of A chromosome control of B chromosome transmission in maize and rye are provided by Puertas et al. (2000) and Puertas (2002). The existence of such a complex set of genetic interactions makes maize one of the best examples of the coevolution between A and B chromosomes.

In the grasshopper E. plorans, recent evidence has suggested that drive suppression may be quite rapid because, in the Torrox (Málaga, Spain) population, B transmission rate has decreased from 0.696 to 0.523 in only six years (Perfectti et al., 2004). Drive suppression in mitotically unstable B chromosomes might consist of the mitotic stabilization of the B chromosome. A possible example has been reported recently in the Brazilian Mogi-Guaçu population of the fish Prochilodus lineatus, in which a B chromosome showed a sharp increase in frequency paralleled by a decline in mitotic instability between the two observational periods of 1979–1980 and 1987–1989 (Cavallaro et al., 2000). In light of such observations, it is conceivable that many B chromosomes show drive only during the initial invasion, because the A genome response might be rather rapid, presumably because the suppressor genes serve other purposes and therefore already exist at low frequency in the populations (Shaw, 1984).

Host Tolerance

Tolerance is defined as the ability of the host genome to withstand B chromosome attack without suffering serious injury. The level of tolerance in this context can be described by the reaction norm of an A genotype exposed to varying numbers of B chromosomes (Camacho *et al.*, 2003). In this sense, host tolerance probably exists along a continuum from complete intolerance (fitness decreases with increasing B numbers), incomplete tolerance (fitness is unaffected at low B numbers but decreases at high B numbers), complete tolerance (fitness is independent of B number), and sometimes even to overtolerance (fitness increases with B number). Because the adverse phenotypic effects of B chromosomes are usually slight at low numbers (Jones, 1985), it seems that the most common response among host genomes is incomplete tolerance.

The evolution of more tolerant host genomes might result from selection operating on the host genome based on the very presence of B chromosomes. As pointed out by Shaw (1984), alleles in the A chromosome set that reduce the intensity of selection against individuals carrying Bs will be favored. As B chromosomes increase in frequency, the selective differential operating on the alleles favored only in B-carrying individuals will therefore increase, such that these alleles will increase in frequency toward fixation, and the fitness of B-carrying individuals will drift slowly upward over time. Shaw (1984) also noted that despite the plausibility of such a scenario, the process would almost certainly be too slow to observe. This might explain the absence in the existing literature of direct reports of the evolution of tolerance to B chromosomes. However, even when a B chromosome exhibiting significant drive in both sexes is found to exert no observable phenotypic effects (as, for example, in the locust *Locusta migratoria*) (Castro *et al.*, 1998), it is still not possible to determine whether the B has showed low virulence since the time of its first infection or whether a higher tolerance (or perhaps B attenuation, see later section) has since evolved. Interestingly, B chromosomes in rye seem to act as selective factors, causing an increase in the fertility of the 0B progeny of 4B plants (González-Sánchez *et al.*, 2004), which might explain how tolerance to Bs can arise.

INTERACTIONS WITH THE HOST ORGANISM

IMPACTS ON THE CELLULAR AND ORGANISMAL PHENOTYPES

As Jones (1995) pointed out in reference to plants, low B chromosome numbers typically do not have much observable impact on the organismal phenotype of the host, but at high numbers they may be quite harmful and lead to reduced viability and/or fertility. The same is applicable to Bs in animals. Fertility, in particular, seems to be one of the main fitness components depressed by B presence in most kinds of organisms, but there are others that range in scale from cellular to ecological effects. Some of these have been alluded to previously, but will be dealt with in more detail in this section.

As discussed in the context of genome size (see Chapters 1 and 2), increases in DNA content result in larger cell sizes. The same effect has long been observed with respect to DNA added in the form of B chromosomes (e.g., Müntzing and Akdik, 1948). The cell containing large numbers of Bs also faces the burden of having more DNA to replicate, which leads to a lengthening of cell cycle duration (see Evans et al., 1972). In the mealybug P. affinis and the grasshopper M. maculatus, this cell-level effect is expressed at the organismal level as an overall slowing of development (Nur, 1966; Hewitt and East, 1978; Harvey and Hewitt, 1979). Other cellular parameters, including nuclear protein and RNA amounts, are affected by B chromosome presence (Kirk and Jones, 1970; Ayonoadu and Rees, 1971).

Specific B chromosome effects on the external phenotype are relatively rare, but are interesting when they do occur. Examples include the plant *Haplopappus gracilis,* in which B presence changes the color of the achenes (Jackson and Newmark, 1960), and corn, in which plants with Bs develop striped leaves (Staub, 1987). To reiterate a much more extreme example from animals, the PSR B chromosome in the hymenopterans *Nasonia vitripennis* and *Trichogramma kaykai* directly influences sex determination when it transforms diploid (i.e., female) zygotes into haploid males by destroying the entire paternal chromosome set, with the notable exception of the B chromosome itself (Werren and Stouthamer, 2003). This remarkable effect of the PSR chromosome does not seem to depend on any single-copy genes, but rather on repetitive DNAs acting as sinks for binding proteins needed for normal chromosome processing in the fertilized egg (Werren and Stouthamer, 2003). Given the richness of most B chromosomes in repetitive DNA, which might generally play this kind of product sequestering role, it is conceivable that many of the functional associations and regulatory effects of B chromosomes on phenotypic traits could operate through this kind of interference with cell processes that might ultimately modify their reaction norm.

In some cases, B effects are observed at the physiological level, as for resistance to rust conferred by B presence in *Avena sativa* (Dherawattana and Sadanaga, 1973) and resistance to antibiotics attributed to the Bs in the fungus *Nectria haematococca* (Miao et al., 1991a,b). B chromosomes frequently affect processes or traits associated with vigor, fertility, and fecundity (for review, see Jones and Rees, 1982), in most cases being detrimental for individual fitness, which reinforces the parasitic nature of B chromosomes. Although it is apparent that most B chromosomes are harmful, there are some cases in which they might be beneficial and increase host fitness. The best candidates are the aforementioned Bs that increase pathogenicity of the fungus *Nectria haematococca* on its host plant and the Bs in the chive *Allium schoenoprasum* that increase viability from embryo to seedling.

B CHROMOSOMES AND HOST REPRODUCTIVE MODE

Clearly, B chromosomes have the potential to exert significant impacts on host fitness. However, this is not a one-way interaction, and certain host characteristics also have the capacity to influence the success of B chromosomes. The most notable among these is host reproductive mode.

As with any other parasite in nature, B chromosomes need a means of infection, and in this case it is sexual reproduction. Both empirical and theoretical works have shown that outcrossing facilitates the spread of parasitic B chromosomes (Burt and Trivers, 1998). In reviewing data compiled from the literature, these authors found, in a sample of 353 British plants, B chromosome presence in only 5.5% of selfers and 6.8% of species with mixed mating systems, but in a staggering 29% of obligately outcrossed species.

At first glance, there would appear to be a tendency for groups that tolerate polyploidy—which often means parthenogenetic species, at least in animals (see Chapter 8)—to display B chromosomes, especially in flowering plants (see Table 4.1). A similar pattern is found in flatworms, where B chromosomes are very common in pseudogamous parthenogenetic polyploids but rare in sexual diploids (Beukeboom *et al.*, 1998). This does not contradict the association of B frequency and outcrossing in plants, because these polyploid flatworms need sperm for reproduction, although the paternal A chromosome set is eliminated, with the exception of B chromosomes that escape expulsion from the egg (Beukeboom *et al.*, 1998). Aneuploidy is also frequent in polyploid flatworms (see Beukeboom *et al.*, 1998), which could constitute a source for frequent origin of neo-Bs in these organisms.

These particular observations aside, the superficial association with polyploidy does not have any clear statistical support. Notably, Jones and Rees (1982) found no evidence for it in their detailed discussion of B chromosomes. More recently, Trivers *et al.* (2004) found no significant relationship between B chromosome presence and whether a species is polyploid or diploid in their comparative analysis of many British plants. In fact, these latter authors reported that species with B chromosomes have, on average, slightly *lower* ploidy levels than those without Bs, although this association disappeared after correcting for differences in study effort. Interestingly, they also found that B chromosomes are more likely found in species with few A chromosomes, which might be a consequence of a more rigid control of meiosis in species with many chromosomes.

A correlation between B chromosome frequency and genome size has also been reported, at least in plants. For instance, in maize, Rosato *et al.* (1998) and Poggio *et al.* (1998) found a negative correlation between genome size and the mean number of B chromosomes in original indigenous populations, although no correlation was observed by Porter and Rayburn (1990) in Arizona populations. Trivers *et al.* (2004), in their comparative analysis, found a strong positive association between genome size and the presence of B chromosomes. As these authors

argue, the simplest explanations for this association are that relaxed selection
against large genome size also means weaker selection against B chromosomes,
and that larger genomes may donate more new B chromosomes.

POPULATION DYNAMICS

The study of population dynamics of B chromosomes has typically focused on
B chromosome changes, such as in the observed correlation between B frequency
and altitude in maize (Poggio *et al.*, 1998) and the fish *Astyanax scabripinnis* (Néo
et al., 2000), or population temperature, as in the grasshopper *M. maculatus*
(Hewitt and Brown, 1970; Hewitt and Ruscoe, 1971). These associations might
be related to the better tolerance for parasitic B chromosomes in populations with
more favorable environments for the host genome. Much less effort has been
devoted to studying changes in populations and species—that is, in the host
gene pool—that might be attributed to B chromosome presence. In some cases
this is partly obvious, as with the local extinction of populations caused by PSR
in *Nasonia* (Werren and Stouthamer, 2003). Although certainly a challenging
prospect, it would be interesting to evaluate the changes occurring in a population or species that is invaded by B chromosomes, and to investigate the consequences that adaptation to a parasitic B chromosome might have in terms of
adaptive costs with respect to other organisms or genetic elements with which the
host interacts. Some more specific questions may also be of particular interest,
such as whether (and if so, how) B chromosomes affected the domestication of
plants like maize and rye, in which they are common.

PARASITE PRUDENCE

Although they may occasionally move among species, B chromosomes are parasites of vertical transmission, which implies that their fitness is closely linked to
that of their hosts. Unlike horizontal parasites, which can infect new hosts by
being highly contagious and virulent, vertical parasites like B chromosomes might
be expected to become prudent parasites that exert a minimal detrimental effect
on their hosts. This expectation of B chromosome attenuation is indeed met in
some, but not all, cases.

A stable B chromosome frequency in natural populations might indicate such
prudence, but it could also reflect host tolerance (or some combination of the
two). For instance, neutralized B chromosomes in the grasshopper *E. plorans* commonly reach a mean number close to one per individual even after having lost
drive, a frequency that can remain approximately stable for several generations
(Camacho *et al.*, 1997). By contrast, PSR in the wasp *N. vitripennis*, which is the

most virulent B chromosome hitherto known and which reduces to zero the fitness of the host genome, reaches only low frequency (up to 6%) in the wild (Beukeboom and Werren, 2000).

Another possible evolutionary pathway for B chromosome prudence, in addition to reduced drive, is its elimination from somatic tissues. In this case, the B chromosome maintains its evolutionary viability by preserving its vertical transmission, but can prevent the most harmful effects from manifesting in the host's soma. Examples of such germline restricted B chromosomes have been reported in the marsupial mammal *Echymipera kalabu* (Hayman *et al.*, 1969) and the Japanese hagfish *Eptatretus okinoseanus* (Kubota *et al.*, 1992).

THE DYNAMICS OF B CHROMOSOME EVOLUTION

THE B CHROMOSOME LIFE CYCLE

Many B chromosome systems seem to show B frequency stability over time, presumably caused by drive-selection balance. This equilibrium is predicted by the parasitic model (Kimura and Kayano, 1961), but may be broken by the evolution of host genes able to suppress B drive (see earlier section). As suppressor genes increase in frequency in the population, B drive will tend to decrease toward values close to the Mendelian rate (0.5). This makes B destiny more dependent on genetic drift and thus on population size. Because B chromosomes are usually deleterious at high numbers, a neutralized B evolving by selection and drift has only one possible fate: extinction. Unless the B acquires regular meiotic behavior and stabilizes its number to two per diploid genome, in which case the B could integrate into the host genome, such a neutralized B is doubly condemned by the action of selection against high B numbers and drift pushing toward the only truly stable status, the absence of Bs.

The material presented in this chapter indicates that a typical B chromosome life cycle consists of three stages: (1) very rapid invasion, lasting only some tens of generations, (2) drive suppression, lasting tens to hundreds of generations (although this may be highly dependent on the number of suppressor genes and the fitness cost of suppression), and (3) drift-selection extinction, lasting thousands of generations, depending mainly on population size (Camacho *et al.*, 1997).

Thus, the coevolutionary arms race between B and A chromosomes consists of a series of mechanisms of attack and defense (Frank, 2000), in which both use their respective available weapons (Fig. 4.5). The primary weapon for a B chromosome is drive, the unfair transmission pattern that allows them to disregard Mendelian laws of inheritance (Camacho *et al.*, 2000). A chromosome defense,

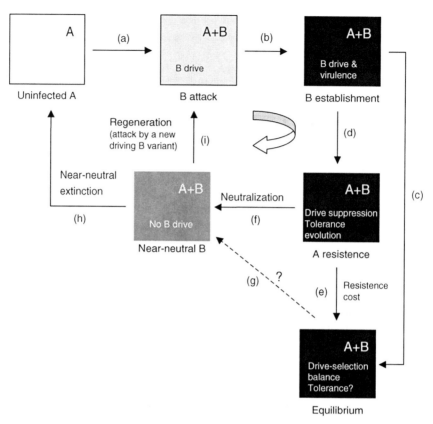

FIGURE 4.5 Evolutionary dynamics of a parasitic B chromosome begin with population invasion by a neo-B, which, by virtue of high drive, would rapidly invade the population (**a**). A beneficial B would also invade quickly but this diagram focuses on parasitic Bs only. As the driving B increases in frequency, it becomes a burden to the host genome because of both virulence (harmful effects) and the increase in B-carrying individuals (**b**). B frequency may reach stability because of drive-selection balance (**c**) or else resistance to Bs may evolve by the increase in frequency of A genes providing suppression of B drive (**d**). Evolution of tolerance to Bs is also conceivable. If resistance has a fitness cost, an equilibrium may also be reached (**e**). In cases where drive suppression shows negligible costs, the B may be neutralized (**f**), which is perhaps also conceivable to occur if the (**c**) situation evolves into drive suppression (**g**). The neutralized B may go to extinction through a near-random walk (**h**) or else may be replaced by a new B variant showing drive (**i**). The last possibility closes a cycle where the B may change from parasitic to near-neutral many times, thereby greatly prolonging the life of the B chromosome polymorphism.

on the other hand, may take place in a number of different ways, ranging from the most direct—suppression of drive—to the most weak—doing nothing and suffering directional selection on tolerance to B chromosomes—and passing through a series of reactions that provoke a multitude of changes in the host individuals, commonly known as "B chromosome effects." It is logical to expect that B chromosome effects may change over time in association with the evolutionary status of the polymorphism, because they include all manner of host responses. These may be very diverse and are dependent on genetic variation in the A chromosomes, which itself may change both in space (among populations) and time (among generations) (see earlier section regarding B effects on A recombination).

But B chromosomes may guard a "secret weapon." Because they are genetically dispensable, they are able to support a high mutation rate with no serious consequences. This high mutation rate is reflected by the large number of B variants found in many species. In the grasshopper *E. plorans,* for example, the estimated mutation rate of B chromosomes in Spain and Morocco is close to 1% (López-León *et al.,* 1993; Bakkali and Camacho, 2004). Some mutations in a parasitic B chromosome may actually lead to drive loss, with the consequent extinction of the B mutant. Conversely, when a parasitic B has been neutralized, some mutations could conceivably occur to restore its capacity for drive. Since the life of a neutralized B may be rather long, there could be ample opportunity for the emergence of mutations that reinstate its parasitic status.

Such a process of "regeneration" has occurred several times in *E. plorans,* in which a neutralized (nondriving) B chromosome is replaced by a parasitic (driving) B derived from it (Camacho *et al.,* 1997). By good fortune, this was witnessed as it occurred in a natural population of *E. plorans* in Torrox (Málaga, Spain) (Zurita *et al.,* 1998). In populations surrounding Torrox, the predominant B chromosome is named B_2, which shows two proximal blocks of a 180 bp sat-DNA and a distal block of rDNA (López-León *et al.,* 1994). In Torrox, a sample collected in 1984 included both B_2 and a second variant, B_{24} (Henriques-Gil and Arana, 1990). This B_{24} variant displays three sat-DNA blocks and a lower amount of rDNA than B_2, and was presumably derived from B_2 (Cabrero *et al.,* 1999). Both B variants showed low initial frequency (0.094 for B_2 and 0.344 for B_{24}) but, eight years later, B_2 had disappeared and B_{24} had sharply increased in frequency (0.975) by virtue of significant drive (0.696) (Zurita *et al.,* 1998). In a 1994 sample, B_{24} frequency was even higher (1.533), suggesting that this B variant was still at the invasion stage. More recent samples have shown that B_{24} is also invading nearby populations around Torrox.

Other examples of regeneration have occurred in *E. plorans.* B_1 is the most widespread B chromosome in the Iberian Peninsula, suggesting that it is the ancestral B, but it has been replaced by B_2 in the Granada and Eastern Málaga provinces, and by B_5 in the Fuengirola (Málaga) region (Henriques-Gil and Arana, 1990). These can be considered primary regenerations because the substituted

B chromosome was the ancestral one. The Torrox regeneration, however, is secondary because the substituted B chromosome, B_2, had already replaced B_1 in a primary regeneration. In light of this process of regeneration, the complete evolutionary pathway of a B chromosome polymorphism is clearly a cycle that can be completed many times, thereby greatly prolonging the life spans of B chromosomes (Camacho *et al.*, 1997).

THE LIFE SPANS OF B CHROMOSOMES

When a B chromosome invades a new population, its destiny depends largely on A chromosome response. Once the B chromosome becomes more established, it can evolve along any of several alternative pathways: (1) conserve its status as a parasitic invader, (2) integrate into the host genome and/or acquire a function, or (3) eventually disappear altogether. The life span of a B chromosome is dependent in part on which pathway it takes, but in all cases it may turn out to be quite long-lived. B chromosomes may be maintained in the population by a drive-selection balance (if parasitic), by balancing selection (if heterotic), or may simply persist fortuitously in large (meta)populations (Camacho *et al.*, 1997).

Perhaps the best evidence that B chromosomes may be long-lived is the presence of similar B chromosomes in closely related species, suggesting that these elements might even survive speciation processes. For instance, Niwa and Sakamoto (1995) suggested a monophyletic origin for the standard B chromosomes in *Secale segetale* and *S. cereale,* on the basis of the indiscriminate meiotic pairing among the Bs coming from both species in an interspecific hybrid. This is a subject in need of additional study, specifically by comparing the DNA constituents of B chromosomes in closely related species. As an example, the probes recently obtained from microdissected B chromosomes in *Apodemus peninsulae* (Trifonov *et al.*, 2002) could be useful for testing homology with Bs in other *Apodemus* species.

Other rodents, such as the black rat *Rattus rattus* and the bush rat *R. fuscipes,* harbor B chromosomes that are very similar in size, morphology, meiotic behavior, and banding patterns (Pretel and Díaz de la Guardia, 1978; Yosida, 1980; Thomson, 1984; Thomson *et al.*, 1984). The A chromosomes of these species differ by only two Robertsonian translocations (Yosida and Sagai, 1973), suggesting that they are indeed closely related. Remarkably, B chromosomes in both species also show a strong mutual resemblance in transmission patterns. In the black rat, data by Yosida (1978) and Stitou *et al.* (2004) indicated that B chromosomes show significant drive when they are transmitted through females but not through males (Table 4.3). In the bush rat, Thompson (1984) showed similar B drive through females. It has been shown that B chromosomes in Moroccan populations of the black rat contain a great deal of inactive rDNA (Stitou *et al.*, 2000). In *R. fuscipes,* Thomson *et al.* (1984) reported that most of the B chromosome

TABLE 4.3 Transmission analysis of B chromosomes in the black rat *Rattus rattus*.
Bs are slightly undertransmitted through males but significantly driven through females.
The test applied here is the Z-test used by López-León (1992), whose critical value is 1.96 for
P = 0.05, the + sign indicating B accumulation and the − sign B elimination. Notice that k_B
(B transmission rate) was high and Z was positive in all crosses where females carried Bs

Species	Author	Bs in parents		Progeny			
		Female	Male	N	Mean Bs	k_B	Z
Rattus rattus	Yosida (1978)	1	1	37	1.54	0.77	3.29
		2	0	17	1.47	0.74	1.94
		3	0	12	1.83	0.61	0.77
	Stitou *et al.* (2004)	0	1	14	0.43	0.43	−0.53
		1	0	42	0.67	0.67	2.16
		1	1	21	1.14	0.57	0.65
		0	2	20	1.10	0.55	0.45
Rattus fuscipes	Thomson (1984)	0	1	20	0.65	0.65	1.34
		1	0	31	0.87	0.87	4.13
		0	2	5	1.00	0.50	0
		2	0	11	1.27	0.64	0.90
		1	1	18	1.39	0.69	1.65
		0	2	5	1.00	0.50	0

is negatively G-banded, the typical pattern shown by rDNA. A simple FISH analysis with an rDNA probe might indicate whether these Bs do in fact share these
DNA sequences. If so, then all these resemblances between B chromosomes in
these two closely related species would suggest the possibility that the Bs are quite
old, perhaps older than the species themselves.

Although no molecular data are yet available in orthopteran insects, the size
and mitotic stability of B chromosomes in 62 species from 21 genera (Table 4.4),
as deduced from the information provided by Hewitt (1979), showed homogeneity of these B characteristics among species in 17 genera and heterogeneity in the
remaining 4, which clearly departs from chance ($\chi^2_{(1)}$ = 8.05, P = 0.0046). It is
therefore quite conceivable that many of the B chromosomes in these congeners
had common origins prior to species differentiation. The extensive dynamism of
B chromosome heterochromatin may speak against this possibility, because rapid
changes could mask or confuse B chromosome ancestry, but the high interest of
the topic merits a more thorough investigation in the future.

TABLE 4.4   Similarity among B chromosomes in two or more species of orthopteran insects for two B properties: size and mitotic stability. Homogeneous pattern means coincidence in the two properties, whereas heterogeneous pattern means difference in at least one of them

Genus	No. species	B size	B stability	Pattern
Acrida	2	small		homogeneous
Acrotylus	3	small		homogeneous
Atractomorpha	3	small	unstable	homogeneous
Chrotogonus	3	small		homogeneous
Circotettix	3	large	stable	homogeneous
Ephippiger	3		unstable	homogeneous
Euscyrtus	2	small		homogeneous
Eyprepocnemis	3	small	unstable	homogeneous
Gryllotalpa	2	small		homogeneous
Gryllus	3	small		homogeneous
Loxoblemmus	2	small		homogeneous
Melanoplus	4			heterogeneous
Oedaleonotus	2	large metacentric		homogeneous
Phaulacridium	2			heterogeneous
Podisma	3			heterogeneous
Scudderia	2	small		homogeneous
Scotussa	4			heterogeneous
Stenobothrus	2	large iso-B	stable	homogeneous
Stetophyma	2	large		homogeneous
Tetrix	2	small		homogeneous
Trimerotropis	10	large	stable	homogeneous

CAN B CHROMOSOMES BECOME BENEFICIAL?

Given that B chromosome polymorphisms may be long-lasting, there should be ample chance for a neutralized (nondriving) B to become beneficial to the host genome. Indeed, Kimura and Kayano (1961) suggested more than 40 years ago that a parasitic B chromosome can become beneficial at low numbers while still being harmful at high numbers. That said, the evidence for heterotic B chromosomes remains scarce. Again, only a few candidates are known, including B chromosomes in *Avena sativa* that confer resistance to rust (Dherawattana and Sadanaga, 1973) and those in the fungus *N. haematococca* that enhance antibiotic resistance (Miao *et al.*, 1991a,b).

The only case in which population analysis has provided evidence for adaptive Bs is in the chive *Allium schoenoprassum*. No drive has been found in this species' Bs (Bougourd and Parker, 1979; Bougourd and Plowman, 1996), but the possibility of rapid drive suppression once the Bs became frequent (perhaps by virtue of initially unnoticed drive) cannot be ruled out. In any case, B presence seems to enhance seed germination in this species, especially under drought conditions (Plowman and Bougourd, 1994), which suggests the possibility that this effect might play a role in B maintenance in natural populations.

CAN B CHROMOSOMES INTEGRATE INTO THE STANDARD GENOME?

Another potential destiny of B chromosomes, and perhaps the most interesting from the perspective of genome evolution, is their integration into the host genome. Dover and Riley (1972) were among the first to point out this possibility by suggesting that the 5B pairing control locus in wheat may have arisen through the incorporation of a critical segment of a supernumerary chromosome from *Aegilops* into the genome of hexaploid wheat. Theoretically, such an integration could follow several different pathways. First, B chromosomes could translocate to an A chromosome to become part of the heterochromatin of the standard genome, in the form of so-called "supernumerary chromosome segments" (SCSs). If these phenomena were frequent, some structural and populational relationships should be apparent between Bs and SCSs. Intriguingly, an SCS located proximally in the smallest autosome in the grasshopper *E. plorans* is composed of the same repetitive DNAs that are the main constituent of Bs in this species (Cabrero *et al.*, 2003b). Likewise, the daisy *Brachycome dichromosomatica* shows a DNA sequence (*Bds1*) in an SCS that is also present in the microBs (Houben *et al.*, 2000). In a recent study of maize, Hsu *et al.* (2003) found a tandem repeat DNA family (TR-1) in the knobs (SCSs) of A chromosomes that shows sequence homology with centromeric DNA sequences in the maize B chromosome. This suggests a common origin of Bs and knobs by some kind of genomic rearrangement. Under conditions in which such a process is common, one would expect a relationship between the frequency of B chromosomes and SCSs in natural populations. Indeed, such an association is observed in the superfamily Acridoidea of grasshoppers. Data provided by Hewitt (1979) indicated the existence of 26 species with Bs and SCSs, 717 species lacking them, 69 with Bs only, and 37 with SCSs only. As showed by López-León *et al.* (1991), there was a significant positive association between B and SCS presence ($\chi^2_{(1)} = 58.7$, $P < 0.001$), pointing to an excess of species bearing both kinds of supernumerary material.

Some possible kinds of translocation by which B chromosome material could integrate into the A chromosomes are transposition, centric fusion, and

B-A interchanges. Transposons seem to be very widespread throughout both A and B chromosomes, and could conceivably provide a vehicle for the movement of DNA sequences from B to A chromosomes and vice versa. Centric fusions seem to occur spontaneously in natural populations of E. plorans (Henriques-Gil et al., 1983; Cabrero et al., 1987), although their absence as a polymorphism suggests that they may rarely prosper. B-A interchanges represent an even more unlikely pathway for B integration because of the high gametic infertility and embryonic mortality engendered by their meiotic behavior (Bakkali et al., 2003).

A second mechanism for B chromosome integration would be the acquisition of a regular behavior during cell division. Because invasion by parasitic Bs is only possible by means of high drive, which is usually based on irregular behavior during cell division, there exists a trade-off between integration and drive, so that integration is only plausible once B frequency is high (when B drive can possibly be lost).

There is at least one case where B chromosomes seem to tend to stabilize their number to one per haploid genome: one in males and two in females in the haplodiploid wasp Trypoxylon albitarse (Araújo et al., 2001), thereby mimicking the situation in the A chromosomes. This seems to occur immediately after a rapid invasion (Araújo et al., 2002), which presumably takes place by means of drive. Haplodiploidy facilitates B regularization, which need only occur in female meiosis. It is hypothesized that the two B chromosomes in the female should pair and segregate to form almost all ovules bearing one B. But pairing itself avoids any preferential segregation of the B toward the viable meiotic pole and thus suppresses drive. The sequential action of preferential segregation of B univalents (when it is still rare in the population) and the regular segregation of B bivalents (when the B is frequent) might explain this observation.

A third pathway of B chromosome integration into the A genome would be through the acquisition of a segregation counterpart with which to pair during cell division. The best candidate is the single X chromosome in many organisms with X0 sex determination. It is well known that B chromosomes sometimes do not segregate randomly with respect to the X univalent in grasshoppers (see López-León et al., 1996). In the Jete population of E. plorans, for instance, the X and B univalents move preferentially to opposite poles during spermatogenesis. Higher regularity in X-B segregation to opposite poles is even shown by two homopteran species, suggesting "the incorporation of the B chromosome into the segregation mechanism with the X chromosome in the place occupied by the Y chromosome in species with [an] XY system" (Nokkala et al., 2000, 2003). This kind of meiotic behavior would render Bs limited to males only, which is tantalizingly suggestive of a possible origin of Y chromosomes as derived B chromosomes. This kind of B integration is among the most interesting aspects of B chromosome evolution, and has indeed been suggested to apply to the Y chromosome in Drosophila (Hackstein et al., 1996; Carvalho, 2002). If this possibility is ultimately confirmed, then B chromosomes would clearly have played an even more remarkable role than previously thought, by serving as a reserve of raw material for the evolution of important genomic novelties.

CONCLUDING REMARKS AND FUTURE PROSPECTS

Based on current estimates, it seems that B chromosomes might be present in about 10–15% of eukaryote species, making them a significant factor in genome evolution. In most cases, they are heterochromatic owing to the high content of repetitive DNA of various types, especially satellite, ribosomal, and mobile element DNA. Their origin has much to do with aneuploidy, and both intra- and interspecific origins seem to occur. The neo-B chromosomes usually degenerate by accumulation of repetitive DNA, following a parallel pathway to sex chromosomes, including the *de novo* evolution of satellite DNA from euchromatic A chromosome sequences.

As was anticipated long ago by Östergren (1945), B chromosomes are best viewed as parasites that infect eukaryote genomes by virtue of their unfair transmission (drive), which allows them to disregard Mendelian laws of inheritance and to persist despite causing some damage to the host. As with most parasites, the interactions with the host are complex and dynamic, and may change considerably over time through an arms race involving a variety of mechanisms of attack and defense on both sides. The primary weapon for a B chromosome is drive, and the primary means of resistance by the host genome is the suppression of drive. However, host resistance does not necessarily instigate the immediate eradication of the Bs, and varying degrees of tolerance may also develop. Incomplete tolerance, in which host fitness is largely unaffected at low B numbers but decreases at high B numbers, seems to be the most common outcome.

Although in general only low B numbers are tolerated, some organisms may harbor very high numbers of B chromosomes, even surpassing the number of A chromosomes. Whatever their level of representation, B chromosomes may interfere with the host at several levels of organization, and each of these represents an exciting avenue for future research. The influence of B chromosomes on A chromosome gene expression, which unfortunately remains poorly studied, represents a prime example. Likewise, the impacts of B chromosomes on the levels of populations and species have scarcely been investigated, and would also provide interesting subject matter for further research.

Additional work is also required regarding the impacts of B chromosomes on the cellular and organismal phenotypes. B effects are rarely manifested overtly in the external phenotype, and low B numbers typically do not have much observable impact on the host phenotype, but at high numbers they may be quite harmful and lead to reduced viability and/or fertility. It is conceivable that many of the functional associations and regulatory effects of B chromosomes on phenotypic traits could operate by a kind of product sequestering role of some of the repetitive DNA contained in them. Again, this is an area awaiting more intensive investigation.

Recent results in the grasshopper E. *plorans* have shown that the effect of parasitic Bs on recombination in the host genome is dependent on the intensity of B chromosome attack, which sheds new light on the contradictory results previously found within some species. Many questions remain to be answered along these lines, such as why some Bs *decrease* chiasma frequency and the mechanism(s) by which chiasma frequency in the host genome is affected in either direction. This latter issue is of particular interest, because it may involve molecular signals similar to those in operation during the response to other kinds of stress.

The few available data suggest that odd–even effects are usually present in the case of B chromosomes whose drive mechanism has something to do with mitotic nondisjunction, and future work should address this possibility.

The theory of centromeric drive (the differential ability of the oocyte poles in capturing centromeres) has recently added an interesting new component to the study of B chromosome drive, by illuminating possible weaknesses of female meiosis on which B chromosomes might set up their drive. Presumably, only a few of the extra chromosomes generated by aneuploidy would possess the ability to take advantage of such oocyte shortcomings, however, or else B chromosomes would be even more frequent in nature. Recent analyses in mammals and grasshoppers indicating that B chromosomes are more frequent in species with asymmetrical karyotypes indirectly support the concept of centromeric drive, but similar tests need to be made in other taxonomic groups.

A typical B chromosome life cycle consists of three stages: (1) very rapid invasion, lasting only some tens of generations, (2) drive suppression, lasting tens to hundreds of generations, and (3) drift-selection extinction, lasting thousands of generations, depending mainly on population size (Camacho et al., 1997). At present, the conclusion that B chromosome invasion is rapid is based primarily on the scarcity of observations of such invasions, a question that can only really be resolved by careful and continued monitoring over many years.

Thus far, the genetic dissection of B chromosome drive and drive-suppression has been performed in only a handful of species (rye, maize, and a few insects). This is, however, one of the most interesting aspects of B chromosome research, and more species should be added to the list in the future.

The presence of similar B chromosomes in closely related species suggests that these elements are long lived, perhaps even persisting through speciation events. This is a subject in need of additional study, specifically by comparing the DNA constituents of B chromosomes in closely related species. When a parasitic B has been neutralized, its long life confers on it ample opportunity to undergo mutations that restore its capacity to drive in parallel to giving it a new appearance, or even allowing it to become beneficial to the host genome. The existence of completely neutralized B chromosomes has only been shown in E. *plorans*, but it should also be examined in some species where drive has not been observed. Investigations into the reasons for maintenance of nondriving Bs, most notably potential functions, would be of significant interest as well.

(Re)integration into the host genome is a particularly interesting fate that may befall a B chromosome, which could occur by translocation to the A chromosomes (e.g., to become part of SCSs), perhaps by transposition, centric fusion, or B–A interchanges. More likely, B chromosome integration could take place via the adoption of regular behavior during cell division (when the Bs reach a high frequency and are thus not so reliant on drive for their persistence), or through the acquisition of a segregation counterpart with which to pair during meiosis (e.g., with the X chromosome in species exhibiting X0 sex determination). This latter scenario illustrates the possible link between B and Y chromosome evolution, and emphasizes the potential role of B chromosomes as sources of new genetic variation.

B chromosomes represent the first elements explicitly identified as "genomic parasites" some 60 years ago and have remained the subject of active study to the present day. Nevertheless, a great deal remains to be learned about the origins, evolutionary histories, impacts, and fates of B chromosomes. As the material presented in this chapter clearly shows, supernumerary chromosomes are common features in many eukaryotic genomes, making a resolution of these issues an integral aspect of genome biology. Perhaps most importantly, the study of B chromosomes reveals the complex coevolutionary interactions that take place not only among organisms within ecosystems, but also among genetic elements within those individual organisms.

REFERENCES

Alfenito MR, Birchler JA. 1993. Molecular characterization of a maize B-chromosome centric sequence. *Genetics* 135: 589–597.

Amos A, Dover G. 1981. The distribution of repetitive DNAs between regular and supernumerary chromosomes in species of *Glossina* (Tsetse): a two-step process in the origin of supernumeraries. *Chromosoma* 81: 673–690.

Araújo SMSR, Pompolo SG, Perfectti F, Camacho JPM. 2001. Integration of a B chromosome into the A genome of a wasp. *Proc R Soc Lond B* 268: 1127–1131.

Araújo SMSR, Pompolo SG, Perfectti F, Camacho JPM. 2002. Integration of a B chromosome into the A genome of a wasp, revisited. *Proc R Soc Lond B* 269: 1475–1478.

Ault JG. 1986. Stable versus unstable orientations of sex-chromosomes in 2 grasshopper species. *Chromosoma* 93: 298–304.

Ayonoadu U, Rees H. 1971. The effects of B-chromosomes on the nuclear phenotype in root meristems of maize. *Heredity* 27: 365–383.

Bakkali M, Camacho JPM. 2004. The B chromosome polymorphism of the grasshopper *Eyprepocnemis plorans* in North Africa: III. Mutation rate of B chromosomes. *Heredity* 92: 428–433.

Bakkali M, Cabrero J, Camacho JPM. 2003. B-A interchanges are an unlikely pathway for B chromosome integration into the standard genome. *Chromosome Res* 11: 115–123.

Bakkali M, Perfectti F, Camacho JPM. 2002. The B chromosome polymorphism of the grasshopper *Eyprepocnemis plorans* in North Africa. II. Parasitic and neutralized B_1 chromosomes. *Heredity* 88: 14–18.

Baverstock PR, Gelder M, Jahnke A. 1982. Cytogenetic studies of the Australian rodent, *Uromys caudimaculatus*, a species showing extensive heterochromatin variation. *Chromosoma* 84: 517–533.

Bell G, Burt A. 1990. B chromosomes: germ-line parasites which induce changes in host recombination. *Parasitology* 100: S19–S26.

Beukeboom LW. 1994. Bewildering Bs: an impression of the 1st B–Chromosome Conference. *Heredity* 73: 328–336.

Beukeboom LW, Werren JH. 2000. The paternal-sex-ratio (PSR) chromosome in natural populations of *Nasonia* (Hymenoptera : Chalcidoidea). *J Evol Biol* 13: 967–975.

Beukeboom LW, Seif M, Plowman AB, et al. 1998. Phenotypic fitness effects of B-chromosomes in the pseudogamous parthenogenetic planarian *Polycelis nigra*. *Heredity* 80: 594–603.

Bidau CJ, Martí DA. 2004. B chromosomes and Robertsonian fusions of *Dichroplus pratensis* (Acrididae): intraspecific support for the centromeric drive theory. *Cytogenet Genome Res* 106: 347–350.

Blunden R, Wilkes TJ, Forster JW, et al. 1993. Identification of the E3900 family, a 2nd family of rye chromosome-B specific repeated sequences. *Genome* 36: 706–711.

Boeskorov GG, Kartavtseva IV, Zagorodnyuk IV, et al. 1995. Nucleolus Organizer Regions and B chromosomes of wood mice (Mammalia, Rodentia, *Apodemus*). *Genetika* 31: 185–192.

Bosemark NO. 1954. On accessory chromosomes in *Festuca pratensis*. II. Inheritance of the standard type of accessory chromosome. *Hereditas* 40: 425–437.

Bougourd SM, Parker JS. 1979. The B chromosome system of *Allium schoenoprasum*. II. Stability, inheritance and phenotypic effects. *Chromosoma* 75: 369–383.

Bougourd SM, Plowman AB. 1996. The inheritance of B chromosomes in *Allium schoenoprasum* L. *Chromosome Res* 4: 151–158.

Bougourd SM, Plowman AB, Ponsford NR, et al. 1995. The case for unselfish B-chromosomes: evidence from *Allium schoenoprasum*. In: Brandham PE, Bennet MD eds. *Kew Chromosome Conference IV*. Kew: Royal Botanic Gardens, 21–34.

Boyes JW, Van Brink JM. 1967. Chromosomes of Syrphidae III. Karyotypes of some species in the tribes Milesiini and Myoleptini. *Chromosoma* 22: 417–455.

Brinkman JN, Sessions SK, Houben A, Green DM. 2000. Structure and evolution of supernumerary chromosomes in the Pacific giant salamander, *Dicamptodon tenebrosus*. *Chromosome Res* 8: 477–485.

Brown SW, Nelson-Rees WA. 1961. Radiation analysis of a lecanoid genetic system. *Genetics* 46: 983–1007.

Burt A, Trivers R. 1998. Selfish DNA and breeding system in flowering plants. *Proc R Soc Lond B* 265: 141–146.

Cabrero J, Camacho JPM. 1987. Population cytogenetics of *Chorthippus vagans*. 2. Reduced meiotic transmission but increased fertilization by males possessing a supernumerary chromosome. *Genome* 29: 285–291.

Cabrero J, Alché JD, Camacho JPM. 1987. Effects of B chromosomes of the grasshopper *Eyprepocnemis plorans* on nucleolar organiser regions activity. Activation of a latent NOR on a B chromosome fused to an autosome. *Genome* 29: 116–121.

Cabrero J, Bakkali M, Bugrov A, et al. 2003a. Multiregional origin of B chromosomes in the grasshopper *Eyprepocnemis plorans*. *Chromosoma* 112: 207–211.

Cabrero J, López-León MD, Bakkali M, Camacho JPM. 1999. Common origin of B chromosome variants in the grasshopper *Eyprepocnemis plorans*. *Heredity* 83: 435–439.

Cabrero J, López-León MD, Gómez R, et al. 1997. Geographical distribution of B chromosomes in the grasshopper *Eyprepocnemis plorans*, along a river basin, is mainly shaped by non-selective historical events. *Chromosome Res* 5: 194–198.

Cabrero J, Perfectti F, Gómez R, et al. 2003b. Population variation in the A chromosome distribution of satellite DNA and ribosomal DNA in the grasshopper *Eyprepocnemis plorans*. *Chromosome Res* 11: 375–381.

Cabrero J, Viseras E, Camacho JPM. 1984. The B-chromosomes of *Locusta migratoria*. I. Detection of negative correlation between mean chiasma frequency and the rate of accumulation of the B's; a reanalysis of the available data about the transmission of these B chromosomes. *Genetica* 64: 155–164.

Camacho JPM, Bakkali M, Corral JM, et al. 2002. Host recombination is dependent on the degree of parasitism. *Proc R Soc Lond B* 269: 2173–2177.

Camacho JPM, Cabrero J, López-León MD, et al. 2003. The B chromosomes of the grasshopper *Eyprepocnemis plorans* and the intragenomic conflict. *Genetica* 117: 77–84.

Camacho JPM, Carballo AR, Cabrero J. 1980. The B-chromosome system of the grasshopper *Eyprepocnemis plorans* subsp. *plorans* (Charpentier). *Chromosoma* 80: 163–166.

Camacho JPM, Díaz de la Guardia R, Ruiz Rejón M. 1981. Polysomy and supernumerary isochromosomes in the grasshopper *Omocestus bolivari* (Chopard). *Heredity* 46: 123–126.

Camacho JPM, Perfectti F, Teruel M, et al. 2004. The odd-even effect in mitotically unstable B chromosomes in grasshoppers. *Cytogenet Genome Res* 106: 325–331.

Camacho JPM, Sharbel TF, Beukeboom LW. 2000. B chromosome evolution. *Phil Trans R Soc Lond B* 355: 163–178.

Camacho JPM, Shaw MW, López–León MD, et al. 1997. Population dynamics of a selfish B chromosome neutralized by the standard genome in the grasshopper *Eyprepocnemis plorans*. *Am Nat* 149: 1030–1050.

Cano MI, Santos JL. 1989. Cytological basis of the B chromosome accumulation mechanism in the grasshopper *Heteracris littoralis* (Ramb). *Heredity* 62: 91–95.

Carlson W. 1969. Factors affecting preferential fertilization in maize. *Genetics* 62: 543–554.

Carlson WR. 1994. Crossover effects of B chromosomes may be 'selfish'. *Heredity* 72: 636–638.

Carlson WR, Roseman RR. 1992. A new property of the maize B chromosome. *Genetics* 131: 211–223.

Carvalho AB. 2002. Origin and evolution of the *Drosophila* Y chromosome. *Curr Opin Genet Dev* 12: 664–668.

Castro AJ, Perfectti F, Pardo MC, et al. 1998. No harmful effects of a selfish B chromosome on several morphological and physiological traits in *Locusta migratoria* (Orthoptera, Acrididae). *Heredity* 80: 753–759.

Cavallaro ZI, Bertollo LAC, Perfectti F, Camacho JPM. 2000. Frequency increase and mitotic stabilization of a B chromosome in the fish *Prochilodus lineatus*. *Chromosome Res* 8: 627–634.

Cebriá A, Navarro ML, Puertas MJ. 1994. Genetic control of B chromosome transmission in *Aegilops speltoides* (Poaceae). *Am J Bot* 81: 1502–1507.

Cheng YM, Lin BY. 2003. Cloning and characterization of maize B chromosome sequences derived from microdissection. *Genetics* 164: 299–310.

Chiavarino AM, González-Sánchez M, Poggio L, et al. 2001. Is maize B chromosome preferential fertilization controlled by a single gene? *Heredity* 86: 743–748.

Chiavarino AM, Rosato M, Rosi P, et al. 1998. Localization of the genes controlling B chromosome transmission rate in maize (*Zea mays* ssp. *mays*, Poaceae). *Am J Bot* 85: 1581–1585.

Chilton MD, MacCarthy BJ. 1973. DNA from maize with and without B chromosomes: a comparative study. *Genetics* 74: 605–614.

Chuma I, Tosa Y, Taga M, et al. 2003. Meiotic behaviour of a supernumerary chromosome in *Magnaporthe oryzae*. *Curr Genet* 43: 191–198.

Covert SF. 1998. Supernumerary chromosomes in filamentous fungi. *Curr Genet* 33: 311–319.

Darlington CD. 1958. *Evolution of Genetic Systems*. Edinburgh and London: Oliver and Boyd.

Darlington CD, Upcott MB. 1941. The activity of inert chromosomes in *Zea mays*. *J Genet* 41: 275–296.

Dawkins R. 1976. *The Selfish Gene*. Oxford: Oxford University Press.

Dearn JM. 1974. Phase transformation and chiasma frequency variation in locusts. II. *Locusta migratoria*. *Chromosoma* 45: 339–352.

Dhar MK, Friebe B, Koul AK, Gill BS. 2002. Origin of an apparent B chromosome by mutation, chromosome fragmentation and specific DNA sequence amplification. *Chromosoma* 111: 332–340.

Dherawattana A, Sadanaga K. 1973. Cytogenetics of a crown rust-resistant hexaploid oat with 42 + 2 fragment chromosomes. *Crop Sci* 13: 591–594.

Dimitri P, Arca B, Berghella L, Mei E. 1997. High genetic instability of heterochromatin after transposition of the LINElike I factor in *Drosophila melanogaster*. *Proc Natl Acad Sci USA* 94: 8052–8057.

Donald TM, Leach CR, Clough A, Timmis JN. 1995. Ribosomal RNA genes and the B chromosome of *Brachycome dichromosomatica*. *Heredity* 74: 556–561.

Doolittle WF, Sapienza C. 1980. Selfish genes, the phenotype paradigm and genome evolution. *Nature* 284: 601–603.

Dover GA. 1975. Heterogeneity of B chromosome DNA. No evidence for a B chromosome specific repetitive DNA correlated with B chromosome effects on meiotic pairing in Triticinae. *Chromosoma* 53: 153–173.

Dover GA, Henderson SA. 1975. No detectable satellite DNA in supernumerary chromosomes of the grasshopper *Myrmeleotettix*. *Nature* 259: 57–59.

Dover GA, Riley R. 1972. Prevention of pairing of homoeologous chromosomes of wheat by an activity of supernumerary chromosomes of *Aegilops*. *Nature* 240: 159–161.

Eickbush DG, Eickbush TH, Werren JH. 1992. Molecular characterization of repetitive DNA sequences from a B chromosome. *Chromosoma* 101: 575–583.

Enkerli J, Bhatt G, Covert SF. 1997. *Nht1*, a transposable element cloned from a dispensable chromosome in *Nectria haematococca*. *Mol Plant-Microbe Interact* 10: 742–749.

Enkerli J, Reed H, Briley A, Bhatt G, Covert SF. 2000. Physical map of a conditionally dispensable chromosome in *Nectria haematococca* mating population VI and location of chromosome breakpoints. *Genetics* 155: 1083–1094.

Evans GM, Rees H, Snell CL, Sun S. 1972. The relationship between nuclear DNA amount and the duration of the mitotic cycle. *Chromosomes Today* 3: 24–31.

Fontana PG, Vickery VR. 1973. Segregation-distortion in the B-chromosome system of *Tettigidea lateralis* (Say) (Orthoptera: Tetrigidae). *Chromosoma* 43: 75–100.

Frank SA. 2000. Polymorphism of attack and defence. *Trends Ecol Evol* 15: 167–171.

Fröst S. 1957. The inheritance of the accessory chromosomes in *Centaurea scabiosa*. *Hereditas* 43: 403–422.

Funnell DL, Matthews PS, VanEtten HD. 2002. Identification of new pisatin demethylase genes (*PDA5* and *PDA7*) in *Nectria haematococca* and non-Mendelian segregation of pisatin demethylating ability and virulence on pea due to loss of chromosomal elements. *Fungal Genet Biol* 37: 121–133.

Ghaffari SM, Bidmeshkipoor A. 2002. Presence and behaviour of B-chromosomes in *Acanthophyllum laxiusculum* (Caryophyllaceae). *Genetica* 115: 319–323.

Gibson I, Hewitt GM. 1972. Isolation of DNA from B chromosomes in grasshoppers. *Nature* 255: 67–68.

González-Sánchez M, Chiavarino M, Jiménez G, *et al.* 2004. The parasitic effects of rye B chromosomes might be beneficial in the long term. *Cytogenet Genome Res* 106: 386–393.

González-Sánchez M, González-González E, Molina F, *et al.* 2003. One gene determines maize B chromosome accumulation by preferential fertilisation; another gene(s) determines their meiotic loss. *Heredity* 90: 122–129.

Gorlov IP, Tsurusaki N. 2000. Analysis of the phenotypic effects of B chromosomes in a natural population of *Metagagrella tenuipes* (Arachnida: Opiliones). *Heredity* 84: 209–217.

Green DM. 1988. Cytogenetics of the endemic New Zealand frog, *Leiopelma hochstetteri*: extraordinary supernumerary chromosome variation and a unique sex-chromosome system. *Chromosoma* 97: 55–70.

Green DM. 1990. Muller's Ratchet and the evolution of supernumerary chromosomes. *Genome* 33: 818–824.

Gutknecht J, Sperlich D, Bachmann L. 1995. A species specific satellite DNA family of *Drosophila subsilvestris* appearing predominantly in B chromosomes. *Chromosoma* 103: 539–544.

Hackstein JHP, Hochstenbach R, Hauschteck-Jungen E, Beukeboom LW. 1996. Is the Y chromosome of *Drosophila* an evolved supernumerary chromosome? *BioEssays* 18: 317–323.

Håkansson A. 1945. Überzählige chromosomen in einer rasse von *Godetia nutans* Hiorth. *Bot Notiser* 2: 1–19.

Håkansson A. 1948. Behaviour of accessory rye chromosomes in the embryo-sac. *Hereditas* 34: 35–59.

Han YN, Liu XG, Benny U, et al. 2001. Genes determining pathogenicity to pea are clustered on a supernumerary chromosome in the fungal plant pathogen *Nectria haematococca*. *Plant J* 25: 305–314.

Harrison A, Olds-Clarke P, King SM. 1998. Identification of the *t* complex-encoded cytoplasmic dynein light chain *tctex1* in inner arm I1 supports the involvement of flagellar dyneins in meiotic drive. *J Cell Biol* 140: 1137–1147.

Harvey AW, Hewitt GM. 1979. B-chromosomes slow development in a grasshopper. *Heredity* 42: 397–401.

Hasegawa N. 1934. A cytological study on 8-chromosome rye. *Cytologia* 6: 68–77.

Hayman DL, Martin PG, Waller PF. 1969. Parallel mosaicism of supernumerary chromosomes and sex chromosomes in *Echymipera kalabu* (Marsupialia). *Chromosoma* 27: 371–380.

Henking H. 1891. Über Spermatogenese und deren Beziehung zur Eientwicklung bei *Pyrrhocoris apterus*. *Z Wiss Zool* 51: 685–736.

Henriques-Gil N, Arana P. 1990. Origin and substitution of B chromosomes in the grasshopper *Eyprepocnemis plorans*. *Evolution* 44: 747–753.

Henriques-Gil N, Arana P, Santos JL. 1983. Spontaneous translocations between B chromosomes and the normal complement in the grasshopper *Eyprepocnemis plorans*. *Chromosoma* 88: 145–148.

Henriques-Gil N, Santos JL, Arana P. 1984. Evolution of a complex polymorphism in the grasshopper *Eyprepocnemis plorans*. *Chromosoma* 89: 290–293.

Henriques-Gil N, Santos JL, Giráldez R. 1982. Genotype-dependent effect of B-chromosomes on chiasma frequency in *Eyprepocnemis plorans* (Acrididae: Orthoptera). *Genetica* 59: 223–227.

Herrera JA, López-León MD, Cabrero J. et al. 1996. Evidence for B chromosome drive suppression in the grasshopper *Eyprepocnemis plorans*. *Heredity* 76: 633–639.

Hewitt GM. 1973a. The integration of supernumerary chromosomes into the orthopteran genome. *Cold Spring Harb Symp Quant Biol* 38: 183–194.

Hewitt GM. 1973b. Variable transmission rates of a B chromosome in *Myrmeleotettix maculatus* (Thunb.). *Chromosoma* 40: 83–106.

Hewitt GM. 1976. Meiotic drive for B chromosomes in the primary oocytes of *Myrmeleotettix maculatus* (Orthoptera: Acrididae). *Chromosoma* 56: 381–391.

Hewitt GM. 1979. Grasshopper and crickets. In: John B ed. *Animal Cytogenetics, vol. 3: Insecta 1 Orthoptera*. Berlin, Stuttgart: Gebruder Borntraeger.

Hewitt GM, Brown FM. 1970. The B-chromosome system of *Myrmeleotettix maculatus* V. A steep cline in East Anglia. *Heredity* 25: 363–371.

Hewitt GM, East TM. 1978. Effects of B chromosomes on development in grasshopper embryos. *Heredity* 41: 347–356.

Hewitt GM, John B. 1967. The B-chromosome system of *Myrmeleotettix maculatus* (Thunb.) III. The statistics. *Chromosoma* 21: 140–162.

Hewitt GM, Ruscoe C. 1971. Changes in microclimate correlated with a cline for B-chromosomes in the grasshopper *Myrmeleotettix maculatus* (Thunb.) (Orthoptera: Acrididae). *J Anim Ecol* 40: 753–765.

Hoffman AA, Parson PA. 1991. *Evolutionary Genetics and Environmental Stress*. Oxford, UK: Oxford University Press.

Houben A, Kynast RG, Heim U, et al. 1996. Molecular cytogenetic characterisation of the terminal heterochromatic segment of the B chromosome of rye (*Secale cereale*). *Chromosoma* 105: 97–103.

Houben A, Leach CR, Verlin D, et al. 1997. A repetitive DNA sequence common to the different B chromosomes of the genus Brachycome. Chromosoma 106: 513–519.

Houben A, Verlin D, Leach CR, Timmis JN. 2001. The genomic complexity of micro B chromosomes of Brachycome dichromosomatica. Chromosoma 110: 451–459.

Houben A, Wanner G, Hanson L, et al. 2000. Cloning and characterisation of polymorphic heterochromatic segments of Brachycome dichromosomatica. Chromosoma 109: 206–213.

Hsiang W. 1958. Cytological studies on migratory locust hybrid, Locusta migratoria migratoria L. × Locusta migratoria malinensis Meyen. Acta Zool Sinica 10: 53–59.

Hsu FC, Wang CJ, Chen CM, et al. 2003. Molecular characterization of a family of tandemly repeated DNA sequences, TR-1, in heterochromatic knobs of maize and its relatives. Genetics 164: 1087–1097.

Itoh H. 1934. Chromosomal variation in the spermatogenesis of a grasshopper, Locusta danica. L. Jap J Genet 10: 115–134.

Jackson RC, Newmark KP. 1960. Effects of supernumerary chromosomes on production of pigment in Haplopappus gracilis. Science 132: 1316–1317.

Jamilena M, Ruiz–Rejón C, Ruiz–Rejón M. 1994. A molecular analysis of the origin of the Crepis capillaris B chromosome. J Cell Sci 107: 703–708.

Jesus CM, de Galetti PM Jr, Valentini SR, Moreira-Filho O. 2003. Molecular characterization and chromosomal localization of two families of satellite DNA in Prochilodus lineatus (Pisces, Prochilodontidae), a species with B chromosomes. Genetica 118: 25–32.

Jiménez G, Manzanero S, Puertas MJ. 2000. Relationship between pachytene synapsis, metaphase I associations, and transmission of 2B and 4B chromosomes in rye. Genome 43: 232–239.

Jiménez MM, Romera F, Gallego A, Puertas MJ. 1995. Genetic control of the rate of transmission of rye B chromosomes. II. 0B x 2B crosses. Heredity 74: 518–523.

John B, Freeman M. 1975. The cytogenetic structure of Tasmanian populations of Phaulacridium vittatum. Chromosoma 53: 283–293.

John B, Hewitt GM. 1965. The B chromosome system of Myrmeleotettix maculatus (Thunb.), I. The mechanics. Chromosoma 16: 548–578.

John UP, Leach CR, Timmis JN. 1991. A sequence specific to B chromosomes of Brachycome dichromosomatica. Genome 34: 739–744.

Jones RN. 1985. Are B chromosomes selfish? In: Cavalier-Smith T ed. The Evolution of Genome Size. London: Wiley, 397–425.

Jones RN. 1991. B-chromosome drive. Am Nat 137: 430–442.

Jones RN. 1995. Tansley review no. 85: B chromosomes in plants. New Phytol 131: 411–434.

Jones RN, Houben A. 2003. B chromosomes in plants: escapees from the A chromosome genome? Trends Plant Sci 8: 417–423.

Jones RN, Rees H. 1967. Genotypic control of chromosome behaviour in rye. XI. The influence of B chromosomes on meiosis. Heredity 22: 333–347.

Jones RN, Rees H. 1969. An anomalous variation due to B chromosomes in rye. Heredity 24: 265–271.

Jones RN, Rees H. 1982. B Chromosomes. New York: Academic Press.

Karamysheva TV, Andreenkova OV, Bochkaerev MN, et al. 2002. B chromosomes of Korean field mouse Apodemus peninsulae (Rodentia, Murinae) analysed by microdissection and FISH. Cytogenet Genome Res 96: 154–160.

Kayano H. 1957. Cytogenetic studies in Lillium callosum. III. Preferential segregation of a supernumerary chromosome in EMCs. Proc Jap Acad 33: 553–558.

Kayano H. 1971. Accumulation of B chromosomes in the germ-line of Locusta migratoria. Heredity 27: 119–123.

Kenton A. 1991. Heterochromatin accumulation, disposition and diversity in Gibasis karwinskyana (Commelinaceae). Chromosoma 100: 467–478.

Keyl HG, Hägele K. 1971. B chromosomen bei Chironomus. Chromosoma 35: 403–417.

Kidwell MG, Lisch DR. 2001. Transposable elements, parasitic DNA, and genome evolution. Evolution 55: 1–24.

Kimura M, Kayano H. 1961. The maintenance of supernumerary chromosomes in wild populations of *Lillium callosum* by preferential segregation. *Genetics* 46: 1699–1712.

King M, John B. 1980. Regularities and restrictions governing C-band variation in acridoid grasshoppers. *Chromosoma* 76: 123–150.

Kirk D, Jones RN. 1970. Nuclear genetic activity in B-chromosome rye, in terms of the quantitative interrelationships between nuclear protein, nuclear RNA and histone. *Chromosoma* 31: 241–254.

Klein AS, Eckhardt RA. 1976. The DNAs of the A and B chromosomes of the mealy bug, *Pseudococcus obscurus*. *Chromosoma* 57: 333–340.

Kovalchuk I, Kovalchuk O, Kalck V, *et al*. 2003. Pathogen-induced systemic plant signal triggers DNA rearrangements. *Nature* 423: 760–762.

Kubota S, Nakai Y, Kuroo M, Kohno S. 1992. Germ line restricted supernumerary (B) chromosomes in *Eptatretus okinoseanus*. *Cytogenet Cell Genet* 60: 224–228.

Kusano A, Staber C, Chan HYE, Ganetzky B. 2003. Closing the (Ran)GAP on segregation distortion in *Drosophila*. *BioEssays* 25: 108–115.

Kuwada Y. 1915. Ueber die chromosomenzhal von *Zea mays* L. *Bot Mag* 29: 83–89.

Langdon T, Seago C, Jones RN, *et al*. 2000. *De novo* evolution of satellite DNA on the rye B chromosome. *Genetics* 154: 869–884.

Leach CR, Donald TM, Franks TK, *et al*. 1995. Organisation and origin of a B chromosome centromeric sequence from *Brachycome dichromosomatica*. *Chromosoma* 103: 708–714.

Leigh EGJ. 1977. How does selection reconcile individual advantage with the good of the group? *Proc Natl Acad Sci USA* 74: 4542–4546.

Lespinasse R. 1973. Comportment des chromosomes surnuméraires et relation avec le taux de mortalité dans une population africaine de *Locusta migratoria migratorioides*. *Chromosoma* 44: 107–122.

Lespinasse R. 1977. Analyse de la transmission des chromosomes surnuméraires chez *Locusta migratoria migratorioides* R et F. *Chromosoma* 59: 307–322.

Lin BY, Chou HP. 1997. Physical mapping of four RAPDs in the B chromosome of maize. *Theor Appl Genet* 94: 534–538.

López-León MD, Cabrero J, Camacho JPM. 1991. Meiotic drive against an autosomal supernumerary segment promoted by the presence of a B chromosome in females of the grasshopper *Eyprepocnemis plorans*. *Chromosoma* 100: 282–287.

López-León MD, Cabrero J, Camacho JPM. 1996. Achiasmate segregation of X and B univalents in males of the grasshopper *Eyprepocnemis plorans* is independent of previous association. *Chromosome Res* 4: 43–48.

López-León MD, Cabrero J, Camacho JPM, *et al*. 1992. A widespread B chromosome polymorphism maintained without apparent drive. *Evolution* 46: 529–539.

López-León MD, Cabrero J, Pardo MC, *et al*. 1993. Generating high variability of B chromosomes in the grasshopper *Eyprepocnemis plorans*. *Heredity* 71: 352–362.

López-León MD, Neves N, Schwarzacher T, *et al*. 1994. Possible origin of a B chromosome deduced from its DNA composition using double FISH technique. *Chromosome Res* 2: 87–92.

López-León MD, Vázquez P, Hewitt GM, Camacho JPM. 1995. Cloning and sequence analysis of an extremely homogeneous tandemly repeated DNA in the grasshopper *Eyprepocnemis plorans*. *Heredity* 75: 370–375.

Lucov Z, Nur U. 1973. Accumulation of B-chromosomes by preferential segregation in females of the grasshopper *Melanoplus femur-rubrum*. *Chromosoma* 42: 289–306.

Maluszynska J, Schweizer D. 1989. Ribosomal RNA genes in B chromosomes of *Crepis capillaris* detected by non-radioactive *in situ* hybridization. *Heredity* 62: 59–65.

McAllister BF, Werren JH. 1997. Hybrid origin of a B chromosome (PSR) in the parasitic wasp *Nasonia vitripennis*. *Chromosoma* 106: 243–253.

McClung CE. 1901. Notes on the accessory chromosome. *Anat Anz* 20: 220–226.

McClung CE. 1902. The accessory chromosome—sex determinant? *Biol Bull* 3: 43–84.

McQuade LR, Hill RJ, Francis D. 1994. B-chromosome systems in the greater glider, *Petauroides volans* (Marsupialia, Pseudocheiridae) 2. Investigation of B-chromosome DNA sequences isolated by micromanipulation and PCR. *Cytogenet Cell Genet* 66: 155–161.

Mendelson D, Zohari D. 1972. Behaviour and transmission of supernumerary chromosomes in *Aegilops speltoides*. *Heredity* 29: 329–339.

Mestriner CA, Galetti PM, Valentini SR, *et al.* 2000. Structural and functional evidence that a B chromosome in the characid fish *Astyanax scabripinnis* is an isochromosome. *Heredity* 85: 1–9.

Miao VP, Covert SF, VanEtten HD. 1991a. A fungal gene for antibiotic resistance on a dispensable ('B') chromosome. *Science* 254: 1773–1776.

Miao VP, Matthews DE, VanEtten HD. 1991b. Identification and chromosomal locations of a family of cytochrome P-450 genes for pisatin detoxification in the fungus *Nectria haematococca*. *Mol Gen Genet* 226: 214–223.

Montgomery EA, Huang SM, Langley CH, Judd BH. 1991. Chromosome rearrangement by ectopic recombination in *Drosophila melanogaster*: genome structure and evolution. *Genetics* 129: 1085–1098.

Morais-Cecílio L, Delgado M, Jones RN, Viegas W. 1997. Interphase arrangement of rye B chromosomes in rye and wheat. *Chromosome Res* 5: 177–181.

Müntzing A. 1944. Cytological studies of extra fragment chromosomes in rye. I. Isofragments produced by misdivision. *Hereditas* 30: 231–248.

Müntzing A. 1948. Accessory chromosomes in *Poa alpina*. *Heredity* 2: 49–61.

Müntzing A. 1954. Cytogenetics of accessory chromosomes (B-chromosomes). Proceedings of the 9th International Congress of Genetics. *Caryologia* S6: 282–301.

Müntzing A, Akdik S. 1948. The effect on cell size of accessory chromosomes in rye. *Hereditas* 34: 248–250.

Néo DM, Moreira-Filho O, Camacho JPM. 2000. Altitudinal variation for B chromosome frequency in the characid fish *Astyanax scabripinnis*. *Heredity* 85: 136–141.

Nicklas RB. 1961. Recurrent pole-to-pole movements of the sex chromosome during prometaphase I in *Melanoplus differentialis* spermatocytes. *Chromosoma* 12: 97–115.

Niwa K, Sakamoto S. 1995. Origin of B-chromosomes in cultivated rye. *Genome* 38: 307–312.

Noda S. 1975. Achiasmate meiosis in the *Fritillaria japonica* group. I. Different modes of bivalent formation in the two sex mother cells. *Heredity* 34: 373–380.

Nokkala S, Grozeva S, Kuznetsova V, Maryanska-Nadachowska A. 2003. The origin of the achiasmatic XY sex chromosome system in *Cacopsylla peregrina* (Frst.) (Psylloidea, Homoptera). *Genetica* 119: 327–332.

Nokkala S, Kuznetsova V, Maryanska-Nadachowska A. 2000. Achiasmate segregation of a B chromosome from the X chromosome in two species of psyllids (Psylloidea, Homoptera). *Genetica* 108: 181–189.

Nur U. 1962. A supernumerary chromosome with an accumulation mechanism in the lecanoid genetic system. *Chromosoma* 13: 249–271.

Nur U. 1963. A mitotically unstable supernumerary chromosome with an accumulation mechanism in a grasshopper. *Chromosoma* 14: 407–422.

Nur U. 1966. The effect of supernumerary chromosomes on the development of mealy bugs. *Genetics* 54: 1239–1249.

Nur U. 1969. Mitotic instability leading to an accumulation of B-chromosomes in grasshoppers. *Chromosoma* 27: 1–19.

Nur U. 1977. Maintenance of a "parasitic" B chromosome in the grasshopper *Melanoplus femur–rubrum*. *Genetics* 87: 499–512.

Nur U, Brett BLH. 1985. Genotypes suppressing meiotic drive of a B chromosome in the mealy bug *Pseudococcus obscurus*. *Genetics* 110: 73–92.

Nur U, Brett BLH. 1987. Control of meiotic drive of B chromosomes in the mealy bug *Pseudococcus affinis (obscurus)*. *Genetics* 115: 499–510.

Nur U, Brett BLH. 1988. Genotypes affecting the condensation and transmission of heterochromatic B chromosomes in the mealy bug *Pseudococcus affinis*. *Chromosoma* 96: 205–212.

Nur U, Werren JH, Eickbush DG, *et al*. 1988. A selfish B chromosome that enhances its transmission by eliminating the paternal genome. *Science* 240: 512–514.

Oliver JL, Posse F, Martínez-Zapater JM, *et al*. 1982. B chromosomes and E1 isoenzyme activity in mosaic bulbs of *Scilla autumnalis*. *Chromosoma* 85: 399–403.

Orgel LE, Crick FH. 1980. Selfish DNA: the ultimate parasite. *Nature* 284: 604–607.

Östergren G. 1945. Parasitic nature of extra fragment chromosomes. *Bot Notiser* 2: 157–163.

Page BT, Wanous MK, Birchler JA. 2001. Characterization of a maize chromosome 4 centromeric sequence: Evidence for an evolutionary relationship with the B chromosome centromere. *Genetics* 159: 291–302.

Palestis BG, Burt A, Jones RN, Trivers R. 2004. B chromosomes are more frequent in mammals with acrocentric karyotypes: support for the theory of centromeric drive. *Proc R Soc Lond B (Suppl)* 271: S22–S24.

Paliwal RL, Hyde BB. 1959. The association of a single B chromosome with male sterility in *Plantago coronopus*. *Am J Bot* 46: 460–466.

Pardo MC, López-León MD, Viseras E, *et al*. 1995. Mitotic instability of B chromosomes during embryo development in *Locusta migratoria*. *Heredity* 74: 164–169.

Pardo-Manuel de Villena F, Sapienza C. 2001. Female meiosis drives karyotypic evolution in mammals. *Genetics* 159: 1179–1189.

Parker JS, Taylor S, Ainsworth CC. 1982. The B chromosome system of *Hypochoeris maculata*. III. Variation in B-chromosome transmission rates. *Chromosoma* 85: 229–310.

Peppers JA, Wiggins LE, Baker RJ. 1997. Nature of B chromosomes in the harvest mouse *Reithrodontomys megalotis* by fluorescence *in situ* hybridization (FISH). *Chromosome Res* 5: 475–479.

Perfectti F, Werren JH. 2001. The interspecific origin of B chromosomes: experimental evidence. *Evolution* 55: 1069–1073.

Perfectti F, Corral JM, Mesa JA, *et al*. 2004. Rapid suppression of drive for a parasitic B chromosome. *Cytogenet Genome Res* 106: 338–343.

Peters GB. 1981. Germ line polysomy in the grasshopper *Atractomorpha similis*. *Chromosoma* 81: 593–617.

Plowman AB, Bougourd SM. 1994. Selectively advantageous effects of B chromosomes on germination behavior in *Allium schoenoprasum* L. *Heredity* 72: 587–593.

Poggio L, Rosato M, Chiavarino AM, Naranjo CA. 1998. Genome size and environmental correlations in maize (*Zea mays* ssp. *mays*, Poaceae). *Ann Bot* 82: 107–115.

Porter HL, Rayburn AL. 1990. B-chromosome and C-band heterochromatin variation in Arizona maize populations adapted to different altitudes. *Genome* 33: 659–662.

Pretel MA, Díaz de la Guardia R. 1978. Chromosomal polymorphism caused by supernumerary chromosomes in *Rattus rattus* ssp. *frugivurus* (Rafinesque, 1814) (Rodentia, Muridae). *Experientia* 34: 325–328.

Puertas MJ. 2002. Nature and evolution of B chromosomes in plants: A non-coding but information-rich part of plant genomes. *Cytogenet Genome Res* 96: 198–205.

Puertas MJ, González-Sánchez M, Manzanero S, *et al*. 1998. Genetic control of the rate of transmission of rye B chromosomes. IV. Localization of the genes controlling B transmission rate. *Heredity* 80: 209–213.

Puertas MJ, Jiménez G, Manzanero S, *et al*. 2000. Genetic control of B chromosome transmission in maize and rye. *Chromosomes Today* 13: 79–92.

Qi ZX, Zeng H, Li XL, *et al*. 2002. The molecular characterization of maize B chromosome specific AFLPs. *Cell Res* 12: 63–68.

Raghuvanshi SS, Kumar G. 1983. No male-sterility gene on B chromosomes in *Plantago coronopus*. *Heredity* 51: 429–433.

Randolph LF. 1928. Types of supernumerary chromosomes in maize. *Anat Rec* 41: 102.

Rebollo E, Arana P. 1995. A comparative study of orientation at behaviour of univalents in living grasshopper spermatocytes. *Chromosoma* 104: 56–67.

Rebollo E, Martin S, Manzanero S, Arana P. 1998. Chromosomal strategies for adaptation to univalency. *Chromosome Res* 6: 515–531.

Rees H, Hutchinson J. 1973. Nuclear DNA variation due to B chromosomes. *Cold Spring Harb Symp Quant Biol* 38: 175–182.

Rees H, Jamieson A. 1954. A supernumerary chromosome in *Locusta*. *Nature* 173: 43–44.

Riera L, Petitpierre E, Juan C, *et al.* 2004. Evolutionary dynamics of a B-chromosome invasion in island populations of the grasshopper *Eyprepocnemis plorans*. *J Evol Biol* 17: 716–719.

Rimpau J, Flavell RB. 1975. Characterisation of rye B chromosome DNA by DNA/DNA hybridisation. *Chromosoma* 52: 207–217.

Roman H. 1948. Directed fertilisation in maize. *Proc Natl Acad Sci USA* 34: 36–42.

Romera F, Jiménez MM, Puertas MJ. 1991. Genetic control of the rate of transmission of rye B chromosomes. I. Effects in 2B x 0B crosses. *Heredity* 66: 61–65.

Rosato M, Chiavarino AM, Naranjo CA, *et al.* 1998. Genome size and numerical polymorphism for the B chromosome in races of maize (*Zea mays* ssp. *mays*, Poaceae). *Am J Bot* 85: 168–174.

Rosato M, Chiavarino AM, Puertas MJ, *et al.* 1996. Genetic control of B chromosome transmission rate in *Zea mays* ssp. *mays* (Poaceae). *Am J Bot* 83: 1107–1112.

Ruíz-Rejón M, Posse F, Oliver JL. 1980. The B chromosome system of *Scilla autumnalis* (Liliaceae): effects at the isozyme level. *Chromosoma* 79: 341–348.

Rutishauser A, Röthlisberger E. 1966. Boosting mechanism of B chromosomes in *Crepis capillaris*. *Chromosomes Today* 1: 28–30.

Sandery MJ, Forster JW, Blunden R, Jones RN. 1990. Identification of a family of repeated sequences on the rye B chromosome. *Genome* 33: 908–913.

Sannomiya M. 1974. Cytogenetic studies on natural populations of grasshoppers with special reference to B-chromosomes I. *Gonista bicolor*. *Heredity* 32: 251–265.

Santos JL, Del Cerro AL, Fernández A, Díez M. 1993. Meiotic behavior of B chromosomes in the grasshopper *Omocestus burri*. A case of drive in females. *Hereditas* 118: 139–143.

Sapre AB, Deshpande DS. 1987. Origin of B chromosomes in *Coix* L. through spontaneous interspecific hybridization. *J Hered* 78: 191–196.

Schartl M, Nanda I, Schlupp I, *et al.* 1995. Incorporation of subgenomic amounts of DNA as compensation for mutational load in a gynogenetic fish. *Nature* 373: 68–71.

Schmid M, Ziegler CG, Steinlein C, *et al.* 2002. Chromosome banding in Amphibia. XXIV. The B chromosomes of *Gastrotheca espeletia* (Anura, Hylidae). *Cytogenet Genome Res* 97: 205–218.

Sharbel TF, Green DM, Houben A. 1998. B chromosome origin in the endemic New Zealand frog *Leiopelma hochstetteri* through sex chromosome devolution. *Genome* 41: 14–22.

Sharbel TF, Pijnacker LP, Beukeboom LW. 1997. Multiple supernumerary chromosomes in the pseudogamous parthenogenetic flatworm, *Polycelis nigra*: lineage markers or remnants of genetic leakage? *Genome* 40: 850–856.

Shaw MW. 1984. The population genetics of the B–chromosome polymorphism of *Myrmeleotettix maculatus* (Orthoptera: Acrididae). *Biol J Linn Soc* 23: 77–100.

Shaw MW, Hewitt GM. 1985. The genetic control of meiotic drive acting on the B chromosome of *Myrmeleotettix maculatus* (Orthoptera: Acrididae). *Heredity* 54: 259–268.

Shaw MW, Hewitt GM, Anderson DA. 1985. Polymorphism in the rates of meiotic drive acting on the chromosome of *Myrmeleotettix maculatus*. *Heredity* 55: 61–68.

Shiflett AM, Enkerli J, Covert SF. 2002. *Nht2*, a *copia* LTR retrotransposon from a conditionally dispensable chromosome in *Nectria haematococca*. *Curr Genet* 41: 99–106.

Silva MJJ, Yonenaga-Yassuda Y. 1998. Heterogeneity and meiotic behaviour of B and sex chromosomes, banding patterns and localization of (TTAGGG)$_n$ sequences by fluorescence *in situ* hybridization in the neotropical water rat *Nectomys* (Rodentia, Cricetidae). *Chromosome Res* 6: 455–462.

Stark EA, Connerton I, Bennet ST, *et al.* 1996. Molecular analysis of the structure of the maize B-chromosome. *Chromosome Res* 4: 15–23.

Staub RW. 1987. Leaf striping correlated with the presence of B chromosomes in maize. *J Hered* 78: 71–74.

Steinemann M, Steinemann S. 1997. The enigma of Y chromosome degeneration: *TRAM*, a novel retro-transposon is preferentially located on the Neo-Y chromosome of *Drosophila miranda*. *Genetics* 145: 261–266.

Stevens NM. 1908. The chromosomes in *Diabrotica vittata, Diabrotica soror* and *Diabrotica 12-punctata*. A contribution to the literature on heterochromosomes and sex determination. *J Exp Zool* 5: 453–470.

Stitou S, Díaz de la Guardia R, Jiménez R, Burgos M. 2000. Inactive ribosomal cistrons are spread throughout the B chromosomes of *Rattus rattus* (Rodentia, Muridae). Implications for their origin and evolution. *Chromosome Res* 8: 305–311.

Stitou S, Zurita F, Díaz de la Guardia R, *et al.* 2004. Transmission analysis of B chromosomes in *Rattus rattus* from Northern Africa. *Cytogenet Genome Res* 106: 344–346.

Szczerbal I, Switonski M. 2003. B chromosomes of the Chinese raccoon dog (*Nyctereutes procyonoides procyonoides* Gray) contain inactive NOR-like sequences. *Caryologia* 56: 213–216.

Taga M, Murata M, VanEtten HD. 1999. Visualization of a conditionally dispensable chromosome in the filamentous ascomycete *Nectria haematococca* by fluorescence in situ hybridisation. *Fungal Genet Biol* 26: 169–177.

Talavera M, López-León MD, Cabrero J, Camacho JPM. 1990. Male germ line polysomy in the grasshopper *Chorthippus binotatus*: extra chromosomes are not transmitted. *Genome* 33: 384–388.

Thomson RL. 1984. B chromosomes in *Rattus fuscipes*. II. The transmission of B chromosomes to off-spring and population studies. Support for the parasitic model. *Heredity* 52: 363–372.

Thomson RL, Westerman M, Murray ND. 1984. B chromosomes in *Rattus fuscipes*. I. Mitotic and mei-otic chromosomes and the effects of B chromosomes on chiasma frequency. *Heredity* 52: 355–362.

Trifonov VA, Perelman PL, Kawada SI, *et al.* 2002. Complex structure of B chromosomes in two mammalian species: *Apodemus peninsulae* (Rodentia) and *Nyctereutes procyonoides* (Carnivora). *Chromosome Res* 10: 109–116.

Trivers R, Burt R, Palestis BG. 2004. B chromosomes and genome size in flowering plants. *Genome* 47: 1–8.

Tsurusaki N. 1993. Geographic variation of the number of B-chromosomes in *Metagagrella tenuipes* (Opiliones, Phalangiidae, Gagrellinae). *Mem Queensl Mus* 33: 659–665.

Uhl CH, Moran R. 1973. The chromosomes of *Pachyphytum* (Crassulaceae). *Am J Bot* 60: 648–656.

Van Valen L. 1973. A new evolutionary law. *Evol Theor* 1: 1–30.

Viseras E, Camacho JPM, Cano MI, Santos JL. 1990. Relationship between mitotic instability and accumulation of B chromosomes in males and females of *Locusta migratoria*. *Genome* 33: 23–29.

Volobujev VT, Timina NZ. 1980. Unusually high number of B-chromosomes and mosaicism by them in *Apodemus peninsulae* (Rodentia, Muridae). *Tsitologiya Genet* 14: 43–45.

Werren JH. 1991. The paternal sex ratio chromosome of *Nasonia*. *Am Nat* 137: 392–402.

Werren JH, Stouthamer R. 2003. PSR (paternal sex ratio) chromosomes: the ultimate selfish genetic elements. *Genetica* 117: 85–101.

Westerman M, Dempsey J. 1977. Population cytology of the genus *Phaulacridium*. VI. Seasonal changes in the frequency of the B-chromosomes in a population of *P. vittatum*. *Austr J Biol Sci* 30: 329–336.

White WJD. 1973. *Animal Cytology and Evolution*, 3rd ed. London: Cambridge University Press.

Wilkes TM, Francki MG, Langidge P, *et al.* 1995. Analysis of rye B-chromosome structure using fluo-rescence *in situ* hybridization (FISH). *Chromosome Res* 3: 466–472.

Williams P. 1970. Genetical effects of B chromosomes in *Lolium perenne*. PhD Thesis, Cardiff, Wales: University of Wales.

Wilson EB. 1906. Studies on chromosomes. V. The chromosomes of *Metapodius*. A contribution to the hypothesis of genetic continuity of chromosomes. *J Exp Zool* 6: 147–205.

Wilson EB. 1907. The supernumerary chromosomes of Hemiptera. *Science* 26: 870–871.

Wolf KW, Mertl HG, Traut W. 1991. Structure, mitotic and meiotic behavior, and stability of centromere-like elements devoid of chromosome arms in the fly *Megaselia scalaris* (Phoridae). *Chromosoma* 101: 99–108.

Wurster-Hill DH, Ward OG, Davis BH, *et al.* 1988. Fragile sites, telomeric DNA sequences, B chromosomes, and DNA content in raccoon dogs, *Nyctereutes procyonides*, with comparative notes on foxes, coyote, wolf, and raccoon. *Cytogenet Cell Genet* 49: 278–281.

Yonenaga-Yassuda Y, De Assis MDL, Kasahara S. 1992. Variability of the nucleolus organizer regions and the presence of the rDNA genes in the supernumerary chromosome of *Akodon* aff. *arviculoides* (Cricetidae, Rodentia). *Caryologia* 45: 163–174.

Yosida TH. 1978. Some genetic analysis of supernumerary chromosomes in the black rat in laboratory matings. *Proc Jap Acad* 54: Ser. B., 440–445.

Yosida TH. 1980. *Cytogenetics of the Black Rat: Karyotype Evolution and Species Differentiation.* Tokyo: University of Tokyo Press.

Yosida TH, Sagai T. 1973. Similarity of Giemsa banding patterns of chromosomes in several species of the genus *Rattus*. *Chromosoma* 41: 3–101.

Zečević L, Paunović D. 1969. The effect of B-chromosomes on chiasma frequency in wild populations of rye. *Chromosoma* 27: 198–200.

Ziegler CG, Lamatsch DK, Steinlein C, *et al.* 2003. The giant B chromosome of the cyprinid fish *Alburnus alburnus* harbours a retrotransposon-derived repetitive DNA sequence. *Chromosome Res* 11: 23–35.

Zurita S, Cabrero J, López-León MD, Camacho JPM. 1998. Polymorphism regeneration for a neutralized selfish B chromosome. *Evolution* 52: 274–277.

PART III

Duplications,
Duplications…

Small-Scale Gene Duplications

John S. Taylor and Jeroen Raes

More than 35 years ago, Susumu Ohno stated that gene duplication was the single most important factor in evolution (Ohno, 1967). This point was reiterated three years later when he proposed that without duplicated genes, many of the major evolutionary transitions of interest to humans would have been impossible. The evolution of complex organisms, he argued, required the creation of new genes by duplication, and not just the modification of existing ones (Ohno, 1970) (see also Chapters 6 and 11). Bold statements such as these, combined with his proposal that at least one whole genome duplication event facilitated the evolution of vertebrates (see Chapter 6), have made Ohno an icon in the genome evolution literature. However, discussion on the occurrence and consequences of gene duplication has a much longer, albeit often neglected, pedigree.

This chapter traces the development of ideas regarding gene (and genome) duplication over the past century, and outlines the information now emerging from detailed genomic analyses. As this discussion and those in the next several chapters show, duplications have indeed been an important factor in the evolution of genes, genomes, and species for a variety of reasons.

The Evolution of the Genome, edited by TR Gregory

THE LONG PEDIGREE OF GENE
DUPLICATION RESEARCH

EARLY CHROMOSOMAL STUDIES

The science of genetics long predates the molecular definition of genes, and most early work in this area necessarily focused on observable changes in organismal phenotypes and chromosome morphology. Despite this obviously limited resolution, many of the key ideas surrounding duplications in genetic material were developed as part of these early investigations.

In 1911, for example, Kuwada proposed that the production of many different varieties of maize (*Zea mays*) was related to an ancient chromosome duplication event. At about the same time, Tischler (1915) noticed morphological differences that were correlated with variation in chromosome number in several closely related plants. This connection between chromosomal and morphological variation, though obvious today, was novel at the time. Notably, Tischler (1915) commented that "even recently, a drawing together of these two disciplines would have been considered absurd." The work of numerous scientists dealing with the importance of chromosome number, including that of Kuwada and Tischler (which was originally published in German), was reviewed by Winge in 1917. In his review, Winge remarked that chromosomes had by that time come to be regarded as being of quite extraordinary importance, the more so since the technical improvements in microscopy had made it possible to penetrate further into their nature. Indeed, many subsequent breakthroughs in the study of genomes, including recent ones, have been linked to comparable technological advances.

As far back as 1918, Calvin Bridges addressed the issue of how gene duplication might contribute to speciation and morphological innovation, as had been proposed by Kuwada and Tischler. Bridges remarked that the main interest in duplications lay in their offering a method for evolutionary increase in the lengths of chromosomes with identical genes, which could subsequently mutate separately and diversify in their effects (quoted in Bridges, 1935). Beginning in the 1920s and 1930s, this connection between gene duplication, speciation, and morphological innovation was addressed experimentally. Blakeslee (1934), for example, used Jimson weeds (*Datura stramonium*) to study the correlation between variation in karyotypes and organismal morphologies. Specifically, a diversity of lineages with unique karyotypes was observed in nature and created in the lab, and by comparing these strains, Blakeslee was able to associate morphological change with the duplication of specific chromosome parts. In one of his experiments, the duplication of a region of the largest chromosome caused the plant capsule to be small and the leaves to be narrow, whereas duplication of a different part of the same chromosome led to a large capsule and relatively broad leaves. Blakeslee made pure-breeding lab strains with duplicated chromosome segments and he considered these to be artificial new species. "Whether nature has used such methods, we do not yet know," he remarked, "but it should be remembered that often, when man has devised a method which he has thought unique, nature has been found to have had the priority in the use of the same method."

Hermann Muller, who, like Bridges, graduated from the famous *Drosophila* genetics lab of Thomas Morgan, produced fruit flies with a small fragment of their X chromosome duplicated and inserted into Chromosome 2 (Muller, 1935). Such duplications, thought Muller, might occur in nature and may be a way of increasing gene number without the typically negative consequences of gaining or losing one or more whole chromosomes. Muller (1935) further proposed that the redundant loci produced by the duplication of chromosome parts could experience divergent mutations and eventually be regarded as unrelated genes.

In 1936, Bridges was able to attribute phenotypic variation in *Drosophila* to the duplication of a particular locus when he concluded that the Bar and Bar-double phenotypes (both of which have reduced eyes) were a consequence of rare, small-scale tandem duplication events observable in giant, multistranded ("polytene") chromosomes (Figs. 5.1 and 5.2). These phenotypes had, in fact, been attributed

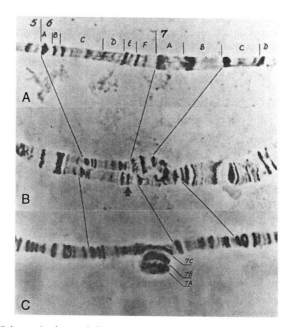

FIGURE 5.1   Polytene (multistranded) chromosomes in *Drosophila,* showing (A) the normal structure of regions 6 and 7 in the middle of the Y chromosome, (B) a heterozygote with a deletion of region 6F–7C in one of the chromosomes (indicated by the arrow), and (C) a reverse tandem duplication of region 6F–7C in the X chromosome. From Snustad *et al.* (1997), reproduced by permission (© John Wiley & Sons).

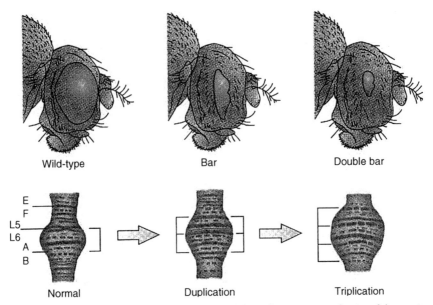

FIGURE 5.2   Effects of duplications for region 16A of the X chromosome on the size of the eyes in *Drosophila*. These phenotypic differences were attributed to duplication events by Sturtevant (1925), but polytene chromosomes were not known at that time, and it was Bridges (1936) who attributed these effects to the duplication of a particular locus. From Snustad *et al.* (1997), reproduced by permission (© John Wiley & Sons).

to duplication events much earlier by Sturtevant (1925), but resolution was poor at that time because polytene chromosomes had not yet been discovered.

Serebrovsky (1938) not only recognized that selection could be relaxed in one of the genes in a duplicate pair, but also that both copies might be modified. Having characterized the roles of *achaete* and *scute,* closely linked genes on the *Drosophila* X chromosome that influence bristle morphology, Serebrovsky concluded that a single gene might influence multiple aspects of the phenotype and that after a duplication event, these multiple functions might be distributed between the duplicates. More specifically, he suggested that the loss of function of one duplicate should result in a specialization of genes, whereby each then fulfills only one function that is strictly limited and important for the life of the organism. This is a theme that will resurface later in the chapter.

Haldane (1932) also suggested that duplication events might be favorable because they produce genes that could be altered without disadvantage to the organism. In addition, he proposed that organisms with multiple copies of genes would be less prone to harmful mutations (Haldane, 1933). This suggestion was based upon Stadler's (1929) demonstration that polyploid species in *Avena* (oats) and *Triticum* (barley) were less sensitive to irradiation than their diploid congeners. The hypothesis that duplicate genes buffer organisms from harmful mutations continues to be tested in modern times (e.g., Wagner, 2000; Gu, 2003).

Goldschmidt (1940), Gulick (1944), and Metz (1947) were among the first to make explicit links between organismal complexity and gene duplication. Goldschmidt, who rejected the notion that organisms as different as humans and amoebae were connected by simple changes in the same set of genes, focused on the role of chromosome repatterning in macroevolution. Gulick argued that the whole history of many-celled organisms must undoubtedly have called for frequent increases in gene count as overall complexity increased, and Metz argued that evolution cannot be explained upon the basis of loss or simple alteration of materials already present in the germ-plasm. As Metz (1947) put it, "New elements must be added, otherwise we would have to assume that the primordial amoeba was endowed with all the germinal components now present throughout the wide range of its descendents, from protozoa to man." Using a style later adopted by Ohno, Metz (1947) argued that the duplication of chromosome parts, together with its consequences, has probably been one of the most important factors in evolution, if not *the* most important.

Twenty years before Ohno, but clearly expanding on ideas developed two decades earlier, Stephens (1951) doubted that evolution took place by the slow accumulation of small genic mutations. Recognizing that mutations were likely to impair original gene function, Stephens proposed that the only way of achieving "evolutionary progress" would be by increasing the number of genetic loci, either by the synthesis of new loci from nongenic material or by the duplication and subsequent differentiation of existing loci via genome duplication or unequal recombination (see later section). Also in 1951, Lewis concluded that numerous traits thought to be based on allele-level variation were actually under the control of closely linked duplicates, or so-called "pseudoalleles." Interestingly, Bailey *et al.* (2002) recently concluded that many single nucleotide polymorphisms (SNPs) in human genomes are, in fact, variants at duplicated loci—that is, pseudoalleles. Lewis (1951) also proposed a model for the creation of multistep biochemical reactions by gene duplication that presaged modern interpretations of duplicate gene evolution (see later section).

In summary, between 1911 and the 1950s, cytology and cytogenetics, with major contributions from *Drosophila* and plant research, produced much evidence for gene (and whole-genome) duplication events. Amazingly, all of this predated the elucidation of the double helix and the conclusive identification of DNA as the molecular basis of heredity. Moreover, many hypotheses concerning the mechanisms by which gene duplication might contribute to important evolutionary phenomena, including speciation and increases in morphological complexity, were developed during this early phase of gene duplication research.

STUDIES AT THE PROTEIN LEVEL: EVIDENCE FOR GENE DUPLICATION AND DIVERGENCE FROM ISOZYMES

In the late 1960s, the development of starch gel electrophoresis allowed variation to be detected in proteins themselves, thereby providing much higher resolution than studies of entire chromosomes or organismal phenotypes. Duplicated loci produce more bands (regions of enzymatic activity) than single-copy genes, and many isozyme electrophoresis studies uncovered gene duplicates in polyploids and in species where no cytological data had predicted their occurrence (see Chapters 7 and 8). In several instances, isozyme studies were designed specifically to test gene and genome duplication hypotheses. For example, the discovery that some duplicated isozymes show parallel linkage in maize (Stuber and Goodman, 1983) supported Kuwada's (1911) hypothesis that this species evolved from a tetraploid ancestor. As discussed in later sections, data from isozyme research not only uncovered new examples of gene duplication, they also shed light on the consequences of these events.

THE ADVENT OF PCR AND DNA SEQUENCING

Although it led to the detection of gene and whole-genome duplication events in several unexpected cases, isozyme research has some drawbacks. For example, most surveys are restricted to a set of approximately 30 enzyme loci. Also, duplicated isozyme-coding genes can only be identified if both copies are expressed in the tissue that is surveyed and if they produce proteins with different electrophoretic mobilities. By contrast, the ability of Polymerase Chain Reaction (PCR)-based DNA sequencing to detect gene duplicates is limited only by the degree of sequence conservation among duplicates because the success of a PCR reaction depends primarily on the annealing of fairly short (e.g., 24 base pair) oligonucleotide primers to a DNA template. When conditions are used that facilitate annealing between imperfectly matched sequences (e.g., comparatively low annealing temperatures), it is not uncommon for PCR to uncover pairs or even larger assemblages of gene duplicates. Fürthauer *et al.* (1999), for example, uncovered three zebrafish *Noggin* genes using a single set of primers.

DNA sequencing has also led to a diversity of hybridization-based tools that have been used to study gene duplication. Southern blotting, for example, involves the hybridization of a DNA probe to genomic DNA that has been cut using restriction enzymes and run on an agarose gel. Colony hybrization is very similar, but involves genomic DNA targets that have been cut and cloned into bacterial vectors. As with the PCR-based approach, conditions may be used in Southern blotting and colony hybridization that facilitate binding between imperfectly matched probes and target DNAs. For example, Nornes *et al.* (1998) used part of the zebrafish *Pax9* gene as a probe to search a zebrafish DNA library (a large set of zebrafish DNA-containing bacteria) for additional *Pax* genes. Among the genes they uncovered were *Pax6* duplicates. That is, a pair of zebrafish genes that, based on subsequent phylogenetic analyses, were shown to be duplicates of human *Pax6*. Another method, known as fluorescence *in situ* hybridization (FISH), involves the hybridization of DNA sequence probes to metaphase or interphase chromosome preparations. This technique allows duplicates to be identified and their chromosomal locations to be visualized. *In situ* hybridizations involving the hybrization of complementary DNA (cDNA) probes to embryo, tissue, or cell preparations have been used to study expression domains of an enormous number of duplicated genes (see later section).

Microarray-based research is another DNA hybridization-based technology for investigating gene expression. A microarray is a glass slide with, typically, cDNA sequences spotted onto it in a grid pattern. The cDNA spots (which may number in the thousands or even tens of thousands) on the microarray slide, obtained by reverse transcription from messenger RNA (i.e., from genes that are being expressed), can come from a diversity of cell types and developmental stages. Alternatively, microarrays that contain particular sets of genes can be constructed. Microarray experiments involve the exposure of two additional sets of cDNAs to the microarray grid. For example, all of the genes that are being expressed (i.e., transcribed into mRNA) in a particular sample are exposed to the grid in the first hybridization step and those in the sample that find complementary DNA on the grid bind to it. Then, a second hybridization is carried out using all of the genes that are expressed in a second sample. The cDNAs from the two test samples (e.g., heart and liver tissue) are labeled with different color tags. By measuring the colors of the spots, it is possible to determine which genes on the grid were being expressed in both, neither, or only one of the samples, and to what degree.

Similar expression data can also be obtained using so-called "promoter plus reporter-gene transgenics." In these experiments, sequences that drive the expression of genes are fused to genes that code for glow-in-the-dark proteins such as green fluorescence protein (gfp). The location of gfp is then used to delimit the expression domains of the sequences being studied. This transgenic technology has been streamlined in the nematode worm *Caenorhabditis elegans* such that very large sets of duplicates can be compared.

Finally, widespread DNA sequencing has led to the formation of large public repositories including the National Center for Biotechnology Information (www.ncbi.nlm.nih.gov) and Ensembl (www.ensembl.org), and it has become possible to study gene duplication using an exclusively bioinformatics-based approach. For example, BLAST (Basic Local Alignment Search Tool; see www.ncbi.nlm.nih.gov/blast) can be used to uncover genes in DNA sequence databases that are homologous to—that is, share a common ancestor with—a given query sequence (Altschul *et al.*, 1990; Pearson and Wood, 2001). BLAST searches combined with phylogenetic analyses have been used to identify sets of orthologous and paralogous[1] genes and even to test hypotheses regarding whole genome duplication events (see Chapter 6). The utility of databases is not limited to sequence comparisons alone, and those dealing with gene function also can be used to characterize retained duplicates. For example, the Gene Ontology (GO) Consortium (www.geneontology.org), has created a searchable database that associates genes with a list of terms related to their biological processes, cellular components, and molecular functions.

GENE DUPLICATIONS IN THE POST-GENOMIC ERA

Interest in gene duplications and the capability to study them have only increased with the emergence of large-scale DNA sequencing. Today, it is not only possible to identify large numbers of gene duplicates, but also to examine their origins and relationships to one another. Indeed, at the broadest level, whole genome sequences now make it theoretically possible to characterize a species' entire set of duplicated genes, or its "paranome" (Friedman and Hughes, 2001). This would involve an "all-against-all" BLAST approach, whereby every gene in a genome is used as a query sequence to search a database containing the same complete set of genes of that species. This way, all paralogous relationships among genes can be determined. In addition, by estimating the time since duplication for each paralogous pair, an overview of the duplication history of that genome can be obtained (as illustrated in Fig. 5.3) and gene birth and death rates can be estimated (as discussed in a later section). More recently, the availability of complete genome sequences from sets of closely related species—so far, two yeasts, two flies, two nematode worms, and several mammals and fishes—means that comparative genomics can be used to determine whether the expansion or contraction of collections of gene duplicates occurred before or after these lineages diverged. In such studies, relationships among orthologs and paralogs are delimited using phylogenetic analyses and gene map (or location) data.

Whole genome sequencing has also allowed the scaling-up of a diversity of functional genomics technologies. By way of example, genomewide microarrays and/or large-scale studies of promoter plus reporter-gene transgenics allow gene

---

[1]Orthologous (ortho = "exact") genes are homologous (homo = "same") genes that occur in different species (i.e., are derived from a gene found in their common ancestor), whereas paralogous (para = "in parallel") genes are homologs in the same species (i.e., formed by within-genome duplication). At the level of entire chromosomes (or major segments thereof), "homologous" chromosomes are those that pair with one another during meiosis. "Homoeologous" chromosomes, on the other hand, are the equivalent chromosomes found in another species. Homoeologous chromosomes may sometimes be found within a single individual, as in polyploids formed through hybridization ("allopolyploids") (see Chapters 6, 7, and 8).

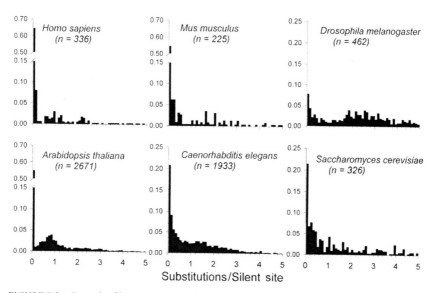

FIGURE 5.3   Example of how paranome analysis can reveal important information about both small-and large-scale duplication events. The different figures show the distribution of the time since duplication (approximated by measuring the evolutionary distance based on the amount of synonymous substitutions) for all pairwise paralogous gene comparisons. For example, while the graph of the *Caenorhabditis elegans* paranome shows a more or less exponential decay, that of *Arabidopsis thaliana* shows a sudden increase at 0.8, caused by the complete genome duplication that happened around that time (see also Chapter 6). From Lynch and Conery (2000), reproduced by permission (© American Association for the Advancement of Science).

expression to be studied at a whole-genome scale. Furthermore, large-scale cDNA cloning into expression vectors, combined with yeast two-hybrid experiments (e.g., Li *et al.,* 2004; Vidalain *et al.,* 2004), and high-throughput mass spectometric protein complex identification (Ho *et al.,* 2002) facilitate the characterization of physical interactions among proteins. Thus for a few species it is now possible to identify all related duplicates within an assemblage and to characterize their expression domains and the proteins they interact with in order to more rigorously assess the occurrence and functional consequences of gene duplication.

Despite these significant advances, some methodological problems do remain. For example, genome annotation and assembly is imprecise (e.g., paralogs can be confused for alleles). Also, BLAST cannot be relied upon to recover all members of a given set of gene duplicates, and phylogenetic analyses are time-consuming and difficult to automate. But success in finding gene duplicates in species with sequenced and assembled genomes is not limited, as it was in pregenomic studies, by such things as PCR primer design, oligonucleotide probe design, or genomic libraries that may or may not contain an entire genome. In addition, given the rapid rate of advance in genome analysis technology and methodology, there is reason to hope that these impediments may not persist for very long.

MECHANISMS OF GENE DUPLICATION

Gene duplication can occur by a variety of mutational mechanisms. Not surprisingly, the first to be observed involved chromosome-level events including the doubling of the entire chromosome complement (polyploidy; discussed in Chapters 6, 7, and 8), the loss or gain of particular chromosomes (aneuploidy), and the duplication of large chromosome parts. Other processes now recognized include "duplicative transposition" and larger-scale "tandem duplication" events such as those observable in polytene chromosomes. Even much smaller-scale and more ancient events whose physical consequences have become obscured have now become discernable with more recent technological innovations including analyses of isozymes, gene sequences, and entire genomes (see also Chapter 6). Some of the most important mechanisms of gene duplication below the level of polyploidy are outlined here.

ANEUPLOIDY

At a level below entire genomes, but still involving large collections of genes, whole chromosomes may be duplicated (or lost) through a process of "nondisjunction." Nondisjunction can occur during meiosis and involves the failure of one or more sets of homologous chromosomes to separate properly and move to opposite poles. The resulting chromosome imbalance, whether involving chromosome loss (monosomy) or gain (polysomy), often has a clear phenotypic effect, as in Blakeslee's (1934) classic study on *Datura stramonium*, where 12 different capsule morphologies were observed as the result of trisomy (three copies) of one of the 12 respective chromosomes, and in the well-known examples of Down (trisomy 21) (Fig. 5.4) and Klinefelter (XXY) (see Fig. 9.2B in Chapter 9) syndromes. Partial polysomy can also be observed, as with some individuals with Down syndrome in whom only part of Chromosome 21 occurs in triplicate (Abeliovich *et al.*, 1985).

DUPLICATIVE TRANSPOSITION

Duplicative transposition involves the duplication and movement within the genome of fragments of DNA ranging in size from 1 to more than 200 kilobases (kb) (for a recent review, see Samonte and Eichler, 2002). Human DNA fragments created by duplicative transposition, which were identified using FISH and bioinformatics tools (e.g., Horvath *et al.*, 2000), have been found to contain repetitive DNA, portions of genes (intron–exon structures), and complete genes. Telomeres and centromeres are especially receptive to such insertions; fragments copied from distant regions of the genome occur side by side at these locations.

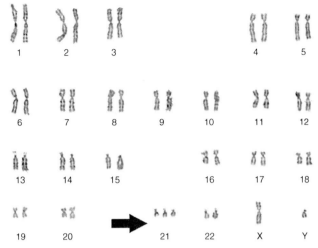

FIGURE 5.4 Aneuploidy, as illustrated by the karyotype of an individual with Down syndrome, which involves trisomy of Chromosome 21. From Bianca (2002), reproduced by permission of *Images in Paediatric Cardiology* and the author.

The most significant evolutionary consequences of duplicative transposition may be the large-scale pericentric chromosomal rearrangements that they appear to facilitate (see Chapter 9). However, there might also be gene-level consequences: some transcripts are patchworks or mosaics comprised of exons originally from distantly located genes.

Accumulating evidence also suggests that reverse transcription plays an important role in gene duplication, as evidenced by the discovery of so-called "processed pseudogenes," which are mRNAs that are transcribed into cDNAs by reverse transcriptase and reinserted in the genome (Marx, 1982; Lewin, 1983; Drouin and Dover, 1987). Like some other transposable elements (i.e., short interspersed nuclear elements, or SINEs) that lack the capacity to carry out their own reverse transcription, processed pseudogenes may be dependent on the molecular machinery of long interspersed nuclear elements (LINEs) for their transposition (e.g., Esnault *et al.*, 2000) (see also Chapters 1 and 3). Interestingly, the most common types of processed pseudogenes in the human genome are ones whose functional progenitors relate to translation (e.g., ribosomal proteins, lamin receptors, translation elongation factor alpha) (Venter *et al.*, 2001).

The absence of upstream regulatory sequences after reinsertion, the inaccuracy of the reverse transcription process, and the high probability of insertion into a genomic region that does not facilitate proper expression suggest that the fate of most processed pseudogenes is degeneration (Graur and Li, 1999). If, however, these cDNAs are inserted through homologous recombination with the original gene, the resulting exonless gene might be expressed by the original promoter and function correctly (Fink, 1987). Alternatively, if the gene is inserted near the promoter of another gene, the transcript can potentially acquire a function in a new transcriptional niche (McCarrey and Thomas, 1987; Brosius, 2003). The insertion of a gene into a retrotransposon, and its subsequent replicative transposition, might also lead to gene duplication. As an example, an *Spm/En*–like transposon containing a complete and expressed MADS-box gene was described by Montag *et al.* (1996).

LOCAL TANDEM DUPLICATION

Duplicated genes that reside next to one another are refered to as "tandem duplicates." Phylogenetic analyses can often be used to uncover the duplication events that have led to the generation of large sets of tandemly duplicated genes, and are therefore useful in testing hypotheses about the mechanisms involved (see the later section on *Hox* genes for an example). Unequal crossing-over (between homologous chromosomes at meiosis) (Fig. 5.5) and possibly unequal sister-chromatid exchange (an event occurring during the mitotic replication of a single chromosome) (e.g., Hu and Worton, 1992) appear to have played roles in the expansion of tandem arrays of gene duplicates.

Tandem duplication by unequal crossing-over was first described for the Bar locus in *Drosophila* (Sturtevant, 1925; Bridges, 1936). Today, one of the most spectacular examples of tandemly duplicated genes known is provided by those encoding ribosomal RNAs, which occur in long tandem arrays of up to thousands

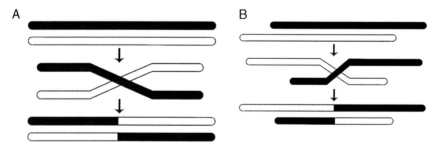

FIGURE 5.5 (A) Normal meiotic recombination, in which homologous chromosomes pair and exchange an equal amount of material, resulting in new genetic combinations. (B) Unequal crossing-over, whereby the homologous chromosomes do not pair correctly at meiosis, resulting in an uneven exchange with one chromosome gaining and one losing material. Depending on how the resulting chromosomes segregate, this could result in either a gain or loss of DNA (see also Chapter 1). This mechanism plays a major role in the generation and expansion of gene families. A similar process of unequal exchange can also occur between sister chromatids during germline mitosis.

of copies in some eukaryotic genomes, depending on the type of rRNA molecule encoded (Brown and Dawid, 1968; Cronn et al., 1996). It is also now clear that tandem duplication is an important process in protein-coding gene duplication. In fact, widespread tandem duplication followed by chromosome rearrangements can even complicate the identification of ancient genome duplications, because these duplication plus rearrangement events create what appear to be entire paralogous chromosomes (see Chapter 6).

THE LIFE AND DEATH OF GENE DUPLICATES IN THE GENOME

THE BIRTH AND DEATH OF GENE DUPLICATES

In a well-known analysis, Lynch and Conery (2000) used data from the human, mouse, chicken, fly, nematode, rice, Arabidopsis, and yeast genome sequencing projects to address a number of questions regarding the evolutionary impacts of gene duplication. Perhaps the most surprising result of their study was the estimate that a gene is as likely to be duplicated as a single nucleotide is to experience a substitution. Revised duplication rate estimates, which included data from three additional unicellular eukaryotes, mosquito, and the pufferfish Takifugu rubripes, were very similar (Lynch and Conery, 2003).

Lynch and Conery (2000) also found that humans and nematodes make new genes faster than Drosophila, Arabidopsis, and yeast. They concluded, from the pattern of nucleotide substitution and from the frequency distribution of gene ages (estimated from mutations at silent sites), that duplicates experience a brief period of relaxed selection and that most become nonfunctional very quickly. This second conclusion seemed to be inconsistent with data from tetraploid lineages (i.e., those that have experienced a recent whole-genome duplication), which retain a large proportion of duplicates (Bailey et al., 1978; Li, 1980). Lynch and Conery (2000) suggested that selection might preferentially retain duplicates produced during whole genome duplication events in order to maintain relative gene dosage. Regarding the evolutionary implications of their results, Lynch and Conery (2000) argued that the high rate of gene duplication has the potential to generate substantial molecular substrate for the origin of evolutionary novelties, but also that the window of time for such evolutionary "exploration" by gene duplicates is narrow. They concluded with the proposal that the most significant consequence of gene duplication might be speciation caused by postreproductive isolation owing to the loss of different duplicates in different populations (a process known as "reciprocal silencing" or "divergent resolution") (see Chapter 6).

In a recent study, an amino acid sequence–based BLAST-plus-phylogeny reconstruction approach was used to survey the human and pufferfish genomes for duplicated genes and to estimate the ages of the duplicates (Vandepoele *et al.*, 2004). Genes that occur once in the urochordate *Ciona intestinalis* and/or *Drosophila*, at least once in pufferfish, and from two and ten times in humans, were identified using BLAST. Phylogenetic trees were reconstructed and the ages of 447 human gene duplication events were estimated. Vandepoele *et al.* (2004) concluded that a large number of these duplicates (360) were formed prior to the divergence of actinopterygians (ray-finned fishes) and sarcopterygians (lobe-finned fishes and their tetrapod relatives). The remaining 87 nodes (174 human paralogs) were formed relatively recently; that is, during the past 50 million years. Most of these young human duplicates were found to be linked and, therefore, probably formed by tandem duplication events.

Recently duplicated human genes (i.e., those with five or fewer synonymous substitutions per 100 synonymous positions) tend to be shorter than older paralogous pairs (those with 34 to 74 synonymous substitutions per 100 synonymous sites) (Nembaware *et al.*, 2002). Nembaware *et al.* (2002) proposed that the probability of a gene being duplicated is correlated with its length, and that gene length data might be useful for settling gene duplication versus whole genome duplication debates because this short gene bias would not be expected in duplicates produced by tetraploidy.

In an effort to determine whether fast- or slow-evolving genes were better at producing duplicates (but not to determine how duplication itself influences rates of molecular evolution) (see later section), Davis and Petrov (2004) compared evolutionary rates at nonsynonymous sites in genes that had been duplicated in the yeast *Saccharomyces cerevisiae* and the nematode *C. elegans* to genes that occurred only once in these species. Each gene from yeast and nematode worm was placed into either the duplicate category or the singleton category according to copy number, and then the rates of nonsynonymous substitutions for orthologs of these two sets of genes were measured in the dipteran insects *Drosophila melanogaster* and *Anopheles gambiae*. The resulting observation was that dipteran orthologs of yeast and worm duplicates have much slower rates of evolution than dipteran orthologs of yeast and worm singletons. That is to say, genes exhibiting low rates of molecular evolution appear more likely to be retained in duplicate. The explanation that seemed most plausible to Davis and Petrov (2004) involved gene expression levels. In yeast, the slowly evolving genes that had been preferentially retained in duplicate appeared to be the genes that were most highly expressed.

THE EVOLUTION OF GENE FAMILIES

By the process of duplication and divergence, certain genes may come to exist as "families" of repeated copies (Dayhoff, 1974; Zuckerkandl, 1975; Ohta, 1990).

In some cases, these are of obvious importance for organismal fitness. Two well-known examples from human medicine include the hemoglobin and immuno-globulin gene families.

Hemoglobin

Hemoglobin, the oxygen-carrying molecule of the blood, is made up of four peptides: two α-globins and two β-globins. There are several different types of α-globins and β-globins (each encoded by a different gene), and at different periods of human development, hemoglobin is comprised of different combinations of these globins. The observation that fetal hemoglobin, for example, has a higher affinity for oxygen than adult hemoglobin, shows that globin gene duplication and divergence has resulted in the production of developmentally specialized hemo-globin. Interestingly, there has been some discussion regarding the reactivation of fetal hemoglobins in adults as a therapeutic treatment for sickle-cell anemia and β-thalassemia (Olivieri and Weatherall, 1998). In humans, the α-globin genes all occur on Chromosome 16 and the β-globins occur on Chromosome 11. These genes are distantly related to a single myoglobin gene on Chromosome 22, showing that a diversity of duplication events led to the structure of this globin superfamily (Fig. 5.6). This system also illustrates that divergence is not always an option for duplicates, given that the loss of one of the two nearly identical α-globin genes leads to α-thalassemia-2 syndrome (Kan *et al.*, 1975).

Other species appear to have exploited globin duplicates in unique ways. High-flying birds, for example, have hemoglobins with especially high affinities for oxygen (see Li, 1997). As a notable example, Rüppell's griffons (*Gyps rueppellii*) have been observed flying more than 10 kilometers above sea level and, unlike other birds, have four hemoglobins circulating in their blood. All four have the same β chain but differ in the α chain, implying the existence of four α loci. Furthermore, the β chain is identical to that found in low-altitude fliers, suggesting that this species' ability to respire at low-oxygen altitudes is due to novel muta-tions in the duplicated α-chain genes (Hiebl *et al.*, 1989).

Immunoglobulins

Humans are able to make antibodies, which are molecules comprised of proteins called immunoglobulins, against millions of potential antigens. Gene duplication plays an important role in this almost unlimited antigen response. Just as a freight train with 100 different cars can be assembled into 100 factorial ($100 \times 99 \times 98 \times 97$, etc.) different 100-car patterns, and an even larger number if train length is allowed to vary, a relatively small number of immunoglobulin genes can be assembled into an effectively infinite number of different patterns or antibodies.

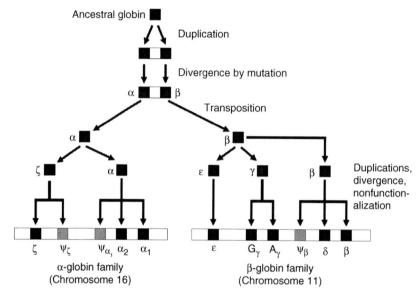

FIGURE 5.6   Duplication and divergence in the hemoglobin gene family. Hemoglobin, which transports oxygen in the blood, is a tetrameric molecule, composed of two α and two β chains. The single-gene ancestors of the α and β gene families were produced by gene duplication approximately 450 million years ago. As a result of subsequent tandem duplication events, mutations, and rearrangements, the human α family now consists of three functional genes (ξ, $α_2$, $α_1$) and two pseudogenes ($ψ_ξ$ and $ψ α_1$), all on Chromosome 16. The human β family includes five functional genes (ε, Gγ, Aγ, δ, β) and one pseudogene ($ψ_β$), all on Chromosome 11. The more distantly related myoglobin gene lies on Chromosome 22. Embryonic ($ξ_2ε_2$, $α_2ε_2$, $ξ_2γ_2$) and fetal ($α_2γ_2$, $α_2β_2$) hemoglobins are composed of different sets of α and β chains from adult ($α_2β_2$, $α_2δ_2$) hemoglobin and have higher affinities for oxygen than adult hemoglobin. Thus, duplication and divergence has led to the production of molecules that are specialized for oxygen transport at different stages of human development. Figure by T.R. Gregory, based on information in the National Center for Biotechnology Information (NCBI) LocusLink database entry for human hemoglobin genes (www.ncbi.nlm.nih.gov/LocusLink/LocRpt.cgi?l=3039 and www.ncbi.nlm.nih.gov/LocusLink/LocRpt.cgi?l=64162).

In the human genome, the "freight train" includes between 149 and 168 functional immunoglobulin genes in four classes: variable sequences (V), leader (or D) sequences, joiner sequences (J), and constant sequences (C) (see the Immunogenetics Gene Database at http://imgt.cines.fr/home.html). In mammalian genomes, the V, D, and J genes are linked, with multiple V genes next to multiple D genes next to multiple J genes. These sets of linked genes are rearranged and assembled by a process called "somatic recombination" in order to produce an enormous number of different antibodies (Tonegawa, 1983).

The Contribution of Gene Duplication to Genome Structure

Genome sequencing projects show that large-scale gene duplication and complete genome duplication events have contributed to gene family expansion and to genome evolution in a great diversity of species. In the pathogenic bacterium *Mycoplasma pneumoniae*, for example, more than 28% of the genome appears to have been produced by lineage-specific duplication events involving about four genes at a time (Jordan *et al.*, 2001). In another bacterial pathogen, *Mycobacterium tuberculosis*, more than 33% of the genome is comprised of recently duplicated genes, but in this species some large clusters of between 20 and 90 genes are also involved (Jordan *et al.*, 2001). Gevers *et al.* (2004) analyzed 106 bacterial genomes in a study designed to determine the extent to which gene duplication contributes to genome structure and to expose strain-specific gene family expansions. Paralogous genes were defined by within-genome all-against-all BLAST surveys. From 7 to 41% of the genes in the genomes surveyed had intragenomic BLAST hits, and the size of a species' or strain's genome was strongly correlated with the number of paralogous genes it contained (see Chapter 10 for a discussion of bacterial genome size evolution).

In the eukaryotic parasite *Plasmodium falciparum*, which is responsible for hundreds of millions of cases of human malaria, and in other *Plasmodium* species, duplication events have produced unlinked duplicated sets or "units" of ribosomal RNA genes (Li *et al.*, 1994; Gardner *et al.*, 2002). These paralogous sets of genes differ mainly in regions that are also variable among other eukaryotic rRNAs. "S-type" rRNA genes are expressed when the parasite is in the mosquito vector and "A-type" rRNA genes are expressed when the parasite occurs in the human host (Waters, 1994). Thus one consequence of gene duplication for *Plasmodium* appears to be different ribosomes for different environments. However, currently there are no clear explanations for these gene expression observations, and in *Plasmodium vivax*, differential expression of rRNA genes correlates better with parasite developmental stage than with host environment (Li *et al.*, 1994).

Detailed analyses show that recent gene duplication events have also played a large role in shaping the human genome. During the sequencing of the human genome, each nucleotide was sequenced approximately five times (Venter *et al.*, 2001). Thus a fragment of DNA that occurs once in human, when used as a BLAST query, would be expected to find, on average, five identical matches in the whole-genome shotgun reads. Recently duplicated sequences would be expected to yield an increase in the number of hits, but with a small decrease in average sequence identity among these hits. Bailey *et al.* (2002) surveyed raw data from the human genome sequencing project (27.3 million reads) using 32,610 clones as queries. This study identified 8595 duplicated regions (defined as sequences having >94% sequence identity over 5000 base pairs) and concluded that 130.5 megabases of the human genome had been recently duplicated. Thus many positions that have been considered to be variable, that is, sites of single nucleotide polymorphisms (SNPs), are in fact paralogous positions/loci. This result is consistent with high estimates for the rate of gene duplication in humans (Lynch and Conery, 2000). From a practical standpoint, it also shows that the algorithms used to assemble the genomes must be fine-tuned so that they do not mistake recently duplicated regions as sequencing errors.

WHAT HAPPENS TO DUPLICATED GENES?

Nonfunctionalization

The fate awaiting most genes appears to be silencing or loss owing to degenerative mutations in coding and/or regulatory modules (Nei, 1969; Li, 1980; Lynch and Conery, 2000) (Fig. 5.7). Lynch and Conery (2000) concluded, for example, that most duplicates become nonfunctional (i.e., become pseudogenes[2]) by the time silent sites have diverged by only a few percent. As noted in Chapter 1, the original definition of "junk DNA" referred to pseudogenes, in particular "classical pseudogenes" (i.e., those produced by the loss of function in a direct genomic duplicate), rather than "processed pseudogenes" (a reinserted, intronless mRNA sequence). While "junk DNA," strictly defined, does not account for the massive variation in genome size among eukaryote species (see Chapters 1 and 2), pseudogenes are nevertheless well represented in many genomes. For example, a recent estimate based on Chromosomes 21 and 22 suggests that the human genome may contain about 9000 processed and 10,000 classical pseudogenes (Harrison et al., 2002). Interestingly, a large portion of the pseudogenes on these chromosomes are located in "pseudogenic hotspots" near the centromeres (Harrison et al., 2002).

Those That Beat the Odds

Although the most common fate of gene duplicates may be loss or pseudogenization, there are clearly important exceptions. With the availability of whole genome sequences, it is possible to characterize these genes that have beaten the odds.

In their survey of 106 bacterial genomes, Gevers et al. (2004) found that a large proportion of retained duplicates were ABC-type transporters, transcription

---

[2]Pseudogenes are indicated by prefixing the name of the relevant functional gene with the Greek letter psi, ψ. For example, the processed Adh pseudogene in Drosophila is labeled as ψAdh.

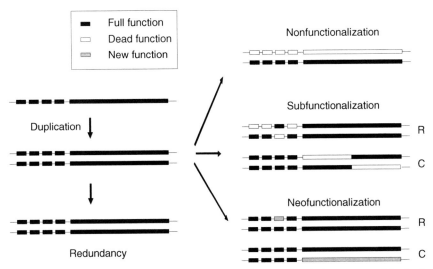

FIGURE 5.7   Overview of the different fates of duplicated genes. After duplication, one of the duplicates can accumulate many degenerative mutations and become nonfunctionalized (i.e., become a
pseudogene). Alternatively, complementary degenerative mutations at the regulatory (R) or coding (C)
levels may cause the duplicates to specialize such that each performs a different subset of the original
functions of the ancestral gene ("subfunctionalization"). Another possibility is that genes can acquire
new functions ("neofunctionalization") based on changes at either the regulatory (R) or coding (C)
levels. Finally, it is also possible that the duplicated genes do not change in function and simply
remain redundant. Of course, combinations of these different processes can occur. Based on a figure
presented by Force *et al.* (1999), reproduced by permission (© Genetics Society of America).

factors, and dehydrogenases. Species with unique gene expansion patterns included
Borrelia burgdorferi (the spirochetal bacterium that causes Lyme disease) with an
excess, relative to other species surveyed, of motility and chemotaxis genes. *Bacteroides
thetaiotaomicron* (a dominant member of the human intestinal microbiota) had the
largest gene family among the species surveyed. Gevers *et al.* (2004) suggested that this
species' 77 outer membrane proteins are involved in nutrient binding.

Conant and Wagner (2002) also investigated functional biases in retained
duplicates in the fully sequenced genomes of the yeasts *S. cerevisiae* and
Schizosaccharomyces pombe, the fly *D. melanogaster*, the nematode *C. elegans*, and
the bacterium *Escherichia coli*. They began by assigning genes to functional categories using the Gene Ontology database. Then they used a BLASTp-based approach
to determine whether each gene was present as a single copy, had one paralog, or
possessed multiple paralogs. Next they asked whether genes with one or more paralogs were over- or underrepresented in each functional class. The ribosomal protein
gene category had few genes with multiple paralogs in all species except *S. pombe*,
suggesting that there was selection against the retention of duplicated ribosomal
proteins. For both yeast species, ribosomal protein genes often had one paralog,

an observation thought to be a consequence of polyploidy. That is, when riboso-mal proteins are duplicated as part of a whole genome duplication event, both copies are retained. Seoighe and Wolfe (1999) had also noted this overrepre-sentation of ribosomal proteins in their analysis of the genes retained after the *S. cerevisiae* genome duplication. In addition, they reported an excess of cyclins and protein classes linked to signal transduction.

Several other biases in the proportion of duplicated genes in functional categories were detected by Conant and Wagner (2002), including an overrepresentation of protein kinases in *D. melanogaster* and histone genes in *C. elegans*. Another interesting observation reported by Conant and Wagner (2002) was that in the yeasts, members of large gene families had a higher proportion of amino acid substitutions than genes in smaller families. This observation is consistent with the suggestion made 70 years earlier by Haldane (1933) that gene duplication may buffer against the effect of harmful mutations.

The duplicated regions of the human genome uncovered by Bailey *et al.* (2002) were defined by size, not gene content. When they looked at the genes within the recently duplicated blocks, they found that some genes were more likely to be duplicated than others. For example, Bailey *et al.* (2002) found that genes associ-ated with immunity and defense, membrane surface interactions, drug detoxifica-tion, and growth and development were particularly common within the recently duplicated segments. In terms of broader comparisons, Jordan *et al.* (2004) pointed out that genes involved in signaling and related cellular processes tend to be commonly found as duplicates in various eukaryotes, whereas genes with poorly characterized functions are much more likely to be present in only one copy.

It is important to note that it is not only the type of gene that can influence the preponderance of duplication, but also the species in which a given gene family finds itself. Indeed, the number of duplicates within a gene family can differ substantially even among closely related organisms. As a prime example, there are 718 putative chemoreceptor proteins in the nematode *C. elegans,* but only 429 in *C. briggsae.* Likewise, whereas 243 cyclin-like F-box genes are found in *C. elegans,* only 98 are found in *C. briggsae* (Stein *et al.,* 2003).

SEQUENCE DIVERGENCE

According to many of the models proposed since the 1930s, gene duplication is followed by a period of relaxed selection, at least for one of the duplicates. It therefore follows that the analysis of changes in rates of molecular evolution after duplication should provide a tool for studying functional divergence of duplicated genes. Several studies have investigated rate variation between duplicates and their single-copy orthologs, or "pro-orthologs" (Sharman, 1999). These studies have provided variable results, ranging from providing no (Hughes and Hughes, 1993;

Cronn *et al.*, 1999), to little (Robinson-Rechavi and Laudet, 2001), to considerable (Van de Peer *et al.*, 2001) evidence for evolutionary rate increase following duplication. The first genome-scale analyses tended to support a "very little increase" hypothesis (Kondrashov *et al.*, 2002; Zhang *et al.*, 2002b), although more recent analyses have yielded larger rate differences (Conant and Wagner, 2003; Jordan *et al.*, 2004; Kellis *et al.*, 2004). For example, a study of closely related gene triplets in four completely sequenced genomes showed that 20–30% of duplicated genes experienced a significant difference in evolutionary rate (Conant and Wagner, 2003). Similarly, a comparison of duplicated *S. cerevisiae* genes with their pro-orthologs in *Kluyveromyces waltii* (another yeast) indicated that in 17% of the cases, one or both duplicates had undergone accelerated evolution since the duplication event (Kellis *et al.*, 2004).

Genome sequence data from humans and mice have been used to investigate the effect that a paralogous gene in humans has on the divergence between its duplicate and the ortholog (or pro-ortholog) in mice. Specifically, Nembaware *et al.* (2002) tested the hypothesis that a given human gene diverges from its mouse ortholog faster when it has a paralog. Interestingly, genes that have distantly related paralogs do evolve more quickly, but genes with closely related paralogs appear not to. This paralog-induced increase in evolutionary rate was most prominant at nonsynonymous positions.

Based on comparisons of several different species pairs, Jordan *et al.* (2004) made the interesting observation that although there is indeed an initial increase in molecular evolutionary rate following gene duplication, in the long run duplicated genes tend to evolve more *slowly* than singletons (i.e., those present in only one copy). For example, Jordan *et al.* (2004) found that in 100 million-year-old gene pairs between humans and mice, those having a paralog evolved significantly more slowly than singletons. In fact, the decrease in evolutionary rate was correlated with the number of paralogs: genes with more paralogs tended to evolve more slowly than genes with fewer paralogs. The observation that duplicated genes evolve more slowly overall than singletons would appear to be at odds with the observation that evolutionary rate often accelerates following a duplication event. However, Jordan *et al.* (2004) note that this could result from the influence of two countervailing forces in the evolution of duplicate genes: (1) acceleration of substitution between paralogs following duplication caused by a relaxation of purifying selection (and/or the active input of positive diversifying selection), and (2) relative reduction of substitution rate for genes with duplicates compared to singletons, based on the stronger functional constraints affecting the types of genes that tend to be found in duplicate versus those that are usually present in one copy (see previous section).

Gribaldo *et al.* (2003) studied duplication and divergence in hemoglobin genes in a test of the hypothesis that site-specific changes in evolutionary rates in a member of a duplicate pair are correlated with functional divergence. Phillipe *et al.* (2003) also investigated the correlation between the evolution of new gene function in paralogs and the rate of DNA sequence evolution. The premise for both studies was that new function is a consequence of a change in protein structure. A significant increase in evolutionary rate appeared not to be an indicator of functional change, but a correlation between "Constant But Different" (CBD) substitutions and functional divergence in proteins was detected (Phillipe et al., 2003). CBDs occur when a typically constant amino acid residue changes once in the phylogenetic tree but not again. Creevy and McInerey (2002) considered such mutations to be evidence for directional adaptive evolution, and they proposed that the discovery of CBDs, or "invariable replacement" (IR) substitutions, was an alternative to the traditional way of detecting positive selection, which involves comparing nonsynonymous and synonymous substitution rates. The idea behind the CBD/IR approach is that during an episode of positive directional selection, advantageous substitutions will occur at positions that then remain invariable at a rate significantly higher than expected from the neutral model.

There are many other methods for investigating site- and branch-specific rate variation within gene families (i.e., after duplication), and for uncovering evidence of postduplication positive selection (reviewed recently by Raes and Van de Peer, 2003). Such methods have been used to identify mutations that have led to functional shifts after gene duplication in a large number of gene families.

CHANGES IN EXPRESSION PATTERNS

Indirect methods of surveying duplicates for functional divergence (e.g., analyses of primary sequences) might one day be used to characterize the consequences of gene duplication at a whole-genome scale. However, the studies summarized in the previous section show that current computational approaches, such as the estimation of rate variation among duplicates, do not tell the whole story. In particular, differences in expression patterns among gene duplicates may also be an important feature of their subsequent evolution.

Differences in duplicate gene expression domains were first revealed by isozyme studies. Avise and Kitto (1975), for example, discovered that duplicated phosphoglucoisomerase (PGI) genes were expressed in different tissues and, by surveying representatives from a diversity of fish species, demonstrated that this expression-level divergence occurred shortly after the genes were duplicated. Ferris and Whitt (1979) compared the expression patterns of duplicated isozymes among species in the tetraploid fish family Catostomidae. In this case, an average of eight duplicated enzymes was studied in 10 different tissues each for 15 species. They concluded that the rates of change in regulatory genes (which might be interpreted as including regulatory elements) and structural genes were uncoupled; values for expression divergence were not correlated with subunit molecular weight or heterozygosity (taken as a proxy for sequence divergence of the coding region). By mapping isozyme expression domains onto a phylogeny of catastomid fishes, Ferris and Whitt (1979) were able to conclude that differential gene expression between some duplicates evolved soon after the 50-million-year-old genome duplication event, whereas other expression domain differences probably arose relatively recently. Divergent expression was usually unidirectional in Ferris and Whitt's study, meaning that one of the two duplicates typically had stronger staining across all of the tissues surveyed. Likewise, Wagner (2002) recently found that the divergence of duplicated genes is often asymmetric in yeast, indicating that one gene loses more functions than the other. Ferris and Whitt (1979) also found variation among tissues with respect to the degree of divergent expression: for eight pairs of duplicates, expression patterns were most similar in the brain and least similar in the liver.

In another early example, Hopkinson *et al.* (1976) surveyed data from 100 human isozyme loci and observed that 20 of these occurred in duplicate. For these 20 loci, the proteins encoded by paralogous genes were very similar with respect to subunit size and subunit number. But, in contrast to this structural similarity, there were several instances of divergent expression among these sets of duplicated human enzymes.

During the last several years, functional genomics data have been used to study gene duplication and divergence. Wagner (2000), for example, analyzed the expression patterns of 124 duplicated pairs of yeast genes using a compilation of 79 microarray experiments. He showed that there was almost no correlation between the divergence in expression pattern and the evolutionary distance of the corresponding proteins, implying a decoupling of the rate of coding sequence evolution and that of expression divergence after duplication. Later Gu *et al.* (2002) showed, also in yeast, that these two rates are coupled, but only for a brief period after duplication. In addition, a significant correlation was found between expression divergence and the number of synonymous substitutions per site between duplicates, which shows that expression divergence increases with evolutionary time. This observation was supported by a negative correlation between the number of conserved regulatory elements between duplicates and time (Papp *et al.*, 2003). A recent microarray-based study of duplicated human genes also found that divergence in expression and in synonymous substitutions was correlated and that expression divergence was more rapid in humans than in yeast (Makova and Li, 2003). Proteins involved in immune response, in particular, appeared to show a more rapid divergence in expression after duplication.

Numerous other studies describing expressional differences between duplicated genes have been published, ranging from detailed studies of single gene duplicates (e.g., Averof *et al.*, 1996; Locascio *et al.*, 2002; Hua *et al.*, 2003) to large-scale analyses of expression of duplicated genes in polyploids or by mining available expression data (Galitski *et al.*, 1999; Gu *et al.*, 2002; Kashkush *et al.*, 2002; Wagner, 2002; Adams *et al.*, 2003). Adams *et al.* (2003), for example, observed several cases of differential, organ-specific expression patterns in polyploid cotton.

NEOFUNCTIONALIZATION

Although the early literature on gene duplication included discussions of specialization on a subset of ancestral functions (Serebrovsky) and protection from harmful mutations (Stadler, Haldane), the primary focus with respect to the role of gene duplication in evolution then, as now, involved the production of redundant genes that have the potential to evolve new functions—that is, "neofunctionalization" (Fig. 5.7). However, it bears noting that despite this long history, the number of examples of the evolution of new, potentially adaptive functions in duplicated genes is still quite small.

One interesting example of neofunctionalization is the duplication of the ribonuclease (*RNAse1*) gene in a leaf-eating colobine monkey. While the diets of most monkeys consist of fruits and insects, colobine monkeys eat leaves. Leaves are fermented in their foregut by symbiotic bacteria, which, when digested, serve as a source of nutrition for the monkeys. Importantly, colobine monkeys have two *RNAse1* genes: *RNAse1a*, which digests double-stranded RNA, and *RNAse1b*, which has undergone several radical amino acid substitutions that appear to allow it to digest bacterial RNA in the acidic foregut. In this sense, duplication and divergence of *RNAse1* genes can be taken to represent an adaptation of these monkeys to a new nutritional niche (Zhang *et al.*, 2002a). In keeping with this, an excess of nonsynonymous substitutions per nonsynonymous site over the number of synonymous substitutions per synonymous site indicates that the driving force behind the acquisition of this new function was positive Darwinian selection (Zhang *et al.*, 2002a).

Another striking example from primates involves the evolution of trichromatic (three-color) vision. Old World primates have trichromatic vision due to the presence of three different visual pigments consisting of the retinal protein bound to one of three different opsin proteins. These pigments have different spectral properties, depending on whether the protein component of the receptor is encoded by the short-wave (SW; autosomal), the middle-wave (MW; X-linked), or long-wave (LW; X-linked) opsin gene. The MW and LW proteins have arisen through a recent gene duplication of an ancestral MW/LW gene. New World monkeys, with only an SW and a MW/LW gene, have dichromatic vision (Dulai *et al.*, 1999). Interestingly, in monkeys there are up to three allelic forms of the MW/LW gene (each with specific spectral properties) and this allows some heterozygous females to develop trichromatic vision from the expression of different alleles in different cone photoreceptors (i.e., owing to a mosaic of cell-specific X inactivation in the retina). A group of New World primates called howler monkeys independently evolved trichromatic vision through a recent gene duplication of the MW/LW gene (Jacobs, 1996; Hunt *et al.,* 1998; Dulai *et al.,* 1999), unambiguously linking duplication of these genes to this evolutionary innovation.

Duplicated *RNAse1* and opsin genes have evolved new functions as a consequence of formerly forbidden amino acid substitutions. By contrast, the origin of the antifreeze glycoprotein (AFGP) of Antarctic notothenioid fishes is an example of extensive postduplication protein remodeling. In this case, a trypsinogen-like protease gene was duplicated. In one copy, a small Thr-Ala-Ala-encoding region expanded through iterative (microsatellite-like) internal duplications. The expanded region codes for the AGFP polyprotein, which is post-translationally cleaved to form the mature AGFPs that bind to growing ice crystals and prevent the fish from freezing. Later, the obsolete exons coding for protease-specific sequences were lost, giving rise to the AGFP gene in its current form (Chen *et al.,* 1997; Cheng and Chen, 1999).

Olfactory receptor (OR) genes provide another important example whose duplication history, genomic organization, and regulation have been recently reviewed (Kratz *et al.,* 2002). Briefly, the proteins coded by these genes are expressed in sensory neurons of the vertebrate nose. There are approximately 100 OR genes in zebrafish and about 1000 in mice and humans, but in contrast to both zebrafish and mice, a large proportion of human OR genes are pseudogenes. Each neuron in the olfactory epithelium expresses only a single allele of a single OR gene locus, meaning that the sensitivity of a given olfactory neuron is limited by the range of odorants to which a particular OR can bind (Kratz *et al.,* 2002). Like the well-known *Hox* genes involved in development (see later section), ORs occur in clusters. In zebrafish, there are two clusters and the most closely related ORs are adjacent to one another, indicating that tandem duplication and/or unequal crossing-over is the major mode by which this family has expanded in this species. In humans and mice, OR clusters are distributed among many chromosomes and members of OR subfamilies are dispersed among clusters and among chromosomes. Thus tandem and duplicative transposition events (or postduplication gene rearrangements) appear to have played roles in the expansion of this family in mammals (Kratz *et al.,* 2002).

In a recent study of human and chimpanzee OR genes, Gilad *et al.* (2003) found a higher rate of gene loss (pseudogene formation) in humans (50%) than in chimps (30%). A reduced need for chemoreception in humans was offered as one explanation, but the observation that intact OR genes appear to have experienced positive selection in humans immediately following the divergence of these two species indicated that the situation is more complex than this. Gilad *et al.* (2003) proposed that the human-specific habit of cooking food might explain both the loss of duplicated OR genes and the occurrence of positive selection in a subset of these genes. Whatever the explanation for this intriguing pattern, this study, like those involving isozymes in catastomids (Ferris and Whitt, 1979) or fish *Hox* genes (Amores *et al.*, 2004), shows that the modification and loss of duplicated genes continue long after the events that produced the duplicates.

SUBFUNCTIONALIZATION

Serebrovsky's (1938) model for the evolution of *achaete* and *scute* has been revived and revised by Jensen (1976), Wistow and Piatigorsky (1987), and Force *et al.* (1999). In all cases, the model involves duplicated genes that share or subdivide the multiple roles of their single-copy ancestor, or pro-ortholog (Fig. 5.7). The attractive feature of these models is that the duplicates are preserved as a consequence of the changes that they are most likely to experience: degenerative mutations.

Jensen's (1976) model involved an ancestral enzyme with very broad specificity being duplicated and the descendants of this molecule specializing on a subset of ancestral functions. Wistow and Piatigorsky's (1987) model, derived from crystallin gene expression data, was called the "gene sharing model." Crystallins, in addition to being enzymes, contribute up to 60% of the protein in the lenses of vertebrate and squid eyes. There are four types of crystallins: α-crystallin belongs to a superfamily of heatshock proteins, β- and γ-crystallins belong to a family of calcium binding proteins, and ε-crystallin is a functional lactate dehydrogenase. Wistow and Piatigorsky (1987) proposed that crystallins might have been recruited as lens proteins because of their especially stable structure. Their model for α-, β-, and γ-crystallin evolution involved enzymatic proteins that gained structural roles as a result of the acquisition of new gene promoter elements, followed by duplication, divergence, and specialization. For ε-crystallin, the product of one gene carries out dual roles. This phenomenon was called "gene sharing," because crystalline enzymes appear to have evolved their structural role prior to duplication. Further evidence from crystallin genes for the gene-sharing model was reported by Piatigorsky and Wistow in 1991. Chickens and ducks have two δ-crystallin genes (δ_1 and δ_2), and a lens-preferred enhancer is present in both genes in both species. δ_2-crystallin codes for argininosuccinate lyase (ASL), and δ_1 appears to play a role only in lens structure. Both δ-crystallin genes are expressed in the lens of ducks, which means that ASL activity is high in duck lenses. However, in chicken lenses, 95% of the δ-crystallin is of δ_1 type—that is to say, the subdivision of roles has proceeded further in chickens.

In 1999, Force and coworkers introduced the "duplication-degeneration-complementation" (DDC) model. In this model, degenerative mutations in regulatory elements controlling the expression of two duplicated genes lead to complementary expression patterns. As a hypothetical example, if the original gene was expressed in both arms and legs, the degeneration of the elements controlling arm expression in one duplicate and the complementary degeneration of the elements controlling leg expression in the other would lead to a partitioning of gene functions between duplicates. Force *et al.* (1999) called this process "subfunctionalization" (Fig. 5.7).

In a notable real-world example of probable subfunctionalization, Force *et al.* (1999) discussed the evolution of the duplicated zebrafish *engrailed* genes *eng1a* and *eng1b,* which were formed very early during the evolution of teleost fishes. *In situ* hybridzation using *eng1a* and *eng1b* probes and 28.5-hour zebrafish embryos showed *eng1b* expression in a specific set of hindbrain and spinal neurons, whereas *eng1a* was expressed in the pectoral appendage bud. *Eng1* in mice and chickens is expressed in both domains, which is what would be predicted if it reflects the ancestral single-copy fish sequence and if *eng1a* and *eng1b* had followed the subfunctionalization model of gene evolution.

Additional models similar to Force *et al.*'s (1999) DDC model were proposed around the same time. Also in 1999, for example, Stoltzfus suggested (as previous authors had) that duplication leads to redundancy (which he called "excess capacity"), and that mutations in one copy that reduced function would not be opposed by purifying selection (i.e., such mutations would be neutral, but would prevent the subsequent loss of the second copy). However, among those mutations that reduced function, some could actually lead to novel functions in a process that Stoltzfus (1999) described as "constructive" neutral evolution. Another similar model was proposed by Hughes (1994) based on his research with the frog *Xenopus laevis.* Hughes (1994) suggested that gene duplication leading to the production of two genes encoding functionally distinct proteins is usually preceded by a period of gene sharing—that is, a period in which a single generalist protein performs two distinct functions. Once gene duplication occurs, he argued, it becomes possible for the products of the two duplicate genes to specialize so that each of them performs only one of the functions performed by the ancestral gene. Following the gene sharing model, specialization can be achieved by a change in the regulation of expression of one or both daughter genes. Thus, in a multicellular organism, each daughter gene might come to be expressed in a more restricted set of tissues than was the ancestral gene. Hughes (1994) added the hypothesis that natural selection may act rapidly to favor certain amino acid replacements that better suit each daughter gene to its specific function.

Shortly after the development of these models, Dermitzakis and Clark (2001) introduced an approach for detecting what might be considered "coding-level subfunctionalization" in duplicated genes using a method based on local evolutionary rate differences between paralogs. Protein-coding subfunctionalization appears to have occurred in fish *Microphthalamia-associated transcription factor* (*Mitf*) and *Synapsin* genes. The duplicates each code for proteins that correspond to isoforms generated by alternative splicing in their human single-copy orthologs (Lister *et al.,* 2001; Altschmied *et al.,* 2002; Yu *et al.,* 2003). In the Fugu *synA* and *synB* genes, Yu *et al.* (2003) showed that divergence was the result of complementary degenerate mutations disabling alternative splicing in each duplicate gene and allowing only one of the transcripts, orthologous to one of the two respective mammalian isoforms, to be expressed. Duplicated *DGCR6* genes appear to be another example of coding-level subfunctionalization. Comparative FISH and genomic sequence analyses suggested that this gene was duplicated in the primate lineage approximately 35 million years ago. Although the function of these genes is not known, Edelmann *et al.* (2001) reported expressed sequence tag (EST)-based evidence that the retention of both copies in humans might be a consequence of asymmetric mutations that decrease the efficacy of each gene and lead to selection for genomes to retain both.

REVERSION

In 1972, Koch proposed a multistep model of enzyme evolution involving gene duplication. According to this model, the evolutionary improvement of enzymatic function by one-step-at-a-time substitutions reaches a plateau, and then only very rare multiple simultaneous mutations or locus duplication can improve enzyme function. Koch's model posits that if evolution takes the duplication route, at some point in the future the advantage of having two genes for one enzyme will diminish. Then one copy is free to experience nonselective (or, as Koch called them, "non-Darwinian") mutational changes. Koch's model involved a second round of growth limitation (selection) but it included "reversion" as a potential solution. Reversion occurs when the degenerated duplicate is revived, i.e., codes for a better enzyme, which will take over in the population by the selective virtue of its superior maximum activity. Thus this model involves a race between degeneration and the evolution of improved function, but only after selection establishes both paralogs in the population.

HOX GENE DUPLICATION AND THE EVOLUTION OF ANIMAL DEVELOPMENT

The study of the *homeobox* gene family has contributed considerably to the understanding of the prevalence and consequences of gene duplication (and equally importantly, of animal development), and therefore warrants special discussion. *Homeobox* genes encode transcription factors that regulate the expression of a diversity of genes early in development. The name of this gene family comes from the possession of a 60 amino acid–long DNA-binding region called the "homeodomain" or "homeobox." This name, in turn, reflects the observation that mutations in some of these genes lead to so-called "homeotic transformations," in which one body structure is replaced by another (e.g., in *bithorax* mutants in *Drosophila*, a segment that normally carries halteres is transformed into one carrying wings, whereas in *antennapedia* mutants, antennae are replaced with legs).

In vertebrates these genes are called *Hox* genes, and typically occur in one or more clusters of up to 13 genes (McGinnis *et al.*, 1984; Gehring, 1998). Different *Hox* genes are expressed in different regions of the developing embryo. Intriguingly, the order of genes within the clusters reflects their order of expression along the anterior to posterior axis. The evolution of these clusters has been driven by a complex history of duplication and divergence that has now been characterized through phylogenetic analyses. These phylogenetic analyses suggest that tandem duplication of a *protoHox* gene produced a four-gene cluster and that this entire cluster was duplicated producing a four-gene *Hox* cluster and, on a different chromosome, a four-gene *ParaHox* cluster (Brooke *et al.*, 1998) (Fig. 5.8). Following the nomenclature of Kourakis and Martindale (2000), the four-gene

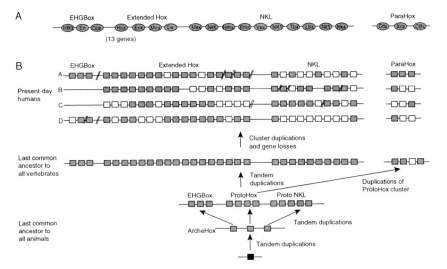

FIGURE 5.8   The evolution of *Hox* genes. A diversity of tandem duplications and two whole-cluster duplication events have led to the current repertoire of human *Hox* and *ParaHox* genes. The cluster duplications were likely linked to whole genome duplications (see Chapter 6), indicating that both small- and large-scale duplication events have played a role in the evolution of these important developmental regulatory genes. (A) Genes thought to have been linked in the common ancestor of vertebrates. Note that *Hox* genes (13 or 14 in total) are represented by a single symbol in this representation. Extended *Hox* includes *Hox* genes, *evenskipped* (*Evx*), *mesenchyme homeobox* (*Mox*), and *distal-less* (*Dlx*). (B) Gene duplication events, beginning at the bottom of the figure with a single *ProtoHox* gene (the middle gene in step two) that have led to the four-cluster complement of *Hox* genes and *ParaHox* genes in humans (see text for details). Homeodomain-containing *NKL* and *EHGBox* genes are also shown (see Pollard and Holland, 2000). Redrawn from Patel and Prince (2000), reproduced by permission (© BioMed Central).

Hox cluster included *anterior Hox, Hox 3, central Hox*, and *posterior Hox*. *Hox1* and *Hox2* are derived from *anterior Hox* and the duplication of *central Hox* has led to the production of *Hox4* through *Hox8*. The remaining *Hox* genes are descendants of *posterior Hox*. Ferrier *et al.* (2000) reported a 14th *Hox* gene in "Amphioxus" (the cephalochordate *Branchiostoma lanceolatum*), and there is evidence for 14-gene *Hox* clusters in sharks and in the coelacanth (Powers and Amemiya, 2004).

The observation that Amphioxus possesses one *Hox* gene cluster (Garcia-Fernandez and Holland, 1994) whereas sarcopterygians (coelacanths, lungfishes, amphibians, reptiles, birds, and mammals) have four, is consistent with the hypothesis that there were two whole-genome duplication events early in vertebrate evolution (Schughart *et al.*, 1989; Holland and Garcia-Fernandez, 1996; Holland, 1997; Spring, 1997) (see Chapter 6).

Inspired by these *Hox* cluster observations, Pébusque *et al.* (1998) combined phylogenetics and gene map data to identify additional large paralogous regions of the human genome. The results supported the hypothesis that large-scale gene duplication occurred before the evolution of bony vertebrates but after the Protostomia–Deuterostoma split. Pébusque *et al.* (1998) did not explicitly link gene duplication to the evolution of vertebrates, but did remark that their study was part of the search for events that have molded animal evolution.

Irvine *et al.* (2002) and Force *et al.* (2002) sequenced *Hox* genes from the sea lamprey (*Petromyzon marinus*), with the hope that counting *Hox* clusters in this species would provide a clearer picture of when *Hox* genes were duplicated, and thus indicate whether or not *Hox* cluster duplication was correlated with the evolution of vertebrates. Both sets of authors concluded that lamprey have four *Hox* gene clusters, which suggests that cluster duplication preceded increases in axial (body plan) complexity.

Further adding to the complexity of the story of *Hox* gene evolution, Amores *et al.* (1998) uncovered seven *Hox* clusters in zebrafish (*Danio rerio*) and, extrapolating from *Hox* clusters to whole genomes, proposed that a fish-specific genome duplication might have been responsible for this gene cluster amplification (see also Meyer and Schartl, 1999). More than four *Hox* clusters have also been described for medaka (*Oryzias latipes*) (Naruse *et al.*, 2000), an African cichlid fish *Oreochromis niloticus* (Málaga-Trillo and Meyer, 2001), the spotted pufferfish *Takifugu rubripes* (Aparicio *et al.*, 1997), and the southern pufferfish *Spheroides nephelus* (Amores *et al.*, 2004). Thus *Hox* clusters, and perhaps the whole genome, duplicated in ray-finned fishes before the divergence of zebrafish, medaka, and pufferfishes.

Postduplication *Hox* gene loss in fishes has been substantial. Zebrafish have only nine more *Hox* genes than humans and mice (48 versus 39). Intriguingly, and in stark contrast to the pattern in mammals where humans and mice possess an identical *Hox* gene complement, the pattern of gene loss differs among fish species. Medaka, the spotted pufferfish, and the southern pufferfish have two

Hoxa, Hoxb, and *Hoxd* clusters, whereas the zebrafish has two *Hoxa, Hoxb,* and *Hoxc* clusters. The *Hox* gene complement even differs among pufferfish species: the *Hoxb7a* gene is absent from the spotted puffer but present in the southern puffer. Both copies of *Hoxb7* must have been retained from the time of duplication (~350 million years ago) to the divergence of these two pufferfish species (~5–35 million years ago), suggesting that the silencing of duplicates is possible long after their duplication.

Hox gene expression studies have uncovered evidence for neofunctionalization, "function shuffling," and subfunctionalization. McClintock *et al.* (2001) studied expression of *Hox1* genes, the so-called "paralogy group (PG) 1." Zebrafish have four PG1 genes—*Hoxa1a, Hoxb1a, Hoxb1b,* and *Hoxc1a*—whereas mice have three PG1 genes—*Hoxa1, Hoxb1,* and *Hoxd1.* Using *in situ* hybridization, McClintock *et al.* (2001) discovered that zebrafish *Hoxa1a* expression is very different from mouse *Hoxa1* expression: in zebrafish, *Hoxa1a* expression in small bilateral cell clusters in the mid- and hindbrain appears to be an example of neofunctionalization. Interestingly, zebrafish *Hoxb1b* appears to perform the same role in zebrafish development that *Hoxa1* performs in mice. This pattern of nonorthologous genes fulfilling equivalent roles was called "function shuffling" (McClintock *et al.,* 2001).

Duplicated *Hoxd4* genes in the southern pufferfish have overlapping but not identical expression domains (Amores *et al.,* 2004). The anterior limit of *Hoxd4a* expression is rhombomere seven (r7), as has been observed in zebrafish and mice. However, *Hoxd4b* expression stops at r8. *S. nephelus Hoxd4b* is expressed more strongly than its paralog in the hindbrain and neural crest (see Fig. 5 in Amores *et al.,* 2004). These *Hoxd4a* and *Hoxd4b* data appear to be an example of subfunctionalization; however, the mutations responsible for these postduplication changes in gene expression have yet to be characterized.

Amores *et al.* (2004) also characterized the expression domains of duplicated *Hoxa2* genes and uncovered what might be an especially interesting example of neofunctionalization. In pufferfish and in zebrafish, *Hoxa2b* is expressed in hindbrain rhombomeres r2–r5. The same pattern has been observed for *Hoxa2* in mice. Pufferfish have a *Hoxa2a* gene and it is expressed in r1. Amores *et al.* (2004) speculated that *Hoxa2a* expression in pufferfish r1 might play a role in the pufferfish-specific invention of the buccal pump, which puffs the stomach with water.

THE GENERAL EVOLUTIONARY IMPORTANCE OF GENE DUPLICATIONS

It is becoming apparent that duplications arise at a surprisingly rapid rate (Lynch and Conery, 2000) and that many gene duplicates are retained. Indeed, large gene families are a feature of all metazoan genomes sequenced to date. As noted earlier in the chapter, some very important genes related to physiology, immunity, and development exist as families of duplicates. It has also become clear that gene family expansion, although punctuated by large-scale episodes of whole genome duplication, is an ongoing phenomenon and involves a great diversity of mutations. Vertebrates have more *Hox* genes than invertebrates, and fishes have duplicates of many tetrapod genes—observations that likely reflect ancient whole genome duplication events (see Chapter 6). On the other hand, smaller-scale duplications are responsible for the fact that old world primates have more opsin genes than most of their new world counterparts, and that mammals have about 10 times more odor receptors than fishes. Even closely related nematode worms and yeasts may differ dramatically in the size of some gene families, again resulting from smaller-scale duplication events.

In light of these findings, it is clear that Ohno's general emphasis on whole-genome duplications was unnecessarily restrictive. It is true that DNA sequencing, which has been especially good at uncovering the enormous contribution of smaller-scale duplication events, was not available when Ohno (1970) published his classic treatise on the subject. But 50 years before DNA sequencing, cytogeneticists had correlated speciation, morphological innovation, and smaller-scale duplication gene events (e.g., Bridges and Sturtevant with the Bar locus, Muller and Blakeslee with partial chromosome duplication, Lewis with pseudoalleles).

Now that complete genome sequence data have been brought to bear on this issue, it is clear that both small- and large-scale duplication events have played major roles. Indeed, Ohno and his predecessors do not appear to have been exaggerating when they proclaimed gene duplication to be a crucial component of the evolutionary process.

CONCLUDING REMARKS AND FUTURE PROSPECTS

A more thorough review clearly shows that the large number of recent papers that begin with "ever since Ohno" statements (including some written by the present authors) fail to acknowledge a long history of inquiry into the occurrence and evolutionary consequences of gene and genome duplication. Chromosome counts, studies of chromosome morphology, and isozyme electrophoresis have all made significant empirical and theoretical contributions to gene and genome duplication research. These early studies, combined with comparatively recent research dealing with gene and whole genome sequences, show that gene duplication is a common occurrence, that the rate of duplication varies among species (humans and nematode worms make new genes faster than *Drosophila, Arabidopsis,* and yeast) and among genes (shorter genes appear to be duplicated more often than long genes, and slowly evolving genes produce duplicates more often than those with a rapid rate of evolution). Whether or not a duplicate is retained depends on its function, its mode of duplication (e.g., whether or not it was duplicated during a whole genome duplication event), the species in which it occurs, and its expression rate. There does not appear to be a consensus with respect to evolutionary rate variation among paralogs and their single-copy orthologs: duplication leads to an increase in evolutionary rate in some cases but not in others. Furthermore, as Philippe *et al.* (2003) pointed out, a change in mutation rate might not be correlated with functional divergence. However, Constant But Different (CBD) or Invariable Replacement (IR) substitutions do appear to occur more frequently following gene duplication. Finally, a diversity of studies has shown that expression divergence and gene loss are phenomena that can occur both shortly and long after gene duplication events.

As this chapter makes clear, the understanding of the impact of gene duplication in evolution has been, and continues to be, enhanced by technological advances in microscopy, biochemistry, and molecular and cell biology. Indeed, although Dyson's (1997) oft-repeated claim that new directions in science are launched by new tools much more often than by new concepts does not accurately describe the study of gene duplication (in which the central ideas have typically been proposed long before the techniques and tools were available to test them), it is very likely that new technologies will continue to advance the field, as they have for more than 100 years. To give just one example, the new method of "high resolution array comparative genomic hybridization (CGH)" provides a novel means of delimiting duplicated fragments (coding and noncoding) in fully sequenced genomes (Ishkanian *et al.*, 2004). So far, this high resolution or "sub-megabase resolution tiling (SMRT) array" CGH has been used with success to look for variation in gene content only between normal and cancerous cell lines, but comparisons between normal cells/individuals or between different species, will, no doubt, soon be reported. Lynch and Conery's (2000) estimation that new human genes arise at a rate of hundreds per generation suggests that SMRT array CGH will turn up interesting among-individual variation in gene content in humans, and Locke *et al.*'s (2003) CGH-based comparison between humans and other apes (chimpanzee, bonobo, gorilla, orangutan), which involved about 12% of the human genome but nonetheless turned up evidence for many lineage-specific gene duplications and deletions (which were later verified using FISH), suggests that this approach has enormous potential for interspecific surveys.

The future of this field, as with others in genome biology, clearly lies in the integration of results from diverse research programs. With the use of broad-based analytical and conceptual approaches, it is now possible (at least in principle) to describe the entire paranome, compare expression domains among all paralogs, and identify the mutations responsible for expression variation among paralogs. It is also possible to compare interaction partners among paralogs, and to correlate this information with ever-increasing knowledge of the pathways in which genes act.

It is very exciting in this sense to consider that a comprehensive description of the occurrence and evolutionary consequences of gene duplication, at least for some species, is now coming within reach after nearly a century of study.

REFERENCES

Abeliovich D, Katz M, Karplus M, Carmi R. 1985. A *de novo* translocation, 14q21q, with a microchromosome-14p21p. *Am J Med Genet* 22: 29–33.

Adams KL, Cronn R, Percifield R, Wendel JF. 2003. Genes duplicated by polyploidy show unequal contributions to the transcriptome and organ-specific silencing. *Proc Natl Acad Sci USA* 100: 4649–4654.

Altschmied J, Delfgaauw J, Wilde B, *et al.* 2002. Subfunctionalization of duplicate *mitf* genes associated with differential degeneration of alternative exons in fish. *Genetics* 161: 259–267.

Altschul SF, Gish W, Miller W, *et al.* 1990. Basic local alignment search tool. *J Mol Biol* 215: 403–410.

Amores A, Force A, Yan YL, *et al.* 1998. Zebrafish *hox* clusters and vertebrate genome evolution. *Science* 282: 1711–1714.

Amores A, Suzuki T, Yan YL, *et al.* 2004. Developmental roles of pufferfish Hox clusters and genome evolution in ray-fin fish. *Genome Res* 14: 1–10.

Aparicio S, Hawker K, Cottage A, *et al.* 1997. Organization of the *Fugu rubripes* Hox clusters: evidence for continuing evolution of vertebrate Hox complexes. *Nat Genet* 16: 79–83.

Averof M, Dawes R, Ferrier D. 1996. Diversification of arthropod *Hox* genes as a paradigm for the evolution of gene functions. *Cell Dev Biol* 7: 539–551.

Avise JC, Kitto GB. 1975. Phosphoglucose isomerase gene duplication in the bony fishes: an evolutionary history. *Biochem Genet* 8: 113–132.

Bailey GS, Poulter RT, Stockwell PA. 1978. Gene duplication in tetraploid fish: models for gene silencing at unlinked duplicated loci. *Proc Natl Acad Sci USA* 11: 5575–5579.

Bailey JA, Gu Z, Clark RA, *et al.* 2002. Recent segmental duplications in the human genome. *Science* 297: 1003–1007.

Bianca, S. 2002. Non congenital heart disease aspects of Down's syndrome. *Images Paediatr Cardiol* 13: 3–11.

Blakeslee AF. 1934. New Jimson weeds from old chromosomes. *J Hered* 25: 81–108.

Bridges CB. 1935. Salivary chromosome maps. *J Hered* 26: 60–64.

Bridges CB. 1936. The bar "gene" a duplication. *Science* 83: 210–211.

Brooke NM, Garcia-Fernadez J, Holland PW. 1998. The ParaHox gene cluster is an evolutionary sister of the Hox gene cluster. *Nature* 392: 920–922.

Brosius J. 2003. Gene duplication and other evolutionary strategies: from the RNA world to the future. *J Struct Funct Genom* 3: 1–17.

Brown DD, Dawid IB. 1968. Specific gene amplification in oocytes. Oocyte nuclei contain extrachromosomal replicas of the genes for ribosomal RNA. *Science* 160: 272–280.

Chen L, DeVries AL, Cheng CH. 1997. Evolution of antifreeze glycoprotein gene from a trypsinogen gene in Antarctic notothenioid fish. *Proc Natl Acad Sci USA* 94: 3811–3816.

Cheng CHC, Chen L. 1999. Evolution of an antifreeze glycoprotein. *Nature* 401: 443–444.

Conant GC, Wagner A. 2002. GenomeHistory: a software tool and its application to fully sequenced genomes. *Nucleic Acids Res* 30: 3378–3386.

Conant GC, Wagner A. 2003. Asymmetric sequence divergence of duplicate genes. *Genome Res* 13: 2052–2058.

Creevy CJ, McInerey JO. 2002. An algorithm for detecting directional and non-directional positive selection, neutrality and negative selection in protein coding DNA sequences. *Gene* 300: 43–51.

Cronn RC, Small RL, Wendel JF. 1999. Duplicated genes evolve independently after polyploid formation in cotton. *Proc Natl Acad Sci USA* 96: 14406–14411.

Cronn RC, Zhao X, Paterson AH, Wendel JF. 1996. Polymorphism and concerted evolution in a tandemly repeated gene family: 5S ribosomal DNA in diploid and allopolyploid cottons. *J Mol Evol* 42: 685–705.

Davis JC, Petrov DA. 2004. Preferential duplication of conserved proteins in eukaryotic genomes. *PLoS Biol* 2: 0318–0326.

Dayhoff MO. 1974. Computer analysis of protein sequences. *Fed Proc* 33: 2314–2316.

Dermitzakis ET, Clark AG. 2001. Differential selection after duplication in mammalian developmental genes. *Mol Biol Evol* 18: 557–562.

Drouin G, Dover GA. 1987. A plant processed pseudogene. *Nature* 328: 557–558.

Dulai KS, von Dornum M, Mollon JD, Hunt DM. 1999. The evolution of trichromatic colour vision by opsin gene duplication in New World and Old World primates. *Genome Res* 9: 629–638.

Dyson, F. 1997. *Imagined Worlds*. Cambridge, MA: Harvard University Press.

Edelmann L, Stankiewicz P, Spiteri E, et al. 2001. Two functional copies of the DGCR6 gene are present on human chromosome 22q11 due to a duplication of an ancestral locus. *Genome Res* 11: 208–217.

Esnault C, Maestre J, Heidmann T. 2000. Human LINE retrotransposons generate processed pseudogenes. *Nat Genet* 24: 363–367.

Ferrier DE, Minguillon C, Holland P, Garcia-Fernandez J. 2000. The amphioxus Hox cluster: deuterostome posterior flexibility and Hox14. *Evol Dev* 2: 284–293.

Ferris S, Whitt GS. 1979. Evolution of the differential regulation of duplicate genes after polyploidization. *J Mol Evol* 12: 267–317.

Fink GR. 1987. Pseudogenes in yeast? *Cell* 49: 5–6.

Force A, Amores A, Postlethwait JH. 2002. Hox cluster organization in the jawless vertebrate *Petromyzon marinus*. *J Exp Zool* 294: 30–46.

Force A, Lynch M, Pickett FB, et al. 1999. The preservation of duplicate genes by complementary, degenerate mutations. *Genetics* 151: 1531–1545.

Friedman R, Hughes AL. 2001. Gene duplication and the structure of eukaryotic genomes. *Genome Res* 11: 373–381.

Fürthauer M, Thisse B, Thisse C. 1999. Three different noggin genes antagonize the activity of bone morphogenetic proteins in the zebrafish embryo. *Dev Biol* 214: 181–196.

Galitski T, Saldanha AJ, Styles CA, et al. 1999. Ploidy regulation of gene expression. *Science* 285: 251–254.

Garcia-Fernandez J, Holland PW. 1994. Archetypal organization of the amphioxus Hox gene cluster. *Nature* 370: 563–566.

Gardner MJ, Shallom SJ, Carlton JM, et al. 2002. Genome sequence of the human malaria parasite *Plasmodium falciparum*. *Nature* 419: 498–511.

Gehring WJ. 1998. *Master Control Genes in Development and Evolution: The Homeobox Story*. New Haven, CT: Yale University Press.

Gevers D, Vandepoele K, Simillion C, Van de Peer Y. 2004. Gene duplication and biased functional retention of paralogs in bacterial genomes. *Trends Microbiol* 12: 148–154.

Gilad Y, Bustamante CD, Lancet D, Pääbo S. 2003. Natural selection on the olfactory receptor gene family in humans and chimpanzees. *Am J Hum Genet* 73: 489–501.

Goldschmidt R. 1940. *The Material Basis of Evolution*. New Haven, CT: Yale University Press.

Graur D, Li WH. 1999. *Fundamentals of Molecular Evolution*. Sunderland, MA: Sinauer.

Gribaldo S, Casane D, Lopez P, Philippe H. 2003. Functional divergence prediction from evolutionary analysis: a case study of vertebrate hemoglobin. *Mol Biol Evol* 20: 1754–1759.

Gu X. 2003. Evolution of duplicate genes versus genetic robustness against null mutations. *Trends Genet* 19: 354–356.

Gu X, Wang Y, Gu J. 2002. Age distribution of human gene families shows significant roles of both large- and small-scale duplications in vertebrate evolution. *Nat Genet* 31: 205–209.

Gu Z, Nicolae D, Lu HHS, Li WH. 2002. Rapid divergence in expression between duplicate genes inferred from microarray data. *Trends Genet* 18: 609–613.

Gulick A. 1944. The chemical formulation of gene structure and gene action. *Adv Enzymol* 4: 1–39.

Haldane JBS. 1932. *The Causes of Evolution*. Ithaca, NY: Cornell University Press.

Haldane JBS. 1933. The part played by recurrent mutation in evolution. *Am Nat* 67: 5–19.

Harrison PM, Hegyi H, Balasubramanian S, *et al*. 2002. Molecular fossils in the human genome: identification and analysis of the pseudogenes in chromosomes 21 and 22. *Genome Res* 12: 272–280.

Hiebl I, Nraunitzer G, Weber RE, Kosters J. 1989. Structural adaptations for high-altitude respiration in bird hemoglobins: Barheaded goose (*Anser indicus, Gorni gus*), Andean goose (*Chloephaga melanoptera*) and Rüppell's griffon (*Gyps rueppellii*). In: Koenig WA, Voelter W. eds. *Proceedings of VIth USSR-FRG Symposium "Chemistry of Peptides and Proteins" vol. 4*, Berlin: Walter de Gruyter & Co.

Ho Y, Gruhler A, Heilbut A, *et al*. 2002. Systematic identification of protein complexes in *Saccharomyces cerevisiae* by mass spectrometry. *Nature* 415: 180–183.

Holland PW. 1997. Vertebrate evolution: something fishy about Hox genes. *Curr Biol* 7: R570–R572.

Holland PW, Garcia-Fernandez J. 1996. Hox genes and chordate evolution. *Dev Biol* 173: 382–395.

Hopkinson DA, Edwards YH, Harris H. 1976. The distribution of subunit numbers and subunit sizes of enzymes: a study of the products of 100 human gene loci. *Ann Hum Genet* 39: 383–411.

Horvath JE, Schwartz S, Eichler EE. 2000. The mosiac structure of human pericentromeric DNA: a strategy of characterizing complex regions of the human genome. *Genome Res* 10: 839–852.

Hu X, Worton RG. 1992. Partial gene duplication as a cause of human disease. *Hum Mutat* 1: 3–12.

Hua LV, Hidaka K, Pesesse X, *et al*. 2003. Paralogous murine *Nudt10* and *Nudt11* genes have differential expression patterns but encode identical proteins that are physiologically competent diphosphoinositol polyphosphate phosphohydrolases. *Biochem J* 373: 81–89.

Hughes AL. 1994. The evolution of functionally novel proteins after gene duplication. *Proc R Soc Lond B* 256: 119–124.

Hughes MK, Hughes AL. 1993. Evolution of duplicate genes in a tetraploid animal, *Xenopus laevis*. *Mol Biol Evol* 10: 1360–1369.

Hunt DM, Dulai KS, Cowing JA, *et al*. 1998. Molecular evolution of trichromacy in primates. *Vision Res* 38: 3299–3306.

Irvine S, Carr JL, Bailey WJ, *et al*. 2002. Genomics analysis of Hox clusters in the sea lamprey *Petromyzon marinus*. *J Exp Zool* 294: 47–62.

Ishkanian AS, Malloff CA, Watson SK, *et al*. 2004. A tiling resolution DNA microarray with complete coverage of the human genome. *Nat Genet* 36: 299–303.

Jacobs GH. 1996. Primate photopigments and primate color vision. *Proc Natl Acad Sci USA* 93: 577–581.

Jensen RA. 1976. Enzyme recruitment in evolution of new function. *Annu Rev Microbiol* 30: 409–425.

Jordan IK, Makarova KS, Spouge JL, *et al*. 2001. Lineage-specific gene expansions in bacterial and archaeal genomes. *Genome Res* 11: 555–565.

Jordan IK, Wolf YI, Koonin EV. 2004. Duplicated genes evolve slower than singletons despite the initial rate increase. *BMC Evol Biol* 4: 22.1–22.11.

Kan YW, Holland JP, Dozy AM, *et al*. 1975 Deletion of the B-globin structure gene in hereditary persistence of foetal haemoglobin. *Nature* 258: 162–163.

Kashkush K, Feldman M, Levy AA. 2002. Gene loss, silencing and activation in a newly synthesized wheat allotetraploid. *Genetics* 160: 1651–1659.

Kellis M, Birren BW, Lander ES. 2004. Proof and evolutionary analysis of ancient genome duplication in the yeast *Saccharomyces cerevisiae*. *Nature* 428:617–624.

Koch AL. 1972. Enzyme evolution: I. The importance of untranslatable intermediates. *Genetics* 72: 297–316.

Kondrashov FA, Rogozin IB, Wolf YI, Koonin EV. 2002. Selection in the evolution of gene duplications. *Genome Biol* 3: research0008.1–0008.9.

Kourakis MJ, Martindale MQ. 2000. Combined-method phylogenetic analysis of Hox and ParaHox genes of the metazoa. *J Exp Zool* 288: 175–191.

Kratz E, Dugas JC, Ngai J. 2002. Odorant receptor gene regulation: implications from genomic organization. *Trends Genet* 18: 29–34.

Kuwada Y. 1911. Meiosis in the pollen mother cells of *Zea Mays L. Bot Mag* 25: 163.

Lewin R. 1983. How mammalian RNA returns to its genome. *Science* 219: 1052–1054.

Lewis EB. 1951. Pseudoallelism and gene evolution. *Cold Spring Harb Symp Quant Biol* 16: 159–174.

Li J, Wirtz RA, McConkey GA, *et al*. 1994. Transition of *Plasmodium vivax* ribosome types corresponds to sporozoite differentiation in the mosquito. *Mol Biochem Parasitol* 65: 283–289.

Li S, Armstrong CM, Bertin N, *et al*. 2004. A map of the interactome network of the metazoan *C. elegans. Science* 303: 540–543.

Li WH. 1980. Rate of gene silencing at duplicated loci: a theoretical study and interpretation of data from tetraploid fishes. *Genetics* 95: 237–258.

Li WH. 1997. *Molecular Evolution*. Sunderland, MA: Sinauer Associates Inc.

Lister JA, Close J, Raible DW. 2001. Duplicate *mitf* genes in zebrafish: complementary expression and conservation of melanogenic potential. *Dev Biol* 237: 333–344.

Locascio A, Manzanares M, Blanco MJ, Nieto MA. 2002. Modularity and reshuffling of *Snail* and *Slug* expression during vertebrate evolution. *Proc Natl Acad Sci USA* 99: 16841–16846.

Locke DP, Segraves R, Carbone L, *et al*. 2003. Large-scale variation among human and great ape genomes determined by array comparative genomic hybridization. *Genome Res* 13: 347–357.

Lynch M, Conery JS. 2000. The evolutionary fate and consequences of duplicate genes. *Science* 290: 1151–1155.

Lynch M, Conery JS. 2003. The origins of genome complexity. *Science* 302: 1401–1404.

Makova KD, Li WH. 2003. Divergence in the spatial pattern of gene expression between human duplicate genes. *Genome Res* 13: 1638–1645.

Málaga-Trillo E, Meyer A. 2001. Genome duplications and accelerated evolution of Hox genes and cluster architecture in teleost fishes. *Am Zool* 41: 676–686.

Marx JL. 1982. Is RNA copied into DNA by mammalian cells? *Science* 216: 969–970.

McCarrey JR, Thomas K. 1987. Human testis-specific PGK gene lacks introns and possesses characteristics of a processed gene. *Nature* 326: 501–505.

McClintock JM, Carlson R, Mann DM, Prince VE. 2001. Consequences of Hox gene duplication in the vertebrates: an investigation of the zebrafish Hox paralogue group 1 genes. *Development* 128: 2471–2484.

McGinnis W, Garber R, Wirz J, *et al*. 1984. A homologous protein-coding sequence in *Drosophila* homeotic genes and its conservation in other metazoans. *Cell* 37: 403–408.

Metz CW. 1947. Duplication of chromosome parts as a factor in evolution. *Am Nat* 81: 81–103.

Meyer A, Schartl M. 1999. Gene and genome duplications in vertebrates: the one-to-four (-to-eight in fish) rule and the evolution of novel gene functions. *Curr Opin Cell Biol* 11: 699–704.

Montag K, Salamini F, Thompson RD. 1996. The ZEM2 family of maize MADS box genes possess features of transposable elements. *Maydica* 41: 241–254.

Muller HJ. 1935. The origination of chromatin deficiencies as minute deletions subject to insertion elsewhere. *Genetica* 17: 237–252.

Naruse K, Fukamachi S, Hirochi M, *et al*. 2000. A detailed linkage map of Medaka, *Oryzias latipes*: comparative genomics and genome evolution. *Genetics* 154: 1773–1784.

Nei M. 1969. Gene duplication and nucleotide substitution in evolution. *Nature* 221: 40–42.

Nembaware V, Crum K, Kelso J, Seoighe C. 2002. Impact of the presence of paralogs on sequence divergence in a set of mouse-human orthologs. *Genome Res* 12: 1370–1376.

Nornes S, Clarkson M, Mikkola I, *et al*. 1998. Zebrafish contains two pax6 genes involved in eye development. *Mech Dev* 77: 185–196.

Ohno S. 1967. *Sex Chromosomes and Sex-linked Genes*. Berlin: Springler-Verlag.

Ohno S. 1970. *Evolution by Gene Duplication*. New York: Springer-Verlag.

Ohta T. 1990. How gene families evolve. *Theor Pop Biol* 37: 213–219.

Olivieri NF, Weatherall DJ. 1998. The therapeutic reactivation of fetal hemoglobin. *Hum Mol Genet* 7: 1655–1658.

Papp B, Pal C, Hurst LD. 2003. Evolution of cis-regulatory elements in duplicated genes of yeast. *Trends Genet* 19: 417–422.

Patel NH, Prince VE. 2000. Beyond the Hox complex. *Genome Biol* 1: reviews 1027.1–1027.4.

Pearson WR, Wood TC. 2001. Statistical significance in biological sequence comparison. In: Balding DJ, Bishop M, Cannings C eds. *Handbook of Statistical Genetics*. Vol. 2, Toronto: John Wiley & Sons Ltd., 39–65.

Pébusque MJ, Coulier F, Birnbaum D, Pontarotti P. 1998. Ancient large-scale genome duplications: phylogenetic and linkage analyses shed light on chordate genome evolution. *Mol Biol Evol* 15: 1145–1159.

Philippe H, Casane D, Gribaldo S, *et al*. 2003. Heterotachy and functional shift in protein evolution. *IUBMB Life* 55: 257–265.

Piatigorsky J, Wistow G. 1991. The recruitment of crystallins: new functions precede gene duplication. *Science* 252: 1078–1079.

Pollard SL, Holland PW. 2000. Evidence for 14 homeobox gene clusters in human genome ancestry. *Curr Biol* 10: 1059–1062.

Powers TP, Amemiya CT. 2004. Evidence for Hox-14 paralog group in vertebrates. *Curr Biol* 14: R183–R184.

Raes J, Van de Peer Y. 2003. Gene duplications, the evolution of novel gene functions, and detecting functional divergence of duplicates in silico. *Appl Bioinformatics* 2: 92–101.

Robinson-Rechavi M, Laudet V. 2001. Evolutionary rates of duplicate genes in fish and mammals. *Mol Biol Evol* 18: 681–683.

Samonte RV, Eichler EE. 2002. Segmental duplications and the evolution of the primate genome. *Nat Rev Genet* 3: 65–72.

Schughart K, Kappen C, Ruddle FH. 1989. Duplication of large genomic regions during the evolution of vertebrate homeobox genes. *Proc Natl Acad Sci USA* 86: 7067–7071.

Seoighe C, Wolfe KH. 1999. Yeast genome evolution in the post-genome era. *Curr Opin Microbiol* 2: 548–554.

Serebrovsky AS. 1938. Genes *scute* and *achaete* in *Drosophila melanogaster* and a hypothesis of gene divergency. *C R Acad Sci URSS* 19: 77–81.

Sharman AC. 1999. Some new terms for duplicated genes. *Cell Dev Biol* 10: 561–563.

Snustad P, Simmons MJ, Jenkins JB. 1997. *Principles of Genetics*. New York: John Wiley & Sons, Inc.

Spring J. 1997. Vertebrate evolution by interspecific hybridization—are we polyploid? *FEBS Lett* 400: 2–8.

Stadler LJ. 1929. Chromosome number and the mutation rate in *Avena* and *Triticum*. *Proc Natl Acad Sci USA* 15: 876–881.

Stein LD, Bao Z, Blasiar D, *et al*. (2003). The genome sequence of *Caenorhabditis briggsae*: a platform for comparative genomics. *PLoS Biol* 1: 166–192.

Stephens SG. 1951. Possible significance of duplications in evolution. *Adv Genet* 4: 247–265.

Stoltzfus A. 1999. On the possibility of constructive neutral evolution. *J Mol Evol* 49: 169–181.

Stuber CW, Goodman MM. 1983. Inheritance, intracellular localization, and genetic variation of phosphoglucomutase isozymes in maize (*Zea mays* L.). *Biochem Genet* 21: 667–689.

Sturtevant AH. 1925. The effects of unequal crossing over at the bar locus in *Drosophila*. *Genetics* 10: 117–147.

Tischler G. 1915. Chromosomenzahl, Form und Individualität in Planzenreiche. *Progr Rei Bot* 5: 164.

Tonegawa S. 1983. Somatic generation of antibody diversity. *Nature* 302: 575–581.

Van de Peer Y, Taylor JS, Braasch I, Meyer A. 2001. The ghost of selection past: rates of evolution and functional divergence of anciently duplicated genes. *J Mol Evol* 53: 436–46.

Vandepoele K, De Vos W, Taylor JS, *et al.* 2004. Major events in the genome evolution of vertebrates: paranome age and size differ considerably between ray-finned fishes and land vertebrates. *Proc Natl Acad Sci USA* 101: 1638–1643.

Vandepoele K, Simillion C, Van de Peer Y. 2004. The quest for genomic homology. *Curr Genomics* 5: 299–308.

Venter JC, Adams MD, Myers EW, *et al.* 2001. The sequence of the human genome. *Science* 291: 1304–1351.

Vidalain PO, Boxem M, Ge H, *et al.* 2004. Increasing specificity in high-throughput yeast two-hybrid experiments. *Methods* 32: 363–370.

Wagner A. 2000. Decoupled evolution of coding region and mRNA expression patterns after gene duplication: implications for the neutralist-selectionist debate. *Proc Natl Acad Sci USA* 97: 6579–6584.

Wagner A. 2002. Asymmetric functional divergence of duplicate genes in yeast. *Mol Biol Evol* 19: 1760–1768.

Waters AP. 1994. The ribosomal RNA genes of *Plasmodium*. *Adv Parasitol* 34: 33–79.

Winge Ö. 1917. The chromosomes: their numbers and general importance. *C R Trav Lab Carlsberg* 13:131–275.

Wistow G, Piatigorsky J. 1987. Recruitment of enzymes as lens structural proteins. *Science* 236: 1554–1556.

Yu WP, Brenner S, Venkatesh B. 2003. Duplication, degeneration and subfunctionalization of the nested *synapsin-Timp* genes in *Fugu*. *Trends Genet* 19:180–183.

Zhang J, Zhang YP, Rosenberg HF. 2002a. Adaptive evolution of a duplicated pancreatic ribonuclease gene in a leaf-eating monkey. *Nat Genet* 30: 411–415.

Zhang L, Vision TJ, Gaut BS. 2002b. Patterns of nucleotide substitution among simultaneously duplicated gene pairs in *Arabidopsis thaliana*. *Mol Biol Evol* 19: 1464–1473.

Zuckerkandl E. 1975. The appearance of new structures and functions in proteins during evolution. *J Mol Evol* 7: 1–57.

CHAPTER 6

Large-Scale Gene and Ancient Genome Duplications

YVES VAN DE PEER AND AXEL MEYER

Duplications of genetic elements can occur by a variety of mechanisms and at different chromosomal and temporal scales. This chapter deals with an important subset of these, namely large-scale gene duplications (versus the small-scale events discussed in Chapter 5) and ancient duplications of whole genomes (versus more recent polyploidy in plants and animals, dealt with in Chapters 7 and 8, respectively). The emphasis in this case is on the techniques used to identify, date, and otherwise investigate such events, as illustrated by some key recent examples. As will be shown, analyses of different eukaryotes clearly indicate that significant portions of their genomes consist of duplicated gene loci, and that many of these gene duplicates have been formed by the duplication of chromosomal blocks and/or entire genomes. The timings of these events, in some cases dating back hundreds of millions of years, suggest that they have played an important role in influencing major patterns of evolutionary diversification.

The Evolution of the Genome, edited by TR Gregory

HISTORICAL PERSPECTIVES ON THE
IMPORTANCE OF LARGE-SCALE DUPLICATIONS

As noted in Chapter 5, Ohno's (1970) book *Evolution by Gene Duplication* has become very influential in the field of genome research. This is a fairly recent phenomenon, with citations of the book tripling between 1990 and 2000, whereas it received only lukewarm reviews at the time of its publication (Wolfe, 2001). In the book Ohno (1970) made the case that not only gene duplications but doublings of entire genomes are the principal forces responsible for generating the genetic raw material necessary for increasing complexity during evolution. As he put it,

> Had evolution been entirely dependent upon natural selection, from a bacterium only numerous forms of bacteria would have emerged. The creation of metazoans, vertebrates, and finally mammals from unicellular organisms would have been quite impossible, for such big leaps in evolution required the creation of new gene loci with previously nonexistent functions.

In other words, Ohno (1970) suggested that "natural selection merely modified, while redundancy created," meaning that gene and genome duplications allowed genes to diversify, take on novel functions, and bring about evolutionary innovation in general. Under Ohno's (1970) preferred interpretation of "neofunctionalization" (see Chapter 5), natural selection would be responsible for the fine-tuning of genes which had, through duplication, the chance to accumulate a sufficiently large number of otherwise "forbidden" mutations. The evidence for Ohno's hypothesis was based mainly on comparative measurements of DNA contents, karyotypic information, and some allozyme data. That is to say, Ohno's tenets were brought forth at a time where the documentation and quantification of genetic variation within populations and between species was largely restricted to scoring allelic variation in enzymes through starch gel electrophoresis and microscopic inspection of karyotypes. Methods to effectively measure genetic variation at the level of the gene or to even sequence DNA had still to be invented.

Like the case with small-scale duplications (see Chapter 5), Ohno (1970) was not the first to notice that the doubling of entire genomes could have been of major importance for evolution. In 1933, for example, Haldane argued that

> Duplications affecting only a few genes would confer only a slight advantage. But duplication of a large section, polysomy of a whole chromosome, or polyploidy, might confer a considerable advantage, provided it caused neither unbalance nor sterility. Whether this advantage is sufficient to be of evolutionary importance is not clear, but the possibility exists.

Haldane (1933) also described another possible mode of making rapid evolutionary jumps, namely by hybridization of the genomes of two different species. He noted that new species formed by hybridization showed heterosis (or "hybrid vigor"), with increased fertility and stability, and therefore higher fitness, relative to their parent species.

As a matter of fact, one of the most important changes in agriculture over the past 50 years has been the improvement of many crops through the production of polyploid hybrids derived from the crossing of highly inbred lines. Whether through allopolyploidization (hybridization) or autopolyploidization (see later section), genome duplication is known to be a very common—perhaps even ubiquitous—occurrence in plant evolution (see Chapter 7). Indeed, many of the most important crop species are recent allopolyploids (e.g., wheat, oat, cotton, and coffee) or recent autopolyploids (e.g., alfalfa and potato), whereas others, such as cabbage, are ancient polyploids (Osborn *et al.*, 2003). Though not as common as in plants, polyploidy is also a widespread phenomenon in the animal kingdom. As discussed in detail in Chapter 8, many species (or sometimes higher taxa, like families) of fishes, amphibians, annelids, molluscs, crustaceans, insects, and other groups have been identified as recent or more ancient polyploids.

With the advent of large-scale genome sequencing, it has become possible to test some of the hypotheses put forth by Ohno and his predecessors by investigating, on a genomewide scale, whether large-scale duplication events of genes or even entire genomes have indeed been important over long evolutionary timescales.

MECHANISMS OF LARGE-SCALE DUPLICATION

In some groups, polyploids appear to form frequently and repeatedly, with a large percentage of species showing signs of recent polyploidization (see Chapters 7 and 8). There is also increasing evidence that many organisms are "paleopolyploids"—that is, ancient polyploids whose genome duplications have been masked by subsequent molecular and chromosomal evolution (see later section). The mechanisms involved in the initial duplications are thought to be the same as those occurring in more recent polyploidization events. Other mechanisms of large-scale duplications, above the level of individual genes but less than entire genomes, are also recognized. The most important of these are described briefly in the following sections. Additional details regarding mechanisms of polyploidization can be found in Chapters 7 and 8, whereas rediploidization, the evolutionary process in which a tetraploid species "decays" to become a diploid, is discussed in more detail by Wolfe (2001).

AUTOPOLYPLOIDY

Polyploidy can occur when an error during meiosis leads to the production of unreduced (i.e., diploid) gametes rather than haploid ones, as shown in Figure 6.1.

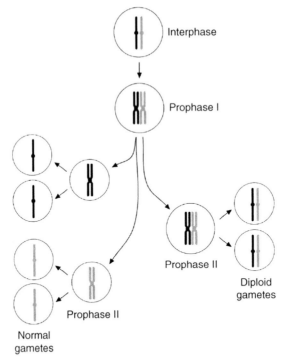

FIGURE 6.1   The basis of autopolyploidization. Autopolyploidization can occur when the pairs of homologous chromosomes have not separated into different nuclei during meiosis. The resulting gametes will be diploid rather than haploid. Based on Brown (1999), reproduced by permission (© BIOS Scientific Publishers).

If two diploid gametes fuse, an autotetraploid will be created whose nucleus contains four copies of each chromosome. Autopolyploids are often viable because each chromosome still has a homologous partner and can therefore form a bivalent during meiosis. This mechanism allows an autopolyploid to reproduce successfully, but prevents interbreeding with the original organism from which it was derived because a cross between a tetraploid and a diploid would give triploid offspring. Unlike tetraploids, triploids are very often sterile because one full set of its chromosomes lacks homologous partners to form the bivalents necessary for segregation (see Chapter 8 for more on the meiotic consequences of polyploidy).

ALLOPOLYPLOIDY

Polyploidy can also result from hybridization of two closely related species, leading to viable hybrids when the genomes are very similar, or to sterile hybrids when

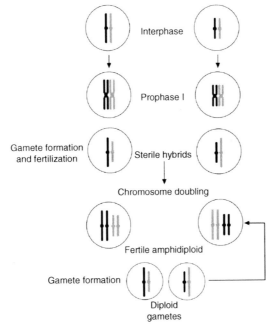

FIGURE 6.2   The basis of allopolyploidization. Allopolyploidy can result from hybridization of two closely related species, possibly leading to sterile hybrids, because chromosomes can not synapse (pair) during meiosis because they are not similar enough. However, when the newly combined genome undergoes a chromosomal doubling, two identical sets of chromosomes are available to pair during meiosis, and a fertile tetraploid is produced.

the chromosomes are insufficiently similar to synapse (pair) during meiosis. However, if the new combined genome undergoes a chromosomal doubling, two identical sets of chromosomes are available to pair during meiosis. As a result, a fertile tetraploid is produced (Fig. 6.2). For the most part, ancient polyploids are assumed to have been formed through allopolyploidy rather than autopolyploidy (e.g., Spring, 2003), although some semiancient polyploids such as the salmonid fishes are believed to be autopolyploid (see Chapter 8).

ANEUPLOIDY

Aneuploids have a chromosome number that differs from an exact multiple of the haploid chromosome set. In such cases, a single chromosome is either lost or added from a normal diploid set of chromosomes. Duplication of individual chromosomes, described extensively for humans (and *Drosophila*), is either lethal or results in serious genetic diseases such as Down syndrome, which is caused by the possession of three copies (trisomy) of Chromosome 21.

BLOCK DUPLICATIONS

Comparative studies have suggested that "block" (or "segmental") duplications—the duplication of large DNA segments—have been a continuing process during evolution. Block duplications are those in which many genes and their upstream regions are duplicated in a single event. However, for a long time it was unclear how such duplications could be generated, whether most of them occur intra- or interchromosomally, whether these tend to be found in direct or inverted orientation, and what sort of sequences are involved at the junctions. Recently, Koszul *et al.* (2004) have used a gene dosage assay for growth recovery in *Saccharomyces cerevisiae* to address these questions. They demonstrated that a majority of the revertant strains resulted from the spontaneous duplication of large DNA segments, both intra- and interchromosomal, ranging from 41 to 655 kilobases (kb) in size. In fact, in many cases dozens of genes were duplicated in a single event. The types of sequences at the breakpoints as well as their superposition with the replication map suggest that spontaneous large segmental duplications mainly result from replication accidents (Koszul *et al.*, 2004).

TANDEM DUPLICATIONS

Tandem duplications are duplications where the two copies of the duplicated region are located immediately adjacent to one another. The process of unequal recombination (or crossing-over) is widely viewed as responsible for the creation of tandem duplications. Well-known examples of gene complexes created by tandem duplications are the *Hox* gene clusters (discussed elsewhere in this chapter) and ribosomal RNA genes (see also Chapter 5).

HOW LARGE-SCALE GENE DUPLICATIONS ARE STUDIED

IDENTIFICATION OF BLOCK DUPLICATIONS

The search for traces of ancient large-scale duplications has received much attention of late, with hypotheses about the number and age of polyploidization events in different eukaryotes a subject of much current debate (Wolfe, 2001; Durand, 2003). Evidence for large-scale gene- or entire genome-duplication events often comes from the detection of block duplications. Identifying duplicated regions at the gene level is usually based on a within-genome comparison that aims to delineate regions of conserved gene content and order (i.e., "colinearity") in different parts of the genome. Disagreement can arise because the detection of colinear regions in genomes is not always straightforward (Gaut, 2001; Vandepoele et al., 2002a).

In general, one attempts to identify a number of homologous gene pairs (typically referred to as "anchor points") in relatively close proximity to each other between two different segments in the genome, either on the same chromosome or on different chromosomes. When such a candidate colinear region has been detected, usually some sort of permutation test is performed in which a large number of randomized datasets is sampled in order to calculate the probability that the observed colinearity could have been generated by chance (Gaut, 2001; Simillion et al., 2002). When the similarity between two genomic segments can be shown to be statistically significant, i.e., unlikely to be the result of chance, the conclusion is that the duplicated genes are the result of a single block duplication.

The statistics that determine colinearity thus depend on two factors: (1) the number of anchor points and (2) their distance from each other. These factors in turn usually depend on the number of "single" genes that interrupt colinearity. The tendency for a high level of gene loss, together with phenomena such as translocations and chromosomal rearrangements (see Chapter 9), often renders it very difficult to find statistically significant homologous regions in the genome, in particular when the duplication events are ancient. Fortunately, techniques are available for dealing with this issue, such as the map-based approach developed by Van de Peer and coworkers described in detail in the following section.

THE MAP-BASED APPROACH

In order to detect chromosomal locations of colinear genes, it is necessary to search for regions that can be paired up because they contain sets of homologous genes. This requires a dataset containing all gene products and their absolute or relative positions in a genomic sequence. The map-based approach to analyzing such data involves only two parameters: (1) G, the "gap size," which specifies the maximum allowable number of intervening, nonhomologous genes between two homologous genes within a colinear segment, and (2) Q, the "diagonal quality" of the colinear regions (see later section).

To detect colinearity in two genomic fragments, a comparative search of all gene products (i.e., the amino acid sequences of the proteins) coded for in the relevant regions is performed using BLASTp (protein–protein Basic Local Alignment Search Tool) (Altschul et al., 1997) (see www.ncbi.nlm.nih.gov/blast). The goal is to detect homologous gene products in the two regions, with two protein sequences considered homologous when they share more than 30% sequence identity over an alignable region of at least 150 amino acids. Homology can still be determined when the matching sequences have an alignable region smaller than 150 amino acids, but this involves more complex analysis to compare the structure, and not just the sequence, of the proteins (using what is called the "homology-derived secondary structure prediction identity cut-off curve") (Rost, 1999).

Once obtained, the information on homologous genes is stored in a so-called "Gene Homology Matrix" (GHM), a hypothetical example of which is illustrated in Figure 6.3. In general terms, such a matrix consists of $m \times n$ elements, with m and n being the total number of genes on each genomic fragment. Pairs of homologous genes ("nonzero elements") in the matrix are identified by the coordinates (x, y). As shown in Figure 6.3A, colinear regions are represented as diagonal lines in the matrix, and tandem duplications are manifested as either horizontal or vertical lines, depending on which genomic segment has the additional copies. Inversions can be detected by looking at the organization of the entries. Gaps in

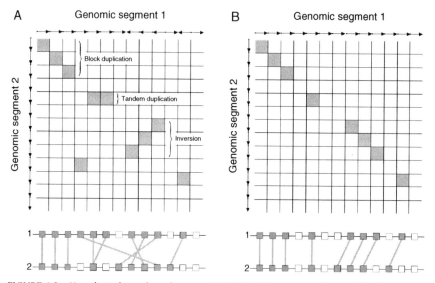

FIGURE 6.3   Hypothetical gene homology matrix (GHM). Each arrow on the axes of both segments represents a gene on the genomic segment. Gray cells illustrate homologous genes (called "anchor points"). (A) The original organization of all genes in their genomic context, with tandem duplications and inversions clearly visible. (B) The same gene homology matrix after tandem remapping and the removal of irrelevant (i.e., not part of a duplication) single data points by the ADHoRe algorithm. In addition, the small inverted colinear segment of three anchor points was restored to its original orientation in order to create a larger colinear region. See text for more details. From Vandepoele et al. (2004b), reproduced by permission (© Bentham Science Publishers Ltd.).

diagonal regions indicate insertions (through translocation, not duplication) or losses of genes in duplicated blocks.

After identification of the homologous genes, irrelevant data points need to be removed by a process referred to as "filtering" (Vandepoele et al., 2002a). The fact that identifying colinearity effectively means finding diagonal series of elements in the matrix reduces the question to what is called a "clustering problem." This allows all elements that are too far away from other elements in the homology matrix to belong to a cluster to be removed during filtering. Next, the vertical and horizontal regions representing tandem duplications are deleted from the matrix. Specifically, these are remapped by collapsing all tandem duplications of a gene with the same orientation and within a distance G into a single element in the matrix. Tandem remapping makes it easier to detect diagonal regions, because then they are no longer interrupted by horizontal or vertical elements. The end result is a matrix in which a duplicated region now appears as a clear diagonal, as illustrated in Figure 6.3B.

In statistical terms, locating the diagonal regions in the matrix involves a special distance function that yields a shorter distance for points that are in diagonally close proximity than for points that are in horizontal or vertical proximity (Vandepoele et al., 2002a). A generalized version of this is depicted in Figure 6.4A, and Figure 6.4B shows the application of this distance function to a hypothetical example. The actual clustering step is conceived as an iterative process, whereby the gap size is gradually increased until the final gap size (G) is reached. During each iteration, the gap size represents the maximum distance between two points in a cluster. Each time the process is repeated, new clusters can be formed and existing clusters can be extended. In the approach described here, by default the initial gap size is set to 3 and is then increased in 10 exponential steps until the final gap size has been reached.

Again, gap size is only one of the two parameters involved in this algorithm. The second is the "quality" of the clusters, meaning that it is important to only join genes to clusters that are assumed to have been created by the same duplication event. This is represented by the second parameter, Q, which determines the extent to which the elements of a cluster actually fit on a diagonal line. This "quality" parameter is estimated by calculating the coefficient of determination (r^2) by linear regression through the points in the clusters. Only clusters with a sufficiently high quality (i.e., higher than the cutoff Q) will be kept. Each addition of a potential gene duplicate to the diagonal line is tested, using the specific distance function described above, to determine the effects on the quality of the line. That is to say, each iteration of the algorithm involves a statistical test of whether the clusters can be enriched by adding single genes ("singletons") or joined with other clusters without badly affecting the cluster's diagonal properties.

Three conditions must be fulfilled for such additions or mergers to be accepted. First, the candidate singleton or cluster must be within a distance

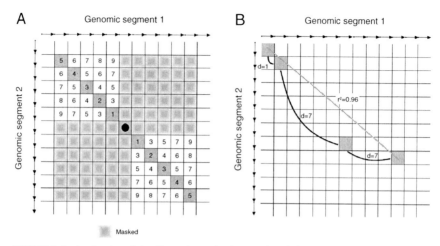

FIGURE 6.4   Application of the diagonal pseudo distance (DPD) function to the detection of elements with diagonal proximity in the gene homology matrix. (A) The DPD for a given cell in the matrix to the central black dot (anchor point). The diagonal pseudo distance is smaller for diagonally orientated elements (gray boxes) than for elements deviating from the diagonal. Shaded boxes represent elements (genes) with an infinite distance to the central dot, because these elements are unlikely to be part of the duplicated segment that contains the black dot. (B) The iterative clustering of elements for a colinear region with positive orientation (i.e., from top left to bottom right) in the homology matrix. All genes lie within a maximum gap distance G (for instance 30) of each other. The best-fit line and its coefficient of determination (r^2) show the quality of the cluster, which is clearly above the predefined Q value cutoff, here set to 0.9. As a result, all four homologous genes are considered to have arisen by a block duplication. From Vandepoele et al. (2004b), reproduced by permission (© Bentham Science Publishers Ltd.).

smaller than or equal to the current gap size in the iteration. Second, the candidate singleton must be positioned within the 99% confidence interval of the cluster (see Fig. 6.5). This confidence interval is computed by considering the best-fit line $y = ax + b$ through all the points in the cluster using the least-squares fit method. Usually, the points in the cluster show a certain degree of deviation from this line, which can be explained by two factors: (1) the error on the calculation of the constants a and b of the regression line, and (2) the error caused by the deviation of the point x_i, y_i from this line. Assuming this deviation is normally distributed, a confidence interval can be calculated that indicates the maximum deviation a candidate singleton can have from the best-fit line. If a singleton or cluster lies within these boundaries, then its effects on the r^2 of the diagonal line will be tested. If adding it does not cause the r^2 to fall below the cutoff (Q), it will be added to the cluster (Fig. 6.5B). The entire process is then repeated, using an increased gap size with this new cluster as the starting point. An example of the real-world application of such a process (using the ADHoRe software tool) to two fragments of the Arabidopsis thaliana genome is shown in Figure 6.6.

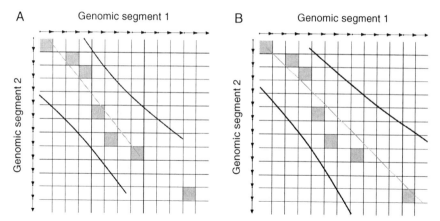

FIGURE 6.5 When adding genes to duplicated segments, it is assessed, using the specific diagonal pseudo distance (DPD) function (see Fig. 6.4), whether the clusters can be enriched with singletons (single genes) (A) or joined with other clusters without badly affecting the cluster's diagonal properties. To this end, the candidate singleton or cluster must be within a distance smaller than or equal to the gap size in the current iteration. Next, the candidate singleton must be positioned within the 99% confidence interval of the cluster. This confidence interval is computed by considering the best-fit line $y = ax + b$ through all the points in the cluster using the least-squares fit method. If these requirements are fulfilled, the segmental block duplication is extended (B).

Compiling a cluster (i.e., identifying a colinear region) is not the end of the procedure, because it is still necessary to remove any clusters that could have arisen by chance. This is accomplished with the use of a permutation test, by sampling a large number of reshuffled datasets and calculating the probability that a colinear region, characterized by a number of conserved genes and an average gap size, can be found by chance. When the similarity between two genomic segments can be shown to be statistically significant in this way, the conclusion is that both

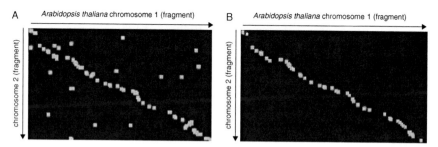

FIGURE 6.6 Example of the application of the ADHoRe software tool to two fragments of the *Arabidopsis thaliana* genome, (A) before and (B) after the filtering process (see Figs. 6.3–6.5).

segments are homologous and have originated by duplication. Permutation tests are very computer intensive, but recently novel, faster statistical methods have been developed to determine the statistical significance of putative homologous segments (Calabrese *et al.*, 2003; Simillion *et al.*, 2004). These methods are based on the observation that a cluster that was generated by chance generally contains fewer anchor points than a truly significant cluster, and that the average distance between these anchor points is also greater. In other words, the more anchor points a cluster contains and the closer these anchor points are located to each other on the diagonal of the GHM, the less likely it is that this cluster has been generated by chance.

Although the identification of block duplications is usually considered strong evidence for large-scale gene duplications, this is not a strict requirement. If many gene duplicates can be shown to have originated at about the same time in evolution, this could also be considered strong evidence that most of these paralogous genes have been created by one single event. Examples of such observations will be discussed later in this chapter.

HIDDEN DUPLICATIONS, GHOST DUPLICATIONS, AND MULTIPLICONS

In addition to the easily recognized "obvious" or "nonhidden" block duplications and tandem duplications (Fig. 6.3), there are also "hidden" and "ghost" duplications that are more difficult to identify (Fig. 6.7). Hidden duplications are heavily degenerated block duplications that cannot be observed by directly comparing both duplicated segments with each other, but only through comparison with a third segment within the genome. Consequently, hidden duplications are important when determining the actual number of duplication events that have occurred over time, as has been demonstrated previously for *Arabidopsis thaliana* (Simillion *et al.*, 2002). An example of such a hidden block duplication in *Arabidopsis* is presented in Figure 6.8.

Ghost duplications are defined as hidden duplications between different genomes. Two genomic segments in the same genome form a ghost duplication when their homology can only be inferred through comparison with the genome of another species (Vandepoele *et al.*, 2002b). In the case of *Arabidopsis* shown in Figure 6.8, for example, if Chromosome 2 proved to be derived from a different parental species than Chromosome 4, then the duplicated segments on Chromosomes 2 and 4 would form a ghost duplication.

As it turns out, a large number of chromosomal segments can often be identified as having been involved in multiple duplications. Such a group of homologous segments is referred to as a "multiplicon." Another way of displaying multiplicons is illustrated in Figure 6.9, which shows a network of colinearity between rice and *Arabidopsis*, including nonhidden, hidden, and ghost duplications.

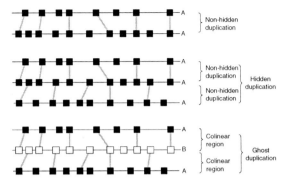

FIGURE 6.7 Schematic representation of nonhidden, hidden, and ghost duplications. Boxes represent the genes on chromosomal segments of genomes A and B, whereas connecting lines indicate the anchor points (i.e., homologous or duplicated genes). Hidden duplications are heavily degenerated block duplications that cannot be observed by directly comparing both duplicated segments, but only through comparison with a third segment from the same genome. Ghost duplications are hidden block duplications that can only be identified through colinearity with the same segment in a different genome. In contrast to hidden duplications, the identification of ghost duplications increases the fraction of the genome involved in a duplication event. From Vandepoele *et al.* (2003), reproduced by permission (© American Society of Plant Biologists).

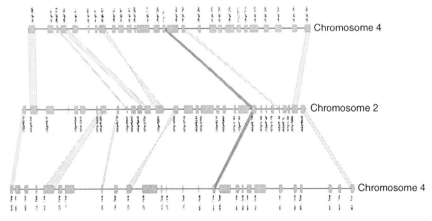

FIGURE 6.8 Example of a multiplicon in *Arabidopsis thaliana*. No duplication can be observed between the two segments on Chromosome 4, because these have only one homologous gene in common (dark gray band). However, both segments still share several, but different, homologous genes with a segment on Chromosome 2. Therefore, both segments on Chromosome 4 form a hidden duplication. If Chromosomes 2 and 4 were found to be derived from two different species, then this would constitute a ghost duplication.

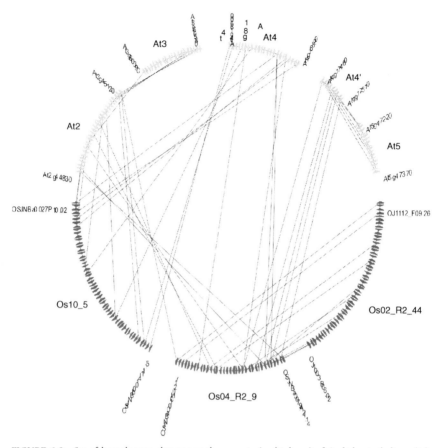

FIGURE 6.9   Set of homologous chromosomal segments (multiplicon) of *Arabidopsis thaliana* (At) and rice (*Oryza sativa*, Os). Arrows represent the genes on the chromosomal segments, whereas connecting lines indicate the anchor points (i.e., homologous or duplicated genes) that are part of a significant colinear region determined by the ADHoRe algorithm. For each genomic segment, the names of the two genes delineating the segment are shown. Chromosomal segments of rice and *Arabidopsis* are shown in dark and light gray, respectively. By considering the colinearity between *Arabidopsis* and rice, a set of at first sight unrelated *Arabidopsis* segments can be joined into a multiplicon with multiplication Level 5, confirming the three duplication events in *Arabidopsis* described earlier (Simillion *et al.*, 2002). Conversely, colinearity between rice and *Arabidopsis* reveals that all three rice segments are linked with each other by two duplication events.

GENOMIC PROFILES: AN EXTENSION TO THE MAP-BASED APPROACH

Although considering transitive homologies such as hidden and ghost duplications allows the identification of many previously undetectable homologous genomic segments, it still requires that these show significant colinearity with at least one other homologous segment. However, it is possible that, within a given multiplicon, one or more segments have diverged so much from the others in gene content and order that they no longer show clear colinearity with any of the other segments. Unfortunately, such segments in the "twilight zone" of genomic homology cannot be detected with any of the currently available methods. New software is being developed (e.g., by Van de Peer and colleagues) to uncover chromosomal segments that are homologous (with respect to having common ancestry) to others but can no longer be identified as such because of extreme gene loss. This is done by aligning clearly colinear segments and using this alignment as a "genomic profile" that combines gene content and order information from multiple segments to detect these heavily degenerated homology relationships (see Fig. 6.10).

After the initial detection of a "Level 2" multiplicon (i.e., a pair of homologous chromosomal segments) with the basic ADHoRe algorithm, an alignment of the two segments that form this multiplicon can be created where the anchor points of the multiplicon are positioned in the same columns. Using this alignment as a "profile," a new type of homology matrix can be constructed in which the gene products of a segment are compared to the gene products of the profile. Once this new GHM is constructed, it is subjected to the basic ADHoRe algorithm, which involves the same statistical validation procedures to detect clusters of anchor points. This time, however, new significant clusters will not reveal homology between two individual segments, but rather between the two segments inside the profile (i.e., the initial Level 2 multiplicon) and a third segment. Because this type of GHM combines gene content and order information of the different segments in the profile, it is possible to detect homology relationships with a third segment that could not be recognized by directly comparing any of the segments of the multiplicon individually with this third segment. If such a third segment is detected, it is added to the multiplicon, thereby increasing its multiplication level, and the corresponding profile is updated by aligning the new segment to it. The entire detection process is then repeated with the newly obtained profile.

By constructing genomic profiles that combine gene content and order information from multiple homologous segments, it becomes possible to detect heavily degenerated homology relationships between segments that no longer show significant colinearity with any of the segments contained in the profile. The strength of this approach is clearly illustrated by the substantial increase in multiplicons it generates in *Arabidopsis* as compared with the traditional approach; indeed, multiplications of Level 5 or greater may be observed in this way (see Simillion *et al.*, 2004).

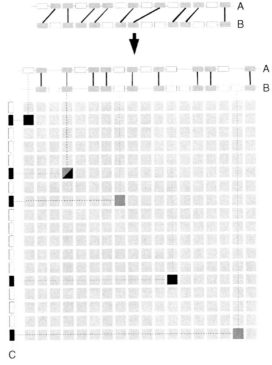

FIGURE 6.10   Detection of homology through a genomic profile. The upper section shows an initially detected Level 2 multiplicon (a pair of homologous chromosomal segments). The gray boxes connected by black lines represent pairs of homologous genes (anchor points) between the two segments. The lower section shows the construction of a homology matrix using this multiplicon as a profile. To accomplish this, the multiplicon is first aligned by inserting gaps at the proper positions (depicted by empty spaces in the alignment). The homology matrix can now be constructed by comparing this profile with the genes of a chromosomal segment C (shown on the left of the matrix). Anchor points in the matrix are detected whenever a gene of this chromosomal segment belongs to the same gene family as one of the genes in any of the segments in the profile. The black squares represent homologs between segments A and C, and the dark gray between B and C. The black/dark-gray square denotes a gene that has a homolog on both segments A and B. Combining segments A and B in a profile thus results in five anchor points with segment C, whereas the individual segments A and B only have three anchor points with segment C, which might be too few to detect statistically significant homology.

DATING DUPLICATION EVENTS

Several methods are commonly being used to date gene duplication events, the most notable of which are (1) absolute dating based on third codon or synonymous substitution rates, (2) absolute dating based on nonsynonymous substitution rates or protein-based distances, and (3) relative and absolute dating by the construction and analysis of phylogenetic trees. These will be discussed in turn.

Absolute Dating Based on Synonymous Substitutions

Because most substitutions in third-codon positions do not result in amino acid replacements (Fig. 6.11), the rate of fixation of these substitutions is expected to be relatively constant in different protein-coding genes (Nei and Kumar, 2000) and to reflect the overall mutation rate (Hughes, 1999a). Time of divergence (T) can be calculated from this as $T = K_S/2\lambda$, where K_S is the fraction of synonymous substitutions per synonymous site and λ is the mean rate of synonymous substitution (Nei and Kumar, 2000). The value for λ differs for various organisms; in *Arabidopsis*, for instance, the estimate is 6.1 synonymous substitutions per 10^9 years, whereas for mammals it is considered to be about 2.5 substitutions per 10^9 years (Lynch and Conery, 2000).

Although silent substitutions have been used extensively to compute duplication events, there is one important caveat, namely that dating based on such substitutions can only be applied when K_S is less than 1. Higher values of K_S point to saturation of synonymous sites and should therefore be used with great caution when drawing any conclusions regarding the date of duplication events. There are different ways to compute the number of synonymous substitutions per synonymous site, depending on which method is used to correct for multiple mutations at these sites. For example, the NTALIGN program in the NTDIFFS software package (Conery and Lynch, 2001) first aligns the DNA sequence of two mRNAs based on their corresponding protein alignment and then calculates K_S by the method of Li (1993). Nei and Gojobori (1986) and Yang and Nielsen (2000) have proposed two alternative methods to compute K_S, both of which are implemented in the PAML phylogenetic analysis package (Yang, 1997).

Protein-Based Distances

Although protein-based distances are known to vary considerably among proteins (Easteal and Collet, 1994) because of different functional constraints, several attempts have been made to use such distances to date duplication events. For example, Vision *et al.* (2000) have used amino acid replacement rates (K_A) to date

```
M    A    L    A    F    D    E    F    G    R    P    F    I    I    L
ATG  GCT  TTG  GCT  TTC  GAT  GAG  TTT  GGC  CGG  CCG  TTC  ATT  ATA  CTA  Duplicate α
ATG  GCG  CTG  GCG  TTC  GAT  GAG  TTC  GGG  CGT  CCG  TTC  ATT  ATA  CTG  Duplicate β
```

FIGURE 6.11 Silent substitutions, indicated in bold, mostly occurring at third codon positions, do not lead to amino acid replacement and are therefore regarded as "neutral," and assumed to follow a clocklike behavior.

block duplication events in *Arabidopsis*. These authors assumed that, whereas the mutation rate of different proteins may vary considerably, the overall distribution of amino acid substitution rates is the same throughout the genome. If that assumption were valid, then any contemporaneously duplicated block containing several homologous pairs would provide a more or less independent sample of the distribution. Furthermore, the average values of K_A for blocks duplicated at the same time must necessarily be much less variable around the true mean than the individual protein values themselves. Unfortunately, there is some evidence from other organisms that rates of protein evolution vary systematically in different regions of the genome. However, for that phenomenon to create problems with dating based on the block averages, the variation among regions would have to be on the same scale as the differences between duplicated blocks of different age classes, and to co-vary among the chromosome pairs in each block (T. Vision, personal communication).

That said, it has been shown that protein distances are not very reliable for dating duplicated blocks containing heterogeneous classes of proteins. For example, different block duplications in *Arabidopsis* estimated to be of similar age based on mean protein distance (Vision *et al.*, 2000) actually turned out to be very heterogeneous in age when compared to dating based on synonymous substitution rates (Raes *et al.*, 2003). The reason is that duplicated blocks that contain a larger fraction of fast-evolving genes will have a relatively high mean protein distance between the paralogous regions and appear older than they actually are. It would therefore seem that the use of synonymous and, consequently, neutral substitutions for evolutionary distance calculations is the more reliable way of estimating duplication events, unless there is no alternative because the duplications are too old.

DATING BY PHYLOGENETIC MEANS

Another way of dating duplication events is by mapping them onto phylogenetic trees. In relative terms, this approach allows a determination of whether duplications have occurred prior to or after a speciation event. For example, in Figure 6.12A, the gene has been duplicated prior to the divergence of zebrafish and pufferfish (~ 150 million years ago), whereas the gene duplications in Figure 6.12B are younger, and have occurred independently in zebrafish and pufferfish after their divergence.

If the timing of a speciation event is known with confidence, gene trees can also be used to infer absolute dates. This is usually performed by the construction of linearized trees (Takezaki *et al.*, 1995), which assumes equal rates of evolution in different lineages of the tree—that is, a molecular clock (see Chapter 9). In order to create such linearized trees, relative rate and branch length tests for rate

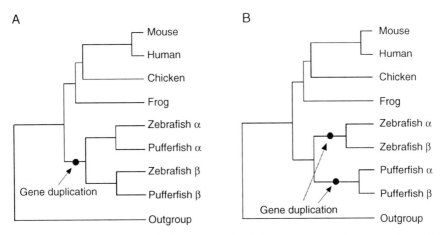

FIGURE 6.12   Relative dating of duplication events by phylogenetic means. Different scenarios—and expected inferred tree topologies—are shown to explain the presence of more genes in fishes. (A) Duplicated fish genes resulting from a gene/genome duplication that preceded the divergence of zebrafish and pufferfish. (B) Duplicated genes formed by independent gene duplications.

heterogeneity are usually applied to these trees to check for deviations from the assumption of a constant molecular clock. Faster or more slowly evolving sequences are then removed so that the dataset contains only sequences evolving at a similar rate. By comparing the divergences of duplicated genes with a fixed calibration point—that is, the date of a particular evolutionary event, such as the divergence between fishes and land vertebrates—the absolute date of origin of paralogous genes can be inferred.

PUTTING THEORY INTO PRACTICE: EVIDENCE FOR LARGE-SCALE GENE DUPLICATION EVENTS

Although there is evidence that individual gene duplications occur frequently and are actually part of a continual process (Lynch and Conery, 2000; Gu et al., 2002) (see Chapter 5), more and more genomic data seem to suggest that many gene duplicates have arisen during major large-scale duplication events. Indeed, ancient duplications of entire genomes have now been documented for members of the three best-studied eukaryotic kingdoms. The first strong evidence for an ancient polyploidy event in eukaryotes came from the yeast *Saccharomyces cerevisiae*. Based on a genomewide analysis, it was postulated that the entire yeast genome had duplicated about 100 million years ago (Wolfe and Shields, 1997), and that as a result, approximately 25% of the yeast genome still consists of duplicated genes

(Seoighe and Wolfe, 1999). Recently, the genome duplication in yeast has been confirmed through comparative analysis with closely related species (Dietrich *et al.*, 2004; Dujon *et al.*, 2004; Kellis *et al.*, 2004). As described in the following sections, some intriguing examples are now also known from animals and plants.

1R/2R: Genome Duplications in Vertebrates

In *Evolution by Gene Duplication*, Ohno (1970) argued that large-scale gene duplication occurred during the evolution of early vertebrates. Although based on rather inaccurate indicators of genome complexity, such as genome size (see Chapter 1) and isozyme patterns, Ohno proposed that two rounds of genome duplications had occurred in the evolutionary past of early vertebrates, one on the shared lineage leading to both cephalochordates and vertebrates, and a second in the fish or amphibian lineage (see also Furlong and Holland, 2002) (Fig. 6.13).

The advent of DNA sequence–based analysis provided more reliable evidence for the hypothesis of two rounds of large-scale gene duplications in the early vertebrates. A prime example of this is the analysis of *Hox* genes (Holland *et al.*, 1994). *Hox* genes encode DNA-binding proteins that specify cell fate along the anterior–posterior axis of bilaterian animal embryos, and occur in one or more clusters of up to 13 genes per cluster (reviewed in Gehring, 1998). The observation that protostome invertebrates, as well as the deuterostome cephalochordate *Branchiostoma lanceolatum* (commonly called "Amphioxus"), possess a single

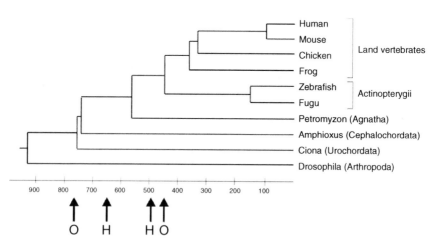

FIGURE 6.13 Phylogenetic tree of major vertebrate groups and their time of divergence. Arrows indicate presumed genome duplications according to (O) Ohno (1970) and (H) Holland *et al.* (1994).

Hox cluster, whereas the lobe-finned fishes (coelacanth and lungfishes), amphibians, reptiles, birds, and mammals have four clusters (Holland and Garcia-Fernandez, 1996; Holland, 1997; Larhammer *et al.*, 2002), supports the hypothesis of two rounds (2R) of entire genome duplications early in vertebrate evolution. Holland *et al.* (1994) proposed that a first duplication occurred on the vertebrate lineage after the divergence of the cephalochordates, and a second one after the divergence of the jawless vertebrates (Fig. 6.13). Although some support can be found for this hypothesis (see, for example, Escriva *et al.*, 2002), the recent discovery of three, and most probably four *Hox* clusters in the lamprey *Petromyzon marinus* suggests that the two rounds of (*Hox* cluster) duplications occurred before the divergence of lampreys and hagfishes (Irvine *et al.*, 2002; Vandepoele *et al.*, 2004a). By contrast, some other authors assume an independent duplication history of the lamprey *Hox* clusters, and therefore do not consider this evidence for two rounds of large-scale gene duplication events prior to the divergence of lampreys and hagfishes from the other vertebrates (Fried *et al.*, 2003).

Spring (1997) uncovered an average of three paralogs in humans for each of 52 *Drosophila* genes and proposed that the additional human genes were produced during two allopolyploidization events in the early vertebrate lineage. The presence in four copies of various segments in vertebrate genomes has been reported in subsequent studies, which likewise is suggestive of two large-scale duplications (Abi-Rached *et al.*, 2002; Lundin *et al.*, 2003). Additional support for 1R or 2R of genome duplication comes from the detection and dating of duplicated blocks in the human genome and from large-scale phylogenetic analyses of gene families (Abi-Rached *et al.*, 2002; Gu *et al.*, 2002). Recently, McLysaght *et al.* (2002) described an extensive gene duplication during early chordate evolution. They suggested that at least one (maybe two) round(s) of polyploidization occurred in the early history of vertebrates, and concluded that humans, like yeast and *Arabidopsis* (see later section), are ancient polyploids. Gu *et al.* (2002) showed that both large- and small-scale duplications are required to explain the age distribution of duplicated human gene families.

Although a consensus seems to be emerging that large-scale gene or even entire genome duplication events have occurred in the evolution of early vertebrates, rediploidization and degeneration of duplicate genes generally makes strong evidence in support of 2R hard to find. As a consequence, the 1R/2R hypothesis of vertebrate genome evolution is still hotly debated, with opinions ranging from strongly in favor (e.g., Holland *et al.*, 1994; Furlong and Holland, 2002; Larhammer *et al.*, 2002; Panopoulou *et al.*, 2003; Spring, 2003; Vandepoele *et al.*, 2004a) to highly skeptical (e.g., Hughes *et al.*, 2001; Martin, 2001; Friedman and Hughes, 2003).

Much of this confusion may stem from the nature of the duplication events themselves, in particular their timing relative to each other. For example, some advocates of the 2R hypothesis believe that the two rounds of genome duplications occurred in very short succession (Larhammer *et al.*, 2003). This would explain why it is generally hard to infer phylogenetic trees of the form ((A,B)(C,D)) using gene duplicates, which in principle should be easy to do if two tetraploidization events had occurred (Skrabanek and Wolfe, 1998; Hughes, 1999b; Martin, 2001). If both genome doublings indeed occurred almost contemporaneously, it is not surprising that they cannot easily be distinguished based on age differences between genes or the topology of gene family trees. However, as more large-scale genome sequence data become available, it should be possible to improve the resolution of such analyses and perhaps to answer this question conclusively.

3R: An Additional Round of Genome Duplication in Teleost Fishes

A few years ago, it was proposed that an additional (3R) genome duplication had occurred in ray-finned fishes (Aparicio *et al.*, 1997; Amores *et al.*, 1998; Wittbrodt *et al.*, 1998). As with the proposed duplication event(s) shared by all vertebrates, the first indications for a fish-specific genome duplication came from studies of *Hox* genes. Extra *Hox* gene clusters have been discovered in the zebrafish (*Danio rerio*), medaka (*Oryzias latipes*), the African cichlid *Oreochromis niloticus*, and the pufferfish *Takifugu rubripes*. The observation that such distantly related species all share this feature suggested the occurrence of an additional genome duplication event in the ray-finned fish lineage (Actinopterygii) before the divergence of most teleost species (Amores *et al.*, 1998; Wittbrodt *et al.*, 1998; Meyer and Schartl, 1999; Naruse *et al.*, 2000; Málaga-Trillo and Meyer, 2001).

More recent comparative genomic studies have turned up many more genes and gene clusters for which two copies exist in fishes but not in other vertebrates (e.g., Postlethwait *et al.*, 2000; Woods *et al.*, 2000; Robinson-Rechavi *et al.*, 2001a; Taylor *et al.*, 2001, 2003; Van de Peer *et al.*, 2001). The findings that different paralogous pairs appear to have originated at about the same time (Taylor *et al.*, 2001), that different fish species seem to share ancient gene duplications (Taylor *et al.*, 2003), and that different paralogs are found on different linkage groups in the same order (i.e., show synteny) with other duplicated genes (Gates *et al.*, 1999; Postlethwait *et al.*, 2000; Woods *et al.*, 2000), all support the hypothesis that these genes arose through a complete genome duplication event. However, it bears noting that some authors have argued that an ancestral whole-genome duplication event was not responsible for the abundance of duplicated fish genes. For example, Robinson-Rechavi *et al.* (2001a,b) counted orthologous genes in fishes and mice and, where extra genes were found in fishes, compared the number of gene duplications occurring in a single fish lineage with that shared by more than one lineage. Most mouse genes surveyed were also found as single copies in fishes. Duplicated fish genes were detected, but most were interpreted as the products of lineage-specific duplication events in fishes and not as a single ancient duplication event.

In order to find further evidence for or against large-scale gene duplication events in early vertebrate evolution, Vandepoele *et al.* (2004a) recently analyzed the complete genomes of the pufferfish *Takifugu rubripes* ("Fugu") and human. Phylogenetic trees were constructed for all (i.e., 3077) gene families containing two to 10 duplicated Fugu genes, and relative dating of duplication events was performed to test whether gene duplications occurred before (1R/2R) or after (3R) the divergence of the lineages that led to ray-finned fishes and land vertebrates (Fig. 6.14). This analysis showed that most paralogous genes in pufferfish are the result of at least two, probably three, complete genome duplications.

Absolute dating of duplication events was performed through the inference from linearized trees (Takezaki *et al.*, 1995). In these linearized trees—where branch-length is drawn directly proportional to time—the split between ray-finned fishes and land vertebrates (dated at 450 million years ago) (Carroll, 1988; Benton, 1990; Zhu *et al.*, 1999) was used as a calibration point for the dating of gene duplication events. The removal of trees with insufficient statistical support left 595 nodes, based on the analysis of 488 gene families, for which an absolute date could be inferred. Combining the results of relative and absolute dating, these 565 duplications could be subdivided into 166 3R and 399 1R/2R duplications (Figs. 6.14 and 6.15).

Put another way, these results indicate that a major fraction (30%) of the Fugu paralogs is younger than the split between ray-finned fishes and land vertebrates, probably arising somewhere between 225 and 425 million years ago. The most plausible and parsimonious explanation for this observation would be a large-scale

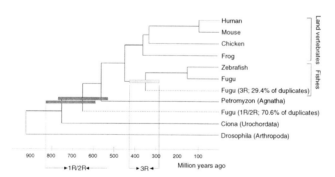

FIGURE 6.14 Phylogenetic tree of major vertebrate groups and superimposed Fugu gene duplication events. Black and gray bars denote large-scale gene duplication events observed in the Fugu genome based on absolute and relative dating and the detection of segmental duplications (see text for details). The time of divergence for the lamprey *Petromyzon*, as a representative of the Agnatha, was taken from Shu *et al.* (1999). From Vandepoele *et al.* (2004a), reproduced by permission (© National Academy of Sciences USA).

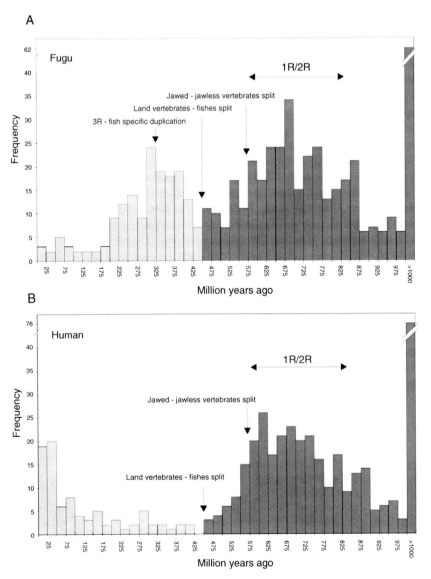

FIGURE 6.15   Age distribution of duplicated genes in the (A) Fugu (pufferfish) and (B) human genomes. Dark bins correspond to duplications prior to the divergence of ray-finned fishes and land vertebrates; light bins correspond to duplications after the split between ray-finned fishes and land vertebrates (see text for details). From Vandepoele *et al.* (2004a), reproduced by permission (© National Academy of Sciences USA).

gene or entire genome duplication. To test whether the sudden increase in the number of duplicated genes in the Fugu genome is the result of an entire genome duplication rather than an increased rate of independent tandem duplication events, Vandepoele et al. (2004a) investigated whether these duplicated genes appear in duplicated blocks on chromosomes (again, the identification of duplicated blocks is usually considered strong evidence for large-scale gene duplication events).

Using the map-based approach outlined above, Vandepoele et al. (2004a) identified statistically significant regions of microcolinearity (showing the same gene content and gene order) within the complete Fugu genome. All genes within such a region are presumed to have been duplicated at the same time, and hence to be of identical age, because it is unlikely that these colinear regions would be created independently on different chromosomes. By applying the ADHoRe algorithm to scaffolds of the available pufferfish genome sequence, and using phylogenetic methods to date the duplicated blocks so identified, Vandepoele et al. (2004a) were able to conclude that the 3R blocks of duplicated genes all arose at approximately the same time, namely about 320 million years ago, with a standard deviation of 67 million years.

Of course, it might be argued that a standard deviation of 67 million years is rather large and could indicate the occurrence of several independent block duplications rather than a single genome duplication event. However, when using an absolute dating approach, such a variance on estimated duplication times is to be expected even when the duplicates are of the same age (Vandepoele et al., 2004a). In particular, within the same block duplication, homologous genes that have been duplicated at the same time can exhibit a considerable difference in estimated duplication time owing to deviations of the molecular clock. The simultaneous duplication for these genes is supported statistically (Vandepoele et al., 2004a), even with the very fragmented nature of the Fugu scaffold dataset used in the analysis. In fact, the number of duplicated blocks is probably much higher, and is expected to rise considerably once better assemblies of the Fugu genome become available. This suggests that the wide distribution of duplicated Fugu genes already observed is in perfect agreement with the hypothesis of a single complete genome duplication event. Overall, such considerations provide very strong support for a complete genome duplication event in the early stages of fish evolution, predating the origin of most modern ray-finned fish species that are believed to have (started to) diverge(d) from each other more than 200 million years ago.

Additional evidence for a fish-specific 3R duplication event comes from analyzing nonfish genomes. Using the same methods as for Fugu, Vandepoele et al. (2004a) performed an analysis of gene duplicates in the human genome. Of the 447 duplication events identified in the human genome sequence, absolute dating suggested that 360 can be attributed to 1R/2R whereas 87 were specific to humans (see Fig. 6.15B). The distribution of inferred ages of duplicated genes shows a similar increase in the number of duplication events around 675 million years ago, as observed in Fugu. Not only does this support the 1R/2R hypothesis, but it also confirms the expectation that no evidence of the hypothesized fish-specific 3R genome duplication event is found in the human genome (Fig. 6.15B).

To summarize, the relative and absolute dating of hundreds of gene families, together with the detection of many duplicated blocks that have originated at about the same time, provides strong support for the hypothesis of a fish-specific genome duplication ~320 million years ago that was not experienced in the lineage of vertebrates leading to humans. This 3R genome duplication event accounts for the large majority of gene duplicates found in the Fugu genome, in contrast to the situation in the human genome, where many more recent tandem and segmental duplication events account for the majority of duplicated genes (Bailey *et al.*, 2002) (Fig. 6.15). Most of the remaining paralogs seem to have been created by one or two much older large-scale duplication events, predating the split between ray-finned fishes and terrestrial vertebrates. Indeed, using the fish-specific genome duplication as a benchmark, and assuming equal rates of gene loss throughout vertebrate evolution, two genome duplications rather than one seem to have occurred—as proposed by Ohno in 1970.

ANCIENT GENOME DUPLICATIONS IN PLANTS

As discussed in Chapter 7, estimates of the prevalence of polyploidy in flowering plants have been increasing over time, beginning at about 30–50% in the 1930s and 1950s, to 70–80% in the 1980s and 1990s. Today, it is becoming more common to suggest that 100% of angiosperms have polyploidy in their ancestry. Much of this new view is based on the discovery of ancient polyploidy even in plants in which it was not at all expected.

As a most notable example, although initial sequencing of the tiny genome of *Arabidopsis thaliana* revealed numerous duplicated segments (Paterson *et al.*, 1996; Lin *et al.*, 1999; Mayer *et al.*, 1999; Terryn *et al.*,1999), this plant was long believed to be a clear example of a diploid organism. However, after bacterial artificial chromosome (BAC) sequences representing approximately 80% of the genome had been analyzed, almost 60% of the genome was found to contain duplicated genes and regions (Blanc *et al.*, 2000). This phenomenon could only be explained by a complete genome duplication event, an opinion shared by the *Arabidopsis* Genome Initiative (2000).

Comparative studies of BACs between *Arabidopsis* and soybean (Grant *et al.*, 2000), and between *Arabidopsis* and tomato (Ku *et al.*, 2000), led to similar conclusions. In the latter case, two complete genome duplications were proposed: one 112 million years ago and another 180 million years ago. After dating duplicated blocks through a molecular clock analysis, Vision *et al.* (2000) also rejected the single-genome duplication hypothesis put forward by the *Arabidopsis* Genome

Initiative (2000). Several different age classes among the duplicated blocks were found, ranging from 50 to 220 million years, and at least four rounds of large-scale duplications were postulated. One of these classes, dated to approximately 100 million years ago, grouped nearly 50% of all the duplicated blocks, suggesting a complete genome duplication at that time (Vision *et al.*, 2000). However, the dating methods used for these gene duplications were based on averaging evolutionary rates of different proteins, which was later criticized because of their high sensitivity to rate differences (Wolfe, 2001; Raes *et al.*, 2003). Nevertheless, Vision *et al.* (2000) had discovered multiplicons of greater than Level 2, which can only be explained by multiple duplication events. By applying the novel techniques described earlier to detect heavily degenerated block duplications in *Arabidopsis*, Simillion *et al.* (2002) showed that the genome of this species had been reshaped by not one, but three entire genome duplication events. Recently, this result has also been confirmed through the construction and dating of evolutionary trees using genes from *Arabidopsis* and other plants (see Bowers *et al.*, 2003).

In stark contrast to *Arabidopsis*, where initial sequencing of the genome quickly revealed numerous duplicated segments, no clear evidence for ancient genome duplications had been reported for rice (*Oryza sativa*) until very recently, even though a paleopolyploid origin had been suggested for this species on several occasions (e.g., Goff *et al.*, 2002; Levy and Feldman 2002). Because the rice genome has now been completely sequenced (Goff *et al.*, 2002; Yu *et al.*, 2002), it is possible to apply the same approaches used in *Arabidopsis*. Based on a BAC assembly covering more than 70% of the genome sequence of *O. sativa,* the ADHoRe algorithm was applied to detect block duplications at the gene level. In addition to the detection of a large number of duplicated segments by direct comparison of all rice genomic scaffolds, a comparative approach using the genome sequence of *Arabidopsis* also yielded a set of ghost duplications, reflecting heavily degenerated duplicated segments. Of the 43 large block duplications (i.e., those with more than five anchor points), 34% of the total number of genes in these segments are retained as duplicates. When taking into account the estimated time of duplication, this fraction of retained gene duplicates is very similar to what has been observed in *Arabidopsis* and yeast (28% and 25%, respectively; Wolfe and Shields 1997; Simillion *et al.*, 2002), which seems to indicate similar rates of gene loss after duplication events.

When examining all multiplicons present in the rice genome through nonhidden, hidden, and ghost duplications, it is apparent that approximately 1.3% of the genome resides in multiplicons higher than Level 2. This implies that, given the quality of the current rice genomic data, a very small number of chromosomal regions have been involved in multiple duplication events. Again, this is very different from the situation in *Arabidopsis*, where the majority of chromosomal regions have been involved in multiple duplication events (Vision *et al.*, 2000; Simillion *et al.*, 2002; Bowers *et al.*, 2003).

In order to answer the question of whether rice is an ancient polyploid, Vandepoele et al. (2003) compared the duplication history of *Arabidopis* and rice by plotting the total number of gene pairs in both species against their genetic distance inferred from the nucleotide substitutions at silent sites. When all duplicated gene pairs in *Arabidopsis* and rice are plotted as a function of K_s, the shape and height of the two curves are quite different (Vandepoele et al., 2003). In *Arabidopsis*, the number of duplicates with K_s values between 0.6 and 0.9 increases dramatically, which corresponds with a genome duplication about 40 to 75 million years ago, as previously reported (Lynch and Conery, 2000; Simillion et al., 2002; Blanc et al., 2003; Bowers et al., 2003). A small but significant increase can also be observed for rice duplicates with K_s values between 0.6 and 1.1. Because the relative increase in the number of duplicates is much smaller in rice than in *Arabidopsis*, a complete genome duplication in rice was considered highly unlikely, with aneuploidy given as the preferred explanation. However, recent analysis of a better assembly of the rice genome does seem to provide evidence for the occurrence of a whole genome duplication in rice about 70 million years ago (Guyot and Keller, 2004; Paterson et al., 2004).

LARGE-SCALE DUPLICATIONS IN THE EVOLUTIONARY PROCESS

THE MAINTENANCE OF DUPLICATED GENES

Before considering the role of large-scale duplications in influencing patterns of evolution, it is important to briefly review what happens to the genes themselves after duplication. Specifically, it is useful to consider why duplicated genes might be preserved in the genome over long evolutionary time periods. Some of these concepts were covered in more detail in Chapter 5 with reference to smaller duplications, but they also apply to genes duplicated *en masse*. The possibilities described here include "neofunctionalization," "subfunctionalization," and functional shift owing to positive selection.

After duplication, the two copies of the gene are redundant, meaning that they perform the same function and that inactivation of one gene should have little or no effect on the biological phenotype (Nowak et al., 1997; Gibson and Spring, 1998; Lynch and Conery, 2000; Gu et al., 2003). Therefore, because one of the copies is freed from functional constraint, mutations in this gene will be selectively neutral and will most often turn the gene into a nonfunctional pseudogene. As discussed in Chapter 5, the hypothesis presented by Ohno (1970) and several of his predecessors that gen(om)e duplications are vital for evolutionary diversification was often based on the notion of "neofunctionalization." That is, instead of being rendered inactive, on rare occasions one of the copies may be converted to a novel gene with a new function by a fortuitous series of nondeleterious mutations (Ohno, 1970, 1973). Although this model has been widely adopted to explain the evolution of functionally novel genes, little evidence has actually been found to support this mechanism. Moreover, under Ohno's model, one might consider it unlikely that anciently duplicated genes still perform completely redundant functions, yet redundancy has been shown to be widespread in the genomes of complex organisms (Nowak *et al.*, 1997, and references therein; Gibson and Spring, 1998; Li *et al.*, 2003).

The more recent alternative "duplication-degeneration-complementation" (DDC) model provides some explanation as to why duplicate genes might be retained (Force *et al.*, 1999; Lynch and Force, 2000). As noted in the previous chapter, this model starts from the assumption that a gene can perform several different functions; for instance, genes are expressed in different tissues and at different times during development, which may be controlled by different DNA regulatory elements. When duplicated genes lose different regulatory subfunctions, each affecting different spatial and/or temporal expression patterns, they must complement each other by jointly retaining the full set of subfunctions that were present in the single ancestral gene. Therefore, degenerative mutations facilitate the retention of duplicate functional genes, where both duplicates now perform different but necessary subfunctions. Therefore, the DDC model predicts that the sum of the retained duplicates is equal to the total number of subfunctions performed by the ancestral gene.

In short, according to the DDC model of Force *et al.* (1999), degenerative mutations preserve rather than destroy duplicated genes, but also change, or at least restrict, their functions to make them more specialized. Such a mechanism may prove to apply to the retention of many different gene duplicates, and indeed an increasing number of genes expected to have been subfunctionalized is being described (e.g., Prince and Pickett, 2002; Van de Peer *et al.*, 2003).

It is not only genes expressed in different tissues or at different times that can be subfunctionalized. For example, Gibson and Spring (1998) argued that selection can prevent the loss of redundant genes (i.e., duplicates) if these genes code for components of multidomain/multimer proteins. This is because inferior copies of these genes (or rather, their gene products) might inhibit the proper working of the "original" gene product. This hypothesis might explain why many transcription factors (TFs), which often form dimers, in gene families of plants contain so many members, many of which are probably redundant (De Bodt *et al.*, 2003; J. Spring, personal communication).

Positive Darwinian selection can also be responsible for functional divergence between duplicated genes (e.g., Zhang *et al.*, 1998; Duda and Palumbi, 1999; Hughes *et al.*, 2000). Most studies that look for evidence of positive Darwinian selection[1]

---

[1]For a review of the computational methods used to detect positive selection in duplicated genes, see Raes and Van de Peer (2003).

compare the ratio of nonsynonymous (p_N) and synonymous (p_S) substitutions (Hughes, 1999a; Nei and Kumar, 2000). In most genes, synonymous substitutions occur at a higher rate than nonsynonymous ones, because purifying selection prevents amino acid sequence changes (which are mostly disadvantageous). Under neutral evolution, the rates of synonymous and nonsynonymous substitutions are expected to be equal (Kimura, 1983). However, under positive Darwinian selection, amino acid replacements (i.e., nonsynonymous mutations) are favored. As a result, nonsynonymous mutations occur at a faster rate than synonymous mutations, as has been shown previously for genes and proteins such as primate lysozymes (Messier and Stewart, 1997), pregnancy-associated glycoproteins (Hughes *et al.*, 2000), primate ribonuclease genes (Zhang and Nei, 2000), conotoxins (Duda and Palumbi, 1999), opsins (Yokoyama *et al.*, 2000; Terai *et al.*, 2002), MYB DNA binding proteins (Jia *et al.*, 2003), and many others (see Endo *et al.*, 1996, and references therein).

Overall, the number of examples of the evolution of new and potentially adaptive functions in duplicated genes is, although growing, still quite small. Some of the more notable examples are the antifreeze proteins in Antarctic fishes (Cheng and Chen, 1999), color vision in new-world monkeys (Dulai *et al.*, 1999), thermal adaptation in *Escherichia coli* (Riehle *et al.*, 2001), and RNA digestion in colobine monkeys (Zhang *et al.*, 2002) (see also Chapter 5). Of course, large-scale gene or complete genome duplications, by whatever means, would provide an enormous number of "extra" genes with the potential to evolve new functions.

WHICH GENES ARE MAINTAINED, AND WHY?

The recent analyses of complete genome sequences have indicated that large-scale gene duplication has probably been rampant during the evolution of plants, fungi, and animals. It is tempting to speculate on the importance of such events for the biological evolution of these organisms. Indeed, as discussed earlier, it is to be expected that such major duplication events have been responsible for important evolutionary transitions and/or adaptive radiations of species (see also Chapters 5 and 11). However, providing hard evidence for direct correlations between large-scale gene duplication events and major leaps in evolution is not straightforward. An important first step in demonstrating that gene duplication events have indeed been of major importance for biological evolution would be to show which (kinds of) genes have generally been retained after gene duplication events.

For bacteria, this is relatively easy to do. With many complete genomes at hand, as well as more reliable genome annotations, it is possible to study which functional classes of genes show an excess of retained genes after duplication. Recent analysis of the functional classification of duplicated genes in bacteria, mainly created by small-scale duplication events such as tandem and operon duplications, revealed a preferential enrichment in functional classes that are involved in transcription, metabolism, and defense mechanisms (Gevers *et al.*, 2004).

Based on such analyses, it is also possible to consider links between gene retention and specific observations regarding the evolution and adaptation of organisms. For example, in the paranome of mycobacteria, two functional classes with an excess of retained duplicated genes are prominent, namely "lipid transport and metabolism" and "secondary metabolites biosynthesis, transport and catabolism" (Gevers *et al.*, 2004). Regarding the fatty acid metabolism, this is in agreement with the complex nature of the *Mycobacterium* cell wall and might reflect adaptive evolution of the bacterial cell surface. The case of *Borrelia burgdorferi* (the Lyme disease spirochete) is also informative in this context. In this species, the biased retention of duplicated motility genes and chemotaxis genes, together comprising more than 6% of its proteome, also appears to be biologically significant. Because *B. burgdorferi* lacks recognizable virulence factors, its ability to migrate to distant sites in its tick and mammalian hosts is probably dependent on a robust chemotaxis response (Fraser *et al.*, 1997). It has been suggested that multiple chemotaxis genes can be differentially expressed under varied physiological conditions or that different flagellar systems exist, requiring different chemotaxis systems (Fraser *et al.*, 1997).

Unfortunately, such analyses are much less straightforward in eukaryote genomes, in particular when the goal is to link gene retention with large-scale gene or entire genome duplication. For *Arabidopsis,* mathematical models are under development that will describe and simulate the retention of gene duplicates through time. Such models assume a constant "background" birth rate of new duplicates on which the three genome duplication events inferred for the *Arabidopsis* genome can be superimposed (Simillion *et al.*, 2002). Furthermore, this can allow different large-scale gene duplication events to have different decay rates with respect to each other and with respect to the continual background duplication process. Modeling both the continual mode of gene duplication as well as large-scale gene duplication events will also allow a comparison of the retention (and decay) of duplicates following large-scale duplication events for different functional categories of genes. It is hoped that this will provide a list of genes or gene categories that have been most important in driving evolution after duplication.

THE MAINTENANCE OF DUPLICATED GENOMES

If Ohno's proposition were true—that redundant genes, produced during large-scale gene duplication events, evolve previously nonexistent functions important for the evolution of phenotypic "complexity"—then traces of such events should be uncovered when the genomes of "complex" organisms are analyzed. Thanks to recent advances in genome sequencing and bioinformatics, it is now recognized that many eukaryotes have undergone large-scale duplications of chromosomal segments and/or entire genomes (Wolfe and Shields, 1997; *Arabidopsis* Genome Initiative, 2000; Wolfe, 2001; Simillion *et al.*, 2002; Blanc *et al.*, 2003; Bowers *et al.*, 2003; Vandepoele *et al.*, 2003, 2004a). However, although duplicated genes and genomes may provide the raw material for evolutionary diversification, and functional divergence of duplicated genes (by several possible mechanisms) might offer a selective advantage to polyploids over a long time period, it is not yet clear how a partially or fully duplicated genome proves beneficial for an organism shortly after the duplication event. In other words, if a new genome doubling is to survive long enough to exert its long-term evolutionary effects, it must provide an immediate selective advantage that allows it to become established. There are several ways in which newly duplicated genomes might fulfill this requirement.

An important characteristic of duplicated genes is that they can buffer the genome against environmental perturbations and mutations, because when one copy of the gene is somehow inactivated, another with the same or similar function can be used instead. For example, Gu *et al.* (2003) have studied the effects of duplicated genes on the "fitness" of individuals of the budding yeast *Saccharomyces cerevisiae*. Based on functional data at the whole-genome level, the knocking out of single-copy genes was shown to generally reduce fitness more severely than deleting one gene of a pair of duplicates. As expected, duplicated genes that are highly similar in sequence are better at compensating for each other than duplicates whose sequences have diverged further. In conclusion, the study of Gu *et al.* (2003) demonstrates that duplicated genes may play an important role in genetic robustness against null mutations.

In plants, polyploidy often has immediate phenotypic effects with potential consequences for fitness, such as increased cell and organ size, faster growth, and increased capacity for invading new habitats (e.g., Osborn *et al.*, 2003) (see Chapter 7). In many cases, such differences in phenotype are probably caused by increased variation in dosage-regulated gene expression (Guo *et al.*, 1996). The fact that most ancient polyploids are thought to have been formed through allopolyploidy rather than autopolyploidy (Spring, 2003) is also relevant in this regard. Specifically, the combination of different genomes can lead to "hybrid vigor," placing the newly formed polyploid at a selective advantage compared to closely related diploid organisms.

Certainly, the prominence of polyploidy in flowering plants (see Chapter 7) implies that it has some adaptive significance, and hybridization has long been considered to be a significant evolutionary force that creates opportunities for adaptive evolution and speciation (Anderson, 1949; Ehrendorfer, 1980; Arnold, 1997; Ramsey and Schemske, 2002; Osborn *et al.*, 2003). Recently, Rieseberg *et al.* (2003) provided evidence that hybridization can play a key role in adaptation. These authors have employed several approaches to study the role of hybridization in ecological adaptation and speciation in sunflowers, and showed that hybridization facilitated ecological divergence. In accordance with Spring (2003), they suggested that hybridization provides a mechanism for large and rapid adaptive transitions, made possible by the genetic variation at hundreds or thousands of genes in a single generation. Amores *et al.* (1998) and Wittbrodt *et al.* (1998) have also suggested that the potentially more complex genomic architecture of fishes resulting from an additional genome duplication might have permitted them to adapt and speciate more quickly in response to changing environments. Many studies have indeed shown that speciation can occur very rapidly in fishes, with the best known case being that of the African cichlids (Meyer *et al.*, 1990; Sturmbauer and Meyer, 1992; Meyer, 1993; Stiassny and Meyer, 1999; Wilson *et al.*, 2000).

In short, genome duplications may offer short-term selective advantages at each of the molecular, phenotypic, and ecological levels, in addition to influencing the long-term diversification of lineages.

SPECIATION AND DIVERGENT RESOLUTION

Based on isozyme studies in ferns, Werth and Windham (1991) developed a model in which the "reciprocal silencing" of genes in geographically separated (allopatric) populations would promote speciation. Recently, this idea was revived in a model called "divergent resolution" (Lynch and Conery, 2000; Lynch and Force, 2000), in which the loss or silencing of gene duplicates may be even more important to the evolution of species diversity than the acquisition of new functions by the duplicated genes.

Divergent resolution occurs when different copies (on different chromosomes) of a duplicated gene are lost in allopatric populations, thereby creating genetic barriers to reproduction between them. Specifically, hybridization between such allopatric populations would produce an F_1 generation with one functional allele and one pseudogene at each of the duplicated loci (see Fig. 6.16). This in itself would not be problematic, but subsequent crosses between F_1 individuals would produce individuals with between zero and four alleles at the duplicated loci (Werth and Windham, 1991; Lynch and Force, 2000; Taylor *et al.*, 2001). Selection against F_2 individuals with more or fewer than two alleles per locus might provide a genetic environment in which speciation alleles (e.g., alleles for assortative mating) would be favored. Therefore, large-scale gene duplications might bring about rapid divergence because natural selection would favor speciation over hybridization in populations fixed for different copies of a duplicated locus.

Genome duplications produce an enormous number of gene duplicates that could be divergently resolved, with such genes potentially playing a prominent role in the generation of biodiversity by promoting the origin of postmating reproductive barriers (Fig. 6.16). In this respect, it is noteworthy that in both ray-finned fishes and flowering plants there is a strong indication for a polyploidization event that seems to coincide with a massive diversification of novel lineages (Bowers *et al.*, 2003; Simillion *et al.*, 2003; Taylor *et al.*, 2003). On the other hand, whereas several studies have shown variation among populations in the retention of

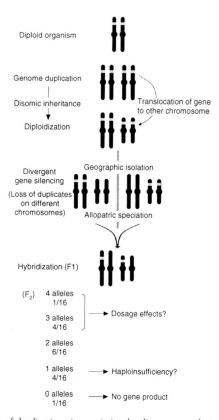

FIGURE 6.16   The role of duplications in speciation by divergent resolution. Gray bands represent a locus that is duplicated (along with all other loci) during a tetraploidization event. In this hypothetical example, diploidization is driven by a reciprocal translocation (to a different chromosome, depicted by a change in chromosome size). If individuals from isolated populations mate, their hybrid progeny would be heterozygous, possessing a functional allele (gray) and a pseudogene (black) at each locus of the duplicated gene. Crosses between the F₁ individuals produce some (approximately 6%) F₂ individuals with only pseudogenes at both of the loci in question, and therefore lacking viability and/or fertility. Others would receive from one allele, which might reduce functionality when the gene product from one functional allele is inadequate to support normal function (a phenomenon called "haploinsufficiency"), to three or four functional alleles, which might have a negative dosage effect. All these might lead to postmating reproductive isolation (Lynch and Force, 2000).

duplicated loci (reviewed by Taylor *et al.*, 2001), none has uncovered the pattern of gene loss predicted by the model. The tetraploid fish family Salmonidae (e.g., trout, salmon, char), which has many more species than its diploid sister group Esocidae (pike, pickerel, mudminnows), would be one good group in which to look for evidence of speciation owing to divergent resolution.

CONCLUDING REMARKS AND FUTURE PROSPECTS

It is becoming increasingly apparent that large-scale duplication events have featured prominently in many taxa, even those with small genomes. As the pace of complete genome sequencing continues to quicken, detailed investigations of this issue will become possible in an ever-widening array of species.

One important challenge for the next wave of studies will obviously be to identify additional duplication events themselves. This will be facilitated by the continued refinement of existing analytical techniques, as well as the development of new ones. It will also be important to discern the mechanisms responsible for these large-scale genomic events and to provide accurate estimates of the timings at which they have taken place. An understanding of the process of rediploidization, in which major duplication events are functionally undone, is also an area of considerable interest. The nature of the gene duplicates that persist, and the process by which duplicate pairs may diverge, is likewise a subject that is only beginning to be understood.

Perhaps most important of all will be the gaining of new insights, from the comparative study of many different genomes, regarding the evolutionary implications of large-scale gene and genome duplications. This involves considerations at several different levels, including impacts at the level of individual genes and entire genomes, the phenotypic and population-level consequences for the first organisms to exhibit the newly duplicated configuration, the possible input into the speciation process, and the long-term implications for major patterns of diversification.

This is indeed an exciting time in the study of genome biology, and one that will undoubtedly continue to alter the understanding of the evolutionary process at both genomic and geological scales.

REFERENCES

Abi-Rached L, Gilles A, Shiina T, et al. 2002. Evidence of en bloc duplication in vertebrate genomes. *Nat Genet* 31: 100–105.

Altschul SF, Madden TL, Schaffer AA, et al. 1997. Gapped BLAST and PSI-BLAST: a new generation of protein database search programs. *Nucleic Acids Res* 25: 3389–3402.

Amores A, Force A, Yan YL, *et al.* 1998. Zebrafish hox clusters and vertebrate genome evolution. *Science* 282: 1711–1714.

Anderson E. 1949. *Introgressive Hybridization.* New York: Wiley.

Aparicio S, Chapman J, Stupka E, *et al.* 2002. Whole-genome shotgun assembly and analysis of the genome of *Fugu rubripes. Science* 297: 1301–1310.

Aparicio S, Hawker K, Cottage A, *et al.* 1997. Organization of the *Fugu rubripes* Hox clusters: evidence for continuing evolution of vertebrate Hox complexes. *Nat Genet* 16: 79–83.

Arabidopsis Genome Initiative. 2000. Analysis of the genome sequence of the flowering plant *Arabidopsis thaliana. Nature* 408: 796–815.

Arnold ML. 1997. *Natural Hybridization and Evolution.* Oxford: Oxford University Press.

Bailey JA, Gu Z, Clark RA, *et al.* 2002. Recent segmental duplications in the human genome. *Science* 297: 1003–1007.

Benton MJ. 1990. Phylogeny of the major tetrapod groups: morphological data and divergence dates. *J Mol Evol* 30: 409–424.

Blanc G, Barakat A, Guyot R, *et al.* 2000. Extensive duplication and reshuffling in the *Arabidopsis* genome. *Plant Cell* 12, 1093–1101.

Blanc G, Hokamp K, Wolfe KH. 2003. A recent polyploidy superimposed on older large-scale duplications in the *Arabidopsis* genome. *Genome Res* 13: 137–144.

Bowers JE, Chapman BA, Rong J, Paterson AH. 2003. Unravelling angiosperm genome evolution by phylogenetic analysis of chromosomal duplication events. *Nature* 422: 433–438.

Brown TA. 1999. *Genomes.* Oxford: BIOS Scientific Publishers.

Calabrese PP, Chakravarty S, Vision TJ. 2003. Fast identification and statistical evaluation of segmental homologies in comparative maps. *Bioinformatics* 19: S174–S180.

Carroll RL. 1988. *Vertebrate Paleontology and Evolution.* New York: W. H. Freeman and Co.

Cheng CHC, Chen L. 1999. Evolution of an antifreeze glycoprotein. *Nature* 401: 443–444.

Conery JS, Lynch M. 2001. Nucleotide substitutions and the evolution of duplicate genes. *Pac Symp Biocomput*, 167–178.

De Bodt S, Raes J, Florquin K, *et al.* 2003. Structural annotation and evolutionary analysis of type I MADS box transcription factors in plants. *J Mol Evol* 56, 573–586.

Dietrich FS, Voegeli S, Brachat S, *et al.* 2004. The *Ashbya gossypii* genome as a tool for mapping the ancient *Saccharomyces cerevisiae* genome. *Science* 304: 304–307.

Duda TF, Palumbi SR. 1999. Molecular genetics of ecological diversification: duplication and rapid evolution of toxin genes of the venomous gastropod *Conus. Proc Natl Acad Sci USA* 96: 6820–6823.

Dujon B, Sherman D, Fisher G. 2004. Genome evolution in yeasts. *Nature* 430: 35–44.

Dulai KS, von Dornum M, Mollon JD, Hunt DM. 1999. The evolution of trichromatic colour vision by opsin gene duplication in new world and old world primates. *Genome Res* 9: 629–638.

Durand D. 2003. Vertebrate evolution: doubling and shuffling with a full deck. *Trends Genet* 19: 2–5.

Easteal S, Collet C. 1994. Consistent variation in amino-acid substitution rate, despite uniformity of mutation rate: protein evolution in mammals is not neutral. *Mol Biol Evol* 11: 643–647.

Ehrendorfer F. 1980. Polyploidy and distribution. In: Lewis WH ed. *Polyploidy—Biological Relevance.* New York: Plenum Press, pp. 45–60.

Endo T, Ikeo K, Gojobori T. 1996. Large-scale search for genes on which positive selection may operate. *Mol Biol Evol* 13: 685–690.

Escriva H, Manzon L, Youson J, Laudet V. 2002. Analysis of lamprey and hagfish genes reveals a complex history of gene duplications during early vertebrate evolution. *Mol Biol Evol* 9: 1440–1450.

Force A, Lynch M, Pickett FB, *et al.* 1999. Preservation of duplicate genes by complementary, degenerative mutations. *Genetics* 151: 1531–1545.

Fraser CM, Casjens S, Huang WM, *et al.* 1997. Genomic sequence of a Lyme disease spirochaete, *Borrelia burgdorferi. Nature* 390: 580–586.

Fried C, Prohaska SJ, Stadler PF. 2003. Independent *Hox*-cluster duplications in lampreys. *J Exp Zool (Mol Dev Evol)* 299: 18–25.

Friedman R, Hughes AL. 2003. The temporal distribution of gene duplication events in a set of highly conserved human gene families. *Mol Biol Evol* 20: 154–161.

Furlong RF, Holland PWH. 2002. Were vertebrates octoploids? *Philos Trans R Soc Lond* B 357: 531–544.

Gates MA, Kim L, Cardozo T, *et al.* 1999. A genetic linkage map for zebrafish: comparative analysis and localization of genes and expressed sequences. *Genome Res* 9: 334–347.

Gaut BS. 2001. Patterns of chromosomal duplication in maize and their implications for comparative maps of the grasses. *Genome Res* 11: 55–66.

Gehring WJ. 1998. *Master Control Genes in Development and Evolution: The Homeobox Story.* New Haven, CT: Yale University Press.

Gevers D, Vandepoele K, Simillion C, Van de Peer Y. 2004. Gene duplication and biased functional retention of paralogs in bacterial genomes. *Trends Microbiol* 12: 148–154.

Gibson TJ, Spring J. 1998. Genetic redundancy in vertebrates: polyploidy and persistence of genes encoding multidomain proteins. *Trends Genet* 14: 46–49.

Goff SA, Ricke D, Lan TH, *et al.* 2002. A draft sequence of the rice genome (*Oryza sativa* L. ssp. *japonica*). *Science* 296: 92–100.

Grant D, Cregan P, Shoemaker RC. 2000. Genome organization in dicots: genome duplication in *Arabidopsis* and synteny between soybean and *Arabidopsis*. *Proc Natl Acad Sci USA* 97: 4168–4173.

Gu X, Wang Y, Gu J. 2002. Age distribution of human gene families shows significant roles of both large- and small-scale duplications in vertebrate evolution. *Nat Genet* 31: 205–209.

Gu Z, Steinmetz LM, Gu X, *et al.* 2003. Role of duplicate genes in genetic robustness against null mutations. *Nature* 421: 63–66.

Guo M, Davis D, Birchler JA, *et al.* 1996. Dosage effect on gene expression in a maize ploidy series. *Genetics* 142: 1349–1355.

Guyot R, Keller B. 2004. Ancestral genome duplication in rice. *Genome* 47: 610–614.

Haldane JBS. 1933. The part played by recurrent mutation in evolution. *Am Nat* 67: 5–19.

Holland PW. 1997. Vertebrate evolution: something fishy about *Hox* genes. *Curr Biol* 7: R570–R572.

Holland PW, Garcia-Fernandez J. 1996. *Hox* genes and chordate evolution. *Dev Biol* 173: 382–395.

Holland PW, Garcia-Fernandez J, Williams NA, Sidow A. 1994. Gene duplications and the origins of vertebrate development. *Development* 120 (Suppl.): 125–133.

Hughes AL. 1999a. *Adaptive Evolution of Genes and Genomes.* Oxford: Oxford University Press.

Hughes AL. 1999b. Phylogenies of developmentally important proteins do not support the hypothesis of two rounds of genome duplication early in vertebrate history. *J Mol Evol* 48: 565–576.

Hughes AL, da Silva J, Friedman R. 2001. Ancient genome duplications did not structure the human *Hox*-bearing chromosomes. *Genome Res* 11: 771–780.

Hughes AL, Green JA, Garbayo JM, Roberts RM. 2000. Adaptive diversification within a large family of recently duplicated, placentally expressed genes. *Proc Natl Acad Sci USA* 97: 3319–3323.

Irvine S, Carr JL, Bailey WJ, *et al.* 2002. Genomics analysis of *Hox* clusters in the sea lamprey *Petromyzon marinus*. *J Exp Zool* 249: 47–62.

Jia L, Clegg MT, Jiang T. 2003. Excess of non-synonymous substitutions suggest that positive selection episodes occurred during the evolution of DNA-binding domains in the *Arabidopsis* R2R3-MYB gene family. *Plant Mol Biol* 52: 627–642.

Kellis M, Birren BW, Lander ES. 2004. Proof and evolutionary analysis of ancient genome duplication in the yeast *Saccharomyces cerevisiae*. *Nature* 428: 617–624.

Kimura M. 1983. *The Neutral Theory of Molecular Evolution.* Cambridge, UK: Cambridge University Press.

Koszul R, Caburet S, Dujon B, Fischer G. 2004. Eukaryotic genome evolution through the spontaneous duplication of large chromosomal segments. *EMBO J* 23: 234–243.

Ku HM, Vision T, Liu J, Tanksley SD. 2000. Comparing sequenced segments of the tomato and *Arabidopsis* genomes: large-scale duplication followed by selective gene loss creates a network of synteny. *Proc Natl Acad Sci USA* 97, 9121–9126.

Larhammar D, Lundin LG, Hallbook F. 2002. The human *Hox*-bearing chromosome regions did arise by block or chromosome (or even genome) duplications. *Genome Res* 12: 1910–1920.

Levy A, Feldman M. 2002. The impact of polyploidy on grass genome evolution. *Plant Physiol* 130: 1587–1593.

Li WH. 1993. Unbiased estimation of the rates of synonymous and nonsynonymous substitution. *J Mol Evol* 36: 96–99.

Li WH, Gu Z, Cavalcanti ARO, Nekrutenko A. 2003. Detection of gene duplications and block duplication in eukaryotic genomes. *J Struct Funct Genomics* 3: 27–34.

Lin X, Kaul S, Rounsley S, *et al.* 1999. Sequence and analysis of chromosome 2 of the plant *Arabidopsis thaliana*. *Nature* 402: 761–768.

Lundin LG, Larhammar D, Hallbook F. 2003. Numerous groups of chromosomal regional paralogies strongly indicate two genome doublings at the root of the vertebrates. *J Struct Funct Genomics* 3: 53–63.

Lynch M, Conery JS. 2000. The evolutionary fate and consequences of duplicate genes. *Science* 290: 1151–1155.

Lynch M, Force A. 2000. The probability of duplicate gene preservation by subfunctionalization. *Genetics* 154: 459–473.

Málaga-Trillo E, Meyer A. 2001. Genome duplications and accelerated evolution of *Hox* genes and cluster architecture in teleost fishes. *Am Zool* 41: 676–686.

Martin A. 2001. Is tetralogy true? Lack of support for the "one-to-four rule." *Mol Biol Evol* 18: 89–93.

Mayer K, Schüller C, Wambutt R, *et al.* 1999. Sequence and analysis of chromosome 4 of the plant *Arabidopsis thaliana*. *Nature* 402: 769–777.

McLysaght A, Hokamp K, Wolfe KH. 2002. Extensive genomic duplication during early chordate evolution. *Nat Genet* 31: 200–204.

Messier W, Stewart CB. 1997. Episodic adaptive evolution of primate lysosymes. *Nature* 385: 151–154.

Meyer A. 1993. Phylogenetic relationships and evolutionary processes in East African cichlids. *Trends Ecol Evol* 8: 279–284.

Meyer A, Kocher TD, Basasibwaki P, Wilson A. 1990. Monophyletic origin of Lake Victoria Africa cichlid fishes suggested by mitochondrial DNA sequences. *Nature* 347: 550–663.

Meyer A, Schartl M. 1999. Gene and genome duplications in vertebrates: the one-to-four (-to-eight in fish) rule and the evolution of novel gene functions. *Curr Opin Cell Biol* 11: 699–704.

Naruse K, Fukamachi S, Mitani H, *et al.* 2000. A detailed linkage map of medaka, *Oryzias latipes*: comparative genomics and genome evolution. *Genetics* 154: 1773–1784.

Nei M, Gojobori T. 1986. Simple methods for estimating the numbers of synonymous and nonsynonymous nucleotide substitutions. *Mol Biol Evol* 3: 418–426.

Nei M, Kumar S. 2000. *Molecular Evolution and Phylogenetics*. Oxford: Oxford University Press.

Nowak MA, Boerlijst MC, Cooke J, Maynard Smith J. 1997. Evolution of genetic redundancy. *Nature* 388: 167–171.

Ohno S. 1970. *Evolution by Gene Duplication*. New York: Springer Verlag.

Ohno S. 1973. Ancient linkage groups and frozen accidents. *Nature* 244: 259–262.

Osborn TC, Pires JC, Birchler JA, *et al.* 2003. Understanding mechanisms of novel gene expression in polyploids. *Trends Genet* 19: 141–147.

Panopoulou G, Hennig S, Groth D, *et al.* 2003. New evidence for genome-wide duplications at the origin of vertebrates using an Amphioxus gene set and completed animal genomes. *Genome Res* 13: 1056–1066.

Paterson AH, Bowers JE, Chapman BA. 2004. Ancient polyploidization predating divergence of the cereals, and its consequences for comparative genomics. *Proc Natl Acad Sci USA* 101: 9903–9908.

Paterson AH, Lan TH, Reischmann KP, *et al.* 1996. Toward a unified genetic map of higher plants, transcending the monocot-dicot divergence. *Nat Genet* 14: 380–382.

Postlethwait JH, Woods IG, Ngo-Hazelett P, *et al.* 2000. Zebrafish comparative genomics and the origins of vertebrate chromosomes. *Genome Res* 10: 1890–1902.

Prince VE, Pickett FB. 2002. Splitting pairs: the diverging fates of duplicated genes. *Nat Rev Genet* 3: 827–837.

Raes J, Van de Peer Y. 2003. Gene duplications, the evolution of novel gene functions, and detecting functional divergence of duplicates *in silico*. *Appl Bioinformatics* 2: 92–101.

Raes J, Vandepoele K, Simillion C, *et al.* 2003. Investigating ancient duplication events in the *Arabidopsis* genome. *J Struct Func Genomics* 3: 117–129.

Ramsey J, Schemske DW. 2002. Neopolyploidy in flowering plants. *Annu Rev Ecol Syst* 33: 589–639.

Riehle MM, Bennette AF, Long AD. 2001. Genetic architecture of thermal adaptation in *Escherichia coli*. *Proc Natl Acad Sci USA* 98: 525–530.

Rieseberg LH, Raymond O, Rosenthal DM, *et al.* 2003. Major ecological transitions in wild sunflowers facilitated by hybridization. *Science* 301: 1211–1216.

Robinson-Rechavi M, Marchand O, Escriva H, *et al.* 2001a. Euteleost fish genomes are characterized by expansion of gene families. *Genome Res* 11: 781–788.

Robinson-Rechavi M, Marchand O, Escriva H, Laudet V. 2001b. An ancestral whole-genome duplication may not have been responsible for the abundance of duplicated fish genes. *Curr Biol* 11: R458–R459.

Rost B. 1999. Twilight zone of protein sequence alignments. *Protein Eng* 12: 85–94.

Seoighe C, Wolfe KH. 1999. Yeast genome evolution in the post-genome era. *Curr Opin Microbiol* 2: 548–554.

Shu DG, Luo HL, Conway Morris S, *et al.* 1999. Lower Cambrian vertebrates from south China. *Nature* 402: 42–46.

Simillion C, Vandepoele K, Saeys Y, Van de Peer Y. 2004. Building genomic profiles for uncovering segmental homology in the twilight zone. *Genome Res* 14: 1095–1106.

Simillion C, Vandepoele K, Van Montagu M, *et al.* 2002. The hidden duplication past of *Arabidopsis thaliana*. *Proc Natl Acad Sci USA* 99: 13627–13632.

Skrabanek L, Wolfe KH. 1998. Eukaryote genome duplication: where's the evidence? *Curr Opin Genet Dev* 8: 694–700.

Spring J. 1997. Vertebrate evolution by interspecific hybridisations: are we polyploid? *FEBS Lett* 400: 2–8.

Spring J. 2003. Major transitions in evolution by genome fusions: from prokaryotes to eukaryotes, metazoans, bilaterians and vertebrates. *J Struct Funct Genomics* 3: 19–25.

Stephens SG. 1951. Possible significance of duplication in evolution. *Adv Genet* 4: 247–265.

Stiassny MLJ, Meyer A. 1999. Cichlids of the African Rift Lakes. *Sci Am* February: 64–69.

Sturmbauer C, Meyer A. 1992. Genetic divergence, speciation and morphological stasis in a lineage of African cichlid fishes. *Nature* 358: 578–581.

Takezaki N, Rzhetsky A, Nei M. 1995. Phylogenetic test of the molecular clock and linearized trees. *Mol Biol Evol* 12: 823–833.

Taylor J, Braasch I, Frickey T, *et al.* 2003. Genome duplication, a trait shared by 22,000 species of ray-finned fish. *Genome Res* 13: 382–390.

Taylor JS, Van de Peer Y, Braasch I, Meyer A. 2001. Comparative genomics provides evidence for an ancient genome duplication event in fish. *Philos Trans R Soc Lond B* 356: 1661–1679.

Terai Y, Mayer WE, Klein J, *et al.* 2002. The effect of selection on a long wavelength-sensitive (LWS) opsin gene of Lake Victoria cichlid fishes. *Proc Natl Acad Sci USA* 99: 15501–15506.

Terryn N, Heijnen L, De Keyser A, *et al.* 1999. Evidence for an ancient chromosomal duplication in *Arabidopsis thaliana* by sequencing and analyzing a 400-kb contig at the APETALA2 locus on chromosome 4. *FEBS Lett* 445: 237–245.

Van de Peer Y, Taylor JS, Braasch I, Meyer A. 2001. The ghost of selection past: rates of evolution and functional divergence of anciently duplicated genes. *J Mol Evol* 53: 436–446.

Van de Peer Y, Taylor JS, Meyer A. 2003. Are all fishes ancient polyploids? *J Struct Funct Genomics* 3: 65–73.

Vandepoele K, De Vos W, Taylor JS, *et al*. 2004a. Major events in the genome evolution of vertebrates: paranome age and size differs considerably between fishes and land vertebrates. *Proc Natl Acad Sci USA* 101: 1638–1643.

Vandepoele K, Saeys Y, Simillion C, *et al*. 2002a. A new tool for the automatic detection of homologous regions (ADHoRe) and its application to microcolinearity between *Arabidopsis* and rice. *Genome Res* 12: 1792–1801.

Vandepoele K, Simillion C, Van de Peer Y. 2002b. Detecting the undetectable: uncovering duplicated segments in *Arabidopsis* by comparison with rice. *Trends Genet* 18: 606–608.

Vandepoele K, Simillion C, Van de Peer Y. 2003. Evidence that rice, and other cereals, are ancient aneuploids. *Plant Cell* 15: 2192–2202.

Vandepoele K, Simillion C, Van de Peer Y. 2004b. The quest for genomic homology. *Curr Genomics* 5: 299–308.

Vision TJ, Brown DG, Tanksley SD. 2000. The origins of genomic duplications in *Arabidopsis*. *Science* 290: 2114–2117.

Werth CR, Windham MD. 1991. A model for divergent, allopatric speciation of polyploidy pteridophytes resulting from silencing of duplicate-gene expression. *Am Nat* 137: 515–526.

Wilson AB, Noack-Kunnmann K, Meyer A. 2000. Incipient speciation in sympatric Nicaraguan crater lake cichlid fishes: sexual selection versus ecological diversification. *Proc R Soc Lond B* 267: 2133–2141.

Wittbrodt J, Meyer A, Schartl M. 1998. More genes in fish? *BioEssays* 20: 511–512.

Wolfe KH. 2001. Yesterday's polyploids and the mystery of diploidization. *Nat Rev Genet* 2: 333–341.

Wolfe KH, Shields DC. 1997. Molecular evidence for an ancient duplication of the entire yeast genome. *Nature* 387: 708–713.

Woods IG, Kelly PD, Chu F, *et al*. 2000. A comparative map of the zebrafish genome. *Genome Res* 10: 1903–1914.

Yang Z. 1997. PAML: a program package for phylogenetic analysis by maximum likelihood. *Comput Appl Biosci* 13: 555–556 (available at http://abacus.gene.ucl.ac.uk/software/paml.html).

Yang Z, Nielsen R. 2000. Estimating synonymous and nonsynonymous substitution rates under realistic evolutionary models. *Mol Biol Evol* 17: 32–43.

Yokoyama S, Blow NS, Radlwimmer FB. 2000. Molecular evolution of color vision of zebra finch. *Gene* 259: 17–24.

Yu J, Hu S, Wang J, *et al*. 2002. A draft sequence of the rice genome (*Oryza sativa* L. ssp. *indica*). *Science* 296: 79–92.

Zhang J, Nei M. 2000. Positive selection in the evolution of mammalian interleukin-2 genes. *Mol Biol Evol* 17: 1413–1416.

Zhang J, Rosenberg HF, Nei M. 1998. Positive Darwinian selection after gene duplication in primate ribonuclease genes. *Proc Natl Acad Sci USA* 95: 3708–3713.

Zhang J, Zhang YP, Rosenberg HF. 2002. Adaptive evolution of a duplicated pancreatic ribonuclease gene in a leaf-eating monkey. *Nat Genet* 30: 411–415.

Zhu M, Yu X. 2002. A primitive fish close to the common ancestor of tetrapods and lungfish. *Nature* 418: 767–770.

Zhu M, Yu X, Janvier P. 1999. A primitive fossil fish sheds light on the origin of bony fishes. *Nature* 397: 607–610.

PART IV

...And More Duplications

CHAPTER 7

Polyploidy in Plants

JENNIFER A. TATE, DOUGLAS E. SOLTIS, AND PAMELA S. SOLTIS

According to the classic definition of Grant (1981, p. 283), which actually traces back to research conducted in the early 1900s (e.g., Winkler, 1916; Winge, 1917), polyploidy represents "the formation of a higher chromosome number . . . by the addition of extra whole chromosome sets present in one or more ancestral organisms. In short, polyploidy is the presence of three or more chromosome sets in an organism." While seemingly simple in concept, understanding the evolution of polyploidy involves several complex questions, some of which are only now beginning to be addressed.

Polyploidy has long been recognized as playing an especially important role in plant evolution. However, the extent to which this is true has been largely underestimated in terms of the commonality of polyploidy, the processes by which it occurs, and the genomic and phenotypic consequences it engenders. Major recent advances in genomic analysis have made it possible to reexamine some of these issues in a new light, and have drastically altered the view of polyploid evolution. This chapter provides a discussion of these recent advancements, put in context by reviewing the classic literature on the study of plant polyploidy.

The Evolution of the Genome, edited by TR Gregory

HISTORY OF THE STUDY OF POLYPLOIDY
IN PLANTS

Plant polyploidy has been studied for nearly a century, dating from the early genetic work of Hugo de Vries on *Oenothera lamarckiana* mut. *Gigas* (Onagraceae), which was discovered to be a tetraploid (Lutz, 1907; Gates, 1909), and from Kuwada's (1911) hypothesis regarding an ancient chromosome duplication in maize (*Zea mays*). The artificial production of a tetraploid form of *Solanum nigrum* (Solanaceae) through decapitation and the regeneration of callus tissue (Winkler, 1916) is considered the first example of the production of a polyploid in the laboratory (Stebbins, 1950). Experimental and exploratory research in the early 1900s demonstrated mechanisms by which polyploids might be formed and highlighted the frequency of polyploids in nature. This information led to the early development of hypotheses on the role of polyploidy in speciation. For example, Winge (1917) observed regular arithmetic series of chromosome numbers[1] in several collections of congeneric species. In *Chenopodium* (Amaranthaceae), some species have 2n = 18, and others have 2n = 36. Similarly, within *Chrysanthemum* (Asteraceae), different species have 2n = 18, 36, 54, 72, and 90. To explain these observed numbers, Winge (1917) formulated a hypothesis of hybridization followed by doubling of the chromosomes (i.e., polyploidy) as a method for the origin of new species. Research on *Rumex* (Polygonaceae) indicated that hybridization and chromosome doubling were operating in natural populations (Kihara and Ono, 1926). Artificial hybridizations in *Nicotiana* (Solanaceae), *Galeopsis* (Lamiaceae), and *Raphanobrassica* (Brassicaceae) confirmed Winge's hypotheses on the origins of polyploidy (Clausen and Goodspeed, 1925; Müntzing, 1930), although it soon became clear that several different mechanisms can generate polyploids (see later section). During this same period, other investigators had determined that many of the major crops—including wheat, oats, cotton, tobacco, potato, and coffee—are polyploid. In addition, evidence was provided for the parentage of some of these polyploid crops, including wheat (McFadden and Sears, 1946), cotton (Beasley, 1940), and tobacco (Goodspeed and Clausen, 1928).

Early influential reviews of polyploidy in plants include those of Müntzing (1936), Darlington (1937), Clausen *et al.* (1945), Löve and Löve (1949),

---

[1]To make use of the definition of polyploidy as the presence of three or more chromosome sets, it is necessary to distinguish ploidal level from the total number of chromosomes found in a sporophytic (i.e., diploid phase) nucleus ("2n"), which is done by indicating the number of basic chromosome sets ("x") included therein. Therefore, a diploid nuclear complement contains 2n = 2x chromosomes, whereas triploids have 2n = 3x, tetraploids have 2n = 4x, and so on through all higher levels (e.g., hexaploidy, octoploidy). This convention can be used in reference to animal polyploidy as well (Chapter 8).

and especially Stebbins (1947, 1950). Following Stebbins's (1940, 1947, 1950) treatments of polyploids, much attention was focused on polyploid complexes and the intricate relationships among groups of hybridizing species, with the result that polyploidy occupied a significant sector of biosystematic research. Studies of *Solanum* (Solanaceae), *Crepis* (Asteraceae), *Paeonia* (Paeoniaceae), *Delphinium* (Ranunculaceae), *Lotus* (Fabaceae), and many other genera revealed complex patterns of morphological and cytogenetic variation (Stebbins, 1950). However, during the latter decades of the 20th century, the study of polyploidy began to wane as plant systematists focused greater attention on phylogeny reconstruction.

Recent years have seen a resurgence in the study of polyploidy, with renewed interest in the mechanisms of polyploid formation and establishment (Ramsey and Schemske, 1998, 2002; Husband, 2004), the frequency of recurrent polyploidization (Werth *et al.*, 1985a,b; Soltis and Soltis, 1999, 2000), the population genetics of polyploids (Husband and Schemske, 1997; Cook and Soltis, 1999, 2000), the ecological effects of plant polyploidy (Levin, 1983; Bayer *et al.*, 1991; Segraves and Thompson, 1999; Thompson *et al.*, 2004), and the genetic, chromosomal, and genomic consequences of polyploidization (Leitch and Bennett, 1997; Bowers *et al.*, 2003; Liu and Wendel, 2003; Osborn *et al.*, 2003). Notably, Mable (2003) recently compiled the number of polyploidy-related papers published during the last decade, based on listings in the PubMed (www.ncbi.nlm.nih.gov/PubMed) library database. Since the late 1990s, the number of such publications has more than doubled, from approximately 60 in 1999 to 140 in 2003. The study of the genetic and genomic aspects of polyploidy has united plant biologists from diverse disciplines, from molecular genetics to evolutionary biology and ecology, and promises to reveal many new insights regarding the success of polyploids in nature.

TYPES OF POLYPLOIDS

Two general types of polyploids have long been recognized: those involving the multiplication of one chromosome set and those resulting from the merger of structurally different chromosome sets. Kihara and Ono (1926) used the terms *autopolyploidy* (auto = "same") and *allopolyploidy* (allo = "different"), respectively, to distinguish between these two types. This convention was adopted in subsequent discussions (Müntzing, 1936; Darlington, 1937; Clausen *et al.*, 1945) and, according to Grant (1981), represented a fundamental distinction between polyploid types. However, despite ongoing attempts to categorize polyploids as either auto- or allopolyploids, nature clearly has produced a continuum of polyploid types that have arisen via an array of processes and which defy placement in either of these two groups (Stebbins, 1950; Lewis, 1980b). Hence there has been debate

TABLE 7.1 Summary of the traditional types of polyploids. A and B represent the different parental genomes and subscripts indicate distinct genomes from the same species. In spite of the name, Stebbins (1950) considered segmental allopolyploids to fall under the broad category of autopolyploidy (because they are derived from distinct genotypes of the same species), whereas Grant (1981) considered these more similar to true allopolyploids (because they result from hybridization of genetically distinct genomes).

Type of polyploidy	Category	Mode of formation	Chromosome pairing and inheritance	Example of genomic composition
Strict auto-polyploidy	Autopolyploidy	Within a species, either from genome doubling in a single individual or fusion of unreduced gametes from genetically similar individuals	Multivalents, polysomic	$AAAA$
Interracial autopolyploidy	Autopolyploidy	Within a species, genetically distinct, but structurally similar chromosomes	Multivalents, polysomic	$AAAA$
Segmental allopolyploidy	Autopolyploidy (Stebbins)/ Allopolyploidy (Grant)	Within a species, but from parental genomes that differ from each other in a large number of genes or chromosomal segments; these are unstable polyploids that evolve toward auto- or true allopolyploidy	Bivalents, disomic or multivalents, polysomic	$A_sA_sA_lA_l$
True (genomic) allopolyploidy	Allopolyploidy	Hybridization between distantly related species	Bivalents, disomic	$AABB$
Autoallo-polyploidy	Allo- and autopolyploidy	Genome doubling following allopolyploid formation	Multivalents and bivalents; some loci are disomic, others polysomic	$AAAABB$

for more than 70 years as to the types of polyploids that should be recognized in nature and the proper definitions of autopolyploidy and allopolyploidy.

Stebbins (1947, 1950) recognized four types of polyploids based on genetic and cytogenetic criteria: *autopolyploids, segmental allopolyploids, true* or *genomic polyploids,* and *autoallopolyploids* (Table 7.1). The first three of these were considered major categories, whereas autoallopolyploidy is a combination of both auto- and allopolyploidy. Stebbins's (1947) classification of polyploid types incorporates more specific information about the polyploid's formation into its definition. Thus autopolyploids are formed by genome duplication within a species and consequently exhibit little morphological or cytogenetic divergence from the diploid progenitor (Fig. 7.1) (see Chapters 6 and 8). In many cases, autopolyploidy has conspicuous effects on chromosome pairing during gamete formation, which can sometimes be used to identify it as the mechanism of polyploidization. In a "normal" diploid cell, homologous chromosomes pair during prophase I of meiosis (also known as synapsis) to form "bivalents" (Fig. 7.2). However, in autopolyploids, which harbor more than two copies of each chromosome, "multivalents" may form during meiosis (Fig. 7.2). In triploids (and other odd-numbered polyploids), a bivalent plus a univalent may form for each set of homologous chromosomes, leading to instability of chromosome complements in the gametes and possibly resulting in sterility (see Chapter 8). On the other hand, despite some

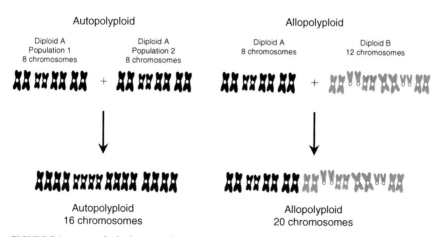

FIGURE 7.1   Autopolyploid versus allopolyploid formation. Autopolyploids are formed by intraspecific hybridization of diploid populations, whereas allopolyploids are formed by interspecific hybridization. It bears noting that the classification into these two categories is not always simple: polyploid evolution can be a complex process, sometimes involving both intra- and interspecific mechanisms. Modified from Leitch and Bennett (1997), reproduced by permission (© Elsevier Inc.). See also Chapter 6.

previous expectations that their meiotic chromosome pairing would be chaotic, even-numbered polyploids may exhibit stable multivalents (e.g., four paired chromosomes in a tetraploid) or regular bivalents, thereby allowing them to produce functional and chromosomally balanced gametes (Fig. 7.2).

Figure 7.3 depicts another frequently observed consequence of autopolyploidy, namely "polysomic inheritance," in which multiple alleles (i.e., more than two) of a given gene can be detected in the offspring. In a mating between two diploid individuals (or following self-fertilization), the offspring are expected to exhibit one of three possible combinations in a predictable ratio of two heterozygotes for every one of each type of homozygote (Fig. 7.3). Two autopolyploids, by contrast, each possess four copies of a given gene, which increases both the number of possible combinations and the complexity of their predicted ratios (Fig. 7.3). Indeed, the segregation ratios illustrated in Figure 7.3 are a simplification and will be true only for genes adjacent to the centromere. Because of "double reduction" (i.e., when two sister chromatids segregate to the same gamete), an even more complicated pattern can result (e.g., Ronfort *et al.*, 1998; Butruille and Boiteux, 2000).

Stebbins also included segmental allopolyploids under the broad category of autopolyploidy. Unlike strict autopolyploids, which result from genome doubling within a single individual or by the production and merger of unreduced (diploid) gametes from genetically similar individuals, segmental allopolyploid formation involves hybridization between genetically distinct races or populations of a single species (Stebbins, 1950, 1985; Soltis and Rieseberg, 1986; Soltis and Soltis, 1993) (Table 7.1). These genomes typically differ from one another in a large number of genes or chromosomal segments. Stebbins (1947, 1950) asserted that

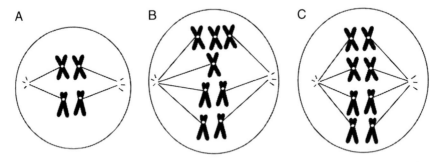

FIGURE 7.2   Chromosome pairing at meiosis (prophase I). (A) Homologous chromosomes align during prophase I in a diploid, with only bivalents forming. (B) Previous hypotheses of chromosome pairing in autopolyploids invoked chaotic pairing of homologs during meiosis, with univalents, bivalents, trivalents, and even quadrivalents possibly forming. (C) Observations of naturally occurring autopolyploids indicate that homologous pairing can be orderly (i.e., bivalents form) and may operate on a first-come, first-served basis.

<table>
<tr><td align="center">Diploid</td><td align="center">Autotetraploid</td></tr>
<tr><td align="center">Aa × Aa</td><td align="center">AAaa × AAaa</td></tr>
<tr><td align="center">AA Aa aa
1 : 2 : 1</td><td align="center">AAAA AAAa AAaa Aaaa aaaa
1 : 8 : 18 : 8 : 1
34</td></tr>
<tr><td align="center">Disomic inheritance</td><td align="center">Tetrasomic inheritance</td></tr>
</table>

FIGURE 7.3 Disomic versus tetrasomic inheritance in autopolyploids. Crosses between two heterozygous individuals are expected to yield these ratios of heterozygous genotypes. This pattern is the one expected in the absence of double reduction, which is the segregation of sister chromatids to the same gamete in an autotetraploid.

segmental allopolyploids are not stable, and will tend to evolve toward the characteristics of either autopolyploids or true allopolyploids. True (or genomic) allopolyploids are typically derived from hybridization between two or more distantly related species, and therefore combine divergent genomes that are unable to pair with each other (Fig. 7.1). Thus an autotetraploid will contain four copies of each chromosome (all four are homologs), whereas an allotetraploid will contain two of each pair of the counterpart chromosomes derived from two different species (called "homoeologous" chromosomes). In (most) allopolyploids, inheritance patterns are therefore disomic (Fig. 7.3).

Autoallopolyploids are hexaploids or higher ploidies that combine the processes of auto- and allopolyploid formation, as for example when an allopolyploid forms by hybridization and then undergoes a within-species (i.e., autopolyploid) duplication. Both homologous and homoeologous chromosome sets are present in such a case, and therefore, the segregation of some loci will be disomic, whereas for others it will be polysomic.

The term *amphiploid,* which is still in use, was introduced to cover all types of polyploids that have formed via hybridization (Clausen *et al.,* 1945). Using this criterion, "amphiploids" (sometimes also called "amphidiploids") could refer to true allopolyploids, interracial autopolyploids (see following), segmental allopolyploids (following Grant), or autoallopolyploids. However, adding to the nomenclatural confusion is the fact that "amphiploid" is most commonly used as a synonym for "allopolyploid" (Grant, 1981).

Grant (1981, p. 300) stated that the "principal criteria for distinguishing between autopolyploids and amphiploids (allopolyploids) are chromosome behavior, fertility, segregation ratios, and morphology," also noting that "these criteria will all break down in individual cases." Indeed, he noted (p. 298) that autopolyploidy and allopolyploidy are "the extreme members of a graded series." In contrast to Stebbins, Grant (1981) included segmental allopolyploidy as a type of amphiploidy (allopolyploidy), rather than a type of autopolyploidy. His summary of the principal types of polyploids is useful, and can be illustrated as follows (with A and B representing the different parental genomes and subscripts indicating distinct genomes from the same species):

I. Autopolyploids
   1. Strict autopolyploid $AAAA$
   2. Interracial autopolyploid $AAAA$

II. Amphiploids
   1. Segmental allopolyploid $A_sA_sA_tA_t$
   2. Genomic allopolyploid $AABB$
   3. Autoallopolyploid $AAAABB$

Distinguishing among categories of polyploids is not always straightforward. Morphological, cytogenetic, and genetic data, coupled with information on fertility, may all be necessary to determine if a given polyploid is an auto-, segmental allo-, true allo-, or autoallopolyploid (Grant, 1981) (see also Chapter 8). Cytogenetic criteria, such as observations of multivalent versus bivalent formation and estimation of chiasma (meiotic crossing over) frequency, have been incorporated into models for inferring auto- versus allopolyploidy (Jackson, 1982; Jackson and Casey, 1982; Jackson and Hauber, 1982). However, some autotetraploids form bivalents while also exhibiting tetrasomic inheritance, suggesting that pairing among the four homologs is random and may occur on a "first-come, first-served" basis (Fig. 7.2). Furthermore, multivalent formation may precede the formation of bivalents, as in the autotetraploid *Tolmiea menziesii* (Saxifragaceae) (Hauber and Soltis, 1989). Alternatively, some species that are considered allotetraploids, such as *Gossypium arboreum* and *G. herbaceum* (Malvaceae), may have formed through hybridization of close relatives and thus exhibit tetrasomic inheritance at some loci. Likewise, although polysomic inheritance is common in autopolyploids and disomic inheritance is generally characteristic of allopolyploids, genetic data alone (especially if based on only one or a few loci) may be misleading. For example, inheritance patterns may differ among loci in segmental allopolyploids, with some loci exhibiting polysomic inheritance and others disomic. Inheritance patterns in autoallopolyploids are likely to be particularly complex (Grant, 1981), which may in itself be an indication that a plant is an autoallopolyploid.

Because complex genetic and morphological combinations can result from successive rounds of hybridization, polyploidization, and back-crossing, Stebbins (1971) advocated "polyploid complex" as a useful term to discuss these plants as a unit. Rather than focusing on a strict categorization of the individuals involved, the polyploid complex encompasses both the progenitors and derivative species, which may belong to the same or different taxonomic species. Other investigators have classified polyploids using taxonomic rank as a criterion. This method was partially employed by Clausen *et al.* (1945), who attempted to combine information from taxonomy, ecology, genetics, and cytology. Lewis (1980b) used a strictly taxonomic approach, with autopolyploidy and allopolyploidy paralleling intraspecific and interspecific polyploidy, respectively. Debate has continued as to the best definitions of autopolyploids and allopolyploids (Soltis and Rieseberg, 1986), but the taxonomic approach is perhaps the most widespread one in use today. This is the view taken in the present chapter, whereby allopolyploids are considered to form through hybridization between species whereas autopolyploids form within species (in particular usually involving crossing between individuals, rather than somatic doubling of a single diploid plant) (Soltis and Rieseberg, 1986).

FREQUENCY OF POLYPLOIDS

HOW COMMON IS POLYPLOIDY IN PLANTS?

It has long been recognized that polyploidy is a major evolutionary factor in plants (Müntzing, 1936; Darlington, 1937; Clausen *et al.*, 1945; Löve and Löve, 1949; Stebbins, 1950; Lewis, 1980b; Grant, 1981), but it has proved difficult to determine the actual frequency of the process in various plant lineages, despite numerous attempts to estimate this over the past 70 years. The angiosperms (flowering plants), in particular, have received much attention regarding the occurrence of polyploidy. Among the earliest estimates, Müntzing (1936) and Darlington (1937) speculated that about one half of all angiosperm species were polyploid, whereas Stebbins (1950) later estimated that this was true for 30–35% of angiosperm species. Grant (1963, 1981), basing his estimate of the frequency of polyploidy in angiosperms on chromosome numbers for 17,138 species that were available in 1955, hypothesized that flowering plants with haploid chromosome numbers of $n = 14$ or higher were of polyploid origin. Using this cutoff point, he inferred that 47% of all flowering plants were of polyploid origin, and proposed that 58% of monocots and 43% of dicots were polyploid. Using additional chromosome counts and the same methods and cutoff as Grant, Goldblatt (1980) recalculated the frequency of polyploidy in the monocots to be 55%. Goldblatt also suggested that Grant's (1963) estimate was too conservative; he thought that taxa with chromosome numbers above $n = 9$ or 10 probably have polyploidy in their evolutionary history. Using these lower numbers, he calculated that at least 70%, and perhaps 80%, of monocots are of polyploid origin. Lewis (1980a) applied an approach similar to Goldblatt's to dicots, and estimated that 70–80%

were polyploid. Most recently, Masterson (1994) used the novel approach of comparing leaf guard cell size in fossil and extant taxa from a few angiosperm families (Platanaceae, Lauraceae, Magnoliaceae) to estimate polyploid occurrence through time. Because guard cell size is often considerably larger in polyploids than in diploids (Fig. 7.4), this provided a reasonable method for estimating whether the fossil taxa were diploid (smaller guard cells than extant taxa) or polyploid (the same or larger guard cell sizes versus extant species). From these comparisons, Masterson (1994) estimated that 70% of all angiosperms had experienced one or more episodes of polyploidy in their ancestry.

The ferns have some of the highest reported chromosome numbers in plants (Manton, 1950; Löve et al., 1977). Vida (1976), assuming an ancestral base number as Stebbins and others had done, estimated that 43.5% of ferns alone were of polyploid origin. Using a similar approach and chromosome number cutoff (n = 14) that he used for the angiosperms, Grant (1981) estimated that 95% of the ferns and their allies were polyploid. Both paleopolyploidy (ancient polyploidy) and neopolyploidy (recent polyploidy) are considered important in the evolutionary history of the fern lineages (Wagner and Wagner, 1980). Because

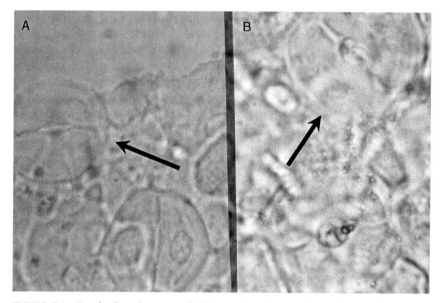

FIGURE 7.4　Guard cells and stomata of *Galax urceolata*. Arrows indicate the paired guard cells. (A) Diploid. (B) Autotetraploid. Cell sizes are generally larger in polyploids than in diploids (though there are exceptions, especially when dealing with gametes). Guard cell sizes, in particular, have been used to identify polyploids in the fossil record and thus to estimate the prevalence of polyploidy in both living and extinct taxa (Masterson, 1994).

homosporous fern gametophytes are hermaphroditic, and thus unreduced male and female gametes produced by a single gametophyte may be in close proximity, the occurrence of polyploidy seems to be facilitated in these species as compared to heterosporous ferns, which have separate male and female gametophytes (Walker, 1979; Haufler, 2002). In keeping with this view, homosporous ferns have higher base chromosome numbers ($x = 20$–70) than heterosporous ferns ($x = 7$–11) (Wagner and Wagner, 1980).

Otto and Whitton (2000) developed an innovative approach for estimating the incidence of polyploidy based on the distribution of haploid chromosome numbers. In examining the distribution of haploid chromosome numbers in various lineages (e.g., ferns, monocots, mammals), they observed a large excess of even over odd haploid chromosome numbers. They argued that this "saw-toothed" distributional pattern is difficult to explain by any mechanism other than frequent polyploidy. That is, this saw-toothed "signature" arises because the haploid number is even following autotetraploidy, or in the case of allopolyploidy when the diploid progenitors have the same number of chromosomes (called "monobasic allotetraploidy"). Some allopolyploids will have odd haploid numbers (e.g., if the parents have $x = 7$ and $x = 8$), so not all polyploids have this signature. Nevertheless, using this signature and a time frame measured in speciation events, Otto and Whitton (2000) were able to estimate that roughly 2–4% of such events in angiosperms and approximately 7% in ferns may have involved polyploidy. Based on this, they suggested that "polyploidization may be the single most common mechanism of sympatric speciation in plants."

In contrast to its obvious abundance in angiosperms and ferns, polyploidy appears to be less common in other plant lineages. Although bryophytes (liverworts, hornworts, and mosses) have not received a great deal of attention regarding the frequency of polyploidy, the mosses contain many polyploid species (Crosby, 1980; Wyatt *et al.*, 1988, 1993; Przywara and Kuta, 1995). However, polyploidy is rare in the liverworts and virtually unknown in the hornworts (Crosby, 1980; Przywara and Kuta, 1995). Similarly, in gymnosperms (nonflowering seed plants), many species are $2n = 12$ and only 1.5% of the species are estimated to be polyploid (Khoshoo, 1959). Redwood, *Sequoia sempervirens* (Taxodiaceae), is the only known hexaploid conifer (Stebbins, 1948). Polyploidy is also found in the Gnetales; some species of *Ephedra* and *Gnetum* are polyploid (Delevoryas, 1980). In the cycads, there are no recorded polyploids (Khoshoo, 1959).

Recent investigations of entire genomes have dramatically altered the polyploidy paradigm. Genomic investigations reveal that flowering plants, and perhaps all eukaryotes, possess genomes with considerable gene redundancy, much of which could be the result of (ancient) whole genome duplications. For example, complete sequencing of the genome of *Arabidopsis thaliana*, which has a very small genome size (157 megabases [Mb]) (Bennett *et al.*, 2003), revealed numerous duplicate genes and suggested two or three rounds of genome duplication (Vision *et al.*, 2000;

Bowers *et al.*, 2003), as reviewed in detail in Chapter 6. Because *Arabidopsis* has only five chromosomes, it was not previously classified as a polyploid using the common cutoff criteria. Genomic data suggest a recent round of duplication, perhaps during the early evolution of the mustard family, Brassicaceae, with a much earlier round of duplication during the early diversification of angiosperms (Blanc *et al.*, 2003; Bowers *et al.*, 2003) (Fig. 7.5). Similarly, diploid members of *Brassica* may be ancient hexaploids based on analyses of linkage maps, with a number of genes clearly represented multiple times (Lagercrantz and Lydiate, 1996; Lukens *et al.*, 2004).

Genome doubling is now known to be widespread, perhaps characterizing most groups of organisms (Leipoldt and Schmidtke, 1982). In addition to the examples from plants, both ancient and more recent genome duplications are found in numerous groups of animals, as discussed in Chapters 6 and 8. Even the yeast genome is now known to have been anciently duplicated (Wolfe and Schields, 1997). Indeed, genomic studies raise the question of whether there really are any true diploids, at least in a historical sense (Soltis *et al.*, 2004a,b). Furthermore, the classic question of what percentage of angiosperms and ferns is of polyploid origin is almost certainly moot. Perhaps the more appropriate question for angiosperms (and many other groups of organisms) is not whether any, but rather *how many*, genome duplication events have occurred in various lineages. Otto and Whitton's (2000) approach provided some estimate for the number of gene duplication events, but sophisticated genomics-based methods are needed to identify ancient episodes of genome duplications like those recently revealed in *Arabidopsis thaliana* and *Oryza sativa* (see Chapter 6).

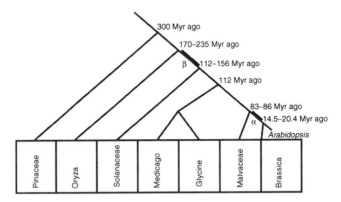

FIGURE 7.5 Multiple genome duplication events in the evolutionary history of *Arabidopsis*. The symbols α and β indicate two genome-wide duplication events inferred and the approximate timing of each. See Chapter 6 for details on how such ancient duplication events are uncovered. Modified from Bowers *et al.* (2003), reproduced by permission (© Nature Publishing Group).

The Frequency of Allopolyploidy versus Autopolyploidy

Allopolyploidy has long been considered much more common than autopolyploidy. In fact, autopolyploids were traditionally thought to be very rare in natural populations of plants. Stebbins (1950) suggested that *Galax aphylla* (now *Galax urceolata*, Diapensiaceae) was one unambiguous example of autopolyploidy. He also offered *Sedum ternatum* and *S. pulchellum* (Crassulaceae) as examples, with *Fritillaria camschatcensis* (Liliaceae) representing "a probable autotriploid." Stebbins (1950, p. 318) went on to discuss several examples of what he termed "intervarietal autopolyploids." A variety is a rank below that of subspecies and is usually used for geographical races and well-marked ecotypes within morphologically variable species. Stebbins stated, in reference to intervarietal polyploids, that "this sort may be found to be not uncommon when more polyploids are analyzed with this possibility in mind." Intervarietal autopolyploids *sensu* Stebbins (1950) included *Biscutella laevigatum* (Brassicaceae), *Dactylis glomerata* (Poaceae), *Allium schoenoprasum* (Alliaceae), *Polygonatum commutatum* (Ruscaceae), *Cuthbertia graminea* (Commelinaceae), *Eriogonum fasciculatum* (Polygonaceae), and "some of the various polyploids of *Vaccinium*" (Ericaceae). Grant (1981) also suggested that autopolyploids were extremely rare in nature. Grant's list of "clear-cut" autopolyploids included *Galax aphylla*, *Biscutella laevigatum*, *Dactylis glomerata*, and *Solanum tuberosum* (Solanaceae). He also noted several "probable" autopolyploids: *Vaccinium uliginosum* (Ericaceae), *Eragrostis pallescens* (Poaceae), and *Galium mollugo* and *G. verum* (Rubiaceae).

The perceived extreme rarity of natural autopolyploids was attributed to concerns about chromosome pairing. Geneticists such as Stebbins maintained that in an autotetraploid, with every chromosome represented four times, normal chromosome pairing at meiosis would be difficult, and multivalent formation would lead to reduced fertility (Fig. 7.2). This prompted harsh statements such as "autopolyploidy is not a help but a hindrance" in natural populations (Stebbins, 1971, p. 126). As noted earlier, however, chromosome pairing in even-numbered polyploids may be stable, such that it does not negate their ability to reproduce.

The inability to detect autopolyploidy probably also influenced early assessments of its natural rarity. Although early investigators realized that an autotetraploid should exhibit tetrasomic inheritance (e.g., Muller, 1914; Haldane, 1930), there was no simple means for determining the inheritance patterns of most genes until the last decades of the 1900s. Prior to that time, researchers were required to make crosses and follow individual traits (presumed to be under simple genetic control) for several generations. The use of enzyme electrophoresis revolutionized plant and animal genetics by allowing multiple molecular forms of individual enzymes ("isozymes") to be detected. Indeed, isozymes have played a critical role in many areas of biology since their discovery by Hunter and Markert (1957), and for several decades they were the most widely used link between the organism and its genes. Allozymes, which are multiple forms of an enzyme encoded by different alleles at a locus, permit an individual to be genotyped. Because allozymic variation is typically present at multiple loci within a species, these loci represent a large pool of potential data for studies of inheritance. By examining the segregation patterns of allozymes, it became possible to determine if a tetraploid plant exhibited disomic or tetrasomic inheritance. Specifically, two plants possessing different allozymes (i.e., alleles) at a gene locus could be crossed (producing a heterozygous F_1 plant), and seeds from this plant could be grown to maturity and either self-pollinated or cross-pollinated among each other to produce F_2 plants. The array of allozymic genotypes in numerous F_2 plants could then be scored to determine whether a $1:2:1$ ratio of homozygotes to heterozygotes to homozygotes was present (disomic inheritance), or if the segregation ratios matched those expected for an autotetraploid with tetrasomic inheritance (Fig. 7.3). Another implication of tetrasomic and higher-level polysomic inheritance in autopolyploids is a concomitant increase in levels of genetic variation in populations of autopolyploids compared to diploids (Moody *et al.*, 1993; Soltis and Soltis, 1993).

Using the allozyme electrophoresis technique, studies of natural populations from diverse plant groups revealed a number of previously unrecognized autopolyploids, including *Tolmiea menziesii* (Saxifragaceae) (Soltis and Soltis, 1989b), *Heuchera micrantha* (Saxifragaceae) (Soltis *et al.*, 1989), *Heuchera grossulariifolia* (Saxifragaceae) (Wolf *et al.*, 1989), *Turnera ulmifolia* (Turneraceae) (Shore, 1991), and *Allium nevii* (Alliaceae) (Rieseberg and Doyle, 1989). Although autopolyploidy is certainly not as common as allopolyploidy, it has probably been an important evolutionary force in plants whose prevalence had been greatly underestimated in the past. In a general sense, some genera and species with chromosome counts existing in prominent polyploid series may represent autopolyploids. Of course, autopolyploidy may be more prevalent in some plant groups than others; for example, several autopolyploids have been documented in the Saxifragaceae, whereas no unambiguous allotetraploids have yet been found in this group (Soltis, 2004).

POLYPLOID FORMATION AND ESTABLISHMENT

MECHANISMS AND CHANCES OF FORMATION

Although enormous strides have been made in the study of some aspects of the genetics and evolution of polyploids, their exact mechanisms of origin remain poorly understood. Polyploidy often is described as "chromosome doubling," which implies a somatic (nonreproductive) event during formation. A somatic chromosome doubling event could occur in a zygote or in developing seedlings or even in active apical meristematic tissues. Both zygotic and meristematic chromosome doubling would immediately result in polyploidy. Importantly, however, it has been noted that critical evidence in support of chromosome doubling in the zygote is lacking (de Wet, 1980), and spontaneous somatic (meristematic) chromosome doubling is considered very rare (e.g., Nasrallah *et al.*, 2000; Grant, 2002). Thus, although "chromosome doubling" in a strict somatic sense may be one method of polyploid formation, it is not considered the most common mechanism. Instead, Harlan and de Wet (1975) and de Wet (1980) proposed that a more common method of polyploidization is via the formation and fusion of unreduced gametes.

The production of pollen and egg cells is a complex process, and the failure of chromosome reduction during meiosis has been observed in many plant species. That is, as a result of several possible meiotic mishaps, a pollen or egg cell can be produced that is not haploid, but rather has the same "unreduced" chromosome number as the parent cell. As reviewed by several authors (de Wet, 1980; Bretagnolle and Thompson, 1995; Ramsey and Schemske, 1998), unreduced gametes are probably produced by most individuals. Obviously, the union of two such unreduced gametes would result in an instantaneous polyploid. However, the rate of production of unreduced gametes is quite low (at most, just a few percent of all gametes produced), such that the probability of an unreduced egg being fertilized by an unreduced sperm (a process termed "bilateral polyploidization") (Stebbins, 1971) is relatively small. It is therefore more likely that the formation of higher polyploids proceeds via the intermediate of triploidy (Bretagnolle and Thompson, 1995; Ramsey and Schemske, 1998). Thus the first step would be fertilization involving haploid pollen and an unreduced (diploid) egg (or, less frequently, vice versa) to yield a triploid zygote ("unilateral polyploidization"). If this triploid is viable and fertile—that is, it survives to maturity and produces functional reproductive structures—it may generate a number of triploid eggs. If these, in turn, back-cross to a normal diploid plant (and hence are fertilized by a haploid pollen grain), the result would be a tetraploid. This triploid step toward tetraploid formation is sometimes referred to as a "triploid bridge," and there is increasing evidence that this is an important stage in polyploid formation (Bretagnolle and Thompson, 1995; Ramsey and Schemske, 1998; Husband, 2004).

Both de Wet (1980) and Ramsey and Schemske (1998) pointed out that adverse growing conditions may increase the frequency with which unreduced gametes are produced (e.g., Clausen *et al.*, 1940; Grant, 1952). The genotype of the individual also seems to play a role, with some individuals and populations more prone to generating unreduced gametes than others. Hybridization itself (particularly among species) may also promote the formation of unreduced gametes: even 30 years ago, Harlan and de Wet (1975) listed 68 genera of flowering plants in which functional unreduced gametes were produced by diploid F_1 hybrids that resulted from the fertilization of a normal haploid egg (from species A) and normal haploid sperm (from species B).

An important challenge to the evolution of higher polyploidy via a triploid bridge is the fact that triploidy may result in reduced fertility, primarily because of meiotic irregularities resulting in aneuploid gametes (i.e., having a chromosome set with more or less than a multiple of a basic number). One other contributing factor, known as "triploid block," stems from the partial or complete failure of the endosperm tissue (the nutritive tissue for developing embryos) following certain interploidal or interspecific crosses (Bretagnolle and Thompson, 1995; Ramsey and Schemske, 1998). In "normal" diploid–diploid fertilization in angiosperms, one sperm nucleus from the pollen grain fertilizes the egg cell, while a second sperm nucleus fuses with two other haploid (polar) nuclei in the central cell of the female gametophyte and forms a triploid nutritive endosperm tissue. In a diploid × diploid cross, the triploid endosperm tissue contains two sets of maternal chromosomes (from the two polar nuclei) and one set of paternal chromosomes (from the sperm nucleus), whereas the embryo is diploid with a 1:1 ratio of maternal and paternal genomes. It was initially thought that this 2:1 ratio of the maternal to paternal ploidies in the endosperm was required for proper embryo development, and that seed abortion following interploidal (e.g., diploid × tetraploid) crosses occurred because of an imbalance in this ploidal ratio (see Johnston et al., 1980; Johnston and Hanneman, 1982). However, later empirical studies showed that the ploidal levels of the maternal and paternal tissues did not necessarily determine whether a seed would abort (Johnston et al., 1980; Johnston and Hanneman, 1982).

More recently, the basic notion of a 2:1 maternal to paternal endosperm genome ratio has been incorporated into the so-called "Endosperm Balance Number (EBN) hypothesis" (Johnston et al., 1980; Johnston and Hanneman, 1982). Essentially, the genome of each species is assigned an EBN, which is based on the endosperm viability (i.e., production of viable seed) following a cross to a reference species. If the cross between the reference and an "unknown" species or cytotype results in the production of viable seed, then it is assumed that the 2:1 maternal to paternal ratio is maintained in the endosperm. The crossed species is then assigned the EBN of the reference species (see Table 7.2). As an example, an arbitrary EBN of two could be assigned to 2x *Species A*. If a cross between 2x *Species A* and 4x *Species B* yields viable 3x offspring, then the 4x *Species B* will also have an EBN of two. Crosses can then be made using 4x *Species B* as a reference to determine the EBN for other species and cytotypes. Higher ploidal levels of the reference species may also be useful to determine the EBN for certain cytotypes or species. For example, 4x *Species A* (i.e., autopolyploid of 2x *Species A*, therefore it has an EBN of four) does not form viable seed when crossed with

TABLE 7.2 Examples of assigning an Endosperm Balance Number (EBN) through crosses with a reference species, which is first arbitrarily assigned an EBN. Crosses are made between the reference species and a different cytotype or species (or both). It is assumed that crosses will produce viable seed when the ratio of maternal to paternal genomes is at the required 2 : 1 in the endosperm. In this case, the crossed species is assigned the same EBN as the reference species, and it may then serve as a reference species for other crosses. Based on information provided by Johnston et al. (1980).

Reference species	Reference EBN	Crossed species	Viable seed produced?	EBN of crossed species
2x Species A	2	4x Species B	Yes	2
4x Species A	4	2x Species A	No	—
4x Species A	4	4x Species B	No	—
4x Species A	4	4x Species C	Yes	4
4x Species B	2	2x Species C	Yes	
4x Species B	2	4x Species C	No	—
8x Species B	4	4x Species C	Yes	4

2x *Species A* or 4x *Species B*, but it does produce viable seed when crossed with 4x *Species C*, which is then assigned an EBN of four. Importantly, differing cytotypes of the same species may not necessarily have the same EBN; therefore, crosses among cytotypes and species must be conducted in order to assign the EBN. Once determined for each cytotype and each species being crossed, however, the EBN can have predictive value in determining which interploidal or interspecific crosses should produce viable offspring. In summary, the EBN, but not necessarily the ploidal levels, should be in a 2:1 maternal to paternal ratio for proper endosperm function. Studies in *Chamerion angustifolium* (Burton and Husband, 2000), *Impatiens* (Arisumi, 1982), *Solanum* (Johnston et al., 1980; Johnston and Hanneman, 1982), and *Trifolium* (Parrott and Smith, 1986) support this hypothesis.

Despite the potential for triploid block, the triploid bridge remains a plausible mode for polyploid formation. As Ramsey and Schemske (1998) pointed out, triploids generate some euploid gametes (i.e., with an exact multiple of the basic chromosome number) and may be semifertile. Once a few tetraploids are produced, an array of intercytotype crosses (backcrosses to the diploids, selfing, or crossing with other triploids) may create more tetraploids or higher ploidal levels. In studies of crosses among diploid, triploid, and tetraploid *Chamerion angustifolium* (Onagraceae), Burton and Husband (2001) found that although the fecundity of the progeny derived from triploid crosses was low, the gametes produced (1x, 2x, and 3x) could function in fertilization and sire progeny of various ploidal levels.

Further support for the triploid bridge as a pathway to tetraploids and higher ploidies should be sought in cytotype contact zones for other groups (Petit *et al.*, 1999).

LIKELIHOOD OF ESTABLISHMENT

Even though polyploidy may have major effects on the evolution of plant lineages, such as providing a mechanism for instantaneous speciation, the conditions that favor the establishment and persistence of recently formed polyploids are still not well understood (e.g., Thompson and Lumaret, 1992; Ramsey and Schemske, 2002). Anecdotal data and broad floristic correlations have suggested numerous relationships between polyploidy and various measures of ecological "success" (e.g., Stebbins, 1947, 1950; Ehrendorfer, 1980; Lewis, 1980b), but these hypotheses have seldom been tested with specific case studies, and the factors that contribute to the long-term success of polyploids have rarely been identified. On the other hand, there is a growing body of both theoretical and empirical work dealing with the more immediate issue of locating suitable mates, and broader aspects such as dispersal to new habitats and competition with (and/or divergence from) related diploid species.

While new polyploids may have reduced fertility, the fertility of early-generation polyploids increases rapidly (Ramsey and Schemske, 2002), meaning that a more significant problem for sexually reproducing neopolyploids involves finding fellow polyploids with which to mate amidst their more numerous diploid counterparts. Thus the "minority cytotype exclusion principle" (Levin, 1975) holds that a newly formed polyploid will be subjected to frequency-dependent selection, with the rare genotype or cytotype at a disadvantage. That is, the tetraploid cytotype will be unlikely to establish if there are proportionately more diploids than tetraploids with which to mate, primarily owing to lower fitness of the triploid progeny compared to "within cytotype" progeny. Recently, Husband (2000) conducted a test of this principle in artificially mixed cytotype ($2x$ and $4x$) populations of *Chamerion angustifolium* (Onagraceae) and, interestingly, found evidence for frequency-dependent fitness among the diploids but not the tetraploids. Assortative mating (within the same phenotype) mediated by pollinators (in this case different types of bees) consistently pollinating similar phenotypes was suggested to contribute to the within-cytotype seed set of the tetraploids. This study indicates that complex population-level processes will be involved in the establishment of polyploids. Factors affecting the reproductive biology of the polyploids themselves may also play an important role in the persistence of polyploids (as discussed later in the chapter).

A few theoretical discussions have focused on the establishment and persistence of polyploids among their diploid progenitors (e.g., Fowler and Levin, 1984; Felber, 1991; Thompson and Lumaret, 1992; Rodriguez, 1996). Two hypotheses describe the conditions under which a new polyploid, in competition with one or both of its parental species, is likely to become established and persist (Fowler and Levin, 1984; Felber, 1991; Rodriguez, 1996). The first suggests that a new polyploid may persist by *replacing* its diploid parent(s), either as a result of an unstable equilibrium between the cytotypes and the stochastic loss of a small diploid population, or by outcompeting it (Fowler and Levin, 1984; Rodriguez, 1996). The second hypothesis suggests that a new polyploid may *coexist* with its diploid parent(s) as a result of habitat differentiation immediately following the origin of the polyploid. Importantly, replacement and habitat differentiation should not be considered mutually exclusive. For example, habitat differentiation could contribute to the initial establishment of a new polyploid, but because of higher fitness, this neopolyploid may be able to replace one or both of its diploid progenitors, at least in some areas.

Views on the fitness of diploid hybrids have also focused on the availability of novel habitats for hybrid derivatives (e.g., Anderson, 1949; Stebbins, 1959). Whereas some models hold that hybrids are consistently less fit than their parents (e.g., Howard, 1982, 1986; Harrison, 1986), the "bounded hybrid superiority model" (Moore, 1977) views hybrids as less fit than their parents in the parental habitats but more fit than either parent in other habitats. Biologists have long recognized the role of disturbance in providing new habitats for hybrids (the "hybridized habitat" of Anderson, 1949), but undisturbed areas may also have open habitats for hybrids (Arnold, 1997). Arnold's (1997) "evolutionary novelty" model of hybrid zone dynamics seems applicable to allopolyploids, which through the combination of the genomes of their parents, represent new entities with their own evolutionary tendencies. Just as different genotypic classes of hybrids may be less fit, equally fit, intermediate, or more fit than their parents (e.g., Cruzan and Arnold, 1993; Arnold, 1997; Emms and Arnold, 1997), populations of an allo-polyploid of independent origin (see later section) may have variable fitnesses relative to their diploid parents.

Several studies suggest evidence of habitat differentiation among cytotypes in species representing a variety of families, including *Claytonia virginica* (Portulacaceae) (Lewis, 1976; Lewis and Suda, 1976), *Festuca apennina* (Poaceae) (Tyler *et al.*, 1978), *Fragaria* spp. (Rosaceae) (Hancock and Bringhurst, 1981), *Dactylis glomerata* (Poaceae) (Lumaret, 1984; Lumaret *et al.*, 1987), *Clarkia* spp. (Onagraceae) (Smith-Huerta, 1984), *Anthoxanthum odoratum* (Poaceae) (Felber, 1988), the *Antennaria rosea* complex (Asteraceae) (Bayer *et al.*, 1991), *Lotus corniculatus* (Fabaceae) (Reynaud *et al.*, 1991), and *Galax urceolata* (Diapensiaceae) (Johnson *et al.*, 2003). In general, greater variability in polyploids for morphological, demographic, and phenotypic traits relative to their diploid progenitors is believed to contribute to habitat differentiation. For example, octoploid *Fragaria chiloensis* and *F. virginiana* display more heterozygosity and morphological variation, have greater physiological homeostasis, and/or occur across a wider range of environments than one of their putative diploid parents, *F. vesca* (Hancock and Bringhurst, 1981). Polyploidy can also lead to more intense partitioning of the habitat, as in the *Antennaria rosea* species complex (Bayer *et al.*, 1991), in which most polyploid populations occupy habitats intermediate to those of the diploid parental species. Thus polyploids may not necessarily occupy harsher habitats than their diploid parents but can be considered "fill-in" taxa that occupy habitats intermediate to those of their progenitors (Ehrendorfer, 1980).

In some cases, the range of the diploids may actually represent a subset of the range of the polyploids; for example, in *Deschampsia cespitosa* (Poaceae) (Rothera and Davy, 1986), *Anthoxanthum odoratum* (Felber, 1988), *Solidago nemoralis* (Asteraceae) (Brammal and Semple, 1990), and *Plantago media* (Plantaginaceae) (Van Dijk *et al.*, 1992, 1997). In more extreme instances, habitat differentiation may be so great that polyploids and diploids come to have very different geographic ranges, with little or no overlap at all. Although the ecological factors at work in most of these cases have not been examined, variation in the geographic distribution of diploid and tetraploid cytotypes of *Anthoxanthum alpinum* appears to relate to differences in flowering phenology (Felber, 1988). Whatever the specific explanation, this presence of only polyploid cytotypes in large geographic areas is a good indicator of the importance of habitat differentiation (Thompson and Lumaret, 1992) and/or superior colonizing abilities (e.g., Clausen *et al.*, 1945; Levin, 1983; Barrett and Richardson, 1986; Lumaret *et al.*, 1987; Barrett and Shore, 1989) in allowing polyploids to persist on a broad ecological scale.

MULTIPLE ORIGINS OF POLYPLOID SPECIES

THE RULE, NOT THE EXCEPTION

The traditional view of speciation in plants mirrors that for other groups of organisms, namely that individual species are taken to form only once in one of several possible modes (e.g., allopatric, sympatric, or saltational speciation) (reviewed in Grant, 1981; Futuyma, 1998; Levin, 2000). Until fairly recently, this standard view of diploid speciation had carried over to treatments of polyploids. In fact, the possibility that a single polyploid species might form more than once does not appear to have been discussed in any of the previous major reviews of polyploidy, including the highly influential discussions by Stebbins (1950, 1971) and Grant (1981). It is only during the past 20 years that a fundamental change has occurred in the understanding of the frequency and importance of multiple origins of polyploid species. Today, it is well established, based on several lines of evidence, that individual polyploid plant species typically form multiple times (Fig. 7.6).

A

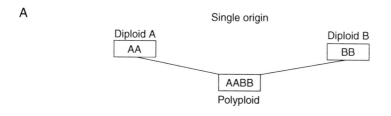

Single origin

B

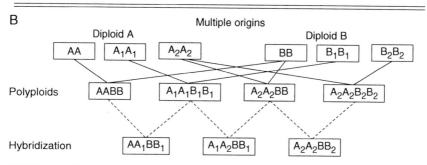

Multiple origins

FIGURE 7.6 (A) Previous view of polyploid origins in which each polyploid species formed once, resulting in a genetically uniform species. (B) New view of recurrent formation from different parental genotypes, generating an array of polyploid genotypes. Crossing and recombination result in additional variability (disomic inheritance assumed). Redrawn from Soltis and Soltis (1999), reproduced by permission (© Elsevier Inc.).

This is not to say that evidence of multiple formation was totally lacking in early discussions, and indeed a few investigations from the premolecular era did suggest that a single polyploid species may form more than once. For example, although Grant (1981) did not mention recurrent polyploidization, he was aware that it occurred based on some of his own research in *Gilia* (Polemoniaceae) (Grant, 2002). Ownbey (1950) and Ownbey and McCollum (1953) reported that the recently formed allotetraploids *Tragopogon mirus* and *T. miscellus* (Asteraceae) had likely formed at least twice. Other examples of recurrent polyploidy from the premolecular literature include species of *Madia* (Asteraceae) (Clausen *et al.,* 1945), *Gutierrezia* (Asteraceae) (Solbrig, 1971), *Mimulus* (Scrophulariaceae) (Mia *et al.,* 1964), and *Rubus* (Rosaceae) (Rozanova, 1938; see Mavrodiev and Soltis, 2001).

Tragopogon (Asteraceae) provides a particularly interesting example of the demonstration of multiple origins prior to the use of molecular markers (Fig. 7.7). In extensive crossing studies, Ownbey and McCollum (1953) produced F_1 hybrids between *T. dubius* and *T. pratensis,* the parents of the allotetraploid *T. miscellus,* and found that the F_1s differed dramatically in the morphology of

the head of flowers based on which species was the maternal parent. When
T. dubius was the maternal parent, the F_1s had "long ligules," whereas when
T. pratensis was the maternal parent, the F_1 plants had "short ligules" (see Fig. 7.7).
These same two long- and short-ligule morphologies are also observed in natural
populations of *T. miscellus*, leading Ownbey and McCollum (1953) to argue that
T. miscellus had formed twice with reciprocal parentage. That is, in some cases
T. pratensis was the maternal parent, and in other cases *T. dubius* served as the
maternal parent. Unfortunately, Ownbey and McCollum (1953) could not induce
polyploids to form from the F_1 hybrids that they had produced, which would have
provided the final piece of evidence. Instead, confirmation of their hypothesis
only came later, with the application of molecular methodologies.

Such investigations based on experimental hybridization and cytogenetics were
essential in earlier times (and still remain useful) for determining what was taking
place in polyploid formation. However, they are very time-consuming and lack
the efficiency of molecular techniques, which can supply more information in
a shorter period. When molecular markers were applied in the study of polyploid
evolution, they revealed that multiple origins were commonplace, and that they

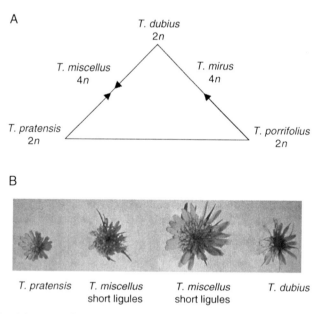

FIGURE 7.7   (A) Origins of *Tragopogon* allopolyploids in the New World. (B) Reciprocally formed
Tragopogon miscellus populations differ in their floral morphologies. When *T. pratensis* is the maternal
parent, the ligules are short, and when *T. dubius* is the maternal parent, the ligules are long.

occurred frequently in some species. In fact, numerous studies have demonstrated that polyploid species may form recurrently from hybridization events between the same diploid parents (Werth et al., 1985a,b; Wyatt et al., 1988; Soltis and Soltis, 1989a, 1991; Brochmann et al., 1992a,b; Soltis et al., 1995; Cook et al., 1998; Doyle et al., 1999; Segraves et al., 1999; Sharbel and Mitchell-Olds, 2001; Doyle et al., 2002). Moreover, this pattern of multiple origins applies to both allo- and autopolyploids.

Molecular markers, including isozymes (Werth et al., 1985a,b; Wyatt et al., 1988; Werth and Windham, 1991; Brochmann et al., 1992b; Soltis et al., 1995; Allen and Eccleston, 1998; Arft and Ranker, 1998), random amplified polymorphic DNA (RAPDs) (Cook et al., 1998), restriction site profiles (Doyle et al., 1990; Soltis and Soltis, 1990a; Brochmann et al., 1992a), and DNA sequences (Segraves et al., 1999; Popp and Oxelman, 2001; Sharbel and Mitchell-Olds, 2001; Doyle et al., 2002) have all been used to document the fact that recurrent formation of polyploid species may be frequent.

In Tragopogon miscellus, for example, chloroplast DNA (cpDNA) evidence was used to confirm that this tetraploid had formed at least twice. Chloroplasts are inherited maternally in most angiosperms, so this is an excellent marker for determining a putative maternal parent of a polyploid. A comparison of cpDNA restriction site patterns from populations of the polyploid T. miscellus and multiple populations of the diploid parents, T. dubius and T. pratensis, demonstrated that plants of T. miscellus with long ligules exhibited the same pattern as plants of T. dubius; conversely, plants of T. miscellus with short ligules had the cpDNA profile of T. pratensis (Soltis and Soltis, 1989a). These data confirmed the hypothesis presented more than 35 years earlier by Ownbey and McCollum (1953) based on less direct methods involving crossing data and comparisons of morphology.

In the fern genus Hemionitis, Ranker et al. (1989) used the diversity of allozyme patterns to demonstrate that the allotetraploid H. pinnatifida formed at least five times. Most of the alleles in the tetraploid matched those found in various populations of the diploid progenitors, supporting the hypothesis of multiple origins from genetically differentiated diploid populations. Intriguingly, some allozymes were "missing" from the sample of alleles detected in the diploid parents, but Ranker et al. (1989) were able to reconstruct the hypothetical parental genotypes (presumably now extinct, or at least not sampled) that could have contributed these alleles.

Very few polyploids that have been investigated with molecular markers are thought to have originated only once,[2] and, therefore, multiple origins are now considered to be the rule in polyploid evolution (Soltis and Soltis, 1993, 1999).

---

[2]Notable exceptions in which polyploidy appears to have occurred only once include salt marsh grass, Spartina anglica (Poaceae) (Huskins, 1931; Marchant, 1968; Raybould et al., 1991; Baumel et al., 2001), and peanut, Arachis hypogaea (Fabaceae) (Kochert et al., 1996).

The number of recurrent origins varies considerably, and may be as few as two (Soltis and Soltis, 1993) or as many as 21 (or possibly even more) for *Tragopogon miscellus* (Asteraceae) (Soltis *et al.*, 1995). As data are gathered for additional polyploid populations and from multiple markers, the number of estimated independent polyploidization events typically increases. For example, autotetraploid *Heuchera grossulariifolia* (Saxifragaceae) was originally suggested to have formed at least three times (Wolf *et al.*, 1990), but additional data suggested as many as five origins (Segraves *et al.*, 1999). Allohexaploid *Draba norvegica* (Brassicaceae) has formed at least 13 times in a small area of Scandinavia (Brochmann *et al.*, 1992a). *Glycine tabacina* (Fabaceae), an allopolyploid, has formed at least six times in Australia (Doyle *et al.*, 2004). *Tragopogon miscellus* and *T. mirus*, two allotetraploid species of goatsbeard, may have formed as many as 21 and 12 times, respectively, in eastern Washington and adjacent Idaho (United States) in the past 80 years—multiple polyploidizations have even occurred within a single small town (Soltis *et al.*, 1995; Cook *et al.*, 1998). Evidently, the recurrent formation of polyploids occurs not only in a diversity of species, but also at several different temporal and geographic scales.

Polyploids constitute a significant proportion of both angiosperm and fern species, and the fact that recurrent origins typify those taxa investigated thus far has important implications. Again, Otto and Whitton (2000) estimated that polyploidization may represent 2–4% and 7% of all speciation events in angiosperms and ferns, respectively. If these estimates are close to correct, then for many land plants, most of which are polyploid, it might be more appropriate to think not in terms of Darwin's (1859) simpler "Origin of Species," but rather of the *origins* of species (Soltis and Soltis, 1999; Soltis *et al.*, 2004b). Moreover, if monophyly is considered a necessary criterion for recognizing species, then the recurrently formed polyploid lineages should be regarded as cryptic phylogenetic species. However, the consensus view among systematists is that these separately formed lineages may produce a tokogenetic (that is, reticulating genealogical) network from which a single "lineage" emerges, although experimental evidence for this is generally lacking (Brochmann and Elven, 1992). The implications of recurrent polyploidization for species concepts (especially the phylogenetic concept) are important and deserve additional attention, but will be left for future authors to consider because they are beyond the scope of this discussion.

GENOTYPIC AND PHENOTYPIC CONSEQUENCES OF MULTIPLE ORIGINS

The realization that recurrent polyploidization is the rule, not the exception, has shattered earlier perceptions of polyploids as genetically depauperate species of uniform genotype that represent evolutionary dead-ends (Wagner, 1970) (Fig. 7.6).

Because of multiple origins, polyploid species can maintain high levels of segregating genetic variation through the incorporation of genetic diversity from multiple populations of their diploid progenitors (Soltis and Soltis, 1999). Polyploid genotypes of separate origin may come into contact and hybridize, resulting in subsequent segregation and recombination, which generates additional genetic variation (Fig. 7.6). Molecular studies suggest that this does indeed occur in nature. In species of *Draba,* distinct genotypes of separate polyploid origin co-occur in the same populations, along with putative recombinants (Brochmann and Elven, 1992). Autopolyploid populations of *Heuchera grossulariifolia* often form a mosaic of genotypes representing separate origins (Segraves *et al.,* 1999). Application of RAPD markers to the allotetraploid *Tragopogon* species reveals that populations of separate origin come into contact (Cook *et al.,* 1998). Importantly, the short time since the origin of *Tragopogon* allotetraploids indicates the rapidity with which polyploid genotypes can come into contact after their formation.

Polyploid species of independent origin may also differ morphologically (Soltis *et al.,* 1995, 2004a; Lowe and Abbott, 1996). For example, Ownbey (1950) first suggested that the dramatic difference in inflorescence morphology between populations of *Tragopogon miscellus* resulted from recurrent formation involving reciprocal parentage (Fig. 7.7). Populations of *T. mirus* of separate origin differ in floral coloration (Ownbey, 1950; Soltis *et al.,* 1995), and other examples of differences in floral morphology among recurrently formed polyploids can be found in *Heuchera grossulariifolia* (Segraves and Thompson, 1999) and *Draba* (Brochmann, 1993).

A CASE OF PARTICULAR INTEREST: THE ARCTIC FLORA

The Arctic flora is of particular interest because it can be used to illustrate the process of recurrent polyploid formation on local, intermediate, and very broad geographical scales. In fact, this flora may be the best one in which to observe recurrent polyploidization on a broad geographical scale, because here diploids and polyploid derivatives exhibit overlapping circumboreal distributions.

On a relatively small scale, the polyploid *Draba norvegica* (Brassicaceae) has formed at least 13 times in a small area of Scandinavia, as revealed by allozyme studies (Brochmann and Elven, 1992; Brochmann *et al.,* 1992a). In the grass *Dupontia,* a similar approach demonstrated that an allopolyploid had formed recurrently across the Arctic of North America (Brysting *et al.,* 2003, 2004; Brochmann *et al.,* 2004). On a large, circumboreal scale, complex patterns of recurrent formation and subsequent interbreeding of polyploid genotypes can be seen in species of *Saxifraga,* where progenitor species come into contact throughout broad areas of the Arctic (Steen *et al.,* 2000; Brochmann and Hapnes, 2001).

A more complex example of recurring tetraploidy in the Arctic is represented by the bird-dispersed Arctic blueberry (or bilberry), *Vaccinium uliginosum* (Ericaceae) (Brochmann *et al.*, 2004). Polyploidy in this complex involves events that probably have taken place at different scales in time and space. The complex has a complete circumpolar/boreal distribution with extension to more southern mountain ranges. About 30 different taxa have been described in this complex, some of them at the species level, but a single, polymorphic species with a variable number of sub-species is now usually recognized. Recent molecular, ploidal, and morphological analyses of populations from the entire geographic range have provided strong evidence for at least three independent origins of tetraploids (reviewed in Alsos *et al.*, 2001, 2002; Alsos, 2003; Brochmann *et al.*, 2004). Three major clades were identified in phylogenetic analyses of DNA sequences: a circumpolar "Arctic-Alpine clade," the boreal "Amphi-Atlantic clade," and the boreal/Beringian "Amphi-Pacific clade," estimated to have diverged between 0.7 and 3.0 million years ago. The Amphi-Pacific clade was diverse and formed a central group in a network analysis, with the chloroplast haplotype of a Beringian diploid in the most central position, suggesting that the entire complex may have originated in Beringia. The Amphi-Atlantic clade was exclusively tetraploid, suggesting that this lineage was derived from an old tetraploidization event, possibly dating back to the onset of the Pleistocene glaciations. In contrast, the tetraploids in the Amphi-Pacific clade were confined to tip haplotypes in the network analysis, suggesting much more recent origins. In the Arctic-Alpine clade, most populations were diploid, but the most widespread (entirely circumpolar) haplotype C was also found in tetraploid populations, suggesting that tetraploidy has also originated independently in this clade.

Traditionally, each Arctic polyploid was thought to have formed only once, with subsequent migration considered responsible for the establishment of a broad geographic distribution. Genetic data suggest instead a very different evolutionary scenario, in which progenitor species that co-occur on a circumboreal scale hybridize repeatedly and subsequently form polyploids. This phenomenon evidently occurs repeatedly over long periods of time and over large geographic areas, and leads to the generation of complexes of different polyploid genotypes and morphotypes. These genotypes themselves may ultimately come into contact and hybridize, generating even more genetic and morphological complexity through subsequent segregation and recombination. This scenario, repeated in numerous lineages, would help to explain the infamous taxonomic uncertainty surrounding polyploid complexes in the Arctic.

IMPACTS OF POLYPLOIDIZATION AT THE CELLULAR AND ORGANISMAL LEVELS

Stebbins (1950) made an important point when he stated that "the only generalization which may be safely made about the morphological and physiological effects of polyploidy is that these depend greatly on the nature of the original genotype." Certainly, some of the effects of polyploidization will differ in auto- versus allopolyploids, and polyploids derived from hybridization between distantly related species may also differ fundamentally from those derived from more closely related species. Although it is important to keep in mind that the conse- quences of polyploidization are not universal, several important attributes of poly- ploids have commonly been discussed in the literature. Levin (1983) provided a very thorough review of this topic, and for the most part, the present review is meant to highlight some of the more recent advances, in addition to discussing the classical literature.

CELL SIZE

One of the most immediate effects of genome doubling is an increase in cell size in polyploids relative to their diploid progenitors, with the resulting decrease in surface area to volume ratio (Stebbins, 1950, 1971) (see Chapter 8). In plants, guard cell and pollen grain sizes are commonly used as proxies for ploidal level in closely related diploid and polyploid species, because both are often significantly larger in polyploids than in the diploid progenitors (Fig. 7.4). Again, Masterson (1994) compared guard cell measurements in fossil and extant angiosperms to infer high estimates of polyploidy within the flowering plants. However, in some groups (e.g., some Saxifragaceae), guard cell and pollen sizes may not increase with ploidy.

Butterfass (1987) compiled pollen grain and guard cell measurements from the literature and found that natural tetraploids tended to have relatively smaller cell sizes than artificial tetraploids representing the same species. This reduction was suggested to result from natural selection for smaller cell size or a "physiological self-adjustment." No significant differences in cell size were detected between induced allopolyploids and autopolyploids (Butterfass, 1987). Occasionally, pollen grain sizes in diploid and polyploid races of a single species may overlap (Lewis, 1980b), making accurate inferences of ploidal levels difficult (e.g., Small, 1983). In one extreme case, the tetraploid species of *Tarasa* (Malvaceae) have smaller pollen grains than the putative diploid progenitor species (Tressens, 1970; Tate and Simpson, 2004). In this case, selection for a smaller floral morphology (including pollen grains) appears to have followed a switch to an autogamous breeding system (i.e., involving self-fertilization) in the polyploids (Tate and Simpson, 2004). Like other polyploids, however, the guard cells in the *Tarasa* tetraploids examined were larger than those of the diploids (J. Tate, J. McDill, B. Simpson, unpublished data). In most cases, guard cell size (or stomatal length) may be a more reliable indicator of ploidal level among closely related taxa than pollen grain size. Occasionally, the effects of polyploidization on somatic cells may result in a generalized "gigas effect," whereby the entire polyploid plant is larger than its diploid relatives (Stebbins, 1950). For example, when Ownbey (1950) first described the New World *Tragopogon* allopolyploids, he characterized them as being "larger in every way" than the diploid parents.

REPRODUCTIVE BIOLOGY

As discussed in previous sections, the success of polyploids involves more than mechanisms of formation—the establishment and persistence of a newly formed polyploid are also critical considerations. Central to this issue is the reproductive biology of the polyploid species. In essence, a new polyploid may be reproductively isolated from its diploid progenitors, not only by virtue of its multiplied chromosome number, but also because the physiological and morphological changes that accompany or follow polyploidization may alter its reproductive biology (Thompson and Lumaret, 1992). These reproductive barriers may be prezygotic (e.g., geographic isolation, flowering phenology differences, pollinator consistency) or postzygotic (e.g., triploid hybrid inviability, inbreeding depression). Whereas isolation from diploids can help in the maintenance of a distinct polyploid lineage, these factors also present challenges to be overcome if newly formed polyploids are to reproduce successfully. Importantly, polyploids have a number of means to persist even when similar cytotypes are lacking from the population, with both sexual and asexual reproductive modes playing an important role in polyploid persistence.

Apomixis, a general term describing asexual reproduction in plants, is a prominent feature in many polyploid groups. The two main forms of apomixis are *vegetative reproduction*, such as stolon or plantlet formation, and *agamospermy*, the production of fertile seeds without fusion between gametes ("seeds without sex") (Bierzychudek, 1990; Asker and Jerling, 1992; Richards, 1997). The form of agamospermy most commonly associated with polyploidy is *gametophytic apomixis*, in which a female gametophyte (embryo sac) forms with the sporophytic chromosome number (Pessino *et al.*, 1999). Two complex pathways (apospory and diplospory) have been described for this process, but essentially both involve an avoidance or a breakdown in female meiosis that ultimately leads to the formation of a mature seed (Asker and Jerling, 1992; Richards, 1997).

Nogler (1984) suggested that dominant genes for apomixis might be linked to recessive lethals that would be expressed normally in the gametophyte (haploid) generation. In a polyploid, these lethal recessives would not be expressed by virtue of the "diploid" state of the gametophyte. Recent research supports the hypothesis that apomixis is inherited as a monogenic dominant trait in most plants, for example, in *Hieracium* (Bicknell *et al.*, 2000), *Panicum* (Savidan, 1981), *Pennisetum* (Ozias-Akins *et al.*, 1993), and *Ranunculus* (Nogler, 1984), although Noyes and Rieseberg (2000) found evidence for two loci controlling agamospermy in triploid *Erigeron annuus* (Asteraceae). It is interesting to note that agamospermy occurs in several unrelated flowering plant families (e.g., Poaceae, Asteraceae, Rosaceae, Ranunculaceae). As more is learned about the underlying genetics controlling agamospermy, particularly in polyploids, potential practical applications (especially for the agricultural industry) may come to light.

While most polyploid species are perennial in habit and/or possess a means of asexual reproduction, polyploidy in annual plants is restricted to species that are self-compatible (Stebbins, 1950). Some examples include members of *Clarkia* (Onagraceae), *Galeopsis* (Lamiaceae), *Gilia* (Polemoniaceae), *Madia* (Asteraceae), *Microseris* (Asteraceae), and *Mentzelia* (Loasaceae) (Grant, 1981). A breakdown of the genetic self-incompatibility mechanisms frequently accompanies polyploidization in annuals or perennials (Stebbins, 1950; Grant, 1981; Richards, 1997). This change occurs in species with a single-locus gametophytic self-incompatibility (GSI) system and is not known to occur in species with multigenic GSI or sporophytic self-incompatibility (SSI) systems (de Nettancourt, 2001; Ramsey and Schemske, 2002; but see Mable 2004a). Similarly, in homosporous ferns, intragametophytic selfing rates are higher in tetraploids than in diploids (Masuyama and Watano, 1990; Soltis and Soltis, 1990b). Because a newly formed polyploid may be reproductively isolated from its progenitors, a switch to self-compatibility greatly enhances the possibility of polyploid establishment (Fowler and Levin, 1984; Ramsey and Schemske, 2002). Moreover, because of their duplicated genomes, polyploids are expected to be buffered from the deleterious effects of inbreeding depression that otherwise occur when normally outcrossing plant species self-fertilize (Stebbins, 1957; Charlesworth and Charlesworth, 1987; Hedrick, 1987; Husband and Schemske, 1997; Cook and Soltis, 1999, 2000).

Miller and Venable (2000) proposed that polyploidy is a trigger for the evolution of gender dimorphism in plants. In this case, dioecy (male and female flowers produced on separate plants) or gynodioecy (female and bisexual [i.e., "perfect"] flowers on separate plants) evolved in polyploid, self-compatible species of *Lycium* (Solanaceae) whose closest relatives were bisexual, self-incompatible diploids. This pathway to gender dimorphism apparently occurred in 12 genera involving at least 20 independent evolutionary events. Miller and Venable's (2000) hypothesis purports that self-incompatibility breaks down following polyploidization, but that these polyploids then suffer from inbreeding depression. The evolution of male sterility follows and ultimately leads to a system of stabilized gender dimorphism in these species. It is interesting to note that members of the Solanaceae have a gametophytic self-incompatibility system, in which polyploidy often causes a breakdown of the incompability reaction. The consequences of this breakdown (e.g., level of inbreeding depression) should be investigated further.

Morphological changes resulting from polyploidization can also have an effect on the reproductive biology of a neopolyploid species. For example, Segraves and

Thompson (1999) found that natural autopolyploid populations of *Heuchera grossulariifolia* (Saxifragaceae) not only have larger flowers than the diploid populations, but also different floral shapes and flowering times. The suites of pollinators visiting sympatric diploid and tetraploid plants differed proportionately when the flowering time of the two cytotypes was synchronous. Moreover, the independently generated autotetraploid populations differed in a number of floral characters, including mean scape length and mean number of flowers on the first inflorescence (Segraves and Thompson, 1999).

Husband and Sabara (2003) recently compiled data for multiple prezygotic and postzygotic reproductive isolating mechanisms in diploid and autotetraploid individuals of *Chamerion angustifolium* (Onagraceae). Their data showed that the autotetraploid individuals were reproductively isolated from the diploids by geographic distance, flowering asynchrony, pollinator fidelity, self-pollination, and gametic selection, with geographic isolation (41%) and pollinator fidelity (44%) representing the greatest proportions of these mechanisms. Importantly, the authors proposed that the emphasis previously placed on postzygotic factors, such as triploid sterility/inviability, may actually be secondary to prezygotic isolating mechansisms.

In summary, a number of features may contribute to the reproductive success of a neopolyploid, and hence its establishment and persistence. These characteristics may include perenniality or a propensity toward apomixis or self-compatibility. Additionally, changes in the morphological features (namely in floral characteristics) following polyploidization may reinforce the prezygotic reproductive barriers that prevent mating between cytotypes.

PHYSIOLOGY AND DEVELOPMENT

Levin (1983) suggested that polyploidy could propel a population into a new adaptive sphere given the myriad changes that accompany a multiplication of the chromosome set. Many physiological and developmental processes are affected by increases in ploidy, including carbon dioxide exchange rates, gene activity, hormone levels, photosynthetic rates, and water balance (Levin, 1983, 2002; Warner and Edwards, 1993). Because of the increased genomic content and cell size in polyploids, many physiological processes are thought to proceed more slowly (Tal, 1980). In particular, meiotic and mitotic rates are slowed in polyploids owing to the longer times required for replication of the genome prior to division. However, polyploids may have larger seeds relative to their diploid progenitors, which can result in faster germination and early development (Levin, 1983).

Despite the prevalence and importance of polyploidy in plants, there have actually been few recent investigations regarding its physiological and developmental consequences (Tal, 1980). Pegtel (1999) measured a number of morphological characters associated with reproduction and early development in a polyploid complex (diploid, tetraploid, hexaploid, and octaploid) of scurvy grass (*Cochlearia officinalis*, Brassicaceae). In this group, seed weight and ploidal level were positively correlated, although no significant differences in growth rate were found. In a comparative study of tetraploid (*Triticum durum*) and hexaploid (*Triticum aestivum*) species of wheat, Von Well and Fossey (1998) found that the hexaploid species had faster metabolic and growth rates than the tetraploid species. In the hexaploid *T. aestivum*, the processes that were more rapid than in the tetraploid *T. durum* included mobilization of food reserves, root growth, and mitotic rate over a 24-hour period. Watanabe *et al.* (1997) examined the effects of the genome donor on photosynthetic rate, chlorophyll content, and chlorophyll a:b ratio in hexaploid and synthetic pentaploid wheat. In this study, the photosynthetic rate of the polyploid was not dependent on the rates exhibited by the genome donors and did not necessarily increase with greater ploidal level. Villar *et al.* (1998) examined relative growth rates and biomass allocation for 20 species of *Aegilops* (Poaceae), including $2x$, $4x$, and $6x$ cytotypes. The results did not demonstrate an effect of ploidal level on relative growth rate or photosynthetic rate per unit leaf area for the cytotypes under study.

It is clear from the few studies conducted to date that polyploidization can and does have variable effects on physiological and developmental processes. As more diverse taxonomic groups at varying ploidal levels are investigated, more general patterns may surface.

GEOGRAPHIC DISTRIBUTION

Polyploids are ubiquitous in floras worldwide, but are generally regarded as more common at high latitudes and altitudes (Tischler, 1935; Löve and Löve, 1949; Ehrendorfer, 1980). Early authors hypothesized that polyploids were concentrated in these areas because they are more hardy and cold-tolerant than their diploid relatives (e.g., Tischler, 1935; Löve and Löve, 1949). On the other hand, Stebbins (1950, 1985) and others contended that increased heterozygosity resulting from hybridization (in allopolyploids) allowed the polyploids to colonize areas not previously occupied by their diploid relatives. Notably, many of the world's worst weeds are polyploids, which further attests to their potential colonizing ability (Brown and Marshall, 1981). Of course, not all polyploids necessarily invade marginal habitats following their formation, but rather may often be formed *in situ*. It has been pointed out that the production of unreduced gametes, and thus the likelihood of polyploid formation, may be affected by cold temperature in such a way as to promote polyploidization at high latitudes and altitudes (e.g., Ramsey and Schemske, 1998; Otto and Whitton, 2000; Mable, 2004b) (see Chapter 8). Similarly, other environmental stresses, such as nutrient deficiencies, herbivory, and disease, may also promote the formation of unreduced gametes (Ramsey and Schemske, 1998). Thus it is not surprising that polyploids are commonly found in disturbed habitats or previously glaciated areas (Ehrendorfer, 1980). Finally, the divergence and subsequent hybridization of populations restricted to refugia during ice ages may also promote the formation of polyploids in the north. As discussed in detail in Chapter 8 with regard to similar distributional patterns reported for many animal polyploids, several (or perhaps all) of these proposed factors probably come into play.

Certainly, patterns will differ among groups and, therefore, knowledge of the distribution of each polyploid species relative to those of its diploid progenitors is crucial. When compared to their diploid relatives, polyploids may occupy different ecological niches (Stebbins, 1950; Ehrendorfer, 1980), and this may be important, if not required, for their establishment (Fowler and Levin, 1984). A detailed study of Arctic *Draba* by Brochmann and Elven (1992) showed that the polyploid species occupy wider ecological niches than the diploid species, and that the geographic partitioning of the ploidal levels is related to breeding system and consequent genetic variation. The diploid species are predominantly autogamous and form populations composed of long-lived individuals found in stressful but stable environments. Ruderal polyploids of *Draba* are autogamous and short-lived and occupy unstable environments, whereas a second category of polyploids (stress-tolerant and competitive) has a mixed breeding system of autogamy/allogamy and is distributed in stable habitats (Brochmann and Elven, 1992).

As noted previously, it has been suggested that polyploidy plays an important role in established floras by generating "fill-in" taxa that occupy niches between their diploid parents, which leads to a greater partitioning of niche space (Ehrendorfer, 1980). This is predicted to be true particularly in "established" floras that are already "saturated." Polyploids that are able to exploit a novel habitat or niche space will be better able to establish and persist in an area where diploids predominate. On a broad scale, Petit and Thompson (1999) assessed species diversity and ecological range related to ploidal level in the Pyrenees for unrelated genera from diverse angiosperm families. Their results indicated that polyploidization led to taxonomic diversity in some genera, which in turn led to a greater exploitation of the habitats by various ploidal levels. A similar example comes from the allopolyploid *Antennaria rosea*, which has formed multiple times from hybridization between eight different sexual species of *Antennaria*. In this case, some niches occupied by the polyploid are similar to those of the diploid parents, whereas others are distinct in several characteristics, including soil composition, precipitation, temperature, and nutrients (Bayer *et al.*, 1991).

PLANT–ANIMAL INTERACTIONS

Recent studies indicate that plant polyploidy can have profound effects on interactions with animal herbivores and pollinators (e.g., Thompson *et al.*, 1997,

2004; Segraves and Thompson, 1999; Husband, 2000; Nuismer and Thompson, 2001; Lou and Baldwin, 2003). Despite the small number of studies on the evolutionary ecology of plant–animal interactions in diploids and polyploids, current data clearly show marked differences in patterns of attack by herbivores on diploids versus polyploids (Thompson *et al.*, 1997; Nuismer and Thompson, 2001; Lou and Baldwin, 2003) and strong differentiation in the use of diploids and polyploids by pollinators (Segraves and Thompson, 1999; Husband, 2000).

Studies of *Heuchera grossulariifolia* (Saxifragaceae) from the northern Rocky Mountains of the United States have demonstrated that autotetraploid lineages of independent origin (Wolf *et al.*, 1989, 1990; Segraves *et al.*, 1999) all experience higher levels of attack by the moth *Greya politella* than sympatric or parapatric diploids (Nuismer and Thompson, 2001). In addition, the geometrid moth, *Eupithecia misturata*, oviposits significantly more eggs onto tetraploid plants than onto diploid plants (Nuismer and Thompson, 2001). However, the stem borer *Greya piperella* attacks diploid plants more frequently than tetraploid plants (Nuismer and Thompson, 2001). Thus even within a single complex of diploid-tetraploid populations, polyploidy has differential effects on herbivores.

Nicotiana attenuata (2n = 24) is a diploid progenitor of two North American allopolyploid species, *N. bigelovii* (2n = 48) and *N. clevelandii* (2n = 48). The other parent of these two polyploids belongs to an n = 12 lineage and is thought to be extinct (Goodspeed, 1954). Recent studies on the response of *N. attenuata* to herbivory by the tobacco hornworm (*Manduca sexta*) indicate that the plants respond differently to manual disturbance or damage and herbivore attack (Baldwin, 2001). That is, specific patterns of hormone signaling, secondary metabolite accumulation, and gene transcript accumulation are elicited when a true insectivorous attack occurs (Baldwin, 2001). When the responses of diploid *N. attenuata* were compared to the derivative polyploid species, *N. bigelovii* and *N. clevelandii*, some differences were found (Lou and Baldwin, 2003). For example, a significant increase in jasmonate production was detected in all three species after wound initiation, but the response in *N. attenuata* and *N. bigelovii* was twice that of *N. clevelandii*. Other responses, such as volatile compound release, were also decreased in *N. clevelandii* relative to the other two species, suggesting that the herbivore recognition system was maintained following allopolyploid speciation of *N. bigelovii* but not *N. clevelandii* (Lou and Baldwin, 2003).

Heuchera grossulariifolia is visited by at least 25 insect species across its range (Thompson *et al.*, 2004), and at least 15 of these species visit the flowers of these plants along the Salmon River of Idaho in the United States (Segraves and Thompson, 1999). The proportions of insect visitors to plants of varying ploidy differ significantly. Bees of the genus *Lasioglossum* and workers of *Bombus centralis* visit diploids more frequently than tetraploids, but *Greya politella*, the bee-fly *Bombyllius major*, and queens of *B. centralis* all visit tetraploids more frequently than diploids (Segraves and Thompson, 1999). However, no major visitor is restricted to plants of either ploidy, allowing for possible gene flow between cytotypes of *H. grossulariifolia*.

These coordinated studies of the effects of polyploidy on interactions with insect herbivores and pollinators indicate that polyploidy certainly affects patterns of attack and visitation, respectively, and suggest that these effects may differ even among closely related species of herbivores (e.g., *Greya politella* and *G. piperella*). However, many questions remain unanswered (Thompson *et al.*, 2004). Most notable are: (1) Are herbivores and pathogens more or less likely to colonize polyploid plants than diploid plants?, and (2) Does polyploidy in plants have predictable effects on the evolution of plant–animal interactions? Answers to these and related questions may clarify the role that plant polyploidy plays in shaping the evolution of interspecific interactions and the organization of terrestrial biodiversity.

IMPACTS OF POLYPLOIDIZATION AT THE GENOME LEVEL

In the last several years, polyploid genome evolution has become a prevalent research topic in molecular biology, particularly as technological advances increasingly permit detailed and high throughput studies of the genome (Wendel, 2000; Liu and Wendel, 2003; Osborn *et al.*, 2003). As mentioned previously, molecular data have advanced the understanding of the widespread nature of polyploidy throughout plant and animal lineages, and have helped to determine parentage for many polyploid species. Despite these insights, however, researchers are only just beginning to understand the genomic complexities that follow polyploid formation. Most studies of plant polyploid genomes (particularly gene expression studies) have focused on the model organism *Arabidopsis thaliana* and its allotetraploid derivative *A. suecica*, or crop plants such as tobacco (*Nicotiana*), wheat (*Triticum*), and cotton (*Gossypium*). Consequently, these groups will represent the majority of the examples presented here. Ongoing studies in natural polyploid populations and complexes, including *Tragopogon* and *Glycine*, should provide additional insights into the evolution of these polyploids in their natural environments.

Changes in the polyploid genome may involve genetic or epigenetic mechanisms (Comai, 2000; Wendel, 2000; Liu and Wendel, 2003). Genetic changes are based on alteration of the DNA sequence itself, resulting in permanent alterations or losses of genes. Possible sequence or chromosomal changes include unequal crossing-over, homoeologous recombination, aneuploidy, gene conversion, insertions and deletions, duplications, and point mutations (Leitch and Bennett, 1997; Soltis and Soltis, 1999, 2000; Wendel, 2000). In addition, the activation of transposable elements (TEs) may be an important component of polyploid genome evolution (Matzke and Matzke, 1998) (see Chapter 3 for a discussion of TEs).

Epigenetic changes, such as DNA methylation, histone modification, RNA interference, and dosage compensation, are heritable and may alter gene expression without a change in DNA sequence (Wolffe and Matzke, 1999; Liu and Wendel, 2003; Osborn et al., 2003), but with dramatic phenotypic effects. Both genetic and epigenetic changes have now been documented in polyploids and an important recent development is the widespread appreciation for the role that epigenetic changes may play in polyploid evolution (Wendel, 2000; Liu and Wendel, 2003).

GENOMIC REARRANGEMENTS

A major discovery of the past decade relates to the extent and rapidity of genome reorganization that may occur in polyploids. Modification of parental diploid genomes, once in a common polyploid nucleus, was traditionally considered minimal. However, a diversity of molecular approaches, including chromosome painting methods like genomic and fluorescence *in situ* hybridizations (GISH and FISH), genetic mapping, and comparative genetics, provide evidence for both intra- and intergenomic reorganization of polyploid genomes (reviewed in Soltis and Soltis, 1993, 1999; Leitch and Bennett, 1997; Wendel, 2000; Raina and Rani, 2001).

Using chromosome painting techniques, for example, nine intergenomic chromosomal rearrangements have been detected in allotetraploid tobacco (*Nicotiana tabacum*) (Fig. 7.8), five intergenomic translocations in allotetraploid oats (*Avena maroccana*), and approximately 18 such rearrangements in allohexaploid *Avena sativa* (Leitch and Bennett, 1997; Soltis and Soltis, 1999; Wendel, 2000). Particularly notable genomic changes have occurred in *Brassica,* in which

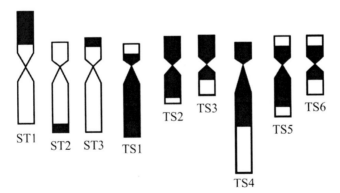

FIGURE 7.8   Genomic rearrangements in allopolyploid tobacco, *Nicotiana tabacum*. Nine intergenomic translocations were identified using genomic *in situ* hybridization (GISH) by Kenton *et al.* (1993). The T parental genome is shown in black and the S parental genome is in white. Redrawn from Leitch and Bennett (1997), reproduced by permission (© Elsevier Inc.).

chromosome mapping suggests that the naturally occurring allopolyploid genomes exhibit extensive reorganization compared to their diploid progenitors (Wendel, 2000). Amazingly, the synthetic allopolyploid *Brassica napus* recreated by Song *et al.* (1995) was found to experience major genomic rearrangements within just five generations. Using these same synthetic allopolyploid lines of *B. napus*, Schranz and Osborn (2000) found heritable differences in flowering time within and among sublineages of seventh generation progeny, suggesting that phenotypic variation may also occur very quickly. Synthetic autotetraploids of *Brassica*'s relative *Arabidopsis thaliana* (Wilna ecotype) likewise showed rearrangement of the 45S rDNA locus relative to the diploid ecotype (Weiss and Maluszynska, 2000).

Results for some cereals (e.g., maize, rice, and wheat) also suggest that rapid genomic change can occur in newly synthesized allopolyploids (Feldman *et al.*, 1997; Liu *et al.*, 1998; Ozkan *et al.*, 2001). In the wheat group (*Aegilops, Triticum*), elimination of noncoding chromosome- and genome-specific sequences began as early as the first generation and was completed by the second or third allopolyploid generation (Feldman *et al.*, 1997; Ozkan *et al.*, 2001). Moreover, the loss of these repetitive elements was not random and was reproducible among the synthetic lines, suggesting a directed basis for genome restructuring (Ozkan *et al.*, 2001), a process that may also continue over evolutionary time. For example, following an ancient polyploidization event in maize, considerable genomic restructuring has occurred (Gaut and Doebley, 1997; Gaut *et al.*, 2000), such that it is approaching a more diploid-like state (Ilic *et al.*, 2003).

However, chromosomal changes do not always accompany polyploidization. Although *Nicotiana tabacum* has undergone considerable rearrangement, the allotetraploids *N. rustica* and *N. arentsii* have not experienced any detectable chromosomal change based on analyses using GISH (Lim *et al.*, 2004). Similarly, using four tandem repetitive sequences as probes with FISH, Pires *et al.* (2004) found that the recent allopolyploids *Tragopogon mirus* and *T. miscellus* have not undergone major chromosomal rearrangements since their formation. Genomic changes also have apparently not occurred in amphiploid *Brassica juncea* (Axelsson *et al.*, 2000), nor in allopolyploid cotton (Liu *et al.*, 2001).

Another noteworthy recent discovery is "intergenomic invasion," or the movement of genetic material from one parental diploid genome to another after these are combined into an allopolyploid nucleus (Wendel, 2000). In allotetraploid cotton, for example, the two parental diploid genomes (designated by the letters A and D) have distinct repetitive elements. Following polyploidy, there has been substantial colonization of the D-genome by A-genome repetitive elements (Zhao *et al.*, 1998).

Data for *Brassica* and cereals also suggest that the extent of genomic change can be influenced by cytoplasmic–nuclear interactions (Wendel, 2000), which is not unexpected. Plant development involves coordination between the expression of nuclear as well as mitochondrial and plastid genes. Following polyploidization, selective pressures operate to stabilize and fine-tune these interactions (e.g., Wendel, 2000), although little is known about the fine-tuning process itself.

GENOMIC DOWNSIZING AND "DIPLOIDIZATION"

Elimination of chromosome- or genome-specific sequences may occur during polyploid formation, which, in some cases, may eventually lead to effectively "diploid" individuals. Intuitively, the expectation is that polyploids should have larger C-values (amounts of DNA in the unreplicated gametic nucleus) than diploids, with the C-values of polyploids increasing in direct proportion to ploidal level (see Chapter 2). This expectation has been shown to hold for some synthetic polyploids and newly formed polyploids but not others. For example, the allotetraploids *Tragopogon miscellus* and *T. mirus*, which are less than 80 years old, have C-values that are additive of their diploid progenitors (Pires *et al.*, 2004), whereas nonadditive changes have been reported in wheat (Ozkan *et al.*, 2003). On a much broader scale, Leitch and Bennett (2004) used the large dataset of C-values in the *Plant DNA C-values Database* (www.rbgkew.org.uk/cval/homepage.html) to make comparisons of diploids and polyploids and obtained important results at odds with the expectation of C-value additivity. Specifically, Leitch and Bennett (2004) found that mean DNA amount per genome within the nucleus (i.e., total nuclear DNA content divided by ploidy) tended to decrease with increasing ploidal level across a wide range of taxa. They concluded that loss of DNA following polyploid formation, which had been reported for a few species, may be a widespread phenomenon (see also Chapter 2). For example, rapid sequence elimination in newly formed wheat polyploids (Liu *et al.*, 1998; Ozkan *et al.*, 2001; Shaked *et al.*, 2001) suggested that differential elimination of genome-specific sequences may facilitate homologous chromosome pairing.

Mechanisms of genome contraction remain poorly understood, but there have been important recent improvements in the understanding of how decreases in genome size could take place (e.g., Vicient *et al.*, 1999; Kirik *et al.*, 2000; Petrov, 2001; Bennetzen, 2002; Frank *et al.*, 2002; Hancock, 2002) (see Chapters 1 and 2). Current data indicate that unequal recombination can slow the increase in genome size and that illegitimate recombination and other deletion processes may be the major mechanisms for decreases in genome size (Bennetzen, 2002; Devos *et al.*, 2002). Studies of microbial genomes suggest that downsizing of some genomes may be the result of homologous recombination at repeated genes, leading to the loss of large blocks of DNA as well as repeated sequences (Frank *et al.*, 2002). Differences in double-stranded break repair may also be responsible for some genome size variation (Kirik *et al.*, 2000), and exonucleolytic degradation of DNA ends might be a significant factor in the evolution of genome size (Orel and Puchta, 2003). Several hypotheses have been proposed to explain the reductions in relative DNA content observed in many polyploids, including (1) reduction of the nucleotypic effects of increased DNA amounts, (2) reduction of the biochemical costs associated with additional DNA amounts, and (3) enhancement of polyploid stability (I.J. Leitch and M.D. Bennett, in preparation).

The Fates of Plant Genes Duplicated by Polyploidy

A fundamental question in polyploid evolution centers on the fates of genes duplicated by polyploidy (homoeologs) and their expression. The topics of recent small-scale and ancient large-scale gene duplications were discussed in Chapters 5 and 6, respectively, but a more detailed discussion of recent plant polyploids seems warranted here. Recent studies have investigated both genetic and epigenetic phenomena. At the primary (sequence) level, rates of molecular evolution and the degree of sequence diversity between polyploids and their diploid progenitors have been at the forefront of polyploid gene evolution studies. Expression studies have applied techniques such as complementary DNA (cDNA)-amplified fragment length polymorphism (cDNA-AFLP) display to analyze genomewide expression differences between polyploids and their diploid progenitors (Comai et al., 2000; Lee and Chen, 2001; Kashkush et al., 2002), or cDNA-single-stranded conformational polymorphism (cDNA-SSCP) analysis to investigate expression levels of particular homoeologous transcripts (Adams et al., 2003). Many of the studies conducted to date illustrate that polyploid genomes are "dynamic" (Soltis and Soltis, 1995). The more that is learned about different polyploid groups and their genomes, the more complex the overall picture of polyploid genome evolution becomes. Nonetheless, regarding the fates of homoeologous loci in polyploids, theory predicts three potential outcomes for these duplicated genes: (1) both copies remain functional, (2) one copy degenerates and becomes silenced or lost while the other copy retains the original function, or (3) the two copies may diverge in function (Ohno, 1970; Lynch and Conery, 2000; Wendel, 2000; Prince and Pickett, 2002) (see Chapters 5 and 6). Each of these possibilities will be explored in the following sections.

Homoeologous Loci: Rates of Molecular Evolution and Sequence Diversity

In a polyploid, every gene is duplicated. By comparing patterns of nucleotide substitutions (both synonymous and nonsynonymous) within a given gene among diploid and tetraploid species, an understanding of the basic evolutionary rates at the DNA level can be gained. Many studies suggest that homoeologous loci evolve independently in the polyploid and accumulate mutations at different rates. For example, in a series of studies on the alcohol dehydrogenase (Adh) gene family, Small et al. (1999) and Small and Wendel (2000a,b, 2002) examined the rates of evolution between members of the gene family in allotetraploid and diploid species of cotton (*Gossypium*). Variable rates of evolution between the loci (*AdhA–AdhE*) were found in the diploid and allotetraploid species (Small and Wendel, 2000a). Furthermore, at two loci (*AdhA* and *AhdC*), the allotetraploids showed higher levels of sequence diversity in the D-genome homoeolog compared to the A-genome homoeolog (Small et al., 1998, 1999; Small and Wendel, 1999, 2000a, 2002). Cronn et al. (1999) examined 16 duplicated genes in allotetraploid cotton (*Gossypium hirsutum*) and found that they evolved independently of each other and that the homoeologous genes in the polyploids did not evolve more rapidly than the diploid genes. Similarly, MYB genes, which are transcription factors, also appear to evolve independently and at relatively equal rates in allotetraploid *G. hirsutum* and the diploid relatives, *G. raimondii* and *G. arboreum* (Cedroni et al., 2003). Recently, Senchina et al. (2003) evaluated 48 nuclear genes in allotetraploid cotton species (*G. barbadense* or *G. hirsutum*) and their respective diploid progenitors and found that the evolutionary rates of the genes differed within species, and that the rates were only slightly higher in the polyploids than in the diploids. Studies of four nuclear loci (*Adh1*, *glb1*, *c1*, and *waxy*) in the tetraploid *Zea perennis* and its closest diploid relative *Z. diploperennis* showed a similar pattern in that sequence diversity was not greater in the tetraploid than the diploid species (Tiffin and Gaut, 2001).

In contrast, Barrier et al. (2001) found higher rates of nonsynonymous to synonymous substitutions in the polyploid species of the Hawaiian silversword alliance relative to their continental North American diploid tarweed relatives. Specifically, the floral regulatory genes APETALA1 (ASAP1) and APETALA3 (ASAP3/TM6) are evolving much faster in the polyploid species than in the diploids. Although a structural gene, chlorophyll a/b binding protein9 (ASCAB9), is evolving more rapidly in the silversword alliance than in the diploid progenitor lineage, the rate is not as high as in the regulatory genes examined (Barrier et al., 2001). Lawton-Rauh et al. (2003) recently conducted a detailed analysis of ASAP1 and ASAP3/TM6 in two polyploid species from the Hawaiian silversword alliance (*Argyroxiphium sandwicense* ssp. *macrocephalum* and *Dubautia ciliolata* ssp. *glutinosa*) and found that homoeologous copies of both regulatory genes show different levels and patterns of nucleotide polymorphisms in each polyploid species.

Nuclear ribosomal RNA (rRNA) genes, particularly the intervening spacer regions between 18S–26S, are among the most commonly used loci for phylogeny reconstruction in plants. As a result, substantial information on the complex evolution of this region in polyploids is available (Alvarez and Wendel, 2003). In plants, hundreds or thousands of rRNA genes are tandemly arrayed at one or more chromosomal loci, and, in theory, these repeats are maintained as homogeneous units by concerted evolution (Zimmer *et al.*, 1980). However, in hybrids or polyploids, the rRNA gene repeats may not be homogenized if the species were formed recently, if the repeats are located at different loci (and therefore intralocus concerted evolution has failed), or if the polyploid or hybrid is asexual (Baldwin *et al.*, 1995). Polyploids may possess both parental types, or only one type if concerted evolution has acted to essentially remove one of the two homoeologs. For example, Wendel *et al.* (1995) found that several different allotetraploid cotton species, each containing the same progenitor AD genomes, possessed rRNA gene copies that were homogenized toward one parental repeat type or the other. That is, four species were more similar to the D genome lineage, and a fifth species was more similar to the A parental genome. In a detailed study of the *Glycine tomentella* allopolyploid complex, Rauscher *et al.* (2004) found that multiple accessions of six different races contained both homoeologs, but that copy number was biased and not always toward the same parental repeat type. Interestingly, the authors were able to detect the rare homoeolog in the allopolyploids using repeat-specific primers that were designed to amplify preferentially a single rDNA repeat (Rauscher *et al.*, 2002).

These studies demonstrate that the evolution of homoeologous loci in polyploids is quite dynamic and varies not only among taxonomic groups but also among types of genes examined (e.g., structural versus regulatory sequences). Although information on sequence diversity and rates of sequence evolution is interesting, more studies of the expression of homoeologous loci are needed (Adams *et al.*, 2003).

Gene Silencing and Gene Loss: Genetic and Epigenetic Mechanisms

Gene silencing via genetic and/or epigenetic mechanisms is predicted to play an important role in regulating duplicate gene expression (Ohno, 1970; Comai, 2000; Wendel, 2000; Prince and Pickett, 2002). The accumulation of genetic changes in coding or regulatory regions can potentially lead to gene silencing (pseudogenization) or gene loss (Ohno, 1970; Lynch and Conery, 2000; Prince and Pickett, 2002) (see Chapter 5). In polyploids, both gene silencing and gene loss have been widely documented (Wendel, 2000). At the extreme, polyploids may experience genomewide gene silencing or loss, such that they appear to be diploids (Ohno, 1970). For example, blocks of duplicated genes have been lost in *Arabidopsis thaliana*, a fact that had previously masked its true polyploid history (Vision *et al.*, 2000). Levels of gene silencing may be high in homosporous ferns, which have high chromosome numbers and may be ancient polyploids, because they show allozyme profiles typical of diploids (Haufler, 1987, 2002). Although it is not certain whether or not homosporous ferns are truly polyploid, numerous degenerative, and therefore nonfunctional, copies of some gene families have been reported, consistent with the hypothesis of ancient polyploidy and extensive gene silencing (Pichersky *et al.*, 1990). In two recently formed allotetraploid species of *Clarkia* (Onagraceae), single mutations have silenced one of the two *PgiC* genes inherited from their diploid progenitors (Ford and Gottlieb, 2002). In *C. similis*, a single nucleotide mutation in exon 5 produces a premature stop codon, whereas in *C. delicata*, a defective allele that contains a deletion in an intron splice junction produces a nonfunctional protein. Interestingly, both *Clarkia* allotetraploids silence the *PgiC2* copy, but they do it through different mechanisms.

Recent studies have shown that gene loss can occur rapidly. In synthetic wheat allopolyploids, gene loss occurred as early as the F_1 hybrid or the first amphiploid generation (Kashkush *et al.*, 2002). Previous studies on wheat polyploids also demonstrated rapid and reproducible elimination of DNA sequences that were not homologous to known genes and likely represented noncoding regions (Feldman *et al.*, 1997; Ozkan *et al.*, 2001; Shaked *et al.*, 2001). However, in hexaploid wheat, down-regulation of transcripts, rather than gene loss, was responsible for the differential expression of genes in allopolyploids compared to the diploid progenitors (He *et al.*, 2003).

Epigenetic mechanisms also contribute to the silencing of duplicated genes in polyploids (Comai, 2000; Wendel, 2000). In both plants and animals, epigenetic regulation of gene expression occurs via nongenetic changes, including cytosine methylation and chromatin remodeling (Finnegan *et al.*, 1993, 1998; Wolffe and Matzke, 1999; Aalfs and Kingston, 2000; Finnegan, 2002; Bender, 2004). Chromatin modifications may involve the repositioning or conformational changes of the nucleosomes, the recruitment of other proteins (including specific histones), or altering the structure of the DNA to change accessibility for regulatory proteins (Aalfs and Kingston, 2000; Finnegan, 2002). Importantly, these changes are heritable, but may be reversible (unlike genetic changes, which are heritable but essentially permanent) (Bender, 2004). Thus, epigenetic modifications may represent another source of genomic diversity on which selection may act. Studies have demonstrated that these epigenetic changes occur rapidly in polyploid evolution, similar to the structural changes discussed previously (Matzke *et al.*, 1999). In an early study on the effects of polyploidization, Mittelsten Scheid *et al.* (1996) showed that a change in ploidal level (from diploid to triploid) in *Arabidopsis thaliana* epigenetically silenced a transgene that was expressed in both parental cytotypes (diploid and tetraploid).

One of the best examples of epigenetic gene silencing is nucleolar dominance, in which one set of parental rRNA genes is silenced in a hybrid or an allopolyploid (Pikaard and Chen, 1998; Pikaard, 1999, 2000). In a study of *A. suecica*, Chen *et al.* (1998) found that the *A. thaliana* rRNA genes were always silenced, but that at least two generations were required for complete silencing. *Brassica* allopolyploids have active rRNA genes from only one parent in the vegetative tissues, whereas floral tissue contains both parental copies, suggesting that the epigenetic modifications are developmentally regulated (Chen and Pikaard, 1997). Chen and Pikaard (1997) also found that nucleolar dominance in these polyploids occurred independently of maternal effects or ploidal level (hexaploids versus tetraploids).

Lee and Chen (2001) showed that approximately 2.5% of the *Arabidopsis suecica* transcriptome was silenced. The silenced genes could be reactivated by treatment of the polyploid plants with a methylation inhibitor (5-aza-2'-deoxycytidine), which suggested that in this case gene silencing was the result of epigenetic DNA methylation. In resynthesized *A. suecica*, about 0.4% of the genes were silenced (Comai *et al.*, 2000). Several of these genes were characterized in detail and were also reactivated by methylation inhibitors, suggesting epigenetic regulation of gene silencing in the new allopolyploids as well. The synthetic *A. suecica* polyploids also showed high levels of phenotypic instability. In synthetic wheat allopolyploids, gene silencing, some of which was attributed to cytosine methylation (Kashkush *et al.*, 2002), was estimated to occur in about 1% of the transcripts.

Recently, Adams *et al.* (2003) examined expression patterns of 40 homoeologous loci in natural and synthetic polyploids of *Gossypium* and found evidence for organ-specific reciprocal silencing. Expression of one parental homoeolog in the ovules was demonstrated, whereas the other parental homoeolog was expressed in different tissues. Reciprocal silencing of the homoeologous loci was predicted to be epigenetically regulated in these polyploids.

Functional Diversification: Neofunctionalization and Subfunctionalization

In addition to gene silencing, genetic changes may lead to functional diversification of the homoeologs, creating genes of new function (neofunctionalization) or resulting in a partitioning of gene function (subfunctionalization) (Ohno, 1970; Lynch and Conery, 2000; Lynch and Force, 2000; Prince and Pickett, 2002) (see Chapters 5 and 6). Functional diversification of duplicated genes can be difficult to demonstrate empirically even in diploids, and many authors have predicted that the phenomenon will be rare (Prince and Pickett, 2002). Some evidence for functional diversification in polyploid plants comes from maize, which has an anciently duplicated genome (Wendel, 2000). Two duplicated genes, *R* and *B*, which encode basic helix-loop-helix transcription activators and are involved in the anthocyanin pathway, showed higher replacement substitution rates when compared to other duplicated gene pairs in maize. Furthermore, *R* and *B* differed mainly in their expression patterns (and not protein function), indicating the importance of mutations in coding regions of regulatory genes, as well as regulatory elements that control *R* and *B* (Wendel, 2000).

Neofunctionalization has yet to be established empirically for polyploids. However, novel phenotypic variation in polyploids may arise through changes in gene expression, including dosage effects, multiple origins, or chromatin remodeling (Osborn *et al.*, 2003). Genomic studies have found evidence for novel expression in polyploids compared to their diploid progenitors, and in most cases, these novel transcripts appear to be transposons (Comai *et al.*, 2000; Kashkush *et al.*, 2002). In allotetraploid *Tragopogon miscellus*, novel transcripts were detected using cDNA-AFLP display and these showed high sequence similarity to genes representing a range of functions, including hypothetical proteins, DNA-helicase-like protein, photosystem II oxygen-evolving complex, and a protein phosphatase-2C (J. Tate, A.-C. Scheen, Z.J. Chen, D. Soltis, P. Soltis, unpublished data).

TRANSPOSABLE ELEMENTS

As discussed in Chapter 3, TEs represent an especially important factor in genome evolution in general, a fact that also applies to polyploids (Matzke and Matzke, 1998; Bennetzen, 2000; Wendel, 2000; Liu and Wendel, 2003). Several studies have shown that TEs are reactivated very early following polyploid formation and that these can contribute to genetic diversity in polyploids (Comai *et al.*, 2000). TEs may be responsible for gene silencing, if inserted into the coding or regulatory regions of a gene (Matzke and Matzke, 1998; Wendel, 2000). However, polyploids are expected to be buffered from the deleterious effects of transposition because they contain duplicated copies of each gene (Matzke and Matzke, 1998). Recently, Kashkush *et al.* (2003) found that genes adjacent to one retrotransposon were silenced or activated by transcriptional activation of the TE. Their work suggested that transposons may play an important role as controlling elements within the genome. At present, an understanding of the role of TEs in polyploid evolution is only just beginning to emerge, and this will undoubtedly be an active area of research in the years to come.

NUCLEAR–CYTOPLASMIC INTERACTIONS

Plants have three genomes (nuclear, mitochondrial, and chloroplastic) that must coordinate complex gene and gene product interactions. Nuclear–cytoplasmic interactions, involving the nuclear genome and the mitochondrial or chloroplastic genome, are predicted to be important in allopolyploid and hybrid evolution (Gill, 1991; Wendel, 2000; Levin, 2003). One of the earliest and most compelling studies was conducted by Song *et al.* (1995) on resynthesized *Brassica* allopolyploids. The reciprocally formed polyploids derived from *B. rapa* (A genome) and *B. nigra* (B genome) showed directional genomic changes in the F_5 generation. In both the AB and BA polyploids, the paternal genome was significantly altered. Other polyploids synthesized from *B. rapa* and *B. oleracea* (C genome) did not show significant genomic changes. Song *et al.* (1995) suggested that this occurred because the A and C cytoplasmic genomes are more closely related and the nuclear–cytoplasmic genomes were more compatible in the AC and CA polyploids. In reciprocally formed populations of *Tragopogon miscellus,* expression differences potentially due to maternal/paternal effects were detected (J. Tate, A.-C. Scheen, Z.J. Chen, D. Soltis, P. Soltis, unpublished data). Furthermore, individuals within the two populations differ in their expression profiles of homoeologous loci. For some genes, individuals preferentially express one homoeolog, whereas for other genes, the individuals express alternative homoeologs. This finding has significant implications not only for nuclear–cytoplasmic interactions, but for multiple origins of polyploidy as well.

CONCLUDING REMARKS AND FUTURE PROSPECTS

There is little doubt that polyploidy has played a very significant role during the evolutionary history of plants, as the high estimates of polyploidy throughout the angiosperm and fern lineages clearly attest. Plant polyploidy is also of substantial importance to humans, given that many of the world's chief agricultural crops have a polyploid origin. Recent genomic studies suggest that even species that were previously assumed to be diploid (most notably, the small-genomed *Arabidopsis thaliana*) experienced one or more ancient rounds of genome duplication, as did vertebrates, yeast, and other eukaryotic lineages (see Chapter 6). In this sense, elucidating the causes and consequences of polyploidy appears fundamental to the study of eukaryotic life forms.

One thing that is already evident is that polyploids are complex, such that generalizations about their evolution provide little more than a starting point for testing hypotheses using various specific polyploid–progenitor pairs. For example, the morphological, physiological, and ecological consequences of polyploidy may vary from group to group, and according to the mode of polyploidization. Important work has been done in these areas for many species, but a great deal more remains to be understood about the patterns and causes of such relationships. This will necessarily involve study at several different levels, from genes and chromosomes to individual organisms and populations to species and entire ecosystems.

The advent of large-scale genome sequencing provides many new opportunities to explore polyploid evolution at the genomic level. Ongoing and future studies will further delve into various aspects of polyploid genome evolution, such as epigenetic regulation of duplicated genes, nuclear–cytoplasmic interactions in polyploids, comparisons of diploid hybrids with polyploids, and genomewide expression changes using microarrays (Wendel, 2000; Osborn *et al.,* 2003; Otto, 2003). It will be particularly interesting to compare polyploids in natural populations to model systems, such as cotton, wheat, and rice, that have been well studied in the past.

Whatever the future directions of plant polyploidy research, it is evident that an improved appreciation of the diversity of genetic and higher-level consequences of polyploidy will greatly advance the understanding of the processes that generate, maintain, and alter polyploid lineages in nature—and, by extension, influence the evolution of the global flora as a whole.

REFERENCES

Aalfs JD, Kingston RE. 2000. What does 'chromatin remodeling' mean? *Trends Biochem Sci* 25: 548–555.

Adams KL, Cronn RC, Percifield R, Wendel JF. 2003. Genes duplicated by polyploidy show unequal contributions to the transcriptome and organ-specific reciprocal silencing. *Proc Natl Acad Sci USA* 100: 4649–4654.

Allen GA, Eccleston CL. 1998. Genetic resemblance of allotetraploid *Aster ascendens* to its diploid progenitors *Aster falcatus* and *Aster occidentalis*. *Can J Bot* 76: 338–344.

Alsos IG. 2003. Conservation biology of the most thermophilous plant species in the Arctic. Genetic variation, recruitment and phylogeography in a changing climate. PhD thesis, University of Tromsø, Tromsø, Norway.

Alsos IG, Engelskjon T, Brochmann C. 2002. Conservation genetics and population history of *Betula nana*, *Vaccinium uliginosum*, and *Campanula rotundifolia* in the arctic archipelago of Svalbard. *Arct Antarct Alpine Res* 34: 408–418.

Alsos IG, Engelskjon T, Taberlet P, Brochmann C. 2001. Circumpolar phylogeography of *Vaccinium uliginosum* inferred from cpDNA sequences. *Bauhinia* 15: 74.

Alvarez I, Wendel JF. 2003. Ribosomal ITS sequences and plant phylogenetic inference. *Mol Phylogenet Evol* 29: 417–434.

Anderson E. 1949. *Introgressive Hybridization*. New York: John Wiley & Sons.

Arft AM, Ranker TA. 1998. Allopolyploid origin and population genetics of the rare orchid *Spiranthes diluvialis*. *Am J Bot* 85: 110–122.

Arisumi T. 1982. Endosperm Balance Number among New Guinea-Indonesian *Impatiens* species. *J Hered* 73: 57–65.

Arnold ML. 1997. *Natural Hybrization and Evolution*. New York: Oxford University Press.

Asker S, Jerling L. 1992. *Apomixis in Plants*. Boca Raton, FL: CRC Press.

Axelsson T, Bowman CM, Sharpe AG, *et al.* 2000. Amphidiploid *Brassica juncea* contains conserved progenitor genomes. *Genome* 43: 679–688.

Baldwin BG, Sanderson MJ, Porter JM, *et al.* 1995. The ITS region of nuclear ribosomal DNA: a valuable source of evidence on angiosperm phylogeny. *Ann Missouri Bot Gard* 82: 247–277.

Baldwin IT. 2001. An ecologically motivated analysis of plant-herbivore interactions in native tobacco. *Plant Physiol* 127: 1449–1458.

Barrett SCH, Richardson BJ. 1986. Genetic attributes of invading species. In Groves RH, Burdon JJ eds. *Ecology of Biological Invasions: An Australian Perspective*. Canberra: Australian Academy of Science, 21–33.

Barrett SCH, Shore JS. 1989. Isozyme variation in colonizing plants. In Soltis DE, Soltis PS eds. *Isozymes in Plant Biology*. Portland, OR: Dioscorides Press, 106–126.

Barrier M, Robichaux RH, Purugganan MD. 2001. Accelerated regulatory gene evolution in an adaptive radiation. *Proc Natl Acad Sci USA* 98: 10208–10213.

Baumel A, Ainouche ML, Levasseur JE. 2001. Molecular investigations in populations of *Spartina anglica* C. E. Hubbard (Poaceae) invading coastal Brittany (France). *Mol Ecol* 10: 1689–1701.

Bayer RJ, Purdy BG, Lebedyx DG. 1991. Niche differentiation among eight sexual species of *Antennaria* Gaertner (Asteraceae: Inuleae) and *A. rosea*, their allopolyploid derivative. *Evol Trends Plants* 5: 109–123.

Beasley JO. 1940. The origin of American tetraploid *Gossypium* species. *Am Nat* 74: 285–286.

Bender J. 2004. DNA methylation and epigenetics. *Annu Rev Plant Physiol Plant Mol Biol* 55: 41–68.

Bennett MD, Leitch IJ, Price HJ, Johnston JS. 2003. Comparisons with *Caenorhabditis* (~100 Mb) and *Drosophila* (~175 Mb) using flow cytometry show genome size in *Arabidopsis* to be ~157 Mb and thus ~25% larger than the Arabidopsis Genome Initiative estimate of ~125 Mb. *Ann Bot* 91: 547–557.

Bennetzen JL. 2000. Transposable element contribution to plant gene and genome evolution. *Plant Mol Biol* 42: 251–269.

Bennetzen JL. 2002. Mechanisms and rates of genome expansion and contraction in flowering plants. *Genetica* 115: 29–36.

Bicknell RA, Borst NK, Koltunow AM. 2000. Monogenic inheritance of apomixis in two *Hieracium* species with distinct developmental mechanisms. *Heredity* 84: 228–237.

Bierzychudek P. 1990. The demographic consequences of sexuality and apomixis in *Antennaria*. In: Kawano S ed. *Biological Approaches and Evolutionary Trends in Plants*. New York: Academic Press, 293–307.

Blanc G, Hokamp K, Wolfe KH. 2003. A recent polyploidy superimposed on older large-scale duplications in the *Arabidopsis* genome. *Genome Res* 13: 137–144.

Bowers JE, Chapman BA, Rong J, Paterson AH. 2003. Unravelling angiosperm genome evolution by phylogenetic analysis of chromosomal duplication events. *Nature* 422: 433–438.

Brammal RA, Semple JC. 1990. The cytology of *Solidago nemoralis* (Compositae: Astereae). *Can J Bot* 68: 2065–2069.

Bretagnolle F, Thompson JD. 1995. Gametes with the somatic chromosome number: mechanisms of their formation and role in the evolution of autopolyploid plants. *New Phytol* 129: 1–22.

Brochmann C. 1993. Reproductive strategies of diploid and polyploid populations of arctic *Draba* (Brassicaceae). *Plant Syst Evol* 185: 55–83.

Brochmann C, Elven R. 1992. Ecological and genetic consequences of polyploidy in Arctic *Draba* (Brassicaceae). *Evol Trends Plants* 6: 111–124.

Brochmann C, Hapnes A. 2001. Reproductive strategies in some arctic *Saxifraga* (Saxifragaceae), with emphasis on the narrow endemic *S. svalbardensis* and its parental species. *Bot J Linn Soc* 137: 31–49.

Brochmann C, Brysting AK, Also IG, et al. 2004. Polyploidy in arctic plants. *Biol J Linn Soc* 82: 521–536.

Brochmann C, Soltis PS, Soltis DE. 1992a. Recurrent formation and polyphyly of Nordic polyploids in *Draba* (Brassicaceae). *Am J Bot* 79: 673–688.

Brochmann C, Soltis PS, Soltis DE. 1992b. Multiple origins of the octoploid Scandinavian endemic *Draba cacuminum*: electrophoretic and morphological evidence. *Nordic J Bot* 12: 257–272.

Brown AHD, Marshall DR. 1981. Evolutionary changes accompanying colonization in plants. In: Schudder GGE, Reveal J eds. *Evolution Today. Proceedings of the Second International Congress of Systematic and Evolutionary Biology*, 351–363.

Brysting A, Aiken S, Lefkovitch L, Boles R. 2003. *Dupontia* (Poaceae) in North America. *Can J Bot* 81: 769–779.

Brysting AK, Fay MF, Leitch IJ, Aiken SG. 2004. One or more species in the genus *Dupontia*?— a contribution to the Panarctic Flora project. *Taxon* 53:365–382.

Burton TL, Husband BC. 2000. Fitness differences among diploids, tetraploids, and their triploid progeny in *Chamerion angustifolium*: mechanisms of inviability and implications for polyploid evolution. *Evolution* 54: 1182–1191.

Burton TL, Husband BC. 2001. Fecundity and offspring ploidy in matings among diploid, triploid and tetraploid *Chamerion angustifolium* (Onagraceae): consequences for tetraploid establishment. *Heredity* 87: 573–582.

Butruille DV, Boiteux LS. 2000. Selection–mutation balance in polysomic tetraploids: impact of double reduction and gametophytic selection on the frequency and subchromosomal localization of deleterious mutations. *Proc Natl Acad Sci USA* 97: 6608–6613.

Butterfass T. 1987. Cell volume ratios of natural and of induced tetraploid and diploid flowering plants. *Cytologia* 52: 309–316.

Cedroni M, Cronn RC, Adams KL, *et al.* 2003. Evolution and expression of MYB genes in diploid and polyploid cotton. *Plant Mol Biol* 51: 313–325.

Charlesworth D, Charlesworth B. 1987. Inbreeding depression and its evolutionary consequences. *Annu Rev Ecol Syst* 18: 237–268.

Chen ZJ, Pikaard CS. 1997. Transcriptional analysis of nucleolar dominance in polyploid plants: biased expression/silencing of progenitor rRNA genes is developmentally regulated in *Brassica*. *Proc Natl Acad Sci USA* 94: 3442–3447.

Chen ZJ, Comai L, Pikaard CS. 1998. Gene dosage and stochastic effects determine the severity and direction of uniparental ribosomal RNA gene silencing (nucleolar dominance) in *Arabidopsis* allopolyploids. *Proc Natl Acad Sci USA* 95: 14891–14896.

Clausen J, Keck DD, Hiesey WM. 1940. *Experimental studies on the nature of species. I. Effect of varied environments on western North American plants.* Washington, DC: Carnegie Institution of Washington Publication No. 520.

Clausen J, Keck DD, Hiesey WM. 1945. *Experimental studies on the nature of species. II. Plant evolution through amphiploidy and autopolyploidy, with examples from the Madiinae.* Washington, DC: Carnegie Institution of Washington Publication No. 564.

Clausen RE, Goodspeed TH. 1925. Interspecific hybridization in *Nicotiana*. II. A tetraploid *glutinosa-tabacum* hybrid, an experimental verification of Winge's hypothesis. *Genetics* 10: 279–284.

Comai L. 2000. Genetic and epigenetic interactions in allopolyploid plants. *Plant Mol Biol* 43: 387–399.

Comai L, Tyagi AP, Winter K, Holmes-Davis R, *et al.* 2000. Phenotypic instability and rapid gene silencing in newly formed *Arabidopsis* allotetraploids. *Plant Cell* 12: 1551–1567.

Cook LM, Soltis PS. 1999. Mating systems of diploid and allotetraploid populations of *Tragopogon* (Asteraceae). I. Natural populations. *Heredity* 82: 237–244.

Cook LM, Soltis PS. 2000. Mating systems of diploid and allotetraploid populations of *Tragopogon* (Asteraceae). II. Artificial populations. *Heredity* 84: 410–415.

Cook LM, Soltis PS, Brunsfeld SJ, Soltis DE. 1998. Multiple independent formations of *Tragopogon* tetraploids (Asteraceae): evidence from RAPD markers. *Mol Ecol* 7: 1293–1302.

Cronn RC, Small RL, Wendel JF. 1999. Duplicated genes evolve independently after polyploid formation in cotton. *Proc Natl Acad Sci USA* 96: 14406–14411.

Crosby MR. 1980. Polyploidy in bryophytes with special emphasis on mosses. In: Lewis WH ed. *Polyploidy: Biological Relevance.* New York: Plenum Press, 193–198.

Cruzan MB, Arnold ML. 1993. Ecological and genetic associations in an *Iris* hybrid zone. *Evolution* 47: 1432–1445.

Darlington CD. 1937. *Recent Advances in Cytology.* 2nd edition. Philadelphia: P. Blakiston's Son & Co.

Darwin C. 1859. *On the Origin of Species by Means of Natural Selection, or the Preservation of Favoured Races in the Struggle for Life.* London: John Murray.

de Nettancourt D. 2001. *Incompatibility and Incongruity in Wild and Cultivated Plants.* 2nd edition. New York: Springer-Verlag.

de Wet JMJ. 1980. Origins of polyploids. In: Lewis WH ed. *Polyploidy: Biological Relevance.* New York, Plenum Press, 3–15.

Delevoryas T. 1980. Polyploidy in gymnosperms. In: Lewis WH ed. *Polyploidy: Biological Relevance.* New York, Plenum Press, 215–218.

Devos KM, Brown JK, Bennetzen JL. 2002. Genome size reduction through illegitimate recombination counteracts genome expansion in *Arabidopsis*. *Genome Res* 12: 1075–1079.

Doyle JJ, Doyle JL, Brown AHD. 1990. Analysis of a polyploid complex in *Glycine* with chloroplast and nuclear DNA. *Austr Syst Bot* 3: 125–136.

Doyle JJ, Doyle JL, Brown AHD. 1999. Origins, colonization, and lineage recombination in a widespread perennial soybean polyploid complex. *Proc Natl Acad Sci USA* 96: 10741–10745.

Doyle JJ, Doyle JL, Brown AHD, Palmer RG. 2002. Genomes, multiple origins, and lineage recombination in the *Glycine tomentella* (Leguminosae) polyploid complex: histone H3-D gene sequences. *Evolution* 56: 1388–1402.

Doyle JJ, Doyle JL, Rauscher JT, Brown AHD. 2004. Diploid and polyploid reticulate evolution throughout the history of the perennial soybeans (*Glycine* subg. *Glycine*). *New Phytol* 161: 121–132.

Ehrendorfer F. 1980. Polyploidy and distribution. In: Lewis WH ed. *Polyploidy: Biological Relevance.* New York: Plenum Press, 45–60.

Emms SK, Arnold ML. 1997. The effect of habitat on parental and hybrid fitness: transplant experiments with Louisiana irises. *Evolution* 51: 1112–1119.

Felber F. 1988. Phenologie de la floraison de populations diploides et tetraploides d'*Anthoxanthum alpinum* et d'*Anthoxanthum odoratum*. *Can J Bot* 66: 2258–2264.

Felber F. 1991. Establishment of a tetraploid cytotype in a diploid population: effect of relative fitness of the cytotypes. *J Evol Biol* 4: 195–207.

Feldman M, Liu B, Segal G, et al. 1997. Rapid elimination of low-copy DNA sequences in polyploid wheat: a possible mechanism for differentiation of homoeologous chromosomes. *Genetics* 147: 1381–1387.

Finnegan EJ. 2002. Epialleles—a source of random variation in times of stress. *Curr Opin Plant Biol* 5: 101–106.

Finnegan EJ, Brettell RI, Dennis ES. 1993. The role of DNA methylation in the regulation of plant gene expression. *Exs* 64: 218–261.

Finnegan EJ, Genger RK, Peacock WJ. 1998. DNA methylation in plants. *Annu Rev Plant Physiol Plant Mol Biol* 49: 223–247.

Ford VS, Gottlieb LD. 2002. Single mutations silence *PgiC2* genes in two very recent allotetraploid species of *Clarkia*. *Evolution* 56: 699–707.

Fowler NL, Levin DL. 1984. Ecological constraints on the establishment of a novel polyploid in competition with its diploid progenitor. *Am Nat* 124: 703–711.

Frank AC, Amiri H, Andersson SGE. 2002. Genome deterioration: loss of repeated sequences and accumulation of junk DNA. *Genetica* 115: 1–12.

Futuyma D. 1998. *Evolutionary Biology.* 3d ed. Sunderland, MA: Sinaur Assoc., Inc.

Gates RR. 1909. The stature and chromosomes of *Oenothera gigas* De Vries. *Arch Zellforsch* 3: 525–552.

Gaut B, Le Thierry d'Ennequin M, Peek AS, Sawkins MC. 2000. Maize as a model for the evolution of plant nuclear genomes. *Proc Natl Acad Sci USA* 97: 7008–7015.

Gaut BS, Doebley JF. 1997. DNA sequence evidence for the segmental allotetraploid origin of maize. *Proc Natl Acad Sci USA* 94: 6809–6814.

Gill BS. 1991. Nucleo-cytoplasmic interaction (NCI) hypothesis of genome evolution and speciation in polyploid plants. In: Sasakuma T, Kinoshita T eds. *Nuclear and Organellar Genomes of Wheat Species.* Proceedings of the Dr. H. Kihara Memorial International Symposium on Cytoplasmic Engineering in Wheat. Yokohama: Kihara Memorial Yokohama Foundation for the Advancement of Life Science, 48–53.

Goldblatt P. 1980. Polyploidy in angiosperms: Monocotyledons. In: Lewis WH ed. *Polyploidy: Biological Relevance.* New York: Plenum Press, 219–239.

Goodspeed TH. 1954. *The Genus Nicotiana.* Waltham, MA: Chronica Botanic.

Goodspeed TH, Clausen RE. 1928. Interspecific hybridization in *Nicotiana*. VIII. The *sylvestris-tomentosa-tabacum* triangle and its bearing on the origin of *Tabacum*. *Univ Cal Publ Bot* 11: 245–256.

Grant V. 1952. Cytogenetics of the hybrid *Gilia millefoliata* × *achilleaefolia*. I. Variations in meiosis and polyploidy rate as affected by nutritional and genetic conditions. *Chromosoma* 5: 372–390.

Grant V. 1963. *The Origin of Adaptations*. New York: Columbia University Press.

Grant V. 1981. *Plant Speciation*. 2nd edition. New York: Columbia University Press.

Grant V. 2002. Frequency of spontaneous amphiploids in *Gilia* (Polemoniaceae) hybrids. *Am J Bot* 89: 1197–1202.

Haldane JBS. 1930. Theoretical genetics of autopolyploids. *J Genet* 22: 359–372.

Hancock JF, Bringhurst RS. 1981. Evolution of California populations of diploid and octoploid *Fragaria* (Rosaceae): a comparison. *Am J Bot* 68: 1–5.

Hancock JM. 2002. Genome size and the accumulation of simple sequence repeats: implications of new data from genome sequencing projects. *Genetica* 115: 93–103.

Harlan JR, de Wet JMJ. 1975. On O. Winge and a prayer: the origins of polyploidy. *Bot Rev* 41: 361–390.

Harrison RG. 1986. Pattern and process in a narrow hybrid zone. *Heredity* 56: 337–349.

Hauber DP, Soltis DE. 1989. Quantitative cytogenetic analysis of tetraploid *Tolmiea menziesii*: a test for autoploid behavior. *Am J Bot* 76: 147.

Haufler CH. 1987. Electrophoresis is modifying our concepts of evolution in homosporous pteridophytes. *Am J Bot* 74: 953–966.

Haufler CH. 2002. Homospory 2002: an odyssey of progress in pteridophyte genetics and evolutionary biology. *BioScience* 52: 1081–1093.

He P, Friebe BR, Gill BS, Zhou JM. 2003. Allopolyploidy alters gene expression in the highly stable hexaploid wheat. *Plant Mol Biol* 52: 401–414.

Hedrick PW. 1987. Genetic load and the mating system in homosporous ferns. *Evolution* 41: 1282–1289.

Howard DJ. 1982. *Speciation and Coexistence in a Group of Closely Related Ground Crickets*. New Haven, CT: Yale University Press.

Howard DJ. 1986. A zone of overlap and hybridization between two ground cricket species. *Evolution* 40: 34–43.

Hunter RL, Markert CL. 1957. Histochemical demonstration of enzymes separated by zone electrophoresis in starch gels. *Science* 125: 1294–1295.

Husband BC. 2000. Constraints on polyploid evolution: a test of the minority cytotype exclusion principle. *Proc R Soc Lond B* 267: 217–223.

Husband BC. 2004. The role of triploid hybrids in the evolutionary dynamics of mixed-ploidy populations. *Biol J Linn Soc* 82: 537–546.

Husband BC, Sabara HA. 2003. Reproductive isolation between autotetraploids and their diploid progenitors in fireweed, *Chamerion angustifolium* (Onagraceae). *New Phytol* 161: 703–713.

Husband BC, Schemske DW. 1997. The effect of inbreeding in diploid and tetraploid *Epilobium angustifolium* (Onagraceae): implications of the genetic basis of inbreeding depression. *Evolution* 51: 737–746.

Huskins CL. 1931. The origin of *Spartina townsendii*. *Genetica* 12: 531.

Ilic K, Sanmigue PJ, Bennetzen JL. 2003. A complex history of rearrangement in an orthologous region of the maize, sorghum, and rice genomes. *Proc Natl Acad Sci USA* 100: 12265–12270.

Jackson RC. 1982. Polyploidy and diploidy: new perspectives on chromosome pairing and its evolutionary implications. *Am J Bot* 69: 1512–1523.

Jackson RC, Casey J. 1982. Cytogenetic analyses of autopolyploids: models and methods for triploids to octoploids. *Am J Bot* 69: 487–501.

Jackson RC, Hauber DP. 1982. Autotriploid and autotetraploid cytogenetic analyses: correction coefficients for proposed binomial models. *Am J Bot* 69: 644–646.

Johnson MTJ, Husband BC, Burton TL. 2003. Habitat differentiation between diploid and tetraploid *Galax urceolata* (Diapensiaceae). *Int J Plant Sci* 164: 703–710.

Johnston SA, Hanneman RE. 1982. Manipulations of endosperm balance number overcome crossing barriers between diploid *Solanum* species. *Science* 217: 446–448.

Johnston SA, Den Nijs TPM, Peloquin SJ, Hanneman Jr. RE. 1980. The significance of genetic balance to endosperm development in interspecific crosses. *Theor Appl Genet* 57: 5–9.

Kashkush K, Feldman M, Levy AA. 2002. Gene loss, silencing and activation in a newly synthesized wheat allotetraploid. *Genetics* 160: 1651–1659.

Kashkush K, Feldman M, Levy AA. 2003. Transcriptional activation of retrotransposons alters the expression of adjacent genes in wheat. *Nat Genet* 33: 102–106.

Kenton A, Parokonny AS, Gleba YY. 1993. Characterization of the *Nicotiana tabacum* L. genome by molecular cytogenetics. *Mol Gen Genet* 240: 159–169.

Khoshoo TN. 1959. Polyploidy in gymnosperms. *Evolution* 13: 24–39.

Kihara H, Ono T. 1926. Chromosomenzahlen und systematische Gruppierung der *Rumex*-Arten. *Z Zellforsch mikr Anatomie* 4: 475–481.

Kirik A, Salomon S, Puchta H. 2000. Species-specific double-strand break repair and genome evolution in plants. *EMBO J* 19: 5562–5566.

Kochert G, Stocker HT, Gimenes M, *et al.* 1996. RFLP and cytogenetic evidence on the origin and evolution of allotetraploid domesticated peanut, *Arachis hypogaea* (Leguminosae). *Am J Bot* 83: 1282–1291.

Kuwada Y. 1911. Meiosis in the pollen mother cells of *Zea Mays* L. *Bot Mag* 25:163

Lagercrantz U, Lydiate DJ. 1996. Comparative genome mapping in *Brassica*. *Genetics* 144: 1903–1910.

Lawton-Rauh A, Robichaux RH, Purugganan MD. 2003. Patterns of nucleotide variation in homoeologous regulatory genes in the allotetraploid Hawaiian silversword alliance (Asteraceae). *Mol Ecol* 12: 1301–1313.

Lee HS, Chen ZJ. 2001. Protein-coding genes are epigenetically regulated in *Arabidopsis* polyploids. *Proc Natl Acad Sci USA* 98: 6753–6758.

Leipoldt M, Schmidtke J. 1982. Gene expression in phylogenetically polyploid organisms. In: Dover G, Flaell R eds. *Genome Evolution*. New York: Academic Press, 219–236.

Leitch IJ, Bennett MD. 1997. Polyploidy in angiosperms. *Trends Plant Sci* 2: 470–476.

Leitch IJ, Bennett MD. 2004. Genome downsizing in polyploid plants. *Biol J Linn Soc* 82: 651–663.

Levin DA. 1975. Minority cytotype exclusion in local plant populations. *Taxon* 24: 35–43.

Levin DA. 1983. Polyploidy and novelty in flowering plants. *Am Nat* 122: 1–25.

Levin DA. 2000. *The Origin, Expansion, and Demise of Plant Species*. New York: Oxford University Press.

Levin DA. 2002. *The Role of Chromosomal Change in Plant Evolution*. New York: Oxford University Press.

Levin DA. 2003. The cytoplasmic factor in plant speciation. *Syst Bot* 28: 5–11.

Lewis WH. 1976. Temporal adaptation correlated with ploidy in *Claytonia virginica*. *Syst Bot* 1: 340–347.

Lewis WH. 1980a. Polyploidy in angiosperms: Dicotyledons. In: Lewis WH ed. *Polyploidy: Biological Relevance*. New York: Plenum Press, 241–268.

Lewis WH. 1980b. Polyploidy in species populations. In: Lewis WH ed. *Polyploidy: Biological Relevance*. New York: Plenum Press, 103–144.

Lewis WH, Suda Y. 1976. Diploids and polyploids from a single population: temporal variation. *J Hered* 67: 391–393.

Lim KY, Matyasek R, Kovařík A, Leitch AR. 2004. Genome evolution in allotetraploid *Nicotiana*. *Biol J Linn Soc* 82: 599–606.

Liu B, Wendel JF. 2003. Epigenetic phenomena and the evolution of plant allopolyploids. *Mol Phylogenet Evol* 29: 365–379.

Liu B, Brubaker CL, Mergeai G, *et al.* 2001. Polyploid formation in cotton is not accompanied by rapid genomic changes. *Genome* 44: 321–330.

Liu B, Vega JM, Feldman M. 1998. Rapid genomic changes in newly synthesized amphiploids of *Triticum* and *Aegilops*. II. Changes in low-copy coding DNA sequences. *Genome* 41: 535–542.

Lou Y, Baldwin IT. 2003. *Manduca sexta* recognition and resistance among allopolyploid *Nicotiana* host plants. *Proc Natl Acad Sci USA* 100: 14581–14586.

Löve A, Löve D. 1949. The geobotanical significance of polyploidy. I. Polyploidy and latitude. *Portugal Acta Biol A*: 273–352.

Löve A, Löve D, Sermolli REGP. 1977. *Cytotaxonomical Atlas of the Pteridophyta*. Vaduz: J. Cramer.

Lowe AJ, Abbott RJ. 1996. Origins of the new allopolyploid species *Senecio cambrensis* (Asteraceae) and its relationship to the Canary Islands endemic *Senecio teneriffae*. *Am J Bot* 83: 1365–1372.

Lukens L, Quijada P, Udall J, *et al.* 2004. Genome redundancy and plasticity within ancient and recent *Brassica* crop species. *Biol J Linn Soc* 82: 665–674.

Lumaret R. 1984. The role of polyploidy in the adaptive significance of polymorphism at the Got-1 locus in the *Dactylis glomerata* complex. *Heredity* 52: 153–169.

Lumaret R, Borrill M. 1988. Cytology, genetics, and evolution in the genus *Dactylis*. *Crit Rev Plant Sci* 7: 55–91.

Lumaret R, Guillerm JL, Delay J, *et al.* 1987. Polyploidy and habitat differentiation in *Dactylis glomerata* L. from Galicia (Spain). *Oecologia* 73: 436–446.

Lutz AM. 1907. A preliminary note on the chromosomes of *Oenothera lamarckiana* and one of its mutants, *O. gigas*. *Science* 26: 151–152.

Lynch M, Conery JS. 2000. The evolutionary fate of duplicated genes. *Science* 290: 1151–1154.

Lynch M, Force A. 2000. The probability of duplicate gene preservation by subfunctionalization. *Genetics* 154: 459–473.

Mable B. 2003. Breaking down taxonomic barriers in polyploidy research. *Trends Plant Sci* 8: 582–590.

Mable B. 2004a. Polyploidy and self-compatibility: is there an association? *New Phytol* 162: 803–811.

Mable B. 2004b. 'Why polyploidy is rarer in animals than in plants': myths and mechanisms. *Biol J Linn Soc* 82: 453–466.

Manton I. 1950. *Problems of Cytology and Evolution in the Pteridophyta*. Cambridge, UK: Cambridge University Press.

Marchant CJ. 1968. Evolution in *Spartina* (Gramineae). II. Chromosomes, basic relationships and the problem of *S.* × *townsendii* agg. *J Linn Soc Bot* 60: 381–409.

Masterson J. 1994. Stomatal size in fossil plants: evidence for polyploidy in majority of angiosperms. *Science* 264: 421–423.

Masuyama S, Watano Y. 1990. Trends for inbreeding in polyploid pteridophytes. *Plant Spec Biol* 5: 13–17.

Matzke MA, Matzke AJM. 1998. Polyploidy and transposons. *Trends Ecol Evol* 13: 241.

Matzke MA, Scheid OM, Matzke AJ. 1999. Rapid structural and epigenetic changes in polyploid and aneuploid genomes. *Bioessays* 21: 761–767.

Mavrodiev EV, Soltis DE. 2001. Recurring polyploid formation: an early account from the Russian literature. *Taxon* 50: 469–474.

McFadden ES, Sears ER. 1946. The origin of *Triticum spelta* and its free-threshing hexaploid relatives. *J Hered* 37: 81–89; 107–116.

Mia MM, Mukherjee BB, Vickery RK. 1964. Chromosome counts in the section Simiolus of the genus *Mimulus* (Scrophulariaceae). VI. New numbers in *M. guttatus*, *M. tigrinus* and *M. glabratus*. *Madroño* 17: 156–160.

Miller JS, Venable DL. 2000. Polyploidy and the evolution of gender dimorphism in plants. *Science* 289: 2335–2338.

Mittelstein Scheid O, Jakovleva L, Afsar K, *et al.* 1996. A change of ploidy can modify epigenetic silencing. *Proc Natl Acad Sci USA* 93: 7114–7119.

Moody ME, Muellert LD, Soltis DE. 1993. Genetic variation and random drift in autotetraploid populations. *Genetics* 154: 649–657.

Moore WS. 1977. An evaluation of narrow hybrid zones in vertebrates. *Quart Rev Biol* 52: 263–277.

Muller HJ. 1914. A new mode of segregation in Gregory's tetraploid primulas. *Am Nat* 48: 508–512.

Müntzing A. 1930. Über Chromosomenvermehrung in *Galeopsis*-Kreuzungen und ihre phylogenetische Bedeutung. *Hereditas* 14: 153–172.

Müntzing A. 1936. The evolutionary significance of autopolyploidy. *Hereditas* 21: 263–378.

Nasrallah ME, Yogeeswaran K, Snyder S, Nasrallah JB. 2000. *Arabidopsis* species hybrids in the study of species differences and evolution of amphiploidy in plants. *Plant Physiol* 124: 1605–1614.

Nogler GA. 1984. Genetics of apospory in apomictic *Ranunculus auricomus*. V. Conclusion. *Bot Helv* 94: 411–422.

Noyes RD, Rieseberg LH. 2000. Two independent loci control agamospermy (apomixis) in the triploid flowering plant *Erigeron annuus*. *Genetics* 155: 379–390.

Nuismer SL, Thompson JN. 2001. Plant polyploidy and non-uniform effects on insect herbivores. *Proc R Soc Lond B* 268: 1937–1940.

Ohno S. 1970. *Evolution by Gene Duplication*. New York: Springer-Verlag.

Orel N, Puchta H. 2003. Differences in the processing of DNA ends in *Arabidopsis thaliana* and tobacco: possible implications for genome evolution. *Plant Mol Biol* 51: 523–531.

Osborn TC, Pires JC, Birchler JA, et al. 2003. Understanding mechanisms of novel gene expression in polyploids. *Trends Genet* 19: 141–147.

Otto SP. 2003. In polyploids, one plus one does not equal two. *Trends Ecol Evol* 18: 431–433.

Otto SP, Whitton J. 2000. Polyploid incidence and evolution. *Annu Rev Genet* 34: 401–437.

Ownbey M. 1950. Natural hybridization and amphiploidy in the genus *Tragopogon*. *Am J Bot* 37: 487–499.

Ownbey M, McCollum GD. 1953. Cytoplasmic inheritance and reciprocal amphiploidy in *Tragopogon*. *Am J Bot* 40: 788–796.

Ozias-Akins PE, Lubbers L, Hanna WW, McNay JW. 1993. Transmission of the apomictic mode of reproduction in *Pennisetum*: co-inheritance of the trait and molecular markers. *Theor Appl Genet* 85: 632–638.

Ozkan H, Levy AA, Feldman M. 2001. Allopolyploidy-induced rapid genome evolution in the wheat (*Aegilops-Triticum*) group. *Plant Cell* 13: 1735–1747.

Ozkan H, Tuna M, Arumuganathan K. 2003. Nonadditive changes in genome size during allopolyploidization in the wheat (*Aegilops-Triticum*) group. *J Hered* 94: 260–264.

Parrott WA, Smith RR. 1986. Evidence for the existence of endosperm balance number in the true clovers (*Trifolium* spp.). *Can J Genet Cytol* 28: 281–286.

Pegtel DM. 1999. Effect of ploidy level on fruit morphology, seed germination and juvenile growth in scurvy grass (*Cochlearia officinalis* L. s.l., Brassicaceae). *Plant Spec Biol* 14: 201–215.

Pessino SC, Ortiz JPA, Hayward MD, Quarìgn CL. 1999. The molecular genetics of gametophytic apomixis. *Hereditas* 130: 1–11.

Petit C, Thompson JD. 1999. Species diversity and ecological range in relation to ploidy level in the flora of the Pyrenees. *Evol Ecol* 13: 45–66.

Petit C, Bretagnolle F, Felber F. 1999. Evolutionary consequences of diploid-polyploid hybrid zones in wild species. *Trends Ecol Evol* 14: 306–311.

Petrov DA. 2001. Evolution of genome size: new approaches to an old problem. *Trends Genet* 17: 23–28.

Pichersky E, Soltis DE, Soltis PS. 1990. Defective chlorophyll a/b-binding protein genes in the genome of a homosporous fern. *Proc Natl Acad Sci USA* 87: 195–199.

Pikaard CS. 1999. Nucleolar dominance and silencing of transcription. *Trends Plant Sci* 4: 478–483.

Pikaard CS. 2000. The epigenetics of nucleolar dominance. *Trends Genet* 16: 495–500.

Pikaard CS, Chen ZJ. 1998. Nucleolar dominance. In: Paule MR ed. *RNA Polymerase I: Transcription of Eukaryotic Ribosomal RNA*. Austin, TX: R.G. Landes Co., 275–293.

Pires JC, Lim KY, Kovařík A, et al. 2004. Molecular cytogenetic analysis of recently evolved *Tragopogon* (Asteraceae) allopolyploids reveal a karyotype that is additive of the diploid progenitors. *Am J Bot* 91: 1022–1035.

Popp M, Oxelman B. 2001. Inferring the history of the polyploid *Silene aegaea* (Caryophyllaceae) using plastid and homoeologous nuclear DNA sequences. *Mol Phylogenet Evol* 20: 474–481.

Prince VE, Pickett FB. 2002. Splitting pairs: the diverging fates of duplicated genes. *Nat Rev Genet* 3: 827–837.

Przywara L, Kuta E. 1995. Karyology of bryophytes. *Pol Bot Stud* 9: 1–83.

Raina SN, Rani V. 2001. GISH technology in plant genome research. *Methods Cell Sci* 23: 83–104.

Ramsey J, Schemske DW. 1998. Pathways, mechanisms, and rates of polyploid formation in flowering plants. *Annu Rev Ecol Syst* 29: 467–501.

Ramsey J, Schemske DW. 2002. Neopolyploidy in flowering plants. *Annu Rev Ecol Syst* 33: 589–639.

Ranker TA, Haufler CH, Soltis PS, Soltis DE. 1989. Genetic evidence for allopolyploidy in the Neotropical fern *Hemionitis pinnatifida* (Adiantaceae) and the reconstruction of an ancestral genome. *Syst Bot* 14: 439–447.

Rauscher JT, Doyle JJ, Brown AHD. 2002. Internal transcribed spacer repeat-specific primers and the analysis of hybridization in the *Glycine tomentella* (Leguminosae) polyploid complex. *Mol Ecol* 11: 2691–2702.

Rauscher JT, Doyle JJ, Brown AHD. 2004. Multiple origins and nrDNA internal transcribed spacer homeologue evolution in the *Glycine tomentella* (Leguminosae) allopolyploid complex. *Genetics* 166: 987–998.

Raybould AF, Gray AJ, Lawrence MJ, Marshall DF. 1991. The evolution of *Spartina anglica* C.E. Hubbard (Gramineae): origin and genetic variability. *Biol J Linn Soc* 43: 111–126.

Reynaud J, Jay M, Blaise S. 1991. Evolution and differentiation of populations of *Lotus corniculatus* s.l. (Fabaceae) from the southern French Alps (Massif du Ventoux and Montagne de Lure). *Can J Bot* 69: 2286–2290.

Richards AJ. 1997. *Plant Breeding Systems*. 2nd edition. New York: Chapman and Hall.

Rieseberg LH, Doyle JJ. 1989. Tetrasomic segregation in the naturally occurring autotetraploid *Allium nevii* alliance. *Hereditas* 111: 31–36.

Rodriguez DJ. 1996. A model for the establishment of polyploidy in plants. *Am Nat* 147: 33–46.

Ronfort J, Jenczewski E, Bataillon T, Rousset F. 1998. Analysis of population structure in autotetraploid species. *Genetics* 150: 921–930.

Rothera SL, Davy AJ. 1986. Polyploidy and habitat differentiation in *Deschampsia cespitosa*. *New Phytol* 102: 449–467.

Rozanova MA. 1938. On polymorphic type of species origin. *C R Acad Sci URSS* 18: 677–680.

Savidan Y. 1981. Genetics and utilisation of apomixis for the improvement of Guineagrass (*Panicum maximum* Jacq.). In: *Proceedings XIV International Grassland Congress*, Smith JA, Hayes VW, eds. Boulder, CO: Westview Press, 182–184.

Schranz ME, Osborn TC. 2000. Novel flowering time variation in resynthesized polyploid *Brassica napus*. *J Hered* 91: 242–246.

Segraves KA, Thompson JN. 1999. Plant polyploidy and pollination: floral traits and insect visits to diploid and tetraploid *Heuchera grossulariifolia*. *Evolution* 53: 1114–1127.

Segraves KA, Thompson JN, Soltis PS, Soltis DE. 1999. Multiple origins of polyploidy and the geographic structure of *Heuchera grossulariifolia*. *Mol Ecol* 8: 253–262.

Senchina DS, Alvarez I, Cronn RC, *et al.* 2003. Rate variation among nuclear genes and the age of polyploidy in *Gossypium*. *Mol Biol Evol* 20: 633–643.

Shaked H, Kashkush K, Ozkan H, *et al.* 2001. Sequence elimination and cytosine methylation are rapid and reproducible responses of the genome to wide hybridization and allopolyploidy in wheat. *Plant Cell* 13: 1749–1759.

Sharbel TF, Mitchell-Olds T. 2001. Recurrent polyploid origins and chloroplast phylogeography in the *Arabis holboellii* complex (Brassicaceae). *Heredity* 87: 59–68.

Shore JS. 1991. Tetrasomic inheritance and isozyme variation in *Turnera ulmifolia* var. *elegans* Urb. and *intermedia* Urb. (Turneraceae). *Heredity* 66: 305–312.

Small E. 1983. Pollen ploidy-prediction in the *Medicago sativa* complex. *Pollen Spores* 25: 305–320.

Small RL, Wendel JF. 1999. The mitochondrial genome of allotetraploid cotton (*Gossypium* L.). *J Hered* 90: 251–253.

Small RL, Wendel JF. 2000a. Copy number lability and evolutionary dynamics of the *Adh* gene family in diploid and tetraploid cotton (*Gossypium*). *Genetics* 155: 1913–1926.

Small RL, Wendel JF. 2000b. Phylogeny, duplication, and intraspecific variation of *Adh* sequences in New World diploid cottons (*Gossypium* L., Malvaceae). *Mol Phylogenet Evol* 16: 73–84.

Small RL, Wendel JF. 2002. Differential evolutionary dynamics of duplicated paralogous *Adh* loci in allotetraploid cotton (*Gossypium*). *Mol Biol Evol* 19: 597–607.

Small RL, Ryburn JA, Cronn RC, *et al.* 1998. The tortoise and the hare: choosing between noncoding plastome and nuclear *Adh* sequences for phylogeny reconstruction in a recently diverged plant group. *Am J Bot* 85: 1301–1315.

Small RL, Ryburn JA, Wendel JF. 1999. Low levels of nucleotide diversity at homoeologous *Adh* loci in allotetraploid cotton (*Gossypium* L.). *Mol Biol Evol* 16: 491–501.

Smith-Huerta NL. 1984. Seed germination in related diploid and allotetraploid *Clarkia* species. *Bot Gaz* 145: 246–252.

Solbrig O. 1971. Polyphyletic origin of tetraploid populations of *Gutierrezia sarothrae* (Compositae). *Madroño* 21: 20–25.

Soltis DE. 2004. Saxifragaceae: a taxonomic treatment. In: Kadereit JW, Jeffrey C eds. *The Families and Genera of Vascular Plants.* Berlin: Springer-Verlag.

Soltis DE, Rieseberg LH. 1986. Autopolyploidy in *Tolmiea menziesii* (Saxifragaceae): evidence from enzyme electrophoresis. *Am J Bot* 73: 310–318.

Soltis DE, Soltis PS. 1989a. Allopolyploid speciation in *Tragopogon*: insights from chloroplast DNA. *Am J Bot* 76: 1119–1124.

Soltis DE, Soltis PS. 1989b. Genetic consequences of autopolyploidy in *Tolmiea* (Saxifragaceae). *Evolution* 43: 586–594.

Soltis DE, Soltis PS. 1990a. Chloroplast DNA and nuclear rDNA variation: insights into autopolyploid and allopolyploid evolution. In: Kawano S ed. *Biological Approaches and Evolutionary Trends in Plants.* San Diego, CA: Academic Press, 97–117.

Soltis DE, Soltis PS. 1993. Molecular data and the dynamic nature of polyploidy. *Crit Rev Plant Sci* 12: 243–273.

Soltis DE, Soltis PS. 1995. The dynamic nature of polyploid genomes. *Proc Natl Acad Sci USA* 92: 8089–8091.

Soltis DE, Soltis PS. 1999. Polyploidy: recurrent formation and genome evolution. *Trends Ecol Evol* 14: 348–352.

Soltis DE, Soltis PS, Ness BD. 1989. Chloroplast DNA variation and multiple origins of autopolyploidy in *Heuchera micrantha* (Saxifragaceae). *Evolution* 43: 650–656.

Soltis DE, Soltis PS, Pires JC, *et al.* 2004a. Recent and recurrent polyploidy in *Tragopogon* (Asteraceae): cytogenetic, genomic and genetic comparisons. *Biol J Linn Soc* 82: 485–501.

Soltis DE, Soltis PS, Tate JA. 2004b. Advances in the study of polyploidy since *Plant Speciation*. *New Phytol* 161: 173–191.

Soltis PS, Soltis DE. 1990b. Evolution of inbreeding and outcrossing in ferns and fern-allies. *Plant Spec Biol* 5: 1–11.

Soltis PS, Soltis DE. 1991. Multiple origins of the allotetraploid *Tragopogon mirus* (Compositae): rDNA evidence. *Syst Bot* 16: 407–413.

Soltis PS, Soltis DE. 2000. The role of genetic and genomic attributes in the success of polyploids. *Proc Natl Acad Sci USA* 97: 7051–7057.

Soltis PS, Plunkett GM, Novak SJ, Soltis DE. 1995. Genetic variation in *Tragopogon* species: additional origins of the allotetraploids *T. mirus* and *T. miscellus* (Compositae). *Am J Bot* 82: 1329–1341.

Song K, Lu P, Tang K, Osborn TC. 1995. Rapid genome change in synthetic polyploids of *Brassica* and its implications for polyploid evolution. *Proc Natl Acad Sci USA* 92: 7719–7723.

Stebbins GL. 1940. The significance of polyploidy in plant evolution. *Am Nat* 74: 54–66.

Stebbins GL. 1947. Types of polyploids: their classification and significance. *Adv Genet* 1: 403–429.

Stebbins GL. 1948. The chromosomes and relationships of *Metasequoia* and *Sequoia*. *Science* 108: 95–98.

Stebbins GL. 1950. *Variation and Evolution in Plants*. New York: Columbia University Press.

Stebbins GL. 1957. Self fertilization and population variability in the higher plants. *Am Nat* 91: 337–354.

Stebbins GL. 1959. The role of hybridization in evolution. *Proc Am Philos Soc* 103: 231–251.

Stebbins GL. 1971. *Chromosomal Evolution in Higher Plants*. London: Addison-Wesley.

Stebbins GL. 1985. Polyploidy, hybridization, and the invasion of new habitats. *Ann Missouri Bot Gard* 72: 824–832.

Steen SW, Gielly L, Taberlet P, Brochmann C. 2000. Same parental species, but different taxa: molecular evidence for hybrid origins of the rare endemics *Saxifraga opdalensis* and *S. svalbardensis* (Saxifragaceae). *Bot J Linn Soc* 132: 153–164.

Tal M. 1980. Physiology of polyploids. In: Lewis WH ed. *Polyploidy: Biological Relevance*. New York: Plenum Press, 61–75.

Tate JA, Simpson BB. 2004. Breeding system evolution in *Tarasa* (Malvaceae) and selection for reduced pollen grain size in the polyploid species. *Am J Bot* 91: 207–213.

Thompson JD, Lumaret R. 1992. The evolutionary dynamics of polyploid plants: origins, establishment, and persistence. *Trends Ecol Evol* 7: 302–307.

Thompson JN, Cunningham BM, Segraves KA, *et al.* 1997. Plant polyploidy and insect/plant interactions. *Am Nat* 150: 730–743.

Thompson JN, Nuismer SL, Merg K. 2004. Plant polyploidy and the evolutionary ecology of plant/animal interactions. *Biol J Linn Soc* 82: 511–519.

Tiffin P, Gaut BS. 2001. Sequence diversity in the tetraploid *Zea perennis* and the closely related diploid *Z. diploperennis*: insights from four nuclear loci. *Genetics* 158: 401–412.

Tischler G. 1935. Die Bedeutung der Polyploidie fur die Verbreitung der Angiospermen, erlautert an den Arten Schleswig-Holstein, mit Ausblicken auf andere Florengebiete. *Bot Jahrb Syst* 67: 1–36.

Tressens SG. 1970. Morfología del polen y evolución en *Tarasa* (Malvaceae). *Bonplandia* 3: 73–100.

Tyler B, Borrill M, Chorlton K. 1978. Studies in *Festuca*. X. Observations on germination and seedling cold tolerance in diploid *Festuca pratensis* and tetraploid *F. pratensis* var. *alpennina* in relation to their altitudinal distribution. *J Appl Ecol* 15: 219–226.

Van Dijk P, Bakx-Schotman T. 1997. Chloroplast DNA phylogeography and cytotype geography in autopolyploid *Plantago media*. *Mol Ecol* 6: 345–352.

Van Dijk P, Hartog M, Van Delden W. 1992. Single cytotype areas in autopolyploid *Plantago media* L. *Biol J Linn Soc* 46: 315–331.

Vicient CM, Suoniemi A, Anamtamat-Jonsson K, *et al.* 1999. Retrotransposon BARE-1 and its role in genome evolution in the genus *Hordeum*. *Plant Cell* 11: 1769–1784.

Vida G. 1976. The role of polyploidy in evolution. In: Novak VJA, Pacltova I eds. *Evolutionary Biology*. Prague: Czechoslovak Academy of Sciences, 267–304.

Villar R, Veneklaas EJ, Jordano P, Lambers H. 1998. Relative growth rate and biomass allocation in 20 *Aegilops* (Poaceae) species. *New Phytol* 140: 425–437.

Vision TJ, Brown DG, Tanksley SD. 2000. The origins of genomic duplications in *Arabidopsis*. *Science* 290: 2114–2117.

Von Well E, Fossey A. 1998. A comparative investigation of seed germination, metabolism and seedling growth between two polyploid *Triticum* species. *Euphytica* 101: 83–89.

Wagner WH. 1970. Biosystematics and evolutionary noise. *Taxon* 19: 146–151.

Wagner WH, Wagner FS. 1980. Polyploidy in pteridophytes. In: Lewis WH ed. *Polyploidy: Biological Relevance*. New York: Plenum Press, 199–214.

Walker TG. 1979. The cytogenetics of ferns. In: Dyer AF ed. *The Experimental Biology of Ferns*. London: Academic Press, 87–132.

Warner DA, Edwards GE. 1993. Effects of polyploidy on photosynthesis. *Photosynth Res* 35: 135–147.

Watanabe N, Kobayashi S, Furuta Y. 1997. Effect of genome and ploidy on photosynthesis of wheat. *Euphytica* 94: 303–309.

Weiss H, Maluszynska J. 2000. Chromosomal rearrangement in autotetraploid plants of *Arabidopsis thaliana*. *Hereditas* 133: 255–261.

Wendel JF. 2000. Genome evolution in polyploids. *Plant Mol Biol* 42: 225–249.

Wendel JF, Schnabel A, Seelanan T. 1995. Bidirectional interlocus concerted evolution following allopolyploid speciation in cotton (*Gossypium*). *Proc Natl Acad Sci USA* 92: 280–284.

Werth CR, Windham MD. 1991. A model for divergent, allopatric speciation of polyploid pteridophytes resulting from silencing of duplicate-gene expression. *Am Nat* 137: 515–526.

Werth CR, Guttman SI, Eshbaugh WH. 1985a. Electrophoretic evidence of reticulate evolution in the Appalachian *Asplenium* complex. *Syst Bot* 10: 184–192.

Werth CR, Guttman SI, Eshbaugh WH. 1985b. Recurring origins of allopolyploid species in *Asplenium*. *Science* 228: 731–733.

Winge O. 1917. The chromosomes: their number and general importance. *C R Trav Lab Carlsberg* 13: 131–275.

Winkler H. 1916. Uber die experimentelle Erzeugung von Pflanzen mit abweichenden chromosomenzahlen. *Z Bot* 8: 417–531.

Wolf PG, Soltis DE, Soltis PS. 1990. Chloroplast DNA and allozymic variation in diploid and autotetraploid *Heuchera grossulariifolia* (Saxifragaceae). *Am J Bot* 77: 232–244.

Wolf PG, Soltis PS, Soltis DE. 1989. Tetrasomic inheritance and chromosome pairing behavior in the naturally occurring autotetraploid *Heuchera grossulariifolia* (Saxifragaceae). *Genome* 32: 655–659.

Wolfe KH, Schields DC. 1997. Molecular evidence for an ancient duplication of the entire yeast genome. *Nature* 387: 708–713.

Wolffe AP, Matzke MA. 1999. Epigenetics: regulation through repression. *Science* 286: 481–486.

Wyatt R, Odrzykoski IJ, Stoneburner A. 1993. Isozyme evidence proves that the moss *Rhizomnium pseudopunctatum* is an allopolyploid of *R. gracile* × *R. magnifolium*. *Mem Torrey Bot Club* 25: 21–35.

Wyatt R, Odrzykoski IJ, Stoneburner A, *et al.* 1988. Allopolyploidy in bryophytes: multiple origins of *Plagiomnium medium*. *Proc Natl Acad Sci USA* 85: 5601–5604.

Zhao XP, Si Y, Hanson RE, *et al.* 1998. Dispersed repetitive DNA has spread to new genomes since polyploid formation in cotton. *Genome Res* 8: 479–492.

Zimmer EA, Martin SL, Beverley SM, *et al.* 1980. Rapid duplication and loss of genes coding for the alpha chains of hemoglobin. *Proc Natl Acad Sci USA* 77: 2158–2162.

CHAPTER 8

Polyploidy in Animals

T. RYAN GREGORY AND BARBARA K. MABLE

The notion that the duplication of entire genomes plays a significant role in evolution has traditionally been reserved for plants (see Chapter 7), but in terms of ancient events, has recently expanded to include groups as diverse as yeast and vertebrates (see Chapter 6). For the most part, however, more recent duplications of complete chromosome sets have been regarded as relatively unimportant in animal evolution. Polyploidy, it is often argued (or quietly assumed), occurs only under remarkably uncommon circumstances among the metazoa and is therefore more a curiosity than an evolutionary force in this group. As the detailed review presented here shows, this conventional assessment is far too limited.

This chapter provides an outline of the methods used for classifying and identifying polyploid animals, followed by a discussion of proposed explanations for its relative rarity as compared to plants (where it may be near-ubiquitous) (see Chapter 7). The known cases of polyploid vertebrates and invertebrates are then reviewed in succession, with notes about the phenotypic and evolutionary consequences interspersed throughout. As will be seen, polyploidy has arisen by a variety of mechanisms in a diverse array of animal taxa, covering nearly every major phylum.

Some recent authors have accurately stressed the need for increased communication between botanists and zoologists interested in the question of polyploidy
(Mable, 2003; Le Comber and Smith, 2004). Because large-scale duplications are
also of substantial import for comparative genomics, improved discourse with
geneticists and genome biologists is likewise required. In this regard, the placement of this chapter following a discussion of plant polyploidy and preceding one
on comparative genomics is not coincidental.

THE ORIGINS AND CLASSIFICATION
OF POLYPLOID ANIMALS

AUTOPOLYPLOIDY AND ALLOPOLYPLOIDY

The classification of polyploids is a somewhat complex issue, with several
different terminological and conceptual options available. Most of the progress in
this debate has come from studies of plants, where it is largely accepted that mode
of origin should be the primary criterion for classifying polyploids. In both plants
and animals, the various mechanisms operating to produce polyploids can be
grouped into two broad categories: *autopolyploidy* and *allopolyploidy* (Thompson
and Lumaret, 1992; Ramsey and Schemske, 1998, 2002) (see Chapter 7).

Autopolyploids can result from any of several processes operating within
a species: (1) simple genomic doubling by a failure of germline cell division
after mitotic replication; (2) the production and fertilization of unreduced
(i.e., diploid) gametes by a failure of cell division after meiosis (in animals, this
occurs primarily in eggs rather than sperm); and (3) polyspermy, or the fertilization of an egg by more than one sperm (Otto and Whitton, 2000). External
factors such as temperature, reproductive mode, and others that may increase the
frequency of such events can influence the extent to which autopolyploids are
generated (Otto and Whitton, 2000).

Allopolyploids, on the other hand, are those that arise through hybridization
and can be further subdivided into *segmental allopolyploids,* which result from
hybridization between partially cross-fertile progenitor species with similar
genomes, and *genomic allopolyploids* (sometimes also referred to as *amphiploids*),
whose parental species are cross-sterile and which therefore essentially contain
two different genomes at their time of origin (Thompson and Lumaret, 1992).
Hybridization is common in plants and often occurs repeatedly between related
species (see Chapter 7). Hybridization is also more frequent in the animal
kingdom than is often assumed (e.g., Allendorf *et al.,* 2001), and it is becoming
apparent that hybridization—in some cases recurrent—is a prevalent mode of
polyploid formation in animals.

IDENTIFYING POLYPLOIDS

Is this animal polyploid? And if so, is it an auto- or an allopolyploid? These are straightforward questions, but ones that are not always easily answered. Although several different criteria have been used to detect and distinguish polyploidy in animals, none has proven entirely reliable. This difficulty in itself probably contributes to the comparatively low number of known cases of polyploidy in animals versus plants.

CHROMOSOME NUMBER AND NUCLEAR DNA CONTENT

In this chapter, animal chromosome numbers will be presented following the conventions used in the botanical literature (Ramsey and Schemske, 1998) (see Chapter 7). Specifically, "2n" indicates the somatic chromosome number and "n" the gametic chromosome number, independent of ploidy level, whereas x is the base number of the species or lineage. As a result, the chromosome complements of diploids are given as $2n = 2x$, triploids as $2n = 3x$, tetraploids as $2n = 4x$, and so on.

Since polyploidy is a multiplication of entire chromosome sets, the simplest and most common method of identifying it is to find chromosome numbers and/or nuclear DNA contents that are multiples of the basal value of the lineage. However, either of these parameters by itself can be misleading. For example, birds typically have high chromosome numbers, but these are not associated with high DNA contents, and instead result from the presence of numerous "microchromosomes." A similar observation holds for butterflies, in which chromosome breakages, and not multiplications, are responsible for the high numbers observed (White, 1978). Conversely, discontinuous variation in DNA content can occur without changes in chromosome number, in a poorly understood process known as "cryptopolyploidy" (Gregory and Hebert, 1999). Moreover, there is often a decrease in total DNA content in polyploids, such that the doubled chromosome set does not include a fully doubled amount of DNA (see Chapters 2 and 7).

These problems aside, karyotyping and DNA quantification remain the most accurate methods for differentiating known polyploids from diploid relatives. Karyotyping is a labor-intensive process, but methods such as flow cytometry and image analysis densitometry allow DNA contents to be assessed rapidly. This can be performed nondestructively for larger animals like vertebrates, which can be sampled for erythrocytes, leukocytes, skin epithelia, or other tissue clippings as dictated by the size and cellular features of the animals under study. With smaller invertebrates, ploidy analysis is less convenient and is often terminal, but can be performed using the same principles.

CELL AND NUCLEUS SIZE

DNA content and cell/nucleus size are strongly linked, irrespective of the source of the DNA (see Chapters 1 and 2). Cell size in itself does not provide convincing evidence of new cases of polyploidy, but it often can be used to distinguish diploids and polyploids in species where both are known to occur. Most commonly, erythrocyte sizes have been used to identify polyploids in both wild and cultured vertebrate species (e.g., Austin and Bogart, 1982; Matson, 1990; Garcia-Abiado et al., 1999; Felip et al., 2001) (Fig. 8.1), but the sizes of toe pad epithelial cells in some frogs (Green, 1980) and nuclear sizes of shed epithelial cells from salamanders (Licht and Bogart, 1987) have also been employed.

It has been suggested by some authors that cell sizes in polyploid animals may eventually shrink to match diploid cell sizes (Pedersen, 1971; Otto and Whitton, 2000), which would challenge the notion that DNA content causally influences cell size (Gregory, 2001a). However, Pedersen's (1971) conclusions, which were based on a three-species comparison of unrelated polyploid and diploid fishes, have been challenged by the detailed recent study of Hardie and Hebert (2003), which included comparisons of related diploid and polyploid fishes. Otto and Whitton's (2000) only reference for this point is Beatty (1957), which in fact deals with attempts to produce artificial polyploid mammals. Overall, convincing evidence for this process is lacking, and it actually seems rather unlikely that polyploid animals could come to possess the same cell sizes as their diploid relatives without a substantial reduction in DNA content.

MEIOTIC CHROMOSOME BEHAVIOR

Aside from comparisons of chromosome numbers and DNA contents, the most commonly cited means of identifying (auto)polyploids is by observing chromosome

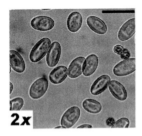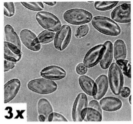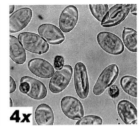

FIGURE 8.1   Photomicrographs of red blood cells from diploid (2x), triploid (3x), and tetraploid (4x) members of the salamander genus *Ambystoma*. Polyploids in this group are formed by hybridization, and can be identified in practice by their enlarged cells. Photos taken at 16x magnification, courtesy of J.P. Bogart. Scale bar equals 50 µm.

FIGURE 8.2 Side view of a first meiotic metaphase from a male triploid hybrid newt resulting from a cross between *Triturus marmoratus* and *T. karelini*. Twelve bivalents are visible in this figure, along with 12 unpaired univalents. The univalents (examples of which are indicated by arrows) will distribute randomly between the resulting gametes, most likely resulting in dysfunctional sperm with unbalanced chromosome complements. From Lantz and Callan (1954), as reprinted in Macgregor (1993). Reproduced by permission of the Indian Academy of Sciences, H.C. Macgregor, and Springer Science and Business Media.

behavior during meiosis. Under the normal conditions of diploidy, equivalent chromosomes pair to form "bivalents," which are then separated such that each gamete receives one copy. In triploids, this is not possible because there will either be a "trivalent" or a bivalent plus a "univalent" (Fig. 8.2), whereas in tetraploids the four chromosome copies may group together to form a "multivalent" (specifically, a "tetravalent").

Univalents and trivalents indicate odd-numbered polyploidy and tetravalents imply even-numbered polyploidy, but the absence of these does not necessarily mean diploidy. For starters, some polyploids do not undergo meiosis at all, making this a moot point. In more ancient polyploids, multiplied chromosomes may diverge and come to segregate without forming multivalents. Finally, the chromosomes of genomic allopolyploids may not be similar enough to pair in this way. In fact, the presence of multivalents is often taken to indicate autopolyploidy, because it is assumed that only chromosomes from the same species (homologs) would be similar enough to allow multivalent associations. However, the structure of at least some chromosomes from different species (homoeologs) may be conserved sufficiently to allow multivalent formation (especially, but not only, in cases of segmental allopolyploidy), and autopolyploids may exhibit mechanisms that suppress multivalent formation. In addition, changes in genomic structure and gene expression following polyploidization are likely to alter meiotic associations. So, although the complete absence of multivalents may indicate an allopolyploid origin and the presence of only multivalents may indicate an autopolyploid origin, it is more common to observe intermediate patterns, which are much more difficult to interpret conclusively. Some phylogenetically based analyses of gene duplicates have been proposed as means for distinguishing auto- and allopolyploidy (Chenuil *et al.*, 1999) as an alternative to relying solely on segregation patterns.

PROTEIN ELECTROPHORESIS

The term "isozyme" refers to different forms of the same enzyme coded at different genetic loci, and the term "allozyme" (from "allelic isozyme") refers to different forms coded at the same locus. By separating different iso/allozymes by gel electrophoresis, it is possible to assess variation in patterns of inheritance, which may be affected by polyploidy. If the functional enzyme consists of one protein subunit (i.e., is a monomer), for example, then one would expect to find one (if homozygous) or two (if heterozygous) bands on the gel for a diploid organism. Three or more bands for a monomer, or an unbalanced staining pattern with one dense band (implying several copies) and one light one (single copy), would indicate that multiple copies of this gene are present (Fig. 8.3). Such evidence can be suggestive of polyploidy, particularly when compared to a related specimen that is known to be diploid. Multimeric enzymes may give complex banding patterns that are more difficult to interpret but can also be used to identify the existence (but not the degree) of polyploidy. It bears noting, however, that even clear examples of duplication could be owing to isolated or tandem gene duplications rather than polyploidy (see Chapter 5). Additionally, if the entire population is homozygous for a given gene, then this will not be informative, and allopolyploidy involving hybridization between two such homozygotes would produce only the expected two bands for a monomeric enzyme. Nevertheless, the discovery of even one duplicated gene may warrant further examination with other methods such as karyotyping and DNA content analysis.

Identification of large numbers of duplicated genes, using allyzome analyses at multiple loci or the methods discussed in detail in Chapters 5 and 6, especially

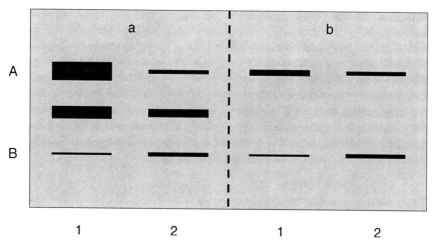

FIGURE 8.3   Zymogram representing expected allozyme banding patterns in tetraploids when gene dosage levels are additive (i.e., staining intensity is proportional to the number of gene copies) for both dimeric (a) and monomeric (b) enzymes. For dimeric enzymes, a heterozygote for two alleles produces two homodimer bands (A and B here) and one heterodimer band (middle band). For each type of enzyme, lane 1 shows the pattern expected for an asymmetrically heterozygous individual (AAAB) and lane 2 shows that expected for a symmetrically heterozygous individual (AABB). Although it is theoretically possible (with perfect additivity) to distinguish triploids and tetraploids with asymmetric patterns from each other (especially using dimeric enzymes where staining intensities are expected to be 4 : 4 : 1 and 9 : 6 : 1 for triploids and tetraploids, respectively) and from diploids using densitometry (e.g., Danzmann and Bogart, 1982), it is not possible to distinguish diploids from tetraploids with symmetric patterns (expected staining intensities are 1 : 2 : 1 for dimeric enzymes for both).

if distributed among several chromosomes, would provide stronger evidence of a complete genome duplication by polyploidy. However, in ancient polyploids it may be difficult to identify duplicate pairs, especially if rediploidization has been extensive (Wolfe, 2001) (see Chapter 6).

WHY IS POLYPLOIDY LESS COMMON IN ANIMALS THAN IN PLANTS?

Polyploidy is much more widely distributed among animal taxa than commonly appreciated (as reviewed in detail later), and there is increasing evidence for ancient genome duplications in a variety of animal groups (see Chapter 6), but it nevertheless remains true that recent polyploidy is far less common in animals than in plants. Since H.J. Muller's classic discussion in 1925, theorists have tried to explain this taxonomic discrepancy. The major proposed explanations of the last 80 years are reviewed here, some of which deal with challenges to the formation of animal polyploids, whereas others have to do with adverse developmental or phenotypic consequences in certain groups. None has proved satisfactory, and indeed the very question may be miscast. As Mable (2004a) has recently pointed out, such an approach based on the rarity of polyploidy has tended to obscure potentially more interesting questions relating to the mechanisms that promote the formation and establishment of polyploidy in animals and plants. It also has the effect of deterring further research into the presence of polyploidy in animals, which confirms by default the supposition of its rarity.

DISRUPTION OF SEX DETERMINATION

The oldest and still most widely cited explanation for the rarity of animal polyploidy is the one proposed by Muller (1925), in which he noted that whereas most plants tend to exhibit both sexes in a single individual (such plants are *monoecious* if they have separate male and female flowers, or *hermaphroditic* if the flowers themselves contain both male and female parts), most animals have two separate sexes (i.e., are *bisexual* or *gonochoristic* or *dioecious*). Muller noted that sex is determined chromosomally in many animals, and that polyploidy was likely to disrupt the development and/or fertility of one of the sexes as a result.

Based on evidence from *Drosophila*, Muller argued that sex determination in animals "depends upon a certain proportion of the genes" present on primary sex chromosomes in relation to those on autosomes. Triploidy, which is taken to be a common intermediate step between diploidy and tetraploidy (see Chapter 7), would disrupt this genetic "ratio" and leave only one fertile—but mateless—sex in the population. Levels higher than triploidy would then become impossible because two triploids could never mate to produce them. On occasion, Muller conceded, tetraploids may form by the fusion of two unreduced gametes, by the fertilization of a triploid female by an anomalous diploid sperm, or by a mitotic irregularity during early cleavage of a diploid zygote—but even in these cases the very rare tetraploid produced would be forced to mate with diploids, leaving only triploid offspring and leading to the same "genetic *cul de sac*" as described earlier.

Under Muller's theory, tetraploids could only be stably maintained if they continued to mate with other tetraploids, or if a tetraploid of the heterogametic sex happened to breed with a triploid. Any interbreeding with the diploid majority would destroy this stability. Even if a tetraploid line could be established, they would still be at a disadvantage relative to diploids, in part because there is no guarantee that the heterogametic sex (e.g., with XXYY sex chromosomes) will pass on either both Xs or neither in such a way as to maintain the necessary ratio.

And, in fact, the majority of the gametes produced would be of the undesirable XY variety. While the odds would seem to be against a newly formed polyploid finding a suitable mate, in plants there is often a shift in flowering time and ecological habitat that could facilitate the reproductive isolation of polyploids (Otto and Whitton, 2000) (see Chapter 7). Such mechanisms may be less common in animals, which could further contribute to the rarity of animal polyploids.

In summary, Muller (1925) argued that "a most remarkable concatenation of events must obtain before a persistent tetraploid line can actually become established, and capable of surviving in a state of nature, in animals having the prevalent type of sex determination." Because plants do not possess the supposedly "prevalent" type of ratio-based sex determination, the problems outlined earlier would not apply to the botanical world. The difficulty for Muller's theory is that this type of sex determination system is actually not very common in animals, either. As it turns out, *Drosophila* is a rather exceptional case (even among Diptera), and the arguments against polyploidy based on its peculiar system of sex determination are applicable to only a few animals. Far more species employ an XX : XY (or ZZ : ZW) system in which the presence or absence of a dominant Y (or W) chromosome determines sex.[1] In these cases the number of X (or Z) chromosomes is often immaterial, thereby freeing potential polyploids from the constraints outlined in Muller's theory (Orr, 1990; Otto and Whitton, 2000).

However, even under the dominant-Y type of system, there would be a problem in maintaining a stable ratio of males and females in the population. As Macgregor (1993) noted, if the initial tetraploid were male, then its genotype would be XXYY, and because the two Xs and two Ys would form bivalents during meiosis, it could be that all of this male's gametes would be XY. As a result, no female-producing gametes would be generated by such a male, and regardless of whether it mated with a diploid or tetraploid female, its offspring would be male (XXY or XXXY). Of course, it is not necessarily the case that the first tetraploid would be male, nor that all of the gametes produced would be XY. Thus, although Otto and Whitton (2000) also noted that the majority of the polyploid offspring would be male (XYYY, XXYY, XXXY) and females (XXXX) would be rare, they did not take such a pessimistic view. Instead, they suggested that this would be only a short-lived difficulty because selection for a balanced sex ratio would quickly take effect to restore the appropriate proportion of females. In the dioecious plant

---

[1]The difference between these sex determination systems is simple. An XX : XY system is one in which females are homogametic (XX) and males are heterogametic (XY), whereas a ZZ : ZW system is the reverse, in which males are homogametic (ZZ) and females are heterogametic (ZW). The former is found in mammals, and the latter in birds, for example.

Melandrium album, for example, there is a fully operational XXXX : XXXY system with
a stable 1:1 sex ratio, and likewise a functional ZZZZ : ZZZW system can be found
in artificially produced tetraploid silkworm moths, *Bombyx mori* (White, 1978).

It should go without saying that one must be wary of broad claims based on
observations in a single model species, but there is still a tendency in some
circles to equate all of biology with *Drosophila.* The old habit of citing Muller's
theory as *the* explanation for the rarity of animal polyploids also seems to die
hard, despite the rather narrow applicability of his particular arguments.

DEGENERATE SEX CHROMOSOMES AND DOSAGE COMPENSATION

Orr (1990) proposed a somewhat more widely applicable theory to explain the
rarity of animal polyploids, also based on the notion of a disruption in the sex
chromosomes. According to Orr (1990), most taxa that possess genetically degenerate sex chromosomes have evolved mechanisms of dosage compensation to preserve the same balance of X to autosomal gene products in both sexes. "Polyploidy
disrupts this delicate balance," he stated, and the resulting unbalanced genotypes
"are almost surely lethal." The balance of gene products between Y chromosomes
and autosomes may also be of some relevance, although this would obviously be
less critical than the case of the far more gene-dense X chromosomes (Orr, 1990).
Unlike Muller's (1925) theory, this approach does not depend on the disruption of
sex determination, although the similarities based on the notion of dosage ratios
are obvious. Under Orr's (1990) theory, polyploidy may be rare among animals
but common in plants because the former often possess degenerate sex chromosomes whereas the latter do not.

In *Drosophila,* dosage compensation involves doubling the transcription rate of
the single X chromosome in males. In the more familiar example of mammals,
compensation is achieved by inactivating one of the X chromosomes in females.
Interestingly, the system in *Drosophila* is probably more amenable to polyploidization than the one in mammals (Orr, 1990), even though it is the one on which
Muller's (1925) theory was based. According to Orr's theory, it is not dioecy per se
that limits the capacity for polyploidization, but dioecy *with the presence of a genetically degenerate sex chromosome* that necessitates a dosage compensation mechanism.
Some important predictions can be derived from this theory.

First, and most obviously, polyploidy should be more common among organisms that lack a system of sex determination based on a dominant but degenerate
sex chromosome. Orr (1990) suggested that this was the case in dioecious plants,
in which the dominant-Y system may be rare. In animals, too, there seems to
be some evidence in support of this contention. For example, in many of the
bisexual polyploid amphibians and reptiles, fully differentiated sex chromosomes

are rare (Bogart, 1980). However, there is evidence of heteromorphic sex chromosomes in several dioecious plants, and in at least one case there are signs of early stages of Y chromosome degeneration (Guttman and Charlesworth, 1998; Charlesworth, 2002). Second, Orr's theory predicts that polyploidy should be more common in groups that lack dosage compensation mechanisms. On this point, the theory suffers from some contradictory observations. For example, some taxa that are commonly believed to lack a dosage compensation mechanism, most notably birds (but see Ellegren, 2002), do not show evidence of polyploidy (Otto and Whitton, 2000).

Overall Orr's theory of X–autosome balance is undoubtedly more widely relevant than Muller's approach based on sex determination, but it too is insufficient to account for the comparative scarcity of polyploidy in animals in general. Chromosomal sex determination may indeed play a role in limiting the occurrence of polyploidy in some animals, either by being disrupted directly (as in the ratio-based system of *Drosophila*) or perhaps by affecting the population sex ratio (as with Y-dominant systems), but this certainly does not provide a complete explanation for the kingdom at large.

IMPEDIMENTS TO MEIOTIC DISJUNCTION

As Macgregor (1993) noted, all sexual species, regardless of sex determination system, require a mechanism to ensure a balanced pairing (bivalent formation) and then separation (disjunction) of sister chromosomes during meiotic cell division. In triploids, the capacity for balanced disjunction may be compromised, because the odd-numbered chromosomes will be unable to pair with a partner. That is, either univalents or trivalents will form, and these will subsequently distribute randomly between the daughter cells. With either trivalents or univalents, some cells will receive two copies of one chromosome whereas others will receive only one, such that the gametes produced will be left with an unbalanced (aneuploid) chromosome set (Fig. 8.2) in terms of both autosomes and sex chromosomes (White, 1973, 1978; Orr, 1990; Macgregor, 1993). Because triploidy is often assumed to be a necessary intermediate step on the way to tetraploidy in both plants and animals (de Wet, 1980) (see Chapter 7), meiotic disjunction-induced sterility could represent a serious barrier to the establishment of balanced ploidy levels (e.g., Jackson, 1976). However, this challenge relates more to why polyploidy should be rare in sexual species in general, rather than in animals versus plants per se. Because dioecy is rare among plants (only about 6% of taxa) (Renner and Ricklefs, 1995) but predominates in animals (Bell, 1982), this could be relevant. Nevertheless, it is notable that a relatively large proportion of the dioecious plant species that do exist are polyploid (Orr, 1990; Mable, 2004b).

438

Interploidy Crosses and Triploid Sterility

The fact that newly arising tetraploid individuals are likely to encounter diploids as potential mating partners much more frequently than other tetraploids poses a problem for their persistence, because matings with diploids are likely to produce sterile triploids. For this reason, an association has often been predicted to exist between selfing and polyploidy in plants (e.g., Stebbins, 1950; Levin, 1975; Lande and Schemske, 1985) (see Chapter 7), and White (1973) argued that the scarcity of self-fertilization in animals was the most important factor keeping the frequency of polyploidy low. However, in plants the causation may also operate in the reverse direction, with polyploidy leading to a breakdown in genetic self-incompatibility mechanisms and allowing selfing to occur more frequently than in diploids (Miller and Venable, 2000; Otto and Whitton, 2000). Moreover, both the strength of the predicted polyploidy-selfing association and the frequency of a breakdown in self-incompatibility are probably not as high as previously suggested (Mable, 2004b). The likelihood of polyploid establishment could be increased even in self-incompatible plants through shifts in flowering time, ecological habitat, or pollinator preference that often accompany polyploidization, which could facilitate the reproductive isolation of different cytotypes (Husband and Schemske, 1998, 2000; Otto and Whitton, 2000; Husband et al., 2002) (see Chapter 7). Such mechanisms may be less common in animals, which could contribute to the relative rarity of animal polyploids, although behavioral isolating mechanisms are known to effectively separate ploidy levels in some groups.

Disruption of Development

Von Wettstein (1927) and Stebbins (1950) suggested long ago that animal development is too complex and sensitive to tolerate a major genomic shift like polyploidy, whereas the development of plants is simpler and therefore more flexible. Obviously, this cannot provide a general explanation for the rarity of animal polyploidy, because polyploids are found in many of the most complex groups. Orr (1990) rejected this view on the basis that parthenogenetic and hermaphroditic animals frequently show polyploidy but do not appear to suffer from excessive developmental abnormalities. In addition, it is relatively easy to experimentally produce polyploids in many animal taxa (Jackson, 1976), with higher ploidy offspring often appearing more vigorous than their progenitors (e.g., Fankhauser, 1945; Bogart and Wasserman, 1972).

Nevertheless, it has been argued that "extensive data from chickens and humans suggest that a generalized disruption of development may explain the near absence of polyploid birds and mammals" (Otto and Whitton, 2000).

In chickens, most triploids and tetraploids are unviable (Bloom, 1972), and the fact that both ZZZZ and ZZWW tetraploids die as embryos indicates that this is a result of a general developmental disruption rather than one involving sex determination (Otto and Whitton, 2000). Nonetheless, a number of adult intersex triploid chickens (3A, ZZW) have been found (Ohno *et al.*, 1963; Abdel-Hameed and Schoffner, 1971). Although very rare, these examples demonstrate that polyploidy is not strictly impossible in birds. In humans, the effects of polyploidy seem even more severe. In fact, about 5% of naturally aborted fetuses are polyploid, although relatively short-lived live births do occasionally occur (Otto and Whitton, 2000). In part, this could relate to the abnormal development of the placenta in polyploids (Creasy *et al.*, 1976), which may be associated with effects on cell size (Otto and Whitton, 2000). Again, the adverse consequences of polyploidy in humans extend beyond the sexual disruption proposed by Muller (1925). As Otto and Whitton (2000) noted: "Although polyploidy can disrupt normal genital formation in humans, the fact that polyploid survival is much lower than that of trisomics involving the sex chromosomes (XYY, XXY, and XXX) and is associated with much more severe abnormalities and early mortality regardless of genotype suggest[s] that a general disruption of development and not a disruption of sexual development, *per se*, explains the absence of polyploidy in adult humans."

In both chickens and humans, these impacts on development presumably result from effects on the balance between maternally and paternally imprinted genes and/or the consequences of the influence of DNA content on cell size (Otto and Whitton, 2000). In terms of this latter possibility, it is quite interesting to note that neither birds nor mammals show any correlation between DNA content (specifically, genome size) and developmental rate, whereas such a relationship is readily discernible in groups like amphibians and crustaceans where polyploidy is reasonably common (see Chapter 1). One must be cautious when extrapolating from one species to an entire class, but the evidence for the adverse developmental effects of polyploidy does seem compelling for at least these two species.

NOT ENOUGH HYBRIDIZATION ... OR MAYBE TOO MUCH?

To the extent that the generation of animal polyploidy is by allopolyploidization, a relevant factor in their formation will be the opportunities for hybridization. Based on their study of cladoceran crustceans, in which allopolyploidy appears predominant, Dufresne and Hebert (1994) argued that "the low frequency of hybridization events in animals may be a primary factor constraining the origin of polyploidy in this kingdom." To be sure, it is logical that hybridization would be more common in plants that broadcast pollen over a wide area than in animals that must actively locate mates. Nevertheless, some groups of animals are known to hybridize rather frequently, and yet polyploidy remains relatively rare. For example, more than 3000 cases of interspecific hybridization are known from fishes (Schultz, 1980), but only a few of these exhibit polyploidy. Moreover, the cladocerans studied by Dufresne and Hebert (1994) are cyclic parthenogens and lack a chromosomal sex determination system, so this neither represents a typical group of animals nor controls for the other theories of polyploid animal rarity described above. More recently, Otto and Whitton (2000) have pointed out that just the opposite problem may actually pertain, with hybridization between newly formed polyploids and their diploid relatives being *too common* in animals. In either case, this explanation is clearly relevant to only one mechanism of polyploid formation and does not touch on the parallel scarcity of autopolyploid animals.

NUCLEOTYPIC CONSTRAINTS

A final (albeit only partial) explanation for the rarity of animal polyploids is one based on the nucleotypic effects of increased DNA content on cellular and organismal phenotypes (see Chapter 1). Such a possibility was hinted at by Muller (1925), who noted that "certain peculiar physiological conditions—e.g., the previous attainment of an optimum surface-volume ratio—might render polyploidy disadvantageous." However, in general this has been overlooked, even in cases where it may most clearly be applicable. For example, Otto and Whitton (2000) pointed out that "polyploidy is more often found in perennial plants and insects, probably because having a long life span increases the chances that rare events will occur (e.g., polyploidization following hybridization), and allows for mating between polyploids and their offspring." This explanation seems insufficient, because other mechanisms of polyploid production might benefit from shorter generation times involving more frequent mating events. On the other hand, it is thought that only perennial plants can tolerate large amounts of nuclear DNA because of the nucleotypic effects on development (see Chapter 2), and similar constraints probably operate in certain groups of insects (see Chapter 1). Cell size–related constraints on DNA content based on metabolic considerations may be especially important in mammals and birds, but far less so in amphibians (see Chapter 1).

POLYPLOIDY AND UNISEXUALITY

One of the most prevalent patterns in animal polyploidy is an association with alterations in reproductive mode (which is also seen in plants).

Unfortunately, literature associated with reproductive modes is filled with complicated and extensive jargon that often differs between plant and animal researchers (Mable, 2003). Prior to expanding on the link between polyploidy and unisexuality, an overview of these definitions is warranted.

DEFINITIONS

The term *parthenogenesis* comes from the Greek *parthenos* (virgin) and *genesis* (origin), and is defined as development from an unfertilized egg, spore, or seed. The various mechanisms of generation and the evolutionary consequences of parthenogenesis have been reviewed in great detail elsewhere (e.g., White, 1973, 1978; Bell, 1982; Suomalainen *et al.*, 1987), and in this chapter only their relevance to the issue of polyploidy will be discussed. Parthenogenesis can take several forms in animals. The most straightforward of these is *thyletoky*, in which unfertilized eggs develop to produce female clones. Other forms include *arrhenotoky*, in which unfertilized eggs develop into males and fertilized eggs develop into females (as in haplodiploid hymenoptera and spider mites), and *amphitoky*, in which either sex may originate from an unfertilized egg. In this chapter, "parthenogenesis" refers only to thyletoky.

According to some authors, "parthenogenesis" should not be considered equivalent to asexual reproduction, because new individuals still develop from sex cells, albeit unfertilized ones. The opposite of parthenogenesis would therefore not be sexual reproduction per se, but sexual reproduction involving the fusion of gametes (*syngamy*), or more importantly of nuclei and genomes (*karyogamy*), from different individuals (Suomalainen *et al.*, 1987). In order to side-step this minor controversy, the term "unisexual" (rather than "asexual") will be used here when referring to all-female populations or species. Unisexuality should not be confused with *hermaphroditism*, which is the possession of both male and female sex organs in a single individual, nor with *self-fertilization*, which is a process undertaken by some hermaphrodites.

Under normal sexual reproduction, or *amphimixis*, haploid male and female gametes are produced by reductive meiosis and then unite to form a diploid zygote. The term *apomixis* is sometimes used interchangeably with thyletoky, particularly in the botanical literature. However, apomixis (also known as *ameiotic thyletoky*) is actually a particular type of parthenogenesis in which meiosis is entirely suppressed, with the maturation divisions in the oocyte occurring instead by mitosis. *Automixis* (or *meiotic thyletoky*), by contrast, involves a meiotic division in the developing oocyte that is later compensated for by a doubling of the chromosome number at some stage in the life cycle (White, 1973).

In another form of unisexuality known as *gynogenesis*, penetration of the egg by sperm is necessary to trigger embryonic development, but the paternal chromosome set is subsequently eliminated or degenerates in the egg and makes no genetic contribution. Because gynogenetic species are unisexual, the eggs must be activated by a male from a related bisexual species. In a somewhat similar process known as *hybridogenesis*, the eggs of a unisexual species are actually fertilized by a male from a related bisexual species, but only the maternal genome is passed to the subsequent set of eggs (without recombination), such that the paternal genome must be replaced by a new hybridization event in each generation (Schultz, 1969). In this way, hybridogens are essentially "perpetual F_1 hybrids," with the interesting consequence that "all paternal traits are derived from the father and none from the grandfather" (Schultz, 1980). As only half the genes are actually passed on in each generation, hybridogens are sometimes called *hemiclonal* rather than clonal. Because they involve the investment of sperm from males of related species but provide no genetic benefit to these "host" males, gynogenesis and hybridogenesis are often considered to represent a type of "sexual parasitism."

WHY ARE POLYPLOIDY AND UNISEXUALITY LINKED?

Constraints on polyploidization caused by generalized developmental disruptions or nucleotypic effects are applicable to both bisexual and unisexual animals. However, other theories of animal polyploid scarcity may be taken to predict a high prevalence of unisexual polyploids. The most obvious benefit of thyletoky (and self-fertilizing hermaphroditism) is that there is no need to locate a mate. Among rare tetraploids, this can be an important benefit to avoid the risk of producing sterile triploid offspring. For this reason, the likelihood of establishing a stable polyploid line may be much higher in parthenogens and selfing hermaphrodites than in sexually reproducing populations (White, 1973).

In terms of both Muller's (1925) and Orr's (1990) theories based on sex chromosome ratios, thyletoky and hermaphroditism are relevant in that they eliminate the production of separate males and females and hence no sex chromosomes are involved at all, degenerate or otherwise. Muller (1925) clearly understood the importance of reproductive mode in influencing the prevalence of polyploidy according to his theory, although he emphasized hermaphroditism rather than parthenogenesis. Thus he predicted with some confidence that "amongst groups of animals like earthworms and freshwater snails, which are normally hermaphroditic, tetraploidy or even higher forms of polyploidy might occur as readily as amongst plants."

The third major barrier to polyploidy that is alleviated by thyletoky is the problem of univalent and/or multivalent formation during meiosis and the unstable aneuploid gametes thereby generated. This does not apply to all potential polyploids, however. Because homologous polyploid chromosomes can pair with one another (Moore, 2002), even-numbered polyploidy (e.g., $4x$, $6x$) may still be compatible with meiotic division, whereas thyletoky has often been assumed to be necessary for odd-numbered polyploidy (e.g., $3x$, $5x$) (Viktorov, 1997). Neither does this apply to all types of thyletoky. Specifically, apomixis (which involves no meiosis) alleviates the problem of uni- and multivalent formation in odd-numbered polyploids, but automixis (which includes meiotic division) provides no such benefit.

To summarize, several of the theories of polyploid rarity in animals predict a much higher incidence of polyploidy in unisexual taxa, although this may not apply to all forms of polyploidy, nor to all types of unisexuality. It is also obvious that unisexuality itself is neither sufficient to bring about polyploidy, because diploid parthenogens are relatively common in some groups (Bell, 1982), nor necessary, as sexual polyploids are known. For example, whereas in weevils, crustaceans, and reptiles, polyploidy is often linked strongly to unisexual reproduction, many polyploid fishes and amphibians, as well as some earthworms and other invertebrates, continue to reproduce by amphimixis.

In the cases in which polyploidy and unisexuality are strongly linked, the question is: Did unisexuality allow the emergence of polyploidy, or did polyploidization necessitate or even instigate a shift to unisexual reproduction? According to Otto and Whitton (2000), polyploidy in plants tends to occur prior to the emergence of apomixis, perhaps because the genes for this mode of reproduction can only be transmitted or expressed in unreduced gametes (e.g., Mogie, 1986; Pessino et al., 1999), whereas in animals unisexuality is often assumed to arise in diploids and therefore to predate polyploidization. Bell (1982, p. 337) described the situation as follows:

> [In animals], it seems unlikely that polyploidy [arises] first, since the first polyploid would be sexually sterile and, having arisen from an amphimictic lineage, would be unlikely to bear alleles directing parthenogenetic reproduction; but if parthenogenesis arose first, we would expect the majority of apomicts to be diploid, which is not the case. It follows that polyploidy and apomixis arise together; or alternatively that parthenogenesis evolves first in diploids, but that only those taxa which succeed in acquiring polyploidy soon thereafter survive for any appreciable length of time.

In this sense, polyploidy may at least be important for the persistence of unisexuality, if not its emergence.

A simultaneous transition to triploidy and unisexuality following the fertilization of a rare diploid egg has been suggested for certain crustaceans and insects (Vandel, 1928; Suomalainen et al., 1987). However, it may be far more common for unisexuality to arise first, followed by the secondary emergence of polyploidy (Bell, 1982; Otto and Whitton, 2000). As Bell (1982) pointed out, a diploid unisexual line of females might become polyploid by several different mechanisms: (1) by an internal chromosomal duplication event, (2) via the fertilization of an unreduced egg by a male of the same species (autopolyploidy), or (3) through hybridization with another species (allopolyploidy). Bell (1982) considered the second possibility to be by far the most likely for unisexuals. The logic behind this assertion is that a newly formed thyletokous female would most likely find itself surrounded by amphimictic conspecifics, such that the unreduced (i.e., diploid) eggs it produces would run a high chance of being fertilized by a nearby male. The offspring of such a mating would be triploid and would carry the genes controlling unisexuality, and would avoid problems with meiotic disjunction because they are apomictic. Additional chance fertilizations by related diploid males would generate higher levels of ploidy. It bears noting, however, that despite these theoretical predictions, few data exist to conclusively determine the sequence of events for most polyploid unisexuals.

POLYPLOIDY IN VERTEBRATES

The ongoing debate as to whether vertebrates in general are paleopolyploids is discussed thoroughly in Chapter 6, and thus will not be treated in any detail here. Rather, much of the remainder of this chapter is devoted to surveying the known cases of relatively recent polyploidy in animals. The discussion begins with the paraphyletic group commonly known as "fishes," broken down into the major taxonomic divisions. Fishes show the most extensive polyploidy among the vertebrates, but numerous other examples are known, especially in amphibians and reptiles.

JAWLESS FISHES

The jawless fishes of the superclass Agnatha are among the most primitive of vertebrates. According to Potter and Robinson's (1973) bold claim, "in the lampreys polyploidy has played a more striking role than in any other vertebrate." Lampreys (class Cephalaspidomorphi) do possess a large number of chromosomes in most cases (2n = 76 to 178), whereas in hagfishes (class Myxini), chromosome numbers are relatively low (2n = 14 to 48). However, there is a negative association between DNA content and chromosome number in this group, with hagfish genomes being about twice as large as those of lampreys (Gregory, 2001b). This is because the chromosomes of lampreys are numerous but very small, perhaps suggesting that they have formed by breakages of formerly larger chromosomes. This is supported by the observation that, within the Cephalaspidomorphi, those species with $2n \approx 76$ chromosomes do not have higher DNA contents than those with $2n \approx 150$ (Gregory, 2001b). It therefore seems very unlikely that polyploidy has been important in this group in the way suggested by Potter and Robinson (1973).

CARTILAGINOUS FISHES

Sharks, rays, and skates (class Chondrychthyes, subclass Elasmobranchii) have not been as well studied karyotypically as bony fishes, but it is clear that in general they possess rather high chromosome numbers (ranging from 2n = 28 to 106) (reviewed by Stingo and Rocco, 2001). Chromosome numbers are generally lower in more recently derived taxa, perhaps suggesting a high initial value in the early ancestors of the group. Although their DNA contents are also high (see Chapter 1), elasmobranchs exhibit no association between DNA content and chromosome number (Stingo and Rocco, 2001). Olmo *et al.* (1982) used reassociation kinetics to assess the genomic composition of several species of elasmobranchs, and concluded that at least some of them gave evidence of being tetraploid or octoploid. By extrapolation from these results and some detailed comparisons of elasmobranch karyotypes, it has been suggested that polyploidy may be of general significance among cartilaginous fishes (Olmo *et al.*, 1982; Schwartz and Maddock, 1986; Stingo and Capriglione, 1986; Stingo and Rocco, 2001).

In order to account for the discrepancy between DNA content and chromosome number, Olmo *et al.* (1982) suggested that polyploidization may have occurred independently in different orders of elasmobranchs, with the preduplication ancestral karyotypes differing among them. Alternatively, the different lineages may have shared a similar karyotype at the beginning, but then underwent very different amounts of deletion and rearrangements since their respective duplication events. Finally, Olmo *et al.* (1982) proposed that polyploidization may have occurred prior to the divergence of the various orders, followed by differing levels of karyotypic change in the time since their evolutionary separation. Schwartz and Maddock (1986), on the other hand, tentatively favored a scenario in which polyploidy played a role very early in elasmobranch evolution, but not in the time since. At least one example of recent triploidy has been documented, in a female specimen of the nurse shark *Ginglymostoma cirratum* (Kendall *et al.*, 1994).

On the basis of currently available data, it is not yet possible to reconstruct the evolution of polyploidy in elasmobranchs. It does seem likely, however, that such a process has played a role in some form in this ancient group, suggesting that further study is well warranted.

LUNGFISHES

Lungfishes, which are taken by many to resemble the earliest ancestor of tetrapods, are members of the class Sarcopterygii (lobe-finned fishes) and the subclass Dipnoi. Dipnoans flourished in the Devonian, but are now represented by a small handful of species in just three genera: *Neoceratodus forsteri* in Australia, *Lepidosiren paradoxa* in South America, and several *Protopterus* species in Africa.

These species have some of the largest genome sizes of any animals (see Chapter 1), but it is not clear that this is related to polyploidy because they do not possess especially high chromosome numbers (2n = 34 in *Protopterus,* 2n = 38 in *Lepidosiren,* and 2n = 58 in *Neoceratodus,* with 20 of the latter being microchromosomes) (Rock *et al.,* 1996; Morescalchi *et al.,* 2002). Similarly, the coelacanth *Latimeria chalumnae* (in the infraclass Coelacanthimorpha within the Sarcopterygii) has a larger genome than most teleosts, but the same basic chromosome number of 2n = 48, and has also been considered to be diploid (Thomson *et al.,* 1973). Nevertheless, at least one species of African lungfish (*Protopterus dolloi*) shows signs of recent tetraploidy, by displaying twice the DNA content and chromosome number of its congeners (2n = 68) (Vervoort, 1980). Unfortunately, few details are available regarding the origin of polyploidy in this species, and it is unknown whether the previously greater diversity of dipnoans included other polyploids.

CHONDROSTEANS

Chondrostean fishes (class Osteichthyes, subclass Actinopterygii, infraclass Chondrostei) were the dominant forms of bony fishes in the Permian but, like the lungfishes, have subsequently declined to a few small genera. Most of the remaining chondrosteans are members of the order Acipenseriformes, which includes the sturgeons (family Acipenseridae, with 23 species in four genera) and the paddlefishes (family Polyodontidae, with two monotypic genera). The North American paddlefish, *Polyodon spathula,* displays a karyotype consisting of 120 chromosomes (many of them microchromosomes) that can easily be arranged into 30 quartets, and on this basis has long been assumed to be tetraploid (Dingerkus and Howell, 1976). Similarly, sturgeons appear to display karyotypes that can be grouped into sets of four homologous chromosomes (Ohno *et al.,* 1969a), although there has been some discrepancy in the estimated number of chromosomes for some sturgeons, reflecting the difficulty in visualizing their numerous microchromosomes. In *Acipenser transmontanus,* for example, estimates have ranged from 230 to 248 to 276 chromosomes (see Ludwig *et al.,* 2001). Nevertheless, it is apparent that species of sturgeon and paddlefish can be placed into three different karyotypic categories: (1) species with ~120 (110 to 130) chromosomes, (2) those with ~250 (220 to 276) chromosomes, and (3) those with ~500 chromosomes (Birstein *et al.,* 1997; Ludwig *et al.,* 2001).

Interestingly, DNA contents in Acipenseriformes are not exceptionally high (ranging from 1.5 to 7.2 picograms [pg]), despite the very large number of (micro)chromosomes. Nevertheless, the species with more chromosomes are those with the most DNA (unlike the situation in agnathans and elasmobranchs), and the distribution of DNA contents is generally discontinuous among species (e.g., Birstein *et al.*, 1993; Blacklidge and Bidwell, 1993). Not surprisingly, it has long been suggested that sturgeons as a group are of tetraploid origin (Ohno *et al.*, 1969a).

There has been considerable disagreement about how to classify the ploidy levels of sturgeons. Some authors consider species with 120 chromosomes to be tetraploid and those with 250 to be octoploid, whereas others view the former group as now behaving cytogenetically as diploids and the latter as only tetraploids (reviewed by Ludwig *et al.*, 2001). This second view was supported by using the number of microsatellite alleles as an indication of ploidy level in 20 species of sturgeons (Ludwig *et al.*, 2001). Some authors have considered *Acipenser mikadoi* and *A. brevirostrum* (which have high DNA contents and upward of 500 chromosomes) to be 16-ploid (e.g., Birstein and DeSalle, 1998), but Ludwig *et al.* (2001) argued that they are octoploid. Blacklidge and Bidwell (1993), on the other hand, took their measurements of nuclear DNA content to indicate that *A. brevirostrum* was 12-ploid, perhaps having formed through hybridization between a tetraploid and an octoploid, followed by an autopolyploid duplication ($4x \times 8x = 6x$, then $6x \times 2 = 12x$). Obviously, there is some work yet to be done in determining the ploidy levels of species in this group.

In any case, the dominant view is that the Acipenseriformes arose by a polyploidization event from a diploid ancestor with $2n = 60$ chromosomes (Birstein *et al.*, 1997), probably with reestablishment of functional diploidy occurring before the major radiation of the order (Fontana, 1994; Ludwig *et al.*, 2001). Additional duplications restricted to members of the genus *Acipenser* have occurred in the time since, and indeed it is believed that higher ploidy levels have evolved several times independently in sturgeons (Ludwig *et al.*, 2001). This process may still be ongoing, because "triploid" individuals can occasionally be found within otherwise "diploid" species (Blacklidge and Bidwell, 1993). Thus the evolution of polyploidy in sturgeons appears to operate at several levels, including an initial tetraploidization event that has subsequently been functionally undone by gene silencing, later elevations to higher ploidy levels in some taxa (up to $8x$ or $16x$, depending on the authority), and even some more recent intraspecific triploidization events.

TELEOSTS

In addition to the proposed genome duplication around the origin of vertebrates, it has been argued that a second round occurred shortly before the diversification of the teleost fishes (see Chapter 6). This group (class Actinopterygii, subclass Neopterygii, superorder Teleostei) is by far the most diverse of all vertebrates and contains roughly half of all vertebrate species. The extent to which an early genome duplication event fostered this extraordinary diversification remains an open question, but it is clear that many groups of teleosts show evidence of more recent polyploidy.

An Entire Family of Polyploids: Salmonidae

The family Salmonidae (order Salmoniformes) is relatively small, including 66 species in 11 genera, grouped into three subfamilies (Froese and Pauly, 2003). Nevertheless, it is of substantial importance from the perspective of this chapter, given that a role for polyploidization in the evolution of the Salmonidae was first proposed nearly 60 years ago (Svärdson, 1945). Svärdson (1945) noted that chromosome numbers in the family tended to exist as multiples of 10 (e.g., Atlantic salmon at 2n ≈ 60, brown trout, brook trout, Arctic char, and whitefish at 2n ≈ 80, and grayling at 2n ≈ 100) and suggested that the basic chromosome number in salmonids was, in fact, $x = 10$. These higher numbers, he reasoned, were produced by polyploidization, such that those species with 2n = 60 were hexaploid (6x), 2n = 80 was 8x, and the grayling (genus *Thymalus*) a remarkable 10x. Similarly, Kupka (1948) suggested that whitefish (*Coregonus* spp.) had a basic chromosome number of $x = 18$, making species with 2n = 70 or 72 tetraploids. In neither case was the evidence for polyploidy particularly convincing, and later observations, most notably the finding of similar nuclear DNA contents among salmonids with different chromosome numbers by Ojima *et al.* (1963) and Rees (1964), ultimately led to the rejection of Svärdson's (1945) and Kupka's (1948) hypotheses (e.g., White, 1973).

More recent surveys of salmonid chromosome numbers have shown that most of them do indeed fall into two main categories: type A karyotypes with 2n ≈ 80 and type B with 2n ≈ 60 (Hartley, 1987; Phillips and Ráb, 2001). In both cases, the number of chromosome arms (called the *nombre fondamental*) is roughly NF = 100. Moreover, because DNA contents do not vary in proportion to these differences in chromosome number, this cannot be explained by a polyploid series within the family (Rees, 1964). Instead, the evidence for polyploidy in salmonids comes from comparisons outside of the Salmonidae.

It was noted very early on that salmonids have both higher DNA contents and chromosome numbers than members of related families (Ohno *et al.*, 1969b; Ohno, 1974). Figure 8.4, for example, shows the result of an early comparison between the diploid anchovy *Engraulis mordax* (family Engraulidae) and the rainbow trout *Oncorhynchus mykiss*, the latter of which contains a significantly higher number of chromosomes (Ohno, 1974). Based on such observations, it was suggested that the entire salmonid family was tetraploid, having undergone some form of genome duplication very early in its evolutionary history. Later, two additional lines of evidence were found to support this conclusion, including the presence of multivalents during meiosis in several species and a high incidence of duplicated enzyme loci (Allendorf and Thorgaard, 1984). Some authors attempted to revive the hypothesis of multiple (indeed, at least *four*) polyploidization events in salmonids on the basis of reassociation kinetic data (e.g., Schmidtke *et al.*, 1979; Schmidtke and Kandt, 1981), but the current view is that the family is tetraploid only (Allendorf and Thorgaard, 1984; Phillips and Ráb, 2001).

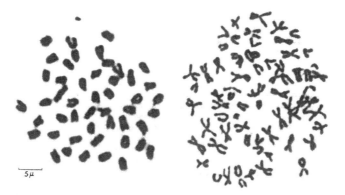

FIGURE 8.4   Early karyotypic evidence for polyploidy in the fish family Salmonidae. At the left, a somatic metaphase spread from the Californian anchovy *Engraulis mordax* (family Engraulidae, order Clupeiformes) with 2n = 48 chromosomes, and at the right, a similar preparation from the rainbow trout *Oncorhynchus mykiss* with 2n ≈ 60. Comparisons such as these, which revealed salmonids to have elevated chromosome numbers as compared to related groups, helped to establish that the entire family is tetraploid. Salmonids are of particular interest because they are in the midst of becoming rediploidized, because they are one of the few major groups to be entirely autopolyploid, and because *O. mykiss* is currently the subject of a complete genome sequencing project. From Ohno (1974), reproduced by permission (© Gebrüder Borntraeger, www.schweizbart.de).

Both the presence of multivalents at meiosis and the observations of tetrasomic inheritance (with little divergence among gene duplicates) in many salmonids suggest that the two contributing genomes that merged during tetraploidization were very similar (Allendorf and Thorgaard, 1984). Although the possibility of segmental allopolyploidy cannot be excluded on this basis, it has been inferred that the salmonid genome was doubled by autopolyploidy (Ohno, 1974; Allendorf and Thorgaard, 1984). The exact mechanisms by which this occurred are not clear, but Allendorf and Thorgaard (1984) suggested the possible intermediate of a gynogenetic triploid female whose diploid eggs were fertilized by the sperm of a related male. However, it does bear noting that modern triploid salmonids are sterile and incapable of reproducing unisexually.

Hartley (1987) suggested two possibilities regarding the karyotype of the first polyploid salmonid. First, if the diploid ancestor was 2n = 48, NF = 48 (as a great many teleosts are), then this would have made the first tetraploid salmonid 4x = 96, NF = 96. However, Hartley (1987) also noted that some of the closest relatives of salmonids (e.g., Osmeridae and Retropinnidae) have 2n = 50 to 56, NF ≈ 60, and on this basis offered the additional possibility that the first polyploid salmonid was actually 4x = 100, NF = 120. Because of the rather large number of subsequent karyotypic changes needed to accommodate the second hypothesis, Phillips and Ráb (2001) have argued that the first of these possibilities is much more likely to be true. In this sense, members of the subfamily Thymallinae have probably maintained the early tetraploid chromosome number (100) while the number of their chromosome arms has increased (146 to 170). The other subfamilies (Coregoninae and Salmoninae) must have experienced a reduction in chromosome numbers (e.g., by centric fusions) while still preserving the ancestral number of chromosome arms (Phillips and Ráb, 2001). That such large-scale chromosomal differences can arise in a family without changes in ploidy is not so surprising when one considers that there can be substantial karyotypic variation even within individual salmonid species (Thorgaard, 1983).

If salmonids did indeed undergo a single autopolyploidization event, one may reasonably ask *when* this key event took place. The scenario outlined by Phillips and Ráb (2001) would indicate that the polyploidization event occurred at or shortly after the origin of the family (because closely related families are clearly diploid) but before the divergence of the three subfamilies (because all three share the elevated chromosome number). Nevertheless, dating this event has proved quite difficult. Allendorf and Thorgaard (1984) described a few different methods that have been used to establish the timing of the event (none of which they considered overwhelmingly convincing) and derived a rather loose (but commonly cited) estimate of 25–100 million years ago. Like many of the questions discussed in this chapter, this is an issue awaiting more conclusive resolution.

Several features of salmonids make them interesting from the perspective of polyploid evolution. For one, they are polyploid but are not parthenogenetic. Indeed, salmonids possess a chromosomal sex determining mechanism, with males being the heterogametic sex. In some species, males are XY, whereas in others they may be XXY or XYY (Phillips and Ráb, 2001), indicating that the sex chromosomes were also doubled without ill effect on reproduction. Another interesting feature of salmonids is that they may continue to polyploidize in modern times. Thus both induced and spontaneous triploids can be found in cultured stocks, and hybridization among related species seems to promote modern polyploidization in some instances (e.g., Capanna *et al.*, 1974; Thorgaard *et al.*, 1982; Allendorf and Thorgaard, 1984).

The formation of multivalents during meiosis substantially increases the risk of producing aneuploid gametes, so for amphimictic tetraploids like salmonids there should be pressure to reduce such pairing and to restore disomic inheritance. In other words, there is a selective advantage to becoming functionally diploid, with the four homologs undergoing structural changes that allow the formation of two separate pairs rather than a single quadrivalent (Allendorf and Thorgaard, 1984). As Wolfe (2001) noted, "diploidization does not necessarily happen simultaneously for all chromosomes or even for all loci on a particular chromosome," and one of the most intriguing features of the Salmonidae is that they are still in the process of rediploidizing. Thus, in some salmonids tetrasomic inheritence patterns can be found with some enzymes, whereas others are now disomic (or intermediate) in this regard (Ohno, 1974; Wright *et al.*, 1983; Allendorf and Thorgaard, 1984; Allendorf and Danzmann, 1997). The legacy of tetraploidy will remain in the form of high DNA contents and chromosome numbers, and possibly duplicate gene expression, even if complete rediploidization is eventually achieved. Indeed, as Allendorf and Thorgaard (1984) pointed out, "all aspects of genetics and evolution in salmonids are affected by their tetraploid ancestry." The fact that the rainbow trout (*Oncorhynchus mykiss*) is currently the subject of a complete genome sequencing project promises to provide important insights along these lines (Thorgaard *et al.*, 2002).

ANOTHER ENTIRELY POLYPLOID FAMILY: CATOSTOMIDAE

There are 68 species of suckers in 13 genera in the family Catostomidae (order Cypriniformes). Like the salmonids, catostomids exhibit higher DNA contents and chromosome numbers than their closest relatives, and have therefore long been presumed to be tetraploid (Uyeno and Smith, 1972) (Fig. 8.5).

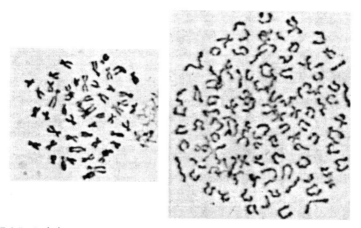

FIGURE 8.5   Early karyotypic evidence for polyploidy in the fish family Catostomidae. At the left, a somatic metaphase spread from the Colorado pikeminnow *Ptychocheilus lucius* (family Cyprinidae, order Cypriniformes) with 2n = 50 chromosomes, and at the right, a similar preparation from the black redhorse *Moxostoma duquesnii* with 2n = 104. As with the salmonids (Fig. 8.4), the catostomids were established as being tetraploid on the basis of evidence such as this. However, unlike the salmonids, the catostomids probably formed by allopolyploidy and are now genetically fully diploidized. Photos were taken at 2500x magnification. From Uyeno and Smith (1972), reproduced by permission (© American Association for the Advancement of Science).

However, unlike salmonids, catostomids appear to be fully functionally diploidized, with only disomic inheritance observed and no multivalents apparent (Uyeno and Smith, 1972; Ferris, 1984), although they retain duplicate gene expression at ~53% of their loci (Ferris and Whitt, 1977). It is believed that they originated about 50 million years ago as an allotetraploid hybrid between two cyprinidlike ancestors (Uyeno and Smith, 1972). Given the uncertainty regarding salmonid origins (25–100 million years ago), it is not possible to say conclusively which family of tetraploids is the older. Interestingly, most catostomids still retain 96–100 chromosomes even though they have progressed much further in the diploidization process than salmonids. This could relate to a difference in genome dynamics between allo- and autopolyploids, but at the very least it shows that polyploidization itself does not automatically induce the extensive chromosomal changes seen in salmonids (Allendorf and Thorgaard, 1984).

SEVERAL SPECIES OF POLYPLOIDS: CYPRINIDAE

Even before he outlined the tetraploid nature of the Salmonidae, Susumu Ohno and his colleagues used comparisons of chromosome numbers, DNA contents, and numbers of gene loci to argue that polyploidy was important in one of the largest fish families in the world, the Cyprinidae (order Cypriniformes) (Ohno *et al.*, 1967; Klose *et al.*, 1968; Wolf *et al.*, 1969). Unlike the salmonid and catostomid cases, however, the cyprinids did not appear to deviate as a whole from the standard teleost karyotype (most have 2n ≈ 50). Instead, only certain species gave signs of being tetraploid (i.e., 2n ≈ 100), including goldfish (*Carassius auratus*) and the common carp (*Cyprinus carpio*) (Ohno *et al.*, 1967; Wolf *et al.*, 1969) (Fig. 8.6). Since that time, more than 50 taxa from several cyprinid subfamilies (mostly Cyprininae) have been identified as polyploids (Buth *et al.*, 1991). This appears to include nearly all species of *Carassius* and the entire genus *Cyprinus* in the subfamily Cyprininae, several members of the genus *Barbus* in the subfamily Barbinae, *Leuciscus alburnoides* in the subfamily Leucisinae, *Phoxinus* species in the subfamily Psilorhynchinae, and some species of *Schizothorax* and *Diptychus* in the subfamily Schizothoracinae (e.g., Zan *et al.*, 1986; Collares-Pereira and Moreira da Costa, 1999; Próspero and Collares-Pereira, 2000; Tsigenopoulos *et al.*, 2002). Moreover, several instances of higher ploidy have been documented among cyprinids, including 6x in *Barbus, Cyprinus,* and *Schizothorax;* 6x and occasionally 8x in *Carassius;* and perhaps even as high as 20x (up to 470 chromosomes) in *Diptychus dipogon* (Zan *et al.*, 1986; Collares-Pereira and Moreira da Costa 1999; Tsigenopoulos *et al.*, 2002).

Some authors have argued that cyprinids may be similar to salmonids in having originally been entirely polyploid, with substantial chromosomal change since their origin. Collares-Pereira and Coelho (1989), for example, suggested that

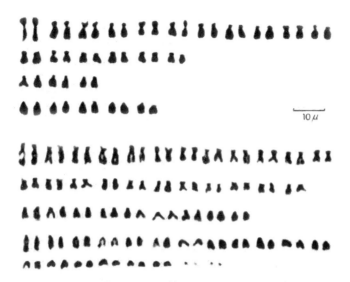

FIGURE 8.6   Early karyotypic evidence for polyploidy in the fish family Cyprinidae. At the top, the karyotype of the diploid barb *Barbus tetrazona* (2n = 50), and at the bottom, that of the carp *Cyprinus carpio* (2n = 104). Unlike the Salmonidae (Fig. 8.4) and Catostomidae (Fig. 8.5), only certain species in the Cyprinidae exhibit polyploidy. From Ohno (1974), reproduced by permission (© Gebrüder Borntraeger, www.schweizbart.de).

2n = 50 is a derived feature in the Cyprinidae, and that the ancestral value for the family was actually 2n = 100. In some ways this is not entirely unrealistic, given that some apparent polyploids have 2n = 100 chromosomes but the same or (rarely) even lower DNA contents than related diploids with 2n = 50 (Collares-Pereira and Moreira da Costa, 1999). This aside, polyploidy is now commonly viewed as the derived feature, occurring only secondarily in certain lineages. Furthermore, polyploid cyprinids show no sign of multivalents at meiosis and most behave genetically as diploids (Schultz, 1980).

In terms of the deeper ancestry of polyploid cyprinids, Zan *et al.* (1986) noted that most of them probably arose sometime in the Oligocene, and that their modern geographic distributions suggested an origin in the region bordering the Qinghai-Xizang plateau in China. Putting these two elements together, Zan *et al.* (1986) proposed that the decrease in temperature engendered by the uplift of the plateau induced polyploid formation in several species—that is, they are autopolyploid. While interesting, this possibility has not found much favor among cyprinid cytogeneticists, and the current view is that cyprinid polyploids have arisen via allopolyploidy (Chenuil *et al.*, 1999; Tsigenopoulos *et al.*, 2002). The lack of multivalents even in fairly recent polyploids and the existence of several unisexual polyploid hybrids both provide evidence in support of allopolyploidy. In the tetraploid carp *Cyprinus carpio*, estimates for the date of the polyploidization event have previously varied from 58 million years ago (Zhang *et al.*, 1995) to less than 16 million years ago (Larhammar and Risinger, 1994), but most recently the time has been placed at as little as 12 million years ago (David *et al.*, 2003). Given this short time, the lack of polysomic inheritance in carp likewise suggests an allotetraploid origin for this species (David *et al.*, 2003).

While there was apparently only one polyploidization event at or near the origin of salmonids, it seems that polyploidy has evolved independently on several occasions in the Cyprinidae, and even sometimes within single genera. As a prime example, the genus *Barbus* contains more than 800 nominal species (although it is taxonomically problematic and almost certainly not monophyletic), which include diploids, tetraploids, and hexaploids (Chenuil *et al.*, 1999; Tsigenopoulos *et al.*, 2002). The polyploids in this case are assumed to have arisen by allopolyploidy, which would not be surprising given the propensity of *Barbus* species to hybridize (Chenuil *et al.*, 1999). A recent phylogenetic study of "*Barbus*" species revealed that tetraploids in southern Africa evolved from local diploids and independently of the Euro-Mediterranean tetraploids (Tsigenopoulos *et al.*, 2002). The origin of the hexaploids of Africa is not resolved, but they do not appear to be derived from either the extant diploids or tetraploids in Africa (Tsigenopoulos *et al.*, 2002). Whatever the ultimate fate of *Barbus* taxonomy, this example provides a good illustration of multiple polyploidization events within a related group of species, as has been commonly found to occur in plants (see Chapter 7).

One of the most complicated polyploid systems known in vertebrates is also found in a cyprinid, namely in the Iberian minnow *Leuciscus alburnoides*. *L. alburnoides* (sometimes also called *Rutilus alburnoides* or *Tropidophoxinellus alburnoides*) exists as a complex of diploid and polyploid forms with varying reproductive modes. This species includes both males and females, but the sex ratio is strongly female-biased. This is because 75–80% of the individuals are triploid females ($2n = 3x = 75$), the remaining 20–25% being a combination of diploid males and females ($2n = 50$), although the proportion varies in different drainages (Alves *et al.*, 2001). The diploid males, diploid females, and triploid females all differ in their respective spatial distributions, which may help to reduce competitive interactions among them (Martins *et al.*, 1998).

Allozyme and genetic analyses have revealed that the entire *L. alburnoides* complex, including both diploids and triploids, arose by hybridization, probably involving the related species *L. pyrenaicus* (Alves *et al.*, 1997, 2001; Carmona *et al.*, 1997). The series of probable hybridization events involved is enormously complex (Fig. 8.7). In diploid hybrid females, reproduction may be by thyletoky or hybridogenesis, depending on the particular combination of parental genomes. Diploid hybrid males may produce either haploid or unreduced (diploid) sperm, with triploid females resulting from either the fertilization of an unreduced egg by a haploid sperm, or of a haploid egg by an unreduced sperm. In some triploid females (again, depending on their parental genome combination), reproduction occurs via a curious process known as "meiotic hybridogenesis" in which the minority genome is excluded, followed by either reductional meiosis to produce haploid eggs or nonreductional meiosis to produce diploid eggs (Alves *et al.*, 1999). Occasionally, fertile tetraploid males and females can result from the fusion of eggs and sperm that are both unreduced (Alves *et al.*, 1999, 2001) (Fig. 8.7).

Hybridization between female finescale dace (*Phoxinus neogaeus*) and male northern redbelly dace (*P. eos*) from the northern Unites States and southern Canada

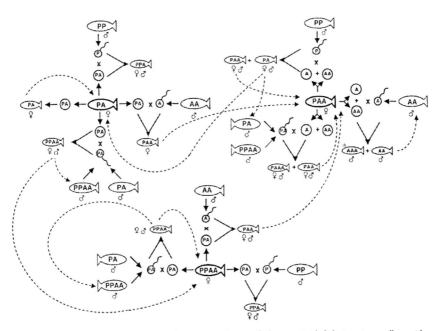

FIGURE 8.7 Relationships among the various forms of the cyprinid fish *Leuciscus alburnoides*, as inferred from experimental crosses. In this figure, adults are represented by fish drawings, eggs by open circles, and sperm by tailed circles. Genomes from *L. pyrenaicus* (P) and another ancestor (A) combine in various ways, including in the production of hybrid polyploids (i.e., those with more than two P and/or A genomes). As noted in the text, several different reproductive modes come into play in this complex. The form marked with an asterisk (AAA ♂) has not been observed in natural populations, but has been generated experimentally. From Alves *et al.* (2001), reproduced by permission (© Kluwer Academic Publishers).

has produced another unisexual "species" of cyprinid, *Phoxinus eos-neogaeus* (Goddard *et al.*, 1998). *P. eos-neogaeus* is remarkable in that it includes diploids, triploids, and diploid–triploid mosaics composed of a mixture of diploid and triploid cells (Dawley and Goddard, 1988; Goddard and Schultz, 1993; Goddard *et al.*, 1998). Several different modes of gametogenesis are known among groups of *P. eos-neogaeus* hybrids, but in at least some diploids and mosaics, reproduction is by a form of "leaky" gynogenesis (see Goddard *et al.*, 1998). One third to one half of the diploid eggs they produce are actually fertilized by sperm from *P. eos* or *P. neogaeus* to produce triploids (i.e., *P. 2 eos-neogaeus* or *P. eos-2 neogaeus*) or mosaics (Goddard *et al.*, 1998). Amazingly, triploid hybrids may reproduce aclonally by maintaining only one copy (usually from *P. eos*) of the parental genomes and producing haploid eggs. When fertilized by sperm from *P. eos*, this generates offspring that are indistinguishable from *P. eos*—except that they may contain the mitochondrial genome from *P. neogaeus* (Goddard and Schultz, 1993; Goddard *et al.*, 1998).

Diploid–triploid, and even diploid–triploid–tetraploid, mosaicism is also found in laboratory populations of the gynogenetic carp *Carassius auratus langsdorfii* (Murayama *et al.*, 1986). As Dawley and Goddard (1988) pointed out (in reference to *P. eos-neogaeus*), such hybrid mosaics could result from one of two processes: "genome loss," whereby a diploid egg is fertilized to produce a triploid in which one copy of the three genomes is later lost from certain cells, or "delayed fusion," in which a sperm enters a diploid egg but remains quiescent and then later merges with a cleavage cell, such that only a subset of the cells ever become triploid. Whatever the cause, this provides an interesting case in which diploidy and polyploidy must be considered the endpoints of a continuum, rather than discrete categories.

Special Cases: Poeciliidae

The family Poeciliidae (order Cyprinodontiformes), which consists primarily of the live-bearing mollies and platties, contains two examples of allopolyploidy and unisexuality that are particularly interesting and which have been the subject of detailed study for decades.

Poecilia

The Amazon molly, *Poecilia formosa*, is not so named because it is native to the Amazon, but rather because it consists almost entirely of females. This species has long been recognized as having originated by hybridization and was the first unisexual vertebrate discovered (Hubbs and Hubbs, 1932). Molecular studies suggest that the relevant event most likely occurred at least 100,000 years ago between a female shortfin molly (*P. mexicana limantouri*) and a male sailfin molly (a member of the *P. latipinna/P. velifera* complex) (Sola *et al.*, 1992; Schartl *et al.*, 1995), although it was initially proposed that one of the progenitors was the black molly, *P. sphenops* (Hubbs and Hubbs, 1932). Throughout its range from southern Texas to northeastern Mexico, *P. formosa* is sympatric with *P. latipinna*, *P. mexicana*, or both.

Gamete production in *P. formosa* is apomictic but reproduction is gynogenetic, such that all descendants of a unisexual female normally will be genetically identical clones. The lack of genetic input by males of the bisexual species is emphasized by the fact that aneuploid sperm from rare triploid males and, in captivity, even males of different related genera (e.g., *Gambusia*, *Lebistes*, and *Limia*) can serve the purpose of triggering gynogenetic egg development in the unisexuals (Hubbs, 1964; Schultz, 1989). This "sexual parasitism" is thought to be facilitated by the hybrid origin of the unisexuals; sexual males of the progenitor species are unable to distinguish hybrid females from those of their own species (Dries, 2003).

P. formosa is predominantly a diploid unisexual species with the same chromosome complement ($2n = 46$) as its parental species, *P. latipinna* and *P. mexicana* (Sola *et al.*, 1992). Occasional triploids ($3x = 69$) have been noted in laboratory populations (Schultz and Kallman, 1968; Nanda *et al.*, 1995), and triploidy has been found to be fairly common in nature (Prehn and Rasch, 1969; Balsano et al., 1972). Laboratory triploids have been sterile, but natural triploids are fertile and continue to produce triploid female offspring (Schlupp *et al.*, 1998). In this case, the initial triploids are formed when a diploid *P. formosa* egg is fertilized by a normal sperm from *P. mexicana* or *P. latipinna* (or, in the lab, *P. sphenops*) following a breakdown of the sperm exclusion mechanism in a process called "paternal introgression." Insofar as there is some question about whether a gynogenetic hybrid including triploids of different origin really exists as a separate "species," these triploids might be more accurately designated as *P. 2 mexicana-latipinna* or *P. mexicana-2 latipinna* (Schultz, 1980), assuming that *P. latipinna* and *P. mexicana* are the progenitor species and depending on which of these was the sympatric host that contributed the additional genome. Diploid–triploid mosaicism may also occur in *P. formosa*, although it has so far been detected in only one individual (Lamatsch *et al.*, 2002).

Poeciliopsis

The second poeciliid of interest, *Poeciliopsis monacha-lucida*, is also a native of northern Mexico of hybrid origin (♀ *P. monacha* × ♂ *P. lucida*), and consists entirely of females. As with *Poecilia formosa*, most individuals of *Poeciliopsis monacha-lucida* are diploid ($2n = 48$), but there is also a relatively common triploid cytotype ($3x = 72$). There is far less confusion regarding the progenitors involved in this case versus that of *Poecilia formosa*, largely because *P. monacha-lucida* can be synthesized by matings in the lab (Schultz, 1973, 1980). Unlike *Poecilia formosa*, this hybrid *Poeciliopsis* has not been given a distinct species name.

Diploid female *P. monacha-lucida* are not gynogenetic, they are hybridogenetic hemiclones, and they produce ova containing only the maternal *monacha* genome. Fertilization by sperm does take place in this hybrid, but the paternal *lucida* genome is not transmitted in the next set of eggs. Mechanistically, this occurs during the last oogonial mitosis by an interesting process of "premeiotic exclusion," which involves the formation of a unipolar spindle that collects only the maternally derived (i.e., *monacha*) chromosomes (Cimino, 1972). As such, each generation represents a new hybridization event in which the paternal *lucida* genome is restored (Ohno, 1974; Dawley *et al.*, 1997). *P. lucida* males derive no genetic benefit from this union, but it is crucial for the maintenance of *P. monacha-lucida* because a mating with a male *P. monacha* would simply produce ordinary *P. monacha* offspring (Ohno, 1974).

Triploidy in *P. monacha-lucida*, as in *Poecilia formosa*, probably arises by the fusion of an unreduced egg from a diploid unisexual hybrid and a haploid sperm from a related bisexual male (Cimino, 1974). The triploid hybrids in *Poeciliopsis*, unlike the diploid hybridogens, are automictic gynogens, and their oogonia undergo endomitosis to reach the hexaploid level prior to a conventional meiosis that produces triploid eggs (Cimino, 1972). This presents an interesting mix of causal factors, with hybridization first producing a diploid hybridogen, and then subsequent hybridization producing a triploid that is gynogenetic. In other words, unisexuality of one form (hybridogenesis) allowed the evolution of polyploidy, which in turn promoted the evolution of a different form of unisexuality (gynogenesis).

Triploidy in *P. monaca-lucida* can occur by the addition of an extra genome from either of the parental species (i.e., *P. 2 monacha-lucida* or *P. monacha-2 lucida*) (Cimino, 1974; Schultz, 1980). Moreover, *P. monaca* and *P. lucida* are not the only species that can hybridize in *Poeciliopsis*, and other documented combinations include the diploids *P. monaca-latidens* and *P. monacha-occidentalis* and the tri-parental triploid *P. monacha-viriosa-lucida* (Schultz, 1980). Diploid–triploid mosaicism has not been observed in *Poeciliopsis* (Dawley *et al.*, 1997), but in the laboratory a triploid female *Poeciliopsis* was found to produce a diploid male off-spring (Cimino and Schultz, 1970). Triploid hybrids may also diploidize, which appears to be in progress in one cytotype of *P. monacha-lucida* with an apparently diploid chromosome complement but showing an elevated DNA content and signs of containing portions of three different genomes (Vrijenhoek and Schultz, 1974). Nevertheless, most hybrids maintain the amount of DNA predicted from the sum of the two parental genomes, suggesting that DNA has not been lost in these cases (Cimino, 1974; Dawley *et al.*, 1997).

Polyploidization has been noted to have some interesting effects in *Poeciliopsis*. Cell sizes are larger in the triploids (Cimino, 1973), but diploids grow to a larger adult body size (Schultz, 1982). And, whereas diploid hybrids also grow faster than triploids, both types of hybrids outgrow the sexual species *P. monacha* and *P. lucida* (Schultz, 1982). Cold tolerance in the triploids is similar to that of diploid hybrids, but tolerance for heat stress is lower in the triploids (Schultz, 1982). Diploids and triploids also differ in their spatial distributions, with triploids preferring areas of higher velocity current (Schenck and Vrijenhoek, 1989), as is also the case in the cyprinid *Leuciscus alburnoides* (Martins *et al.*, 1998).

MISCELLANEOUS POLYPLOID FISHES

South American armored catfishes of the genus *Corydoras* (order Siluriformes, family Callichthyidae) vary greatly in their chromosome numbers and DNA contents, and can generally be classified into five different groups on this basis, some of which may have formed via a combination of polyploidy and other chromosomal rearrangements (Oliveira *et al.*, 1992, 1993). In one species this is particularly clear, because *C. aeneus* is known to exist as a complex of diploid and tetraploid cytotypes (Turner *et al.*, 1992). It has not been determined whether *C. aeneus* is an example of allo- or autotetraploidy, but it is notable in any case that this species may have XY sex chromosomes (Turner *et al.*, 1992), which, once again, introduced no barrier to polyploidization.

Diploid–triploid–tetraploid complexes can also be found in the spined loach genus *Cobitis* (order Cypriniformes, family Cobitidae), which includes both bisexual and gynogenetic forms. Triploids are unisexual in this complex, but both unisexual and bisexual tetraploid forms are known (Vasil'ev *et al.*, 1999). The triploids in this case are of allopolyploid origin, one being *C. 2 sinensis-longicorpus*, and another containing one genome from *C. taenia* and two from an as-yet-unidentified species (Vasil'ev *et al.*, 1999). The tetraploids are also produced by hybridization as a cross between diploid bisexual males and triploid gynogens. As it turns out, the resulting reproductive system of the offspring depends on whether *C. melanoleuca* (bisexual) or *C. taenia* (gynogenetic) provided the paternal genome (Vasil'ev *et al.*, 1999). A recent phylogenetic study of hybrids in the *C. elongatoides-taenia* complex revealed reciprocal hybridizations between parental species, meaning that unisexuality evolved more than once in both groups (Janko *et al.*, 2003). Triploids, in this case, also appear to have arisen from the fertilization of diploid eggs produced by the gynogenetic hybrids (Janko *et al.*, 2003).

The mud loach *Misgurnus fossilis* (family Cobitidae) appears to be a tetraploid ($4x = 100$) (Raicu and Taisescu, 1972), and some clown loaches of the genus *Botia* likewise show evidence for polyploidy on the basis of elevated chromosome numbers. Also, whereas most Japanese populations of the loach *Misgurnus anguillicaudatus* are diploid ($2n = 50$), fully tetraploid ($4x = 100$, and not genetically rediploidized) forms are known from the Hubei Province of China (Zhang *et al.*, 1998). Triploid *M. anguillicaudatus* can be produced by mating tetraploids with diploids, resulting in both sterile males and fertile females. Interestingly, female triploids lay both small haploid and large triploid eggs simultaneously (Zhang et al., 1998). Presumably, the triploid eggs are produced via a premeiotic endomitotic replication (to 6x) in the oogonia, whereas the haploid eggs form without this prior duplication and with the extra set of univalents eliminated during meiosis (Zhang et al., 1998). Pentaploid M. anguillicaudatus have also been induced experimentally by inhibiting the release of the second polar body in triploids using hydrostatic pressure, and both triploids and pentaploids fertilized by normal diploid males produce viable progeny (Arai et al., 1993; Matsubara et al., 1995). Gynogenetic tetraploids can also be formed experimentally by fertilizing triploid females with UV-irradiated spermatozoa (Arai et al., 1993).

A few other cases of natural polyploidy in teleosts deserve mention. For example, natural allotriploidy has been found, albeit at relatively low frequency, in gynogenetic hybrid silversides of the Menidia clarkhubbsi complex (order Atheriniformes, family Atherinidae) along the Texas coast (Echelle et al., 1988). Single natural spontaneous triploid individuals have been reported in the stone loach Neomacheilus barbatulus (order Cypriniformers, family Balitoridae) (Collares-Pereira and Madeira, 1995), the banded knifefish Gymnotus carapo (order Gymnotiformes, family Gymnotidae) (Fernandes-Matioli et al., 1998), and the characin Characidium gomesi (order Characiformes, family Crenuchidae) (Centofante et al., 2001). Spontaneous triploidy has also been reported in a few specimens of Astyanax eigenmanniorum and A. scabripinnis (order Characiformes, family Characidae) (Fauaz et al., 1994; Maistro et al., 1994) and in the aimara Hoplerythrinus unitaeniatus (order Characiformes, family Erythrinidae) (Giuliano-Caetano and Bertollo, 1990). In one rather intriguing recent case, the stinging catfish Heteropneustes fossilis (order Siluriformes, family Heteropneustidae) from India was reported to include naturally occurring haploids, triploids, tetraploids, and a few aneuploid levels in between (Pandian and Koteeswaran, 1999). Hybrids between the killifish Fundulus heteroclitus and F. diaphanus are usually diploid, but rare allotriploids also occur (Dawley, 1992). Finally, though based only on estimates of nuclear DNA contents, Hardie and Hebert (2003) highlighted the possible existence of triploidy in the one-spot squeaker catfish Synodontis notatus (order Siluriformes, family Mochokidae), and tetraploidy in the green jobfish Aprion virescens (order Perciformes, family Lutjanidae). Almost certainly, many additional examples of polyploid fishes await discovery.

Artificially induced triploidy has become very common in teleost fishes since the early work of Swarup (1959) and others, and has found particularly useful applications in aquaculture, where sterile all-female stocks may be preferred for a variety of reasons (Ihssen et al., 1990; Felip et al., 2001). This can be accomplished by heat, cold, or high-pressure shock, or through artificial hybridization (reviewed by Ihssen et al., 1990; Felip et al., 2001). Cell sizes are larger in such triploids (see Garcia-Abiado *et al.*, 1999, and references therein), but it seems that cell number is generally reduced so that body size may not be affected (Felip *et al.*, 2001). Growth rates may be the same as, or higher or lower than, those of diploids, depending on the species (Ihssen *et al.*, 1990; Felip *et al.*, 2001). While triploidization by artificial means may have other implications for the biology of these fishes (see Benfey, 1999), the fact that they are usually sterile means they are of no evolutionary significance and thus are of only peripheral interest to this chapter.

AMPHIBIANS

It has long been known that artificial polyploids can be produced relatively easily in several amphibians by various means, and that spontaneous triploids occur at nontrivial frequencies in many species (reviewed in Fankhauser, 1945; Kawamura, 1984; Lowcock and Licht, 1990; Borkin *et al.*, 1996). In fact, the first reported case of vertebrate polyploidy was in the frog *Rana esculenta* (Hertwig and Hertwig, 1920). However, in general terms polyploidy does not appear to have been a predominant force in amphibian evolution, even though numerous cases are known in this group. Although there is a significant positive correlation between genome size and chromosome number in amphibians as a class ($r = 0.26$, $P < 0.0001$, $n = 313$), this does not hold within either the frogs (order Anura; $P > 0.36$) or salamanders (order Caudata; $P > 0.74$) taken separately. In fact, the majority of frogs have only $2n = 22$ to 26 chromosomes, and most salamanders have $2n = 22$ to 28 chromosomes, so ancestral polyploidy is almost certainly not a shared feature within or between these groups (King, 1990). Indeed, it has been claimed several times that there are no known cases of exclusively polyploid families, genera, or even species assemblages in the Amphibia (Bogart, 1980; Kawamura, 1984; King, 1990). While this does seem to be generally true, two possible exceptions to this rule will be discussed in this chapter: the frog genus *Xenopus* and the salamander family Sirenidae.

Although salamanders invariably have larger genomes than frogs (see Chapter 1), polyploidy is much more common in the latter group. Polyploid frogs are almost exclusively bisexual, whereas unisexual reproduction is more common among polyploid salamanders (Bogart, 1980). Unfortunately, very little is known about caecilians (legless amphibians in the order Apoda), in terms of either genome size or ploidy. It has sometimes been claimed that, among frogs, only members of the primitive family Pipidae are allopolyploids (e.g., Beçak and Beçak, 1998), but evidence for hybrid origins has increased along with new findings of tetraploidy in various groups (e.g., Channing and Bogart, 1996; Martino and Sinsch, 2002; Vences *et al.*, 2002).

DIPLOID–POLYPLOID SPECIES PAIRS IN FROGS

To date, roughly 30 polyploid species of frogs representing eight families have been identified. The total number of cases in nature is difficult to estimate, however, because in many instances diploid–polyploid species pairs are cryptic and therefore difficult to identify (Bogart, 1980). Examples of such pairs include *Bufo regularis* (2x) + *B. amarae* (4x) and *Bufo viridis* (2x) + *B. danatensis* (4x) in the Bufonidae (Bogart, 1980; Tandy *et al.*, 1982), and *Phyllomedusa distincta* (2x) + *P. tetraploidea* (4x) in the Hylidae (Haddad *et al.*, 1994), and there are many more. In some cases, diploid and tetraploid "populations" of certain "species" have been reported, as in *Phyllomedusa burmeisteri* in the Hylidae from South America (Batistic *et al.*, 1975) and *Hoplobatrachus* (formerly *Dicroglossus*) *occipitalis* in the Ranidae from Africa (Bogart and Tandy, 1976). It could be, however, that these actually represent cryptic diploid–tetraploid species pairs, as was found to be the case with the African species *Tomopterna delalandii* (2x) and *T. tandyi* (4x) in the Ranidae, which were first described as two populations of what was then called *Pyxicephalus delalandii* (Bogart and Tandy, 1976). It is now recognized that *T. tandyi* arose as an allotetraploid hybrid of *T. delalandii* × *T. cryptotis* as recently as 1.5 million years ago (Channing and Bogart, 1996). Because most of these polyploids (with the exception of *T. tandyi*) display at least some multivalents during meiosis and have morphologically and karyotypically very similar diploid counterparts, they are usually considered to be recently derived autopolyploids (see Tymowska, 1991). However, chromosome morphology may be quite conserved among closely related species within a group, and convergence in frog body forms is common. Thus it is difficult to conclusively assign autotetraploid status without additional evidence.

Notwithstanding the growing list of diploid–tetraploid species pairs, it remains the case that diploid partner species may not always be readily identified. For example, *Scaphiophryne gottlebei* in the family Microhylidae is likely an allotetraploid (the only one known from Madagascar), but is an ancient species whose diploid partner (if any) has not been identified and may no longer exist (Vences *et al.*, 2002). Other less well studied polyploid frogs include the South American leptodactylid species *Pleurodema bibroni* and *P. kriegi* (both 4x) (Barrio and Rinaldi de Chieri, 1970) and *Ceratophrys aurita* (formerly *C. dorsata*) and *C. ornata* (both 8x) (Beçak *et al.*, 1967a,b; Bogart, 1980), and the African arthroleptid (some say astylosternid) *Astylosternus diadematus* (4x) (Bogart and Tandy, 1981). Another leptodactylid, *Eleutherodactylus binotatus*, has the same number of chromosomes as its diploid congeners (2n = 22), but four times the DNA content and has been proposed to be an ancient octoploid that is now fully diploidized at the karyotypic level (Beçak and Beçak, 1974, 1998).

The fact that many of the diploid–tetraploid species pairs of frogs appear as cryptic species means that they differ very little in outward morphology. Despite the influence of polyploidy on cell size, it is often found that polyploid amphibians reduce their total cell number such that they achieve the same body size as diploids (Fankhauser, 1955; Bogart, 1980). However, because they can genuinely be classified as different species, there must exist some strong barriers to reproduction between them. It is notable in this regard that with one possible exception (Haddad *et al.*, 1994), diploids and polyploids tend to differ in the acoustical properties of their mating calls (e.g., Bogart, 1980; Channing and Bogart, 1996; Stöck, 1998; Martino and Sinsch, 2002). This could relate to a difference in the morphology and/or action of the vocalization organs, perhaps caused by changes in cell size (Bogart, 1980; McLister *et al.*, 1995; Keller and Gerhardt, 2001) or to alterations in neural firing rates (Brenowitz *et al.*, 1985). However, call parameters do not change in a consistent manner with polyploidy in different species, and in at least one study of diploids versus polyploids, call muscle cell characteristics did not appear to be universally influenced by increased ploidy (Tito and Boyd, 1999). It is therefore also possible that strong selection against matings across ploidy types (because triploid hybrids are unviable or sterile) could secondarily generate these differences (Castellano *et al.*, 2002). Whatever the underlying mechanism, it is apparent that this behavioral consequence has been important in the maintenance of newly formed polyploid species (Bogart, 1980; Otto and Whitton, 2000).

Following from the general outline just given, it is worthwhile to consider some specific examples of polyploid frogs from the families Leptodactylidae, Hylidae, Myobatrachidae, Ranidae, and Pipidae that are of particular interest.

Hyla versicolor (Hylidae)

The North American treefrogs *Hyla versicolor* and *H. chrysoscelis* represent a cryptic species pair that was initially diagnosed acoustically on the basis of differences in pulse rates of mating calls (Blair, 1958). Early experimental studies showed that the two species could generate viable hybrids, but that these individuals were completely sterile (Johnson, 1959, 1963, 1966). These observations were later explained by the fact that *H. versicolor* ($4x = 48$, ~ 19 pg per nucleus) is a recently derived tetraploid from the diploid *H. chrysoscelis* ($2n = 24$, ~ 9 pg/N) (Wasserman, 1970; Bogart and Wasserman, 1972).

Based on the ease of inducing diploid eggs in *H. chrysoscelis*, Bogart and Wasserman (1972) suggested that *H. versicolor* could have originated by autopolyploidy when a diploid ovum from a triploid female was fertilized by a normal haploid sperm. They also postulated that newly arising autotetraploids could have been immediately reproductively isolated from their diploid progenitors because the ratio of difference in pulse rates of the mating calls is very similar to the differences in cell size of diploids and tetraploids (~ 1.7x). However, the story may not be quite that simple. The complication arises from the fact that *H. chrysoscelis* is divided into at least two distinct lineages that differ geographically, genetically, and behaviorally, but which cannot be distinguished morphologically (Ptacek *et al.*, 1994), and may therefore in fact represent cryptic species (Bogart, 1980). Of course, even if *H. chrysoscelis* is actually a composite of two cryptic species, *H. versicolor* could still be an autopolyploid if it arose from only one of these, or if it had been generated prior to the divergence of the *H. chrysoscelis* lineages. On the other hand, Maxson *et al.* (1977) suggested that immunological comparison data indicated an origin of *H. versicolor* by hybridization between the two lineages of *H. chrysoscelis* approximately 4 million years ago (although this timing remains controversial) (see Ralin *et al.*, 1983).

Using protein electrophoretic data, Ralin *et al.* (1983) proposed that the tetraploidization event leading to *H. versicolor* occurred only once, coinciding with the onset of the Illinoian glacial stage 575,000 years ago or the beginning of the Sangamon interglacial period 375,000 years ago. Romano *et al.* (1987) reported notable similarities in allozyme frequencies between populations of diploids and tetraploids living sympatrically, but concluded that this was caused by similar selection pressures in the shared environment. More recently, Ptacek *et al.* (1994) used a detailed comparison of mitochondrial DNA sequences to show that *H. versicolor* had, in fact, arisen at least three times independently, and that the two lineages of *H. chrysoscelis* had each served as the progenitor to one of the descendant *H. versicolor* lineages. A potential "hybrid" origin was suggested by the observation that these mitochondrially defined tetraploid lineages were more similar to diploids from the opposite part of the range than to their currently sympatric diploids (Ptacek *et al.*, 1994), whereas nuclear allozyme data showed the opposite pattern (Mable, 1996).

The origin of the third lineage of *H. versicolor* remains uncertain because it is not clearly allied with either of the *H. chrysoscelis* lineages (Ptacek *et al.*, 1994); these tetraploid "orphans" are found in northern regions and mostly occur beyond the current limits of the range of the diploids (Mable, 1996). It is therefore possible that this "lineage" descended from a third line of *H. chrysoscelis* which is now extinct, or it could be that it formed as a hybrid of the other two lineages. Although multivalents have been observed in *H. versicolor* for many chromosomes (Bogart and Wasserman, 1972), it is not yet clear whether any or all three of the polyploid lineages are derived from just one lineage each of *H. chrysoscelis* or from hybridization between lineages (Ptacek *et al.*, 1994). As such, the remarkable possibility exists that, as currently defined, "*Hyla versicolor*" is a polyphyletic "species" that has arisen by both strict autopolyploidy and "intervarietal" autopolyploidy (segmental allopolyploidy) in the sense used in plants, or possibly even by allopolyploidy if the two lineages are actually cryptic species. As if this situation were not complex enough, it also seems that

H. versicolor occasionally hybridizes accidentally with *H. chrysoscelis* to form sterile triploids (Gerhardt *et al.,* 1994). Based on comparisons of nuclear gene frequencies in sympatric and allopatric tetraploids from the same mitochondrial lineages, it has been suggested that the independently arisen lineages of *H. versicolor* may also hybridize with one another in areas where they cohabitate (Espinoza and Noor, 2002). However, the data presented cannot distinguish this hypothesis from multiple paternal ancestors and rely on the assumption of a divergence time of four million years ago (which has not been conclusively demonstrated). Artificial polyploid hybrids have also been created in this genus (but crosses are only successful when the tetraploid is the female parent). For example, vigorous triploid hybrids have been produced by mating *H. versicolor* with *H. arborea,* with subsequent 3x, 4x, and 5x backcrosses to the tetraploid parent (Mable and Bogart, 1995), and hexaploids have been produced by mating *H. versicolor* with the lowland burrowing treefrog, *Pternohyla fodiens* (Green, 1980).

Because of their extensive sympatry, recent divergence, and roughly 2-fold difference in DNA content and cell size, *H. versicolor* and *H. chrysoscelis* have been the subject of several studies to evaluate possible links between polyploidy and organismal features. Thus, possible interspecific variation in body mass, water content, and dehydration tolerance (Ralin, 1981); oxygen consumption rates and postexercise lactic acid concentrations (Kamel *et al.,* 1985); larval life-history traits (Ptacek, 1996); and freeze tolerance (Irwin and Lee, 2003) have all been examined. Intriguingly, for the most part, differences in these parameters between the species have not been found. This suggests that, although polyploidization may be important in the origin of new (cryptic) species in frogs, it has probably contributed very little to morphological evolution.

Odontophrynus americanus (Leptodactylidae)

The South American genus *Odontophrynus* contains fewer than 10 species, most of which are diploids with a basic chromosome number of 2n = 22. However, it was noted early on that *O. americanus* has 4x = 44, representing the very first naturally occurring bisexual polyploid animal found (Beçak *et al.,* 1966). The situation in *O. americanus* was recognized more than 30 years ago by Bogart and Wasserman (1972) to represent the neotropical counterpart of the *Hyla veriscolor* system, complete with a cryptic diploid partner. However, until very recently, the diploids and tetraploids had not been considered separate species, and only after careful analysis of morphometric, allozyme, and acoustical differences has a separate species designation been provided for the diploids, now called *Odontophrynus cordobae* (Martino and Sinsch, 2002).

The chromosomes of *O. americanus* can be classified into 11 groups of four homologs with up to 11 quadrivalents forming during meiosis, suggesting that this is an autopolyploid that has not yet rediploidized (Beçak *et al.,* 1966;

Macgregor, 1993). However, Beçak and Beçak (1998) have noted several significant chromosomal aberrations that suggest a certain degree of genomic instability in this species. Interestingly, although the haploid nuclear DNA content of *O. americanus* is roughly twice that of most of its diploid congeners, there are other frogs with 2n = 22 that have much larger genome sizes (Gregory, 2001b). Evidently, the genome that was duplicated in this species was at the lower end of the frog genome size distribution.

Under laboratory conditions, triploid hybrids between *O. cultripes* (2x) × *O. americanus* (4x) can be formed, and include an even ratio of male and female progeny (Beçak and Beçak, 1970). In this regard, Beçak and Beçak (1970) suggested that hybridization to produce fertile triploids that then mate with each other may be an important step in the generation of higher ploidy levels. Such a process would involve an interesting mixture of auto- and allopolyploidization mechanisms.

Neobatrachus spp. (Myobatrachidae)

The first reports of polyploidy in amphibians from Australia were by Mahony and Robinson (1980), who showed by karyotypic analysis that two species of burrowing frogs in the genus *Neobatrachus* were tetraploids with 4x = 48, whereas four others were all diploids with 2x = 24. The DNA contents and erythrocyte nucleus and cell sizes of the presumed polyploids were also found to be proportionately larger than those of their diploid counterparts (Mahony and Robinson, 1980). Taxonomic revisions have shown that *Neobatrachus* contains four bisexual tetraploid species (known as *N. sudelli, N. centralis, N. kunapalari*, and *N. aquilonius*) (King, 1990; Roberts, 1997) as well as six diploids.

As with *Odontophrynus americanus*, polyploid *Neobatrachus* species exhibit quadrivalents during meiosis (Mahony and Robinson, 1980), which suggests an origin by autopolyploidy but which could also represent hybridization between very closely related species (Roberts, 1997). Like in the other polyploid frogs, mating calls are different between polyploids and diploids in *Neobatrachus* (Roberts, 1997). There is sympatry between some polyploids and diploids, and occasional hybridization occurs (Mable and Roberts, 1997). Attempts to resolve the origins of the polyploids using immunological or electrophoretic techniques have been unsuccessful, and the exact nature and timing of the events remains unclear, but Mable and Roberts (1997) used divergence in mitochondrial sequences to show that at least two independent polyploid origins had occurred in this genus.

Xenopus spp. (Pipidae)

African clawed frogs of the genus *Xenopus* display some of the most variable ploidy levels seen among vertebrates (King, 1990). Traditionally, *Xenopus* species with 20 or 36 chromosomes had been considered as diploids, those with 72 chromosomes as tetraploids, and the 108 chromosomes of *X. ruwenzoriensis* as a hexaploid (e.g., Bogart, 1980). As King (1990) pointed out, this would require some rather unlikely chromosomal changes, making it much more parsi-monious to assume that 36 chromosomes constitute a tetraploid complement. Indeed, it is now thought that *X. laevis* (for example) is, in fact, tetraploid (Kobel and Du Pasquier, 1986; Cannatella and de Sá, 1993), making *X. ruwenzoriensis* a dodecaploid (Kobel and Du Pasquier, 1986; Tymowska, 1991). For this reason, and despite decades of work on *X. laevis*, research in developmental genetics is shifting to the smaller, faster-developing, and genetically much more convenient diploid *Xenopus tropicalis* (Vogel, 1999). Likewise, *X. tropicalis* has been selected over *X. laevis* as the subject of a complete genome sequencing project for the same reason.

In fact, *X. tropicalis* is the only known diploid in the genus (with 2n = 20), but even here there is at least one associated cryptic tetraploid species, *X. epitropicalis* (4x = 40), which differs from *X. tropicalis* primarily with regard to mating call character-istics and some additional karyotypic features (Loumont, 1983; Tymowska, 1991). According to many authors, the 2n = 20 karyotype of *X. tropicalis* forms the basis of the entire polyploid series in the genus (e.g., Tymowska, 1991). However, there is a clear discrepancy between diploid *X. tropicalis* (2n = 20), tetraploid *X. epitropicalis* (4x = 40), and the other *Xenopus* polyploids (all 4x = 36 or multiples thereof). Indeed, the karyotypes of *X. tropicalis/X. epitropicalis* and those of other species like *X. laevis* are so different as to make direct comparisons difficult (Tymowska, 1991).

Interestingly, systematists now consider *X. tropicalis* and *X. epitropicalis* to warrant placement into their own genus (*Silurana*) on the basis of morphological characteristics (e.g., Cannatella and Trueb, 1988; de Sá and Hillis, 1990; Cannatella and de Sá, 1993; Rödel, 2000; Frost, 2004). Molecular data have either supported the separation into two genera (but not two subfamilies, as some mor-phologists suggested) (de Sá and Hillis, 1990) or have favored placement into differ-ent subgenera within *Xenopus* (e.g., Kobel *et al.*, 1998). In either case, the (sub)genus *Xenopus* itself would therefore consist entirely of polyploids, potentially making it the only group of amphibians not directly linked to related diploids. Importantly, such a distinction would clarify the mismatch between 4x = 36 and 4x = 40 tetraploids, and implies that the (extinct) diploid ancestor of the former had a chromosome complement of 2n = 18, and not 2n = 20 as in *X. (S.) tropicalis*.

The lack of multivalents during meiosis (at least in the tetraploids), an absence of closely related diploids, observations of mixed interbreeding, and the relatively common generation of polyploid oocytes in female hybrids all point to allopoly-ploidy in *Xenopus* species (Tymowska, 1991; Kobel, 1996). It is believed that the first duplication/hybridization event occurred at least 30 million years ago, with more recent events perhaps having been stimulated by environmental changes during the Pleistocene (Tymowska, 1991). Since then, most of the polyploids have become functional diploids in terms of gene expression patterns. The dodecaploid *X. ruwenzoriensis* presents a notable exception to this pattern, having arisen more recently (Kobel, 1996; Sammut *et al.*, 2002).

Xenopus are one of the few groups of polyploid frogs with a well-characterized sex-determination system. Researchers have found that the genetic ZZ/ZW sex determination system in this genus (females are heterozygous for dominant female-determining factors and for recessive male-determining factors, although no morphologically distinct sex chromosomes have been found) is disrupted by polyploidy (reviewed in Kobel and Du Pasquier, 1986). Specifically, W dominance is weakened with increasing dosages of male-determining factors, and genetic sex determination may give way to environmental sex determination at ploidies where the dosages are equal. This example emphasizes that much more flexibility may exist in sex determination systems in animals than assumed in the models of Muller (1925) and Orr (1990).

Not only has polyploidy been important in the evolution of the genus *Xenopus* itself, but it also appears to have played a major role in the evolution of species associated with it. *Xenopus laevis* is known to serve as host to at least 29 different genera of metazoan and protozoan parasites, many of which are strictly specific to it and its fellow *Xenopus* species (Tinsley and Jackson, 1998). Most generally, it is very interesting to note that the parasites harbored by *X. tropicalis* and *X. epitropicalis* differ from those found in association with other *Xenopus* species (Tinsley and Jackson, 1998). Even more intriguingly, it appears that speciation in monogenean flatworms of the genus *Protopolystoma* and other parasites has closely followed the speciation by polyploidy of its *Xenopus* hosts (Tinsley and Jackson, 1998; Jackson and Tinsley, 2001). The coevolutionary influence of polyploidy goes a step further in this regard, because it appears that allopolyploidy may confer some resistance to certain species of these parasites (Jackson and Tinsley, 2003).

Rana esculenta (Ranidae)

One of the most unusual forms of polyploidy in amphibians is found in the water frog *Rana esculenta* from Europe and western Asia. As with the fishes *Poeciliopsis* and *Phoxinus*, *Rana esculenta* polyploids are formed by hybridogenesis, in this case by hybridization between the bisexual species *Rana lessonae* and *R. ridibunda* (King, 1990). Unlike the fish cases, however, both male and female hybridogens (diploids and triploids) are produced in *R. esculenta*.

In somatic tissues, *R. esculenta* represents a mosaic of *lessonae* and *ridibunda* genomes. However, in the germline, one of the parental genomes is eliminated before premeiotic DNA synthesis, with hybridization occurring anew in each generation by a backcross with the parental species whose genome was eliminated. When *R. esculenta* cohabitates with *R. lessonae,* the *lessonae* genome is eliminated, and when *R. ridibunda* is the sympatric species, the *ridibunda* genome is eliminated. Triploids contain two copies of one genome and one of the other, and in this case it is the genome in the minority that is eliminated (Vinogradov *et al.,* 1990).

The maintenance of this sexually parasitic system is thought to be mediated by frequency-dependent assortative mating, which explains why sexual *R. lessonae* are not genetically swamped by the hybrids (Som *et al.*, 2000).

Bufo pseudoraddei baturae (Bufonidae)

In the Bufonidae, tetraploidy is found in *Bufo amarae* and *B. danatensis*, which, like many other anurans, are paired with related diploid species (Bogart, 1980; Tandy *et al.*, 1982). More recently, triploidy has been found in *B. pseudoraddei baturae*, a newly described member of the *Bufo viridis* complex from an isolated region in the Karakoram mountain range in central Asia (Stöck *et al.*, 2002). What sets this case apart from all other vertebrate examples is that this toad exhibits odd-numbered ploidy but does not reproduce by thyletoky, gynogenesis, or hybridogenesis—it is both triploid and fully bisexual, something that has long been considered impossible (Stöck *et al.*, 2002). Both 3x males and 3x females are fertile, and their matings produce fertile 3x offspring. In males, spermatogenesis involves the elimination of one genome (3x to 2x) and recombination of the other two during meiosis to produce haploid sperm (2x to 1x). Females produce diploid eggs, probably by duplicating only one genome (3x to 4x), or perhaps by eliminating one genome (3x to 2x) and then undergoing a mitotic doubling (2x to 4x), prior to meiosis (4x to 2x) (Stöck *et al.*, 2002). As Stöck *et al.* (2002) pointed out, "the fact that euploid gametes can be produced regularly on the basis of uneven ploidy in the soma encourages speculation that other bisexually reproducing all-triploid species exist."

As this and the previous examples show, polyploidy has arisen many times and by various routes in frogs. This is not the case with salamanders, however, in which polyploidy is confined to only a few families.

POLYPLOIDY IN SALAMANDERS: AMBYSTOMATIDAE

The family Ambystomatidae (mole salamanders) contains about 30 species exclusive to North America, all of them usually classified into the single genus *Ambystoma* (AmphibiaWeb, 2004; but see Petranka, 1998). This includes the Mexican axolotl (*A. mexicanum*), which has long been an important model organism in developmental biology and in which triploidy is frequently induced as part of this research (Tank *et al.*, 1987). Spontaneous triploidy is known to occur at low frequency in a number of species of urodeles but naturally occurring triploids have only rarely been found (reviewed in Fankhauser, 1945; Lowcock and Licht, 1990), except in the *Ambystoma laterale-jeffersonianum* complex of "unisexual" salamanders.

Forty years ago, Uzzell (1963, 1964) reported two groups of naturally occurring unisexual triploid female salamanders that he considered to represent the species *A. platineum* and *A. tremblayi* (previously described by Cope, 1867, and Comeau, 1943, respectively). Using a variety of cytogenetic techniques, Sessions (1982) deduced that the initial diploid hybrid female was formed by a mating between a male *A. laterale* and a female *A. jeffersonianum*. Presumably this hybrid female produced diploid eggs that were fertilized by one or the other of the diploid progenitors, resulting in one of the triploid species. A mating between this hybrid and diploid males of *A. jeffersonianum* produced *A. platineum*, and a mating with *A. laterale* generated *A. tremblayi*. However, it has been argued that these hybrids do not constitute legitimate species, and that *A. platineum* should simply be identified as *A. 2 jeffersonianum-laterale*, and *A. tremblayi* as *A. jeffersonianum-2 laterale* (Lowcock *et al.*, 1987). Similarly, a unisexual triploid from Kelley Island, Ohio, dubbed *A. nothagenes* by Kraus (1985), was shown to be recreated in each generation as a hybrid of *A. texanum* × *A. laterale* (Bogart *et al.*, 1987).

There has long been considerable uncertainty as to the reproductive mode of these all-female hybrids, with some authors assuming thyletoky, some suggesting hybridogenesis, and still others favoring gynogenesis (see Bogart and Licht, 1986). However, it remains the case that there has never been convincing evidence for thyletoky in salamanders, and indeed matings with diploid males of related species are generally considered necessary for embryogenesis. A gynogenetic system in this case would involve the automictic production of triploid eggs by the female hybrids, for which there was reported to be experimental evidence (King, 1990). Sperm from sexual diploids would be needed to stimulate egg development but there would be no fertilization because these sperm would not be incorporated. However, following a careful study of certain populations of *Ambystoma* hybrids, Bogart and Licht (1986) suggested that probably neither thyletoky nor pure gynogenesis was occurring, but rather that diploid females might produce both haploid and diploid eggs in the same egg mass, the fertilization of which would generate diploids and triploids, respectively. A recent review by Bogart (2003) summarizes the current understanding of this highly complicated system. Although gynogenesis is probably the most frequent mode of reproduction at colder temperatures (at which breeding normally takes place), sperm incorporation can occur at an increased rate at higher temperatures (Bogart *et al.*, 1989). Because females appear capable of producing both reduced and unreduced eggs, both increases and decreases in ploidy would be possible when sperm is incorporated (Bogart, 2003).

Whatever the reproductive system, it is clear that *A. jeffersonianum* and *A. laterale* are not the only species that hybridize to form polyploids. *A. texanum* and *A. tigrinum* may also be involved, and in fact 20 different nuclear combinations of diploids, triploids, tetraploids, and pentaploids have been documented (Lowcock *et al.*, 1987; Bogart, 2003). In some cases, three different genomes may be involved, as in *A. laterale-jeffersonianum-texanum* or *A. laterale-texanum-tigrinum* (Lowcock *et al.*, 1987; Kraus *et al.*, 1991), or genomes from the different contributing species may be found in different numbers of copies, as in *A. 3 laterale-texanum* or *A. laterale-2 jeffersonianum-tigrinum* (Lowcock *et al.*, 1987). Hybrids containing all four genomes have also been identified, as have rare hybrid males (Bogart, 2003).

In a particularly interesting twist on this common intermingling of nuclear genomes, it has been shown that the hybrid lineages maintain their own mitochondrial genomes independent of what happens in terms of the nuclear genome(s). Mitochondrial DNA studies of the unisexual "hybrid" females and putative possible ancestors showed the unisexual lineages to be about four to five million years old, which would make them the most ancient unisexuals known in vertebrates (Hedges *et al.*, 1992; Spolsky *et al.*, 1992). More recent evidence, based on more extensive sampling of mitochondrial DNA sequences that were not restricted to possible hybridizing species, suggests that the original female that gave rise to the unisexual complex is not found among the four species whose genomes might be included in their nuclear genome. Subsequent nuclear genome swapping has probably been an important factor in the evolution of this complex, and the hybrids likely have much more dynamic genomic histories than previously thought (Bogart, 2003).

The sizes of nuclei and cells are influenced by polyploidy in salamanders (e.g., Austin and Bogart, 1982; Licht and Bogart, 1987), and in some cases this can be expressed as slight differences in body size (Licht and Bogart, 1989, 1990; Lowcock *et al.*, 1991). However, as with the frog *Hyla versicolor*, polyploidy does not seem to exert any significant effects on organismal features like rates of oxygen consumption and water loss, embryonic development rate, or temperature tolerance (Licht and Bogart, 1989, 1990). There may, however, be some selective benefit in the form of larger egg size among polyploids (Licht and Bogart, 1990).

A POLYPLOID FAMILY OF SALAMANDERS (?): SIRENIDAE

Polyploidization in the Ambystomatidae is obviously extensive, but it is limited to certain species and is not typical of the family as a whole. Although it contains fewer polyploid species than the Ambystomatidae (because it contains only three extant species in total), the Sirenidae is of particular interest because it may be the only truly polyploid family of amphibians (Morescalchi and Olmo, 1974; Morescalchi *et al.*, 1986).

All three sirenids have higher chromosome numbers per somatic nucleus than most other salamanders: *Pseudobranchus striatus*, 2n = 64; *Siren lacertina*, 2n = 52; and *Siren intermedia*, 2n = 46 (Morescalchi and Olmo, 1974; Morescalchi *et al.*, 1986). The first two species have karyotypes that can be grouped relatively easily into quartets (or possibly sextets in *P. striatus*) of morphologically similar chromosomes, and even though 46 is not divisible by four, *S. intermedia*'s karyotype can also be assembled into quartets with two additional chromosomes lacking partners (Morescalchi *et al.*, 1986). Combined with some of the highest DNA contents in vertebrates, these karyotypic features suggest tetraploidy in *Siren* spp. and possibly hexaploidy in *P. striatus* (Morescalchi *et al.,* 1986). However, if this is so then it must be an ancient condition, because these species behave genetically as diploids.

Other families of salamanders may have very large genome sizes but a fairly typical salamander chromosome number (e.g., Amphiumidae, 2n = 28; Proteidae, 2n = 38), high chromosome numbers but a standard salamander genome size (e.g., Hynobiidae, 2n = 40 to 62), or both high DNA contents and chromosome numbers, but including many microchromosomes (e.g., Cryptobranchidae, 2n = 60) (King, 1990). In none of these cases is polyploidy implicated. More work is needed to determine whether the Sirenidae is indeed a small but fully polyploid family, and if so, when and how the polyploidization event(s) took place and how this may have contributed to the origin and/or evolution of the family.

RARE TRIPLOIDY IN NEWTS

Artificial polyploids have long been studied in the newt genus *Triturus* (e.g., *T. viridescens* and *T. alpestris*) (Fankhauser, 1955). Triploidy can be readily generated by hybridization in this genus (e.g., in crosses between *Triturus marmoratus* and *T. karelini*) (Fig. 8.2), although this clearly leads to meiotic problems owing to the presence of univalents (Macgregor, 1993). Single, naturally occurring autotriploid specimens have been found in each of the Danube newt *T. dobrogicus,* the crested newt *T. cristatus,* and the smooth newt *T. vulgaris* (Borkin *et al.,* 1996; Litvinchuk *et al.,* 1998, 2001). In general, spontaneous autotriploidy is believed to occur in less than 1% of individuals within *Tritutus* species, and at about the same rate in salamanders in general (Lowcock and Licht, 1990; Litvinchuk *et al.,* 2001).

REPTILES

There is a significant positive correlation between chromosome number and nuclear DNA content in reptiles at large (r = 0.31, P < 0.0001, n = 170), but not within any of the turtles (P > 0.36), lizards (P > 0.34), or snakes (P > 0.91). Moreover, turtles display very stable chromosome numbers but the most variable genome sizes among reptiles, whereas other reptilian groups often show notable chromosomal variation but little difference in DNA content (Olmo, 1986). This disconnect between chromosome number and DNA content suggests that polyploidy has not been of general importance in this class, and indeed no exclusively polyploid higher taxa of reptiles are thought to exist (Bickham, 1984).

That is not to say that polyploidy does not occur in reptiles, however. In fact, there is some reason to expect polyploidy in this group, because the class contains about 40 parthenogenetic species, most of them probably of hybrid origin

(Bogart, 1980; Bickham, 1984; Olmo, 1986; Darevsky, 1992). The majority of these are lizards (representing 15 genera and seven families), but a few examples are also known from snakes (Darevsky, 1992). In addition, the sex chromosomes of lizards are in an early stage of differentiation (Bickham, 1984), which suggests that some of the primary barriers to polyploidy should be alleviated in members of this group. Not surprisingly, numerous examples of polyploidy (although usually only triploidy) have been reported in lizards.

THE GENUS CNEMIDOPHORUS

Parthenogenesis does not guarantee the generation of fertile polyploids in lizards, and many examples are known in which the parthenogens remain diploid. In some cases, as with *Archaeolacerta* species (family Lacertidae), the hybrids that form in sympatry are sterile (Darevsky, 1992). Polyploids do commonly result in the best-studied group of parthenogenetic lizards, the whiptail lizards of the new world genus *Cnemidophorus*[2] (family Teiidae). Indeed, at least eight species from this genus are triploid hybrids, with some species (e.g., *C. tesselatus* and *C. lemniscatus*) having both diploid and triploid parthenogenetic populations (Darevsky, 1992). Interestingly, *C. lemniscatus* may actually represent a complex of independently formed hybrids, further complicating the situation in this "species."

The generation of triploidy in these lizards is thought to involve at least two rounds of hybridization, one to produce a diploid parthenogenetic species and a second between these parthenogens and a bisexual relative. In some cases, multiple species are involved in this process, leading to a rather complex network of hybrid relationships (Fig. 8.8). Tetraploidy, the highest level attained by any reptiles, can be produced experimentally by mating a triploid with a diploid, as for example with *C. sonorae* ($3x$) × *C. tigris* ($2x$) (Hardy and Cole, 1998). On occasion, tetraploids may occur naturally by such matings between bisexual diploids and unisexual triploids, but these are sterile (Neaves, 1971; Darevsky, 1992; Hardy and Cole, 1998). Thus, even though they may grow more quickly and achieve a larger adult body size than their relatives of lower ploidy, these tetraploids are of little consequence in evolutionary terms because only diploids and triploids are capable of reproducing (Hardy and Cole, 1998). In the triploid dessert grassland whiptail, *C. uniparens* (= *C. burti-2 inornatus*), reproduction is automictic, involving an endomitotic doubling of the chromosomes prior to meiosis (Cuellar, 1971). Intriguingly, even though females can reproduce without males,

---

[2]It should be noted that most of the triploid *Cnemidophorus* species have recently been classified into the new genus *Aspidoscelis* (Reeder *et al.*, 2002). However, in order to avoid confusion with the existing literature, *Cnemidophorus* is maintained in this discussion.

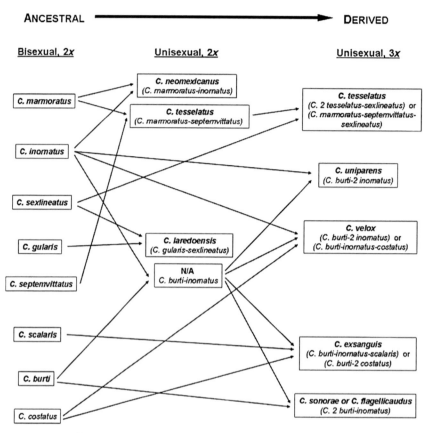

FIGURE 8.8 Schematic representation of the hybrid origins of three diploid and six triploid parthenogenetic whiptail lizards in the genus *Cnemidophorus*. Converging arrows indicate a hybridization between the two source species. Based on Darevsky (1992), which was modified from Dessauer and Cole (1989). Reproduced by permission (© Society for the Study of Amphibians and Reptiles).

female dominance hierarchies are thought to influence mating success, and behavioral stimulation from other females may be required (Grassman and Crews, 1987). It remains to be seen whether this is true in general of parthenogenetic *Cnemidophorus* species.

In some cases, hybridization is a continual process. For example, annual hybridizations have been observed between diploid parthenogens in *C. tesselatus* and diploid males from *C. tigris marmoratus* in the desert grasslands of New Mexico (Taylor *et al.*, 2001). The resulting hybrids are sterile triploids; so, although it is of interest that the hybridization process can be observed as it occurs, it is doubtful that this particular system represents an incipient stage of species formation (Taylor *et al.*, 2001).

OTHER POLYPLOID REPTILES

A few other examples of polyploidy are scattered throughout the lizards. At least two diploid–triploid hybrid complexes in butterfly lizards of the genus *Leiolepis* (family Uromastycidae) have been documented, including the aptly named *L. triploida*. This species was once considered to be an autopolyploid, but is now recognized as being derived from a hybridization between the unisexual *L. boehmei* and a member of the bisexual *L. belliana* group (Darevsky, 1992). The other triploid parthenogen in this genus, *L. guntherpetersi*, is also of hybrid origin (Darevsky, 1992).

The triploid form of the Australian gecko *Heternotia binoei* (family Gekkonidae) is also a product of hybridization, and evidence suggests that it has formed repeatedly in this way (Moritz *et al.*, 1989, 1990). Some triploid clones of unisexual gekkos, such as in *Hemidactylus garnotii*, *H. vietnamensis*, and *H. stejnergeri*, may be direct autopolyploids, whereas others, like those in *Lepidodactylus lugubris*, may have formed by hybridization between species or between cytotypes of the same species (Darevsky, 1992; Moritz *et al.*, 1993), in the latter case also making them autopolyploids under a broad definition of the term. The centralian slider *Lerista frosti* (family Scincidae) was suggested as another possible polyploid, although this was based solely on its possession of an elevated DNA content (MacCulloch *et al.*, 1996).

Only one triploid unisexual snake is known (*Ramphotyphlops braminus*) (Wynn *et al.*, 1987), and there are no known cases of polyploidy in crocodilians (Bickham, 1984; Darevsky, 1992). Polyploidy was thought to be absent in turtles, but Bull and Legler (1980) reported apparent triploidy in the twist-necked turtle *Platemys platycephala* (family Chelidae). As it turns out, this species is not a true triploid, but consists of diploid–triploid or triploid–tetraploid mosaics (Bickham *et al.*, 1985, 1993). The means by which this condition occurs in both males and females of these bisexual turtles is not well understood. Presumably, this involves autopolyploidy (Bickham *et al.*, 1985), in contrast to the mosaicism observed in unisexual hybrid fishes (*Carassius*, *Phoxinus*, and *Poecilia*). Remarkably, the mosaic males appear to produce haploid gametes, regardless of their somatic ploidy level (Bickham *et al.*, 1993). This is the only known example of a bisexual vertebrate species that has mosaics at a high natural frequency (Bickham *et al.*, 1993). Diploid–triploid mosaicism also occurs in lizards of the genus *Lacerta* (Kapriyanova, 1989) but, like the fish examples, these are unisexual and are produced by hybridization.

MAMMALS AND BIRDS

As noted earlier, triploidy associated with intersexuality has occasionally been observed in individual domestic birds (e.g., Ohno *et al.*, 1963; Abdel-Hameed and Schoffner, 1971), and it has been reported in one individual of the blue-and-yellow macaw, *Ara ararauna* (Tiersch *et al.*, 1991). Diploid–triploid mosaics are known in humans, cats, mice, rabbits, cattle, pigs, mink, and chickens, but these are usually malformed or at least sterile (Fechheimer *et al.*, 1983). The 1950s saw concerted attempts to produce parthenogenetic and polyploid mammals for use in agriculture, but this was not met with success (Beatty, 1957). It was thought for some time that the golden hamster (*Mesocricetus auratus*), with 2n = 44 chromosomes, had evolved as a polyploid hybrid of two related species with 2n = 22, but it has since been determined that a range of chromosome numbers from 2n = 22 to 2n = 44 can be found among its relatives (Macgregor, 1993). In short, natural polyploidy is exceedingly rare in mammals and birds, perhaps for several of the reasons outlined at the beginning of the chapter.

However, recent evidence suggests that the conclusion that polyploidy cannot be tolerated in these groups may be premature. Notably, the South American red viscacha rat *Tympanoctomys barrerae* (family Octodontidae), which has twice the DNA content and number of chromosomes (2n = 102) as its closest relatives, provided the first example of a naturally occurring tetraploid mammal (Gallardo *et al.*, 1999, 2003, 2004). Interestingly, two new species of octodontids thought to be sister to *T. barrerae* (*Pipanoctomys aureus* and *Salinoctomys loschalchalerosorum*) (Mares *et al.*, 2000) also appear to have high chromosome numbers and DNA contents (Honeycutt *et al.*, 2003). Comparison of gene copy number in the sex-linked androgen receptor gene of *T. barrerae*, *P. aureus*, and two closely related diploid species (*Octomys mimax* and *O. gliroides*) reinforced the tetraploid status of *T. barrerae* and suggested that *P. aureus* was also a tetraploid (2n = 92) (Gallardo *et al.*, 2004). These data also provided evidence that the sex chromosomes are doubled in *T. barrerae*, with four copies of the X chromosome in females and up to three in males.

Sperm, lymphocytes, and liver cells are larger in *T. barrerae* than in its diploid relatives (Gallardo *et al.*, 1999, 2003). Despite a lack of nuclei, mammalian red blood cell sizes are correlated positively with genome size in mammals (Gregory, 2000) (see Chapter 1), but it has been reported recently that the erythrocytes of *T. barrerae* do not follow this trend (Gallardo *et al.*, 2003). However, it seems that this conclusion may have been based on problematic erythrocyte size measurements by Gallardo *et al.* (2003), who gave a diameter of 7.5 micrometers (μm) for *T. barrerae*, which is smaller than that usually reported for humans (~7.7 μm). Visual comparison of cells from the two species raises doubts about this finding (Fig. 8.9). In addition, Gallardo *et al.* (2003) reported a value for *Rattus rattus* (7.0 μm) that is higher than all previous measurements for *Rattus* (~6.5 μm), and it is questionable whether *Ctenomys medocinus*, with a genome at the high end of the mammalian distribution (4.9 pg), could have by far the smallest erythrocytes (4.8 μm) ever reported for a rodent. As it turns out, an earlier measurement found the diameter of *T. barrerae* erythrocytes to be 8.3 μm, which is larger than the rodent average (7.2 μm) in roughly the same proportion as are its hepatocytes

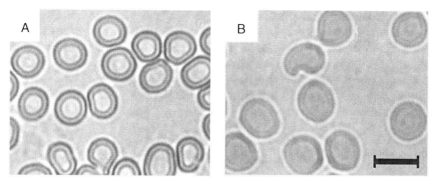

FIGURE 8.9   Photomicrographs of dried erythrocytes from (A) human (diameter = 7.7 μm) and (B) the tetraploid red viscacha rat, *Tympanoctomys barrerae* (diameter = 8.3 μm). Despite a recent report to the contrary (Gallardo *et al.* 2003), *T. barrerae* clearly has enlarged erythrocytes, especially as compared to the rodent average of 7.2 μm (Gregory, 2001c). Both photos taken at 40x magnification. Scale bar equals 10 μm. Blood smears of *T. barrerae* were kindly provided by Milton Gallardo.

(Gregory, 2001c). Based on this, it would seem that erythrocytes are affected by elevated DNA content in the same way as other cells in this polyploid mammal.

Both tetraploid species, *T. barrerae* and *P. aureus,* are presumed to be of hybrid origin based on exclusively bivalent associations during meiosis (Gallardo *et al.,* 2004). It will be interesting to see if *S. loschalchalerosorum* is also found to be polyploid. More generally, despite intuitive objections to the presence of bisexual polyploidy in mammals with an XX : XY sex determination system, it seems that further efforts should be made to assess whether the octodontids are a truly exceptional group with respect to ploidy or whether additional examples might be found in other taxa.

VERTEBRATE POLYPLOIDS: A SUMMARY

As the examples provided here show, the ancient genome duplication of the vertebrate lineage has been followed by many more recent polyploidization events (including in the teleosts at large) (see Chapter 6). Although recent polyploidy is rare in mammals and birds, this is clearly not true of groups like fishes and amphibians. In reptiles, too, there are numerous known instances. In some cases, one or the other of auto- or allopolyploidy may appear to predominate, but overall both mechanisms are generally found in each of the major taxa. As in plants (see Chapter 7), there are signs of repeated polyploid formation in certain lineages, and there may in fact be complex networks of hybridization among some related species. Polyploidy is found in association with all of the major reproductive systems in vertebrates, including bisexuality, thyletoky, gynogenesis, and hybridogenesis. In some cases, polyploidy is a feature shared by an entire family, whereas in others the polyploids are found in cryptic association with a diploid species, or even just represent a minor form within a mostly diploid species. There is clearly substantial diversity in the preponderance, mechanisms of formation, and biological and evolutionary implications of polyploidy among the vertebrates.

POLYPLOIDY IN INVERTEBRATES

Christensen (1980, p. 52) voiced a fairly common sentiment about the occurrence of polyploidy when he argued that "in most animal phyla it is considered either unknown or of extremely rare occurrence." As noted previously, views such as this tend to represent self-fulfilling prophecies, because their primary effect is to stifle further research into animal polyploidy. Polyploidy has already been shown in the previous sections to be relatively common in vertebrates and, in direct contrast to such pessimistic pronouncements, also occurs widely among invertebrates. In fact, examples of naturally occurring polyploidy can be found in nearly all of the major invertebrate phyla, and cases may yet be discovered in the remaining taxa (e.g., Cnidaria, Echinodermata) if interest could be stimulated in such a survey.

CRUSTACEANS

Chromosome numbers vary enormously in the Crustacea, from $2n = 6$ in the copepod *Acanthocyclops robustus* to $2n > 200$ in several decapods, reaching as high as $2n = 254$ in the hermit crab *Pagurus ochotensis*, one of the highest counts yet recorded for an animal (Lécher *et al.*, 1995; Deiana *et al.*, 1999). However, it is far from clear that this variation is the result of polyploidy, and in fact the phenomenon appears to be restricted to certain groups.

In the order Amphipoda within the class Malacostraca, for example, polyploidy has only been found in one bisexually reproducing species (*Pontoporeia affinis*) (Salemma, 1984), although there is some indication that the genus *Gammarus* may also exhibit some speciation by polyploidy (Lécher *et al.*, 1995). More generally, chromosome data are suggestive of polyploidy in some other amphipods (Salemma and Heino, 1990; Coleman, 1994), although karyotypic data alone cannot establish this convincingly. Polyploidy is known in only one member of the malacostracan order Isopoda, the terrestrial woodlouse (or sow bug) *Trichoniscus pusillus*, which contains diploid bisexual and triploid thyletokous forms (Vandel, 1927, 1928; Lécher *et al.*, 1995). By contrast, polyploidy may be of general importance in the order Decapoda, in which chromosome numbers are all high, ranging from 2n = 54 to 254 (Lécher *et al.*, 1995; Deiana *et al.*, 1999). It has been suggested that primitive decapod species tend to have lower chromosome numbers, perhaps indicating that the higher numbers of more recently derived groups are attributable to polyploidy (Lécher *et al.*, 1995). Indeed, polyploidy may be important in each of the infraorders Astacidea (lobsters and crayfishes), Anomura (hermit crabs), and Brachyura (crabs), all of which have high chromosome numbers and DNA contents (Lécher *et al.*, 1995). In the family Scyllaridae (slipper lobsters), at least, there is a positive correlation between chromosome number and DNA content, which supports the hypothesis of polyploidy (Deiana *et al.*, 1999).

Spontaneously arisen unisexual forms are present within the class Ostracoda, and triploid and tetraploid clones can be found in high Arctic populations of *Cyprinotus incongruens*. This appears to include both auto- and allopolyploidy, because some of the polyploids have formed by matings between bisexual and unisexual forms within *C. incongruens* whereas others arose by the incorporation of sperm from the related bisexual species *C. glaucus* (Turgeon and Hebert, 1994). Polyploids are also found in Arctic populations of *Candona rectangulata*, *Cypridopsis vidua*, and *Prinocypris glacialis*, the latter of which may include levels as high as pentaploidy (Havel and Hebert, 1989; Little and Hebert, 1997).

Polyploidy is not found in the class Copepoda, but quantum shifts in genome size ("cryptopolyploidy") are common in this group (see Chapter 1). No polyploid barnacles (class Cirripedia) have been found, but some are hermaphroditic and may yet provide examples of polyploid forms if studied more thoroughly (Lécher *et al.*, 1995). For the time being, the best-studied examples of polyploidy in crustaceans come from water fleas and brine shrimp, both in the class Branchiopoda.

WATER FLEAS (ORDER CLADOCERA)

Water fleas in the order Cladocera are typically unisexual or cyclically parthenogenetic, normally consisting of all-female clones but with sex determined environmentally and males produced during times of stress. In cladocerans, thyletoky is apomictic and interspecific hybridization is common. Although they have some of the smallest genome sizes among the Crustacea (Gregory, 2001b), water fleas provide some of the best-known examples of polyploidy in the subphylum.

As a prime example, several Arctic members of the *Daphnia pulex* species complex may hybridize to generate obligately thyletokous polyploids from formerly diploid cyclic parthenogens (Dufresne and Hebert, 1994). Two polyploid species are recognized in this complex, *D. middendorffiana* (all 4x clones) and *D. tenebrosa* (both 2x and 4x clones) (Dufresne and Hebert, 1997, 1998). Interestingly, the outcome of hybridization is dependent on the directionality of the interspecific cross. In the cross between *D. pulex* and *D. pulicaria*, for example, diploid hybrid clones arise when *D. pulex* is the maternal parent, but polyploids occur when the female parent is *D. pulicaria*. Likewise, interspecific hybridization among cyclically parthenogenetic species accounts for the examples of Arctic polyploid members of the cladoceran genus *Bosmina* (Little *et al.*, 1997). Notably, only even-numbered polyploids (usually 4x) are known from *Daphnia* and *Bosmina* (Lécher *et al.*, 1995).

BRINE SHRIMP (ORDER ANOSTRACA)

Members of the genus *Artemia* inhabit hypersaline inland waters and coastal salt ponds, and are therefore commonly known as brine shrimp. Nearly a century ago, Caesare Artom noted that *"Artemia salina"* contained diploid bisexual, diploid unisexual, and polyploid unisexual forms (Artom, 1906, 1911; Barigozzi, 1974; Lécher *et al.*, 1995). *Artemia* is now recognized to contain several distinct species, the majority of them bisexual and one unisexual. The unisexual species, *Artemia parthenogenetica*, is believed to be derived from a bisexual species, probably *A. tunisiana* (Browne and Bowen, 1991).

A. *parthenogenetica* consists entirely of thyletokous females, but is remarkable in that this includes both automictic and apomictic forms. The automicts, which undergo a form of meiosis but lack syngamy, are invariably diploid, whereas the apomictic cytotypes are polyploids. The polyploids are larger than the diploids in *Artemia* (Artom, 1926; Zhang and King, 1993), which is also the case in triploid versus diploid *Trichoniscus pusillus* isopods (Vandel, 1928). Intriguingly, polyploidy in *A. parthenogenetica* typically consists of odd-numbered ploidy levels (usually 3x and 5x), which emphasizes the significance of their having adopted a system of apomictic reproduction. This pattern of polyploidization clearly cannot be explained by internal duplication events in diploid unisexuals, because this would generate tetraploids and then octoploids. Instead, it has been suggested that polyploidy in *A. parthenogenetica* is generated by the fusion of a pronucleus containing 42 chromosomes with a polar body of 21, resulting in a nucleus with a triploid content of 63 chromosomes (Barigozzi, 1974). The potential role of repeated hybridization should also not be ignored in this case, but whatever the mechanistic basis, *A. parthenogenetica* provides an especially clear picture of the progression from bisexuality to unisexuality to polyploidy.

INSECTS

The diversification of the insects was the most extensive of any multicellular life forms on Earth but, perhaps somewhat surprisingly, appears to have proceeded with only a minimal input from polyploidy. Some insects do exhibit polyploidy, but these generally insist on defying expected trends. For example, polyploidy is extremely rare in aphids (order Hemiptera) even though they reproduce by cyclic thyletoky (Takada *et al.*, 1978). Similarly, although many butterflies (order Lepidoptera) display very high chromosome numbers, these are the result of the fragmentation of existing chromosomes and not of polyploidy (White, 1978). In the order Orthoptera, which have by far the highest DNA contents among insects, only the katydid *Saga pedo* is thought to be polyploid (Lokki and Saura, 1980).

Scattered examples of polyploidy (3x, 4x, and rarely 5x or 6x) can be found in at least nine different insect orders, almost always associated with thyletoky but arising by a variety of mechanisms (Table 8.1). For example, Seiler (1946, 1961) proposed that in the bagworm moth *Solenobia triquetrella* (order Lepidoptera), automictic thyletokous diploids arose first within an amphimictic diploid stock, and then later gave rise directly (e.g., by the fusion of two conspecific diploid gametes) to an automictic autotetraploid race without any triploid intermediate (see also Lokki *et al.*, 1975).

In a very different example from the Lepidoptera, B.L. Astaurov used the diploid silkworm moth *Bombyx mori* and its relative *B. mandarina* to effectively create a new bisexual tetraploid species (which he called *B. allotetraploidus*) via a

TABLE 8.1   Incidence of natural polyploidy in insects. Ploidy level is indicated as multiples of the basic chromosomes number (2x = diploid, 3x = triploid, etc.). Reproductive mode is designated as bisexual (B), thyletokous (T), or gynogenetic (G). Consult the text and Lokki and Saura (1980), Suomalainen *et al.* (1987), and Otto and Whitton (2000) for original references.

Species and taxonomic info	Ploidy	Reproduction
Order Blattaria		
Family Blaberidae		
Pycnoscelus surinamensis	2x, 3x	T (3x)
Order Coleoptera		
Family Chrysomelidae		
Altica lazulina	3x	T
Bromius obscurus	2x, 3x	B (2x), T (3x)
Calligrapha alnicola	4x	T
Calligrapha amator	4x	T
Calligrapha apicalis	4x	T
Calligrapha ostryae	4x	T
Calligrapha scalaris	2x, 4x	B (2x), T (4x)
Calligrapha vicina	4x	T
Calligrapha virginea	4x	T

(Continues)

TABLE 8.1 *(Continued)*

Species and taxonomic info	Ploidy	Reproduction
Family Curculionidae		
Aramigus tessellatus	2x, 3x (and 4x? 5x?)	B (2x), T? (3x)
Barynotus moerens	3x, 5x	T
Barynotus obscurus	4x	T
Barynotus squamosus	3x	T
Blosyrus japonicus	3x, 4x, 5x, 6x	T
Callirhopalus bifaciatus	3x, 4x, 5x	T
Callirhopalus minimus	3x, 4x	T
Callirhopalus obesus	2x, 3x, 4x, 6x	T
Callirhopalus setosus	4x	T
Catapionus gracilicornis	2x, 3x, 4x, 5x, 6x, 10x	T
Cyrtepistomus castaneus	3x	T
Eusomus ovulum	3x	T
Foucartia squamulata	3x	T
Liophloeus tesselatus	3x	T
Lissorhoptrus oryzophilus	3x	T
Listroderes costirostris	3x	T
Macrocorynus griseoides	3x	T
Myllocerus fumosus	3x	T
Myllocerus nipponicus	3x	T
Myosides pyrus	3x	T
Myosides seriaehispidus	3x	T
Naupactus peregrinus	3x	T
Otiorrhynchus anthracinus	5x	T
Otiorrhynchus chrysocomus	3x, 4x	T
Otiorrhynchus dubius	4x	T
Otiorrhynchus gemmatus	3x	T
Otiorrhynchus ligustici	3x	T
Otiorrhynchus niger	3x, 4x	T
Otiorrhynchus ovatus	3x	T
Otiorrhynchus pauxillus	3x	T
Otiorrhynchus proximus	3x	T
Otiorrhynchus pupillatus	4x	T
Otiorrhynchus rugifrons	3x	T

(Continues)

TABLE 8.1   *(Continued)*

Species and taxonomic info	Ploidy	Reproduction
Otiorrhynchus salicis	3x	T
Otiorrhynchus scaber	3x, 4x	T
Otiorrhynchus singularis	3x	T
Otiorrhynchus subcostatus	3x	T
Otiorrhynchus subdentatus	3x, 4x	T
Otiorrhynchus sulcatus	3x	T
Peritelus hirticornis	3x, 4x	T
Polydrusus mollis	2x, 3x	T
Scepticus insularis	2x, 3x, 4x, 5x	T
Sciaphilus asperatus	3x	T
Sciopithes obscurus	4x	T
Strophosomus melanogrammus	3x	T
Trachyphloeus aristatus	3x	T
Trachyphloeus bifoveolatus	3x	T
Trachyphloeus scabriculus	3x	T
Trachyrhinus sp.	3x	T
Trophiphorus carinatus	3x	T
Trophiphorus cucullatus	4x	T
Trophiphorus terricola	4x	T
Family Ptinidae		
Ptinus clavipes f. *mobilis*	3x	G
Family Scolytidae		
Ips acuminatus	2x, 3x	B (2x), G (3x)
Ips borealis	2x, 3x	B (2x), G (3x)
Ips perturbatus	2x, 3x	B (2x), G (3x)
Ips pilifrons	2x, 3x	B (2x), G (3x)
Ips tridens	2x, 3x	B (2x), G (3x)
Order Diptera		
Family Agromyzidae		
Phytomyza crassiseta	2x, 3x	T (3x)
Family Chamaemyiidae		
Ochthiphila polystigma	3x	T
Family Chironomidae		
Limnophyes virgo	3x	T

(Continues)

TABLE 8.1   *(Continued)*

Species and taxonomic info	Ploidy	Reproduction
Lundstroemia parthenogenetica	3x	T
Pseudosmittia sp.	3x	T
Family Psychodidae		
Phytomyza crassiseta	2x, 3x	B or T (2x), T (3x)
Psychoda parthenogenetica	3x	T
Family Simulidae		
Cnephia mutata	2x, 3x	T (3x)
Gymonopais sp.	2x, 3x	T (3x)
Prosimulum macropyga	2x, 3x	T (3x)
Prosimulium ursinum	2x, 3x	B (2x), T (3x)
Order Embiidina		
Family Oligotomidae		
Haploembia solieri	2x, some 3x	T
Order Hemiptera		
Family Coccidae		
Physokermes hemicryphus	2x, 3x	T (3x)
Family Delphacidae		
Muellerianella fairmairei	2x, 3x	G (3x)
Order Hymenoptera		
Family Apidae		
Melipona quinquefasciata	♂2x, ♀4x?	B
Family Diprionidae		
Diprion simile	♂2x, ♀4x	B, T
Order Lepidoptera		
Family Psychidae		
Solenobia fennicella	4x	T
Solenobia lichenella	2x, 4x	T (4x)
Solenobia seileri	4x	T
Solenobia triquetrella	2x, 4x	B or T (2x), T (4x)
Order Orthoptera		
Family Tettigoniidae		
Sago pedo	4x	T
Order Phasmida		
Family Phasmatidae		
Basillus atticus carius	2x, 3x	T
Basillus lynceorum	3x	T

unisexual triploid intermediate (reviewed in Suomalainen *et al.*, 1987). First, apomictic tetraploid thyletokous clones were produced by heat-treating eggs, and then this all-female line was crossed with males of B. *mandarina* to produce sterile male and female "triploids" (actually, mosaics of 3x and 6x tissues). The subsequent fertilization of triploid eggs from the allotriploid females and haploid sperm of B. *mandarina* generated fertile bisexual allotetraploids that were reproductively incompatible with the diploid progenitor species.

Yet another mechanism is revealed in stick insects of the genus *Bacillus* (order Phasmida), in which most species are thyletokous diploids produced by hybridization. Triploid unisexuals have resulted twice from further hybridization, as in the automictic diploid–triploid B. *atticus carius* and the apomictic trihybrid B. *lynceorum* (= B. *rossius-grandii-atticus*) (Mantovani *et al.*, 1992; Manaresi *et al.*, 1993; Marescalchi and Scali, 2003) (Fig. 8.10). Higher levels of polyploidy are not known in stick insects, but presumably they could result from further hybridization involving these allotriploids.

Table 8.1 shows that the majority (>60%) of known polyploid insect species are weevils (order Coleoptera, family Curculionidae), which is perhaps fitting because the Curculionidae is the largest family among all animals. Some 52 species from 24 genera are recognized to be polyploid or to contain polyploid "races," ranging from 3x to 10x (Suomalainen *et al.*, 1987; Saura *et al.*, 1993). There are no exclusively polyploid higher taxa among weevils, and it is thought

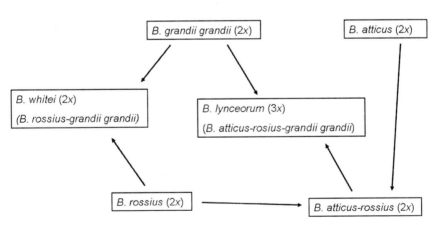

FIGURE 8.10 Schematic representation of the hybrid origins of the stick insect species *Bacillus whitei* (diploid) and B. *lanceorum* (triploid). Converging arrows indicate a hybridization between the two source species. As with the whiptail lizards (Fig. 8.8), there are multiple hybrid combinations in this genus. Based on Mantovani *et al.* (1992), reproduced by permission (© Elsevier Inc.).

that these reflect 52 independent polyploidization events (Saura *et al.*, 1993). Notably, although there are a few unisexual diploid weevils, only apomictic thyletoky is found among the polyploids (Smith and Virkki, 1978; Suomalainen *et al.*, 1987). In some other beetles, such as bark beetles of the genus *Ips* (family Scolytidae) and the brown spider beetle *Ptinus clavipes* (family Ptinidae), triploidy is associated with gynogenesis rather than thyletoky (Lanier and Kirkendall, 1986) (Table 8.1).

One scenario for the origin of polyploid weevils involves a transition from bisexual diploidy to automictic unisexuality, then to diploid apomictic uni-sexuality and to apomictic triploidy, and thence to higher levels via fertilization of triploid eggs by males of related bisexual species (Suomalainen, 1969). However, this explanation encounters the difficulty that diploid unisexuals are only rarely encountered (Saura *et al.*, 1993). Instead, it would seem that apomictic triploidy, which is the dominant ploidy level, may be the first step on the way to higher ploidy. In European populations of *Otiorhynchus scaber*, recent evidence based on high clonal diversity and patterns of distribution suggests that there is a recurrent transition from diploid sexuality to unisexual triploidy then to tetraploidy (Stenberg *et al.*, 2000). Most recently, it has been shown that triploidy has evolved independently at least three times in *O. scaber* (Stenberg *et al.*, 2003). In cases such as this, the diploid automictic unisexual step would be skipped when triploidy occurs by the fertilization of an unreduced egg, likely by interspecific hybridization (Saura *et al.*, 1993). In fact, unisexual reproduction would be con-sidered an escape from sterility in this sense, because these triploids would be unable to reproduce bisexually. As an interesting consequence, this means that the few diploid unisexuals would have been derived secondarily from the triploid apomicts (Saura *et al.*, 1993). However, empirical evidence to distinguish between these possibilities is lacking and it is probable that different mechanisms could operate to generate polyploidy and thyletoky in this family (Tomiuk *et al.*, 1994).

As Lokki and Saura (1980) pointed out, polyploid insects tend to be flightless and to have slow life cycles lasting two or more years. There is also a positive correlation between body size and level of polyploidy in weevils (Suomalainen, 1969; Smith and Virkki, 1978; Suomalainen *et al.*, 1987).

MOLLUSCS

Molluscs represent the most diverse phylum of predominantly aquatic animals and are composed of several very different groups, including the best-known classes Bivalvia (clams, oysters, and their kin), Gastropoda (snails, slugs, limpets), and Cephalopoda (octopuses, squids, cuttlefishes). Polyploidy has not been found in the last of these, perhaps only because it has rarely been sought in this group. By contrast, several examples are known from both bivalves and gas-tropods. That said, it bears noting that in molluscs, "when polyploidy occurs, it usually does so at the species level and probably has been of little significance in the derivation of higher taxa" (Burch and Huber, 1966).

BIVALVES

Triploidy (and more recently, tetraploidy) is often induced by chemical (cytochalasin B or 6-methylaminopurine) treatment in commercial stocks of Pacific oyster, *Crassostrea gigas*, because this disrupts the development of the gonads (and thereby prevents the reduction in meat quality normally associated with this change) and allows them to be marketed year-round (Allen, 1988; Eudeline *et al.*, 2000). In the dwarf surfclam, *Mulinia lateralis*, induced triploidy significantly increases body size, which is also a favorable effect for aquaculture (Guo and Allen, 1994). Triploidy (and sometimes tetraploidy) has also been induced in the blue mussel, *Mytilus edulis*, and more than a dozen other bivalve species (Beaumont and Fairbrother, 1991). However, although this shows that polyploidization is fairly easy to induce and is well tolerated in many bivalves, natural polyploidy is considered extremely rare in this group (Nakamura, 1985).

Of course, there are always exceptions. Natural triploidy occurs in several species of clams in the genus *Corbicula* (Okamoto and Arimoto, 1986; Burch *et al.*, 1998) as well as triploidy, pentaploidy, and hexaploidy in unisexual (some gynogenetic) marine clams of the genus *Lasaea* (Thiriot-Quiévreux *et al.*, 1989; Ó Foighil and Thiriot-Quiévreux, 1991, 1999). All species of *Lasaea* studied so far, with the exception of the bisexual species *L. australis* and *L. colmani*, are polyploid and are probably allopolyploids in particular (Ó Foighil and Thiriot-Quiévreux, 1999). Variable polyploidy of uneven multiples was recently reported in populations of the mussel *Mytilus trossulus* off the west coast of Canada, but this may have been "naturally induced" by infection or pollution (González-Tizón *et al.*, 2000).

One major exception to polyploid rarity in bivalves comes from the freshwater clam family Sphaeriidae, a hermaphroditic group capable of self-fertilization that was long ago suggested as a candidate for extensive polyploidy (Burch and Huber, 1966). In fact, only one diploid species (*Sphaerium corneum*, 2n = 36) is so far known from this family, whereas all the other species studied from the widespread genera *Sphaerium*, *Musculium*, and *Pisidium* display high and variable chromosome numbers (Lee and Ó Foighil, 2002). Assuming a basic haploid chromosome number of $x = 19$ for the family (Lee, 1999), ploidy levels reach as high as $10x = 190$ in *Pisidium casertanum* (Barsiene *et al.*, 1996), $11x = 209$ in *Sphaerium occidentale*, and a staggering $13x = 247$ in *Musculium securis* (Burch *et al.*, 1998). If this is truly the case, it would be very interesting not only because these levels are so high, but also because many of them are odd-numbered. Amazingly, mature sperm appear to be produced not only in even-numbered polyploids like

A   B

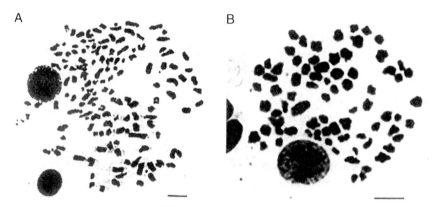

FIGURE 8.11   Photomicrographs of (A) mitotic (2n ≈ 152), and (B) meiotic (n = 76) metaphase chromosomes of the clam *Sphaerium striatinum*, which is believed to be octoploid (8*x*). Other members of
the family Sphaeriidae may reach levels as high as 13*x*. Scale bars equal 5 μm. From Lee (1999),
reproduced by permission (© Japan Mendel Society).

Sphaerium striatinum (8*x*) and *P. casertanum* (10*x*), but also in *S. occidentale* (11*x*)
and *M. securis* (13*x*) (Burch *et al.,* 1998; Lee, 1999) (Fig. 8.11), although there is
no evidence that reproduction is actually sexual in these latter species.
Polyploidization in the Sphaeriidae appears to be an ancient condition, predating
the divergence of *Sphaerium* and *Musculinum* in the Miocene (Lee and Ó Foighil,
2002). It is not clear whether this initial event occurred by auto- or allopolyploidy,
but it has been suggested that subsequent speciation events within *Pisidium* are
autopolyploid in nature (Lee and Ó Foighil, 2002).

GASTROPODS

Freshwater snails (one of the groups predicted by Muller, 1925, to be extensively
polyploid), as well as marine and terrestrial taxa in the subclass Pulmonata, are
hermaphroditic and many are capable of self-fertilization (Burch and Jung, 1993).
Despite this, based on a survey of chromosome numbers, White (1940) argued that
pulmonate snails displayed no evidence of polyploidy. Although he later recognized
a few examples, White (1978) still held that polyploidy was rare in snails and other
molluscs.

    For example, polyploidy has typically been considered absent in pond snails of the
family Lymnaiedae, despite their being hermaphroditic (Burch and Huber, 1966;
White, 1978). Two polyploid specimens of *Lymnaea peregra* have been found in
Spain, but this appears to be a byproduct of ecological stress, to which this species

is cytogenetically sensitive (Barsiene *et al.*, 1996). Other families do contain natural polyploid species, however. The first example of polyploidy in pulmonate snails was found in the family Planorbidae, in the tetraploid *Gyraulus circumstriatus* from North America (Burch, 1960; Burch and Jung, 1993). The two other members of the subgenus to which *G. circumstriatus* belongs (*Torquis*), *G. huronensis* and *G. parvus*, are also tetraploid, so presumably polyploidization occurred just once prior to their divergence (Burch and Jung, 1993).

Chromosome numbers in the family Planorbidae are generally conservative, with most species being diploid with 2n = 36 (Goldman *et al.*, 1984). Nevertheless, about 7% of species in this family are polyploid (Burch and Huber, 1966), including members of the well-studied African genus *Bulinus*, which varies from diploid to octoploid. At one point, snails of different ploidy level in this genus were granted separate species status as *Bulinus tropicus* (2x = 36), *B. truncatus* (4x = 72), *B. hexaploidus* (6x = 108), and *B. octoploidus* (8x = 144) (Wu, 1972). However, Wu (1972) suggested that *B. hexaploidus* should be contained within *B. tropicus* and *B. octoploidus* within *B. truncatus*. Today, this entire system is more generally referred to as the *Bulinus truncatus/tropicus* complex, with *B. tropicus* representing diploids from South Africa and *B. truncatus* comprising tetraploids from Egypt (Goldman *et al.*, 1983). Polyploidy in this genus is believed to be the result of allopolyploidy and possibly of segmental allopolyploidy in particular (Goldman *et al.*, 1983; Städler *et al.*, 1996). A similar allopolyploid mechanism appears to have operated in the European freshwater limpet, *Ancylus fluviatilis* (family Ancylidae), which also includes tetraploids, hexaploids, and octoploids (Städler *et al.*, 1996). In both *Bulinus* and *Ancylus*, polyploidy is linked to a high proportion of self-fertilization (Städler *et al.*, 1996).

Freshwater snails of the genus *Campeloma* (family Viviparidae), consist of both bisexual and unisexual forms. Allotriploidy is found in the unisexual hybrid *C. parthenum* (= *C. limum-2 geniculum*), in this case formed via hybridization between a sexual female *C. limum* and a male *C. geniculum* to produce a unisexual diploid, followed by a backcross with another male *C. geniculum* (Johnson *et al.*, 1999). Interestingly, no higher ploidy levels are found in this genus, which Johnson *et al.* (1999) took to indicate strong selection against levels higher than triploidy, perhaps relating to effects on cell and body size and/or growth rate. However, problems with sterility among triploids could also generate such a limitation.

Tetraploidy has been observed in *Ferrissia parallela* and *F. tarda* in the family Ancyclidae and *Melanoides tuberculata*, *M. lineatus*, and *M. scabra* in the family Thiaridae (Jacob, 1959; Burch and Huber, 1966; Burch and Jung, 1993; Barsiene *et al.*, 1996). The highest level recorded so far occurs in *Physa chukchensis* (family Physidae), which may be as high as 10x or even 12x (Barsiene *et al.*, 1996). In *Melanoides*, the tetraploids are hardier and have higher salinity tolerance than diploids and also produce larger broods, but they do not differ in body size (Jacobs, 1959). By contrast, hexaploids and octoploids are larger than diploids and tetraploids in *Bulinus* (Wu, 1972). It is interesting to note that polyploid *Bulinus* snails appear to be more susceptible to infection from *Schistosoma haematomum* (trematode flatworms that are also parasites of humans) (Natarajan *et al.*, 1965), although *Biomphalaria glabrata*, the predominant planorbid vector of *Schistosoma mansoni*, is diploid (Goldman *et al.*, 1984) and has one of the smallest genomes among gastropods (Gregory, 2003). Polyploid *Bulinus* also appear to grow more slowly and to estivate more poorly under dry conditions than diploids (Goldman *et al.*, 1983). Yet, despite these specific disadvantages of polyploidy, the tetraploid *B. truncatus* is one of the most widespread freshwater molluscs in Africa, a fact that may be attributable to the benefits of hybrid vigor (Goldman *et al.*, 1983).

This propensity for polyploids to be widespread may also have contributed to the success of "*Potamopyrgus jenkensi*" (actually a form of *P. antipodarum*, family Hydrobiidae), which is an invader of Britain and mainland Europe that was accidentally introduced in the early- or mid-19th century from New Zealand (possibly via Australia) (Hughes, 1996). "*P. jenkensi*" is a hardy snail with wide salinity and thermal tolerances and which seems able to colonize new areas quickly (Hughes, 1996). Most populations of *P. antipodarum* in their native New Zealand are bisexual diploids (2n = 34), but a few are apomictic thyletokous triploids (3x = 46–52), including the form initially described separately as *P. jenkensi* in Europe and Australia (Hughes, 1996).

ANNELIDS

Segmented worms of the phylum Annelida are divided into three classes: Polychaeta (marine polychaete worms), Pogonophora (beard worms), and Clitellata (divided into the subclasses Oligochaeta, which includes earthworms and freshwater worms, and Hirudinea, which includes leeches). Polyploid nuclei were produced by potassium chloride treatment in the polychaete *Chaetopterus* as early as 1906 by F.R. Lillie. Although Muller's (1925) prediction that earthworms, which are hermaphroditic, would exhibit polyploidy to the same degree observed in plants has proved overly optimistic, polyploidy has certainly been important in the evolution of some annelid groups.

Indeed, about 40% of cytotypes described in oligochaetes are known to be polyploid, and even this may be an underestimate, because undescribed polyploid forms potentially exist in many species (Christensen, 1980; Casellato, 1987). On the other hand, in polychaetes, which are mostly bisexual, polyploidy seems very rare and is limited to only a few suspected (but unconfirmed) examples (Christensen, 1980). The situation is less clear for leeches, which are typically hermaphroditic but which appear to have both low chromosome numbers (Christensen, 1980) and small genome sizes (Gregory, 2001b), although this is based on a small number of data points. Leeches are sometimes included in lists of animals that exhibit polyploidy, but primary references to this effect are rather difficult to locate. In any case, this is an area in which future work is clearly warranted.

For the most part, autopolyploidy is thought to predominate in oligochaetes, a process that would be facilitated by their hermaphroditic reproductive systems (Christensen, 1961, 1980). Multivalents are rare in polyploid annelids, but Christensen (1980) argued that there are specific chromosome pairing mechanisms at play that suppress such meiotic anomalies in these presumed autopolyploids. In some cases, less well developed sexual organs may be the only morphological difference of note between diploid and polyploid oligochaetes (Christensen, 1980), and this has sometimes been used to assess the distribution of polyploidy in certain species (e.g., Coates, 1995). The lack of other morphological differences is itself sometimes taken as evidence for autopolyploidy, because allopolyploids might be expected to exhibit intermediate or at least divergent morphologies from their parental species because of their mixed gene complements (e.g., Christensen, 1961, 1980). However, one must bear in mind the examples of hybrid polyploid frogs that exist as cryptic species along with their diploid progenitors. Unlike these frogs, however, most polyploid annelids are not granted separate species status and are instead considered different forms of the same species as their diploid counterparts.

Some species of earthworms display no evidence of polyploidy whatsoever (e.g., *Allolobophora chloritica*, *Eisenia fetida*, and the entire genus *Lumbricus*), whereas others may be exclusively polyploid (e.g., *Eiseniella tetraedra*, *Dendrobaena octaedra*, and *Octolasion tyrtaeum*) (Casellato, 1987). Nearly 50% of earthworms in the family Lumbricidae are polyploid (Casellato, 1987; Viktorov, 1997). The maximum level in this group seems to be decaploidy, which is found in at least two species: *Eisenia rosea* ($10x = 160$ to 174) and *Octolasion cyaneum* ($10x = 190$) (Viktorov, 1997) (Fig. 8.12). Reproduction in most (>60%) polyploid earthworms is thyletokous (apomictic in all cases except the automictic *O. cyaneum*); about 25% are bisexual and at least one is gynogenetic, although in several cases the reproductive system has not been determined (Viktorov, 1997). Triploid earthworms are unisexual, whereas even-numbered polyploids may be either unisexual or hermaphroditic, a difference that presumably arises owing to difficulties with meiosis in the former (Casellato, 1987). Notably, *E. rosea* is bisexual whereas *O. cyaneum* is thyletokous (Viktorov, 1997).

In freshwater oligochaetes of the family Tubificidae, which includes the well-known *Tubifex* worm used as aquarium fish food, more than half of the species that have been studied karyotypically are believed to be polyploid, whereas in the related family Naididae, few likely cases have been reported even though the latter are hermaphroditic (Christensen, 1980). According to Christensen (1980, p. 52), this indicates that "hermaphroditism would thus seem to be a necessary but by no means sufficient precondition for the occurrence of polyploidy." It bears noting,

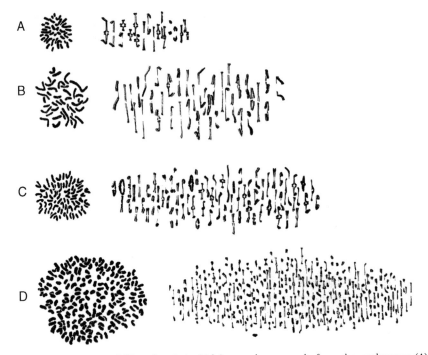

FIGURE 8.12   Mitotic (left) and meiotic (right) metaphase spreads from the earthworms (A) *Lumbricus terrestris* (2n = 36), (B) *Octolasion lacteum* (2n = 36 to 38), (C) *Eiseniella tetraedra* (2n = 72), and (D) *Octolasion cyaneum* (2n ≈ 190). Note that many earthworm species contain both diploid and polyploid forms, but that in some genera (e.g., *Lumbricus*), polyploidy appears to be absent. Based on drawings by Muldal (1952), reproduced by permission (© Nature Publishing Group). All drawings were made at 2500x magnification, except the meiotic figure (right) in (D), which was done at 1260x.

however, that the tubificids are unisexual, whereas the naidids are not, which is probably the more important distinction with respect to polyploid formation (Casellato, 1987).

Like the lumbricids, about 50% of the members of the family Enchytraeidae are polyploid (Christensen, 1980; Coates, 1995). However, in this group the link with unisexual reproduction breaks down, given that up to 75% of the polyploids are bisexual (Christensen, 1980; Bell, 1982). The best-studied species, *Lumbricillus lineatus,* is one of the rare enchytraeid exceptions in which reproduction among polyploids is gynogenetic. This species is common in marine intertidal zones in Europe, North America, and Australia, and is known to frequently include sympatric hermaphroditic diploids along with gynogenetic triploids, tetraploids, and pentaploids (Christensen, 1980; Coates, 1995). According to Coates (1995), the genetic similarity observed between polyploids and diploids implies the repeated generation of polyploids, rather than being explained by gene flow, intrinsic cytogenetic similarities, or responses to similar external pressures. *Lumbricillus* is often found in disturbed areas, for example the sites of oil spills (Suchanek, 1993), and it has been suggested that polyploidy may play a role in their ability to survive this and other environmental stresses (Christensen, 1980; Coates, 1995). Because they are able to feed on oiled organic debris, this may have important practical significance for cleanup efforts (Coates, 1995).

FLATWORMS

The phylum Playthelminthes is generally divided into four classes: the mostly free-living Turbellaria, and the parasitic Cestoda (tapeworms), Trematoda (flukes), and Monogenea (a different type of flukes). Nearly all are potentially self-fertilizing hermaphrodites, and it was noted early on by White (1940) that, at least in the Turbellaria, "there is definite evidence for the occurrence of polyploidy." To be sure, most of the polyploid flatworms now known are turbellarians, mainly in the order Tricladida but also in the Macrostomida and Typhloplanoida (Benazzi and Benazzi Lentati, 1976; Martens *et al.*, 1989).

Good examples of polyploidy at the species level are provided by the turbellarians *Polycelis nigra* and *P. felina*, which include both hermaphroditic diploid and gynogenetic triploid biotypes (Benazzi and Benazzi Lentati, 1976; Beukeboom *et al.*, 1998). In *P. nigra*, these biotypes may exist alone or in sympatry, but the occasional tetraploids that occur are found only in a few otherwise purely triploid populations (Beukeboom *et al.*, 1998). A similar situation of particular interest has recently been revealed in the turbellarian *Schmidtea polychroa*, which are hermaphrodites incapable of self-fertilization. Diploid bisexual and triploid gynogenetic forms also exist in this species, but what sets this case apart is that all biotypes produce functional haploid sperm, so that the gynogens do not need to live in sympatry with related bisexuals (D'Souza *et al.*, 2004). Paternal introgression occurs occasionally as sperm from either a diploid bisexual or a fellow triploid fertilizes a triploid egg to produce tetraploids. Importantly, tetraploids can sometimes produce diploid eggs by meiosis, which, when fertilized by the haploid sperm from a triploid or diploid, produce triploid offspring. In other words, similar to the situation in *Ambystoma* salamanders, there is a continual alternation of ploidy levels occurring in this species, from $3x$ to $4x$ and back again. There can also be exchange of genetic material without changing ploidy level, presumably as one paternal genome is swapped for another (D'Souza *et al.*, 2004). Tetraploids have much lower fecundity than triploids, but they are obviously very important in this system of "occasional sex" because they play a key role in preserving genetic diversity and in mitigating the long-term difficulties of pure unisexuality (D'Souza *et al.*, 2004).

Various reproductive systems can be associated with polyploidy in turbellarians, even within a single genus. *Dugesia lugubris* and *D. benazzii* contain both hermaphroditic ($2x$) and gynogenetic ($3x$, $4x$, and $6x$) biotypes (Benazzi and Benazzi Lentati, 1976; Macgregor, 1993), whereas different forms of *D. tigrina* reproduce either bisexually or by fission, and in the latter case both diploids and triploids are found (Benazzi and Benazzi Lentati, 1976). *Crenobia alpina* also appears to have fissiparous $2x$, $3x$, and $6x$ forms (Benazzi and Benazzi Lentati, 1976). At least two species of turbellarians with triploid thyletokous forms are also known, *Gyratrix hermaphroditus* and *Tetracelis marmorosa* (Benazzi and Benazzi Lentati, 1976; Benazzi, 1982). Finally, comparisons of chromosome numbers also suggest the existence of $3x$, $4x$, and sometimes $8x$ or even $10x$ in species of *Dendrocoelum*, *Macrostomum*, *Mesostoma*, *Phaenocora*, and *Phagocata* (Benazzi and Benazzi Lentati, 1976; White, 1978; Benazzi, 1982). Diploid–polyploid mosaicism is also found in turbellarians (Hoshino *et al.*, 1991). On a higher taxonomic scale, there is good cytogenetic evidence that all of the more recently derived members of the turbellarian order Proseriata formed by polyploidization from a common ancestor with the diploid family Monocelididae (Martens *et al.*, 1989). Notably, it seems that these triploids tend to inhabit more stressful environments than their diploid relatives (Martens *et al.*, 1989).

Triploidy has also been reported in nonturbellarian platyhelminthes, where it is typically associated with unisexuality. Thus, the tapeworms *Atractolytocestus huronensis* and *Diphyllobothrium erinacei* are thyletokous (auto)triploids (Jones and Mackiewicz, 1969; Sasada, 1978). Diploid and (presumed auto)triploid forms are known in the tapeworm species *Caryophyllaeus laticeps*, *Gladiacris catostomi*, and *G. laruei*, and may also occur in *G. oligorchis* (Grey and Mackiewicz, 1980; Mackiewicz, 1982). The trematode fluke *Allocreadium fasciatusi* is likewise a thyletokous triploid whereas its hermaphroditic relative *A. handiai* is diploid (Ramanjaneyulu and Madhavi, 1984). Unisexuality coupled with triploidy has been found within several additional trematode species, including *Bunodera sacculata* (Cannon, 1971), *Fasciola* sp. (Agatsuma *et al.*, 1994a), and *Paragonimus pulmonalis* and *P. westermani* (Terasaki, 1980; Agatsuma *et al.*, 1994b). Tetraploids are also known in *P. westermani*, and they and the triploids are thought to have formed by hybridization among different diploid strains of this species or with a closely related species (Agatsuma *et al.*, 1992, 1994b).

MISCELLANEOUS INVERTEBRATES

NEMATODES

Most nematodes (roundworms) are bisexual, but gynogenesis, thyletoky, and hermaphroditism are also found in some groups (Triantaphyllou, 1983). Both triploids and tetraploids have been induced artificially in *Caenorhabditis elegans* (the latter by heat shock, the former in a $2x \times 4x$ cross) (Nigon, 1951; Madl and Herman, 1979). More importantly, natural polyploidy is known from several genera of nematodes, including the plant parasites *Longidorus* and *Xiphinema*, in which it is associated with automictic thyletoky. The same may hold in species of *Anguina* and some populations of *Ditylenchus dipsaci* (Triantaphyllou, 1983). *Heterodera trifolii*, *H. lespedezae*, and *H. galeopsidis* also contain triploids and tetraploids in association with unisexuality (Triantaphyllou and Hirschmann, 1978), whereas *H. glycines* includes natural (auto)tetraploids but reproduction is amphimictic (Triantaphyllou and Riggs, 1979). Body size is larger in both artificial and natural polyploids relative to diploids in many nematodes (Nigon, 1951; Madl and Herman, 1979; Triantaphyllou and Riggs, 1979; Triantaphyllou and Hirschmann, 1997).

The best-studied examples in this phylum come from thyletokous species of root knot nematodes of the genus *Meloidogyne*. Natural triploids are found in the apomictic species *M. incognita*, *M. arenaria*, and *M. javanica*, and are probably of hybrid origin (Hugall *et al.*, 1999). Tetraploid (and perhaps up to octoploid) forms are found in two additional species, the automictic *M. hapla* and the apomictic *M. microcephala*. These are believed to be autopolyploids formed from automictic populations (Triantaphyllou, 1984, 1991; Triantaphyllou and Hirschmann, 1997), although they do not show evidence of multivalents at meiosis (Goldstein and Triantaphyllou, 1981). Whatever the case, it is believed that unisexuality evolved first in some members of this genus, thereby facilitating the emergence of polyploidy (Triantaphyllou, 1984).

ROTIFERS

Walsh and Zhang (1992) studied two sympatric morphotypes of the bisexual rotifer *Euchlanis dilatata*, and found that the small morphotype males were haploid and females diploid in accordance with the haplodiploid sex determination system in these animals (arrhenotoky). The large morphotype, however, had a triploid chromosome complement, thereby demonstrating the first known case of polyploidy in rotifers. Diploids reached adult body size sooner, but triploids were larger at each stage. Walsh and Zhang (1992) also mentioned the existence of correlations between body size and temperature and latitude in this species, which may be relevant in terms of ploidy differences among populations. Duplicate genes have been found in the ancient asexual bdelloid rotifers, but recent cytogenetic analysis shows that this is not the result of recent polyploidy; it could, however, relate to an ancient polyploidization event early in the history of this lineage (Mark Welch *et al.*, 2004).

TARDIGRADES

Members of the phylum Tardigrada are small aquatic animals known commonly as "water bears" because of their superficial resemblance to tiny ursid mammals.

Bertolani (1975) has suggested that "the lack of males in many populations of tardigrades and some karyological observations indicate that parthenogenesis and polyploidy are frequent events." In particular, diploid bisexual and triploid apomictic thyletokous biotypes are found in *Macrobiotus richtersi*, and bisexual diploids, both auto- and apomictic triploids, and automictic tetraploids occur in *M. hufelandi*, sometimes in sympatry. Bisexual diploids, and apomictic triploid and tetraploid unisexual biotypes are also found in *Hypsibius oberhaeuseri*, and tetraploids occur in *Isohypsibius granulifer*, which may be a self-fertilizing hermaphrodite (Bertolani, 1975; Bell, 1982). Polyploids are generally larger than diploids in these species (Bertolani, 1975).

ARACHNIDS

The subclass Arachnida represents the diverse group of arthropods that includes spiders, scorpions, mites, ticks, and their kin. Thyletoky occurs sporadically in mites and ticks (subclass Acari), and in at least one species, the ioxidid tick *Haemophysalis longicornis*, both diploid bisexual and triploid thyletokous forms are known (Oliver, 1971). More intensive study in this group may yet reveal many other cases.

POLYPLOIDY AND GEOGRAPHIC DISTRIBUTION IN INVERTEBRATES

Some phenotypic correlates of polyploidy, most notably cell size, are essentially universal. Body size, by contrast, is linked to polyploidy in only some groups of invertebrates and generally not in any vertebrates. Correlations between polyploidy and ecological parameters are also easier to discern among invertebrates, the most notable being the tendency for polyploids to be found in harsher environments, and especially at higher latitudes and/or altitudes. Such patterns have been observed in nearly all of the invertebrate groups discussed in the previous sections.

In crustaceans, for example, the amphipod *Pontoporeia affinis* has expanded across a broader geographic range and diversity of habitats following the last glacial period than its diploid relatives (Salemma, 1984). Polyploid brine shrimp are found more often at low and high latitudes rather than in temperate regions (Zhang and Lefcort, 1991), and polyploid water fleas and ostracod crustaceans dominate in the Arctic but are absent from temperate climes in Europe and North America (Beaton and Hebert, 1988; Havel and Hebert, 1989; Turgeon and Hebert, 1994; Ward et al., 1994; Little and Hebert, 1997).

In insects, geographical distribution is the most striking correlate of polyploidy. The wingless tetraploid katydid *Saga pedo*, the only unisexual member of its genus, is more widely distributed north of the Mediterranean than its bisexual diploid counterparts, and polyploid forms of the moth *Solenobia triquetrella* are distributed much more broadly throughout Europe than diploids (Suomalainen *et al.,* 1987). This pattern is particularly clear in weevils, in which unisexual polyploid forms are much more wide-ranging than bisexual diploids, especially in the north or at high altitudes, in both Europe and Japan (Suomalainen, 1969; Takenouchi, 1976; Suomalainen *et al.,* 1987; and numerous references therein). In molluscs, high altitude springs and very small habitats in Spain are inhabited by polyploid species whereas diploids are found at lower altitudes and in larger water bodies (Barsiene *et al.,* 1996).

With regard to annelids, Omodeo (1951) was the first to point out that polyploid earthworms tend to have wider geographical distributions than diploids, and in the following year Muldal (1952) stated quite succinctly that "On the whole it can be safely concluded that within their genera the polyploids are all the most successful and widespread species." However, there are clear exceptions: most notably, the cosmopolitan and archetypal earthworms in the genus *Lumbricus,* in which polyploidy is either absent or very rare (Viktorov, 1997). Nevertheless, in at least one example (*Eisenia nordenskiolodi*), bisexual octoploids are the most widespread form, whereas thyletokous septaploids are less successful (Viktorov, 1997). As with the earthworms, polyploids are geographically widespread in *Lumbricillus lineatus,* but here their gynogenetic mode of reproduction means that polyploids must still obtain sperm from diploids, tetraploids, or pentaploids (triploids cannot produce viable sperm) (Christensen, 1980; Coates, 1995).

Numerous explanations have been offered to account for these patterns. These have not always been mutually exclusive, and in some cases several of the factors discussed in the following sections may be important.

UNISEXUALITY

The first report of the correlation in an invertebrate between ploidy and latitude was provided by Vandel (1928) in his study of the terrestrial isopod *Trichoniscus pusillus*. However, this pattern was not phrased in terms of a correlation with ploidy, but rather as "geographic parthenogenesis" (Vandel, 1928). Because polyploidy and unisexuality are commonly linked in invertebrates, it is very difficult to determine which feature is the most important. Unisexuality has obvious advantages in terms of capabilities for dispersal and rapid reproduction (Bell, 1982), and it has been suggested by some authors that the capacity for colonization by single individuals explains why unisexual, and hence polyploid, earthworms (for example) have wider distributions than diploids (Christensen, 1980; Casellato, 1987; Terhivuo, 1988).

It has also been argued that unisexuality provides an increased tolerance for unstable or disturbed environments. As Wright and Lowe (1968) put it in reference to *Cnemidophorus* lizards, allotriploid unisexuals are essentially "weeds" that inhabit environments considered unsuitable by their bisexual diploid relatives. In *Trichoniscus pusillus,* the original "geographic parthenogen," it now appears that the association is not so much with geography (e.g., latitude) or climate (e.g., temperature), but with the geology. Specifically, unisexual (3*x*) *T. pusillus* are more abundant in areas with calcareous soils and rocks, perhaps because environments with a limestone base contain fluctuating microclimates to which they are better adapted than bisexual (2*x*) forms (Fussey, 1984).

Not all cases of "geographic polyploidy" are potentially "geographic parthenogenesis" in disguise, however. In some plants, for example, it has been possible to disentangle the two features by comparing related diploid and polyploid bisexual species (e.g., Bierzychudek, 1985). It has been possible in crustaceans to perform the converse comparison, namely between diploid and polyploid unisexuals (Adamowicz *et al.,* 2002). Again, in the unisexual ostracod *Prinocypris glacialis,* triploid and pentaploid forms are found in the high Arctic, whereas at lower Arctic latitudes a mixture of diploids and triploids predominates (Little and Hebert, 1997). There is no shift in breeding system at the higher elevation, suggesting that polyploidy itself is of some significance independent of unisexuality. In like fashion, both North American and European high Arctic populations of the *Daphnia pulex* complex are exclusively polyploid, whereas low Arctic environments contain both diploids and polyploids, and temperate populations consist entirely of diploids (Beaton and Hebert, 1988; Ward *et al.,* 1994). All but the more southerly populations are obligate parthenogens, so once again it would seem that polyploidy itself plays a role in determining the patterns of distribution. As such, the question remains: Why should polyploids occur in harsher environments, especially in the cold? Several possible answers exist, as discussed in the following sections.

GENETIC FACTORS

Most simply, both allo- and autopolyploids often exhibit more allelic variation than diploids, which could contribute to their ability to invade new habitats not accessible to their diploid relatives (Suomalainen, 1969; Suomalainen *et al.,* 1987; Otto and Whitton, 2000). Allopolyploids, in particular, may be at an advantage owing to the heterotic benefits of hybridization. Moreover, having multiple copies of the genome may provide a buffer against the accumulation of mutations, and therefore grant a longer lifespan to polyploid versus diploid unisexual lineages. It has been argued that in *Daphnia* this increased clonal longevity allows polyploids to extend farther north of where they are formed than their diploid unisexual counterparts (Ward *et al.,* 1994). Taking an opposite approach, it has also been pointed out that unreduced gametes occur more commonly in the cold, so that perhaps being in a harsh environment promotes the formation of polyploids (Otto and Whitton, 2000; Mable, 2004a). Certainly, heat and cold shock are two of the most common means of inducing polyploidy in the laboratory.

PHYSIOLOGY, DEVELOPMENT, AND ECOLOGY

Another obvious possibility is that polyploidy itself may directly affect parameters relevant for survival in cold climates. For example, polyploid clones in the *Daphnia pulex* complex have been shown under experimental conditions to produce larger (and fewer) eggs and neonates and to display larger body sizes than diploid clones (Dufresne and Hebert, 1998). Body size is correlated with ploidy level in several of the other invertebrate groups discussed previously as well, which almost certainly relates to the effects of DNA content on cell size (Gregory, 2001a,c). This is in keeping with a "few, large offspring strategy" being favored in the cold, where ecological complexity is low and the pressures of predation and competition are relaxed (Yampolsky and Scheiner, 1996).

Polyploid *Daphnia* also develop more quickly at 10°C (but more slowly at 24°C) as compared to related diploid clones (Dufresne and Hebert, 1998), and polyploid *Artemia parthenogenetica* have slower growth rates than diploids under ideal salinity conditions but tend to outgrow diploids under low salinity stress (Zhang and King, 1993). Polyploid *Artemia* also have lower experimental mortality rates than diploids at extreme temperatures (both high and low), and pentaploid *Artemia* voluntarily occupy a much wider range of temperatures in experimental thermal gradients (especially at the colder end) than diploids (Zhang and Lefcort, 1991). There is some indication that polyploid weevils can tolerate lower temperatures than diploids, but more work is needed in this area (Suomalainen *et al.*, 1987).

HISTORY

As Soumalainen (1969) noted, ecological factors such as those described in the previous section do not preclude the action of historical ones, which may be just as important. In the north, the most important historical factor is glaciation, the most recent of which ended with the Pleistocene about 10,000 years ago. Both terrestrial and aquatic organisms were greatly affected by the advance of the ice sheets, and those that survived usually did so by remaining in unfrozen refugia.

There are at least two ways that these refugia could affect current polyploid distributions. The first relates to the higher dispersal capacities of existing polyploid unisexuals versus diploid bisexuals after the retreat of the ice. Unisexual tetraploid *Solenobia triquetrella* moths, for example, have greatly expanded their ranges to cover much of Europe since the end of the last ice age, "while the bisexuals stay where they were left 10,000 years ago [the southeastern border of the Swiss Alps]" (Suomalainen *et al.*, 1987).

In the second scenario, segregation into glacial refugia is seen as actually promoting the generation of polyploids. In *Daphnia*, the Arctic polyploids are of hybrid origin, and it has been argued that the separation, divergence, and subsequent hybridization upon contact of refugial populations was responsible for creating polyploids *in situ* (Dufresne and Hebert, 1994, 1997). That said, it is important to note that no polyploid *Daphnia* have ever been found in temperate regions of either North America or Europe despite rather exhaustive sampling efforts, even though hybridization is known to occur frequently in these regions (Adamowicz *et al.*, 2002). In other words, the opportunity to form polyploids by hybridization clearly exists in temperate regions, yet no polyploidy is observed there. This clearly suggests that there is more to the explanation of polyploid dominance in the Arctic (and absence in more southerly climes) than opportunities for formation, although these historical factors should not be dismissed.

Interestingly, the *Daphnia* fauna of southern Argentina—a temperate region—consists mostly of polyploids, so these are obviously capable of surviving in warmer conditions (Adamowicz *et al.*, 2002). In this case, the historical factor of incumbency seems to play a particularly important role, and "polyploids may dominate the temperate regions of Argentina solely because of a fortuitous priority of arrival" (Adamowicz *et al.*, 2002). Notably, *Daphnia* disperse passively when their resting eggs are carried by birds, and bird migration routes to this region usually involve flights directly from the Arctic, where the source populations are predominantly polyploid (Adamowicz *et al.*, 2002).

THE EVOLUTIONARY FATE OF POLYPLOIDS

Polyploidy is often assumed to represent an evolutionary dead end, largely because of its association with unisexual reproduction. But this tendency to view polyploidy and unisexuality as "tantamount to phylogenetic suicide" has been criticized before (e.g., Bogart, 1980) and it is now becoming apparent that many polyploid lineages are quite successful and in some cases rather long-lived. Vertebrates in general, and teleost fishes in particular, appear to have undergone polyploidization events in their early ancestry, but have certainly not suffered as a consequence (see Chapter 6). Groups containing more recent polyploids like the cyprinid fishes are similarly faring quite well in terms of diversity, and even unisexual polyploid salamander lineages may persist for millions of years (Hedges *et al.*, 1992; Spolsky *et al.*, 1992).

There are several reasons that polyploidy is not necessarily an evolutionarily terminal condition. First, rediploidization may occur, making the group paleopolyploid but functionally diploid (see Chapter 6). Second, polyploidy may persist within a species for a long time through recurrent formation. Third, polyploids may have important advantages in terms of improved diversification ability, increased tolerance to harsh conditions, and wider geographic range, making them more resistant to extinction than their diploid counterparts and thereby providing improved success in lineage-level selection (see Chapter 11). Finally, although genetic heterozygosity is often assumed to be low in (auto)polyploids, in actuality it often equals or surpasses that found in diploids (Otto and Whitton, 2000). Even when unisexuality is involved, there may be mechanisms in place that resupply genetic variation, particularly among gynogens and hybridogens that "mate" with related bisexuals (e.g., Bogart, 2003; D'Souza *et al.*, 2004). The polyploidization event itself may also generate considerable genomic variation, as has been noted in several plants (see Chapter 7).

Overall, there is no reason to expect polyploidy to automatically imply imminent evolutionary doom. Far from it, in fact, because some of the most successful lineages on the planet have a high degree of polyploidy in their ancestry.

CONCLUDING REMARKS AND FUTURE PROSPECTS

Polyploidy is clearly not as common in animals as it is in plants, but neither is it anywhere near as rare as often assumed. In part, the relatively low rate of discovery of polyploidy in animals reflects the low level of effort put into looking for it, which in turn is probably driven by the low expectation of success in finding it. Furthermore, the assessment of polyploid rarity is somewhat subject to the usual bias in zoological research toward the "charismatic megafauna" (i.e., mammals and birds) and model organisms like *Drosophila*, in which polyploidy is indeed rare but which are not necessarily typical of the kingdom at large.

Nevertheless, the study of animal polyploidy does persist among cytogeneticists. In this regard, much active research is currently being done to elucidate the histories and implications of both recent and ancient polyploidization events. As with plants, studies such as these are revealing an increasing number of cases in which polyploids have formed repeatedly (see Chapter 7). There is also a renewed and growing interest in the genetic and evolutionary consequences of polyploidy in eukaryotes in general, stimulated by the recent discovery of paleopolyploidy in numerous unexpected places (e.g., vertebrates, yeast, *Arabidopsis thaliana*) (see Chapter 6). As noted recently, there has been a substantial increase in publications relating to polyploidy since the late 1990s (Mable, 2003), with the number of citations of Ohno's (1970) *Evolution by Gene Duplication* tripling between 1990 and 2000 (Wolfe, 2001).

All of the most fundamental questions about animal polyploidy remain to be answered, including the mechanisms that promote it, the proportion of cases that represent auto- versus allopolyploidy, and the various molecular, cytological, physiological, reproductive, ecological, and evolutionary consequences of genome doubling. Even the most obvious questions of why polyploidy is rarer in animals than in plants, and why it is often associated with unisexuality, have not yet been fully resolved. In more specific instances, molecular techniques can be used to elucidate the parental species that contributed genomes to a known polyploid and to study the structure of the resulting composite genome. The possession of the complete genome sequence of at least one semirecent polyploid animal (the rainbow trout, *Oncorhynchus mykiss*) in the near future will undoubtedly shed a great deal of light on such issues.

One by one, the traditional assumptions about animal polyploidy have faltered in the face of new evidence. For example, it was believed until very recently that polyploidy could not occur in mammals, but then *Tympanoctomys barrerae* was identified as a polyploid in 1999. It was also long assumed that triploidy absolutely required unisexuality to persist, but enter the bisexual triploid frog *Bufo pseudoraddei baturae* in 2002. The supposed rules about animal polyploidy, it would seem, are meant to be broken. It is therefore a virtual guarantee that many additional surprises will be found in the study of animal polyploidy—so long as an effort is made to uncover them.

REFERENCES

Abdel-Hameed F, and Schoffner RN. 1971. Intersexes and sex determination in chickens. *Science* 172: 962–964.

Adamowicz SJ, Gregory TR, Marinone MC, Hebert PDN. 2002. New insights into the distribution of polyploid *Daphnia*: the Holarctic revisited and Argentina explored. *Mol Ecol* 11: 1209–1217.

Agatsuma T, Ho L, Jian H, Habe S, *et al.* 1992. Electrophoretic evidence of a hybrid origin for tetraploid *Paragonimus westermani* discovered in northeastern China. *Parasitol Res* 78: 537–538.

Agatsuma T, Terasaki K, Yang L, Blair D. 1994a. Genetic variation in the triploids of Japanese *Fasciola* species, and relationships with other species in the genus. *J Helminthol* 68: 181–186.

Agatsuma T, Yang L, Kim D, Yonekawa H. 1994b. Mitochondrial DNA differentiation of Japanese diploid and triploid *Paragonimus westermani*. *J Helminthol* 68: 7–11.

Allen SK. 1988. Triploid oysters ensure year-round supply. *Oceanus* 31: 58–63.

Allendorf FW, Danzmann RG. 1997. Secondary tetrasomic segregation of *MDH-B* and preferential pairing of homeologues in rainbow trout. *Genetics* 145: 1083–1092.

Allendorf FW, Thorgaard GH. 1984. Tetraploidy and the evolution of salmonid fishes. In: Turner BJ ed. *Evolutionary Genetics of Fishes*. New York: Plenum Press, 1–53.

Allendorf FW, Leary RF, Spruell P, Wenburg JK. 2001. The problems with hybrids: setting conservation guidelines. *Trends Ecol Evol* 16: 613–622.

Alves MJ, Coelho MM, Collares-Pereira MJ. 1997. The *Rutilus alburnoides* complex (Cyprinidae): evidence for a hybrid origin. *J Zool Syst Evol Res* 35: 1–10.

Alves MJ, Coelho MM, Collares-Pereira MJ. 2001. Evolution in action through hybridisation and polyploidy in an Iberian freshwater fish: a genetic review. *Genetica* 111: 375–385.

Alves MJ, Coelho MM, Prospero MI, Collares-Pereira MJ. 1999. Production of fertile unreduced sperm by hybrid males of the *Rutilus alburnoides* complex (Teleostei, Cyprinidae): an alternative route to genome tetraploidization in unisexuals. *Genetics* 151: 277–283.

AmphibiaWeb. 2004. *AmphibiaWeb: Information on Amphibian Biology and Conservation*. Berkeley, CA. http://amphibiaweb.org.

Arai K, Matsubara K, Suzuki R. 1993. Production of polyploids and viable gynogens using spontaneously occurring tetraploid loach *Misgurnus anguillicaudatus*. *Aquaculture* 117: 227–235.

Artom C. 1906. Ricerche sperimentali sul modo di riprodursi dell'*Artermia salina* Lin. di Cagliari. *Biol Zentralbl* 26: 26–32.

Artom C. 1911. Analisi comparativa della sostanza cromatica nelle mitosi di maturazione e nelle prime mitosi di segmentazione dell'uovo dell'*Artemia* sessuate di Cagliari (*univalens*) e dell'uovo dell'*Artemia partenogenetica* di Capodistria (*bivalens*). *Arch Zellforsch* 7: 277–295.

Artom C. 1926. Tetraploidismo e gigantismo: essame comparativo degli stadi postembrionali dell'*Artemia salina* diploide e tetraploide. *Int Rev gesamten Hydrobiol* 16: 51–80.

Austin NE, Bogart JP. 1982. Erythrocyte area and ploidy determination in the salamanders of the *Ambystoma jeffersonianum* complex. *Copeia* 1982: 485–488.

Balsano JS, Darnell RM, Abramoff P. 1972. Electrophoretic evidence of triploidy associated with populations of the gynogenetic teleost *Poecilia formosa*. *Copeia* 1972: 292–297.

Barigozzi C. 1974. *Artemia*: a survey of its significance in genetic problems. *Evol Biol* 7: 221–252.

Barrio A, Rinaldi de Chieri P. 1970. Estudios citogenétocos sobre el genero *Pleurodema* y sus consecuencias evolutivas (Amphibia, Anura, Leptodactylidae). *Physis* 30: 309–319.

Barsiene J, Tapia G, Barsyte D. 1996. Chromosomes of molluscs inhabiting some mountain springs of eastern Spain. *J Mollusc Stud* 62: 539–543.

Batistic RF, Soma M, Beçak ML, Beçak W. 1975. Further studies on polyploid amphibians: a diploid population of *Phyllomedusa burmeisteri*. *J Hered* 66: 160–162.

Beaton MJ, Hebert PDN. 1988. Geographic parthenogenesis and polyploidy in *Daphnia pulex*. *Am Nat* 132: 837–845.

Beatty RA. 1957. *Parthenogenesis and Polyploidy in Mammalian Development*. Cambridge, UK: Cambridge University Press.

Beaumont A, Fairbrother J. 1991. Ploidy manipulation in molluscan shelfish: a review. *J Shellfish Res* 10: 1–18.

Beçak ML, Beçak W. 1970. Further studies on polyploid amphibians (Ceratophrydidae). III. Meiotic aspects of the interspecific triploid hybrid: *Odontophrynus cultripes* (2n = 22) × *O. americanus* (4n = 44). *Chromosoma* 31: 377–385.

Beçak ML, Beçak W. 1974. Diploidization in *Eleutherodactylus* (Leptodactylidae-Amphibia). *Experientia* 30: 624–625.

Beçak ML, Beçak W. 1998. Evolution by polyploidy in Amphibia: new insights. *Cytogenet Cell Genet* 80: 28–33.

Beçak ML, Beçak W, Rabello MN. 1966. Cytological evidence of constant tetraploidy in the bisexual South American frog *Odontophrynus americanus*. *Chromosoma* 19: 188–193.

Beçak ML, Beçak W, Rabello MN. 1967a. Further studies on polyploid amphibians (Ceratophrydidae). I. Mitotic and meiotic aspects. *Chromosoma* 22: 192–201.

Beçak W, Beçak ML, Lavalle D, Schreiber G. 1967b. Further studies on polyploid amphibians (Ceratophrydidae). II. DNA content and nuclear volume. *Chromosoma* 23: 14–23.

Bell G. 1982. *The Materpiece of Nature*. Berkeley, CA: University of California Press.

Benazzi M. 1982. Speciation events evidenced in Turbellaria. In: Barigozzi C ed. *Mechanisms of Speciation*. New York: Alan R. Liss, 301–344.

Benazzi M, Benazzi Lentati G. 1976. *Animal Cytogenetics, Vol. 1: Platyhelminthes*. Berlin: Gebrüder Borntraeger.

Benfey TJ. 1999. The physiology and behavior of triploid fishes. *Rev Fish Sci* 7: 39–67.

Bertolani R. 1975. Citology and systematics in Tardigrada. *Mem Inst Ital Idrobiol* 32 (Suppl.): 18–35.

Beukeboom LW, Sharbel TF, Michiels NK. 1998. Reproductive modes, ploidy distribution, and supernumerary chromosome frequencies of the flatworm *Polycelis nigra* (Platyhelminthes: Tricladida). *Hydrobiologia* 383: 277–285.

Bickham JW. 1984. Patterns and modes of chromosomal evolution in reptiles. In: Sharma AK, Sharma A eds. *Chromosomes in Evolution of Eukaryotic Groups, Vol. II*. Boca Raton, FL: CRC Press, 13–40.

Bickham JW, Hanks BG, Hale DW, Martin JE. 1993. Ploidy diversity and the production of balanced gametes in male twist-necked turtles (*Platemys platycephala*). *Copeia* 1993: 723–727.

Bickham JW, Tucker PK, Legler JM. 1985. Diploid-triploid mosaicism: an unusual phenomenon in side-necked turtles (*Platemys platycephala*). *Science* 227: 1591–1593.

Bierzychudek P. 1985. Patterns in plant parthenogenesis. *Experientia* 41: 1255–1264.

Birstein VJ, DeSalle R. 1998. Molecular phylogeny of Acipenserinae. *Mol Phylogenet Evol* 9: 141–155.

Birstein VJ, Hanner R, DeSalle R. 1997. Phylogeny of the Acipenseriformes: cytogenetic and molecular approaches. In: Birstein VJ, Waldman JR, Bemis WE eds. *Sturgeon Biodiversity and Conservation.* Dordrecht, The Netherlands: Kluwer Academic Publishers, 127–155.

Birstein VJ, Pletaev AI, Goncharov BF. 1993. DNA content in Eurasian sturgeon species determined by flow cytometry. *Cytometry* 14: 377–383.

Blacklidge KH, Bidwell CA. 1993. Three ploidy levels indicated by genome quantification in Acipenseriformes of North America. *J Hered* 84: 427–430.

Blair WF. 1958. Mating call in the speciation of anuran amphibians. *Am Nat* 92: 27–51.

Bloom SE. 1972. Chromosome abnormalities in chicken (*Gallus domesticus*) embryos: types, frequencies and phenotypic effects. *Chromosoma* 37: 309–326.

Bogart JP. 1980. Evolutionary implications of polyploidy in amphibians and reptiles. In: Lewis WH ed. *Polyploidy: Biological Relevance.* New York: Plenum Press, 341–378.

Bogart JP. 2003. Genetics and systematics of hybrid species. In: Sever DM ed. *Reproductive Biology and Phylogeny of Urodela, Vol. 1.* Enfield, NH: Science Publishers Inc., 109–134.

Bogart JP, Licht LE. 1986. Reproduction and the origin of polyploids in hybrid salamanders of the genus *Ambystoma. Can J Genet Cytol* 28: 605–617.

Bogart JP, Tandy M. 1976. Polyploid amphibians: three more diploid-tetraploid cryptic species of frogs. *Science* 193: 334–335.

Bogart JP, Tandy M. 1981. Chromosome lineages in African ranoid frogs. *Monit Zool Ital* Suppl. 15: 55–91.

Bogart JP, Wasserman AO. 1972. Diploid-polyploid cryptic species pairs: a possible clue to evolution by polyploidization in anuran amphibians. *Cytogenetics* 11: 7–24.

Bogart JP, Elinson RP, Licht LE. 1989. Temperature and sperm incorporation in polyploid salamanders. *Science* 246: 1032–1034.

Bogart JP, Lowcock LA, Zeyl CW, Mable BK. 1987. Genome constitution and reproductive biology of the *Ambystoma* hybrid salamanders on Kelleys Island in Lake Erie. *Can J Zool* 65: 2188–2201.

Borkin LJ, Litvinchuk SN, Rosanov JM. 1996. Spontaneous triploidy in the crested newt, *Triturus cristatus* (Salamandridae). *Russ J Herpetol* 3: 152–156.

Brenowitz EA, Rose G, Capranica RR. 1985. Neural correlates of temperature coupling in the vocal communication system of the gray treefrog (*Hyla versicolor*). *Brain Res* 359: 364–367.

Browne RA, Bowen ST. 1991. Taxonomy and population genetics of *Artemia.* In: Browne RA, Sorgeloos P, Trotman CN eds. *Artemia Biology.* Boca Raton, FL: CRC Press, 221–235.

Bull JJ, Legler JM. 1980. Karyotypes of side-necked turtles (Testudines: Pleurodira). *Can J Zool* 58: 828–841.

Burch JB. 1960. Chromosomes of *Gyraulus circumstriatus*, a freshwater snail. *Nature* 186: 497–498.

Burch JB, Huber JM. 1966. Polyploidy in mollusks. *Malacologia* 5: 41–43.

Burch JB, Jung Y. 1993. Polyploid chromosome numbers in the *Torquis* group of the freshwater snail genus *Gyraulus* (Mollusca: Pulmonata: Planorbidae). *Cytologia* 58: 145–149.

Burch JB, Park G-B, Chung E-Y. 1998. Michigan's polyploid clams. *Mich Acad* 30: 351–352.

Buth DG, Dowling TE, Gold JR. 1991. Molecular and cytological investigations. In: Winfield I, Nelson J eds. *The Biology of Cyprinid Fishes.* London: Chapman & Hall, pp. 83–126.

Cannatella DC, de Sá RO. 1993. *Xenopus laevis* as a model organism. *Syst Biol* 42: 476–507.

Cannatella DC, Trueb L. 1988. Evolution of pipoid frogs: intergeneric relationships of the aquatic frog family Pipidae (Anura). *Zool J Linn Soc* 94: 1–38.

Cannon LRG. 1971. The life cycle of *Bunodera sacculata* and *B. luciopercae* (Trematoda: Allocreadiidae) in Algonquin Park, Ontario. *Can J Zool* 49: 1417–1429.

Capanna E, Cataudella S, Volpe R. 1974. An intergenic hybrid between the rainbow trout and the freshwater char (*Salmo gairdneri* × *Salvelinus fontinalis*). *Boll Pesca Pisci Idrobiol* 29: 101–106.

Carmona JA, Sanjur OI, Doadrio I, et al. 1997. Hybridogenetic reproduction and maternal ancestry of polyploid Iberian fish: the *Tropidophoxinellus alburnoides* complex. *Genetics* 146: 983–993.

Casellato S. 1987. On polyploidy in oligochaetes with particular reference to lumbricids. In: Bonvicini Pagliai AM, Omodeo P eds, *On Earthworms*. Modena, Italy: Mucchi, 75–87.

Castellano S, Tontini L, Giacoma C, et al. 2002. The evolution of release and advertisement calls in green toads (*Bufo viridis* complex). *Biol J Linn Soc* 77: 379–391.

Centofante L, Bertollo LAC, Moreira O. 2001. Comparative cytogenetics among sympatric species of *Characidium* (Pisces, Characiformes). Diversity analysis with the description of a ZW sex chromosome system and natural triploidy. *Caryologia* 54: 253–260.

Channing A, Bogart JP. 1996. Description of a tetraploid *Tomopterna* (Anura: Ranidae) from South Africa. *S Afr J Zool* 31: 80–85.

Charlesworth D. 2002. Plant sex determination and sex chromosomes. *Heredity* 88: 94–101.

Chenuil A, Galtier N, Berrebi P. 1999. A test of the hypothesis of an autopolyploid vs. allopolyploid origin for a tetraploid lineage: application to the genus *Barbus* (Cyprinidae). *Heredity* 82: 373–380.

Christensen B. 1961. Studies on cyto-taxonomy and reproduction in the Enchytraeidae, with notes on parthenogenesis and polyploidy in the animal kingdom. *Hereditas* 47: 387–450.

Christensen B. 1980. *Animal Cytogenetics, Vol. 2: Annelida*. Berlin: Gebrüder Borntraeger.

Cimino MC. 1972. Meiosis in triploid all-female fish (*Poeciliopsis*, Poeciliidae). *Science* 175: 1484–1486.

Cimino MC. 1973. Karyotypes and erythrocyte sizes of some diploid and triploid fishes of the genus *Poeciliopsis*. *J Fish Res Bd Can* 30: 1736–1737.

Cimino MC. 1974. The nuclear DNA content of diploid and triploid *Poeciliopsis* and other poeciliid fishes with reference to the evolution of unisexual forms. *Chromosoma* 47: 297–307.

Cimino MC, Schultz RJ. 1970. Production of a diploid male offspring by a gynogenetic triploid fish of the genus *Poeciliopsis*. *Copeia* 1970: 760–763.

Coates KA. 1995. Widespread polyploid forms of *Lumbricillus lineatus* (Müller) (Enchytraeidae: Oligochaeta): comments on polyploidism in the enchytraeids. *Can J Zool* 73: 1727–1734.

Coleman CC. 1994. Karyological studies in amphipoda (Crustacea). *Ophelia* 9: 93–105.

Collares-Pereira MJ, Coelho MM. 1989. Polyploidy versus diploidy: a new model for the karyological evolution of Cyprinidae. *Arq Mus Bocage, N.S.* 1: 375–383.

Collares-Pereira MJ, Madeira JM. 1995. Spontaneous triploidy in the stone loach *Neomacheilus barbatulus* (Balitoridae). *Copeia* 1995: 483–484.

Collares-Pereira MJ, Moreira da Costa L. 1999. Intraspecific and interspecific genome size variation in Iberian Cyprinidae and the problem of diploidy and polyploidy, with review of genome sizes within the family. *Folia Zool* 48: 61–76.

Comeau NM. 1943. Une ambystome nouvelle. *Ann Assoc Can Fr Adv Sci* 9: 124–125.

Cope ED. 1867. A review of the species of the Amblystomidae. *Proc Acad Nat Sci Philadelphia* 19: 166–211.

Creasy MR, Crolla JA, Alberman ED. 1976. A cytogenetic study of human spontaneous abortions using banding techniques. *Hum Genet* 31: 177–196.

Cuellar O. 1971. Reproduction and the mechanism of meiotic restitution in the parthenogenetic lizard *Cnemidophorus uniparens*. *J Morphol* 133: 139–165.

D'Souza TG, Storhas M, Schulenburg H, et al. 2004. Occasional sex in an "asexual" polyploid hermaphrodite. *Proc R Soc Lond B* 271: 1001–1007.

Danzmann RG, Bogart JP. 1982. Gene dosage effects on MDH isozyme expression in diploid, triploid, and tetraploid treefrogs of the genus *Hyla*. *J Hered* 73: 277–280.

Darevsky IS. 1992. Evolution and ecology of parthenogenesis in reptiles. In: Adler K ed. *Herpetology: Current Research on the Biology of Amphibians and Reptiles. Proceedings of the 1st World Conference of Herpetology*. Oxford, OH: Society for the Study of Amphibians and Reptiles, 21–39.

David L, Blum S, Feldman MW, et al. 2003. Recent duplication of the common carp (*Cyprinus carpio* L.) genome as revealed by analyses of microsatellite loci. *Mol Biol Evol* 20: 1425–1434.

Dawley RM. 1992. Clonal hybrids of the common laboratory fish *Fundulus heteroclitus*. *Proc Natl Acad Sci USA* 89: 2485–2488.

Dawley RM, Goddard KA. 1988. Diploid-triploid mosaics among unisexual hybrids of the minnows *Phoxinus eos* and *Phoxinus neogaeus*. *Evolution* 42: 649–659.

Dawley RM, Rupprecht JD, Schultz RJ. 1997. Genome size of bisexual and unisexual *Poeciliopsis*. *J Hered* 88: 249–252.

de Sá RO, Hillis DM. 1990. Phylogenetic relationships of the pipid frogs *Xenopus* and *Silurana*: an integration of ribosomal DNA and morphology. *Mol Biol Evol* 7: 365–376.

de Wet JMJ. 1980. Origins of polyploids. In: Lewis WH ed. *Polyploidy: Biological Relevance*. New York: Plenum Press, 3–15.

Deiana AM, Cau A, Coluccia E, *et al.* 1999. Genome size and AT-DNA content in thirteen species of decapoda. In: Schram FR, von Vaupel Klein JC eds. *Crustaceans and the Biodiversity Crisis*. Leiden, The Netherlands: Koninklijke Brill NV, 981–985.

Dessauer HC, Cole CJ. 1989. Diversity between and within nominal forms of unisexual teiid lizards. In: Dawley RM, Bogart JP eds. *Evolution and Ecology of Unisexual Vertebrates*. Albany, NY: New York State Museum, 49–70.

Dingerkus G, Howell MW. 1976. Karyotypic analysis and evidence of tetraploidy in the North American paddlefish, *Polyodon spathula*. *Science* 194: 842–844.

Dries LA. 2003. Peering through the looking glass at a sexual parasite: are Amazon mollies red queens? *Evolution* 57: 1387–1396.

Dufresne F, Hebert PDN. 1994. Hybridization and the origins of polyploidy. *Proc R Soc Lond B* 258: 141–146.

Dufresne F, Hebert PDN. 1997. Pleistocene glaciations and polyphyletic origins of polyploidy in an arctic cladoceran. *Proc R Soc Lond B* 264: 201–206.

Dufresne F, Hebert PDN. 1998. Temperature-related differences in life-history characteristics between diploid and polyploid clones of the *Daphnia pulex* complex. *Écoscience* 5: 433–437.

Echelle AA, Echelle AF, DeBault LE, Durham DW. 1988. Ploidy levels in silverside fishes (Atherinidae, *Menidia*) on the Texas coast: flow-cytometric analysis of the occurrence of allotriploidy. *J Fish Biol* 32: 835–844.

Ellegren H. 2002. Dosage compensation: do birds do it as well? *Trends Genet* 18: 25–28.

Espinoza NR, Noor MAF. 2002. Population genetics of a polyploid: is there hybridization between lineages of *Hyla versicolor*? *J Hered* 93: 81–85.

Eudeline B, Allen SK, Guo X. 2000. Optimization of tetraploid induction in Pacific oysters, *Crassostrea gigas*, using first polar body as a natural indicator. *Aquaculture* 187: 73–84.

Fankhauser G. 1945. The effects of changes in chromosome number on amphibian development. *Quart Rev Biol* 20: 20–78.

Fankhauser G. 1955. The role of nucleus and cytoplasm. In: Willier BH, Weiss PA, Hamburger V eds. *Analysis of Development*. Philadelphia: W.B. Saunders Co., 126–150.

Fauaz G, Vicente VE, Moreira O. 1994. Natural triploidy and B-chromosomes in the neotropical fish genus *Astyanax* (Characidae). *Rev Brasil Genet* 17: 157–163.

Fechheimer NS, Isakova GK, Belyaev DK. 1983. Mechanisms involved in the spontaneous occurrence of diploid-triploid chimerism in the mink (*Mustela vison*) and chicken (*Gallus domesticus*). *Cytogenet Cell Genet* 35: 238–243.

Felip A, Zanuy S, Carillo M, Piferrer F. 2001. Induction of triploidy and gynogenesis in teleost fish with emphasis on marine species. *Genetica* 111: 175–195.

Fernandes-Matioli FMC, Almeida-Toledo LF, Toledo SA. 1998. Natural triploidy in the Neotropical species *Gymnotus carapo* (Pisces: Gymnotiformes). *Caryologia* 51: 319–322.

Ferris SD. 1984. Tetraploidy and the evolution of the catostomid fishes. In: Turner BJ ed. *Evolutionary Genetics of Fishes*. New York: Plenum Press, 55–93.

Ferris SD, Whitt GS. 1977. Loss of duplicate gene expression after polyploidization. *Nature* 265: 258–260.

Fontana F. 1994. Chromosomal nucleolar organizer regions in four sturgeon species as markers of karyotype evolution in Acipenseriformes. *Genome* 37: 888–892.

Froese R, Pauly D. 2003. *Fishbase*. www.fishbase.org.

Frost DR. 2004. *Amphibian Species of the World: An Online Reference. Version 3.0.* http://research.amnh.org/herpetology/amphibia.

Fussey GD. 1984. The distribution of the two forms of the woodlouse *Trichoniscus pusillus* Brandt (Isopodae: Oniscoidea) in the British Isles: a reassessment of geographic parthenogenesis. *Biol J Linn Soc* 23: 309–321.

Gallardo MH, Bickham JW, Honeycutt RL, *et al.* 1999. Discovery of tetraploidy in a mammal. *Nature* 401: 341.

Gallardo MH, Bickham JW, Kausel G, *et al.* 2003. Gradual and quantum genome size shifts in the hystricognath rodents. *J Evol Biol* 16: 163–169.

Gallardo MH, Kausel G, Jiménez A, *et al.* 2004. Whole-genome duplications in South American desert rodents (Octodontidae). *Biol J Linn Soc* 82: 443–451.

Garcia-Abiado MAR, Dabrowski K, Christensen JE, Czesny S. 1999. Use of erythrocyte measurements to identify triploid saugeyes. *N Am J Agracult* 61: 319–325.

Gerhardt HC, Ptacek MB, Barnett L, Torke KG. 1994. Hybridization in the diploid-tetraploid treefrogs *Hyla chrysoscelis* and *Hyla versicolor*. *Copeia* 1994: 51–59.

Giuliano-Caetano L, Bertollo LAC. 1990. Karyotype variability in *Hoplerythrinus unitaeniatus* (Pisces, Characiformes, Erythrinidae). II. Occurrence of natural triploidy. *Rev Brasil Genet* 13: 231–237.

Goddard KA, Schultz RJ. 1993. Aclonal reproduction by polyploid members of the clonal hybrid species *Phoxinus eos-neogaeus* (Cyprinidae). *Copeia* 1993: 650–660.

Goddard KA, Megwinoff O, Wessner LL, Giaimo F. 1998. Confirmation of gynogenesis in *Phoxinus eos-neogaeus* (Pisces: Cyprinidae). *J Hered* 89: 151–157.

Goldman MA, LoVerde PT, Chrisman CL. 1983. Hybrid origin of polyploidy in freshwater snails of the genus *Bulinus* (Mollusca: Planorbidae). *Evolution* 37: 592–600.

Goldman MA, LoVerde PT, Chrisman CL, Franklin DA. 1984. Chromosomal evolution in planorbid snails of the genera *Bulinus* and *Biomphalaria*. *Malacologia* 25: 427–446.

Goldstein P, Triantaphyllou AC. 1981. Pachytene karyotype analysis of tetraploid *Meloidogyne hapla* females by electron microscope. *Chromosoma* 84: 405–412.

González-Tizón AM, Martínez-Lage A, Ausió J, Méndez J. 2000. Polyploidy in a natural population of mussel, *Mytilus trossulus*. *Genome* 43: 409–411.

Grassman M, Crews D. 1987. Dominance and reproduction in a parthenogenetic lizard. *Behav Ecol Sociobiol* 21: 141–147.

Green DM. 1980. Size differences in adhesive toe-pad cells of treefrogs of the diploid-polyploid *Hyla versicolor* complex. *J Herpetol* 14: 15–19.

Gregory TR. 2000. Nucleotypic effects without nuclei: genome size and erythrocyte size in mammals. *Genome* 43: 895–901.

Gregory TR. 2001a. Coincidence, coevolution, or causation? DNA content, cell size, and the C-value enigma. *Biol Rev* 76: 65–101.

Gregory TR. 2001b. *Animal Genome Size Database*. www.genomesize.com.

Gregory TR. 2001c. The bigger the C-value, the larger the cell: genome size and red blood cell size in vertebrates. *Blood Cells Mol Dis* 27: 830–843.

Gregory TR. 2003. Genome size estimates for two important freshwater molluscs, the zebra mussel (*Dreissena polymorpha*) and the schistosomiasis vector snail (*Biomphalaria glabrata*). *Genome* 46: 841–844.

Gregory TR, Hebert PDN. 1999. The modulation of DNA content: proximate causes and ultimate consequences. *Genome Res* 9: 317–324.

Grey AJ, Mackiewicz JS. 1980. Chromosomes of caryophyllidean cestodes: diploidy, triploidy, and parthenogenesis in *Glaridacris catostomi*. *Int J Parasitol* 10: 397–407.

508   Gregory and Mable

Guo X, Allen SK. 1994. Sex determination and polyploid gigantism in the dwarf surfclam (*Mulinia lateralis* Say). *Genetics* 138: 1199–1206.

Guttman DH, Charlesworth D. 1998. An X-linked gene with a degenerate Y-linked homologue in a dioecious plant. *Nature* 393: 263–266.

Haddad CFB, Pombal JP, Batistic RF. 1994. Natural hybridization between diploid and tetraploid species of leaf-frogs, genus *Phyllomedusa* (Amphibia). *J Herpetol* 28: 425–430.

Hardie DC, Hebert PDN. 2003. The nucleotypic effects of cellular DNA content in cartilaginous and ray-finned fishes. *Genome* 46: 683–706.

Hardy LM, Cole CJ. 1998. Morphology of a sterile, tetraploid, hybrid whiptail lizard (Squamata: Teiidae: *Cnemidophorus*). *Am Mus Novitates* 3228: 1–16.

Hartley SE. 1987. The chromosomes of salmonid fishes. *Biol Rev* 62: 197–214.

Havel JE, Hebert PDN. 1989. Apomictic parthenogenesis and genotypic diversity in *Cypridopsis vidua* (Ostracoda: Cyprididae). *Heredity* 62: 383–392.

Hedges SB, Bogart JP, Maxson LR. 1992. Ancestry of unisexual salamanders. *Nature* 356: 708–710.

Hertwig G, Hertwig P. 1920. Triploide Froschlarven. *Arch Mikr Anat* 94: 34–54.

Honeycutt RL, Rowe RL, Gallardo MH. 2003. Molecular systematics of the South American caviomorph rodents: relationships among species and genera in the family Octodontidae. *Mol Phylogenet Evol* 26: 476–489.

Hoshino K, Ohnishi K, Yoshida W, Shinozawa T. 1991. Analysis of ploidy in a planarian by flow cytometry. *Hydrobiologia* 227: 175–178.

Hubbs CL. 1964. Interactions between a bisexual fish species and its gynogenetic sexual parasite. *Bull Texas Mem Mus* 8: 1–72.

Hubbs CL, Hubbs LC. 1932. Apparent parthenogenesis in nature, in a form of fish of hybrid origin. *Science* 76: 628–630.

Hugall A, Stanton J, Moritz C. 1999. Reticulate evolution and the origins of ribosomal internal transcribed spacer diversity in apomictic *Meloidogyne*. *Mol Biol Evol* 16: 157–164.

Hughes RN. 1996. Evolutionary ecology of parthenogenetic strains of the prosobranch snail, *Potamopyrgus antipodarum* (Gray) (= *P. Jenkinsi* (Smith)). *Malacol Rev* Suppl. 6 (Mollusc Reprod): 101–113.

Husband BC, Schemske DW. 1998. Cytotype distribution at a diploid-tetraploid contact zone in *Chamerion (Epilobium) angustifolium* (Onagraceae). *Am J Bot* 85: 1688–1694.

Husband BC, Schemske DW. 2000. Ecological mechanisms of reproductive isolation between diploid and tetraploid *Chamerion angustifolium*. *J Ecol* 88: 689–701.

Husband BC, Schemske DW, Burton TL, Goodwillie C. 2002. Pollen competition as a unilateral reproductive barrier between sympatric diploid and tetraploid *Chamerion angustifolium*. *Proc R Soc Lond B* 269: 2565–2571.

Ihssen PE, McKay LR, McMillan I, Phillips RB. 1990. Ploidy manipulation and gynogenesis in fishes: cytogenetic and fisheries applications. *Trans Am Fish Soc* 119: 698–717.

Irwin JT, Lee RE. 2003. Geographic variation in energy storage and physiological responses to freezing in the gray treefrogs *Hyla versicolor* and *H. chrysoscelis*. *J Exp Biol* 206: 2859–2867.

Jackson JA, Tinsley RC. 2001. Host-specificity and distribution of cephalochalmydid cestodes: correlation with allopolyploid evolution of pipid anuran hosts. *J Zool* 254: 405–419.

Jackson JA, Tinsley RC. 2003. Parasite infectivity to hybridising host species: a link between hybrid resistance and allopolyploid speciation? *Int J Parasitol* 33: 137–144.

Jackson RC. 1976. Evolution and systematic significance of polyploidy. *Annu Rev Ecol Syst* 7: 209–234.

Jacob J. 1959. The chromosomes of six melaniid snails (Gastropoda: Prosobranchia). *Cytologia* 24: 487–497.

Janko K, Kotlík P, Ráb P. 2003. Evolutionary history of asexual hybrid loaches (*Cobitis*: Teleostei) inferred from phylogenetic analysis of mitochondrial DNA variation. *J Evol Biol* 16: 1280–1287.

Johnson C. 1959. Genetic incompatibility in the call races of *Hyla versicolor* Le Conte in Texas. *Copeia* 1959: 327–335.

Johnson C. 1963. Additional evidence of sterility between call-types in the *Hyla versicolor* complex. *Copeia* 1963: 139–143.

Johnson C. 1966. Species recognition in the *Hyla versicolor* complex. *Tex J Sci* 18: 361–364.

Johnson SG, Hopkins R, and Goddard K. 1999. Constraints on elevated ploidy in hybrid and nonhybrid parthenogenetic snails. *J Hered* 90: 659–662.

Jones AW, Mackiewicz JS. 1969. Naturally occurring triploidy and parthenogenesis in *Atractolytocestus huronensis* Anthony (Cestoidea: Caryophyllidea) from *Cyprinus carpio* L. in North America. *J Parasitol* 55: 1105–1118.

Kamel S, Marsden JE, Pough FH. 1985. Diploid and tetraploid grey treefrogs (*Hyla chrysoscelis* and *Hyla versicolor*) have similar metabolic rates. *Comp Biochem Physiol* 82A: 217–220.

Kapriyanova LA. 1989. Cytogenetic evidence for genome interaction in hybrid lacertid lizards. In: Dawley RM, Bogart JP eds. *Evolution and Ecology of Unisexual Vertebrates.* Albany, NY: New York State Museum, 236–240.

Kawamura T. 1984. Polyploidy in amphibians. *Zool Sci* 1: 1–15.

Keller MJ, Gerhardt HC. 2001. Polyploidy alters advertisement call structure in gray treefrogs. *Proc R Soc Lond B* 268: 341–345.

Kendall C, Valentino S, Bodine AB, Luer CA. 1994. Triploidy in a nurse shark, *Ginglymostoma cirratum*. *Copeia* 1994: 825–827.

King M. 1990. *Animal Cytogenetics, Vol.4: Chordata, No. 2: Amphibia.* Berlin: Gebrüder Borntraeger.

Klose J, Wolf U, Hitzeroth H, *et al.* 1968. Duplication of the LDH gene loci by polyploidization in the fish order Clupeiformes. *Humangenetik* 5: 190–196.

Kobel HR. 1996. Allopolyploid speciation. In: Tinsley RC, Kobel HR eds. *The Biology of Xenopus.* Oxford, UK: Clarendon Press, 391–401.

Kobel HR, Du Pasquier L. 1986. Genetics of polyploid *Xenopus*. *Trends Genet* 2: 310–315.

Kobel HR, Barandun B, Thiébaud CH. 1998. Mitochondrial rDNA phylogeny in *Xenopus*. *Herpetol J* 8: 13–17.

Kraus F. 1985. A new unisexual salamander from Ohio. *Occas Pap Mus Zool Univ Michigan* 709: 1–24.

Kraus F, Ducey PK, Moler P, Miyamoto MM. 1991. Two new triparental unisexual *Ambystoma* from Ohio and Michigan. *Herpetologica* 47: 429–439.

Kupka E. 1948. Chromosomale Verschiedenheiten bei schweizerischen Coregonen (Felchen). *Rev Suisse Zool* 55: 285–293.

Lamatsch DK, Schmid M, Schartl M. 2002. A somatic mosaic of the gynogenetic Amazon molly. *J Fish Biol* 60: 1417–1422.

Lande R, Schemske DW. 1985. The evolution of self-fertilization and inbreeding depression in plants. I. Genetic models. *Evolution* 39: 24–40.

Lanier GN, Kirkendall LR. 1986. Karyology of pseudogamous *Ips* bark beetles. *Hereditas* 105: 87–96.

Lantz LA, Callan HG. 1954. Phenotypes and spermatogenesis of interspecific hybrids between *Triturus cristatus* and *T. marmoratus*. *J Genet* 52: 165–185.

Larhammar D, Risinger C. 1994. Molecular genetic aspects of tetraploidy in the common carp *Cyprinus carpio*. *Mol Phylogenet Evol* 3: 59–68.

Le Comber SC, Smith C. 2004. Polyploidy in fishes: patterns and processes. *Biol J Linn Soc* 82: 431–442.

Lécher P, DeFaye D, Noel P. 1995. Chromosomes and nuclear DNA of Crustacea. *Invert Reprod Dev* 27: 85–114.

Lee T. 1999. Polyploidy and meiosis in the freshwater clam *Sphaerium striatinum* (Lamarck) and chromosome numbers in the Sphaeriidae (Bivalvia, Veneroida). *Cytologia* 64: 247–252.

Lee T, Ó Foighil D. 2002. 6-Phosphogluconate dehydrogenase (PGD) allele phylogeny is incongruent with a recent origin of polyploidization in some North American Sphaeriidae (Mollusca, Bivalvia). *Mol Phylogenet Evol* 25: 112–124.

Levin DA. 1975. Minority cytotype exclusion in local plant populations. *Taxon* 24: 35–43.

Licht LE, Bogart JP. 1987. Comparative size of epidermal cell nuclei from shed skin of diploid, triploid and tetraploid salamanders (genus *Ambystoma*). *Copeia* 1987: 284–290.

Licht LE, Bogart JP. 1989. Embryonic development and temperature tolerance in diploid and polyploid salamanders (genus *Ambystoma*). *Am Midland Nat* 122: 401–407.

Licht LE, Bogart JP. 1990. Comparative rates of oxygen consumption and water loss in diploid and polyploid salamanders (genus *Ambystoma*). *Comp Biochem Physiol* 97A: 569–572.

Lillie FR. 1906. Observations and experiments concerning the elementary phenomena of embryonic development in *Chaetopterus*. *J Exp Zool* 3: 153–268.

Little TJ, Hebert PDN. 1997. Clonal diversity in high arctic ostracodes. *J Evol Biol* 10: 233–252.

Little TJ, DeMelo R, Taylor DJ, Hebert PDN. 1997. Genetic characterization of an arctic zooplankter: insights into geographic polyploidy. *Proc R Soc Lond B* 264: 1363–1370.

Litvinchuk SN, Rosanov JM, Borkin LJ. 1998. A case of natural triploidy in a smooth newt *Trituris vulgaris* (Linnaeus, 1758), from Russia (Caudata: Salamandridae). *Herpetozoa* 11: 93–95.

Litvinchuk SN, Rosanov JM, Borkin LJ. 2001. Natural autotriploidy in the Danube newt, *Triturus dobrogicus* (Salamandridae). *Russ J Herpetol* 8: 74–76.

Lokki J, Saura A. 1980. Polyploidy in insect evolution. In: Lewis WH ed. *Polyploidy: Biological Relevance*. New York: Plenum Press, 277–312.

Lokki J, Suomalainen E, Saura A, Lankinen P. 1975. Genetic polymorphism and evolution in parthenogenetic animals. II. Diploid and polyploid *Solenobia triquetrella* (Lepidoptera: Psychidae). *Genetics* 79: 513–525.

Loumont C. 1983. Deux espèces nouvelles de *Xenopus* du Cameroun (Amphibia, Pipidae). *Rev Suisse Zool* 90: 169–177.

Lowcock LA, Griffith H, Murphy RW. 1991. The *Ambystoma laterale-jeffersonianum* complex in central Ontario: ploidy structure, sex ratio, and breeding dynamics in a bisexual-unisexual community. *Copeia* 1991: 87–105.

Lowcock LA, Licht LE. 1990. Natural autotriploidy in salamanders. *Genome* 33: 674–678.

Lowcock LA, Licht LE, Bogart JP. 1987. Nomenclature in hybrid complexes of *Ambystoma* (Urodela: Ambystomatidae): no case for the erection of hybrid "species." *Syst Zool* 36: 328–336.

Ludwig A, Belfiore NM, Pitra C, et al. 2001. Genome duplication events and functional reduction of ploidy levels in sturgeon (*Acipenser*, *Huso* and *Scaphirhynchus*). *Genetics* 158: 1203–1215.

Mable BK. 1996. Evolution of polyploidy on two continents: phylogenetic resolution and implications. PhD thesis, University of Texas, Austin, Texas.

Mable BK. 2003. Breaking down taxonomic barriers in polyploidy research. *Trends Plant Sci* 8: 582–590.

Mable BK. 2004a. 'Why polyploidy is rarer in animals than in plants': myths and mechanisms. *Biol J Linn Soc* 82: 453–466.

Mable BK. 2004b. Polyploidy and self-compatibility: is there an association? *New Phytol* 162: 803–811.

Mable BK, Bogart JP. 1995. Hybridization between tetraploid and diploid species of treefrogs (genus *Hyla*). *J Hered* 86: 432–440.

Mable BK, Roberts JD. 1997. Mitochondrial DNA evolution of tetraploids in the genus *Neobatrachus* (Anura: Myobatrachidae). *Copeia* 1997: 680–689.

MacCulloch RD, Upton DE, Murphy RW. 1996. Trends in nuclear DNA content among amphibians and reptiles. *Comp Biochem Physiol* 113B: 601–605.

Macgregor HC. 1993. *An Introduction to Animal Cytogenetics*. London: Chapman & Hall.

Mackiewicz JS. 1982. Caryophyllidea (Cestoidea): perspectives. *Parasitology* 84: 397–417.

Madl JE, Herman RK. 1979. Polyploids and sex determination in *Caenorhabditis elegans*. *Genetics* 93: 393–402.

Mahony MJ, Robinson ES. 1980. Polyploidy in the Australian leptodactylid frog genus *Neobatrachus*. *Chromosoma* 81: 199–212.

Maistro EL, Dias AL, Foresti F, et al. 1994. Natural triploidy in *Astyanax scabripinnis* (Pisces, Characidae) and simultaneous occurrence of macro B-chromosomes. *Caryologia* 47: 233–239.

Manaresi S, Marescalchi O, Scali V. 1993. The trihybrid genome constitution of *Bacillus lyncerorum* (Insecta Phasmatodea) and its karyotypic variations. *Genome* 36: 317–326.

Mantovani B, Scali V, Tinti F. 1992. New morphological and allozymic characterization of *Bacillus whitei* and *B. lynceorum* hybrid complexes (Insecta Phasmatodea). *Biol Zentralbl* 111: 75–91.

Mares MA, Braun JK, Barquez RM, Díaz MM. 2000. Two new genera and species of halophytic desert mammals from isolated salt flats in Argentina. *Occas Pap Mus Tex Tech Univ* 203: 1–27.

Marescalchi O, Scali V. 2003. Automictic parthenogenesis in the diploid-triploid stick insect *Bacillus atticus* and its flexibility leading to heterospecific diploid hybrids. *Invert Reprod Dev* 43: 163–172.

Mark Welch JL, Mark Welch DB, Meselson M. 2004. Cytogenetic evidence for asexual evolution of bdelloid rotifers. *Proc Natl Acad Sci USA* 101: 1618–1621.

Martens PM, Curini-Galletti MC, Van Oostveldt P. 1989. Polyploidy in Proseriata (Platyhelminthes) and its phylogenetical implications. *Evolution* 43: 900–907.

Martino AL, Sinsch U. 2002. Speciation by polyploidy in *Odontophrynus americanus*. *J Zool* 257: 67–81.

Martins MJ, Collares-Pereira MJ, Cowx IG, Coelho MM. 1998. Diploids v. triploids of *Rutilus alburnoides*: spatial segregation and morphological differences. *J Fish Biol* 52: 817–828.

Matson TO. 1990. Erythrocyte size as a taxonomic character in the identification of Ohio *Hyla chrysoscelis* and *H. versicolor*. *Herpetologica* 46: 457–462.

Matsubara K, Arai K, Suzuki R. 1995. Survival potential and chromosomes of progeny of triploid and pentaploid females in the loach, *Misgurnus anguillicaudatus*. *Aquaculture* 131: 37–48.

Maxson L, Pepper E, Maxson RD. 1977. Immunological resolution of a diploid-tetraploid species complex of tree frogs. *Science* 197: 1012–1013.

McLister JD, Stevens ED, Bogart JP. 1995. Comparative contractile dynamics of calling and locomotor muscles in three hylid frogs. *J Exp Biol* 198: 1527–1538.

Miller JS, Venable DL. 2000. Polyploidy and the evolution of gender dimorphism in plants. *Science* 289: 2335–2338.

Mogie M. 1986. On the relationship between asexual reproduction and polyploidy. *J Theor Biol* 122: 493–498.

Moore G. 2002. Meiosis in allopolyploids: the importance of 'Teflon' chromosomes. *Trends Genet* 18: 456–463.

Morescalchi A, Olmo E. 1974. Sirenids: a family of polyploid urodeles? *Experientia* 30: 491–492.

Morescalchi A, Olmo E, Odierna G, Rosati C. 1986. On the polyploidy in the family Sirenidae (Amphibia: Caudata). In: Rocek Z ed. *Studies in Herpetology*. Prague: Charles University, 165–170.

Morescalchi MA, Rocco L, Stingo V. 2002. Cytogenetic and molecular studies in a lungfish, *Protopterus annectens* (Osteichthyes, Dipnoi). *Gene* 295: 279–287.

Moritz C, Brown WM, Densmore LD, et al. 1989. Genetic diversity and the dynamics of hybrid parthenogenesis in *Cnemidophorus* (Teiidae) and *Heteronotia* (Gekkonidae). In: Dawley RM, Bogart JP eds. *Evolution and Ecology of Unisexual Vertebrates*. Albany, NY: New York State Museum, 87–112.

Moritz C, Case TJ, Bolger DT, Donnellan S. 1993. Genetic diversity and the history of pacific island house geckos (*Hemidactylus* and *Lepidodactylus*). *Biol J Linn Soc* 48: 113–133.

Moritz C, Donnellan S, Adams M, Baverstock PR. 1990. The origin and evolution of parthenogenesis in *Heterodontia binoei* (Gekkonidae: Reptilia): extensive clonal diversity in a parthenogenetic vertebrate. *Evolution* 43: 994–1003.

Muldal S. 1952. The chromosomes of the earthworms. I. The evolution of polyploidy. *Heredity* 6: 55–76.

Muller HJ. 1925. Why polyploidy is rarer in animals than in plants. *Am Nat* 59: 346–353.

Murayama Y, Hijikata M, Kojima K, et al. 1986. The appearance of diploid-triploid-tetraploid mosaic individuals in polyploid fish, ginbuna (*Carassius auratus langsdorfii*). *Experientia* 42: 187–188.

Nakamura HK. 1985. A review of molluscan cytogenetic information based on the CISMOCH-Computerized Index System for Molluscan Chromosomes. Bivalvia, Polyplacophora and Cephalopoda. *Venus* 44: 193–225.

Nanda I, Schartl M, Feichtinger W, Schlupp I, et al. 1995. Chromosomal evidence for laboratory syn-thesis of a triploid hybrid between the gynogenetic teleost *Poecilia formosa* and its host species. *J Fish Biol* 47: 619–623.

Natarajan R, Burch JB, Gismann A. 1965. Cytological studies of Planorbidae (Gastropoda: Basommatophora). II. Some African Planorbinae, Planorbininae and Bulininae. *Malacologia* 2: 239–251.

Neaves WB. 1971. Tetraploidy in a hybrid lizard of the genus *Cnemidophorus* (Teiidae). *Breviora* 381: 1–25.

Nigon V. 1951. Polyploidie expérimentale chez un nématode libre, *Rhabditis elegans* Maupas. *Bull Biol Fr Belg* 85: 187–225.

Ó Foighil D, Thiriot-Quiévreux C. 1991. Ploidy and pronuclear interaction in northeastern Pacific *Lasaea* clones (Mollusca: Bivalvia). *Biol Bull* 181: 222-231.

Ó Foighil D, Thiriot-Quiévreux C. 1999. Sympatric Australian *Lasaea* species (Mollusca: Bivalvia) differ in their ploidy levels, reproductive modes and developmental modes. *Zool J Linn Soc* 127: 477–494.

Ohno S. 1974. *Animal Cytogenetics, Vol 4: Chordata, No. 1: Protochordata, Cyclostomata, and Pisces.* Berlin: Gebrüder Borntraeger.

Ohno S, Kittrell WA, Christian LC, et al. 1963. An adult triploid chicken (*Gallus domesticus*) with a left ovotestis. *Cytogenetics* 2: 42–49.

Ohno S, Muramoto J, Christian L, Atkin NB. 1967. Diploid-tetraploid relationship among old-world members of the fish family Cyprinidae. *Chromosoma* 23: 1–9.

Ohno S, Muramoto J, Stenius C, et al. 1969a. Microchromosomes in Holocephalian, Chondrostean and Holostean fishes. *Chromosoma* 26: 35–40.

Ohno S, Muramoto J, Klein J, Atkin NB. 1969b. Diploid-tetraploid relationship in clupeoid and salmonoid fish. *Chromosomes Today* 2: 139–147.

Ojima Y, Maeki K, Takayama S, Nogusa S. 1963. A cytotaxonomic study on the Salmonidae. *Nucleus* 6: 91–98.

Okamoto A, Arimoto B. 1986. Chromosomes of *Corbicula japonica*, *C. sandai* and *C. (Corbiculina) leana* (Bivalvia: Corbiculidae). *Venus* 45: 194–202.

Oliveira C, Almeida-Toledo LF, Mori L, Toledo-Filho SA. 1992. Extensive chromosomal rearrangements and nuclear DNA content changes in the evolution of the armoured catfishes genus *Corydoras* (Pisces, Siluriformes, Callichthyidae). *J Fish Biol* 40: 419–431.

Oliveira C, Almeida-Toledo LF, Mori L, Toledo-Filho SA. 1993. Cytogenetic and DNA content studies of armoured catfishes of the genus *Corydoras* (Pisces, Siluriformes, Callichthyidae) from the south-east coast of Brazil. *Rev Brasil Genet* 16: 617–629.

Oliver JH. 1971. Parthenogenesis in mites and ticks (Arachnida: Acari). *Am Zool* 11: 283–299.

Olmo E. 1986. *Animal Cytogenetics, Vol. 4: Chordata, No. 3A: Reptilia.* Berlin: Gebrüder Borntraeger.

Olmo E, Stingo V, Cobror O, et al. 1982. Repetitive DNA and polyploidy in selachians. *Comp Biochem Physiol* 73B: 739–745.

Omodeo P. 1951. Problemi zoogeografici ed ecologici relativi a lombrichi peregrini, con particolare riguardo al tipo di riproduzione ed alla struttura cariologica. *Bull Zool* 18: 117–122.

Orr HA. 1990. "Why polyploidy is rarer in animals than in plants" revisited. *Am Nat* 136: 759–770.

Otto SP, Whitton J. 2000. Polyploid incidence and evolution. *Annu Rev Genet* 34: 401–437.

Pandian TJ, Koteeswaran R. 1999. Natural occurrence of monoploids and polyploids in the Indian cat-fish, *Heteropneustes fossilis*. *Curr Sci* 76: 1134–1137.

Pedersen RA. 1971. DNA content, ribosomal gene multiplicity, and cell size in fish. *J Exp Zool* 177: 65–79.

Pessino SC, Ortiz JPA, Hayward MD, Quarin CL. 1999. The molecular genetics of gametophytic apomixis. *Hereditas* 130: 1–11.

Petranka JW. 1998. *Salamanders of the United States and Canada.* Washington, DC: Smithsonian Institution Press.

Phillips R, Ráb P. 2001. Chromosome evolution in the Salmonidae (Pisces): an update. *Biol Rev* 76: 1–25.

Potter IC, Robinson ES. 1973. The chromosomes of the cyclostomes. In: Chiarelli AB, Capanna E eds. *Cytotaxonomy and Vertebrate Evolution.* New York: Academic Press, 179–203.

Prehn LM, Rasch EM. 1969. Cytogenetic studies of *Poecilia* (Pisces). I. Chromosome numbers of naturally occurring poeciliid species and their hybrids from eastern Mexico. *Can J Genet Cytol* 11: 880–895.

Próspero MI, Collares-Pereira MJ. 2000. Nuclear DNA content variation in the diploid-polyploid *Leuciscus alburnoides* complex (Teleostei, Cyprinidae) assessed by flow cytometry. *Folia Zool* 49: 53–58.

Ptacek MB. 1996. Interspecific similarity in life-history traits in sympatric populations of gray treefrogs, *Hyla chrysoscelis* and *Hyla versicolor*. *Herpetologica* 52: 323–332.

Ptacek MB, Gerhardt HC, Sage RD. 1994. Speciation by polyploidy in treefrogs: multiple origins of the tetraploid, *Hyla versicolor*. *Evolution* 48: 898–908.

Raicu P, Taisescu E. 1972. *Misgurnus fossilis*, a tetraploid fish species. *J Hered* 63: 92–94.

Ralin DB. 1981. Ecophysiological adaptation in a diploid-tetraploid complex of treefrogs (Hylidae). *Comp Biochem Physiol* 68A: 175–179.

Ralin DB, Romano MA, Kilpatrick CW. 1983. The tetraploid treefrog *Hyla versicolor*: evidence for a single origin from the diploid *H. chrysoscelis*. *Herpetologica* 39: 212–225.

Ramanjaneyulu JV, Madhavi R. 1984. Cytological investigations on two species of aliocreadiid trematodes with special reference to the occurrence of triploidy and parthenogenesis in *Allocreadium fasciatusi*. *Int J Parasitol* 14: 309–316.

Ramsey J, Schemske DW. 1998. Pathways, mechanisms, and rates of polyploid formation in flowering plants. *Annu Rev Ecol Syst* 29: 467–501.

Ramsey J, Schemske DW. 2002. Neopolyploidy in flowering plants. *Annu Rev Ecol Syst* 33: 589–639.

Reeder TW, Cole CJ, Dessauer HC. 2002. Phylogenetic relationships of whiptail lizards of the genus *Cnemidophorus* (Squamata: Teiidae): a test of monophyly, reevaluation of karyotypic evolution, and review of hybrid origins. *Am Mus Novitates* 3365: 1–61.

Rees H. 1964. The question of polyploidy in the Salmonidae. *Chromosoma* 15: 275–279.

Renner SS, Ricklefs RE. 1995. Dioecy and its correlates in the flowering plants. *Am J Bot* 82: 596–606.

Roberts JD. 1997. Call evolution in *Neobatrachus* (Anura: Myobatrachidae): speculations on tetraploid origins. *Copeia* 1997: 791–801.

Rock J, Eldridge M, Champion A, *et al.* 1996. Karyotype and nuclear DNA content of the Australian lungfish, *Neoceratodus forsteri* (Ceratodidae: Dipnoi). *Cytogenet Cell Genet* 73: 187–189.

Rödel M-O. 2000. *Herpetofauna of West Africa, Vol. I. Amphibians of the West African Savanna.* Frankfurt: Edition Chimaira.

Romano MA, Ralin DB, Guttman DH, Skillings JH. 1987. Parallel electromorph variation in the diploid-tetraploid gray treefrog complex. *Am Nat* 130: 864–878.

Salemma H. 1984. Polyploidy in the evolution of the glacial relict *Pontoporeia* spp. (Amphipoda, Crustacea). *Heriditas* 100: 53–60.

Salemma H, Heino T. 1990. Chromosome-numbers of Fennoscandian glacial relict Crustacea. *Ann Zool Fenn* 27: 207–210.

Sammut B, Marcuz A, Du Pasquier L. 2002. The fate of duplicated major histocompatibility complex class Ia genes in a dodecaploid amphibian, *Xenopus ruwenzoriensis*. *Eur J Immunol* 32: 2698–2709.

Sasada K. 1978. Studies on the chromosomes of parasitic helminths. II. Triploidy and cytological mechanism of parthenogenesis in *Diphyllobothrium erinacei* (Cestoda: Diphyllobothriidae) [in Japanese]. *Jap J Parasitol* 27: 547–560.

Saura A, Lokki J, Suomalainen E. 1993. Origin of polyploidy in parthenogenetic weevils. *J Theor Biol* 163: 449–456.

Schartl M, Wilde B, Schlupp I, Parzefall J. 1995. Evolutionary origin of a parthenoform, the Amazon molly *Poecilia formosa*, on the basis of a molecular genealogy. *Evolution* 49: 827–835.

Schenck RA, Vrijenhoek RC. 1989. Coexistence among sexual and asexual forms of *Poeciliopsis*: foraging behaviour and microhabitat selection. In: Dawley RM, Bogart JP eds. *Evolution and Ecology of Unisexual Vertebrates.* Albany: New York State Museum, 39–48.

Schlupp I, Nanda I, Döbler M, et al. 1998. Dispensible and indispensible genes in an ameiotic fish, the Amazon molly *Poecilia formosa*. *Cytogenet Cell Genet* 80: 193–198.

Schmidtke J, Kandt I. 1981. Single-copy DNA relationships between diploid and tetraploid teleostean fish species. *Chromosoma* 83: 191–197.

Schmidtke J, Schmitt E, Matzke E, Engel W. 1979. Non-repetitive DNA sequence divergence in phylogenetically diploid and tetraploid teleostan species of the family Cyprinidae and the order Isospondyli. *Chromosoma* 75: 185–198.

Schultz RJ 1969. Hybridization, unisexuality, and polyploidy in the teleost *Poeciliopsis* (Poeciliidae) and other vertebrates. *Am Nat* 103: 605–619.

Schultz RJ. 1973. Unisexual fish: laboratory synthesis of a "species." *Science* 179: 180–181.

Schultz RJ. 1980. Role of polyploidy in the evolution of fishes. In: Lewis WH ed. *Polyploidy: Biological Relevance*. New York: Plenum Press, 313–340.

Schultz RJ. 1982. Competition and adaptation among diploid and polyploid clones of unisexual fishes. In: Dingle H, Hegmann JP eds. *Evolution and Genetics of Life Histories*. New York: Springer-Verlag, 103–119.

Schultz RJ. 1989. Origins and relationships of unisexual poeciliids. In: Meffe GK, Snelson FF eds. *Ecology and Evolution of Livebearing Fishes (Poeciliidae)*. Englewood Cliffs, NJ: Prentice Hall, 69–87.

Schultz RJ, Kallman KD. 1968. Triploid hybrids between the all-female teleost *Poecilia formosa* and *Poecilia sphenops*. *Nature* 219: 280–282.

Schwartz FJ, Maddock MB. 1986. Comparisons of karyotypes and cellular DNA contents within and between major lines of elasmobranchs. In: Uyeno T, Arai R, Taniuchi T, Matsuura K eds. *Indo-Pacific Fish Biology*. Tokyo: Ichthyological Society of Japan, 148–157.

Seiler J. 1946. Die Verbreitungsgebiete der verschiedenen Rassen von *Solenobia triquetrella* (Psychidae) in der Schweiz. *Rev Suisse Zool* 53: 529–533.

Seiler J. 1961. Untersuchungen über die Entstehung der Parthenogenese bei *Solenobia triquetrella* F.R. (Lepidoptera, Psychidae). III. Die geographische Verbeitung der drei Rassen von *Solenobia triquetrella* (Bisexuell, diploid und tetraploid parthenogenetisch) in der Schweiz und in den angrenzenden Ländern und die Beziehung zur Eiszeit. *Z Vererbungsl* 92: 261–316.

Sessions SK. 1982. Cytogenetics of diploid and triploid salamanders of the *Ambystoma jeffersonianum* complex. *Chromosoma* 84: 599–621.

Smith SG, Virkki N. 1978. *Animal Cytogenetics, Vol. 3: Insecta, No. 5: Coleoptera*. Berlin: Gebrüder Borntraeger.

Sola L, Iaselli V, Rossi AR, et al. 1992. Cytogenetics of bisexual/unisexual species of *Poecilia*. III. The karyotype of *Poecilia formosa*, a gynogenetic species of hybrid origin. *Cytogenet Cell Genet* 60: 236–240.

Som C, Anholt BR, Reyer HU. 2000. The effect of assortative mating on the coexistence of a hybridogenetic waterfrog and its sexual host. *Am Nat* 156: 34–46.

Spolsky CM, Phillips CA, Uzzell T. 1992. Antiquity of clonal salamander lineage revealed by mitochondrial DNA. *Nature* 356: 706–707.

Städler T, Loew M, Streit B. 1996. Genetics and mating systems of polyploid freshwater hermaphrodite snails. *Malacol Rev* Suppl. 6 (Mollusc Reprod): 121–127.

Stebbins GL. 1950. *Variation and Evolution in Plants*. New York: Columbia University Press.

Stenberg P, Lundmark M, Knutelski S, Saura A. 2003. Evolution of clonality and polyploidy in a weevil system. *Mol Biol Evol* 20: 1626–1632.

Stenberg P, Terhivuo J, Lokki J, Saura A. 2000. Clone diversity in the polyploid weevil *Otiorhynchus scaber*. *Hereditas* 132: 137–142.

Stingo V, Capriglione T. 1986. DNA and chromosomal evolution in cartilaginous fish. In: Uyeno T, Arai R, Taniuchi T, Matsuura K eds. *Indo-Pacific Fish Biology*. Tokyo: Ichthyological Society of Japan, 140–147.

Stingo V, Rocco L. 2001. Selachian cytogenetics: a review. *Genetica* 111: 329–347.

Stöck M. 1998. Mating call differences between diploid and tetraploid green toads (*Bufo viridis* complex) in middle Asia. *Amphibia-Reptilia* 19: 29–42.

Stöck M, Lamatsch DK, Steinlein C, et al. 2002. A bisexually reproducing all-triploid vertebrate. *Nat Genet* 30: 325–328.

Suchanek TH. 1993. Oil impacts on marine invertebrate populations and communities. *Am Zool* 33: 510–523.

Suomalainen E. 1969. Evolution in parthenogenetic Curculionidae. *Evol Biol* 3: 261–296.

Suomalainen E, Saura A, Lokki J. 1987. *Cytology and Evolution in Parthenogenesis.* Boca Raton, FL: CRC Press.

Svärdson G. 1945. Chromosome studies on Salmonidae. *Medd St undersökn försökanst sotvattensfisket* 23: 1–151.

Swarup H. 1959. Effect of triploidy on the body size, general organization and cellular structure in *Gasterosteus aculeatus* (L.). *J Genet* 56: 143–155.

Takada H, Blackman RL, Miyasaki M. 1978. Cytological, morphological, and biological studies on a laboratory-reared triploid clone of *Myzus persicae* (Sulzer). *Kontyu* 46: 557–573.

Takenouchi Y. 1976. A study of polyploidy in races of Japanese weevils (Coleoptera: Curculionidae). *Genetica* 46: 327–334.

Tandy M, Bogart JP, Largen MJ, Feener DJ. 1982. A tetraploid species of *Bufo* (Anura Bufonidae) from Ethiopia. *Monit Zool Ital* Suppl. 17: 1–79.

Tank PW, Charlton RK, Burns ER. 1987. Flow cytometric analysis of ploidy in the axolotl, *Ambystoma mexicanum*. *J Exp Zool* 243: 423–433.

Taylor HL, Cole CJ, Hardy M, et al. 2001. Natural hybridization between the teiid lizards *Cnemidophorus tesselatus* (parthenogenetic) and *C. tigris marmoratus* (bisexual): assessment of evolutionary alternatives. *Am Mus Novitates* 3345: 1–65.

Terasaki K. 1980. Comparative studies on the karyotypes of *Paragonimus westermani* (s. str.) and *P. pulmonalis*. *Jap J Parasitol* 29: 239–243.

Terhivuo, J. 1988. The Finnish Lumbricidae (Oligochaeta) fauna and its formation. *Ann Zool Fenn* 25: 229–247.

Thiriot-Quiévreux C, Insua Pombo AM, Albert P. 1989. Polyploïdie chez un bivlave incubant *Lasaea rubra* (Montagu). *C R Acad Sci* 308: 115–120.

Thompson JD, Lumaret R. 1992. The evolutionary dynamics of polyploid plants: origins, establishment and persistence. *Trends Ecol Evol* 7: 302–307.

Thomson KS, Gall JG, Coggins LW. 1973. Nuclear DNA contents of coelacanth erythrocytes. *Nature* 241: 126.

Thorgaard GH. 1983. Chromosomal differences among rainbow trout populations. *Copeia* 1983: 650–662.

Thorgaard GH, Bailey GS, Williams D, et al. 2002. Status and opportunities for genomics research with rainbow trout. *Comp Biochem Physiol* 133B: 609–646.

Thorgaard GH, Rabinovitch PS, Shen MW, et al. 1982. Triploid rainbow trout identified by flow cytometry. *Aquaculture* 29: 305–309.

Tiersch TR, Beck ML, Douglass M. 1991. ZZW autotriploidy in a blue-and-yellow macaw. *Genetica* 84: 209–212.

Tinsley RC, Jackson JA. 1998. Correlation of parasite speciation and specificity with host evolutionary relationships. *Int J Parasitol* 28: 1573–1582.

Tito MB, Boyd SK. 1999. Morphometric differences in calling muscles of the gray treefrogs, *Hyla chrysoscelis* and *Hyla versicolor*. *Am Zool* 39: 112A.

Tomiuk J, Loeschcke V, Schneider M. 1994. On the origin of polyploid parthenogenetic races in the weevil *Polydrusus mollis* (Coleoptera: Curculionidae). *J Theor Biol* 167: 89–92.

Triantaphyllou AC. 1983. Cytogenetic aspects of nematode evolution. In: Stone AR, Platt HM, Khalil LF eds. *Concepts in Nematode Systematics*. London: Academic Press, 55–71.

Triantaphyllou AC. 1984. Polyploidy in meiotic parthenogenetic populations of *Meloidogyne hapla* and a mechanism of conversion to diploidy. *Rev Nématol* 7: 65–72.

Triantaphyllou AC. 1991. Further studies on the role of polyploidy in the evolution of *Meloidogyne*. *J Nematol* 23: 249–253.

Triantaphyllou AC, Hirschmann H. 1978. Cytology of the *Heterodera trifolii* parthenogenetic species complex. *Nematologica* 24: 418–424.

Triantaphyllou AC, Hirschmann H. 1997. Evidence of direct polyploidization in the mitotic partheno-genetic *Meloidogyne microcephala*, through doubling of its somatic chromosome number. *Fund Appl Nematol* 20: 385–391.

Triantaphyllou AC, Riggs RD. 1979. Polyploidy in an amphimictic population of *Heterodera glycines*. *J Nematol* 11: 371–376.

Tsigenopoulos CS, Ráb P, Naran D, Berrebi P. 2002. Multiple origins of polyploidy in the phylogeny of southern Asian barbs (Cyprinidae) as inferred from mtDNA markers. *Heredity* 88: 466–473.

Turgeon J, Hebert PDN. 1994. Evolutionary interactions between sexual and all-female taxa of *Cyprinotus* (Ostracoda: Cyprididae). *Evolution* 48: 1855–1865.

Turner BJ, Diffoot N, Rasch EM. 1992. The callichthyid catfish *Corydoras aeneus* is an unresolved diploid-tetraploid sibling species complex. *Ichthyol Expl Freshwaters* 3: 17–23.

Tymowska J. 1991. Polyploidy and cytogenetic variation in frogs of the genus *Xenopus*. In: Green DM, Sessions SK eds. *Amphibian Cytogenetics and Evolution*. San Diego, CA: Academic Press, 259–297.

Uyeno T, Smith GR. 1972. Tetraploid origin of the karyotype of catostomid fishes. *Science* 175: 644–646.

Uzzell TM. 1963. Natural triploidy in salamanders related to *Ambystoma jeffersonianum*. *Science* 139: 113–115.

Uzzell TM. 1964. Relations of the diploid and triploid species of the *Ambystoma jeffersonianum* complex (Amphibia, Caudata). *Copeia* 1964: 257–300.

Vandel A. 1927. Triploïdie et parthénogenèse chez l'Isopode, *Trichoniscus (Spiloniscus) provisorius* Racovitza. *C R Acad Sci* 183: 158–160.

Vandel A. 1928. La parthénogenèse géographique. Contribution à l'étude biologique et cytoloqigue de la parthénogenèse naturelle. *Bull Biol Fr Belg* 62: 164–281.

Vasil'ev VP, Vinogradov AE, Rozanov YM, Vasil'eva ED. 1999. Cellular DNA content in different forms of the bisexual-unisexual complex of spined loaches of the genus *Cobitis* and in Luther's spined loach *C. lutheri* (Cobitidae). *J Ichthyol* 39: 377–383.

Vences M, Aprea G, Capriglione T, *et al.* 2002. Ancient tetraploidy and slow molecular evolution in *Scaphiophryne*: ecological correlates of speciation mode in Malagasy relict amphibians. *Chromosome Res* 10: 127–136.

Vervoort A. 1980. Tetraploidy in *Protopterus* (Dipnoi). *Experientia* 36: 294–296.

Viktorov AG. 1997. Diversity of polyploid races in the family Lumbricidae. *Soil Biol Biochem* 29: 217–221.

Vinogradov AE, Borkin LJ, Günther R, Rosanov JM. 1990. Genome elimination in diploid and triploid *Rana esculenta* males: cytological evidence from DNA flow cytometry. *Genome* 33: 619–627.

Vogel G. 1999. Frog is a prince of a new model organism. *Science* 285: 25.

von Wettstein F. 1927. Die Erscheinung der Heteroploidie, besonders im Pflanzenreich. *Ergeb Biol* 2: 311–356.

Vrijenhoek RC, Schultz JH. 1974. Evolution of a trihybrid unisexual fish (*Poeciliopsis*, Poeciliidae). *Evolution* 28: 306–319.

Walsh EJ, Zhang L. 1992. Polyploidy and body size variation in a natural population of the rotifer *Euchlanis dilatata*. *J Evol Biol* 5: 345–353.

Ward RD, Bickerton MA, Finston T, Hebert PDN. 1994. Geographical cline in breeding systems and ploidy levels in European populations of *Daphnia pulex*. *Heredity* 73: 532–543.

Wasserman AO. 1970. Polyploidy in the common tree toad *Hyla versicolor* Le Conte. *Science* 167: 385–386.

White MJD. 1940. Evidence for polyploidy in the hermaphrodite groups of animals. *Nature* 146: 132–133.

White MJD. 1973. *Animal Cytology and Evolution*. Cambridge, UK: Cambridge University Press.

White MJD. 1978. *Modes of Speciation*. San Francisco, CA: W.H. Freeman and Co.

Wolf U, Ritter H, Atkin NB, Ohno S. 1969. Polyploidization in the fish family Cyprinidae, Order Cypriniformes. I. DNA-content and chromosome sets in various species of Cyprinidae. *Humangenetik* 7: 240–244.

Wolfe KH. 2001. Yesterday's polyploids and the mystery of diploidization. *Nat Rev Genet* 2: 333–341.

Wright JE, Johnson K, Hollister A, May B. 1983. Meiotic models to explain classical linkage, pseudolinkage and chromosome pairing in tetraploid derivative salmonids. In: Rattazzi MC, Scandalios JG, Whitt GS eds. *Isozymes: Current Topics in Biological and Medical Research*. New York: Alan R. Liss, 239–260.

Wright JW, Lowe CH. 1968. Weeds, polyploids, parthenogenesis, and the geographical and ecological distribution of all-female species of *Cnemidophorus*. *Copeia* 1968: 128–138.

Wu S-K. 1972. Comparative studies on a polyploid series of the African snail genus *Bulinus* (Bassommatophora: Planorbidae). *Malacol Rev* 5: 95–164.

Wynn AH, Cole CJ, Gardner AL. 1987. Apparent triploidy in the unisexual Brahminy blind snake, *Ramphotyphlops braminus*. *Am Mus Novitates* 2868: 1–7.

Yampolsky LY, Scheiner SM. 1996. Why larger offspring at lower temperatures? A demographic approach. *Am Nat* 147: 86–100.

Zan R, Song Z, Liu W. 1986. Studies on karyotypes and nuclear DNA contents of some cyprinoid fishes, with notes on fish polyploids in China. In: Uyeno T, Arai R, Taniuchi T, Matsuura K eds. *Indo-Pacific Fish Biology*. Tokyo: Ichthyological Society of Japan, 877–885.

Zhang H, Okamoto N, Ikeda Y. 1995. Two *c-myc* genes from a tetraploid fish, the common carp (*Cyprinus carpio*). *Gene* 153: 231–236.

Zhang L, King CE. 1993. Life history divergence of sympatric diploid and polyploid populations of brine shrimp *Artemia parthenogenetica*. *Oecologia* 93: 177–183.

Zhang L, Lefcort H. 1991. The effects of ploidy level on the thermal distributions of brine shrimp *Artemia parthenogenetica* and its ecological implications. *Heredity* 66: 445–452.

Zhang Q, Arai K, Yamashita M. 1998. Cytogenetic mechanisms for triploid and haploid egg formation in the triploid loach *Misgurnus anguillicaudatus*. *J Exp Zool* 281: 608–619.

PART V

Sequence and Structure

Comparative Genomics in Eukaryotes

ALAN FILIPSKI AND SUDHIR KUMAR

Although the word "genome," meaning the total hereditary material of an organism, was coined in 1920 (see Chapter 1), the general concept goes back at least as far as the 4th century BCE, when Aristotle implicated blood as the heredity substance. The blood of the mother, it was thought, supplied matter to the developing fetus whereas the semen (a purified form of blood) of the father conveyed form (Aristotle, 1953). Ironically, although the notions of "blood relations" and characteristics being "in one's blood" persist, it is now known that the blood of mammals actually contains very little genetic material because their erythrocytes contain neither nuclei nor mitochondria (see Chapter 1). As scientific method and technique advanced, heredity eventually came to be associated with bodies called chromosomes in the nuclei of cells (late 19th and early 20th centuries) and finally with the long double-stranded nucleotide polymers called DNA molecules that are wound up within those chromosomes (mid-20th century).

This chapter outlines the development and current status of comparative eukaryotic genomics, from the earliest studies of basic chromosome structure to the sequencing of entire genomes. In the process, a review is provided of the structure, organization, and composition of the primary eukaryotic genomes that have been sequenced thus far. This is a truly exciting time for the biological sciences, with avenues of research now opening up that had not even been conceived only a few decades ago. Some of the vast possibilities that are already apparent are discussed at the end of the chapter, but this is necessarily a highly truncated list owing to the ever-accelerating rapidity with which the field is advancing.

THE EARLY HISTORY OF COMPARATIVE EUKARYOTIC GENOMICS

THE BASICS OF EUKARYOTIC CHROMOSOME STRUCTURE

Figure 9.1A depicts a typical eukaryotic chromosome in the unreplicated form. Photographic representations often show the chromosome while it is replicating during mitosis, because chromosomes are easier to photograph in this stage. In this latter case (Fig. 9.1B), each chromosome looks more like the letter X, with each arm in two replicates (sister chromatids) emanating from the centromere. The shorter arm is generally depicted at the top of the image and (in humans) is designated as the "p" arm, from the French *petit bras* ("little arm"); the longer arm is labeled the "q" arm, after the French *queue* ("tail"). Chemical stains, described in more detail later, bring out characteristic light and dark banding patterns. Conventions for designating individual chromosomes with numbers or letters are historical (and often idiosyncratic) and reflect usage that evolved within specific communities of investigators.

Before the advent of techniques for reading the sequence of nucleotide bases of a DNA molecule, the chromosome provided the most detailed view available of

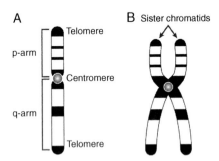

FIGURE 9.1   Schematic representation of a typical human chromosome. In this example, a submetacentric chromosome is shown as it might appear after chemical staining by the Giemsa method. Darkly stained regions are heterochromatic (condensed); lighter regions are euchromatic (uncondensed).

the physical eukaryotic genome. The term "karyotype" refers to a description or depiction (karyogram) of the set of all chromosomes in an organism. It is customary to depict the autosomal (nonsex) chromosomes arranged in homologous pairs in a standard order, usually from largest to smallest, with the shorter arm of each one oriented toward the top of the picture. The sex chromosomes typically are placed last. Sometimes only a haploid chromosome set is depicted. Figure 9.2 shows two examples of eukaryotic karyotypes (from African elephant and Siberian tiger).

Each eukaryotic chromosome is linear with a constriction somewhere along its length called the centromere, and is capped by condensed regions called telomeres. A long DNA double helix molecule stretches from one telomere to the other. Chromosomes consist of a tightly coiled complex of DNA and of proteins such as

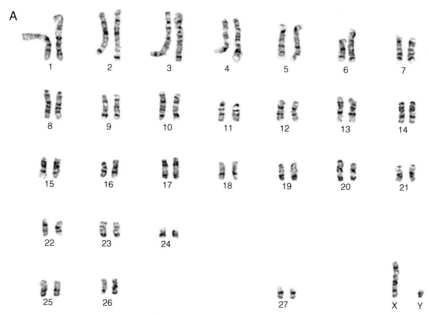

FIGURE 9.2   Representative eukaryote karyotypes from (A) the African elephant *Loxodonta africana*, with 2n = 58 (Houck *et al.* 2001), and (B) the Siberian tiger *Panthera tigris altaica* (Suedmeyer *et al.* 2003). The normal karyotype for the Siberian tiger is 2n = 38, but this individual has a sex chromosome set of XXY and thus exhibits Klinefelter syndrome. Notice that the chromosome naming and ordering conventions for the two species differ. Felid karyotypes follow the standard established for the common cat in which the chromosomes are labeled with a combination of letters and numbers. The elephant karyotype is arranged with acrocentric/telocentric pairs first, followed by the two metacentric pairs, followed by the pair that distinguishes African from Asian species. Reproduced by kind permission of the Zoological Society of San Diego's Center for Reproduction of Endangered Species Genetics Division and the Kansas City Zoo.

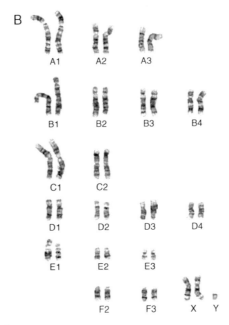

B

A1   A2   A3

B1   B2   B3   B4

C1   C2

D1   D2   D3   D4

E1   E2   E3

F2   F3   X   Y

FIGURE 9.2 *(Continued)*.

histones, which means that the DNA molecule, if extended, would be several orders of magnitude longer than the chromosome—in fact, compared to an average mammalian chromosome length of only a few micrometers (μm) during mitosis, the DNA would be a few centimeters long. The DNA, as well as the protein, in and around the centromeres and telomeres has characteristic properties. For example, DNA near the centromere of a human chromosome contains hundreds of thousands of repeats of a characteristic 171-base pair (bp) sequence called an alpha satellite sequence. Telomeres and nearby regions likewise have characteristic repeat sequences. Centromeres and telomeres are heterochromatic (condensed), whereas other regions in the chromosome can be either heterochromatic or euchromatic (uncondensed), as indicated by the banding pattern. Most genes and other single-copy DNA are found in the euchromatic portions.

KARYOTYPING: THE BEGINNING OF COMPARATIVE GENOMICS

Until the 1970s, banding techniques to consistently reveal the fine structure (chromatin patterns) of chromosomes were unavailable, and the only genome comparisons were based on the number, relative sizes, and shapes of the chromosomes. Even this rough characterization required nontrivial laboratory techniques and generated potentially ambiguous results. For example, the correct chromosome count for humans was not established until 1956 (Ford and Hamerton, 1956; Tjio and Levan, 1956). Even at this level of detail, a great deal of variation among karyotypes of different organisms was apparent. In some organisms, the chromosomes all have the same morphology; for example, all mouse chromosomes are acrocentric (centromere near one end). Other organisms, such as humans, have a mixture of different chromosome morphological types. Chromosomes also vary considerably in size, both within and among genomes. Some chromosomes of fungi and green algae are 1 μm or less in length, whereas some animal and plant chromosomes are more than 30 μm long. Some birds and lizards have a mixture of small and large chromosomes. In terms of numbers of chromosomes, the male of the ant *Myrmecia pilosula* has just one, whereas the fern *Ophioglossum reticulatum* has a diploid chromosome number of 1260. It is remarkable that even closely related and phenotypically similar species such as the Indian muntjac (*Muntiacus muntjak*) and Chinese muntjac (*Muntiacus reevesi*) can differ greatly in chromosome number ("n" represents the haploid number), with 2n = 6 (females) and 2n = 7 (males) for the Indian species as compared to a more typical mammalian value of 2n = 46 for the Chinese species. Even with this extreme karyotypic difference, viable hybrid offspring are known, indicating a high degree of sequence conservation in spite of the radical difference in chromosome number (Levy et al., 1992). Except for frequent evidence of polyploidy in plants (see Chapter 7), there seems to be little phylogenetic pattern or overall trend to chromosome number among major eukaryotic taxa.

Once banding techniques became available in the 1970s (Caspersson et al., 1970; Pardue and Gall, 1970; Seabright, 1971), finer aspects of genomic relationships became visible. Several types of banding can be produced by different dyes and treatments. The most common is G-banding, which produces a characteristic pattern of alternating light and dark regions (note that in plants G-banding does not produce good results). In the human genome up to 850 bands are visible. This method uses trypsin to partially digest the histones of the chromosome prior to staining with the DNA-binding dye called Giemsa, which preferentially stains heterochromatic regions of the chromosome to produce dark bands. The differential staining effect is strongly correlated to locally low G+C content of the DNA, but is not completely explained by the G+C content (Niimura and Gojobori, 2002). Other banding methods in use are R-banding, which gives essentially the reverse of the G-band pattern; Q-banding, which uses fluorescent dye and identifies much the same regions as G-banding; and C-banding, which primarily stains the constitutive heterochromatin of the centromeres.

Following the advent of these powerful techniques for determining chromosomal homology, the genomes of certain groups, perhaps most notably mammals, began to be studied intensively. With a few exceptions, the content of the mammalian genome is more highly conserved than its karyotype might suggest. For example, although the diploid chromosome number ranges from 2n = 6 or 7 in some varieties of muntjac to 2n = 84 in the black rhino *Diceros bicornis* (Hungerford *et al.*, 1967), haploid genome size varies only from less than 2 billion base pairs (gigabases, Gb) in some bats to more than 8 Gb in one species of (polyploid) rat. Indeed, most groups (such as primates, artiodactyls, marsupials, and monotremes) have much smaller ranges, with sizes close to 3 Gb (Gregory, 2001; Hedges and Kumar, 2002) (see Chapter 1).

With the availability of denser genetic maps and the use of more genomic markers and powerful chromosome painting techniques such as ZOO-FISH (Scherthan *et al.*, 1994; Wienberg and Stanyon, 1995), researchers obtained a more refined picture of relative organization within and among mammalian orders. ZOO-FISH (a modified version of fluorescent *in situ* hybridization [FISH] techniques), for example, is based on interspecies chromosome painting in which DNA from fluorescent-labeled individual chromosomes of one species is hybridized *in situ* to the genome of another species. This method has greatly facilitated the identification of evolutionarily conserved chromosomes, chromosome arms, and segments (Raudsepp *et al.*, 1996; Iannuzzi *et al.*, 1998; Richard *et al.*, 2003).

Comparison of homologous markers in different species showed several common patterns of gross genomic change. Besides duplications and chromosomal fission and fusion, homologous chromosome segments could be identified in different relative locations in different genomes. Chromosomal rearrangement is the transfer of chromosome segments either to other chromosomes (interchromosomal rearrangement) or within a chromosome (intrachromosomal rearrangement). The most common form of interchromosomal rearrangement is the *reciprocal translocation,* in which two chromosomes exchange terminal (end) segments. Other forms of interchromosomal rearrangements are *simple translocation,* in which a terminal segment of one chromosome breaks off and attaches itself to the end of another chromosome, and *intercalary transposition,* in which an internal segment of one chromosome moves to a nonterminal position on another chromosome. Common forms of intrachromosomal rearrangements are *simple transpositions* (a segment moves from one part of a chromosome to another part of the same chromosome) and *in-place inversion* (a genomic segment remains in the same place but its direction is reversed). These are depicted in Figure 9.3.

A comparison of human (2n = 46) and chimpanzee (2n = 48), for example, reveals an almost perfect correspondence between respective pairs of chromosomes, with metacentric human Chromosome 2 being divided into two acrocentric chimp Chromosomes 12 and 13. This homology is clearly apparent even in banding patterns (Fig. 9.4). Comparison to other primates shows that the divided form is likely the ancestral state, indicating that the human–chimpanzee difference arose by a fusion of the telomeres of the two acrocentric precursors in the

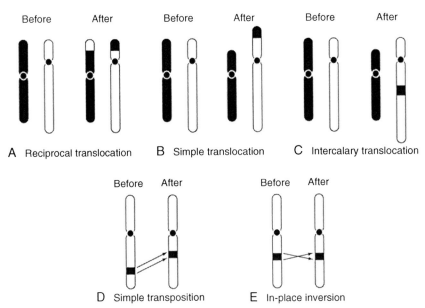

FIGURE 9.3   Diagrams of the most common kinds of chromosomal rearrangements. In each of (A), (B), and (C), an original (nonhomologous) pair of chromosomes is shown on the left and the result after the rearrangement is shown on the right. In (D) and (E), a single chromosome is shown before and after rearrangement. Reciprocal translocations (A) are the most common type of interchromosomal exchanges and involve a swapping of terminal ends between two chromosomes. Simple translocations (B) involve the breakage of a terminal end from one chromosome and its fusion to the terminal end of another chromosome, whereas intercalary translocations (C) involve the transfer of a part of one chromosome to a nonterminal part of another. Chromosomal segments may also change position within chromosomes, as by simple transposition (D), or may stay in the same place but be reversed in direction, as by in-place inversion (E).

human line after divergence from chimpanzees (Yunis and Prakash, 1982). Further evidence of this fusion is the presence of remnants of the extra centromere (Avarello *et al.*, 1992) and the extra telomeres (Ijdo *et al.*, 1991). The gibbon (2n = 58) contains many rearrangements with respect to the human genome. For example, the contents of human Chromosome 2 are now dispersed among gibbon Chromosomes 1a, 14, 17, 19, 20, and 22b. If, however, one considers macaques—the immediate outgroup species to the human, chimpanzee, and gibbon clade—it is evident that the chimpanzee pattern that involves separate chromosomal homologs to each arm of human Chromosome 2 again holds. The conclusion is that extensive rearrangements have taken place along the lineage leading to gibbons from their common ancestor with humans and chimps. Similarly, human Chromosome 21 appears to have evolved from two ancestral

528

Filipski and Kumar

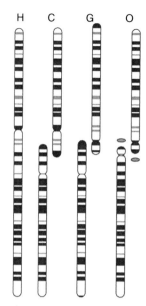

FIGURE 9.4   Schematic representation of evidence for chromosomal evolution. The homologs of the p and q arms of human (H) Chromosome 2 are separate acrocentric chromosomes in other primates (C = chimpanzee, G = gorilla, O = orangutan), indicating a fusion in the human lineage after descent from the most recent common ancestor of human and chimpanzee. Further evidence of this transformation is given by the presence of two inverted arrays of the characteristic vertebrate telomeric repeat in a head-to-head arrangement in the human chromosome at the apparent point of fusion, and the fact that *in situ* hybridization signals the presence of alphoid patterns typical of centromeres at the point in the human chromosome corresponding to the position of the homologous centromere in the other primates. Adapted from Yunis and Prakash (1982), reproduced by permission (© American Association for the Advancement of Science).

blocks present in marsupials and monotremes (Graves, 1996). In this way fissions and fusions can be inferred in a group of closely related lineages. A similar example involves the relation of the nucleoside phosphorylase (NP) gene with the gene complex *PKM2-MP1-HEXA*. In humans, *NP* is on Chromosome 14 and the *PKM2-MP1-HEXA* group lies on Chromosome 15. This separation is preserved in chimpanzees, but the two groups are syntenic (together on the same chromosome) in macaques, rhesus monkeys (Estop *et al.*, 1983), baboons (Thiessen and Lalley, 1986; Thiessen and Lalley, 1987), and pigs (Gellin *et al.*, 1981), indicating general conservation except for a fission in a recent ancestor of chimps and humans (Murphy *et al.*, 2001).

The general pattern of widespread chromosomal conservation with interspersed rapidly evolving lineages is found in many parts of the mammalian phylogeny.

For instance, lemurs (2n = 60), which are basal primates, show an actively evolving genome structure. This time the primary mode is fission, with Chromosome 1 of humans, chimpanzees, and macaques being present as three separate chromosomes (2, 22, and 23) in lemurs. A primatewide comparison of conserved chromosomal regions allows inference of a most parsimonious primate ancestor with 2n = 50 (O'Brien and Stanyon, 1999). The technique is to find large, universally conserved segments and treat these as units that are rearranged through time, while attempting to minimize the number of rearrangements. The reconstruction of chromosome evolution requires about seven major translocation rearrangements to get from the common ancestor of all living primates (60–80 million years ago) (Tavare et al., 2002) to modern humans. This rate of one large-scale rearrangement per 10 million years seems to be a characteristic of the primate lineage. For most extant primate species, fewer than 20 major rearrangements are needed to reconstruct evolution from the common primate ancestor (O'Brien and Stanyon, 1999; Hedges and Kumar, 2003), although many more minor rearrangements are inferred to have taken place (Kumar et al., 2001).

Mouse and human chromosomal homology has been mapped in greater detail owing to the availability of complete genomes. Mice show extensive rearrangements as compared to humans, which are thought to have taken place primarily within the rodent lineage. It has been estimated that chromosomal rearrangements between mouse and rat proceed ten times faster than between far less closely related species such as humans and cats (Stanyon et al., 1999). Indeed, the differences between the cat (2n = 38) and human genomes are not extensive (Nash and O'Brien, 1982; O'Brien and Nash, 1982), and can be accounted for by some 13 translocations and fissions/fusions involving large blocks of genes. This implies a roughly equivalent rate of rearrangements in carnivores as in primates, and indicates that the common mammalian ancestral genome was probably something between that of humans and cats. Dogs (2n = 78) and some bears (Nash et al., 1998) exhibit a somewhat more rapidly evolving genomic architecture with a greater number of karyotypic changes and rearrangements. An ancestral carnivore genome with 2n = 42 has been reconstructed using the same methods discussed previously (Murphy et al., 2001). Among the cetartiodactyls, cows (2n = 60) exhibit a high number of conserved segments with respect to humans, with many of these resulting from intrachromosomal movements such as inversion. Most of the genomic distance (in terms of chromosomal rearrangements) between humans and bovines may be accounted for by 40 to 50 interchromosomal translocations and a similar number of intrachromosomal rearrangements (Band et al., 2000; Jiang et al., 2002).

The banding pattern, morphology (except among ruminants), and gene content of the X chromosome are very highly conserved among eutherian mammals (Chowdhary et al., 1998). A portion called XCR is even identifiable among marsupials as well as eutherians, whereas a more recently added XAR portion has been created from autosomal material prior to eutherian diversification. A study involving 25 markers revealed complete conservation of order between humans and cats, whereas mice showed seven conserved segments with respect to the others (Murphy *et al.,* 1999). Other studies show more intrachromosomal rearrangement of the X chromosomes of some mammals, however (Nadeau, 1989; Farr and Goodfellow, 1992). Wakefield and Graves (1996) found only one of 42 markers on the human X chromosome with an autosomal homolog in a eutherian (*AMD2* on Chromosome 20 of rat). Thus the X chromosome exhibits the same general pattern in this respect as the autosomes, although at a much slower pace. The evolution of the mammalian Y chromosome is also anomalous in several ways—although homologs of human Y-chromosome genes may sometimes appear on X chromosomes of eutherian mammals, it is very rare to find them in autosomes. But the Y chromosome does tend to have a high degree of activity in terms of both content and organization, especially in primates (Archidiacono *et al.,* 1998; Skaletsky *et al.,* 2003). For these reasons, the mammalian sex chromosomes are usually excluded from generalizations based on the autosomes.

Human Chromosome 17 is conserved as an entire chromosome in chimpanzees, macaques, lemurs, tree shrews, cats, horses, pigs, dolphins, cows, Chinese (but not Indian) muntjacs, and sheep, and as an arm in minks, bats, harbor seals, spectacled bears, and giant pandas. Human Chromosome 20 is conserved as an entire chromosome in chimps, lemurs, horses, and pigs and as an arm in gibbons, macaques, tree shrews, cats, minks, dolphins, bats, spectacled bears, and giant pandas. Murid rodents (mice and rats), as usual, form an exception in which fragmentation prevails, but even there, human Chromosome 20 forms a conserved unit in both rats and mice. Note that the human Chromosome 20 homolog appears in both cows and Indian muntjacs as two segments separated by material from human Chromosome 10, indicating that an inversion took place prior to the divergence of the cervids and bovids. Similar arguments based on distribution of synteny led one group (Chowdhary *et al.,* 1998) to postulate a primordial eutherian karyotype of 2n = 48 consisting of human chromosome segments 1p, 1q, 2pter-q13, 2q13-qter, 3+21, 4, 5, 6, 7, 8, 9, 10, 11, 12+22a, 13, 14+15, 16q+19q, 16p, 17, 18, 19p, 20, 22b, X and Y (where "ter" refers to the terminus of the respective arm, and "a" and "b" refer to portions of Chromosome 22).

The genomes of marsupials and monotremes appear to be more conserved than those of eutherian mammals. Among marsupials (2n ranging from 14 to 22), a primitive genome with 2n = 14 appears to best account for the existing diversity. The rock wallabies, however, have a more actively evolving genome, with some 20 different karyotypes described. Similar activity has been noted in some mouse populations, with six different karyotypic races resulting from multiple Robertsonian (centric) fusions being noted on the island of Madeira from a founding population only 500 years old. Amazingly, some of these races have diploid chromosome numbers as low as the 20s, compared with a more typical

2n = 40 (Britton-Davidian *et al.*, 2000). The monotremes, platypus and echidna, have very similar karyotypes (Graves, 1996), despite having been separated for as long as, or longer than, the major orders of eutherians.

In summary, gene order and synteny on the mammalian genome tends to be, with a few notable exceptions such as murid rodents, rather conserved, even when karyotypic change is rampant. It is difficult to infer an ancestral karyotype because fission and fusion are both fairly frequent, but gene order at a coarse level is probably not very different from what is observed today in humans or cats. Mammals generally display a slow rate of chromosome exchange (one or two major exchanges in 10 million years) punctuated in certain lineages by episodes of radical genome reorganization. The reason for these episodes remains unknown (O'Brien *et al.*, 1999; Kumar *et al.*, 2001).

Chromosome painting results have been used to confirm some phylogenetic hypotheses, such as the close relationship of carnivores with perissodactyls and artiodactyls in the hypothesized superordinal clade ferungulata (O'Brien *et al.*, 1999; Murphy *et al.*, 2001). On the whole, however, there appears to be limited phylogenetic information to be obtained from comparative genomics at the karyotype level. As will be seen, this is not the case with sequence-based comparisons.

Besides chromosome number and structure, the other crude descriptor of genomes is haploid DNA content, or "C-value." By the mid-20th century, techniques had been developed to measure this parameter (see Chapters 1 and 2). It became increasingly clear that (1) genome size varies enormously among species, and (2) except at a very basic level (e.g., prokaryotes versus eukaryotes) there is no correlation between genome size and notions of organismal complexity. For example, genome DNA content is now known to vary over several orders of magnitude among eukaryotes and by a factor of 350 even among vertebrates. On the other hand, some groups of eukaryotes, such as birds, mammals, and teleost fishes, show relatively little variation. The factors correlating with genome content are numerous and diverse and their interactions complex. Despite early hopes that C-value might prove to be a simple characterization of an organism's complexity, it has instead raised many more biological questions than it answered (see Chapters 1 and 2).

Genome Architecture

In the last half of the 20th century, a more detailed, sequence-oriented picture of the overall architecture of the eukaryotic genome began to take shape. In the late 1970s, it was discovered that eukaryotic messenger RNA (mRNA) was shorter than the genomic DNA from which it was transcribed and that sections, then known as intervening sequences, were spliced out in eukaryotes and their viruses after initial transcription (Berget *et al.*, 1977; Chow *et al.*, 1977). It was also found that a large fraction of eukaryotic DNA was repetitive and did not appear to code for proteins or to have any of the functions known by analogy to prokaryotic DNA (Britten and Kohne, 1968). Masatoshi Nei called such apparently useless DNA "nonsense DNA" (Nei, 1969), whereas Susumu Ohno called it "junk DNA" (Ohno, 1972). One thing was clear: the eukaryotic genome was much messier to deal with than the compact prokaryotic genome. Although a number of theories have since been proposed to explain the presence of noncoding DNA and the interrupted coding sequences, much uncertainty remains about the functional and evolutionary significance of these features (see Chapters 1 and 2). Certain repetitive elements may simply be parasitic or "selfish" DNA (Doolittle and Sapienza, 1980; Orgel and Crick, 1980). Scientists have at least now created a taxonomy of the repetitive elements and know something about the means by which they replicate (see Chapter 3). Figure 9.5 shows a breakdown of different DNA types in the human genome; other eukaryotes have similar classes of elements, but in different proportions.

The structure of a typical eukaryotic gene is depicted in Figure 9.6. A segment of DNA that does not contain any stop codons when interpreted according to its implied reading frame (with the first nucleotide being the first position of the first codon) is called an Open Reading Frame (ORF) (Doolittle, 1986). An ORF is not necessarily part of a protein-coding gene, but it may be. A eukaryotic protein-coding gene may contain several noncontiguous ORFs, possibly in different reading frames from each other. The problem of genes being interrupted by so-called "intervening sequences" is now well known as the intron–exon distinction, and in general the mechanisms by which this occurs are known. These characteristics, interrupted coding sequences and large amounts of DNA whose function is unclear, characterize the eukaryotic genome.

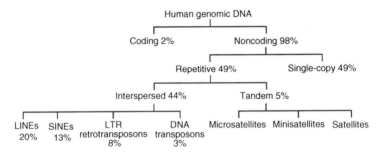

FIGURE 9.5   Composition of the human genome in terms of DNA classes. Percentages are approximate; in particular, exact figures for tandem repeats are not well known because they are most common in difficult-to-sequence constitutive heterochromatic regions (e.g., centromeres). Compositions of other eukaryotes may be radically different. For example, the housefly *Musca domestica* has about 90% single-copy DNA, whereas the toad *Bufo bufo* has only about 20% single-copy DNA (John and Miklos, 1988).

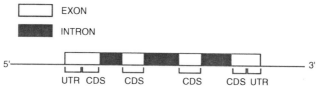

FIGURE 9.6   Architecture of a typical eukaryotic gene. The region from the beginning of the first exon to the end of the last exon is transcribed. Later, the introns are removed during splicing. The Coding DNA Sequences (CDS) are finally translated into a polypeptide. Enhancer and promoter sequences (not indicated) that control transcription may be located near the gene, upstream or downstream, or even within introns. Untranslated regions (UTRs) are not themselves translated into polypeptides, but they control translation in various ways. Lengths and numbers of introns vary widely among different groups of eukaryotes. In humans, the mean intron length is 3400 bp, but the most common (modal) intron length in human genes is around 100 bp. The average number of introns per human gene is about nine.

Another property of the eukaryotic genome is lack of uniformity. Chromosomes, and regions within chromosomes, vary a great deal in almost any parameter one can think of, including gene density, G + C content, and so on; these variations cannot be explained merely by sampling error. Even at a coarse level, the bands revealed by Giemsa staining suggest differences in regions within a chromosome. Bernardi and others have identified five families of regions called *isochores* in the human genome (Bernardi *et al.*, 1985, 1988). Each of these regions is at least 300 kilobases (kb) in length and has a characteristic G + C content. The regions of low G + C content are designated L1 and L2 regions and are gene-poor, whereas regions with high G + C content are designated H1, H2, and H3 and are gene-rich. For example, the H1 regions make up only 3% of the human genome but contain 25% of the genes. It also appears that long genes are less likely to appear in the G + C–rich isochores (but see Duret *et al.*, 1995). The isochores are correlated with G-bands, with the dark G-bands tending to be made up of L1 and L2 isochores, with some contribution from the H1 family (Saccone *et al.*, 1993). Other vertebrates and plants also seem to have an isochore structure. Some controversy, discussed in a later section, arose after the human genome sequence was available about the uniformity of G + C content within isochores, but the concept remains a useful tool for dividing the genome into identifiable regions.

WORKING WITH EUKARYOTIC GENOMES

Today, technology appears to drive the biological sciences as much as hypothesis does (Galison, 1997; Volti, 2001). In contemporary biology, there are already many terabytes of molecular sequence data available and the rate at which it is accumulating is rapidly accelerating. Fortunately software (including databases and specialized search and analytical tools) has been able to keep up with the data explosion and runs on common, inexpensive hardware in most cases. Turning a eukaryotic genome into grist for this mill usually involves the following steps: mapping, sequencing, and annotation.

MAPPING: GENETIC AND PHYSICAL

Mapping involves determining the position of recognizable markers in the genome of interest. The markers may be genes or other sequence features, and the positions may be reckoned crudely in terms of the chromosome or arm on which the marker resides, of its genetic linkage with other markers measured in centiMorgans (linkage map), or of its position on the chromosome specified in terms of base pairs (physical map). Mapping technology began with genetic linkage mapping of *Drosophila melanogaster* and other model organisms in the early 20th century (Sturtevant, 1913), followed by the first rudimentary physical maps two decades later (Bridges, 1935). By the 1970s and 1980s, physical genome mapping underwent great advances. Markers such as Sequence Tagged Sites (STSs) (Olson *et al.*, 1989) based on the Polymerase Chain Reaction (PCR) (Mullis *et al.*, 1986) allowed the accurate mapping of a large number of DNA segments, typically a few hundred base pairs in length, to physical addresses in the genome.

SEQUENCING: THE HOLY GRAIL OF COMPARATIVE GENOMICS

It is the ability to read and assemble sequences of nucleotide bases that has enabled the emergence of comparative genomics as it is currently known. Most DNA sequencing has been based on the Sanger method (Sanger *et al.*, 1977). In this technique, the DNA to be sequenced is first denatured into single strands using heat and then a labeled primer sequence is annealed to the strands near the 3′ end of the region of interest. At this point, the solution is divided into four batches corresponding to the four nucleotide bases. Nucleotides and DNA polymerase are added to each batch, and each is additionally given a solution of one of four different kinds of dideoxynucleotides corresponding to the four bases. Dideoxynucleotides are essentially the same as nucleotides except they contain a hydrogen group on the 3′ end instead of a hydroxyl group. These specially modified nucleotides terminate any DNA chain into which they are incorporated because a phosphodiester bond cannot form between the dideoxynucleotide and the next potential nucleotide. DNA synthesis in each batch thus results in a collection of strands of different lengths, each terminating in the same nucleotide.

When these strands are separated by gel electrophoresis, their lengths indicate positions of that nucleotide. Modern automated sequencing methods (Strauss *et al.*, 1986) have streamlined this technique but the same basic principle is used. For example, all four reactions are run together with distinctively labeled dideoxynucleotides so that the sequence can be automatically read from a single lane using gel or capillary methods.

The problem with the basic Sanger chain termination method is that reads are limited to several hundred or at best a few thousand bases, whereas DNA sequences of interest are often far larger. This problem is addressed using the so-called "shotgun method" of obtaining overlapping random sequence reads from the larger sequence and assembling these on the basis of matching overlapping areas into larger contiguous segments called "contigs." This works well except in the presence of low-complexity or repetitive DNA, where matching may not indicate actual overlap. Eukaryotic whole-genome sequencing projects have taken either of two approaches: hierarchical shotgun or whole-genome shotgun. The public National Human Genome Research Institute (NHGRI) Human Genome Project was an example of the former, whereas Celera used the latter approach in its human genome analysis (see later section for more on this). In the hierarchical shotgun approach, the genome is first broken down into a library of cloned regions (e.g., bacterial artificial chromosomes) whose relationship to the entire genome is known through mapping. Each of these clones is then sequenced by the shotgun method, and the resulting sequences are assembled. The generation and mapping of the clone library is a large part of the effort. The whole-genome shotgun method dispenses with the mapped clone library step, and reads are obtained directly from the target genome (see also Chapter 10). This latter method is much more cost-effective, but is more error-prone and requires more sophisticated assembly methods because the maps are not available as a top-down guide.

Other, radically different, sequencing methods are on the horizon. An example is "nanopore sequencing" (Deamer and Branton, 2002). This technique analyzes individual strands of DNA by applying an electric current as they pass through a tiny membrane channel or pore. As charged bases pass through the pores in single file, they block the flow of current in a manner characteristic of the polymer's sequence.

ANNOTATION: MAKING BIOLOGICAL SENSE OF THE LETTERS

Once the sequence of a genome has been determined, the job of interpreting the lengthy string of A's, C's, G's, and T's can begin. At a minimum, the goal is to identify the locations of all the functional units such as genes, transposable elements, and regulatory regions in the sequence, and ultimately to determine the functional relationship of these elements to each other and to expression data, proteins, phenotype, and disease. All of these kinds of information, when attached to sequences, constitute annotation. Generally, annotation is expert-labor intensive, but automated tools for comparing and parsing sequence data are indispensable. All major sequence databases provide record fields for annotating sequence entries.

THE GENESIS OF LARGE-SCALE SEQUENCING PROJECTS FOR EUKARYOTES

SEQUENCING THE HUMAN GENOME: THE MOST AMBITIOUS IDEA

By 1985 the idea of sequencing the human genome began to be discussed. It was an extremely ambitious notion. The first tiny viral genome had been sequenced barely 10 years before, and the completion of even the first prokaryote genome sequence lay a decade in the future (see Chapter 10). Even physical mapping of eukaryotic genomes had been done only for yeast and simple animals. Nevertheless, buoyed by the mounting wave of successful "big science" projects, from the Manhattan project to the Apollo program to the recent Keck telescope, Robert Sinsheimer of the University of California at Santa Cruz and later Walter Gilbert became early proponents of the idea (Gilbert and Bodmer, 1986). It was a controversial as well as a bold idea. Some biologists were appalled by the notion that "assembly line science" might replace the small research group or at least compete with it for funding (Chargaff, 1980). Gilbert countered that the human genome sequence would be the "raw material for the science of the 21st century" (Gruskin and Smith, 1987).

Who would fund the project, even if it were seen as feasible? Private sources did not seem enthusiastic. The most interested agency of the U.S. federal government seemed to be the Department of Energy's (DOE) Office of Health and Environmental Research, which had been studying the genetic consequences of the use of atomic energy. By 1987 the National Institutes of Health (NIH) had also joined the bandwagon and funded a small feasibility study (Roberts, 1987). By 1988 the NIH and DOE had signed a memorandum of cooperation and famed DNA pioneer James Watson was named Associate Director of Human Genome Research at NIH. The project was under way, with high visibility (Goujon, 2001).

Funding for the Human Genome Project (HGP) formally began in 1990. Gilbert estimated that the overall project would cost about $1 per base and would require 15 years, although the cost at the time was closer to $10 per finished base (Collins et al., 2003). The first five-year plan proposed the creation of complete genetic and physical (STS) maps of the human genome, with sequencing to commence when costs declined to less than $0.50 per base. Ultimate project goals included 100,000 mapped single nucleotide polymorphisms (SNPs, or "snips") as well as the sequencing of 95% of the euchromatic portion of the genome with 99.99% accuracy (Collins *et al.*, 2003). Around the same time, an international coordinating committee, the Human Genome Organization (HUGO), was formed to coordinate international funding and to iron out the anticipated disputes over such issues as intellectual property rights. By 1994, Robert Waterston declared, accurately, that the technology was then available to complete sequencing by 2001, four years ahead of schedule (Boguski, 1995).

In the early 1990s it began to be appreciated that complementary DNA (cDNA) libraries based on expressed sequences (mRNA) would be essential for the discovery and annotation of human genes. To address this, the Expressed Sequence Tag (EST) method was developed (Adams *et al.*, 1991, 1992; Okubo *et al.*, 1992). This allows rapid generation and sequencing of partial mRNA sequences found in a cell. Although this method does not give genomic DNA sequences, it provides an efficient way to characterize expressed genes. Currently the National Center for Biotechnology Information (NCBI) EST Database (www.ncbi.nlm.nih.gov/dbEST) contains nearly 23 million sequences from hundreds of organisms and is a valuable resource for gene prospecting and gene expression studies.

Other organisms were not being neglected. The HGP funded a subsidiary sequencing project for the common gut bacterium *Escherichia coli* as a test bed, and two more eukaryote sequencing projects were also undertaken: the yeast (*Saccharomyces cerevisiae*) genome project (1989) in Europe and the nematode (*Caenorhabditis elegans*) project spearheaded by Sulston and others in the United Kingdom (1990). A sequencing project for thale cress (also called mustard weed), *Arabidopsis thaliana*, was also put forward to the U.S. funding agencies in 1989. All of these were model organisms upon which a great deal of work had already been done, and all were known to have very compact genomes. In a way, the HGP acted as an umbrella to shelter these far more modest projects: if it were indeed feasible to sequence the entire human genome, so the logic went, then surely these more diminutive projects would be relatively simple and would provide a valuable place to develop new techniques.

PRIVATE VERSUS PUBLIC EFFORTS

In May 1998, scientist and entrepreneur J. Craig Venter announced plans to form a new private company named Celera that would sequence a large portion of the human genome within three years for a cost of around $300 million, using whole-genome shotgun methods that were faster and less labor-intensive. The investment was to return a profit by selling access to a database of high-quality well-annotated sequence data. The Celera project ran in parallel to the public NHGRI effort and both announced completion of first drafts at the same time (International Human Genome Sequencing Consortium, 2001; Venter *et al.*, 2001). Some controversy swirled around the manner in which the Celera group released its data, however. Only limited amounts could be freely downloaded without signed nondisclosure agreements. Many felt that that was an unacceptable compromise between scientific openness and commercial interest and that the paper should not have been published in an academic journal. Eric Lander, director of the Whitehead Center for Genome Research and a key figure in the publicly funded International Human Genome Sequencing Consortium, was quoted as saying, "This is the first time in history that a paper reports a scientific result, but tells readers that to see it, they must sign a contract." Further questions were raised about the ability of Celera's whole-genome shotgun to have succeeded at all without building upon the public project's scaffolding (Russo, 2001).

Despite such friction, the overall results of the two projects seemed to be in fairly good accord (Aach *et al.*, 2001), and the two efforts complemented and stimulated each other's progress. From there, Celera went on to involvement in the sequencing of other organisms. Although the private sector has had significant involvement in many sequencing efforts, questions remain about the profitability of the work in the face of publicly funded competition, as well as legal and ethical issues about ownership and restriction of the use of scientific data. It had been argued that, by its nature, the human genome sequence belongs to humanity as a whole (Macer, 1991). In January 2002, Venter left Celera to undertake other projects, including genetic engineering of life forms (see Chapter 10). In April 2002, Venter revealed that the genome sequenced by Celera was not that of a randomly chosen subject, but of Venter himself. One of Venter's announced projects is writing a book analyzing his own genome (Wade, 2002).

So far almost all sequencing projects have operated under the guidelines that sequence data, if not annotations and analyses, were to be released to the public with minimal delay. In 1996 this was formalized as the so-called "Bermuda Principles" (named after the location of a meeting convened by the Wellcome trust), which call for automatic, rapid (within 24 hours) release of sequence assemblies to the public domain and which discourage the patenting of genes by sequencing labs (Collins *et al.*, 2003).

The following sections describe whole-genome sequencing efforts, in roughly chronological order, amended somewhat to discuss related projects or organisms together. In each case, the key events in the project are described and the findings discussed in the context of previously sequenced organisms. Table 9.1 lists a few properties of completely sequenced eukaryotic genomes, and Figure 9.7 shows the progress of some major sequencing projects.

TABLE 9.1   Some basic data about eukaryotic organisms that have been fully sequenced (as of spring 2003). In most cases, data were obtained from the publication announcing completion of the initial release of the respective sequencing project. Chr (n) refers to the haploid chromosome number of the organism.

Species	Common name	Taxon	C-value (Mb)	Chr (n)	Genes	G + C	Exons/ gene	Year
Anopheles gambiae	Mosquito	Insect	278	3	13,700	0.35	3.7	2002
Arabidopsis thaliana	Thale cress	Dicot plant	157	5	25,500	0.35	5.2	2000
Caenorhabditis elegans	Roundworm	Nematode	100	6	19,820	0.36	6	1998
Ciona intestinalis	Sea squirt	Chordate	160	14	16,000	0.35	6.8	2002
Drosophila melanogaster	Fruit fly	Insect	180	4	13,600	0.41	4	2000
Encephalitozoon cuniculi	Parasitic microsporidian	Protist	2.9	11	2000	0.50	1.01	2001
Takifugu rubripes	Pufferfish	Fish	400	22	35,000	0.48	9	2002
Homo sapiens	Human	Mammal	3400	23	35,000	0.41	9	2001
Magnaporthe grisea	Rice blast fungus	Fungus	40	7	11,000	0.52	3	2002
Mus musculus	Mouse	Mammal	3250	20	35,000	0.42	9	2002
Neurospora crassa	Bread mold	Fungus	40	7	10,000	0.50	2.7	2003
Oryza sativa	Rice	Monocot plant	490	12	50,000	0.43	5.1	2002
Plasmodium falciparum	Human malaria pathogen	Protist	23	14	5300	0.19	2.4	2002
Plasmodium yoelii	Rodent malaria pathogen	Protist	23	14	5900	0.23	2	2002
Rattus norvegicus	Brown rat	Mammal	3100	21	35,000	0.42	9	2004
Saccharomyces cerevisiae	Yeast	Fungus	12.5	16	6128	0.38	1.04	1996
Schizosaccharomyces pombe	Fission yeast	Fungus	14	3	4824	0.36	2	2002

540

Filipski and Kumar

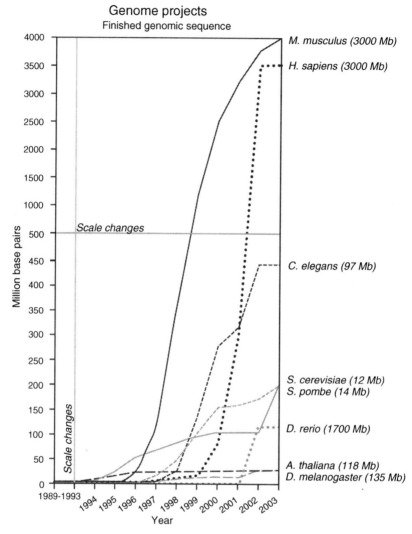

FIGURE 9.7 Cumulative sequencing progress in millions of base pairs for several eukaryotic organisms. Adapted from the European Molecular Biology Laboratory's European Bioinformatics Institute (EMBL-EBI) Genome Monitoring Table at www2.ebi.ac.uk/genomes/mot/.

GENOME SEQUENCING IN FUNGI

SACCHAROMYCES CEREVISIAE: THE FIRST EUKARYOTE TO BE SEQUENCED

As discussed in the previous section, the *S. cerevisiae* project was undertaken around the same time as, and was inspired by, the HGP discussions. At an estimated length of 13 million base pairs (megabases, Mb), the yeast genome is nearly an order of magnitude smaller than other prominent model genomes (e.g., *C. elegans* at 100 Mb and *Arabidopsis* at 157 Mb) (Bennett *et al.*, 2003). Also, a great deal was known about this yeast. For millennia, it has been used in food production for its ability to ferment glucose to ethanol and carbon dioxide. It has been a model organism since the 1970s, and the knowledge of its biochemistry and genetics is highly advanced, in part owing to the early economic importance of the organism to bakers, brewers, and vintners. Moreover, a clone library had already been constructed in the late 1980s (Link and Olson, 1991). The *S. cerevisiae* sequencing project was funded by a mixture of public and private sources primarily centered in Europe and driven by André Goffeau of the Université Catholique de Louvain in Belgium. The philosophy was to supply expertise and coordination among many established laboratories instead of constructing massive new sequencing centers. In 1989, 35 laboratories formed a consortium for the purpose.

In 1992 the complete sequence of Chromosome 3 was published in *Nature* (Oliver *et al.*, 1992) and completion of the entire genome sequence was announced in April 1996 (Goffeau *et al.*, 1996). Soon thereafter, all putative genes (at least those longer than 100 codons) were identified, with the result that for the first time the list of approximately 6000 genes necessary for the functioning of a complete, free-living, eukaryotic organism was known. Most of the yeast genome, about 72%, consists of open reading frames (ORFs), in contrast to the human genome where the figure is less than 2%. Introns tend to be short and almost always located near the 5′ end of the gene; sometimes they occur just after, or even within, the ATG initiation codon. Intron–exon splice sequences are highly conserved. This fortuitous combination of features confirmed the good choice of *S. cerevisiae* as the "practice" model organism with which to begin before being forced to develop more sophisticated methods of exon prediction. Protein-coding genes seem to be randomly oriented on both strands. $G + C$ content varies at many scales, with higher gene density in the broad peaks and a general $G + C$ deficit in subtelomeric and pericentromeric regions. Because so much of the yeast genome codes for proteins, repeats are correspondingly rarer. Ty elements (a kind of LTR retrotransposon) account for less than 3% of the genome, whereas short tandem repeats inhabit small regions around the centromeres and telomeres. Thus the *S. cerevisiae* genome has all the major elements of the larger eukaryotic genomes but in different proportions (Dujon, 1996).

The yeast genome provided an opportunity to face the challenge of functional analysis on a small scale. Perhaps the most striking (if not downright surprising) finding of the yeast genome sequencing effort was that at least a third of the putative genes, as identified from ORFs, had no clear-cut previously known homologs. These orphan genes had escaped the notice of traditional genetic methods, indicating that the state of ignorance of genetics was far greater than was realized. Furthermore, alteration of many of these genes had no apparent effect on the phenotype. A systematic program, dubbed EUROFAN, was begun to test each gene by knocking it out, or disrupting its expression. It appeared that about ⅔ of the disrupted genes on Chromosome 3 led to no obvious difference in phenotype (Goujon, 2001). Something about the function of some genes could be inferred by structural clues, such as the presence of transmembrane helices, but clearly significant advancements were needed to elucidate the function of all known genes.

OTHER FUNGAL SEQUENCING PROJECTS

Other relatives of *S. cerevisiae* were also sequenced with a view to comparison. The fungi are a large and diverse kingdom, with at least 100,000 species, including mushrooms, yeasts, and molds (Hawksworth, 1991). Sequencing was begun on another important fungus, the fission yeast *Schizosaccharomyces pombe,* in 1995 and completed in 2002 (Wood *et al.,* 2002), making it the sixth eukaryote to be sequenced. This model organism, first isolated from pombe (an East African beer), is only distantly related to *S. cerevisiae.* Its genome is about 10% larger, but contains ∼20% fewer genes and only three chromosomes as opposed to 16 for *S. cerevisiae. S. pombe* has on average one intron per gene, much more than *S. cerevisiae.* A pilot deletion project was conducted to determine the apparently essential genes in this organism (Decottignies *et al.,* 2003). The evidence suggests that about 18% of the *S. pombe* genes are in this essential category and that the more phylogenetically widespread a gene is the more likely it is to be required, such that many of the essential eukaryotic genes appeared with the first eukaryotic cell some two billion years ago and have remained strongly conserved.

Genome sequencing work among the fungi has continued and in 2003 the first genome of a filamentous fungus, the intensively studied model mold *Neurospora crassa,* was announced (Galagan *et al.,* 2003) and preliminarily annotated (Mannhaupt *et al.,* 2003). This project has been an important advance in spanning the range of fungal genomes, because the two previously sequenced fungi, *S. cerevisiae* and *S. pombe,* have restricted metabolic and developmental capabilities owing to their specific environmental niches and therefore do not provide a general paradigm for the fungi (Bennett, 1997). An interesting feature of the *N. crassa* genome is the anomalously low fraction of its genes that are members of multigene families (Fig. 9.8). This is thought to be a result of a process called

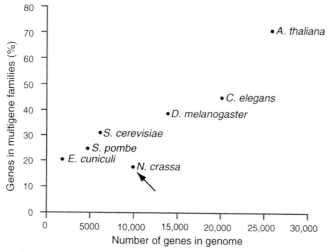

FIGURE 9.8   The proportion of genes in multigene families as a function of the number of genes in the genomes of selected sequenced eukaryotic organisms. The arrow indicates the mold *Neurospora crassa*, which has an especially low proportion of genes in multigene families. Adapted from Galagan *et al.* (2003), reproduced by permission (© Nature Publishing Group).

repeat-induced-point mutation (RIP), which is apparently unique to fungi and mutates and epigenetically silences repetitive DNA. It has been suggested that RIP may be a defense against selfish or mobile DNA (Selker, 1990) (see Chapter 3). As a consequence, *N. crassa* has very little repetitive DNA except for short or highly diverged segments (Krumlauf and Marzluf, 1980). Surprisingly, of the 10,000 predicted protein products of this genome, 41% have no significant matches to known proteins. The fact that new proteins are still being discovered at this rate suggests that the universal proteome is large indeed (Hynes, 2003). It is also interesting that for many *N. crassa* genes, the only known homologs are in prokaryotes (Mannhaupt *et al.*, 2003).

In 2003, three close relatives of *S. cerevisiae* were sequenced (Kellis *et al.*, 2003). These organisms, *Saccharomyces paradoxus, S. bayanus,* and *S. mikatae,* were selected not so much for their intrinsic interest (although two of the three are used in winemaking) but because of their evolutionary similarity to *S. cerevisiae.* In such a case, the whole-genome shotgun method becomes even more efficient because the known genome can be used to help assemble the newly sequenced ones. The real benefit, however, is comparative. When several similar sequences are available, it becomes possible to identify regions that are more conserved than would be expected by chance. Such regions are candidates for small genes or hitherto unknown regulatory regions. Conversely, putative functional genes that cannot be found in the close relatives are suspected to actually be nonfunctional.

Such considerations led to a revision of the gene count for *S. cerevisiae* by an addition of 43 small genes and a deletion of about 500 putative nonconserved genes. Also, many known and newly discovered regulatory motifs were identified. This kind of near-neighbor comparative genomics will be essential for fully parsing any genome, including that of humans.

CAENORHABDITIS ELEGANS AND *DROSOPHILA MELANOGASTER*: THE FIRST ANIMAL GENOMES TO BE SEQUENCED

THE WORM PROJECT

The *C. elegans* sequencing project was initiated by two groups: John Sulston and Alan Coulson at the Medical Research Council (MRC) Laboratory of Molecular Biology in Cambridge in the United Kingdom, and Robert Waterston at the Washington University School of Medicine in St. Louis in the United States. The genome of this organism is much larger than that of *S. cerevisiae* (100 Mb as opposed to 13 Mb) and contains the genes needed for development of a multicellular animal, in addition to genes for the housekeeping functions common to all eukaryotic cells. An advantage of using a nematode as a model organism, other than its simplicity and the knowledge of all developmental cell lineages (Sulston *et al.*, 1983), is that nematodes are thought to have diverged early among the animals, so that genes with homologs in both *C. elegans* and another animal are likely to be ancestral to the entire kingdom.

Results of the *C. elegans* sequencing project were reported in 1998 (*C. elegans* Sequencing Consortium, 1998). This provided the first opportunity for genome comparison between two species from different eukaryotic kingdoms: the metazoan *C. elegans* and the unicellular yeast *S. cerevisiae*. Again, the *C. elegans* genome is about eight times as large as the yeast genome and contains about three times as many genes (20,000 vs. 6000). It was anticipated that the two organisms would contain a common core of genes associated with basic eukaryotic cell maintenance, and this was indeed the case. About 20% of *C. elegans* proteins and 40% of yeast proteins had a very similar homolog (BLASTp P-value $< 10^{-10}$; see www.ncbi.nlm.nih.gov/blast) in the other organism (Chervitz *et al.*, 1998). Figure 9.9 shows the distribution of functions of genes that had identifiable homologs in both species. In most cases, the classification was obtained from yeast annotations. The numbers in each section represent the ratio of worm to yeast genes in each category. These data support the idea that core eukaryotic cellular functions are performed by a highly conserved group of genes without many paralogs, even in larger genomes. It is encouraging for the process of understanding genomes that annotations of core functions seem to be transferable between disparate organisms.

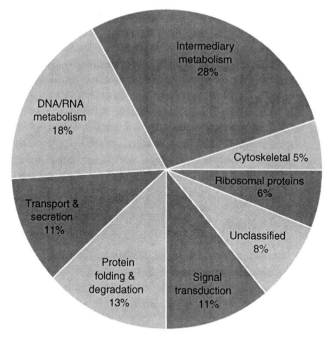

FIGURE 9.9   Distribution of core biological function genes conserved in both yeast (*Saccharomyces cerevisiae*) and nematode worm (*Caenorhabditis elegans*). Yeast and worm protein sequences were clustered into closely related groups. Each sequence group (including groups with two or more sequences) was assigned into a single functional category, relying primarily on the functional annotations for the yeast genes when available. The unclassified category contains groups of sequences without annotation. The number (in %) within each category reflects the ratio of worm to yeast proteins for that category. Adapted from Chervitz *et al.* (1998), reproduced by permission (© American Association for the Advancement of Science).

The complementary aspect of the comparison is to look for genes or families of genes that do not appear to have homologs in both organisms. Although there are some regulatory and signal transduction domains in *S. cerevisiae* genes that do not appear in *C. elegans*, it is primarily the other way around, as expected. Gene types present in *C. elegans* and not in *S. cerevisiae* include those involved in extracellular signaling and adhesion, such as epidermal growth factor (EGF) and factors involved in programmed cell death. In general, comparisons between *C. elegans* and *S. cerevisiae* are consistent with the understanding that common core eukaryotic functions are preserved whereas most disparities can be ascribed to obvious differences in organismal requirements, in this case multicellularity.

A surprising feature of the *C. elegans* genome is that about 15% of its genes are grouped into operons containing from two to eight genes each (Spieth *et al.*, 1993).

In the manner of prokaryotes, all genes in an operon are controlled by a single promoter and produce a single pre-mRNA transcript. However, in contrast to the situation in prokaryotes, genes from a single operon need not be functionally related and are translated separately. They are apparently not ancestrally related to prokaryote operons, but are evolutionarily conserved, as most also appear in *Caenorhabditis briggsae*, a species that last shared a common ancestor with *C. elegans* around 50–100 million years ago. Operon organization in *C. elegans* is thought to be facilitated by a type of mRNA splicing called "trans-splicing" (Nilsen, 1989). Trans-splicing is known to occur in other animals, such as flatworms and hydra, but the extent of occurrence of operons within or beyond the nematodes is not yet known (Blumenthal and Gleason, 2003).

THE FRUIT FLY PROJECT

Despite its early lead as a genetic model organism and the wealth of genetic mapping and functional studies that had been done on it, *D. melanogaster* was not the first choice for an animal to be sequenced. There are a number of reasons for this. For one thing, a large part of the *D. melanogaster* genome is heterochromatic, making it more difficult to map.

A large portion of the *D. melanogaster* genome was sequenced in 2000 by a consortium of private and public research groups lead by Celera Genomics (Adams *et al.*, 2000). The project was notable in that it was the first eukaryote project to use the whole-genome shotgun sequencing method that Celera would later use on the human genome. The initial draft sequence contained many gaps and regions of low sequence quality, but these deficiencies were rectified by the third release (Celniker *et al.*, 2002). Although only about 120 Mb of the 180 Mb genome was sequenced (i.e., the euchromatic portion), it is thought that this includes the vast majority of the protein-coding genes. This result was built upon by a sequencing project for the euchromatic portion of the sister species *D. pseudoobscura*, using a comparative sequence approach. The *Drosophila* genome will be discussed in more detail in relevant later sections.

THE HUMAN GENOME PROJECT

The "completion" of the human genome sequencing project was announced in February 2001 by simultaneous publication of special issues of the journals *Nature* and *Science* describing results of the publicly and privately funded efforts, respectively (International Human Genome Sequencing Consortium, 2001; Venter *et al.*, 2001). This accomplishment surely ranks with the moon landing as a major achievement of the "big science" paradigm. Although the precise point chosen as the completion date was, unlike setting foot on the moon, somewhat arbitrary, at the time of the big announcement more than 90% of the genome was sequenced, the remainder being mostly highly repetitive heterochromatic DNA, which is difficult to sequence and thought not to be very informative. At that time, draft sequences were publicly available, and, although the job of annotation had hardly begun (some 40% of open reading frames were of unknown function), enough was known to paint a broad picture of the human genomic landscape and to compare it to the only truly completely sequenced animal at the time, *C. elegans*.

Annotation of the human genome was greatly facilitated by homology searches for genes known to exist in other vertebrates such as the pufferfishes (genera *Tetraodon* and *Takifugu* [Fugu]) (Brenner *et al.,* 1993; Roest Crollius *et al.,* 2000a,b, 2002). Although the Fugu sequencing project was not officially completed until October of 2001, extensive clone libraries were available.

Even without annotation, however, many simple statistical tests could be performed. One controversial question relates to variation in $G+C$ content and its correlation with other local properties of the genome such as gene density. Bernardi and coworkers (Bernardi *et al.,* 1985; Bernardi, 1995) had postulated that the genome is a mosaic of five different types of compositionally homogeneous regions known as isochores. More recent examination showed that most of the $G+C$ content variance among small (20 kb) regions can be explained by the average $G+C$ content of larger (300 kb) windows that contain them (International Human Genome Sequencing Consortium, 2001), meaning that the hypothesis of strict homogeneity among the small regions was not supportable, and leading to the conclusion that isochores are not as strict or as homogeneous as some expected. On the other hand, further analyses showed that for other choices of region size, the hypothesis of homogeneity may not be rejected (Li, 2002). In any case, regions of distinctive composition certainly exist in the human genome, and Bernardi (2001) added that the original description of isochores did not specify strict statistical homogeneity.

Another interesting feature of the human genomic landscape is the density of CpG islands (Bird, 1986). The notation "CpG" refers to a guanine nucleotide immediately following a cytosine in a DNA strand: 5′ . . . CG . . . 3′. (The "p" in CpG refers to the phosphodiester bond that connects adjacent nucleotides in a strand as distinguished from the hydrogen bonds between the C and G in complementary strands.) A CpG island is a DNA region, usually a few hundred nucleotides in length, with a higher-than-usual $G+C$ content and much higher density of the usually underrepresented CpG dinucleotide. About 30,000 CpG islands were detected in the human genome and it was noted that CpG island density correlates with gene density (International Human Genome Sequencing Consortium, 2001). Other studies showed that about half of human and mouse genes are associated with upstream CpG islands, so the feature becomes important for gene detection (Antequera and Bird, 1993).

A strikingly large portion of the human genome consists of transposable elements. In humans, identified transposable elements make up about 44% of the genome, as compared to about 7% for *C. elegans,* 10 to 22% for *D. melanogaster,* and 14% for *A. thaliana* (see Chapter 3). Short interspersed nuclear elements (SINEs), long interspersed nuclear elements (LINEs), LTR retrotransposons, and DNA transposon copies make up approximately 13%, 20%, 8%, and 3% of the sequence, respectively, in the human genome (see Chapters 1 and 3). More interesting is the comparative age distribution of these elements. Figure 9.10 shows the distribution of ages of these elements for both mice and humans, revealing a marked decline in all transposon activity for the human lineage, going back at least as far as the eutherian radiation, to the extent that only *Alu* and *LINE1* elements show any recent activity; interestingly, mice show no similar decline in TE activity (International Human Genome Sequencing Consortium, 2001).

Figure 9.11 shows a similar comparison extended to cover humans, *D. melanogaster, C. elegans,* and *A. thaliana.* The interspersed repeats in the human

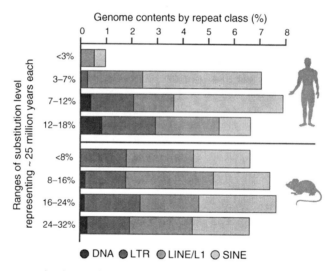

FIGURE 9.10   Age distribution of interspersed repeats in the human and mouse genomes. Bases covered by interspersed repeats were sorted by their divergence from their consensus sequence (which approximates the repeat's original sequence at the time of insertion). CpG dinucleotides in the consensus were excluded from the substitution level calculations because the CT transition rate in CpG pairs is about 10-fold higher than other transitions and causes distortions in comparing transposable elements with high and low CpG content. The data are grouped into bins representing roughly equal time periods of 25 million years. There is a different correspondence between substitution levels and time periods owing to different rates of nucleotide substitution in the two species. Adapted from the International Human Genome Sequencing Consortium (2001), reproduced by permission (© Nature Publishing Group).

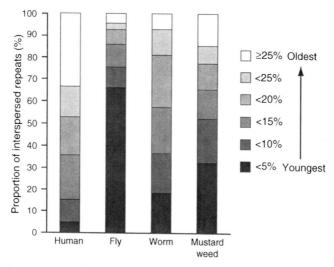

FIGURE 9.11   Comparison of the age of interspersed repeats in four eukaryotic genomes. The copies of repeats were pooled by their nucleotide substitution level from the consensus. Adapted from the International Human Genome Sequencing Consortium (2001), reproduced by permission (© Nature Publishing Group).

genome tend to be older than those of *C. elegans* and *A. thaliana,* and much older than those of *D. melanogaster*. One reason for this may be the increased rate of genome "cleaning" in flies owing to small deletions (Petrov *et al.,* 1996). In humans, most interspersed repeats belong to one of two families, *Alu* and *LINE1*. In *D. melanogaster, C. elegans,* and *A. thaliana,* on the other hand, there are no such dominant families, but a greater diversity of repeat types. This may be because of the much higher fraction of the shorter-lived DNA transposons in *D. melanogaster, A. thaliana,* and *C. elegans* (25%, 49%, and 87%, respectively) than in the human genome. This relative lack of DNA transposons in humans (and probably mammals as a whole) may be related to the improved immune system of this group.

One of the simplest-sounding questions to ask about the human genome is "how many genes does it encode?" It turns out that this is still a very difficult question to answer, even with the complete sequence in hand. In contrast to bacteria, for which precise gene counts can usually be determined (see Chapter 10), there is still extensive uncertainty about the exact number of protein-coding genes in the genomes of humans and other mammals. The reason is that, in the human genome, less than 2% of the DNA codes for proteins (Fig. 9.5). Small exons (see Fig. 9.6 and Table 9.2) are easily lost in the noncoding material. Conversely, some of the nongenic DNA has features in common with protein-coding genes. About 0.5% of the human genome, for example, is thought to consist of pseudogenes,

TABLE 9.2   Some characteristics of human genes. The sample was obtained by aligning genomic DNA from the Human Genome Project with a curated set of full-length mRNA sequences called RefSeq (Pruitt and Maglott, 2001). Some lengths may be underestimated, particularly for the untranslated regions (UTR), because of the currently poor ability in detecting these. Adapted from the International Human Genome Sequencing Consortium (2001), reproduced by permission (© Nature Publishing Group).

	Median	Mean	Sample size
Internal exon size	122 bp	145 bp	43,317 exons
Number of exons	7	8.8	3501 genes
Intron size	1023 bp	3365 bp	27,238 introns
3′ UTR	400 bp	770 bp	689
5′ UTR	240 bp	300 bp	463
Coding sequence	1100 bp (367 aa)	1340 bp (447 aa)	1804
Genomic extent	14 kb	27 kb	1804

remnants of genes that are no longer functional. One study suggests that $1/5$ of *C. elegans* annotated genes may in fact be pseudogenes (Mounsey *et al.*, 2002). Thus any count may be either too high or too low and it is difficult even to establish a tight upper or lower bound. Protein-coding genes were identified in the human genome sequence by comparisons with expression libraries, genes known from other organisms, and the use of gene-finding programs to detect ORFs. To the surprise of most experts, the best informed estimates of the total number of protein-coding genes in the human genome settled in the neighborhood of 30,000 to 35,000, much lower than most previous estimates, many of which favored a figure at least twice as large and which ranged up to 140,000.[1]

The exact number of genes in the human genome is still unknown because it remains possible that some small genes have been missed by gene-finding programs and/or that some identified genes are really only pseudogenes. That said, a total of 30,000 to 35,000 currently seems to be a fair number. This means that the human genome, although about 30 times as large as that of *C. elegans*, contains only around twice as many genes. Part of the explanation for this disparity is that human genes are much more extended by introns. Although the most common intron length in humans (around 90 bp) is only around twice

---

[1]A wager conducted between 2000 and 2003 among participants at the Cold Spring Harbor Genome meetings yielded several hundred guesses ranging from 25,947 to over 150,000 with a mean of around 60,000. All submitted guesses were higher than the official provisional total of 24,847 based on the Ensembl database, so the lowest entry was declared the winner on 30 May, 2003 (http://www.ensembl.org/Genesweep/). See the June 2000 editorial in *Nature Genetics*, "The Nature of the Number" (vol. 28, p.127–128) and Pearson (2003) for details.

as great as the corresponding figure for *C. elegans*, the mean value for humans (3300 bp) is more than 10 times the mean length for *C. elegans*, indicating that some human genes are very extended indeed. Typically, these sprawling genes are found in G + C-poor regions of the human genome.

Another issue raised by this low gene count relates to complexity. Humans have only two or three times as many genes as *D. melanogaster* or *C. elegans*—are not humans more than two or three times more complex? In reference to the old "C-value paradox" (Thomas, 1971), which expressed similar concern about raw genome size, Hahn and Wray (2002) called this the "G-value paradox" and Claverie (2001) dubbed it the "N-value paradox." Although "paradox" is perhaps too strong a word to express a subjective discomfort of this sort (see Chapter 1), mammals are known to be quite complex in some areas, such as the immune system, number of cell types, nervous system, and so on. One solution to this apparent discrepancy lies in the use of alternative splicing and alternative polyadenylation (Edwalds-Gilbert *et al.*, 1997). Alternative splicing allows a single form of pre-mRNA transcript to be spliced into a number of different forms by skipping exons or by recognizing alternative splice sites (see Fig. 9.12). The old idea of "one gene, one protein" is long dead, but the extent to which a gene can produce different products is not easy to estimate. Early methods based on EST alignments suggested that at least 35% of human genes may be involved in alternative splicing (Hanke *et al.*, 1999; Mironov *et al.*, 1999; Brett *et al.*, 2000). Refined estimates based on the complete sequence of several human chromosomes put the fraction at closer to 60%, with an average of at least two or three transcripts per gene. This is much higher than estimates for *C. elegans* of around 22% alternatively spliced, with an average of less than two transcripts per gene. Thus the human transcriptome may be several times larger, in comparison to invertebrates, than the gene count would suggest. It would seem that the initially high estimates of gene number arose, at least in part, by a failure to appreciate this. The interaction of genes through chains of transcriptional regulation may also allow great complexity

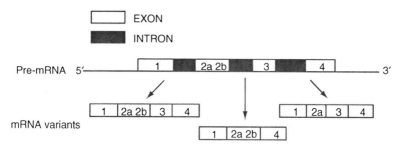

FIGURE 9.12   An illustration of alternative splicing. The same transcribed pre-mRNA strand can be spliced into several variant mature mRNAs. Exons can be included or excluded as units, or alternative splice sites can be used within a single exon (exon 2 in this case).

to arise from a limited number of basic forms (Huang *et al.*, 1999; Fickett and Wasserman, 2000). About 10% of human genes code for transcription factors (proteins that bind to DNA and affect how it is transcribed) whereas only about 5% of yeast genes do. Such a difference, coupled with a more complex network of transcription enhancers and promoters, can result in a much larger set of gene expression patterns leading to a nonlinear increase in organismal complexity (Levine and Tjian, 2003). Unfortunately, the identification and understanding of transcriptional control regions in the human genome lags behind the ability to identify ORFs. In short, the resolution of the G- or N-value paradox may be simply that the relation of gene number to complexity should have not been expected to be linear. Of course, the same was true with genome size and the C-value paradox (see Chapter 1).

Some have argued, both before and after the announcement of the estimates based on the human draft sequence, that there are fundamental limits on the number of genes. George (2002) suggested that the number of genes is limited in organisms with an adaptive immune system by the burden of self-recognition. Pal and Hurst (2000) argued that increase in gene number may be limited by increasing probability of error, both heritable (accumulation of deleterious mutations) and especially nonheritable (e.g., regulatory failure). Another important aspect of comparative genomics is the identification of new genes in humans (taken as a representative vertebrate) that do not have homologs in other sequenced species. It appears that less than 10% of the proteome is in this category, which includes immune and nervous system proteins. Figure 9.13 shows a distribution of where homologs to human genes have been found.

GENOME VARIATION IN HUMAN POPULATIONS

Another aspect of comparative genomics relates to sequence differences within a species or population. The most common variation of this kind is the single nucleotide polymorphism (SNP), defined as occurring when different nucleotide bases (single nucleotide alleles) appear at a homologous site in a population. Usually, a less frequent allele must occur at an arbitrarily specified frequency, say 1% of the population, to qualify a site as polymorphic, but disease-causing alleles are obviously also of interest at much lower levels of frequency. With the initial draft of the human genome sequence it became possible to assess SNP distribution in a comprehensive manner. Data of this kind are important not only for studies in population genetics and the history of the human species, but also for medical applications, as many known SNPs are associated with heritable diseases (Taylor *et al.*, 2001). In an initial analysis of 1.42 million SNPs, mostly collected by the Human Genome Project and a nonprofit consortium called TSC ("The SNP Consortium"), it was found that two homologous chromosomes randomly selected from the population can be expected to differ in one site out of 1331

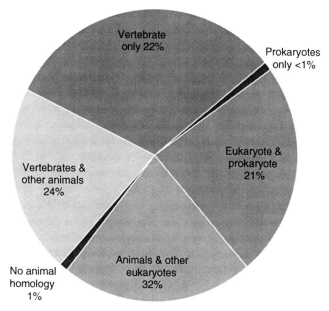

FIGURE 9.13 Distribution of the homologs of the predicted human proteins. For each protein, a homolog to a phylogenetic lineage was considered present if a search of the NCBI nonredundant protein sequence database, using the gapped BLASTp program (www.ncbi.nlm.nih.gov/blast), gave a random expectation (*E*) value of less than or equal to 0.001. Additional searches for probable homologs with lower sequence conservation were performed using the PSI-BLAST program, run for three iterations using the same cutoff for inclusion of sequences into the profile. Adapted from the International Human Genome Sequencing Consortium (2001), reproduced by permission (© Nature Publishing Group).

(Sachidanandam *et al.*, 2001). What is the total number of SNPs (at the 1% level) in the human population? Using classical neutral population genetic methods, Kruglyak (2001) placed the figure at 11 million sites, so that less than 15% have been identified. Thus increased depth (number of individuals assayed) and breadth (genome coverage) will be needed before the catalog of human genotype variation can be said to be complete. Of course, most SNPs are believed to be neutral, so only functionally relevant changes need be considered for many biomedical applications (Kruglyak and Nickerson, 2001). The number of these is expected to be much smaller than the total given above, but it is difficult to separate them out, especially those that may lie in unidentified control regions of the genome. Another simplification is to use not raw SNPs in association studies, but linked groups of alleles called haplotypes (Seltman *et al.*, 2003). As the database of human SNPs and haplotypes grows, the hope is that it may be used to help develop somatic gene therapies for specific diseases and to predict an individual's reaction to therapeutic drugs ("pharmacogenetics" or "pharmacogenomics") (Stephens, 1999; March, 2000; McCarthy and Hilfiker, 2000). Finding these associations is statistically challenging because although there is a great deal of data (Huang et al., 2003), many of the associations are quite weak (Ioannidis, 2003).

Other potentially important forms of genomic variation in human populations have been described recently. Notably, Sebat et al. (2004) found evidence that copy-number polymorphisms involving large segments (more than 100 kb) of DNA contribute substantially to genomic variation between normal humans. However, the significance of such polymorphisms for human health, for example through gene dosage variation, remains largely unknown.

The genome sequence of the chimpanzee, the closest relative of humans, is seen as an important key to understanding exactly what "makes us human" and is therefore of great interest to evolutionary biologists and to the public at large. The mouse–human comparison provides a broad perspective on the mammalian genome, but the high number of rearrangements between the two species requires intensive searching for homologies. The human–chimp comparison will allow researchers to focus on differences. It had been reported (Sibley and Ahlquist, 1984, 1987; Ebersberger et al., 2002) that sequence divergence between chimp and human was less than a few percent, although a recent study has found that aligned regions of human and chimp genomic DNA differ by around 5% when indels are accounted for (Britten, 2002). On the other hand, when attention is restricted to a sampling of genes themselves, the percent nonsynonymous DNA difference was found to be just 0.6% (Wildman et al., 2003). It will be of great interest to isolate these differences, not only in protein-coding exons, but also in regulatory regions such as promoters and enhancers, and relate them to function. The prevailing hypothesis is that differences between human and chimpanzee are primarily owing to differences in gene expression during development, so expression studies are also essential. For example, preliminary mRNA studies show that central nervous system expression patterns diverge more between humans and primate relatives than do patterns for other organs (Normile, 2001; Enard et al., 2002). Large-scale chimpanzee sequencing is already under way in the United States and in Japan by a group called the International Chimpanzee Genome Sequencing Project. In terms of biomedical research, however, there are stronger arguments for sequencing the rhesus macaque (Macaca mulatta) than the chimp (Cyranoski, 2002). Although differences between chimp and human pathologies are of significant interest, the chimp is no longer a common laboratory animal, compared to the rhesus macaque, which provides models for many human diseases.

PUFFERFISH SYNERGY

The human genome project was aided by other concurrent sequencing projects. Although not a traditional model organism for genetics, the pufferfish

Takifugu rubripes ("Fugu") was an ideal sequencing subject because of its remarkably compact genome and so was the second vertebrate to be completely sequenced. Fugu appears to have approximately the same number of genes as humans and a very similar exon–intron pattern, but has much shorter introns and intergenic regions, resulting in a genome about ⅛ as big as that of humans (Brenner *et al.*, 1993; Hedges and Kumar, 2002). The combined factors of compact genome and improved sequencing methods (whole-genome shotgun, in this case) allowed the Fugu genome to be completed for a cost of only about 12 million dollars, less than 1% of the total spent on the human genome project (Aparicio *et al.*, 2002). Besides the evolutionary insights to be gained from comparison of two distantly related vertebrates, the Fugu genome was used to help find functional elements and annotate the human genome (Aparicio *et al.*, 2002). The Fugu draft sequence was produced by a consortium led by the U.S. Department of Energy's Joint Genome Institute (JGI) in Walnut Creek, California, and the Singapore Biomedical Research Council's Institute for Molecular and Cell Biology (IMCB). The consortium's sequencing efforts were aided by two U.S. companies, Celera Genomics, Rockville, Maryland, and Myriad Genetics, Inc., Salt Lake City, Utah. Completion was announced at the 13th International Genome Sequencing and Analysis Conference in San Diego, California, on October 26, 2001.

Sequencing of the freshwater, nonpoisonous pufferfish *Tetraodon nigroviridis* was also announced by Genoscope (The French National Sequencing Center) in Paris, and the Whitehead Institute Center for Genome Research in Cambridge, Massachusetts, at around the same time as the report of the pufferfish genome. These two bony fishes with similarly compact genomes provide a useful contrast of vertebrate genome divergence. The pufferfish and human lineages have been separated for more than 400 million years, whereas *Tetraodon* and *Takifugu* are thought to have diverged 20–30 million years ago.

THE MOUSE AND RAT GENOMES: THE RISE OF MODERN MAMMALIAN COMPARATIVE GENOMICS

Approximately a year after the "completion" of the human genome project, when about 95% of the euchromatic sequence was available in finished form, the first draft sequence of the mouse genome was released (Mouse Genome Sequencing Consortium, 2002). Although the mouse genome is perhaps atypical of mammalian genomes in some ways, it is among the most valuable to use to shed light on human biology. This is largely owing to the status of mouse as the preeminent model mammalian organism in genetics. A great deal is known about the function of many mouse genes and many more can be elucidated using knockout studies that would be impossible to conduct in humans. Homologies between mouse and human genes are readily determined by sequence comparisons and provide an initial key for functional studies. Further clues are provided by relative conservation and divergence of different genes and DNA stretches that lie outside of known genes. This latter category of conserved extragenic DNA provides valuable pointers to the location of promoters, enhancers, and other hard-to-detect but extremely important functional elements. Sensitive methods have been developed for determining genomewide homology mapping, even covering regions apparently not under selection (Schwartz *et al.*, 2003).

In many such applications of comparative genomics, it is necessary to distinguish carefully between homology and orthology of sequences. For example, if a pair of sequences is taken, one from mouse and one from human, homology (common ancestry) can be inferred based on a high degree of similarity at protein or DNA sequence level. However, it is not known whether the two sequences first diverged from their common ancestral sequence at the time of the rodent-primate split, or earlier. In the first case, when the sequences first diverge via a speciation event, they are called orthologous; in the second case, they must have first diverged by a gene duplication event and are said to be paralogous (see Chapter 5). The importance of the difference is that orthologous sequences can be used as proxies for their respective species in phylogenetic and timing analyses, whereas paralogous sequences cannot. This is illustrated in Figure 9.14.

For this reason, care was taken to identify mouse–human orthologs as quickly as possible. Although the orthology of two sequences cannot be determined with absolute certainty without extensive species and genome sampling and the use of a phylogenetic approach (Zmasek and Eddy, 2001), some useful methods are being employed to compare completely sequenced species like mouse and human. Specifically, a pair of sequences, h from human and m from mouse, is considered orthologous if h's closest match in the mouse genome is m and m's closest match in the human genome is h. In this way, human orthologs can be found for about 80% of mouse genes (Mouse Genome Sequencing Consortium, 2002).

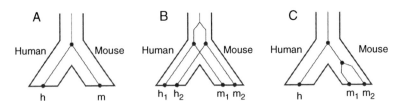

FIGURE 9.14 Orthology versus paralogy. In (A), the sequences h and m are orthologous because their most recent ancestral sequence coincides with the divergence of the mouse and human lineages. In (B), a gene duplication occurred before the divergence of these lineages. In this case, h_1 and m_1 are orthologs, as are h_2 and m_2 whereas h_1 and m_2 are paralogs, as are h_2 and m_1. Note that orthology is not necessarily a one-to-one relationship. In (C), h is orthologous to both m_1 and m_2.

Although the mouse genome is somewhat smaller than that of humans (2.5 Gb euchromatic DNA for mouse vs. 2.9 Gb for human) and has a slight but statistically significant difference in G+C content (Fig. 9.15), it appears to contain about the same number of genes. The main resource for mammalian gene detection and verification is the use of expression data such as cDNAs and ESTs (Hubbard *et al.,* 2002). Results from such searches are then integrated with results from *de novo* gene prediction, producing a final catalog. This process is far from clear-cut, as it relies on complete transcript libraries and uses gene-finding programs that have difficulties separating noise from data when genes are spread out as much as they are in mammals. This process can be facilitated when two genomes are available, because conserved regions within homologies may lead to discovery and validation of splice sites and other genomic elements (Korf *et al.,* 2001; Wiehe *et al.,* 2001).

The fraction of genes in mice or humans that do not appear to have homologs (orthologs or paralogs) in the other is less than 2%. The expansions of certain gene families are readily apparent in the mouse lineage, such as those involving immunity and olfaction, relative to their presence in humans, suggesting either mouse duplications or human losses in these functional areas. An example is the oligo-adenylate synthetase (OAS) gene family involved in interferon-induced antiviral response, which shows many recent murine gene duplications (Kumar *et al.,* 2000). Although the gene sequences have been generally well conserved, their positions in the genome have not—that is, genes that are syntenic in one genome

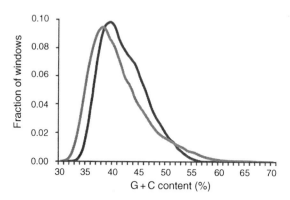

FIGURE 9.15 Distribution of G+C content of the human (gray line) and mouse (black line) genomes. Mice have a slightly higher mean G+C content than humans (42% versus 41%), but humans have a larger fraction of windows with either high or low G+C content. The distribution was determined using the unmasked genomes in 20 kb nonoverlapping windows, with the fraction of windows (y-axis) in each percentage bin (x-axis) plotted for both humans and mice. Adapted from the Mouse Genome Sequencing Consortium (2002), reproduced by permission (© Nature Publishing Group).

are not so in the other. Detailed analysis shows that chromosomal rearrangements have moved segments of DNA both within and among chromosomes, breaking many syntenies. The exact number of relocated segments is difficult to observe because some may be very small, particularly near the centromeres and telomeres, but the number of large segments conserved between humans and mice is at least several hundred (Kumar et al., 2001; Eichler and Sankoff, 2003). This means that the average length of conserved segments between human and mouse genomes is rather small.

On a finer level, gene structure seems to be highly conserved. Analysis of 1506 pairs of genes for which there is strong evidence of orthology shows that 86% genes have the same number of exons and 91% of orthologous exon pairs have the same length in humans and mice. Only about 1% of introns, however, have identical length, and the average length of the mouse introns in this set was 3888 bp compared to 4661 bp for human. This reflects the smaller euchromatic C-value for mice (Mouse Genome Sequencing Consortium, 2002).

In April 2004, the complete genome sequence was published for another important (and indeed, the first) mammal used in medical research, the brown rat *Rattus norvegicus*. The rat genome sequence was obtained using a new approach that combines aspects of the traditional mapping and whole-genome shotgun methods used in the public and private human genome sequencing projects, respectively. Unlike previous mammalian analyses (i.e., between mice and humans), the mouse-rat comparison allows inferences to be made regarding genome evolution over a relatively short time scale (i.e., only 12–40 million years) (Kumar and Hedges, 1998; Rat Genome Sequencing Project Consortium, 2004). The euchromatic portion of the rat genome appears to be intermediate in size (2.75 Gb) relative to that of mice (2.5 Gb) and humans (2.9 Gb), and contains a similar number of genes. The preponderance of segmental duplications (which occur primarily in pericentromeric regions) is also intermediate in rats, and there are signs that some gene families have expanded by duplication in rats but not in mice (Rat Genome Sequencing Project Consortium, 2004). The activity of *L1* transposable elements (a LINE) also seems to be higher in rats than in mice, although a roughly similar number of SINE copies (~ 300,000) appears to have been inserted into the genomes of both rodents after the divergence of their respective lineages.

Sequences comprising about one billion nucleotides (39% of the rat genome) appear to be common to all three mammals, representing an "ancestral eutherian core," which includes around 95% of the known protein-coding and regulatory regions. About 28% of the rat sequence aligns only with mice, not humans, and another 29% aligns with neither of the two mammals. As compared with primates, rodents appear to have much more dynamic genomes, experiencing a faster rate of both molecular (base substitutions) and chromosomal (rearrangements) evolution (Kumar et al., 2001; Kumar and Subramanian, 2002; Rat Genome Sequencing Project Consortium, 2004).

GENOME SEQUENCING IN PLANTS
AND THEIR PATHOGENS

COMPARATIVE GENOMICS OF *ARABIDOPSIS*

Arabidopsis thaliana was a natural choice for the first plant genome to be sequenced. It had the smallest known genome and highest gene density of any plant, plus it was already extensively studied, is easy to grow, and has a short life cycle. Although the genome is only 157 Mb, it contains homologs to nearly all genes found in flowering plants but with much less repetitive DNA than most.

The *A. thaliana* project began in 1990 and involved researchers from many countries. Sequencing itself had begun by 1993, funded primarily by the European Union. By 1996, funding agencies from the United States, as well as Europe and Japan, also contributed to the work, and the *Arabidopsis* Genome Initiative (AGI) was set up with the intention of completing the sequencing (clone libraries were already available) by 2004. The agreement was a model of international scientific cooperation. A coordinating committee was to assign different portions of the genome as needed to prevent duplication of work. No sequence information was to be withheld to benefit any private group, and partial sequences were to be released as soon as available to one of the major databanks. As with all genome projects, it was becoming increasingly clear that ongoing annotation was vital to the value of the data and an organization was formed to help curate annotations. Another important part of the AGI project was a parallel effort to sequence gene expression data in the form of cDNA. Because of advancing sequencing technology, the project was completed well ahead of schedule, with the first report released in 2000 (*Arabidopsis* Genome Initiative, 2000).

The genome size, gene content, and gene family diversity of *A. thaliana* are comparable to that of *C. elegans,* but differences of gene content in different functional classes are significant. For example, less than 20% of *A. thaliana* proteins involved in transcription have strict homologs within *C. elegans* (BLASTp E-value less than 10^{-30}). In contrast, more than 40% of proteins involved in protein synthesis and signal transduction have such homologs, suggesting common ancestry. It is particularly interesting that relatively high proportions (15–30%) of proteins in the energy and metabolism categories have close matches in *E. coli*. This may result from lateral transfers or unusually extreme conservation. About 35% of genes in *A. thaliana* are apparently unique, or at least are not present in the animal and yeast genomes sequenced thus far. About 150 families of genes, including structural proteins and enzymes, appear to be unique to plants (*Arabidopsis* Genome Initiative, 2000).

The proportion of proteins belonging to families of more than five members is substantially higher in *A. thaliana* (37.4%) than in *C. elegans* (24.0%), as is the proportion of gene families with more than two members. These features of

A. *thaliana,* and presumably other plant genomes, may indicate less constraint on genome size in plants and/or a higher propensity for genome duplication. In fact, most (58%) of the A. *thaliana* genome is in the form of large (at least 100 kb) duplicated segments with more than 50% sequence identity. This indicates that the genome was structured by a past polyploidy event (see Chapter 6), as is known to be very common in plant evolution (see Chapter 7). Tandem segment duplications also appear to be common in A. *thaliana,* with about 1500 tandemly duplicated arrays of genes, containing an average of around three genes each, but up to a maximum of 23. This suggests that unequal crossing over may be an important factor in plant genome evolution as well.

THE RICE GENOME

Rice (*Oryza sativa*) is the most important food crop in the world, providing staple nourishment for half the world's population, and is also the cereal crop with the smallest genome, about 490 Mb in size. Rice genetics has been intensively studied and comprehensive genetic and physical maps have been available for some time. For many mapped traits (Gale and Devos, 1998), the rice genome exhibits a strong colinearity, or preservation of genetic linkage relationships, with the much larger genomes of other grain crops such as wheat, barley, oats, and corn (Fig. 9.16). In the same way, A. *thaliana* serves as a genomic key to the Brassica group of crop plants (cabbage, cauliflower, mustard, rape, rutabaga, and turnip, among others). These qualities made rice a practically ideal next choice for sequencing after the tiny-genomed model plant A. *thaliana.* Two strains of rice, the *japonica* and *indica* varieties, were sequenced by different groups (Goff *et al.,* 2002; Yu *et al.,* 2002). The *japonica* announcement by the private company Syngenta raised controversy because the full results were not deposited in a public data bank such as Genbank, although the public did have limited access to a private database (Brickley, 2002). This marked the second time the journal *Science* allowed authors of a scientific paper to withhold full access to sequence data described in a publication, the other instance being the Celera human genome paper.

Although estimates of the number of genes in rice are subject to the uncertainties that apply to all large-genomed eukaryotes, plus some unique problems, there seem to be many more genes than in A. *thaliana,* and quite probably more than in mammals such as *H. sapiens.* This high gene number was not anticipated (Messing, 2001). A remarkable feature of the rice genome is the presence of a strong G + C content gradient within genes. Often, the 5' end is up to 25% richer in G + C content than the 3' end (Wong *et al.,* 2002). No comparable gradient is seen in A. *thaliana.* A consequence of this gradient is that codon usage also varies from one end of a gene to the other, complicating the work of

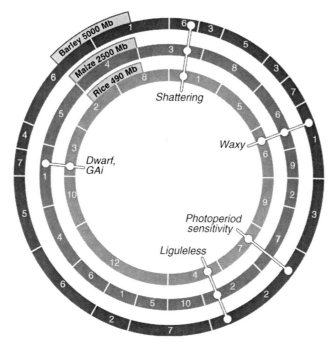

FIGURE 9.16   Genome colinearity among the grasses. The rice genome is the smallest among the grasses most commonly grown as crops. However, enough genomic similarity exists that the genomes of these grain species can be approximated as concentric circles, and the information from the smaller genome of rice provides insight into where to find genes of interest in the larger genomes of the other grasses. Each circle represents a single grain species, with its chromosomes collected end-to-end in a parsimonious manner to best match the structure of the rice genome. A few examples of genetic loci associated with particular traits are shown to illustrate that similar genes occur in similar portions of the genomes across these grasses. Other grain crops that show colinearity with rice include wheat, oats, pearl millet, sorghum, sugar cane, and foxtail millet. Adapted from www.ipw.agrl.ethz.ch/ ~ mbucher/useful/riceposter.pdf, reproduced by permission (© American Association for the Advancement of Science).

gene-detection programs. Therefore gene predictions and counts have an unusual amount of uncertainty in this case.

It appears that about 80% of *A. thaliana* genes have a homolog in rice, whereas only 50% of predicted rice genes have a homolog in *A. thaliana*. The latter figure extends to other sequenced organisms as well, so that about half of rice's genes seem to be novel, without known homologs or functions. Some of this figure may be an effect of the G + C content gradient in rice making homology determination difficult, but it still suggests a great deal of innovation in rice, perhaps amplified by duplications, since the monocot–eudicot divergence.

The Rice Blast Fungus: *Magnaporthe grisea*

In addition to the sequencing of this most important cereal crop, rice's major fungal pathogen, the rice blast fungus, *Magnaporthe grisea*, has also been sequenced (Martin *et al.*, 2002). *M. grisea* represents an excellent model organism for studying fungal phytopathogenicity and host–parasite interactions. Like many other fungal pathogens, *M. grisea* is a haploid, filamentous Ascomycete. It has a fairly small genome of around 40 Mb partitioned into seven chromosomes. *M. grisea* is also closely related to the nonpathogen *Neurospora crassa,* an important model organism for the study of eukaryotic genetics and biology. The main line of defense against this fungus has recently been genetic, via host resistance, although this entails a constant arms race against a rapidly evolving pathogen. Having both sequenced genomes should provide valuable insight for understanding questions of fungal–host interactions. For example: What genes come into play in both species during infection? How does the pathogen recognize when it is on a suitable surface to initiate the infection process? Which genes control host specificity? Mitigating the effects of this fungus could directly help feed tens of millions of people.

OTHER INVERTEBRATE ANIMAL GENOMES

The Mosquito Genome

Anopheles gambiae is the major vector of the human malaria pathogen *Plasmodium falciparum* in Africa. Although malaria has been eliminated in Europe, new malaria control techniques are urgently needed in sub-Saharan Africa, and improved understanding of the ecological relationship of the pathogen and its hosts may provide a key to its elimination. *A. gambiae* is also of interest in that it provides a comparison to *D. melanogaster*. The lineages diverged more than 250 million years ago (Zdobnov *et al.*, 2002) and initial studies, prior to full sequencing, showed considerable divergence in terms of genome rearrangements, although broad conservation of synteny on chromosomal arms was noted (Bolshakov *et al.*, 2002).

As shown in Table 9.1, the mosquito genome is more than twice the size of the *D. melanogaster* genome, although they contain a very similar number of genes (Holt *et al.*, 2002). The difference is mostly owing to a greater amount of intergenic material in the mosquito. For example, the transposable element content of *A. gambiae* is approximately 16% and 60% of euchromatin and heterochromatin, respectively (Rizzon *et al.*, 2002). The fact that most dipterans, including many *Drosophila* species, have genomes closer in size to the mosquito's, suggests that the lineage containing *D. melanogaster* experienced a reduction during recent evolutionary times (Petrov *et al.,* 1996). Some mechanisms for this loss of non-coding material have been proposed (Hartl, 2000; Petrov, 2001).

An interesting contrast between arthropod and vertebrate genomes is provided by the frequency of large duplicated blocks. The number of blocks in *A. gambiae* containing at least three genes that also appear elsewhere was only about 100, compared to more than 1000 for the human genome (Holt *et al.,* 2002). The presence of such repeats in the mouse genome seem comparable to that for human (Mural *et al.,* 2002), whereas the *C. elegans* and Fugu genomes show little evidence of such duplications (*C. elegans* Sequencing Consortium, 1998; Aparicio *et al.,* 2002).

Preliminary analyses of the *A. gambiae* genome and expressed proteome suggest several strategies for reducing human disease associated with this animal. For example, comparisons of gene expression profiles before and after the female's blood meal reveal that certain products (lipid synthesis and transport proteins, egg melanization factors, lysosomal enzymes) are up-regulated, whereas others (involving cytoskeletal and muscle contractile machinery, glycolysis, and proteins associated with vision) are down-regulated. Understanding these changes may provide opportunities for intervention to disrupt reproduction. Other approaches involve disrupting the mechanism by which the mosquito finds the human (e.g., odor receptors [Hill *et al.,* 2002]), or by interfering at some point in *P. falciparum*'s complex life cycle within its host, possibly using the *A. gambiae* immune response (Dimopoulos *et al.,* 2001; Christophides *et al.,* 2002).

THE SEA SQUIRT: A PRIMITIVE CHORDATE

Ciona intestinalis is a urochordate, the most basal branching group of chordates, and therefore was considered an important target for complete genome sequencing. The adult is a sessile filter feeder, but the tadpole has a notochord. This invertebrate chordate has approximately half as many genes as sequenced vertebrates and gives a perspective on the evolution of this group (Dehal *et al.,* 2002). The difference in gene content is thought to result from the proliferation of gene families involved in vertebrate development, so that the *C. intestinalis* genome provides a view into vertebrate ancestry. On the other hand, some genes known to be in both insects and vertebrates are missing in *C. intestinalis*. The *Hox* gene family provides an interesting example. Invertebrates have a single cluster of up to 13 homologs, whereas vertebrates have several clusters. *C. intestinalis* is in the invertebrate camp in that it has a single (albeit widely spread) cluster of nine *Hox* genes, although *Hox* 7, 8, 9, and 11 are apparently absent (Gionti *et al.,* 1998). A remarkable example of an apparent *C. intestinalis* innovation is a set of genes related to the substance tunicin (Krishnan, 1975). Tunicin is a cellulose-like carbohydrate present in the urochordate (or, "tunicate") body-casing. *C. intestinalis* contains at least one potential cellulose synthase and several endoglucanases related to the synthesis and degradation of this material. Homologs are found in *A. thaliana* and in termites and wood-eating cockroaches, although horizontal gene transfer from symbionts may be involved (Lo *et al.*, 2000; Nakashima *et al.*, 2004).

GENOMEWIDE DUPLICATIONS IN VERTEBRATES?

As discussed in detail in Chapter 6, comparisons of vertebrate and invertebrate genomes have also shed much light on the debate about genomewide duplication events in the early history of vertebrates. In addition to the *Hox* genes mentioned previously, other gene families have been found to suggest two rounds of whole genome duplication in early vertebrates (Wolfe, 2001; Gu *et al.*, 2002) (see Chapter 6). Ohno (1970) argued several decades ago that such events may have given vertebrates a sudden leap in body-plan complexity, which, if true, would have significant implications for the understanding of large-scale vertebrate evolution (see Chapter 11).

PROTIST GENOMES

One should not forget that animals, plants, and fungi form only a fraction of total eukaryote diversity. The "Protista" form a polyphyletic collection of unicellular eukaryotes whose genomes are also interesting and important for several reasons. First, many are pathogens that cause an enormous amount of human disease throughout the world, particularly in the tropics. Their early divergence relative to other eukaryotes (Baldauf, 2003) (Fig. 9.17) also makes them important for major evolutionary questions such as those concerning the evolution of organelles. Some, such as *Dictyostelium* (discussed in a later section), are important as simple models of intercellular communication.

ENCEPHALITOZOON CUNICULI: A PARASITIC EUKARYOTE WITH A TINY GENOME

Encephalitozoon cuniculi is a member of the microsporidia, a group of obligate parasites that infest many animal hosts, including rabbits and immunocompromised humans. Because they lack mitochondria, the microsporidia were at first thought to have originated from a deep eukaryotic branch before the organelles were acquired, but closer inspection showed that their nuclear genomes contain some mitochondrial-type genes. This fact, along with further phylogenetic analysis, indicates that they are more properly classified as fungi that have lost their mitochondria (Katinka *et al.*, 2001).

Most obligate parasites are degenerate in various ways, and *E. cuniculi* is no exception. Its tiny genome (2.9 Mb) is smaller than that of many prokaryotes, including *E. coli*. Sequencing announced in 2001 showed that this small size results from a rarity of introns, lack of transposable elements, reduced metabolic and synthetic activities, reduced intergenic spacers, and even reduced protein lengths relative to eukaryotic orthologs. The mean protein length is only 363 amino acids (aa), compared to 472, 435, 543, and 461 aa for *S. cerevisiae*, *C. elegans*, *D. melanogaster*, and human, respectively, and is more comparable to that of prokaryotes such as *E. coli* (315 aa) and *Mycoplasma pneumoniae* (348 aa) (figures from www.ebi.ac.uk/proteome). In a comparison of 350 proteins with *S. cerevisiae* homologs, the *E. cuniculi* sequence was shorter in 85% of the cases, with an average difference of 14.6% (Katinka *et al.*, 2001). Zhang (2000) has suggested that the longer proteins of higher eukaryotes allow more sophisticated gene regulation networks.

PLASMODIUM: THE MALARIA PATHOGEN

In 1996, the International Malaria Genome Sequencing Consortium was formed to sequence the genome of the protist *Plasmodium falciparum*. This organism is of great importance because of its devastating effect as a pathogen—it is responsible for more than 2.5 million deaths each year, a large proportion of them children—but also presents special problems in sequencing. Its genome is twice the size of that of yeast, and its extremely low G + C content (~ 20%) creates technical problems for sequencing. Despite these complications, the complete sequence was announced in 2002 (Gardner *et al.*, 2002). The sequence for the related pathogen in mice, *Plasmodium yoelii*, was announced at the same time for comparative analysis (Carlton *et al.*, 2002).

Compared to free-living protists, *P. falciparum* has fewer genes for enzymes and membrane transporters, but contains an extensive apparatus for evading host defenses. More than 60% of its proteins do not appear to have homologs in previously sequenced eukaryotes. It is not known whether this high figure is a result of *Plasmodium*'s phylogenetic position, high A + T content, or parasite status. Comparison with other eukaryotic genomes reveals that, in terms of overall genome content, *P. falciparum* is slightly more similar to *Arabidopsis thaliana* than to other taxa (Gardner *et al.*, 2002). However, the implied affinity with the plant kingdom may be related to horizontal gene transfer.

Now that the genomes of *P. falciparum*, *Anopheles gambiae*, and *Homo sapiens* have been completed, the raw data are available to break or control this devastating parasitic cycle. The *P. yoelii* sequence is of great importance here as well, because it has been used as a proxy in laboratory studies for the human parasite, whose life cycle cannot be completed *in vitro*.

DICTYOSTELIUM: THE "SLIME MOLD"

Dictyostelium discoideum is a haploid protist that has been intensively studied, primarily because of its social life. A group of free-living cells can aggregate into a motile mass that exhibits morphological and biochemical development for the purpose of common reproduction. For this reason, *Dictyostelium* is an excellent model organism for studying intercellular signaling, specialization, and cooperation in a simple context. The *Dictyostelium* genome has six chromosomes totaling about 34 Mb. Like *Plasmodium,* its low G+C content (30%, down to 10% in some regions) challenges conventional sequencing methods. Currently *Dictyostelium* is being sequenced through a collaborative effort among American, British, French, and German laboratories.

In general, most protist genomes can now be sequenced quickly and relatively cheaply. Given the immense impact of these organisms on human health throughout the world, fruits of this research hold the promise of great medical benefits.

COMPARATIVE GENOMICS AND PHYLOGENETICS IN EUKARYOTES

The availability of more sequence data is also refining ideas regarding within-eukaryote relationships and their times of divergence (see review in Hedges, 2002). The tree in Figure 9.17 reveals several important aspects of the current understanding of early eukaryote evolution. It shows the close relationship between animals and fungi, the relationships of the several forms of algae with plants, and that the "protists" are indeed a group with very diverse origins (Baldauf, 2003).

Figure 9.18 shows typical positions where eukaryotic proteins are found in phylogenetic reconstruction in relation to their prokaryotic homologs. Eukaryotic proteins involved in transcription and translation often cluster with archael homologs, whereas metabolic proteins often cluster with bacteria (Rivera *et al.,* 1998). This pattern is thought to result from the symbiotic origin of eukaryotes and horizontal gene transfer from organelles to the nucleus (Margulis, 1996; Gupta, 1998; Doolittle, 1999).

Associated with the phylogeny question is the issue of timing: How long ago did certain lineages diverge from one another? Traditionally, this was settled by the dating of fossils, giving lower bounds on the age of divergences. The molecular

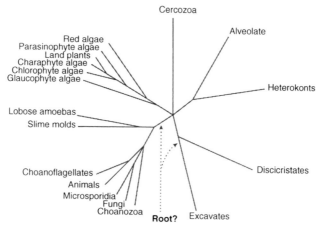

FIGURE 9.17 A consensus phylogeny of eukaryotes based on both molecular and ultrastructural data. The vast majority of characterized eukaryotes, with the notable exception of major subgroups of amoebae, can now be assigned to one of eight major groups. Opisthokonts ("basal flagellum") have a single basal flagellum on reproductive cells and flat mitochondrial cristae (most eukaryotes have tubular ones). Eukaryotic photosynthesis originated in plants; theirs are the only plastids with just two outer membranes. Heterokonts ("different flagellae") have a unique flagellum decorated with hollow tripartite hairs (stramenopiles) and, usually, a second plain one. Cercozoans are amoebae with filose pseudopodia, often living within tests (hard outer shells), some very elaborate (foraminiferans). Amoebozoa are mostly naked amoebae (lacking tests), often with lobose pseudopodia for at least part of their life cycle. Alveolates have systems of cortical alveoli directly beneath their plasma membranes. Discicristates have discoid mitochondrial cristae and, in some cases, a deep (excavated) ventral feeding groove. Amitochondrial excavates lack substantial molecular phylogenetic support, but most have an excavated ventral feeding groove, and all lack mitochondria. Adapted from Baldauf (2003), reproduced by permission (© American Association for the Advancement of Science).

clock hypothesis—that the rate of molecular sequence divergence is often constant for a particular gene over multiple lineages—promised to shed new light on the entire question. After initial controversy about the universality of the clock, tests were devised to reject genes that violated rate-constancy. In theory, phylogenetic trees based on clocklike genes would have branch lengths proportional to elapsed time. If properly calibrated against at least one well-established divergence time, the differences among gene sequences could be used to extrapolate the timings of all phylogenetic events in the tree. Of course, things are never so easy, and one of the main shortcomings of the plan was the relatively large variance of estimates based on too few sites/genes (Benton and Ayala, 2003; Hedges and Kumar, 2003). Not surprisingly, estimates inferred using many genes or proteins are more robust (Hedges and Kumar, 2003). It is also important to consider differences

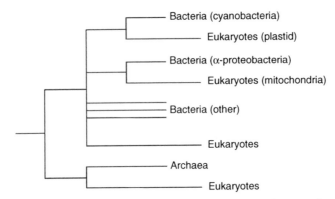

FIGURE 9.18   Eukaryotes consistently evolve faster than prokaryotes. This tree indicates the four
general locations where eukaryotic protein sequences typically cluster in the evolutionary tree of
prokaryotes. The rate of evolution on each lineage (branch) is indicated diagrammatically by relative
branch length (long branch signifies fast rate; branch length is not meant to be proportional to time
or actual rate). Adapted from Hedges *et al.* (2001), reproduced by permission of the author.

among major taxa; for example, Hedges *et al.* (2001) found that protein products
of eukaryotic genes evolve, on the average, 1.2 to 1.6 times faster than their
prokaryotic homologs.

A related question in comparative genomics involves the rate of neutral evolution (essentially equal to mutation rate) (Kimura, 1983). Does it vary from lineage
to lineage and among genomic regions? These questions have been intensively
studied in mammals. Early studies (Wu and Li, 1985; Li and Tanimura, 1987)
reported that the neutral rate for rodents was up to several times greater than
those for primates and artiodactyls. Other studies also reported significant
differences among lineages or regions (Britten, 1986; Wolfe *et al.*, 1989;
Mouchiroud *et al.*, 1995; Matassi *et al.*, 1999). Generation time differences (Li and
Tanimura, 1987) and differences in DNA repair mechanisms (Britten, 1986)
have been suggested as factors to account for this variation. However, the numbers of genes used in these studies were small and estimation errors may have
been made because of incorrect fossil dates or inappropriate outgroups (Easteal
et al., 1995). A large-scale study using more than 5000 genes and representatives
of a broad range of placental mammal groups concluded that there is little significant variation in neutral evolution rate among lineages, and that the rate is
approximately 2.2×10^{-9} per base pair per year (Kumar and Subramanian, 2002).
This improves prospects for accurate phylogenetic timing using neutral (4-fold
degenerate) DNA sites. Ellegren *et al.* (2003) presented a detailed account of the
current debate on rates of mutation and neutral evolution in mammalian
genomes.

CONCLUDING REMARKS AND
FUTURE PROSPECTS

COMPLETE GENOME SEQUENCING

Table 9.1 lists 18 eukaryotes as "completely" sequenced, including two plants, nine animals, four fungi, and three protists. Because of decreasing costs (now at a few cents per base for large-scale shotgun sequencing), this list is expected to grow rapidly, perhaps reaching 100 in only four or five years. Of course, many factors have influenced the choice of organisms to be sequenced. Traditional model organisms have been favored, as have organisms with small genomes. From this admittedly biased sample, a picture of phylogenetics and gene relationships is nevertheless coming to light. One of the surprises in the picture is how much all the sampled kingdoms have in common. Figure 9.19 shows the relative number of shared and unique protein domains among 14 of these species (Chothia *et al.*, 2003). The overwhelming fraction of domains in each organism is shared by all

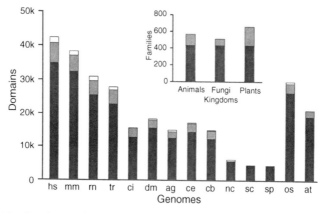

FIGURE 9.19   Contribution of common families to the protein repertoires of various eukaryotes. All or part of about 50% of eukaryote sequences are homologous to domains in proteins of known structure. The numbers of domains that belong to the 429 families common to all 14 eukaryotes studied are shown in black. Additional contributions of families common to the genomes in only one kingdom are shown in gray. For the animal genomes—human (*Homo sapiens*, hs), mouse (*Mus musculus*, mm), rat (*Rattus norvegicus*, rn), pufferfish (*Takifugu rubripes*, tr), sea squirt (*Ciona intestinalis*, ci), fruit fly (*Drosophila melanogaster*, dm), mosquito (*Anopheles gambiae*, ag), and nematodes (*Caenorhabditis elegans*, ce, and *C. briggsae*, cb)—there are 136 additional common families. For the three fungi—bread mold (*Neurospora crassa*, nc), budding yeast (*Saccharomyces cerevisiae*, sc), and fission yeast (*Schizosaccharomyces pombe*, sp)—there are 75 additional common families. For the two plants—rice (*Oryza sativa*, os) and thale cress (*Arabidopsis thaliana*, at)—there are 229 additional common families. Adapted from Chothia *et al.* (2003), reproduced by permission (© American Association for the Advancement of Science).

eukaryotes, a smaller fraction is kingdom-specific, and a mere sliver is unique to the organism, even though, because of the sparse sampling, some organisms on the list represent a broad group such as dicot plants or fishes. An emerging theme of the entire eukaryote sequencing effort is the common basis of eukaryotic life. Although ancestral characters, such as mitochondria, are occasionally lost, and duplicated protein genes acquire novel functions, there is still a great common core of protein domains that constitute the machinery of eukaryotic life, whether animal or plant, fungus or protist, unicellular or multicellular.

The NHGRI maintains a list of "high priority sequencing targets." Large NHGRI-sponsored sequencing centers that have excess capacity resulting from completion of human, mouse, and rat sequencing may use the released capacity for the sequencing of these organisms (Olson and Varki, 2003). As of May 22, 2002, this particular list included the chimpanzee *P. troglodytes,* the chicken, the honeybee, the sea urchin *S. purpuratus,* a protozoan *T. thermophila,* and 15 fungi (Check, 2002). Plants are excluded from this list, because those are handled in the United States by the National Science Foundation and the United States Department of Agriculture. The NHGRI's priority list is being reviewed and updated three times a year and will no doubt soon be expanded to include other model eukaryotic organisms.

In terms of economic impact and potential direct value to a large portion of the world's population, comparative plant genomics, especially with respect to cereal plants, is of immense importance. The factors of population growth (conservatively predicted at 2 billion additional mouths to feed over the next 50 years), the requirement to better manage and preserve world ecosystems, and the as-yet not fully understood effects of global climate change present a formidable challenge to agriculture. Although cereal production uses about half of available farmland and accounts, directly or indirectly, for $2/3$ of all human caloric intake, per capita cereal production has been declining (Dyson, 1999). Extracting more food value from cereal crops is key to feeding a hungry world for the foreseeable future and this requires a deeper understanding of the organisms. Comparative plant genomics has been instrumental in building a better understanding of the complete spectrum of cereal plant characteristics and the genetic bases of the differences among crops. Because cereal genomes tend to be large with an abundance of repetitive sequences (common hexaploid wheat *T. aestivum* has a 17 Gb haploid genome size), the availability of high-throughput sequencing methods was essential to this program.

In April 2001 a meeting was held at the International Maize and Wheat Improvement Center (CIMMYT) in Mexico that outlined five prongs of cereal research: (1) alleviating abiotic stress, (2) alleviating biotic stress, (3) adding value to cereals, (4) improving the yield potential of some cereals, especially by modifying photosynthesis, and (5) coordinating the development of comprehensive, freely available genomic tools and databases for improving cereals. An international Cereals Comparative Genomics Initiative (CCGI) program was sponsored by the Rockefeller Foundation and USAID to help advance these goals.

After basic research, the next steps toward crop enhancement using results from comparative genomics would be the actual production, testing, and distribution of genetically modified crop seeds. This raises complex political and ethical issues. The process is not unprecedented, however. Commercially produced transgenic Bt (insecticidal) corn, cotton, and soybeans and Roundup-Ready (antiherbicidal) corn and soybeans are now common in some countries, including the United States. Calgene's transgenic Flavr Savr tomato was approved for use in the United States in 1994, but was not a commercial success because of problems with characteristics of the crop related to its growth and harvesting (Nicholl, 2002). In 1999, on the noncommercial, humanitarian front, Ingo Potrykus and collaborators developed a method of inserting beta-carotene, a precursor of vitamin A, into rice endosperm where it does not normally occur (Ye *et al.*, 2000). Such rice was dubbed "Golden Rice" and had great potential of improving the nutrition of hundreds of millions of people, including hundreds of thousands of children who go blind from vitamin A deficiency. Even though it was essentially publicly funded and altruistically motivated, this project has encountered many obstacles of a bureaucratic, legal, and political nature (Potrykus, 2001) and is progressing slowly.

However, basic research and high-throughput sequencing continue at an increasingly rapid and economical pace. Sequencing efforts are in progress for various amphibians and fishes, several more insects, turkey, sea urchin, cow, dog, horse, kangaroo, pig, numerous fungi, algae, oat, coffee, soybean, cotton, barley, banana, corn, and protist parasites such as *Giardia* and *Leishmania*. Sequencing of the honeybee (*Apis mellifera*), jointly funded by the NHGRI and the U.S. Department of Agriculture (USDA), was recently completed in draft form. Draft sequences have also recently been completed for the chicken (*Gallus domesticus*) by the Washington University Genome Sequencing Center, and for the chimpanzee (*Pan troglodytes*) by the Broad Institute and Washington University Genome Sequencing Center. Updates of past, present, and future projects are available from the Genomes OnLine Database (GOLD) (www.genomesonline.org).

Genomic studies of eukaryotes have been biased toward vertebrates (mammals in particular), economically important plants, traditional model organisms, and pathogens. The range of eukaryote diversity is very broad, however, and knowledge of it is increasing every year. For example, recent studies show an unexpected abundance and diversity of the so-called picoeukaryotes (eukaryotes less than a few micrometers in size) (Moon-van der Staay *et al.*, 2001). The existence of very small eukaryotes, similar in size to, and even smaller than, many bacteria, has been known for some time. The currently smallest known autotrophic eukaryote is *Ostreococcus tauri* (Courties *et al.*, 1998). This planktonic organism isolated from the Mediterranean is less than 1 μm in size and has a 10.2 Mb genome consisting of 14 linear chromosomes, plus several mitochondria and one chloroplast. It now appears that such miniature eukaryotes may be a more significant part of the planktonic community than formerly appreciated (Lopez-Garcia *et al.*, 2001) and may be represented in at least five of the eight major divisions of eukaryotes shown in Figure 9.17. Such eukaryotes tend to have small genomes and may provide an inexpensive way of exploring eukaryote genome diversity (Baldauf, 2003).

Partial-Genome Comparisons

On April 14, 2003, the International Human Genome Sequencing Consortium announced (final) "completion" of the stated goals of the Human Genome Project, to the specified degree of coverage, accuracy (at least 99.99%), and annotation. This was accomplished approximately two and a half years ahead of the projected completion date of 2005, and at a cost of about 10% less than the projected $3 billion. In addition, a great deal of extra data were produced, including more than 3 million SNPs, and cDNAs for more than 70% of human and mouse genes. All the data have been quickly deposited in public databases, with no restriction on usage. But is a sequencing project ever really completed? The real value of sequence data lies in its annotation. This aspect of the job is never-ending, but pointing the way is the NHGRI Encyclopedia of DNA Elements (ENCODE) initiative, which seeks to develop efficient, comprehensive, high-throughput technologies for the identification and verification of all types of sequence-based functional elements, particularly those other than coding sequences, such as non-protein-coding genes, transcriptional regulatory elements, and determinants of chromosome structure and function. Development of these technologies will allow not only comprehensive annotation of the immediate target, the human genome, but will lead to efficient analysis of homologous regions in a range of mammals and other vertebrates for broad comparative purposes.

A similar comparative approach can be taken using partial genomic sequences. As a complement to whole-genome sequencing efforts, researchers in the NIH Intramural Sequencing Center Comparative Sequencing Program sequenced segments of genomic DNA from 12 vertebrate species, all orthologous to a segment of about 1.8 Mb on human Chromosome 7, containing 10 genes, including the gene mutated in cystic fibrosis (Thomas *et al.*, 2003). From these sequences they identified more than 1000 "multispecies conserved sequences" (MCSs), most of which did not involve known coding sequences. These MCSs were found to be overrepresented in regions immediately upstream of transcription start sites and in introns, and most do not correspond to known regulatory elements. This work thus provides a rich supply of candidates for future functional studies.

THE TREE OF LIFE

One of the principal ongoing projects of modern biology is the construction of a comprehensive tree of life showing evolutionary relationships. Early attempts at finding the relationships among taxa were based on morphological characters and had many successes, as in sorting out the relationships among major groups such as birds, mammals, reptiles, amphibians, and fishes. Some finer divisions of the tree of life remained controversial, such as the relation among mammalian orders, and the deepest divisions among major groups of organisms remained murky. Introduction of molecular methods eventually resolved some questions, such as placement of the cetaceans, and revised the view of the deepest divisions of the tree of life, but the use of only a few genes proved yet unable to resolve, for example, basic questions about the radiation of mammals.

A recent important project umbrella relating to comparative genomics is the National Science Foundation's Tree of Life Initiative. The NSF has solicited proposals for establishing a phylogenetic framework, Darwin's "Great Tree of Life," for the approximately 1.7 million described species of organisms. Many phylogenetic studies have been done by small teams of researchers to elucidate relationships within taxa of interest using molecular or morphological methods, but the Tree of Life project is aimed at funding larger multidisciplinary teams that are able to apply as much evidence as possible to create a definitive large-scale structure to which smaller specialized studies will eventually contribute. Teams of investigators also will be supported for projects in data acquisition, analysis, algorithm development, and dissemination in computational phylogenetics and phyloinformatics. Because it summarizes biological diversity, such a great tree would be useful in many fields, such as tracking the origin and spread of emerging diseases and their vectors, bioprospecting for pharmaceutical and agrochemical products, targeting biological control of invasive species, and evaluating risk factors for species conservation and ecosystem restoration. This project is timely in that, on one hand, a great flood of new molecular characters in the form of sequences is now available, whereas on the other hand, there appears to be a major extinction event induced by human activity, the result of which is that information about species, many of them as-yet unknown, is being lost. The first round of funding has supported methodological studies as well as taxon-specific studies on roundworms, spiders, and birds. In support of such large-scale phylogenetic efforts, it has recently been shown that current methods for constructing phylogenetic trees can be scaled up to infer very large phylogenies (Tamura et al., 2004).

THE CHARTER OF GENOMICS

The availability of sequence data on a large scale has already begun transforming the modern view of biological patterns and processes at the taxon, population, individual, and biochemical levels. Although the Human Genome Project was the quintessential "big science" project, data from it are having the effect of democratizing and globalizing science. Sequence data are now freely available everywhere and require only inexpensive consumer-grade computing equipment to process. Laboratories throughout the world can now add this genomic view to their biological investigations with little additional investment in terms of equipment. Potential applications of these kinds of knowledge to problems in medicine and agriculture range from safer use of pharmaceuticals (pharmacogenomics) to developing effective countermeasures to devastating crop diseases such as rice blast. Of course, knowledge is always a two-edged sword and many questions relating to privacy, safety, and other ethical concerns will arise. These applications and concerns have been laid out by the United States National Human Genome Research Institute, which defined its continuing mission and its vision for the future of the genomics community in terms of three broad areas (Collins *et al.*, 2003).

The first area deals with the application of genomics to biology, which is concerned with elucidating the structure and function of genomes. Goals consistent with this area are: (1) comprehensive identification of the structural and functional components encoded in the human genome, (2) elucidation of the organization of genetic networks and protein pathways in establishing phenotypes, (3) development of a detailed understanding of heritable variation in the human genome, and (4) understanding evolutionary variation among species and the underlying mechanisms.

The second area deals with the application of genomics to human health, which is focused on translating genome-based knowledge into health benefits. The goals of this area are to develop (1) robust strategies for identifying genetic contributions to disease and drug response, (2) strategies to identify gene variants that contribute to good health and resistance to disease, (3) genome-based approaches to prediction of disease susceptibility, drug response, and detection of illness, (4) genome-based approaches to molecular taxonomy of disease states, and (5) new understanding of genes and pathways to develop new therapeutic approaches to disease.

The third area is the application of genomics to society, which is focused on promoting the use of genomics to maximize benefits and minimize harm to society. The goals of this area are to (1) develop policy options for the uses of genomics in both medical and nonmedical settings, (2) understand the relationships between genomics, race, and ethnicity, and the consequences of uncovering these relationships, (3) understand the consequences of deciphering the genomic contributions to human traits and behaviors, and (4) assess how to define the ethical boundaries for uses of genomic information. Unlike the original set of goals for the NHGRI, these are quite broad and open-ended and can be viewed as a comprehensive charter for genomics in the coming century.

Although its roots can be traced back to the earliest chromosomal work, comparative genomics involving (nearly) complete genome sequencing is a science still in its infancy. Fast-growing and full of potential, its maturation is likely to influence an increasingly broad array of biological disciplines. Already, widespread implications can be envisioned for evolutionary biology, medicine, and agriculture; in some cases, these have already become reality. The large-scale comparison, and perhaps even manipulation, of genomes is a complex undertaking involving numerous empirical, analytical, and ethical issues. Undoubtedly, both important challenges and exciting discoveries lie ahead for genome biology.

REFERENCES

Aach J, Bulyk ML, Church GM, et al. 2001. Computational comparison of two draft sequences of the human genome. *Nature* 409: 856–859.

Adams MD, Celniker SE, Holt RA, et al. 2000. The genome sequence of *Drosophila melanogaster*. *Science* 287: 2185–2195.

Adams MD, Dubnick M, Kerlavage AR, et al. 1992. Sequence identification of 2,375 human brain genes. *Nature* 355: 632– 634.

Adams MD, Kelley JM, Gocayne JD, et al. 1991. Complementary DNA sequencing: expressed sequence tags and human genome project. *Science* 252: 1651– 1656.

Antequera F, Bird A. 1993. Number of CpG islands and genes in human and mouse. *Proc Natl Acad Sci USA* 90: 11995–11999.

Aparicio S, Chapman J, Stupka E, et al. 2002. Whole-genome shotgun assembly and analysis of the genome of *Fugu rubripes*. *Science* 297: 1301–1310.

Arabidopsis Genome Initiative. 2000. Analysis of the genome sequence of the flowering plant *Arabidopsis thaliana*. *Nature* 408: 796–815.

Archidiacono N, Storlazzi CT, Spalluto C, et al. 1998. Evolution of chromosome Y in primates. *Chromosoma* 107: 241–246.

Aristotle. 1953. *Generation of Animals*. Cambridge, MA: Harvard University Press; W. Heinemann Ltd.

Avarello R, Pedicini A, Caiulo A, et al. 1992. Evidence for an ancestral alphoid domain on the long arm of human chromosome 2. *Hum Gene* 89: 247–249.

Baldauf SL. 2003. The deep roots of eukaryotes. *Science* 300: 1703–1706.

Band MR, Larson JH, Rebeiz M, et al. 2000. An ordered comparative map of the cattle and human genomes. *Genome Res* 10: 1359–1368.

Bennett JW. 1997. White Paper: genomics for filamentous fungi. *Fungal Genet Biol* 21: 3–7.

Bennett MD, Leitch IJ, Price HJ, Johnston JS. 2003. Comparisons with *Caenorhabditis* (~ 100 Mb) and *Drosophila* (~ 175 Mb) using flow cytometry show genome size in *Arabidopsis* to be ~ 157 Mb and thus ~ 25% larger than the Arabidopsis Genome Initiative estimate of ~ 125 Mb. *Ann Bot* 91: 547–557.

Benton MJ, Ayala FJ. 2003. Dating the tree of life. *Science* 300: 1698–1700.

Berget SM, Moore C, Sharp PA. 1977. Spliced segments at the 5′ terminus of adenovirus 2 late mRNA. *Proc Natl Acad Sci USA* 74: 3171–3175.

Bernardi G. 1995. The human genome: organization and evolutionary history. *Annu Rev Genet* 29: 445–476.

Bernardi G. 2001. Misunderstandings about isochores. Part 1. *Gene* 276: 3–13.

Bernardi G, Mouchiroud D, Gautier C. 1988. Compositional patterns in vertebrate genomes: conservation and change in evolution. *J Mol Evol* 28: 7–18.

Bernardi G, Olofsson B, Filipski J, et al. 1985. The mosaic genome of warm-blooded vertebrates. *Science* 228: 953–958.

Bird AP. 1986. CpG-rich islands and the function of DNA methylation. *Nature* 321: 209–213.

Blumenthal T, Gleason KS. 2003. *Caenorhabditis elegans* operons: form and function. *Nat Rev Genet* 4: 112–120.

Boguski MS. 1995. The turning point in genome research. *Trends Biochem Sci* 20: 295–296.

Bolshakov VN, Topalis P, Blass C, et al. 2002. A comparative genomic analysis of two distant diptera, the fruit fly, *Drosophila melanogaster*, and the malaria mosquito, *Anopheles gambiae*. *Genome Res* 12: 57–66.

Brenner S, Elgar G, Sandford R, et al. 1993. Characterization of the pufferfish (*Fugu*) genome as a compact model vertebrate genome. *Nature* 366: 265–268.

Brett D, Hanke J, Lehmann G, et al. 2000. EST comparison indicates 38% of human mRNAs contain possible alternative splice forms. *FEBS Lett* 474: 83–86.

Brickley P. 2002. A scrap over sequences, take two. *The Scientist* 16 (May 13): 55.

Bridges CB. 1935. Salivary chromosome maps with a key to the banding of the chromosomes of *Drosophila melanogaster*. *J Hered* 26: 60–64.

Britten RJ. 1986. Rates of DNA sequence evolution differ between taxonomic groups. *Science* 231: 1393–1398.

Britten RJ. 2002. Divergence between samples of chimpanzee and human DNA sequences is 5%, counting indels. *Proc Natl Acad Sci USA* 99: 13633–13635.

Britten RJ, Kohne DE. 1968. Hundreds of thousands of copies of DNA sequences have been incorporated into the genomes of higher organisms. *Science* 161: 529–540.

Britton-Davidian J, Catalan J, da Graca Ramalhinho M, et al. 2000. Rapid chromosomal evolution in island mice. *Nature* 403: 158.

C. elegans Sequencing Consortium. 1998. Genome sequence of the nematode *C. elegans*: a platform for investigating biology. *Science* 282: 2012–2018.

Carlton JM, Angiuoli SV, Suh BB, et al. 2002. Genome sequence and comparative analysis of the model rodent malaria parasite *Plasmodium yoelii yoelii*. *Nature* 419: 512–519.

Caspersson T, Zech L, Johansson C. 1970. Differential binding of alkylating fluorochromes in human chromosomes. *Exp Cell Res* 60: 315–319.

Celniker SE, Wheeler DA, Kronmiller B, et al., 2002. Finishing a whole-genome shotgun: release 3 of the *Drosophila melanogaster* euchromatic genome sequence. *Genome Biol* 3: research0079.1–0079.14.

Chargaff E. 1980. In praise of smallness: how we can return to small science? *Perspect Biol Med* 23: 370–385.

Check E. 2002. Priorities for genome sequencing leave macaques out in the cold. *Nature* 417: 473–474.

Chervitz SA, Aravind L, Sherlock G, et al. 1998. Comparison of the complete protein sets of worm and yeast: orthology and divergence. *Science* 282: 2022–2028.

Chothia C, Gough J, Vogel C, Teichmann SA. 2003. Evolution of the protein repertoire. *Science* 300: 1701–1703.

Chow LT, Gelinas RE, Broker TR, Roberts RJ. 1977. An amazing sequence arrangement at the 5' ends of adenovirus 2 messenger RNA. *Cell* 12: 1–8.

Chowdhary BP, Raudsepp T, Fronicke L, and Scherthan H. 1998. Emerging patterns of comparative genome organization in some mammalian species as revealed by Zoo-FISH. *Genome Res* 8: 577–589.

Christophides GK, Zdobnov E, Barillas-Mury C, *et al.* 2002. Immunity-related genes and gene families in *Anopheles gambiae. Science* 298: 159–165.

Claverie JM. 2001. What if there are only 30,000 human genes? *Science* 291: 1255–1257.

Collins FS, Green ED, Guttmacher AE, Guyer MS. 2003. A vision for the future of genomics research. *Nature* 422: 835–847.

Collins FS, Morgan M, Patrinos A. 2003. The Human Genome Project: lessons from large-scale biology. *Science* 300: 286–290.

Courties C, Perasso R, Chrétiennot-Dinet M-J, *et al.* 1998. Phylogenetic analysis and genome size of *Ostreococcus tauri* (Chlorophyta, Prasinophyceae). *J Phycol* 34: 844–849.

Cyranoski D. 2002. Almost human. *Nature* 418: 910–912.

Deamer DW, Branton D. 2002. Characterization of nucleic acids by nanopore analysis. *Acc Chem Res* 35: 817–825.

Decottignies A, Sanchez-Perez I, Nurse P. 2003. *Schizosaccharomyces pombe* essential genes: a pilot study. *Genome Res* 13: 399–406.

Dehal P, Satou Y, Campbell RK, *et al.* 2002. The draft genome of *Ciona intestinalis*: insights into chordate and vertebrate origins. *Science* 298: 2157–2167.

Dimopoulos G, Muller HM, Levashina EA, Kafatos FC. 2001. Innate immune defense against malaria infection in the mosquito. *Curr Opin Immunol* 13: 79–88.

Doolittle RF. 1986. *Of URFs and ORFs: A Primer on How to Analyze Derived Amino Acid Sequences.* Mill Valley, CA: University Science Books.

Doolittle WF. 1999. Phylogenetic classification and the universal tree. *Science* 284: 2124–2129.

Doolittle WF, Sapienza C. 1980. Selfish genes, the phenotype paradigm and genome evolution. *Nature* 284: 601–603.

Dujon B. 1996. The yeast genome project: what did we learn? *Trends Genet* 12: 263–270.

Duret L, Mouchiroud D, Gautier C. 1995. Statistical analysis of vertebrate sequences reveals that long genes are scarce in GC-rich isochores. *J Mol Evol* 40: 308–317.

Dyson T. 1999. World food trends and prospects to 2025. *Proc Natl Acad Sci USA* 96: 5929–5936.

Easteal S, Collet C, Betty D. 1995. *The Mammalian Molecular Clock.* New York: R.G. Landes.

Ebersberger I, Metzler D, Schwarz C, Paabo S. 2002. Genomewide comparison of DNA sequences between humans and chimpanzees. *Am J Hum Genet* 70: 1490–1497.

Edwalds-Gilbert G, Veraldi KL, Milcarek C. 1997. Alternative poly(A) site selection in complex transcription units: means to an end? *Nucleic Acids Res* 25: 2547–2561.

Eichler EE, Sankoff D. 2003. Structural dynamics of eukaryotic chromosome evolution. *Science* 301: 793–797.

Ellegren H, Smith NG, Webster MT. 2003. Mutation rate variation in the mammalian genome. *Curr Opin Genet Dev* 13: 562–568.

Enard W, Khaitovich P, Klose J, *et al.* 2002. Intra- and interspecific variation in primate gene expression patterns. *Science* 296: 340–343.

Estop AM, Garver JJ, Egozcue J, *et al.* 1983. Complex chromosome homologies between the rhesus monkey (*Macaca mulatta*) and man. *Cytogenet Cell Genet* 35: 46–50.

Farr CJ, Goodfellow PN. 1992. Hidden messages in genetic maps. *Science* 258: 49.

Fickett JW, Wasserman WW. 2000. Discovery and modeling of transcriptional regulatory regions. *Curr Opin Biotechnol* 11: 19–24.

Ford CE, Hamerton JL. 1956. Chromosomes of man. *Nature* 178: 1020–1023.

Galagan JE, Calvo SE, Borkovich KA, *et al.* 2003. The genome sequence of the filamentous fungus *Neurospora crassa. Nature* 422: 859–868.

Gale MD, Devos KM. 1998. Plant comparative genetics after 10 years. *Science* 282: 656–659.

Galison PL. 1997. *Image and Logic: A Material Culture of Microphysics.* Chicago: University of Chicago Press.

Gardner MJ, Hall N, Fung E, *et al.* 2002. Genome sequence of the human malaria parasite *Plasmodium falciparum. Nature* 419: 498–511.

Gellin J, Echard G, Benne F, Gillois M. 1981. Pig gene mapping: PKM2-MPI-NP synteny. *Cytogenet Cell Genet* 30: 59–62.

George AJ. 2002. Is the number of genes we possess limited by the presence of an adaptive immune system? *Trends Immunol* 23: 351–355.

Gilbert W, Bodmer WF. 1986. Two cheers for human gene sequencing. *The Scientist* 1 (Oct. 20): 11.

Gionti M, Ristoratore F, Di Gregorio A, *et al.* 1998. Cihox5, a new *Ciona intestinalis* Hox-related gene, is involved in regionalization of the spinal cord. *Dev Genes Evol* 207: 515–523.

Goff SA, Ricke D, Lan TH, *et al.* 2002. A draft sequence of the rice genome (*Oryza sativa* L. ssp. *japonica*). *Science* 296: 92–100.

Goffeau A, Barrell BG, Bussey H, *et al.,* 1996. Life with 6000 genes. *Science* 274: 546, 563–567.

Goujon P. 2001. *From Biotechnology to Genomes*. River Edge, NJ: World Scientific Pub. Co.

Graves JA. 1996. Mammals that break the rules: genetics of marsupials and monotremes. *Annu Rev Genet* 30: 233–260.

Gregory TR. 2001. *Animal Genome Size Database*. www.genomesize.com.

Gruskin KD, Smith TF. 1987. Molecular genetics and computer analyses. *Comput Appl Biosci* 3: 167–170.

Gu X, Wang Y, Gu J. 2002. Age distribution of human gene families shows significant roles of both large- and small-scale duplications in vertebrate evolution. *Nat Genet* 31: 205–209.

Gupta RS. 1998. Protein phylogenies and signature sequences: a reappraisal of evolutionary relationships among archaebacteria, eubacteria, and eukaryotes. *Microbiol Mol Biol Rev* 62: 1435–1491.

Hahn MW, Wray GA. 2002. The g-value paradox. *Evol Dev* 4: 73–75.

Hanke J, Brett D, Zastrow I, *et al.* 1999. Alternative splicing of human genes: more the rule than the exception? *Trends Genet* 15: 389–390.

Hartl DL. 2000. Molecular melodies in high and low C. *Nat Rev Genet* 1: 145–149.

Hawksworth D. 1991. The fungal dimension of biodiversity: magnitude, significance and conservation. *Mycol Res* 95: 641–655.

Hedges SB. 2002. The origin and evolution of model organisms. *Nat Rev Genet* 3: 838–849.

Hedges SB, Kumar S. 2002. Vertebrate genomes compared. *Science* 297: 1283–1285.

Hedges SB, Kumar S. 2003. Genomic clocks and evolutionary timescales. *Trends Genet* 19: 200–206.

Hedges SB, Chen H, Kumar S, *et al.* 2001. A genomic timescale for the origin of eukaryotes. *BMC Evol Biol* 1: 4.1–4.10.

Hill CA, Fox AN, Pitts RJ, *et al.* 2002. G protein-coupled receptors in *Anopheles gambiae*. *Science* 298: 176–178.

Holt RA, Subramanian GM, Halpern A, *et al.* 2002. The genome sequence of the malaria mosquito *Anopheles gambiae*. *Science* 298: 129–149.

Houck ML, Kumamoto AT, Gallagher DS, Benirschke K. 2001. Comparative cytogenetics of the African elephant (*Loxodonta africana*) and Asiatic elephant (*Elephas maximus*). *Cytogenet Cell Genet* 93: 249–252.

Huang L, Guan RJ, Pardee AB. 1999. Evolution of transcriptional control from prokaryotic beginnings to eukaryotic complexities. *Crit Rev Eukaryot Gene Expr* 9: 175–182.

Huang Q, Fu YX, Boerwinkle E. 2003. Comparison of strategies for selecting single nucleotide polymorphisms for case/control association studies. *Hum Genet* 113: 253–257.

Hubbard T, Barker D, Birney E, *et al.* 2002. The Ensembl genome database project. *Nucleic Acids Res* 30: 38–41.

Hungerford DA, Chandra HS, Snyder RL. 1967. Somatic chromosomes of a black rhinoceros (*Diceros bicornis* Gray 1821). *Am Nat* 101: 357–358.

Hynes M. 2003. The *Neurospora crassa* genome opens up the world of filamentous fungi. *Genome Biol* 4: 271.1–271.4.

Iannuzzi L, Di Meo GP, Perucatti A, Bardaro T. 1998. ZOO-FISH and R-banding reveal extensive conservation of human chromosome regions in euchromatic regions of river buffalo chromosomes. *Cytogenet Cell Genet* 82: 210–214.

Ijdo J, Baldini A, Ward DC, *et al.* 1991. Origin of human chromosome 2: an ancestral telomere-telomere fusion. *Proc Natl Acad Sci USA* 88: 9051–9055.

International Human Genome Sequencing Consortium. 2001. Initial sequencing and analysis of the human genome. *Nature* 409: 860–921.

Ioannidis JP. 2003. Genetic associations: false or true? *Trends Mol Med* 9: 135–138.

Jiang Z, Melville JS, Cao H, *et al.* 2002. Measuring conservation of contiguous sets of autosomal markers on bovine and porcine genomes in relation to the map of the human genome. *Genome* 45: 769–776.

John B, Miklos GLG. 1988. *The Eukaryote Genome in Development and Evolution.* London: Allen & Unwin.

Katinka MD, Duprat S, Cornillot E, *et al.* 2001. Genome sequence and gene compaction of the eukaryote parasite *Encephalitozoon cuniculi. Nature* 414: 450–453.

Kellis M, Patterson N, Endrizzi M, *et al.* 2003. Sequencing and comparison of yeast species to identify genes and regulatory elements. *Nature* 423: 241–254.

Kimura M. 1983. *The Neutral Theory of Molecular Evolution.* Cambridge: Cambridge University Press.

Korf I, Flicek P, Duan D, Brent MR. 2001. Integrating genomic homology into gene structure prediction. *Bioinformatics* 17 (Suppl. 1): S140–S148.

Krishnan G. 1975. Nature of tunicin and its interaction with other chemical components of the tunic of the ascidian, *Polyclinum madrasensis* Sebastian. *Indian J Exp Biol* 13: 172–176.

Kruglyak L, Nickerson DA. 2001. Variation is the spice of life. *Nat Genet* 27: 234–236.

Krumlauf R, Marzluf GA. 1980. Genome organization and characterization of the repetitive and inverted repeat DNA sequences in *Neurospora crassa. J Biol Chem* 255: 1138–1145.

Kumar S, Hedges SB. 1998. A molecular timescale for vertebrate evolution. *Nature* 392: 917–920.

Kumar S, Subramanian S. 2002. Mutation rates in mammalian genomes. *Proc Natl Acad Sci USA* 99: 803–808.

Kumar S, Gadagkar SR, Filipski A, Gu X. 2001. Determination of the number of conserved chromosomal segments between species. *Genetics* 157: 1387–1395.

Kumar S, Mitnik C, Valente G, Floyd-Smith G. 2000. Expansion and molecular evolution of the interferon-induced 2′-5′ oligoadenylate synthetase gene family. *Mol Biol Evol* 17: 738–750.

Levine M, Tjian R. 2003. Transcription regulation and animal diversity. *Nature* 424: 147–151.

Levy HP, Schultz RA, Cohen MM. 1992. Comparative gene mapping in the species *Muntiacus muntjac. Cytogenet Cell Genet* 61: 276–281.

Li W-H. 2002. Are isochore sequences homogeneous? *Gene* 300: 129–139.

Li W-H, Tanimura M. 1987. The molecular clock runs more slowly in man than in apes and monkeys. *Nature* 326: 93–96.

Link AJ, Olson MV. 1991. Physical map of the *Saccharomyces cerevisiae* genome at 110–kilobase resolution. *Genetics* 127: 681–698.

Lo N, Tokuda G, Watanabe H, *et al.* 2000. Evidence from multiple gene sequences indicates that termites evolved from wood-feeding cockroaches. *Curr Biol* 10: 801–804.

Lopez-Garcia P, Rodriguez-Valera F, Pedros-Alio C, Moreira D. 2001. Unexpected diversity of small eukaryotes in deep-sea Antarctic plankton. *Nature* 409: 603–607.

Macer D. 1991. Whose genome project? *Bioethics* 5: 183–211.

Mannhaupt G, Montrone C, Haase D, *et al.* 2003. What's in the genome of a filamentous fungus? Analysis of the *Neurospora* genome sequence. *Nucleic Acids Res* 31: 1944–1954.

March R. 2000. Pharmacogenomics: the genomics of drug response. *Yeast* 17: 16–21.

Margulis L. 1996. Archaeal-eubacterial mergers in the origin of Eukarya: phylogenetic classification of life. *Proc Natl Acad Sci USA* 93: 1071–1076.

Martin SL, Blackmon BP, Rajagopalan R, *et al.* 2002. MagnaportheDB: a federated solution for integrating physical and genetic map data with BAC end derived sequences for the rice blast fungus *Magnaporthe grisea. Nucleic Acids Res* 30: 121–124.

Matassi G, Sharp PM, Gautier C. 1999. Chromosomal location effects on gene sequence evolution in mammals. *Curr Biol* 9: 786–791.

McCarthy JJ, Hilfiker R. 2000. The use of single-nucleotide polymorphism maps in pharmacogenomics. *Nat Biotechnol* 18: 505–508.

Messing J. 2001. Do plants have more genes than humans? *Trends Plant Sci* 6: 195–196.

Mironov AA, Fickett JW, Gelfand MS. 1999. Frequent alternative splicing of human genes. *Genome Res* 9: 1288–1293.

Moon-van der Staay SY, De Wachter R, Vaulot D. 2001. Oceanic 18S rDNA sequences from picoplankton reveal unsuspected eukaryotic diversity. *Nature* 409: 607–610.

Mouchiroud D, Gautier C, Bernardi G. 1995. Frequencies of synonymous substitutions in mammals are gene-specific and correlated with frequencies of nonsynonymous substitutions. *J Mol Evol* 40: 107–113.

Mounsey A, Bauer P, Hope IA. 2002. Evidence suggesting that a fifth of annotated *Caenorhabditis elegans* genes may be pseudogenes. *Genome Res* 12: 770–775.

Mouse Genome Sequencing Consortium. 2002. Initial sequencing and comparative analysis of the mouse genome. *Nature* 420: 520–562.

Mullis K, Faloona F, Scharf S, *et al.* 1986. Specific enzymatic amplification of DNA in vitro: the polymerase chain reaction. *Cold Spring Harb Symp Quant Biol* 51 (Pt. 1): 263–273.

Mural RJ, Adams MD, Myers EW, *et al.* 2002. A comparison of whole-genome shotgun-derived mouse chromosome 16 and the human genome. *Science* 296: 1661–1671.

Murphy WJ, Stanyon R, O'Brien SJ. 2001. Evolution of mammalian genome organization inferred from comparative gene mapping. *Genome Biol* 2: reviews0005.1–0005.8.

Murphy WJ, Sun S, Chen ZQ, *et al.* 1999. Extensive conservation of sex chromosome organization between cat and human revealed by parallel radiation hybrid mapping. *Genome Res* 9: 1223–1230.

Nadeau JH. 1989. Maps of linkage and synteny homologies between mouse and man. *Trends Genet* 5: 82–86.

Nakashima K, Yamada L, Satou Y, *et al.* 2004. The evolutionary origin of animal cellulose synthase. *Dev Genes Evol* 214: 81–88.

Nash WG, O'Brien SJ. 1982. Conserved regions of homologous G-banded chromosomes between orders in mammalian evolution: carnivores and primates. *Proc Natl Acad Sci USA* 79: 6631–6635.

Nash WG, Wienberg J, Ferguson-Smith MA, *et al.* 1998. Comparative genomics: tracking chromosome evolution in the family ursidae using reciprocal chromosome painting. *Cytogenet Cell Genet* 83: 182–192.

Nei M. 1969. Gene duplication and nucleotide substitution in evolution. *Nature* 221: 40–42.

Nicholl DST. 2002. *An Introduction to Genetic Engineering.* 2nd edition. Cambridge: Cambridge University Press.

Niimura Y, Gojobori T. 2002. *In silico* chromosome staining: reconstruction of Giemsa bands from the whole human genome sequence. *Proc Natl Acad Sci USA* 99: 797–802.

Nilsen TW. 1989. Trans-splicing in nematodes. *Exp Parasitol* 69: 413–416.

Normile D. 2001. Gene expression differs in human and chimp brains. *Science* 292: 44–45.

O'Brien SJ, Nash WG. 1982. Genetic mapping in mammals: chromosome map of domestic cat. *Science* 216: 257–265.

O'Brien SJ, Stanyon R. 1999. Phylogenomics: ancestral primate viewed. *Nature* 402: 365–366.

O'Brien SJ, Eisenberg JF, Miyamoto M, *et al.* 1999. Genome maps 10. Comparative genomics. Mammalian radiations. Wall chart. *Science* 286: 463–478.

O'Brien SJ, Menotti-Raymond M, Murphy WJ, *et al.* 1999. The promise of comparative genomics in mammals. *Science* 286: 458–462, 479–481.

Ohno S. 1970. *Evolution by Gene Duplication*. Berlin: Springer-Verlag.

Ohno S. 1972. So much "junk" DNA in our genome. In: Smith HH ed. *Evolution of Genetic Systems*. New York: Gordon and Breach, 366–370.

Okubo K, Hori N, Matoba R, *et al.* 1992. Large scale cDNA sequencing for analysis of quantitative and qualitative aspects of gene expression. *Nat Genet* 2: 173–179.

Oliver SG, van der Aart QJ, Agostoni-Carbone ML, *et al.* 1992. The complete DNA sequence of yeast chromosome III. *Nature* 357: 38–46.

Olson M, Hood L, Cantor C, Botstein D. 1989. A common language for physical mapping of the human genome. *Science* 245: 1434–1435.

Olson MV, Varki A. 2003. Sequencing the chimpanzee genome: insights into human evolution and disease. *Nat Rev Genet* 4: 20–28.

Orgel LE, Crick FH. 1980. Selfish DNA: the ultimate parasite. *Nature* 284: 604–607.

Pal C, Hurst LD. 2000. The evolution of gene number: are heritable and non-heritable errors equally important? *Heredity* 84: 393–400.

Pardue ML, Gall JG. 1970. Chromosomal localization of mouse satellite DNA. *Science* 168: 1356–1358.

Pearson H. 2003. Geneticists play the numbers game in vain. *Nature* 423: 576.

Petrov DA. 2001. Evolution of genome size: new approaches to an old problem. *Trends Genet* 17: 23–28.

Petrov DA, Lozovskaya ER, Hartl DL. 1996. High intrinsic rate of DNA loss in *Drosophila*. *Nature* 384: 346–349.

Potrykus I. 2001. Golden rice and beyond. *Plant Physiol* 125: 1157–1161.

Pruitt KD, Maglott DR. 2001. RefSeq and LocusLink: NCBI gene-centered resources. *Nucleic Acids Res* 29: 137–140.

Rat Genome Sequencing Project Consortium. 2004. Genome sequence of the Brown Norway rat yields insights into mammalian evolution. *Nature* 428: 493–521.

Raudsepp T, Fronicke L, Scherthan H, *et al.* 1996. Zoo-FISH delineates conserved chromosomal segments in horse and man. *Chromosome Res* 4: 218–225.

Richard F, Messaoudi C, Bonnet-Garnier A, *et al.* 2003. Highly conserved chromosomes in an Asian squirrel (*Menetes berdmorei*, Rodentia: Sciuridae) as demonstrated by ZOO-FISH with human probes. *Chromosome Res* 11: 597–603.

Rivera MC, Jain R, Moore JE, Lake JA. 1998. Genomic evidence for two functionally distinct gene classes. *Proc Natl Acad Sci USA* 95: 6239–6244.

Rizzon C, Marais G, Gouy M, Biemont C. 2002. Recombination rate and the distribution of transposable elements in the *Drosophila melanogaster* genome. *Genome Res* 12: 400–407.

Roberts L. 1987. Agencies vie over Human Genome Project. *Science* 237: 486–488.

Roest Crollius H, Jaillon O, Bernot A, *et al.* 2000a. Estimate of human gene number provided by genome-wide analysis using *Tetraodon nigroviridis* DNA sequence. *Nat Genet* 25: 235–238.

Roest Crollius H, Jaillon O, Bernot A, *et al.* 2002. Genome-wide comparisons between human and *Tetraodon*. *Ernst Schering Res Found Workshop*: 11–29.

Roest Crollius H, Jaillon O, Dasilva C, *et al.* 2000b. Characterization and repeat analysis of the compact genome of the freshwater pufferfish *Tetraodon nigroviridis*. *Genome Res* 10: 939–949.

Russo E. 2001. Behind the sequence. *The Scientist* 15 (Mar. 5): 1.

Saccone S, de Sario A, Wiegant J, *et al.* 1993. Correlations between isochores and chromosomal bands in the human genome. *Proc Natl Acad Sci USA* 90: 11929–11933.

Sachidanandam R, Weissman D, Schmidt SC, *et al.* 2001. A map of human genome sequence variation containing 1.42 million single nucleotide polymorphisms. *Nature* 409: 928–933.

Sanger F, Air GM, Barrell BG, *et al.* 1977. Nucleotide-sequence of bacteriophage Phi X174 DNA. *Nature* 265: 687–695.

Scherthan H, Cremer T, Arnason U, *et al.* 1994. Comparative chromosome painting discloses homologous segments in distantly related mammals. *Nat Genet* 6: 342–347.

Schwartz S, Kent WJ, Smit A, *et al.* 2003. Human-mouse alignments with BLASTZ. *Genome Res* 13: 103–107.

Seabright M. 1971. A rapid banding technique for human chromosomes. *Lancet* 2: 971–972.

Sebat J, Lakshmi B, Troge J, *et al.* 2004. Large-scale copy number polymorphism in the human genome. *Science* 305: 525–528.

Selker EU. 1990. Premeiotic instability of repeated sequences in *Neurospora crassa*. *Annu Rev Genet* 24: 579–613.

Seltman H, Roeder K, Devlin B. 2003. Evolutionary-based association analysis using haplotype data. *Genet Epidemiol* 25: 48–58.

Sibley CG, Ahlquist JE. 1984. The phylogeny of the hominoid primates, as indicated by DNA-DNA hybridization. *J Mol Evol* 20: 2–15.

Sibley CG, Ahlquist JE. 1987. DNA hybridization evidence of hominoid phylogeny: results from an expanded data set. *J Mol Evol* 26: 99–121.

Skaletsky H, Kuroda-Kawaguchi T, Minx PJ, *et al.* 2003. The male-specific region of the human Y chromosome is a mosaic of discrete sequence classes. *Nature* 423: 825–837.

Spieth J, Brooke G, Kuersten S, *et al.* 1993. Operons in *C. elegans*: polycistronic mRNA precursors are processed by trans-splicing of SL2 to downstream coding regions. *Cell* 73: 521–532.

Stanyon R, Yang F, Cavagna P, *et al.* 1999. Reciprocal chromosome painting shows that genomic rearrangement between rat and mouse proceeds ten times faster than between humans and cats. *Cytogenet Cell Genet* 84: 150–155.

Stephens JC. 1999. Single-nucleotide polymorphisms, haplotypes, and their relevance to pharmaco-genetics. *Mol Diagn* 4: 309–317.

Strauss EC, Kobori JA, Siu G, Hood LE. 1986. Specific-primer-directed DNA sequencing. *Anal Biochem* 154: 353–360.

Sturtevant AH. 1913. The linear arrangement of six sex-linked factors in *Drosophila*, as shown by their mode of association. *J Exp Zool* 14: 43–59.

Suedmeyer WK, Houck ML, Kreeger J. 2003. Klinefelter syndrome (39 XXY) in an adult Siberian tiger (*Panthera tigris altaica*). *J Zoo Wildl Med* 34: 96–99.

Sulston JE, Schierenberg E, White JG, Thomson JN. 1983. The embryonic cell lineage of the nematode *Caenorhabditis elegans*. *Dev Biol* 100: 64–119.

Tamura K, Nei M, Kumar S. 2004. Prospects for inferring very large phylogenies by using the neighbor-joining method. *Proc Natl Acad Sci USA* 101: 11030–11035.

Tavare S, Marshall CR, Will O, *et al.* 2002. Using the fossil record to estimate the age of the last common ancestor of extant primates. *Nature* 416: 726–729.

Taylor JG, Choi EH, Foster CB, Chanock SJ. 2001. Using genetic variation to study human disease. *Trends Mol Med* 7: 507–512.

Thiessen KM, Lalley PA. 1986. New gene assignments and syntenic groups in the baboon (*Papio papio*). *Cytogenet Cell Genet* 42: 19–23.

Thiessen KM, Lalley PA. 1987. Gene assignments and syntenic groups in the sacred baboon (*Papio hamadryas*). *Cytogenet Cell Genet* 44: 82–88.

Thomas CA. 1971. The genetic organization of chromosomes. *Annu Rev Genet* 5: 237–256.

Thomas JW, Touchman JW, Blakesley RW, *et al.* 2003. Comparative analyses of multi-species sequences from targeted genomic regions. *Nature* 424: 788–793.

Tjio JH, Levan A. 1956. The chromosome number of Man. *Hereditas* 42: U1–6.

Venter JC, Adams MD, Myers EW, *et al.* 2001. The sequence of the human genome. *Science* 291: 1304–1351.

Volti R. 2001. *Society and Technological Change*. New York: Worth Publishers.

Wade N. 2002. Thrown aside, genome pioneer plots a rebound. *New York Times* April 30.

Wakefield MJ, Graves JA. 1996. Comparative maps of vertebrates. *Mamm Genome* 7: 715–716.

Wiehe T, Gebauer-Jung S, Mitchell-Olds T, Guigo R. 2001. SGP-1: prediction and validation of homologous genes based on sequence alignments. *Genome Res* 11: 1574–1583.

Wienberg J, Stanyon R. 1995. Chromosome painting in mammals as an approach to comparative genomics. *Curr Opin Genet Dev* 5: 792–797.

Wildman DE, Uddin M, Liu G, *et al.* 2003. Implications of natural selection in shaping 99.4% non-synonymous DNA identity between humans and chimpanzees: enlarging genus *Homo*. *Proc Natl Acad Sci USA* 100: 7181–7188.

Wolfe KH. 2001. Yesterday's polyploids and the mystery of diploidization. *Nat Rev Genet* 2: 333–341.

Wolfe KH, Sharp PM, Li WH. 1989. Mutation rates differ among regions of the mammalian genome. *Nature* 337: 283–285.

Wong GK, Wang J, Tao L, *et al.* 2002. Compositional gradients in Gramineae genes. *Genome Res* 12: 851–856.

Wood VR, Gwilliam MA, Rajandream M, *et al.* 2002. The genome sequence of *Schizosaccharomyces pombe*. *Nature* 415: 871–880.

Wu CI, Li WH. 1985. Evidence for higher rates of nucleotide substitution in rodents than in man. *Proc Natl Acad Sci USA* 82: 1741–1745.

Ye X, Al-Babili S, Kloti A, *et al.* 2000. Engineering the provitamin A (beta-carotene) biosynthetic pathway into (carotenoid-free) rice endosperm. *Science* 287: 303–305.

Yu J, Hu S, Wang J, *et al.* 2002. A draft sequence of the rice genome (*Oryza sativa* L. ssp. *indica*). *Science* 296: 79–92.

Yunis JJ, Prakash O. 1982. The origin of man: a chromosomal pictorial legacy. *Science* 215: 1525–1530.

Zdobnov EM, von Mering C, Letunic I, *et al.* 2002. Comparative genome and proteome analysis of *Anopheles gambiae* and *Drosophila melanogaster*. *Science* 298: 149–159.

Zhang J. 2000. Protein-length distributions for the three domains of life. *Trends Genet* 16: 107–109.

Zmasek CM, Eddy SR. 2001. A simple algorithm to infer gene duplication and speciation events on a gene tree. *Bioinformatics* 17: 821–828.

CHAPTER 10

Comparative Genomics in Prokaryotes

T. Ryan Gregory and Rob DeSalle

As a group, prokaryotes demonstrate the most diverse niche occupancies, metabolisms, and geographic distributions of any living organisms. Prokaryote species may be free-living or exist in pathogenic or mutualistic associations with hosts, either intra- or extracellularly. They may exhibit aerobic or anaerobic metabolisms, and many acquire their energy from chemical sources that are thoroughly toxic to other living things. They can be found from the depths of the ocean to high in the atmosphere and everywhere in between, and have played a fundamental role in shaping and maintaining both the biotic and abiotic characteristics of the planet (Oren, 2004). Their resiliency in the face of environmental change is phenomenal, allowing them to have survived and reproduced in much the same form for nearly four billion years. As Gould (1996, p. 176) noted, "On any possible, reasonable, or fair criterion, bacteria are—and always have been—the dominant forms of life on earth."

A major contributor to this unrivaled evolutionary success lies in the variability and dynamic nature of prokaryotic genomes. It should therefore come as little surprise that the rapid expansion of prokaryote genomics over the last decade has shed much light on the biology of these organisms, including their diversity, abundance, classification, physiology, and evolution. Unfortunately, the extraordinary rate at which new information is becoming available has had the consequence of making it somewhat difficult to access and interpret. In this regard, the present chapter provides an outline of some of the key insights being gleaned from the detailed analysis of prokaryotic genome sequences, as well as discussing the past and possible future of this fast-evolving field of inquiry.

WHAT IS A PROKARYOTE?

CLASSIFYING PROKARYOTES THE OLD-FASHIONED WAY

The term "prokaryote" is a designation based on exclusion. That is, an organism qualifies as a prokaryote based on something it does not possess, namely a nucleus. Unfortunately, such "negative" classifications have a tendency to lump unrelated taxa into the same category, creating paraphyletic assemblages that obscure true evolutionary relationships. However, "positive" unifying characteristics of prokaryotes are not easy to come by, as their size is miniscule (0.2–10 micrometers [μm]), their morphological characters misleading, and their chemical properties extremely diverse. These impediments to detailed classification aside, it has been fairly well accepted for some time that prokaryotes can be divided at the most fundamental level into two large monophyletic groups ("domains"): the Bacteria and the Archaea.

About 6500 species of prokaryotes have been described and named, but this undoubtedly represents only a tiny fraction of their real diversity (Curtis *et al.*, 2002; Oren, 2004). Most of the described species are Bacteria, which can be further classified according to several commonly recognized characteristics, most notably the specific properties (or absence) of a cell wall. The traditional view of prokaryotic classification groups species according to whether they lack a cell wall altogether (Tenericutes), or are either thin-walled and "Gram-negative" (Gracilicutes) or thick-walled and "Gram-positive" (Firmicutes), as revealed by treatment with the classic Gram stain.[1] Gram-positive bacteria may be further subdivided according to specific genomic features.

---

[1]Gram-staining has been among the most common methods of assessing certain features of prokaryote morphology for well over a century, having been first developed in 1884 by the Danish physician Hans Christian Joachim Gram. The procedure involves staining first with crystal (or gentian) violet and then with iodine, followed by a decolorizing wash with ethanol and then a counterstaining with safranin. The violet-iodine stain is not washed out in Gram-positive bacteria primarily because of the presence of a thick layer of peptidoglycan surrounding the cell, so these appear dark purple or blue-black. Gram-negative species have a second membrane rather than the peptidoglycan layer, and so the violet-iodine stain washes out of their cells, and these appear pink or red because of the counterstaining with safranin.

According to the current understanding of prokaryotic diversity and classification, the largest and most physiologically variable taxon within the prokaryotes are the Gram-negative "proteobacteria," which are divided into five major groups labeled with the Greek letters alpha through epsilon (α, β, γ, δ, ϵ). This group contains "chemoautotrophic" forms, which are free-living and include the nitrogen-fixing bacteria found in mutualistic association with legume roots, "chemoheterotrophic" forms, most of which can employ either aerobic or anaerobic metabolism and may be pathogenic (e.g., *Salmonella enterica*) or mostly harmless (e.g., *Escherichia coli*), and "purple bacteria," which undergo a type of photosynthesis that uses an electron donor other than water (e.g., hydrogen sulphide) and therefore releases no oxygen. Based on more recent molecular sequence analyses, the proteobacteria appear to be a well-supported monophyletic group. The same is true of the cyanobacteria (once called "blue-green algae"), which are photosynthetic and were responsible for producing the oxygen-rich atmosphere upon which animal life now depends. Examples of other well-known groupings based on the traditional approach to classification include the Spirochetes, which are chemoheterotrophs that may be either free-living or parasitic, and the Chlamydiae, comprising strictly parasitic species, some of which cause disease in humans.[2]

The domain Archaea represents a diverse group of organisms consisting predominantly of extremophiles that inhabit environments characterized by extreme temperature, salinity, and/or pH. As such, archaeal species are classified into groups with names such as the Sulfolobales, Methanobacteriales, and Thermoplasmales, and are organized into functional categories such as "halophiles," "methanogens," or "thermophiles" (Forterre *et al.*, 2002) (Fig. 10.1). (The "Mendosicutes," a group previously defined by classical microbial taxonomic approaches, are now included in this domain). Not all Archaea are extremophiles, however, and some even inhabit the human body (Miller and Wolin, 1982; Belay *et al.*, 1990; Kulik *et al.*, 2001). Until recently, the Archaea were thought not to be involved in pathogenesis in humans, but recent work suggests that some species of methanogenic Archaea are associated with periodontitis (Lepp *et al.*, 2004).

The Archaea are currently categorized into three major groups—the Crenarchaeota, Euryarchaeota, and Korarchaeota. Only the first two of these have

---

[2]Useful listings of prokaryote taxonomy include the Deutsche Sammlung von Mikroorganismen und Zellkulturen (DSMZ) GmbH (www.dsmz.de), the National Center for Biotechnology Information (NCBI) Taxonomy Browser (www.ncbi.nlm.nih.gov/Taxonomy/taxonomyhome.html), the List of Bacterial Names and Standing in Nomenclature (www.bacterio.cict.fr) (Euzéby, 1997), the online-only publication *The Prokaryotes: An Evolving Electronic Resource for the Microbiological Community* (141.150.157.117:8080/prokPUB/index.htm), and *Bergey's Manual of Systematic Bacteriology* and *Bergey's Manual of Determinative Bacteriology* (see www.cme.msu.edu/bergeys).

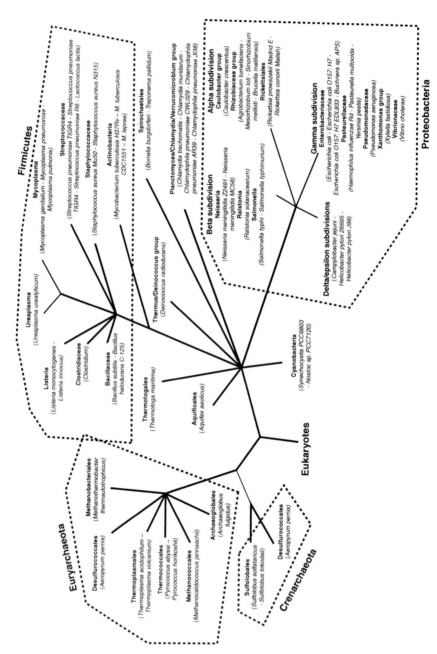

For legend, see opposite page.

been relatively well characterized in a phylogenetic context (Forterre *et al.*, 2002) (Fig. 10.1). The Crenarchaeota contain organisms that can thrive under extreme temperatures; *Pyralobus fumaris*, the species with the highest known growth temperature at 113°C, is a member of this group. The Euryarchaeota contain the methanogens and circumscribe a diverse array of species from varied and often extreme habitats (DeLong, 1992; Kjems *et al.*, 1992; Barns *et al.*, 1994). The existence of a fourth distinct group, the Nanoarchaeota, has recently been suggested, and so far consists of only one known species, *Nanoarchaeum equitans* (Huber *et al.*, 2002).

THE DEEPEST SPLIT OF ALL?

Over the past 30 years, microbial classification has been changed drastically by the use of molecular approaches. Many different techniques have been employed to distinguish different strains and species of prokaryotes, including whole genome DNA–DNA hybridization (e.g., Wayne *et al.*, 1987; Cho and Tiedje, 2001), restriction enzyme cutting patterns (e.g., Lucier and Brubaker, 1992), and phylogenetic analyses based on single gene sequences such as the ribosomal 16S subunit (16S rRNA) (e.g., Stackebrandt and Goebel, 1994) and more recently based on large numbers of protein-coding genes from genome-level studies (e.g., Palys *et al.*, 1997, 2000; Brochier *et al.*, 2002; Forterre *et al.*, 2002; Savva *et al.*, 2003). A recent proposal for the evolutionary relationships among major prokaryotic groups based on these types of analyses is shown in Figure 10.1.

One of the key realizations of such molecular studies was that classification by exclusion had indeed been misleading. Carl Woese and colleagues, using 16S rRNA gene sequences, showed that the Bacteria and Archaea (previously "Eubacteria" and "Archaebacteria"), traditionally lumped together as "prokaryotes," are actually very deeply divergent (Woese and Fox, 1977; Woese and Gupta, 1981; Woese, 1987, 1998, 2000). The most surprising result of these phylogenetic studies was that the root of the tree of life appeared to split the Archaea and the Bacteria in such a way that they do not share a common ancestor to the exclusion of eukaryotes (Iwabe *et al.*, 1989; Gogarten and Taiz, 1992) (Fig. 10.1). The topology of this deep split remains a subject of debate (e.g., Cavalier-Smith, 1998, 2002; Gupta, 1998a,b; Mayr, 1998; Charlebois, 1999), but if accurate would mean that the term "prokaryote" holds little meaning in an evolutionary sense—it is used in this chapter only for the sake of convenience, with the necessary caveats implied throughout.

FIGURE 10.1   Proposed classification and evolutionary relationships among prokaryotes based on a subset of species whose genomes have been completely sequenced. The main groups are clearly evident from this, although of course the tree is constantly growing and sprouting new branches as more genome sequence information accrues. From Saccone and Pesole (2003), reproduced by permission (© John Wiley & Sons).

THE RISE OF COMPLETE PROKARYOTIC
GENOME SEQUENCING

FROM VIRUSES TO VENTER

Although the first successes with DNA sequencing involved the analysis of individual genes, it was not long before the minds of geneticists and molecular biologists were opened to the prospect of sequencing complete genomes of viruses and independently living organisms. Indeed, the first entire genome sequence—the 5375 base pair (bp) genome of the phage φX174 ("phi X"), comprised of nine genes—was made available nearly 30 years ago (Sanger *et al.,* 1977). Even with the simplistic nature of this viral genome and the small number of genes contained within it, surprising discoveries regarding genomic structure were made as a result of this sequencing effort. In particular, this first sequenced genome showed the existence of the φX174 B gene overlapping with the A gene (Smith *et al.,* 1977). Subsequent sequencing of viral genomes has resulted in the discovery of myriad genetic and genomic novelties (Canchya *et al.,* 2003; Casjens, 2003).

These early forays into viral genome sequencing emboldened researchers to suggest that the genomes of more complex organisms could be sequenced as well. By the late 1980s, genome projects were proposed and subsequently initiated for humans and a range of model organisms, including the bacteriologists' favorite, the gut bacterium *E. coli* (see Chapter 9). In the early 1990s, DNA sequencing techniques finally came of age, and with the development of dye terminator chemistry and the invention of automated capillary sequencing machines, the efficiency of DNA sequencing was increased several fold. These high-throughput methods suggested that the completion of whole bacterial genomes was coming within reach, but such projects were stalled because of the time-consuming nature of the approach used to produce the templates for sequencing. Specifically, almost all whole genome projects at the time were being accomplished by first cloning the genomic DNA of the target organism (i.e., by inserting large fragments, from 20 kilobases (kb) to a little over 100 kb, into a bacterium that can be easily cultured) and then constructing a physical map of those large cloned fragments. These cloned fragments were then systematically subcloned into fragments of the size optimal for DNA sequencing (between 0.5 and 2 kb) (see Frangeul *et al.,* 1999, Sensen, 1999, and Fraser *et al.* 2000 for reviews of bacterial genome sequencing methods).

In the mid-1990s, Hamilton Smith and J. Craig Venter decided that what was needed was a high-throughput method of obtaining the cloned templates for sequencing, and they developed what is known today as the "shotgun cloning and sequencing" approach (see Fig. 10.2). In their technique, the cumbersome task of obtaining the large cloned fragments and a physical map were sidestepped by simply shredding the genomic DNA of an organism into fragments of specific

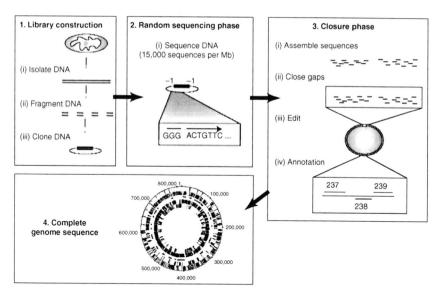

FIGURE 10.2   Diagrammatic representation of the steps involved in a whole-genome shotgun sequencing project. The advent of the shotgun approach, first used by Fleischmann *et al.* (1995) in the sequencing of the *Haemophilus influenzae* genome, greatly increased the efficiency of whole genome sequencing in prokaryotes and remains the standard method in use today. Adapted from Fraser *et al.* (2000), reproduced by permission (© Nature Publishing Group).

sizes (typically 2, 5, and 10 kb) and then cloning these fragments into sequencing vectors. This made many thousands of clones for an organism's genome ready for sequencing in very short order. The assembly of a complete genome from the sequences of these clones was then contingent upon finding overlapping ranges from clone to clone. Although this may seem a daunting task, Smith and Venter reasoned that with clever computer programming and substantial computing power, the assembly of whole genomes from these cloned sequences could be accomplished by producing an excess of sequenced clones.

In practice, it is common to sequence each base in an organism's genome six to 10 times (i.e., to generate 6x to 10x coverage), which provides enough overlapping regions to permit assembly of the fragments. Often there are parts of sequenced genomes represented 20 or more times in the overall assembly, whereas other areas are represented only a few times. Several computer programs are now available to accomplish the difficult task of assembling the random sequence fragments, including PHRED, PHRAP, and CONSED (www.phrap.org). These programs search the sequenced fragments for perfectly overlapping regions and piece them together into contiguous stretches, or "contigs." Once the contigs are constructed, and if enough coverage of the genome has been generated, the full chromosome can be assembled by joining the overlapping regions among the various contigs. If there are gaps in the assembled genome after the initial shotgun stage, these are usually filled with other rounds of shotgun sequencing or a process of "targeted PCR walking" through the gapped regions using the sequences of the assembled genome flanking the gaps.

With early versions of these methods in hand, Smith and Venter proposed in 1994 to sequence the genome of one of Smith's favorite organisms, *Haemophilus influenzae* (affectionately known as "H. flu"), which has a genome about half the size of the *E. coli* genome being sequenced by other researchers using the traditional cloning and mapping approach. The project was met with rejection at the National Institutes of Health (NIH) because reviewers felt that the shotgun method simply would not work. What the reviewers did not know was that Smith and Venter's group from The Institute for Genomic Research (TIGR) had already sequenced nearly 90% of H. flu's genome and were in the process of "closing" the other 10%. Shortly after the proposal was rejected, the complete genome sequence of H. flu was published in *Science* (Fleischmann *et al.*, 1995). In the end, a total of 26,708 sequences was generated and assembled to cover the 1,830,137 base pairs of the *H. influenzae* genome.

THE PROKARYOTE GENOME SEQUENCING EXPLOSION

It is amazing to consider that the first published genome sequence from a cellular organism came less than a decade ago (Fleischmann *et al.*, 1995). Within five years, 30 genomes had been completed, and another 90 were in the works (Fraser *et al.*, 2000; Nierman *et al.*, 2000). As of the spring of 2004, nearly 170 prokaryote genome sequences have been published, with roughly 370 additional projects underway, not counting multiple strains per species (Fig. 10.3). Of course, such summaries are out of date almost as soon as they are written, so it is fortunate that updated online databases are available for consultation. The Genomes OnLine Database (GOLD) (www.genomesonline.org) (Kyrpides, 1999; Bernal *et al.*, 2001) is the most comprehensive, and provides basic details and links to other information regarding finished and ongoing prokaryotic and eukaryotic genome projects. TIGR also maintains databases of completed genome projects (www.tigr.org/tdb/mdb/mdbcomplete.html), as does the Genome News Network (www.genomenewsnetwork.org).

As Fraser and colleagues noted recently, "the random sequencing phase of a [prokaryotic] genome sequencing project, representing more than 99% of the genome sequence, can easily be completed in just a few days at a cost of approximately 3 to 4 cents per bp" (Fraser *et al.*, 2002). The remaining closure and annotation phases generally take a few months (Fraser *et al.*, 2002)—though obviously all aspects of the sequencing process are likely to become increasingly

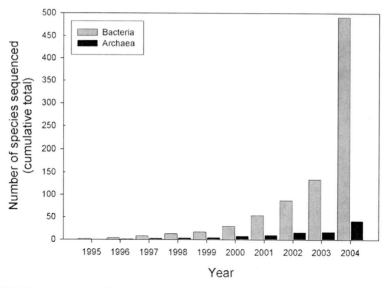

FIGURE 10.3   A summary of the number of bacterial and archaeal genome sequencing projects carried out between 1995 and 2004, the first decade of the complete genome sequencing era. The value for 2004 includes both published and ongoing projects as of the spring of 2004. Data were tabulated from the list provided in the Genomes OnLine Database (GOLD) (www.genomesonline.org). Only one strain and/or sequencing project was included per species, though in many cases multiple independent projects have been carried out for the same species.

efficient in the future. In total, a completely sequenced and assembled microbial genome sequence may cost 8 or 9 cents per bp at present, an order of magnitude cheaper than the initial cost of the *H. influenzae* project less than 10 years ago. This increasing cost-effectiveness of large-scale sequencing helps to explain the accelerating pace with which new projects are being initiated (Fig. 10.3).

GENERAL INSIGHTS ABOUT PROKARYOTE GENOMES

The ability to perform broad comparisons is one of the most important benefits of having such a large (and growing) dataset of prokaryote genome sequences. These in turn allow generalizations about prokaryotes and their genomes to be formulated (or rejected, as the case may be). The following sections provide a review of some of the most notable general insights that have emerged from the study of a large number of prokaryote genomes. Some specific findings from individual species of particular interest are provided later in the chapter.

GENOME ORGANIZATION: ASSUMPTIONS AND EXCEPTIONS

The traditional description of prokaryotic genomes is that they are very small, are contained in a single circular chromosome comprising only one replicon, and are found in one copy per cell. In many cases, this basic view is accurate, as for example with the classic image of the circular genome of *Escherichia coli* shown in Figure 10.4. In the relatively small number of Archaea studied to date, these assumptions do appear to hold (Saccone and Pesole, 2003). However, as with many aspects of prokaryotic biology, the increased breadth of taxonomic sampling in genomic analyses has revealed some intriguing exceptions to all of these assumptions among the Bacteria (Casjens, 1998; Cole and Saint-Girons, 1999; Bendich and Drlica, 2000).

The topic of prokaryotic genome size is discussed in some detail in a later section, but it bears mentioning here that though they do indeed tend to be small relative to most eukaryotes, there is nevertheless some overlap between certain

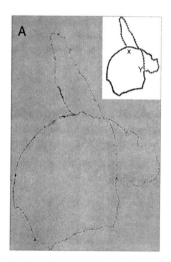 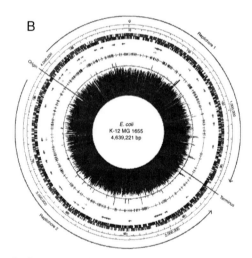

FIGURE 10.4 (A) The classic autoradiograph of an intact replicating chromosome of the gut bacterium *Escherichia coli* presented by Cairns in 1963 (reproduced with modifications to improve the clarity). A diagram of the replicating circular bacterial chromosome (with two replication forks, X and Y) is shown in the inset. This image is often cited to illustrate the circular nature of prokaryotic chromosomes, although examples of linear chromosomes have since been identified in several bacteria (Casjens, 1998; Bendich and Drlica, 2000; Bendich, 2001). Reproduced by permission (© Cold Spring Harbor Laboratory Press). (B) The complete circular genome map of the K12-MG1655 non-pathogenic laboratory strain of *E. coli*. Modified from Blattner *et al.* (1997), reproduced by permission (© American Association for the Advancement of Science).

bacteria and some parasitic protists with highly reduced genomes (Bendich and Drlica, 2000) (see Chapter 1). With regard to genome structure, the exceptions are even more pronounced. For instance, linear chromosomes are found in all species so far examined in the genera *Borrelia* (Spirochetes) and *Streptomyces* (Actinomycetes) (Chen, 1996; Casjens, 1998). In the latter case, this may involve a natural interchange between linear and circular geometry (Volff and Altenbuchner, 1998). Close relatives of both *Borrelia* and *Streptomyces* exhibit the more typical circular configuration, and the dispersed distribution of chromosome linearity on the bacterial phylogenetic tree is indicative of multiple evolutionary origins of the linear geometry (Casjens, 1998). Some proteobacteria also have been suggested to possess linear replicons as part of their genomes, as for example with *Agrobacterium tumefaciens,* which has been reported to have both a 3000 kb circular and a 2100 kb linear replicon. Small linear elements have also been noted to occur rarely in the genomes of *Klebsiella, Thiobacillus,* and even *Escherichia* (Casjens, 1998).

The example of *Agrobacterium* contradicts yet another common notion regarding prokaryotic genomes, namely that they are contained within a single chromosome. In fact, a great many bacteria possess secondary replicon structures (plasmids) in addition to their primary chromosomes. Although plasmids have often been defined as "extrachromosomal" material, in numerous cases these are large and contain important housekeeping genes, indicating that they probably warrant classification as chromosomes in their own right (Casjens, 1998; Bendich and Drlica, 2000). This applies to Gram-negative proteobacteria such as *Agrobacterium, Burkholderia, Brucella, Rhizobium,* and *Rhodobacter,* the Gram-positive firmicute *Bacillus thuringiensis,* and the spirochete *Leptospira* (Casjens, 1998). In *Deinococcus radiodurans,* there are four genetic elements, including two main circular chromosomes, a megaplasmid, and a smaller plasmid. The genomic organization of the Lyme disease spirochete *Borrelia burgdorferi* is perhaps the most exceptional of all, consisting of 12 linear chromosomes/plasmids and nine circular ones ranging from 5 to 54 kb in length (Casjens, 1998; Casjens *et al.,* 2000).

Finally, the assumption that all prokaryotes possess only one copy of the genome per cell (i.e., are haploid) has also been shown not to hold across all species (Casjens, 1998; Bendich and Drlica, 2000). As Casjens (1998) pointed out, this is an oversimplification, given that during periods of exponential growth, "bacteria, especially fast-growing bacteria, contain on average four or more times as many copies of sequences near the origin as near the terminus of replication." Moreover, in species such as *Azotobacter vinelandii* and *Borrelia hermsii,* several complete copies of the genome may be present (Casjens, 1998). In *Deinococus radiodurans,* four copies of its genome appear to be present even during the stationary growth phase. Homologous recombination among these copies can act to regenerate an intact chromosome following severe damage, a fact that may help to explain this species' extraordinary tolerance for radiation (Casjens, 1998).

Individuals of *Buchnera aphidicola* typically contain more than 100 copies of the genome per cell (Komaki and Ishikawa, 1999; Shigenobu *et al.*, 2000). Some specimens of the giant endosymbiotic bacterium *Epulopiscium fishelsoni* (which lives in the gut of the surgeonfish *Acanthurus nigrofuscus*) recently have been reported to be very highly polyploid, apparently containing a staggering 150 times more DNA than a diploid human lymphocyte (Bresler and Fishelson, 2003). An exception to the assumption of universal haploidy is also found in at least one species of Archaea, namely *Methanocaldococcus* (formerly *Methanococcus*) *jannaschii*, which contains from three to more than 10 copies of its genome per cell during exponential growth, and one to five during stationary growth (Bernander, 2000).

STRUCTURE OF PROKARYOTIC CHROMOSOMAL DNA

Although many distinctions between bacterial, archaeal, and eukaryotic genomes that were once thought absolute have since become blurred, there are clearly some important differences among these groups. This goes beyond the notion that eukaryote genomes are linear and contained within a membrane-bound nucleus whereas those of prokaryotes are (mostly) circular and interact with the cytoplasm (albeit while still forming an observable "nucleoid"). At a finer resolution, it is apparent that differences can generally be found in the structure of the DNA molecules at a level below the individual chromosome in these three major taxa.

In (most) eukaryotes, the highly compact structure of chromosomal DNA is achieved primarily by wrapping the DNA molecule around a series of histone proteins to form "nucleosomes." Bacteria lack histones, and thus their genome is not structured into nucleosomes. Nevertheless, the level of compaction of bacterial DNA (1000- to 10,000-fold; i.e., to fit ~ 1 mm of DNA into a ~ 1 μm prokaryotic cell) is reasonably comparable to the extensive compaction characteristic of eukaryotic chromatin (White and Bell, 2002; Sherratt, 2003). In part, this is accomplished by the action of negative supercoiling and structural maintenance of chromosomes (SMC) proteins, which are found in all three branches of cellular life (Sherratt, 2003; Case *et al.*, 2004). Bacteria also possess several small basic proteins (e.g., HU and FIS) that can compact DNA to varying degrees (White and Bell, 2002).

Interestingly, some members of the Archaea do possess histones, and indeed "remarkable parallels can be drawn between the structure and modification of chromatin components in the archaeal and the eukaryotic domains of life" (White and Bell, 2002). Specifically, histone homologs have been found in many of the Euryarchaeota sequenced thus far, but not in any of the Crenarchaeota (White and Bell, 2002). When they occur, the histones of archaea are shorter than those of eukaryotes because they lack the N- and C-terminal tails that play an important role in the regulation of compaction and accessibility of nucleosomes in eukaryotes (White and Bell, 2002). Several other DNA-binding proteins can be found in the Archaea that may play a role in regulating DNA compaction (reviewed in White and Bell, 2002).

DNA REPLICATION IN PROKARYOTES

Whereas much of what is known about bacterial chromosome structure, organization, and replication has traditionally come from the study of a handful of species (e.g., *E. coli, Bacillus subtilis,* and *Caulobacter crescentus*), the availability of a wide range of genome sequences has made it clear that "all bacteria use the same principles for chromosome organization and segregation, replication, repair and recombination" (Sherratt, 2003). Information is also emerging regarding DNA replication in the Archaea (see Bernander, 2000), which appears to involve a single replicon origin as in the Bacteria, even though the proteins involved in archaeal replication are most similar to those of eukaryotes (Vas and Leatherwood, 2000).

It has typically been assumed that only eukaryotic cell division involves mitosis, commonly understood to involve the condensation, alignment at a metaphase plate, and movement of chromosomes to the poles along mitotic spindles. However, it turns out that these features of cell division are not ubiquitous among eukaryotes, and that for each one there are examples of taxa in which it does not occur (Bendich and Drlica, 2000). More importantly, some mitosis-like phenomena have now been documented in certain bacterial species (Bendich and Drlica, 2000).

One interesting result of the detailed examination of bacterial chromosome replication is that a single origin of replication (Messer, 2002) and termination (Postow *et al.,* 2001) exist in conserved positions in bacterial species. Even among bacterial species with linear chromosomes, the mechanisms of replication initiation, termination, and chromosome location during replication appear to be conserved to a certain degree (Kobryn and Chaconas, 2001). This conservation of sites is suggestive of the interesting possibility that bacterial initiation and termination sites are the functional equivalent of eukaryotic centromeres (Sherratt, 2003). The physical positioning of the bacterial chromosomes with the initiation site away from the division plane is also reminiscent of eukaryotic chromosomal positioning during cell division.

GENE CONTENT

Only a small fraction of many eukaryotic genomes is composed of protein-coding genes (e.g., less than 2% of the human genome), whereas the genic portion of both archaeal and bacterial genomes is typically around 85–95% (Saccone and Pesole, 2003). As a result, the study of prokaryotic genome evolution predominantly involves the analysis of gene sequences, rather than transposable elements (see Chapter 3), pseudogenes (see Chapter 5), introns, and the other types of noncoding DNA that make up the majority of eukaryotic genomes (see Chapters 1 and 2).

Determining the content and characteristics of genes in fully sequenced prokaryote genomes can reveal much about their biology and evolution. The first step in this process lies in locating potential genes, a task that is facilitated in prokaryotes by the general lack of introns. One method of finding genes is to search the sequence for sufficiently lengthy open reading frames (ORFs), which represent sequences located in between start and stop codons. However, this does not provide any insight into the function of the putative gene, and a more reliable and informative technique is to search the genome sequence for homologs of genes from other species (i.e, orthologs) that have previously been identified and characterized. Sequences with a relatively high similarity to known genes/proteins can be located in sequence databases using algorithms such as the Basic Local Alignment Search Tool (BLAST) (www.ncbi.nlm.nih.gov/BLAST) (see Pertsemlidis and Fondon, 2001, for review) or FASTA (www.ebi.ac.uk/fasta33). If these methods do not reveal sufficient similarity with known sequences, then it is still possible to search for protein signatures specific to one or more protein families using the ProSite, ProDom, Pfam, and TIGRfam databases (see Saccone and Pesole, 2003, for a detailed overview of database search methods). When no similarity can be found to any known proteins with these searches, it is necessary to use computational methods based on compositional heterogeneity at the three codon positions or a measure of "linguistic complexity" (which is higher in coding versus noncoding regions) to identify potential genes (see Saccone and Pesole, 2003). In total, these identification methods have already revealed nearly 450,000 genes in the various prokaryote genomes that have been sequenced over the past 10 years (Fig. 10.5).

Thus far, the genomes of Bacteria have been found to contain as few as 480 genes in *Mycoplasma genitalium* to as many as 8317 in *Bradyrhizobium japonicum,* with an average of about 3100 per genome. Variation in gene number among bacteria typically reflects differences in lifestyle, with the lowest gene contents found in specialized parasites and symbionts and the largest in species with more complicated life cycles (e.g., including a sporulation phase or living in complex association with legumes) (Saccone and Pesole, 2003). In Archaea, both the average number of genes per genome (~ 2220) and the overall range, from 563 in *Nanoarchaeum equitans* to 4540 in *Methanosarcina acetivorans*, are smaller than in Bacteria.

Roughly 15% of the genes identified in both Archaea and Bacteria remain unclassified or of unknown function, and the largest fraction in both groups remains genes coding for uncharacterized hypothetical proteins that are not conserved among species, followed by those coding for conserved hypothetical proteins (see Fig. 10.5). In general, specific functional identifications of genes have proceeded further in bacterial than in archaeal species (Saccone and Pesole, 2003) (Fig. 10.5). Those genes that have been characterized can be assigned to

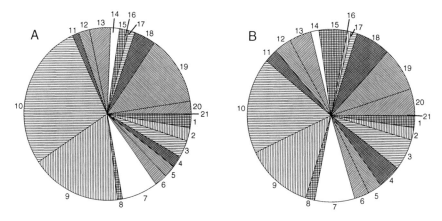

Category	Gene role	A) Archaea Genes	%	B) Bacteria Genes	%
1	Amino acid biosynthesis	1172	2.52	8592	2.15
2	Biosynthesis of cofactors, prosthetic groups, and carriers	1154	2.49	9340	2.34
3	Cell envelope	1516	3.26	23,423	5.87
4	Cellular processes	1006	2.17	18,094	4.53
5	Central intermediary metabolism	1048	2.26	10,795	2.70
6	DNA metabolism	1115	2.40	11,030	2.76
7	Energy metabolism	3193	6.88	30,789	7.71
8	Fatty acid and phospholipid metabolism	465	1.00	6963	1.74
9	Hypothetical proteins (conserved)	8078	17.40	51,595	12.92
10	Hypothetical proteins (not conserved)	12,859	27.69	72,770	18.22
11	Other categories	604	1.30	10,379	2.60
12	Protein fate	991	2.13	12,797	3.20
13	Protein synthesis	1915	4.12	17,009	4.26
14	Purines, pyrimidines, nucleosides, and nucleotides	779	1.68	6233	1.56
15	Regulatory functions	829	1.79	22,048	5.52
16	Signal transduction	1	<0.01	2845	0.71
17	Transcription	556	1.20	4476	1.12
18	Transport and binding proteins	1934	4.16	26,745	6.70
19	Unclassified	6031	12.99	32,426	8.12
20	Unknown function	1132	2.44	20,450	5.12
21	Viral functions	59	0.13	568	0.14

FIGURE 10.5   Classification of identified gene regions by role category based on combined data from complete genome sequences from (A) Archaea (16 species; 46,437 genes) and (B) Bacteria (106 species; 399,367 genes). Data were compiled from the TIGR Comprehensive Microbial Resource (www.tigr.org/tigr-scripts/CMR2/CMRHomePage.spl) in the spring of 2004. Only one strain per species was included. Disrupted reading frames and the very small number of pathogen response genes found in some species were omitted. Updates and data for each individual species are available online as part of the database.

one of 15 specific functional categories, relating primarily to amino acid and protein synthesis and transport, various metabolic pathways, and other functions involving cell maintenance and growth. As shown in Figure 10.5, the proportions in these categories vary slightly between Archaea and Bacteria, although the overall distributions are similar. Of course, the genic makeup of species within these two groups may vary considerably according to differences in biology. Information regarding the patterns of gene content of each individual sequenced species is available from the TIGR Comprehensive Microbial Resource (www.tigr.org/tigr-scripts/CMR2/CMRHomePage.spl) (Peterson *et al.*, 2001).

Comparisons among the first 16 genome sequences from species of Archaea have indicated the presence of a "gene core" of 313 conserved sequences common to all of the species examined to date, along with a "variable shell" that is more susceptible to lineage-specific gene loss or movement among genomes (Makarova *et al.*, 1999; Makarova and Koonin, 2003). Most of the genes in the archaeal gene core are universal to all three domains of life (i.e., orthologs can be found in both bacteria and eukaryotes as well). However, 16 are exclusive to the Archaea, and 61 are restricted to the Archaea-Eukaryota grouping; most of these appear to function in an "information processing" capacity, especially relating to translation and RNA processing (Makarova and Koonin, 2003).

Genes shared among prokaryote genomes by virtue of common evolutionary descent are frequently evaluated in terms of "clustered orthologous groups" (COGs), in which each COG consists of individual orthologous genes, or sets of duplicated genes (paralogs) that as a group are orthologous, from three or more phylogenetic lineages (Tatusov *et al.*, 1997). It has been argued that this is superior to evaluating individual shared genes in phylogenetic analyses, especially across large evolutionary distances, because duplication (i.e., the presence of paralogs instead of orthologs) can be a confounding factor unless a large number of taxa are evaluated simultaneously (Tatusov *et al.*, 1997). (On the other hand, this method has some shortcomings in that it is based only on similarity and does not include phylogenetic information, which itself can lead to confusion regarding paralogy versus orthology.) In the first application of this approach, Tatusov *et al.* (1997) compared the complete genome sequences from seven species representing five major lineages (including members of three bacterial taxa, one archaeon, and one unicellular eukaryote) and identified 720 COGs shared by at least three of the seven species. Since this initial analysis, an entire database has been assembled, which now contains COGs for more than 60 prokaryotes and several eukaryotes (www.ncbi.nlm.nih.gov/COG) (Tatusov *et al.*, 2001, 2003). The dataset includes almost 5000 COGs, covering roughly 140,000 protein-coding genes, with each of the prokaryotic species analyzed being included in about 360 to 2250 individual COGs (Tatusov *et al.*, 2003). This dataset is expected to be of considerable utility in the annotation and functional analysis of newly sequenced genomes (Tatusov *et al.*, 2001, 2003). It has also proved useful in the reconstruction of genome phylogenies, as for example using the online Shared Ortholog and Gene Order Tree Reconstruction Tool (SHOT) (www.bork. embl-heidelberg.de/~korbel/SHOT_v2) (Korbel *et al.*, 2002).

GENE ORDER: PLASTICITY AND STABILITY

In general, the genomes of prokaryotes are very dynamic, with insertions, deletions, inversions, and translocations commonly observed among related species or even between different strains of the same species (Hughes, 2000). The net result is that the particular complement of genes and their order along the chromosome are not typically conserved over evolutionary time (Fig. 10.6A). In some cases, genes that are grouped into operons in one species may be dispersed throughout the genome in others, with only a few exceptions (e.g., ribosomal protein operons) remaining fairly consistent (Saccone and Pesole, 2003). There are nevertheless some common themes within this system of genome plasticity, such as the tendency for large inversions to be symmetric around the axis of replication (see Hughes, 2000), presumably so there is no collision between transcription and replication along the (usually) single circular chromosome (Casjens, 1998). These large symmetric inversions around the origin and terminus of replication are common in bacteria, generating so-called "X-alignments" when viewed on gene position plots (see Eisen *et al.*, 2000) (Fig. 10.6B,C). Although rearrangements within species tend to be less common than among species, these increase greatly in frequency in pathogens and appear to be associated with the ability to infect eukaryotes, perhaps reflecting a mechanism for evading host immune defenses (Hughes, 2000).

In bacteria, chromosome breaks are repaired through the action of the RecA protein, which uses a duplicate copy of the damaged sequence as a template for repair. Under normal conditions, the homologous sequence on a sister chromosome is used for this purpose, but on occasion RecA can promote recombination among gene duplicates (paralogs), which can result in a rearrangement of gene order along the chromosome or the duplication or deletion of intervening sequences (Hughes, 2000). *Buchnera aphidicola,* an obligate mutualistic endosymbiont of aphids, lacks the *recA* gene and appears to compensate for the resulting inability to repair chromosome breakage by maintaining more than 100 copies of the genome in each cell (Hughes, 2000). *B. aphidicola* is of particular interest in the present context because it represents an example of extreme genomic stasis. In particular, a comparison of the complete genome sequences of two long-separated strains inhabiting different host species revealed that, despite a large amount of sequence divergence between the two strains, no chromosome

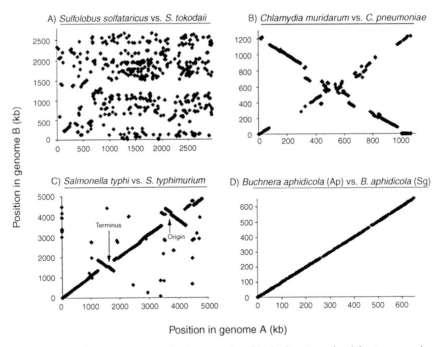

FIGURE 10.6   Gene position plots showing examples of both plasticity and stability in gene order between closely related species of prokaryotes. In these plots, the location of a given gene, measured as its distance from a given starting point in kilobases (kb), is plotted on one axis each for the two species being compared. Unless otherwise indicated, the origin of the axes represents the origin of replication in the chromosomes. (A) The archaeons *Sulfolobus solfataricus* and *S. tokodaii*, whose genomes share very little common gene order and are clearly extremely dynamic. (B) The bacteria *Chlamydia muridarum* and *C. pneumoniae*, which exhibit a clear "X-alignment," indicating a single, large, symmetrical inversion around the origin of replication (see also Eisen *et al.*, 2000; Hughes, 2000). (C) The bacteria *Salmonella typhi* and *S. typhimurium*, which show evidence of two smaller symmetrical inversions, one around the origin of replication and one around the replication terminus. (D) Two strains (or possibly species) of the endosymbiotic bacterium *Buchnera aphidicola* living in distantly related aphid hosts (Ap = *Acyrthosiphon pisum*; Sg = *Schizaphis graminum*). In this case, there has been remarkable stasis in gene order for 50–70 million years, despite considerable sequence divergence (see Tamas *et al.*, 2002). Based on a figure presented by Mira *et al.* (2002), reproduced by permission (© Elsevier Inc.).

rearrangements or gene acquisitions, and only a few gene losses, have occurred in this species over a period of 50–70 million years (Fig. 10.6D). This makes the genome of *B. aphidicola* more than 2000 times less dynamic than those of *E. coli* and *Salmonella* spp., its closest free-living relatives (Tamas *et al.*, 2002). Similarly stable genomes are found in *Corynebacterium* species, which also lack certain recombinational DNA repair genes (Nakamura *et al.*, 2003).

BASE PAIR COMPOSITION

As discussed in Chapter 9, the genomes of eukaryotes often differ substantially in composition with regard to the total percentage of nucleotides consisting of G+C (or, conversely, A+T) pair bonds. Even within a eukaryotic genome there may be particular regions of high or low G+C content (also written as "GC content") known as "isochores." There is also tremendous variation in overall G+C content among the genomes of the Bacteria, ranging from a low of 22.4% in *Wigglesworthia glossinidia brevipalpis*[3] to 72.1% in *Streptomyces coelicolor* (Fig. 10.7). The average for 125 bacterial genomes so far available is roughly 46.8% G+C. In some cases, this variation in G+C content has proved useful for taxonomic purposes, for example in further subdividing the Gram-positive bacteria into low- and high-GC groups (e.g., Nelson *et al.*, 2000). In the Archaea, as with gene numbers, both the average (44.3%) and range (31.4% in *Methanocaldococcus jannaschii* to 67.9% in *Halobacterium* sp.) in G + C content are lower than in the Bacteria.

Overall, G+C content is positively correlated with genome size in the Bacteria, but not in the Archaea (Moran, 2002; Wernegreen, 2002) (Fig. 10.8). In eukaryotes, genome size and G+C content are associated in a triangular relationship, whereby G+C content is rather variable in smaller genomes but stabilizes at about 46% in the larger genomes of salamanders (Vinogradov, 1998) (see Chapter 1). Some prokaryotes show evidence of within-genome regions of low or high G+C content (though this is not necessarily the same phenomenon as eukaryote isochores), which may arise by several mechanisms, including chance mutational biases, selection (e.g., for the higher chemical stability of GC bonds), and insertion of foreign genetic elements (Liò, 2002; Daubin and Perrière, 2003; Saccone and Pesole, 2003) (Fig. 10.7). Position in the genome may also exert an important influence on the resulting G+C content of a given region. As one example of this, in many bacteria there is an enrichment of A+T content near the replication terminus (Daubin and Perrière, 2003).

While "G+C content" provides information regarding the overall number of GC base pairs across the two strands in a DNA molecule, it does not specify whether there are more Gs or Cs in one strand or the other. This second compositional character is expressed by the parameter known as "GC skew," which is used to indicate any biases in the relative contribution of G or C within an individual DNA strand. This is calculated as GC skew = $(G - C) \div (G + C)$, using the values for G and C amount within a single strand rather than the entire double-stranded

---

[3]An average G+C content of only 19.9% has been reported for *Carsonella ruddii*, an obligate intracellular γ-proteobacterial endosymbiont of psyllid insects, but this estimate was based on the sequencing of a small portion of the genome (Clark *et al.*, 2001).

A) *Streptomyces coelicolor*

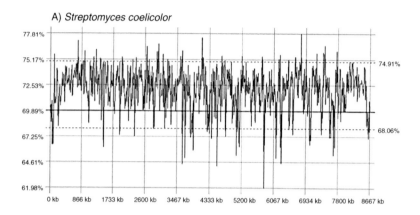

B) *Escherichia coli*

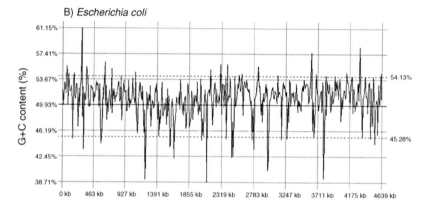

C) *Wigglesworthia glossinidia brevipalpis*

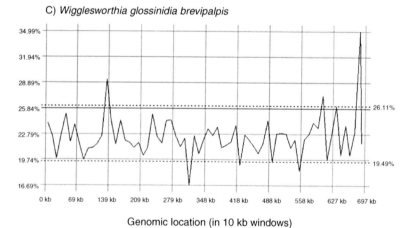

Genomic location (in 10 kb windows)

For legend, see opposite page.

DNA molecule (AT skew can be calculated by substituting A and T into this formula). A GC skew of zero indicates that there are no biases, and that G and C bases are present in equal proportions within each strand.

Very few exceptions are known from prokaryotes in which the genome as a whole exhibits significant GC skew (e.g., *Xylella fastidiosa*) (Saccone and Pesole, 2003). On the other hand, certain regions of the genome frequently exhibit GC skew, especially the origin and terminus of replication. Indeed, GC skew is often used to identify these regions in a newly sequenced genome. Because the origin of replication is commonly labeled as the "first" base pair in a circular prokaryote chromosome, this subsequently affects the labeling of all other features of interest along the genome.

REPEAT CONTENT

Repetitive elements have long been well known to represent a major fraction of eukaryotic genomes, but repetition in the genomes of prokaryotes has only become recognized fairly recently. Several different categories of repeats have been characterized in bacterial genomes, with considerable variation according to size (from a few bp to several kb), arrangement (direct or inverted), and location (clustered or dispersed). Prominent examples include various families of palindromic repeats (e.g., Bachellier *et al.*, 1999), tandem repeats (e.g., Yeramian and Buc, 1999), more widely dispersed gene-sized repeats (e.g., rRNA or *rrn* operons, *Rhs* elements) (Hill, 1999), duplicated genes (e.g., Jordan *et al.*, 2001; Gevers *et al.*, 2004) (see Chapter 5), and even long strict repeats (e.g., Rocha *et al.*, 1999). Repeated elements play a role in packaging DNA into nucleosomes in the Archaea, as they do in eukaryotes (e.g., Bailey and Reeve, 1999), and may also increase the rate at which gene order is lost among species (Mira *et al.*, 2002; Rocha, 2003). Inverted repeats tend to be associated with genomic instability caused by intrachromosomal homologous recombination in bacteria, and hence are rarer than direct repeats, especially in highly stable genomes (Achaz *et al.*, 2003).

By constructing detailed "genome atlases," it has been possible to provide a general visual overview of base pair composition, repeat structure, and structural

---

FIGURE 10.7   G+C content displays for three bacterial species whose genomes differ greatly in base pair composition. (A) *Streptomyces coelicolor*, which has the highest average G + C content known among bacteria at 72.1%. (B) *Escherichia coli* (K12-MG1655 strain), which has a G + C content of 50.7% (i.e., near the bacterial average of 46.8%). (C) *Wigglesworthia glossinidia brevipalpis*, which has the lowest G+C content among sequenced bacteria at 22.4%. The plot shows the G + C% in each 10 kb window at the genomic locations indicated on the X-axis (note the differences in the values on the three Y- and X-axes, reflecting the substantial divergence in G+C% and genome size among these three species). The solid central line on each plot represents the median G+C content of the genome, and the dashed lines represent the 5% lower and 95% upper limits (such that 90% of the genome falls in between these lines). Figures were generated using the GC plot feature of the TIGR Comprehensive Microbial Resource (www.tigr.org/tigr-scripts/CMR2/CMRHomePage.spl).

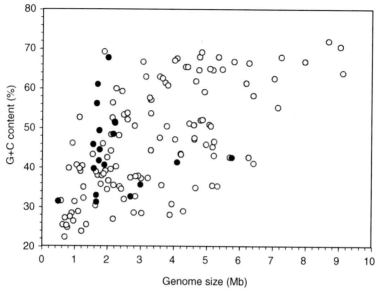

FIGURE 10.8   The relationship between G+C content and genome size in prokaryotes, as revealed by 143 completely sequenced genomes. There is a significant positive correlation in Bacteria (○) but not in Archaea (●). The regression statistics were as follows, Archaea: $r^2 = 0.001$, $P = 0.90$, $n = 18$; Bacteria: $r^2 = 0.38$, $P < 0.0001$, $n = 125$; all prokaryotes: $r^2 = 0.34$, $P < 0.0001$, $n = 143$. The regressions were very slightly stronger following log-transformation, but not substantially different. Data were taken from the Center for Biological Sequence Analysis (CBS) Genome Atlas Database (www.cbs.dtu.dk/services/GenomeAtlas/) in the spring of 2004.

parameters of sequenced genomes (Jensen *et al.*, 1999; Hallin and Ussery, 2004). To construct a genome atlas, each of these three general categories of features is visualized in the form of color-coded circles, which are then combined into a single figure of concentric colored maps showing regions of high and low G+C content, repetitive sequence locations, and so on. Atlases for the various sequenced genomes, along with basic information regarding genome size, gene number, and base pair composition, are available as part of the Center for Biological Sequence Analysis (CBS) Genome Atlas Database (www.cbs.dtu.dk/services/GenomeAtlas).

HORIZONTAL TRANSFER OF GENETIC MATERIAL

The inheritance of genetic information is usually considered as being "vertical"— that is, from parent down to offspring, with the result that evolution is also considered an essentially vertical process of "descent with modification." However, in prokaryotes in particular there is growing evidence for the movement of genetic material from one unrelated genome to another, including both "recombination" between conspecifics and "horizontal (or lateral) gene transfer" (HGT) among different strains and/or species. Overall, the frequency, importance, and evolutionary implications of HGT remain subjects of vigorous debate (e.g., Eisen, 2000a; Gogarten et al., 2002; Jain et al., 2002; Kurland et al., 2003; Lawrence and Hendrickson, 2003), but at the very least it is clear that numerous mechanisms of transfer operate in prokaryotes, and that gene exchange can and does occur both within prokaryote species and across higher taxonomic boundaries.

IDENTIFYING AND CHARACTERIZING HORIZONTAL TRANSFERS

Several criteria can be used to identify and classify horizontal transfer events. A common first step involves finding an unexpected strong similarity in gene or protein sequences (e.g., by BLAST search) between two unrelated species. The origin of this similarity by HGT then can be evaluated by examining the topology of a phylogenetic tree to determine if the genes from the two unrelated taxa group together, or if the anomalous gene is highly restricted in distribution and is not shared by close relatives (Eisen, 2000a; Ochman et al., 2000; Koonin et al., 2001; Gogarten et al., 2002; Brown, 2003). Other potential identifiers include the conservation of gene order in regions of the genome between unrelated taxa, which may imply the transfer of entire functional clusters (operons), or by noting anomalous $G + C$ contents or codon usage biases in certain genes that do not match the composition of the rest of the genome (Eisen, 2000a; Ochman et al., 2000; Koonin et al., 2001; Gogarten et al., 2002; Brown, 2003; Philippe and Douady, 2003). These identification methods make different assumptions and may emphasize different sets of potentially transferred genes (Ragan, 2001; Lawrence and Ochman, 2002).

Identifying HGT events can be challenging, but it is even more difficult to determine the direction in which genetic material was transferred. For this, it is usually necessary to obtain some estimate of how common the gene in question is among the two species and their relatives. The assumption is that the gene is likely to have moved from a clade in which it is common to one in which it remains rare (Koonin et al., 2001; Brown, 2003), although obviously this is neither easy to gauge nor conclusive. In a more recent approach, Planet et al. (2003) used "tree reconciliation," which reconstructs the coevolution of parasites and hosts or genes and organisms, to trace the complex history of horizontal transfers, duplications, and losses of the tad ("tight adherence") locus in the human pathogen Actinobacillus actinomycetemcomitans and other Bacteria and Archaea. This not only provides the most parsimonious solution to the question of direction of transfer, but also reveals patterns that can mimic horizontal transfer, such as duplication with subsequent differential loss.

Finally, the nature of an HGT event can be characterized by cross-species comparisons in terms of the relationship between the horizontally acquired gene and its preexisting homolog (if any) in the recipient's genome. Thus an HGT event may involve the acquisition of an entirely new gene, the acquisition of a distantly related paralog (duplicate copy) of an existing gene, or the replacement of an existing gene by a distant homolog (Koonin *et al.*, 2001).

TRANSFER FROM VIRUSES TO BACTERIA: PROPHAGES IN BACTERIAL GENOMES

Just as the bodies (and sometimes the very cells) of eukaryotes play host to a wide range of bacteria, prokaryotes are often subject to infiltration by viruses (bacteriophages) that exploit their cellular machinery to make copies of themselves. "Lytic" phages are destructive viruses that infect the bacterial host and program the synthesis of viral progeny, which then are released from the dead host cell (Casjens, 2003). "Temperate" phages, by contrast, exist in a stable relationship with the host bacteria, simply integrating their genetic material into the host prokaryote's genome (usually by physical incorporation into the bacterial chromosome, but in some instances as separate plasmids) to be replicated along with the host's DNA, largely without causing harm to the cell (Casjens, 2003). As part of this process, a significant amount of phage DNA (called the "prophage") may become permanently integrated into bacterial genomes.

The first complete genome ever sequenced was from a phage (φX174), and since then phages have remained the subject of a substantial sequencing effort, with hundreds of viral genomes now sequenced[4] and thousands expected within a few years (Brüssow and Hendrix, 2002; Casjens, 2003). However, from the present perspective their relevance is based on the remarkable abundance and diversity of prophages found in bacterial genomes. Of the first 82 bacterial genome sequences completed, 51 (that is, nearly two thirds) were found to contain one or more (up to 20) recognizable representatives of more than 230 different kinds of prophages (Casjens, 2003). About 12% of the genomes of some strains of *Streptococcus pyogenes* and 16% of that in pathogenic *E. coli* are composed of prophages (Banks *et al.*, 2002; Canchaya *et al.*, 2003) (Fig. 10.9). In *Borrelia burgdorferi*, prophages may constitute as much as 20% of the total sequence (Casjens, 2003). Once integrated into the bacterial genome, prophages rarely retain the ability to function as phages and tend to undergo

---

[4]Information regarding viral genome sequences can be obtained from the European Molecular Biology Laboratory's European Bioinformatics Institute database (EMBL-EBI) (www.ebi.ac.uk/genomes), the NCBI Taxonomy Browser (www.ncbi.nlm.nih.gov/Taxonomy/taxonomyhome.html), and the CBS Genome Atlas Database (www.cbs.dtu.dk/services/GenomeAtlas).

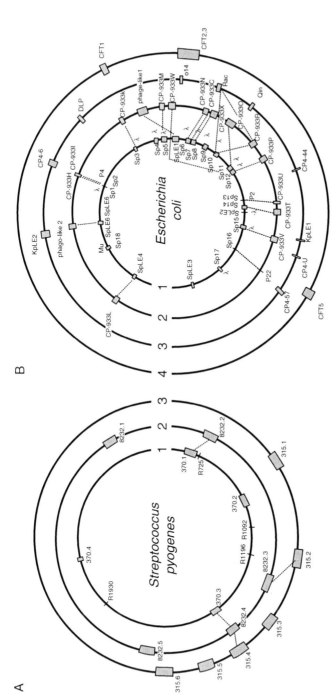

FIGURE 10.9 Prophage content and locations (gray boxes) in several strains of two species of bacteria. (A) *Streptococcus pyogenes*, also known as "group A *Streptococcus*" (GAS), which causes a wide range of infections. The numbered rings represent the genomes of three different serotypes: (1) M1, (2) M18, (3) M3. (B) *Escherichia coli*, a normally benign gut bacterium that includes some enterohemorrhagic and uropathogenic strains. The numbered rings represent the genomes of four different strains: (1) O157:H7 VT2-Sakai, (2) O157:H7 EDL933, (3) K12-MG1655, (4) CFT073. Prophages account for about 12% and 16% of the *S. pyogenes* and pathogenic *E. coli* genomes, respectively (Canchaya et al., 2003). Note that the circumferences of these schematic circular drawings are not to scale and therefore do not reflect the real relative lengths of the chromosomes depicted. Adapted from Canchaya et al. (2003), reproduced by permission (© American Society for Microbiology).

a complex mutational decay process. However, some of the genes they encode remain functional in the bacterial chromosome (Casjens, 2003).

TRANSFER AMONG BACTERIA

From a medical standpoint, gene transfer among bacteria is of significant concern, because this can involve the movement of individual antibiotic resistance genes as well as entire clusters of genes involved in virulence ("pathogenicity islands"). Indeed, antibiotic resistance has been suggested to most commonly evolve by the acquisition of foreign DNA, rather than by the mutation of existing genes (Rowe-Magnus and Mazel, 1999; Rowe-Magnus et al., 1999; but see Kurland et al., 2003). The potential movement of genes from bioengineered strains of bacteria into wild strains or unrelated species is likewise an important consideration (Brown, 2003). The transfer of DNA among bacteria is facilitated by specialized regions within the genome known as "integrons" or "gene capture systems," which act to incorporate foreign open reading frames and convert them to functional genes (see Rowe-Magnus et al., 1999, for review). The practical challenges involved in HGT are further exacerbated by the fact that several different mechanisms and sequence types are involved in the transfer of DNA sequences among bacteria (Dutta and Pan, 2002).

Transformation: The Uptake of Free DNA

In one of the most straightforward mechanisms of horizontal transfer, bacterial cells may take up free DNA from their surroundings and incorporate it into their genome. This process of "transformation" has long been known to permit harmless forms of bacteria to become pathogenic (e.g., Avery et al., 1944). In general, bacteria that can take up DNA in this way are considered "competent," with competency varying across species and also according to environmental conditions such as population density, temperature, or pH.

Conjugation Involving Plasmids and Other Elements

As noted previously, the extrachromosomal elements known as plasmids come in different sizes and shapes, ranging from a few to several hundred kb, and taking either circular or linear forms. These elements replicate autonomously, and several different plasmids, each potentially in multiple copies, may be present within a given cell. Plasmids often contain genes enabling bacterial virulence or antibiotic resistance. Plasmids also play a role in the process of "conjugation" (also known as "bacterial sex"), whereby a "fertility" or "sex" plasmid encoding genes for the synthesis of a "pilus" (a thin, hollow protein tube that can join two cells and allow the transfer of genetic material) is passed from a "male" (or "F+") to a "female" ("F–") cell.

Other DNA sequences can also be passed as part of conjugation, by being spliced from one genome by enzymes, passed through the pilus, and added to the second genome by recombination. Conjugation can occur not only among species of bacteria, including only distantly related ones, but occasionally may also take place across domains (e.g., between the bacterium *E. coli* and the yeast *Saccharomyces cerevisiae*) (Heinemann and Sprague, 1989; Jain *et al.*, 2002).

Transduction by Phages

It is well recognized that bacterial genes, including some involved in pathogenicity, are often transferred among bacteria via phages by the process known as "transduction" (for detailed recent reviews, see Boyd *et al.*, 2001; Boyd and Brüssow, 2002; Wagner and Waldor, 2002). There are essentially two mechanisms by which this can take place: "generalized transduction" and "specialized transduction."

Generalized transduction occurs when a viral particle is mispackaged to include host genomic DNA rather than, or in addition to, that of the phage. In other cases, the encapsulation of host DNA can also be mediated by specific genes within the host genome. Thus some bacterial genomes encode "gene transfer agents" (GTAs), which are tailed phagelike particles that encapsulate random fragments of the bacterial genome. In either instance, the encapsulated fragments are then delivered, virus-like, to other individuals of the same species, and replace existing counterpart fragments by homologous recombination (Casjens, 2003).

Under specialized transduction, host genomic DNA sequences are integrated into the phage genome, which then is transferred to new hosts either via the normal viral mode or through the influence of "helper phages." This latter process may involve complex interactions among unrelated bacteriophages to facilitate the transfer and expression of phage-encoded virulence genes (Boyd *et al.*, 2001). Interestingly, interactions may also take place between the hosts of pathogenic bacteria and the prophages present in the bacterial genome. For example, when *Streptococcus pyogenes* bacteria (a common cause of pharyngitis) are cultured together with human pharyngeal cells, they up-regulate and secrete streptococcal pyrogenic exotoxin C, which is known to be encoded by a prophage gene. This up-regulation of prophage genes within the bacteria appears to be stimulated by a chemical compound produced by the pharyngeal cells (Broudy *et al.*, 2001).

The Movement of Transposable Elements

At least 500 different transposable elements (TEs) have been identified in bacteria, divided into several major types: Class I includes *insertion sequences* (IS), which are usually 0.7 to 2 kb in length and include a transposase enzyme gene flanked by a 9 to 40 bp set of inverted repeats; Class II consists of a few IS elements along with the *transposons,* which are larger than 2 kb and contain genes unrelated to transposition; Class III is represented by *mutator phages,* such as *Mu,* which, though viral in origin, replicate in the host genome by a transposition mechanism (Chalmers and Blot, 1999). This classification scheme differs from that used in eukaryotes, which is discussed in detail in Chapter 3. In some cases, TEs may comprise a significant fraction of the genome, as with the 1.2 Mb sequence of *Mycoplasma mycoides,* more than 13% of which consists of three kinds of IS elements (Westberg *et al.,* 2004). However, in general TEs are not nearly as abundant in the genomes of prokaryotes, in either absolute or relative terms, as they are in eukaryotes; notably, TEs comprise ~ 45% of the human genome but only ~ 1% of the *E. coli* genome. Nevertheless, TEs still have the capacity to promote mutations and genomic rearrangements in prokaryotes. Of relevance to the present discussion is the fact that transposons may encode genes for antibiotic resistance and therefore provide a mechanism of transferring these genes among strains (Chalmers and Blot, 1999).

TRANSFER ACROSS DOMAINS

The notion that genes may be transferred from members of one domain to another is far more controversial than the question of HGT within the Bacteria or Archaea, and is the subject of ongoing debate. Some mechanisms of transfer may be common to both prokaryotic domains, which could allow the sharing of genes between them. On the other hand, no plasmids, phages, or TEs are known to commonly cross from prokaryotes to eukaryotes, and eukaryotic cells are not competent and therefore cannot pick up free DNA in the way that many bacteria can (Brown, 2003). Nonetheless, the intriguing observation of conjugation between bacteria and yeast (Heinemann and Sprague, 1989) shows that HGT is possible across even the deepest of taxonomic chasms.

HGT between Bacteria and Archaea

Although far more emphasis has been placed on the study of horizontal transfer in Bacteria, there is evidence that the process also occurs among species of Archaea (Boucher *et al.,* 2003). Transfer across these domains might be considered unlikely given the common (but incomplete) impression of Archaea as extremophiles experiencing little contact with bacteria, but there nevertheless are signs that such events do take place when bacteria and archaea share a common environment (e.g., Snel *et al.,* 2002; Boucher *et al.,* 2003; Brown, 2003).

As one commonly cited example, completion of the genome sequence of *Thermotoga maritima,* a thermophile and member of one of the deepest branching and most slowly evolving lineages of bacteria, revealed 24% of its genome to be more similar to representatives of the Archaea than to other bacteria, a much higher percentage than in most bacterial species (Logsdon and Faguy, 1999; Nelson *et al.*, 1999). The preferred explanation for this unusual similarity was that there had been extensive horizontal gene transfer between the *T. maritima* and archaeon lineages (Nelson *et al.*, 1999). Similarly, Aravind *et al.* (1998) suggested that at least 10% of the genes of the thermophilic bacterium *Aquifex aeolicus* had been transferred from extremophilic archaea, although a revised estimate was closer to 4% (Aravind *et al.*, 1999; Logsdon and Faguy, 1999). It bears noting that several alternative explanations (some of which were considered and rejected by Nelson *et al.*, 1999, and Aravind *et al.*, 1999) have been offered for these patterns, including differential gene loss and/or divergence in thermophilic versus non-thermophilic bacteria, the expansion and resulting overrepresentation of a few gene families related to archaeal sequences, an ancient shared ancestry between the hyperthermophilic bacteria and archaeons (each of which is among the oldest lineages in its respective domain), or simply a current lack of orthologs from other bacteria in the existing gene databases (Kyrpides and Olsen, 1999; Logsdon and Faguy, 1999; Brown, 2003).

Bacteria and Archaea may also cohabitate, and thus encounter opportunities to exchange genes, in nonextreme environments. Methanogenic archaea are very common in mammalian digestive tracts, for example, and therefore come into contact with well-known bacteria like *E. coli*. Importantly, it seems that the transfer of archaeal genes may have played a role in the emergence of pathogenicity in virulent strains of *E. coli* (Faguy, 2003). As Faguy (2003) noted, if transfers can occur across these two (presumably) most distantly related domains, "then potentially any organism is a reservoir of virulence genes for pathogens." There is also evidence that genes may pass from bacteria to methanogenic archaea, as for example with *Methanosarcina mazei,* about 30% of whose open reading frames appear to have closer affinities to bacterial sequences than to those of its fellow archaeons (Deppenmeier *et al.*, 2002).

HGT between Prokaryotes and Eukaryotes

It is well appreciated that there has been extensive (and still ongoing) transfer of genes from mitochondria and chloroplasts to the nuclear genomes of their eukaryotic host cells, and several hypotheses have been presented to explain this (see, e.g., Berg and Kurland, 2000; Blanchard and Lynch, 2000; Mourier *et al.*, 2001; Selosse *et al.*, 2001; Douglas and Raven, 2002; Howe *et al.*, 2003). This marks an important part of eukaryotic genome evolution, even though a large number of the genes transferred end up as pseudogenes (in the case of mitochondrial sequences, known as "nuclear pseudogenes of mitochondrial origin," or "numts") (Bensasson *et al.*, 2001). However, because these events occurred as part of the transition of formerly free-living bacteria into organelles, this represents a relatively isolated and ancient event. Of greater interest in the context of this chapter is the possibility of more recent HGT between prokaryotes and eukaryotes that still live as separate species.

Along these lines, one of the more intriguing (but subsequently controversial) findings reported from the completion of the human genome sequence was that more than 100 human genes had apparently been acquired from bacteria via horizontal transfer at some point during vertebrate evolution (International Human Genome Sequencing Consortium, 2001). As exciting as this possibility was, several reports quickly appeared in print to provide refutation of the claim of extensive HGT made in the initial human genome sequence report (Andersson *et al.*, 2001; Roelofs and Van Haastert, 2001; Salzburg *et al.*, 2001; Stanhope *et al.*, 2001). More careful examination of the issue indicated that the presence of bacteria-like genes in the human genome can be explained as a product of differential loss of certain genetic elements in bacteria and eukaryotes. As with reports of HGT between Bacteria and Archaea, this could potentially also represent methodological artifacts based on the incorrect assignment of certain sequences to orthologous groups due to the relatively imprecise nature of BLAST searches (Kurland *et al.*, 2003). As such, the possibility of recent HGT from bacteria to vertebrate genomes has been rejected. On the other hand, there is some indication that bacterial and fungal genes may be present in the genome of the nonvertebrate chordate *Ciona intestinalis* (Dehal *et al.*, 2002). More broadly, evidence has recently been presented to suggest that relatively late HGT events have provided animals with genes involved in the production of signaling chemicals important in neural, neuroendocrine, and immune systems (Lakshminarayan *et al.*, 2004).

In keeping with the notion of gene transfer from cellular endosymbionts (i.e., organelles) to host nuclei, the intriguing possibility was raised that genes from intracellular parasites like *Wolbachia* may likewise have migrated to the genomes of their eukaryote hosts (e.g., Kondo *et al.*, 2002). The availability of complete genome sequences for both *W. pipientis* and its host *Drosophila melanogaster* made it possible to test this idea explicitly, but detailed comparisons did not reveal any evidence for such gene transfer (Wu *et al.*, 2004). On the other hand, when Klotz and Loewen (2003) examined the evolution of the gene family encoding the enzymes known as catalytic hydroperoxidases, they reported not only frequent transfer among bacterial genomes, but also several HGT events from bacteria to fungi and the protist ancestor of algae and plants.

Some of the better examples of prokaryote–eukaryote HGT involve shared genes between pathogenic bacteria and their eukaryotic hosts. In such cases, it is generally thought that the direction of transfer is primarily from eukaryotes to prokaryotes, rather than the reverse as with organelles (Budd *et al.*, 2004). For example, Gamieldien *et al.* (2002) identified 19 genes of eukaryotic origin in *Mycobacterium tuberculosis* that contribute to its pathogenicity. While this is among the highest number of eukaryote-to-prokaryote transfers documented, examples have also been noted in both animal and plant pathogens such as *Chlamydia pneumoniae, Pseudomonas aeruginosa,* and *Xylella fastidiosa,* as well as in some archaeons like *Halobacterium* spp. (Koonin *et al.,* 2001). Most recently, Budd *et al.* (2004) have suggested that α_2-macroglobulin genes (which encode a protease inhibitor used in host defense) were transferred from animal hosts to invasive bacteria, followed by widespread sharing of the genes among distantly related bacterial species.

IMPLICATIONS FOR PROKARYOTE EVOLUTION AND THE STUDY THEREOF

Like many other issues in evolutionary biology, the question is not whether HGT occurs, but rather how frequently and with what implications. In some species, HGT has been suggested to be extensive, for example accounting for nearly $1/4$ of the *Thermotoga maritima* genome (Nelson *et al.,* 1999). Using $G+C$ content and codon bias information, Lawrence and Ochman (1998) estimated that about 18% of the *E. coli* genome had been derived from HGT events. According to their calculations, the *E. coli* genome has experienced at least 243 HGT events representing about 1600 kb of DNA (with ~550 kb still remaining after subsequent losses) since diverging from its common ancestor with *Salmonella enterica* about 100 million years ago— a mean transfer rate of 16 kb per million years. In a more general survey of $G+C$ content distributions, codon and amino acid usage biases, and gene positions in 17 Bacteria and seven Archaea, Garcia-Vallvé *et al.* (2000) estimated that 1.5 to 14.5% represented a more typical contribution of HGT to prokaryote genomes. More recently, Koski *et al.* (2001) argued that codon bias and base pair composition are poor indicators of HGT, and provided additional evidence that horizontal transfer had indeed been very important in the evolution of the *E. coli* genome.

Although the overall rate of HGT remains unclear, there are indications that not all genes are equally likely to be transferred horizontally. Kunin and Ouzounis (2003) estimated that only about 25–39% of protein families are involved in HGT. In particular, "operational" genes (i.e., those involved in housekeeping) are much more likely to move via HGT than are "informational" genes (i.e., those involved in transcription, translation, etc.) (Jain *et al.,* 1999). One possible explanation for this is the "complexity hypothesis," which postulates that informational genes are less amenable to HGT because they are often part of complex networks of genes, whereas operational genes are not (Jain *et al.,* 1999). However, this question, like many others in the study of HGT, still awaits a conclusive resolution (see also Doolittle, 1999; Lawrence, 1999; Eisen, 2000a).

Most contentious of all is the question regarding the significance of HGT for understanding the evolutionary histories of prokaryote lineages. Some authors acknowledge the occurrence and relevance of HGT, but rank it well below other processes such as gene losses, *de novo* origins, and duplications as a factor in modern prokaryote genome evolution (e.g., Snel *et al.*, 2002; Kunin and Ouzounis, 2003; Kurland *et al.*, 2003). Intriguingly, it seems that genes transferred horizontally are more likely to be duplicated, which indicates that these processes are not entirely independent (Hooper and Berg, 2003). In any case, where losses and non-HGT gains are considered to predominate, the main phylogenetic signal in genome evolution is seen as vertical and Darwinian, with HGT representing only minor (albeit sometimes rather distracting) "noise" (Kurland *et al.*, 2003). Other authors point out that HGT appears to have contributed to the evolution of numerous important traits and to the large-scale diversification of lineages, making it a major force in prokaryotic genome evolution (e.g., Ochman *et al.*, 2000; Boucher *et al.*, 2003; Martin *et al.*, 2004). Still others have gone so far as to claim that HGT has been sufficiently pervasive as to make the reconstruction of strictly branching phylogenetic trees a misguided procedure (Doolittle, 1999; Philippe and Douady, 2003), and perhaps that a "circle of life" best represents the deepest evolutionary pattern (Rivera and Lake, 2004). Some have presented intermediate views on this latter question, suggesting that the construction of a global "tree of life" is rendered difficult, but not impossible, by the occurrence of extensive HGT and gene loss (e.g., Wolf *et al.*, 2002).

Generally, there is more agreement regarding the contribution of HGT when considering the earliest stages of cellular evolution (Brown, 2003; Kurland *et al.*, 2003). Woese (1998, 2000, 2002) has suggested that at the time of the universal ancestor of cellular life (i.e., before the divergence of the three domains), cells were simple, replication was inaccurate, and HGT was so rampant that this ancestor must be considered a diverse community of cells rather than a single type of organism. However, when cellular systems became more refined, the frequency of HGT declined and allowed the divergence of (mostly) separate evolutionary lineages. This transition from simple cells that primarily exchanged DNA horizontally to more sophisticated ones that transmit their genetic information vertically has been called the "Darwinian threshold" (Woese, 2002). Such a scenario would not preclude the reconstruction of the subsequent evolutionary histories of the three domains, but it would have major implications for how the base of the tree of life must be envisioned (Rivera and Lake, 2004). On the other hand, some authors insist that this threshold has never been reached in prokaryotes, and favor the extreme view that there still exists but one giant, genetically interconnected bacterial "species" (Margulis and Sagan, 2002).

Whatever the ultimate contribution of HGT to prokaryote genome evolution proves to be, it has already clearly exerted a significant influence on the field of prokaryote genomics.

HIGHLIGHTS FROM SPECIFIC PROKARYOTE GENOME SEQUENCING PROJECTS

With several hundred prokaryote genome sequencing projects now either completed or under way, it would be impossible to review more than a small minority of them in a chapter of this format. As a compromise, the following sections provide only one or two highlights chosen from each year of the decade since complete sequencing began. These are not meant to serve as exemplars for prokaryotes at large, but rather to illustrate the kinds of insights that can be gleaned from the detailed study of individual genomes, in addition to the broader comparative analyses described earlier. More detailed information is available regarding these and other species from the original publications and the various databases listed elsewhere in the chapter.

HAEMOPHILUS INFLUENZAE (1995)

Although much subsequent work in prokaryote genomics has relied on comparisons among species, the whole genome sequence of the Gram-negative bacterium *H. influenzae*, the first to be sequenced among cellular organisms (Fleischmann *et al.*, 1995), allowed for some interesting and important advances largely on its own when first completed. (Comparisons with close relatives will soon be possible as well, because four additional strains of *H. influenzae*, as well as *H. ducreyi* and *H. somnus*, are currently being sequenced). Two areas, DNA uptake and the biochemistry of virulence, provide good examples of useful investigations made on the heels of the groundbreaking publication of this first complete genome sequence.

Studies of H. flu provided some of the first genome-level insights into how the uptake of foreign DNA can be regulated among bacteria. Specifically, the Rd strain of H. flu recognizes a 9-bp sequence (5'-AAGTGCGGT-3') in the genomes of members of its own species and preferentially takes up DNA from those individuals with this recognition sequence. The whole genome sequence of H. flu revealed that there were 1465 copies of this 9-bp uptake site in the genome. Whereas the 9-bp sequence was thought to be the only factor involved in DNA recognition for uptake, genomewide comparison of these 1465 copies revealed further similarities flanking the 9-bp region—in particular, a 29-bp consensus region in conjunction with two 6-bp A+T–rich regions separated by about one turn of the DNA helix. In addition, 17% of these uptake recognition sites are found in inverted repeats that are capable of forming stem-loop structures in mRNA that are hypothesized to serve as signals for transcription termination (Smith *et al.*, 1995).

Lipopolysaccharide (LPS) is a complex molecule comprised of core polysaccharides, polysaccharide side chains, and lipid A, and is an important contributor to the structural integrity of Gram-negative bacteria. LPS also serves to protect pathogenic forms from host immune defenses (it is also called "endotoxin"). Given that virulent strains of H. flu cause numerous ailments (e.g., bacterial meningitis caused by the type b strain; the sequenced Rd laboratory strain is benign), an understanding of the biochemistry of its virulence may have substantial medical benefits. One of the first functional analyses of pathogenesis using the

H. flu genome sequence involved a study of LPS and its involvement in disease. In particular, the whole genome sequence of H. flu allowed researchers to compare H. flu genomic sequences to the LPS-related biogenesis gene sequences from other organisms. These comparisons led to the determination of 25 candidate genes in the H. flu genome that were examined for their involvement in LPS biosynthesis. Subsequent screening of these genes using a battery of biological and animal model approaches revealed that the majority of these candidate genes are indeed involved in LPS biosynthesis. More importantly, by using the LPS candidate gene information and an animal model system, researchers were able to determine the minimal LPS structure involved in potential pathogenesis (Hood *et al.*, 1996).

METHANOCALDOCOCCUS JANNASCHII (1996)

The genome sequence of the deep-sea, high pressure- and heat-tolerant methanogenic euryarchaeote *Methanocaldococcus* (*Methanococcus*) *jannaschii*, the first among the Archaea, was published not long after the beginning of the complete sequencing era (Bult *et al.*, 1996). In very general terms, it was similar to the genome of H. flu, for example in both size (1.66 versus 1.83 Mb) and gene number (both ~ 1740 in the original publications; the estimate for *M. jannaschii* was increased slightly not long after) (Kyrpides *et al.*, 1996). Unfortunately, the completion of the *M. jannaschii* genome did not in itself resolve the issue of relatedness among the three domains of life, because genes from *M. jannaschii* related to metabolism and cell division were more similar to those of bacteria, whereas genes involved in translation, transcription, and replication were more like those of eukaryotes (Bult *et al.*, 1996; Graham *et al.*, 2001).

Reflecting both the deep divergence between Archaea and Bacteria and the general lack of information regarding archaeal biology, Bult *et al.* (1996) noted that only 38% of the predicted gene regions identified could be initially assigned a function. Moreover, only 11% of genes from *H. influenzae* and 17% from *Mycoplasma genitalium* (also sequenced the previous year) (Fraser *et al.*, 1995) found matches in the *M. jannaschii* genome, whereas 83% of gene regions matched between these two bacteria (Bult *et al.*, 1996). More detailed analyses have since improved the characterization of archaeal genes considerably (e.g., now about 70% have been assigned probable functions in *M. jannaschii*) (Makarova and Koonin, 2003).

Whole genome comparisons within the Archaea have already revealed numerous important insights about their biology and evolution (e.g., Gaasterland, 1999; Makarova and Koonin, 2003), even though archaeal sequencing projects remain far fewer in number than those dealing with bacteria (Fig. 10.3). For example, an early comparison with fellow methanogen *Methanobacterium thermoautotrophicum*, whose genome sequence was published the following year, revealed that about

19% of the genes were shared between them (Smith *et al.*, 1997). It is clear that gene order has not been preserved between *M. jannaschii* and *M. thermoautotrophicum*, which has also been found in pairs of nonmethanogenic Archaea (e.g., *Pyrococcus* spp., *Sulfolobus* spp.) (Gaasterland, 1999; Mira *et al.*, 2002) (Fig. 10.6).

ESCHERICHIA COLI (1997)

E. coli is one of the best-known species of bacteria, not only because of its close association with humans as a facultative anaerobe inhabiting the digestive tract and its role as a preeminent model of bacterial genetics, but also because contamination of food and water sources with a pathogenic strain can cause outbreaks of hemorrhagic colitis and potentially fatal hemolytic uremic syndrome. Not surprisingly, the K12-MG1655 nonpathogenic laboratory strain of *E. coli* was the earliest organism to be suggested as a candidate for a full sequencing project, although it took nearly six years to complete using the slower pre-shotgun sequencing methods (Blattner *et al.*, 1997). The VT2-Sakai and EDL933 strains of enterohemorrhagic *E. coli* O157:H7 (Hayashi *et al.*, 2001; Perna *et al.*, 2001) and the uropathogenic strain CTF073 (Welch *et al.*, 2002) have been sequenced since, thereby allowing much opportunity for comparison among *E. coli* genomes. This will increase considerably in the near future because eight additional strains of *E. coli* are currently being sequenced.

The *E. coli* strains sequenced thus far vary considerably in the sizes and gene contents of their genomes, with the pathogenic O157:H7 and CTF073 strains containing a remarkable 600 to 900 kb more DNA, and between 300 and over 1100 more genes, than the nonpathogenic K12 strain (Table 10.1). The differences between the O157:H7 and the K12 strain are even more striking when a detailed comparison is made. Sequences in the genomes of K12 and O157:H7 can be easily aligned over a 4,100,000 base pair stretch showing a high degree of

TABLE 10.1   Genome sizes and gene contents of one benign strain and three pathogenic strains of the common gut bacterium *Escherichia coli*. The pathogenic strains have significantly larger genomes and more genes than the nonpathogenic strain. Data from the TIGR Comprehensive Microbial Resource (www.tigr.org/tigr-scripts/CMR2/CMRHomePage.spl).

Strain	Status	Genome size (bp)	Gene number
K12-MG1655	Nonpathogenic	4,639,221	5288
CFT073	Uropathogenic	5,231,428	5605
O157:H7 EDL933	Enterohemorrhagic	5,528,445	5680
O157:H7 VT2-Sakai	Enterohemorrhagic	5,498,450	6423

apparent colinearity (common gene order) for the genomes of the two strains (Hayashi et al., 2001; Perna et al., 2001). However, scattered throughout the length of this aligned stretch are hundreds of sections of DNA that are unique to either O157:H7 or K12. Perna et al. (2001) dubbed these unique stretches "O-islands" and "K-islands," respectively. O-islands account for 1.34 Mb and 1387 protein coding genes, whereas K-islands account for only 0.53 Mb and 528 genes (Eisen, 2001; Perna et al., 2001). A similar comparison of CFT073 and K12 revealed 1.30 Mb of CFT073-specific islands and 0.72 Mb unique to K12 (Welch et al., 2002). In an interesting study, Kolisnychenko et al. (2002) used targeted deletions to remove 12 of the K-islands and 24 of 44 transposable elements from the model K12-MG1655 strain in order to generate a new strain of E. coli called MDS12 with a "cleaner" genome 8.1% smaller and with 9.3% fewer genes than the original strain. Ultimately, this deletion method may be used to create a strain of E. coli with a fully streamlined genome containing no extraneous genetic material, which could be useful as a model for functional genomics and in other applications (Kolisnychenko et al., 2002).

As a further demonstration of the remarkable genomic diversity within E. coli, larger three-way comparisons of the predicted protein overlap of the CFT073, O157:H7, and K12 strains showed that only 39.2% of their protein-coding genes are common to all three (Welch et al., 2002). In fact, such comparisons have revealed that O157:H7 and CFT073 are as dissimilar to each other as either is to the nonpathogenic K12 strain. Furthermore, the pathogenic CFT073 strain lacks the Type III secretion factor genes and phage- and plasmid-encoded virulence genes that are found commonly in O157:H7 isolates (Welch et al., 2002). So, although the different E. coli strains appear to have a common backbone of genes in similar order that has been preserved during the vertical evolution of the strains, overall their genomes represent very different mosaics, probably resulting from the differential uptake of DNA (Eisen, 2001; Hayashi et al., 2001; Perna et al., 2001; Welch et al., 2002).

One of the aspects of pathogenicity in E. coli O157:H7 involves the production of a Shiga-like toxin very similar to that generated by pathogenic species of Shigella, which themselves cause some 160 million (1.1 million of them fatal) cases of bacillary dysentery each year (Jin et al., 2002). Interestingly, genetic analyses have indicated that Shigella may not be a valid genus because the various "species" appear to have been derived independently from strains of E. coli as recently as 35,000 to 270,000 years ago (Pupo et al., 2000; Wang et al., 2001; Jin et al., 2002). The O157:H7 and K12 strains of E. coli, by contrast, may have diverged from each other as long as 4.5 million years ago (Reid et al., 2000).

The complete sequence of the 2a serotype of Shigella flexneri has indicated that it shares a roughly 3.9 Mb backbone with the O157:H7 and K12 strains of E. coli (Jin et al., 2002; Wei et al., 2003). Interestingly, S. flexneri appears to be more closely related to the nonpathogenic K12 strain of E. coli than to the O157:H7 strain (Jin *et al.*, 2002). In part, *Shigella* pathogenicity is dependent on the presence of a large plasmid containing genes required for virulence. However, the mere presence of this plasmid would not have been enough to transform benign strains of *E. coli* into highly pathogenic *Shigella*; instead, this involves extensive regulatory "dialog" between the chromosome and the virulence plasmid, as well as large symmetric inversions, gains and losses of genes, pseudogenizations, and other genomic events that occurred after the acquisition of the plasmid (Venkatesan *et al.*, 2001; Jin *et al.*, 2002; Wei *et al.*, 2003). Because no vaccine against *Shigella* is available, a determination of the genomic underpinnings of its pathogenicity is particularly important (Jin *et al.*, 2002; Wei *et al.*, 2003) and represents an area of active current research.

MYCOBACTERIUM TUBERCULOSIS (1998)

Most members of the Gram-positive bacterial genus *Mycobacterium* are harmless soil- or water-dwelling species, but those that have become obligate intracellular pathogens have had a major impact on human health. Notably, *M. tuberculosis* has resulted in countless millions of deaths as the primary causative agent of tuberculosis, *M. leprae* causes leprosy (Hansen's disease) and was the first bacterium to be identified as a disease-causing agent, and *M. bovis* infects both livestock and humans and results in a loss of about $3 billion annually to the agriculture industry. The H37Rv laboratory strain of *M. tuberculosis* was the first member of this important genus to have its genome completely sequenced (Cole *et al.*, 1998), followed by *M. leprae* (Cole *et al.*, 2001), the CDC1551 clinical strain of *M. tuberculosis* (Fleischmann *et al.*, 2002), and *M. bovis* (Garnier *et al.*, 2003). *M. avium*, *M. marinum*, *M. microti*, *M. smegmatis*, and *M. ulcerans* projects are also under way at present.

Obviously, it is expected that the availability of complete genome sequence information will improve the success in combating these pathogens (Cole *et al.*, 1998; Young, 1998). However, medical implications aside, *Mycobacterium* species are also of considerable interest with regard to genome evolution. For example, the *M. tuberculosis* genome is known to contain about 4000 genes, over half of which have arisen by gene duplication (Brosch *et al.*, 2000; Domenech *et al.*, 2001). *M. tuberculosis* has an inordinate number of genes involved in fatty acid metabolism (more than 250, versus only 50 in *E. coli*) (Domenech *et al.*, 2001). Also of interest is the fact that the genome sequences of *M. tuberculosis* and *M. bovis* are more than 99.95% identical, although deletions in *M. bovis* have resulted in a slightly smaller genome size (Garnier *et al.*, 2003). It was long thought that the human tubercle had descended from the bovine form after the domestication of cattle, but it now seems that "*M. bovis* has evolved from a progenitor of the *M. tuberculosis* complex [which includes *M. africanum*, *M. bovis*,

M. microti, and *M. tuberculosis*] as a clone showing distinct host preference" (Garnier *et al.,* 2003).

   M. leprae, which is biologically unique for having the longest doubling time of any known bacterium (~14 days), is also very peculiar at the genomic level (Cole *et al.,* 2001). For example, its genome is more than 1 Mb smaller than that of *M. tuberculosis.* More strikingly, only about 49.5% of the *M. leprae* genome is accounted for by the 1604 protein-coding genes, whereas a further 27% is composed of 1116 pseudogenes (the remaining 23.5% may be former genes rendered unrecognizable by mutation) (Cole *et al.,* 2001). The net result is that, since its divergence from the common ancestor with *M. tuberculosis, M. leprae* appears to have lost more than 2000 genes through a process of "massive gene decay" (Cole *et al.,* 2001).

DEINOCOCCUS RADIODURANS (1999)

As its name implies, the Gram-positive bacterium *Deinococcus radiodurans* is the most radiation-resistant organism identified to date, being 200 times more tolerant of ionizing radiation and 20 times more resistant to ultraviolet radiation than *E. coli* (White *et al.,* 1999). In addition to being of basic biological interest, this feature makes *D. radiodurans* a promising candidate for bioremediation of contaminated sites (White *et al.,* 1999). As noted previously, the genome of *D. radiodurans* is also of interest because it is comprised of four distinct elements, including two circular chromosomes (2,648,638 bp and 412,348 bp), a megaplasmid (177,466 bp), and a small plasmid (45,704 bp), with a high average G+C content of about 67% and coding for a total of 3187 genes (White *et al.,* 1999).

   The complete genome sequence of *D. radiodurans* revealed a number of genes involved in the prevention of DNA damage (e.g., by scavenging oxygen radicals). However, by far the dominant strategy in this species for surviving radiation is to use effective mechanisms to repair even extensive DNA damage, which it accomplishes without DNA rearrangement or increased mutation rates (White *et al.,* 1999; see also Makarova *et al.,* 2001). Most of the DNA repair genes identified in the *D. radiodurans* genome have homologs in other species, but it is apparent that this species places a high emphasis on redundancy in these sequences. Moreover, many of the extra copies are the result of very recent duplication events (White *et al.,* 1999). Polyploidization (i.e., the duplication of the entire genome) and the presence of repeats that induce recombination in *D. radiodurans* also certainly play a role in its tolerance for radiation, although it is evident that homologous recombination is not the means by which this species repairs extensive double-strand breaks (Levin-Zaidman *et al.,* 2003).

   Other intriguing mechanisms of radiation tolerance discovered through the detailed study of *D. radiodurans* include the ability to export damaged nucleotides out of the cell (White *et al.*, 1999), the presence of a specialized single-stranded DNA binding protein (SSB) (Eggington *et al.*, 2004), a pattern of DNA strand exchange induced by the RecA protein that is the exact opposite of the typical bacterial system (Kim and Cox, 2002), the separation of the multiple genome copies into four distinct cellular compartments, and the formation of a ringlike chromatin structure following irradiation (Levin-Zaidman *et al.*, 2003). However, despite all of these recent insights, the means by which *D. radiodurans* is able to persist in its remarkable lifestyle remains a subject of ongoing investigation.

VIBRIO CHOLERAE (2000)

Species of Gram-negative γ-proteobacteria in the genus *Vibrio* make up a significant fraction of the culturable heterotrophic bacteria of oceans and estuaries, and exert a substantial influence on oceanic nutrient cycling (Heidelberg *et al.*, 2000). Some species are devastating pathogens of marine invertebrates and vertebrates, but the best known is *V. cholerae*, pathogenic strains of which are the agents responsible for cholera, a severe and potentially fatal diarrheal disease that persists in areas with poor sanitation and which has so far resulted in seven pandemics (Dziejman *et al.*, 2002). *V. cholerae* (El Tor N16961 strain, cause of the most recent cholera pandemic in 1961) was the first member of the genus to be sequenced (Heidelberg *et al.*, 2000), although the genomes of the marine pathogens *V. parahaemolyticus* (Makino *et al.*, 2003) and *V. vulnificus* (Chen *et al.*, 2003) have also been published since, and *V. fischeri* and *V. salmonicida* are due to be completed soon.

The *V. cholerae* genome is composed of two circular chromosomes (2,961,146 bp and 1,072,314 bp), with most of the genes with recognizable roles in growth, viability, and pathogenicity located on the larger chromosome (Heidelberg *et al.*, 2000). Some genes essential for survival are located on the smaller chromosome, including many coding for intermediaries of metabolic pathways. Overall, the smaller chromosome contains a greater fraction of hypothetical genes and sequences of foreign origin, and appears to have been descended from a large megaplasmid which had itself previously acquired genes from several different bacterial lineages (Heidelberg *et al.*, 2000). Indeed, the transfer of genetic material across species is believed to have played an important role in the transformation of this formerly free-living organism into a major human pathogen. Genomewide comparisons (e.g., using microarrays) of the pathogenic El Tor N16961 and closely related pathogenic and nonpathogenic strains have already begun to shed light on the specific genes involved in allowing pandemic outbreaks (Dziejman *et al.*, 2002).

Like *V. cholerae*, *V. parahaemolyticus* and *V. vulnificus* both have two circular chromosomes, one large and one small. Also as with *V. cholerae*, the majority of genes involved in growth and viability in *V. parahaemolyticus* are located on Chromosome 1, and genes related to environmental adaptation and pathogenicity are more common on Chromosome 2 (Makino *et al.*, 2003). However, the sizes of the chromosomes vary among the three species, with the large chromosomes of *V. cholerae*, *V. parahaemolyticus*, and *V. vulnificus* being ~3.0, 3.4, and 3.3 Mb, respectively, and the small chromosomes being ~1.0, 1.9, and 1.9 Mb. The larger genomes of *V. vulnificus* and *V. parahaemolyticus* as compared to *V. cholerae* have presumably resulted from more frequent gene duplication events in the former, but could also reflect more gene loss in the latter (Chen *et al.*, 2003; Makino *et al.*, 2003).

Comparisons among the three species of *Vibrio* also revealed extensive shuffling of genes both within and among chromosomes (Fig. 10.10), with greater conservation in the large chromosome and a closer similarity in gene order between *V. parahaemolyticus* and *V. vulnificus* versus *V. cholerae* (Chen *et al.*, 2003; Makino *et al.*, 2003). As a prime example, the large superintegron is located on Chromosome 2 in *V. cholerae*, but on Chromosome 1 in *V. parahaemolyticus* and *V. vulnificus*. However, the genes involved are very different even between *V. parahaemolyticus* and *V. vulnificus*, indicating that this region evolves rapidly (Chen *et al.*, 2003). Likewise with regard to pathogenicity, it seems that *V. cholerae* and *V. parahaemolyticus* use different mechanisms for infection even though they ultimately cause similar symptoms (Makino *et al.*, 2003).

STREPTOCOCCUS SPP. (2001)

Although nearly 40 prokaryote genomes had been published by then, most attention quickly shifted to eukaryotes in mid-February of 2001, for the obvious reason that this marked the publication of the draft human genome sequence (see Chapter 9). Nevertheless, 2001 also saw the completion of some bacterial sequences that are of substantial importance in their own right. Appropriately enough, this included *Streptococcus pneumoniae* ("pneumococcus"), the species used in the famous "transformation experiments" of Avery *et al.* (1944) that first helped to identify DNA as the hereditary material. *Streptococcus* is a low G+C Gram-positive bacterium, and indeed was used in the initial development of Gram stain. Most importantly, *S. pneumoniae* is a causative agent of several serious diseases, including penumonia, meningitis, and otitis media, leading to millions of deaths each year, particularly among children and the elderly (Tettelin *et al.*, 2001). There is therefore great hope that comparative genomics will allow an identification of the genes responsible for its virulence and thus the development of effective measures of treatment and control (Hollingshead and Briles, 2001).

Two strains of *S. pneumoniae*, the nonpathogenic R6 (Hoskins *et al.*, 2001) and the virulent TIGR4 (Tettelin *et al.*, 2001), were sequenced in 2001, along with a

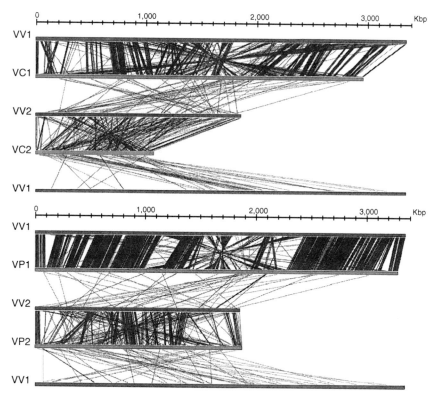

FIGURE 10.10 Extensive intra- and interchromosomal gene shuffling in the genomes of three pathogenic *Vibrio* species. Homologous blocks of genes among genomes are indicated by the lines between chromosomes, and crossed lines represent recombination events. All three of the *Vibrio* species shown here have two circular chromosomes, one large and one small, though the sizes of the respective chromosomes vary somewhat among species as a result of differential rates of gene duplication and/or loss (Chen *et al.*, 2003; Makino *et al.*, 2003). The smaller chromosome is believed to have been descended from a former megaplasmid, and contains genes from several bacterial lineages (Heidelberg *et al.*, 2000). VV1 = large Chromosome 1 of *V. vulnificus*, VV2 = small Chromosome 2 of *V. vulnificus*, VC1 = large Chromosome 1 of *V. cholerae*, VC2 = small Chromosome 2 of *V. cholerae*, VP1 = large Chromosome 1 of *V. parahaemolyticus*, VP2 = small Chromosome 2 of *V. parahaemolyticus*. Modified from Chen *et al.* (2003), reproduced by permission (© Cold Spring Harbor Laboratory Press).

representative of one serotype of *Streptococcus pyogenes* (also called "group A *Streptococcus*," or GAS), which causes a wide variety of infections ranging from mild to severe (Ferretti *et al.*, 2001). These were followed a short time later by the sequencing of two strains of *Streptococcus agalactiae* ("group B *Streptococcus*," or GBS, a cause of several invasive bacterial infections in newborns) (Glaser *et al.*, 2002; Tettelin *et al.*, 2002), one of *S. mutans* (the leading cause of tooth decay)

(Ajdić *et al.*, 2002), and additional serotypes of *S. pyogenes* (Beres *et al.*, 2002; Smoot *et al.*, 2002; Nakagawa *et al.*, 2003).

The genome of the "smooth" R6 strain of *S. pneumoniae*, which lacks the polysaccharide capsule required for virulence, appears to be slightly smaller than that of the encapsulated and virulent TIGR4 strain (2,038,615 bp versus 2,160,837 bp) and to code for fewer genes (2043 versus 2236) (Hoskins *et al.*, 2001; Tettelin *et al.*, 2001). More than 3% of the R6 strain genome consists of repetitive elements from three classes (BOX, RUP, and IS), and in the TIGR4 strain the total is around 5%, which is the highest known repetitive element content among bacteria (Hoskins *et al.*, 2001; Tettelin *et al.*, 2001). Although R6 maintains a large number of genes identified as being involved in virulence, it has undergone a conspicuous 7.5-kb deletion within the region containing capsule biosynthesis genes (Hoskins *et al.*, 2001). As the early experiments involving the transformation of nonencapsulated to encapsulated forms indicated, gene transfer is common in *S. pneumoniae* (see Claverys *et al.*, 2000), and indeed "its genome is littered with genes that are apparently derived from other bacteria," including some from distantly related Gram-negative species that cohabitate in the human host (Hoskins *et al.*, 2001). As Claverys *et al.* (2000) noted, this provides cells of *S. pneumoniae* with access to a "global genome" that is larger than any one individual genome. Given that there are some 90 different capsule types, and making the reasonable assumption that the genes encoding them differ by ~5 kb from one another, then this would mean that there is a pool of 450 kb, or nearly 25% of another entire genome, available for uptake by individuals of *S. pneumoniae* (Claverys *et al.*, 2000).

In total, four strains of *S. pyogenes* have now been sequenced: SF370 (M1 serotype), MGAS315 and SSI-1 (M3 serotype), and MGAS8232 (M18 serotype). The genomes of MGAS315, MGAS8232, and SF370 are very similar, and share a large "backbone" or "core" of genes in common order, covering 90% of the genome, or ~1.7 Mb (Beres *et al.*, 2002; Smoot *et al.*, 2002). However, the SSI-1 strain appears to have undergone extensive translocations and symmetrical rearrangements relative to the others, resulting in an X-alignment on gene location plots of this strain versus other *S. pyogenes* and *S. pneumoniae* genomes (Nakagawa *et al.*, 2003). Importantly, an increase in X-rearranged strains appears to coincide with the recent resurgence of severe *S. pyogenes* infections in Japan (Nakagawa *et al.*, 2003).

In broader terms, *S. agalactiae* (GBS) and *S. pyogenes* (GAS) share a genic backbone covering about 55% of their genomes, with only 36 breakpoints interrupting colinearity (Glaser *et al.*, 2002). By contrast, very little conservation of gene order is observed between *S. agalactiae* and *S. pneumoniae* (Glaser *et al.*, 2002). Some 50% of the genes in *S. agalactiae* are shared by both *S. pyogenes* and *S. pneumoniae* (Tettelin *et al.*, 2002). Detailed comparisons of the gene complements of different species and strains of *Streptococcus* is expected to reveal a great deal regarding the evolution of pathogenicity in this genus. Indeed, *Streptococcus* is set to become one of the most heavily sequenced genera of all, with projects now in progress to sequence various strains of S. agalactiae, S. equi, S. gordonii, S. mitis, S. pneumoniae, S. pyogenes, S. sanguinis, S. sobrinus, S. suis, S. thermophilus, and S. uberis. Not all of these are pathogenic, but they could still be of medical relevance. For example, genome subtraction studies between pathogenic and nonpathogenic species can help to reveal the genes involved in pathogenesis (Hollingshead and Briles, 2001). Furthermore, it is known that genetic exchange between S. mitis or S. oralis (nonpathogenic oral bacteria) and their close relative S. pneumoniae can contribute to penicillin resistance in the latter (Claverys et al., 2000; Hollingshead and Briles, 2001). Given the rapid increase in penicillin and other antibiotic resistance in S. pneumoniae over the recent past, this is a major concern (Claverys et al., 2000; Tettelin et al., 2001). Fortunately, even before the final publication of the S. pneumoniae and S. pyogenes sequences, large-scale genomic data were being used to identify new antibiotic and vaccine candidates (see Graham et al., 2001; Hollingshead and Briles, 2001).

STREPTOMYCES COELICOLOR (2002)

The filamentous Gram-positive soil-dwelling bacteria in the genus Streptomyces are important from a variety of perspectives. They are well recognized as the most prolific manufacturers of natural pharmaceutical compounds, and are responsible for producing more than 2/3 of the known natural antibiotics and a diverse array of other useful chemicals (Bentley et al., 2002; Thompson et al., 2002). These bacteria accomplish the synthesis of the medically important compounds by means of secondary metabolic pathways, making knowledge of the genomes of these organisms an important roadmap to the production of antibiotics. Moreover, comparative genomic analyses promise to shed considerable light on both the remarkable antibiotic producing abilities of Streptomyces and the pathogenic qualities of its actinomycete relatives, Mycobacterium and Corynebacterium. Streptomyces also represents a useful model for developmental biology because this organism develops different "tissues" as part of a filamentous pattern of growth that branches and produces hyphae with reproductive spores.

Streptomyces coelicolor is no less interesting from a genomic perspective. Notably, its genome is arranged in a linear chromosome of 8,667,507 bp plus two linear plasmids of 356,023 bp and 31,317 bp, making it both unique in organization and one of the largest so far identified in prokaryotes. S. coelicolor also has the highest known G+C content (72.1%) and one of the largest predicted gene counts (7825) among prokaryotes, a remarkable 12% of which are involved in regulation (Bentley et al., 2002). About 55% of the identified gene regions are thought to be pseudogenes or otherwise inactive reading frames (Bentley et al., 2002).

As would be expected, a good proportion of the *S. coelicolor* genome is devoted to secondary metabolism, with more than 20 gene clusters of varying size dedicated to these functions. Nearly half of the essential genes required for cell division, translation, transcription, and DNA replication are located as part of a large "central core," which is flanked on either side by "contingency" genes that code for nonessential functions (Bentley *et al.*, 2002). This central core (but not the outer regions) shares a perceptible degree of gene order with the genomes of *Mycobacterium tuberculosis* and *Corynebacterium diphtheriae*, albeit with several inversions apparent. The interesting implication of this is that the central core is ancestral in *S. coelicolor*, with the chromosome arms consisting of subsequently acquired DNA (Bentley *et al.*, 2002).

The recently completed *S. avermitilis* genome exhibits several similarities to that of *S. coelicolor*, including linear organization, large size (9,025,608 bp, plus a linear plasmid of 94,287 bp), high G+C content (70.7%) and number of genes (7547), 30 clusters of genes involved in secondary metabolism, a highly conserved central gene core of ~6.5 Mb containing essential genes, and more variable chromosomal arms containing nonessential genes (Ikeda *et al.*, 2003). It will soon be possible to determine whether these features are common to several other members of the genus, including *S. ambofaciens*, *S. diversa*, *S. noursei*, *S. peucetius*, and *S. scabies*, which are currently being sequenced. To be sure, comparisons among these species should reveal much information regarding the mechanisms that have made these organisms so important to human pharmacology.

BACILLUS ANTHRACIS (2003)

Members of the genus *Bacillus* are low G+C Gram-positive, spore-forming, rod-shaped, aerobic or facultatively anaerobic bacteria. The sequence of the well-studied and industrially beneficial *B. subtilis*, published in 1997, marked the first for a Gram-positive species (Kunst *et al.*, 1997). The genome sequence of the alkaliphilic *B. halodurans*, which grows well at a pH above 9.5, was sequenced a few years later (Takami and Horikoshi, 2000). The expansion of gene families by duplication and several unexpected repetitions proved to be notable features of the *B. subtilis* genome (Kunst *et al.*, 1997). Interestingly, although these two genomes are almost identical in size (~4.2 Mb), gene count (~4100), and mean G+C content (~43.5%), it was noted that they differ almost completely in gene order (Takami and Horikoshi, 2000).

Although these early findings are of interest, without question the complete *Bacillus* genome sequence that has attracted the most attention is that of *B. anthracis*, the causative agent of inhalational anthrax (Read *et al.*, 2003). Two strains of *B. cereus*, a cause of food poisoning and a close relative of *B. anthracis*, have also been sequenced for comparative purposes (Ivanova *et al.*, 2003; Rasko *et al.*, 2004), and projects for at least 10 more strains of *B. anthracis* are now ongoing, along with programs for B. *alcalophilus* (an obligate alkaliphile that grows at pH 11), B. *amyloliquefaciens* (formerly a subspecies of B. *subtilis*), B. *clausii* (another alkaliphile), B. *licheniformis* (which is used in industry for fermentation and enzyme production), B. *stearothermophilus* (a thermophile), and B. *thuringiensis* (an insect pathogen used as a biological pesticide).

Fully pathogenic strains of B. *anthracis* carry two virulence plasmids: pXO1, which is 182 kb in size and contains toxin and other genes, and pXO2, which is 96 kb and carries genes involved in capsule synthesis and degradation (for reviews, see Baillie and Read, 2001; Mock and Fouet, 2001). The "Porton" isolate of the Ames strain sequenced by Read *et al.* (2003) does not contain these plasmids, but the plasmids of the A2012 ("Florida") isolate of the Ames strain had been sequenced the year before, following the bioterrorist attacks in Florida, New York, and Washington, D.C., in the autumn of 2001 (Read *et al.*, 2002; Fraser 2004a). Sequences were also previously available for the Sterne strain pXO1 and Pasteur strain pXO2 plasmids (Okinaka *et al.*, 1999a,b).

The early comparative analysis of the genome from the A2012 isolate of B. *anthracis* versus existing plasmid and chromosomal sequence data revealed several single nucleotide polymorphisms (SNPs, or "snips") among isolates. Along with previously determined sets of markers, such as variable number tandem repeats (VNTRs) and amplified fragment length polymorphisms (AFLPs), these allow specific isolates to be identified, and thus the sources of outbreaks, be they natural or bioterrorist in origin, to be tracked (e.g., Jackson *et al.*, 1997; Keim *et al.*, 1997; Ramisse *et al.*, 1999; Enserink, 2001; Read *et al.*, 2002; Fraser, 2004a). In fact, it was genetic "fingerprinting" analyses like these that identified the 2001 anthrax attack strain as having come from an American biodefense laboratory.

In B. *anthracis,* as in other related endospore-forming Gram-positive bacteria, the rRNA, tRNA, and ribosomal protein genes are concentrated near the origin of replication, a feature that may maximize protein synthesis during early rounds of DNA replication after exiting the dormant endospore phase (Read *et al.*, 2003). There is a high degree of colinearity between the B. *cereus* ATCC 10987 and ATCC 14579 strains and the B. *anthracis* Ames strain (Rasko *et al.*, 2004), with B. *anthracis* and B. *cereus* ATCC 14579 sharing a common core of about 75–80% of their genes (Ivanova *et al.*, 2003). Interestingly, sequence comparisons indicate that B. *cereus* ATCC 10987 is more closely related to B. *anthracis* Ames than to B. *cereus* ATCC 14579 (Rasko *et al.*, 2004). Moreover, the 208 kb pBc10987 plasmid of B. *cereus* ATCC 10987 bears some intriguing parallels to the pXO1 plasmid of B. *anthracis,* such that it may provide a convenient nonlethal model for the study of the B. *anthracis* virulence plasmid (Rasko *et al.*, 2004). Overall, comparisons with B. *subtilis* and B. *cereus* have revealed a large number of genes potentially associated with pathogenesis in members of the B. *cereus* group (which includes B. *anthracis*), but not specific to anthrax per se (Read *et al.*, 2003).

The nature of the conserved genes suggests that the group may have evolved from an opportunistic insect pathogen rather than a benign soil bacterium (Ivanova *et al.*, 2003). The subsequent intentional weaponization of *B. anthracis* and other pathogens is another story.

WOLBACHIA PIPIENTIS (2004)

Wolbachia spp. are Gram-negative α-proteobacteria found in obligate intracellular associations (some mutualistic, some parasitic) with possibly millions of different arthropod and nematode host species, perhaps making them the most common infectious bacteria on Earth. *Wolbachia* species are inherited maternally from hosts to their offspring, and several different forms of reproductive parasitism have emerged in the genus to enhance their own transmission in the host's female line. These include inducing infected females to reproduce without mating (partheno-genesis; e.g., in haplodiploid wasps), feminization of males such that they become functional phenotypic females (e.g., in isopod crustaceans), selective killing of male embryos, and cytoplasmic incompatibility which prevents infected males from producing offspring with uninfected females (Wu *et al.*, 2004).

The first genome sequence to be published from this intriguing group of bacteria was that of *Wolbachia pipientis* wMel, a parasite of the fruit fly *Drosophila melanogaster* (Wu *et al.*, 2004). Species of *Wolbachia* from eight additional hosts are currently being sequenced as well, such that detailed comparisons among close relatives will soon become possible. In the first *Wolbachia* genome analysis, most comparisons were made with *Rickettsia* spp., which are grouped with *Wolbachia* in the order Rickettsiales and which are obligate intracellular parasites usually associated with arthropods, but some of which may also be pathogens of humans. The two *Rickettsia* genomes sequenced thus far are *R. prowazekii*, the cause of epidemic typhus (Andersson *et al.*, 1998), and *R. conorii*, which causes Mediterranean spotted fever (Ogata *et al.*, 2001).

The genomes of *Wolbachia* and *Rickettsia* are very interesting in that although they are small (0.95 to 1.6 Mb) and contain a low number of genes (only 834 in *R. prowazekii*), they include a large amount of repetitive DNA (24% in *R. prowazekii*) and many pseudogenes and transposable elements, which are not normally found in prokaryote genomes, especially those of intracellular endosym-bionts (Andersson *et al.*, 1998; Andersson and Andersson, 2001; Ogata *et al.*, 2001; Wu *et al.*, 2004). There is no large-scale conservation of gene order in *W. pipientis* as compared with *Rickettsia* genomes, indicating a high degree of rearrangement, probably associated with the presence of so much repetitive DNA (Wu *et al.*, 2004). It is evident that *Wolbachia* has lost genes since it last shared a common ancestor with *Rickettsia*, although it is likely that this ancestor also had a streamlined genome (Wu *et al.*, 2004). Interestingly, some of the same genes have been lost in *Buchnera*, the γ-proteobacterial endosymbiont of aphids, although these two genera are only distantly related (Wu *et al.*, 2004). Yet despite this genome downsizing, there is evidence of continued tandem gene duplication in *Wolbachia*, with more than 50 gene families expanded in this lineage (Wu *et al.*, 2004).

It was suggested as part of the *R. prowazekii* sequence report that *Rickettsia* may represent the closest living relative of the α-proteobacterial ancestor to modern-day mitochondria (Andersson *et al.*, 1998, 2003; Gray, 1998). However, more recent analyses including the sequence of *W. pipientis* showed evidence in favor of a grouping of *Wolbachia* and *Rickettsia* to the exclusion of mitochondria, and did not support the specific placement of mitochondria in the order Rickettsiales within the α-proteobacteria lineage (Wu *et al.*, 2004).

THE EVOLUTION OF GENOME SIZE IN PROKARYOTES

As with eukaryotes, the earliest studies of prokaryotic genome size predated the advent of complete sequencing by several decades.[5] Pre-genomics measures of genome size in Bacteria and Archaea have been accomplished by use of reassociation kinetics (especially in older studies), sometimes by fluorometry or image analysis densitometry (e.g., Loferer-Krössbacher *et al.*, 1999), and especially by pulse-field gel electrophoresis (PFGE), which allows chromosome-sized DNA fragments (which would otherwise diffuse too much) to be resolved on agarose gel systems by applying the electric current in a timed, alternating fashion from two different angles. Complete sequencing is clearly the most accurate method of genome size determination in prokaryotes (though not in eukaryotes; see Chapters 1 and 2), but PFGE remains an efficient means of acquiring large genome size datasets (e.g., Bauman *et al.*, 1998; Redenbach *et al.*, 2000; Sun *et al.*, 2001; Islas *et al.*, 2004).

In stark contrast to the eukaryotes (see Chapters 1 and 2), there was never a "C-value paradox" during the early stages of genome size study in prokaryotes,

---

[5]Until very recently, no comprehensive prokaryote genome size database equivalent to those for animals (www.genomesize.com) or plants (www.rbgkew.org.uk/cval/homepage.html) had been made available online. However, the *Prokaryote Genome Size Database* has recently been launched (www.genomesize.com/prokaryotes) and currently includes nearly 700 estimates made by pulse-field gel electrophoresis (based on the dataset of Islas *et al.*, 2004). Older compilations of nonsequencing estimates can be found in Wallace and Morowitz (1973), Herdman (1985), Trevors (1996), Baumann *et al.* (1998), Casjens (1998), and Cole and Saint-Girons (1999), and updated lists of sequenced genomes that include genome size data are available from the Genomes OnLine Database (www.genomesonline.org), the TIGR Comprehensive Microbial Resource (www.tigr.org/tigr-scripts/CMR2/CMRHomePage.spl), and the CBS Genome Atlas Database (www.cbs.dtu.dk/services/GenomeAtlas).

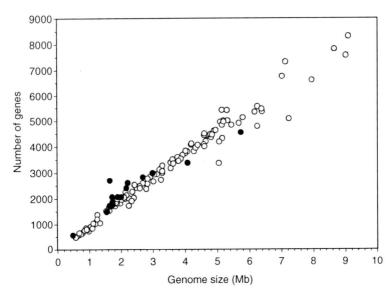

FIGURE 10.11   The relationship between gene (i.e., open reading frame) number and genome size in prokaryotes, as revealed by data from 140 completely sequenced genomes. Unlike in eukaryotes, gene number is strongly positively correlated with genome size in both Archaea (●) and Bacteria (○). The regression statistics were as follows, Archaea: $r^2 = 0.88$, $P < 0.0001$, $n = 18$; Bacteria: $r^2 = 0.97$, $P < 0.0001$, $n = 122$; all prokaryotes: $r^2 = 0.97$, $P < 0.0001$, $n = 140$. The regressions were very slightly stronger following log-transformation, but not substantially different. It has been reported that the archaeon *Aeropyrum pernix* and the bacterium *Mycobacterium leprae* represent exceptions to this trend, with the former having more than the expected number of genes and the latter exhibiting fewer than expected (Doolittle, 2002; Tanaka *et al.*, 2003). However, that these two species are distinct outliers is not so readily apparent with the large dataset used here, in which the relationship generally becomes slightly looser at the higher end of the distribution. Moreover, if the large number of pseudogenes in the *M. leprae* genome are included, this species falls on the line as well (see Mira *et al.*, 2001). Data were taken from the Center for Biological Sequence Analysis (CBS) Genome Atlas Database (www.cbs.dtu.dk/services/GenomeAtlas) in the spring of 2004.

given that genome sizes are invariably small and directly correlated with gene number in these organisms (Fig. 10.11). Nevertheless, the limited variation in prokaryote genome sizes, despite considerable metabolic and ecological diversity, presents an important puzzle in its own right, albeit for the opposite reason as compared to eukaryotes. Indeed, this conspicuous *lack* of variation among prokaryotes is just as important a question as the enormous genome size variation underlying the eukaryotic "C-value enigma" discussed in detail in Chapters 1 and 2.

Overall, prokaryote genome sizes range about 20-fold in size, from 490 kb to more than 9 Mb. Thus prokaryotes possess the smallest genomes among cellular organisms, but also overlap with the low end of the eukaryotic genome size range (see Chapter 1). In Bacteria, the sizes of sequenced genomes range from 580,074 bp in *Mycoplasma genitalium* (Fraser *et al.*, 1995) to 9,105,828 bp in *Bradyrhizobium japonicum* (Kaneko *et al.*, 2002). Based on nonsequencing methods in bacteria, a genome as small as 450 kb has been reported for intracellular γ-proteobacterial symbionts in the genus *Buchnera* (Gil *et al.*, 2002), and genomes may reach sizes as large as 9.7 Mb in the nitrogen-fixing α-proteobacterium *Azospirillum lipoferum* (Islas *et al.*, 2004) and 9.9 Mb in species of δ-proteobacteria in the genus *Stigmatella* (Casjens, 1998). The mean for all bacterial genome sizes measured by either sequencing or PFGE is 3.10 ± 0.09 Mb (Fig. 10.12). Archaeal genome sizes appear in general to be smaller and more constrained than those of Bacteria (though note that these have also been much less well studied),

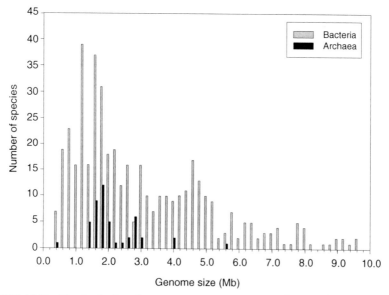

FIGURE 10.12 The distribution of genome size variation among prokaryotes based on complete genome sequencing and pulse-field gel electrophoresis (PFGE) estimates. This includes 18 complete sequences and 29 PFGE measurements for Archaea (black bars), and 125 complete sequences and 323 PFGE estimates for Bacteria (gray bars). Based on this combined dataset, the mean genome size for the Archaea is 2.22 ± 0.13 Mb, and for the Bacteria is 3.10 ± 0.09 Mb. For sequenced genomes alone, the means are 2.19 ± 0.27 Mb for Archaea and 3.40 ± 0.17 Mb for Bacteria. Values from multiple strains per species were averaged, and complete sequencing data were used preferentially where measurements had been made by both methods. Complete genome sequence data were taken from the Center for Biological Sequence Analysis (CBS) Genome Atlas Database (www.cbs.dtu.dk/services/GenomeAtlas) in the spring of 2004, and the PFGE estimates were taken from the dataset compiled by Islas *et al.* (2004), now available as part of the *Prokaryote Genome Size Database* (www.genomesize.com/prokaryotes).

ranging from 490,885 bp in *Nanoarchaeum equitans* (Waters *et al.*, 2003) to 5,751,492 bp in *Methanosarcina acetivorans* (Galagan *et al.*, 2002), with a mean of 2.22 ± 0.13 Mb (Fig. 10.12). In light of these data, two main questions emerge regarding prokaryote genome size evolution: (1) Why are all prokaryote genomes small, especially when there are obvious mechanisms like HGT that can add significant quantities of DNA?, and (2) What accounts for the variation in genome size that does occur among species of prokaryotes?

Factors that Limit Prokaryote Genome Size

The most significant factor that keeps prokaryote genomes small, and strongly distinguishes them from those of most eukaryotes in this sense, is a very low amount of noncoding DNA, be it in the form of transposable elements, introns, or pseudogenes. To be sure, plasmids, prophages, and other foreign sequences do contribute a significant percentage of DNA to many prokaryote cells, but these often contain coding genes and pale in comparison to the amount of DNA represented by mobile elements in eukaryote genomes (see Chapters 1 and 2). For example, whereas some strains of *Shigella* may contain hundreds of copies of the insertion sequence *IS1* (Lawrence *et al.*, 2001), a very high number for bacteria, this is as nothing compared to the more than 3,000,000 transposable element sequences present in each copy of the human genome (International Human Genome Sequencing Consortium, 2001). Also, whereas TE number and genome size appear positively correlated in eukaryotes (Kidwell, 2002; Lynch and Conery, 2003) (see Chapter 3), there is, if anything, only a very weak association between prophage content and genome size in prokaryotes, despite the significant contribution of these elements to some genomes (Casjens, 2003).

Several explanations have been offered to account for the paucity of noncoding elements in prokaryotic genomes. As in eukaryotes, there may be constraints associated with cell size and division rate. In keeping with this, selection for rapid replication is thought to have played a role in the reduction of genome size among organelles (Selosse *et al.*, 2001), and in bacteria there is a strong positive association between genome size and cell size (Shuter *et al.*, 1983). However, the cell size relationship may actually relate to larger cells requiring more gene products (and thus larger genomes), and it has been shown that genome size is not associated with variation in doubling time among culturable bacteria (Mira *et al.*, 2001). Indeed, given that the slowest generation time known among bacteria is found in the small-genomed *Mycobacterium leprae*, it is obvious that a low DNA content is hardly enough to ensure a rapid cell division rate. So, whereas selection for small cell size or rapid replication does not provide a likely explanation for variation in genome size among prokaryotes (see later section), it is nevertheless plausible that these pressures create a relatively low threshold of tolerable DNA contents in this group as a whole. As Gregory (2004) put it, doubling time may not correlate with genome size over the limited real-world range, "but how quickly could a bacterium with a human-sized genome divide, especially with only one replicon origin?" Likewise, there are clear cases in eukaryotes where the genome of one species could not physically fit inside the cell of another (see Chapter 1), and a similar general constraint must operate with regard to the tiny cells of prokaryotes.

As an alternative, Lawrence *et al.* (2001) suggested that high deletion rates are maintained in prokaryote genomes not because compact chromosomes are themselves favored by selection, but rather because the DNA that is removed is potentially detrimental to the cell. Specifically, Lawrence *et al.* (2001) argued that the most efficient way to neutralize genetic parasites like phages (which may kill the cell) and TEs (which may deactivate important genes by inserting into them) is to delete them through recombination between short regions of identity. Under this scenario, an equilibrium deletion rate would be reached that can effectively eliminate parasitic DNA but is not so overzealous as to remove essential genes. Importantly, such a mechanism would ultimately delete all noncoding elements, not just parasitic phages and TEs. The accumulation of pseudogenes in the genomes of some intracellular pathogens like *Rickettsia* and *Mycobacterium* would be a result of relaxed deletional mechanisms in sheltered environments (i.e., eukaryote hosts) where there is less threat from parasitic elements (Lawrence *et al.*, 2001). Perhaps not surprisingly, the accumulation of pseudogenes in *Rickettsia* has also been interpreted in terms of reduced pressure for rapid replication among intracellular parasites that are not competing with other bacterial species and that may kill their hosts prematurely if they replicate too quickly (Holste *et al.*, 2000).

Taking a population genetics approach, Lynch and Conery (2003) provided yet another possible explanation for the small genome sizes of prokaryotes. Under their model, the massive long-term effective population sizes of prokaryotes result in strong purifying selection against mildly deleterious insertions, which can only accumulate when genetic drift plays a more significant role (i.e., in the smaller populations characteristic of large multicellular eukaryotes). Moreover, the rate of insertion and subsequent half-life of duplicated genes is shorter in prokaryotic genomes, which could also contribute to their small size (Lynch and Conery, 2003). This model must be interpreted with caution, however, because one of the explicit predictions made, namely that genome size should correlate negatively with population size (and by extension positively with body size) in mammals has not withstood testing (Vinogradov, 2004). Furthermore, large population sizes are generally associated with *larger* genomes within prokaryotes, whereas population bottlenecks lead to a loss of genes (and a reduction in genome size) by drift (Moran, 2002; Wernegreen, 2002) (see later section). As such, this mechanism does not provide an explanation for variation in genome size among prokaryotes, even if it is possibly of relevance in the comparison with eukaryotes.

Whatever the reason behind it, the general lack of noncoding DNA means that the major determinant of the size of a given prokaryote genome is the number of genes it contains (Doolittle, 2002; Tanaka *et al.*, 2003) (Fig. 10.11). It is therefore also of relevance in the present context that prokaryotic genomes contain a small number of genes as compared with (most) eukaryotes. One obvious interpretation of the low gene number is that a large number of genes is not required to produce their relatively simple morphologies and to carry out their single-celled functions. To be sure, the microbial lifestyle has proved highly successful over vast expanses of evolutionary time. However, this can also be viewed from the opposite perspective, in which the possession of a singular circular chromosome with one replicon origin strongly constrains the size of the genome that can be obtained, and therefore the number of genes that can be contained within it, and hence the potential complexity of the resulting organism (Maynard Smith and Szathsmáry, 1995; Vellai *et al.*, 1998; Vellai and Vida, 1999). That is to say, prokaryotes may have small genomes because they are simple, and/or prokaryotes may be simple because they are limited to small genome sizes.

MECHANISMS OF GENOME SIZE INCREASE

In angiosperm plants, the distribution of genome sizes is highly right-skewed, implying a small ancestral genome that has been expanded in only a few lineages (see Chapter 2). Whereas it is clear that significantly fewer species of prokaryotes exhibit "large" genomes than small to medium-sized ones, the distribution is not nearly so right-skewed as it is in plants (Fig. 10.12). This pattern is compatible with the notion that there have been both increases and decreases in genome size over the course of prokaryote evolution. Because there is so little noncoding DNA in prokaryotes, the mechanisms responsible for generating this variation must largely involve changes in gene number.

In general, the largest prokaryote genome sizes are found in species that produce a large number of gene products or have complex life cycles, such as free-living soil-dwelling and aerobic aquatic bacteria (e.g., Stepkowski and Legocki, 2001; Islas *et al.*, 2004) (Fig. 10.13). It is also of interest that some of the large-genomed species possess linear and/or multiple chromosomes. Thus the three genera with the largest genomes known among bacteria are *Bradyrhizobium* (linear chromosome), *Streptomyces* (linear chromosome), and *Burkholderia* (multiple circular chromosomes). In cases such as these, genome expansion relates to the acquisition or *de novo* evolution of new coding genes, possibly facilitated by the possession of a unique and more accommodating chromosome structure (but note that some species with large genomes, like *Mesorhizobium*, have single circular chromosomes, whereas some with multiple linear chromosomes, such as *Borrelia*, have small genomes). More generally, gene duplications have clearly

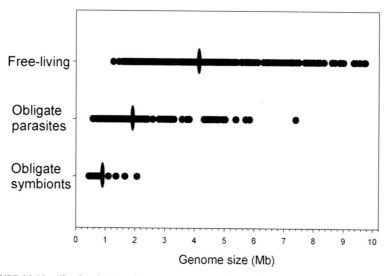

FIGURE 10.13   The distribution of genome sizes according to lifestyle in the Bacteria. Each point represents the genome size (measured by pulse-field gel electrophoresis) of one species or strain of bacteria categorized as either free-living ($n = 398$), obligately parasitic ($n = 227$), or obligately symbiotic ($n = 20$) as in Islas *et al.* (2004). The means for each category are indicated with vertical ellipses. Data were provided by S. Islas and A. Lazcano, Universidad Nacional Autónoma de México.

played a role in shaping prokaryote genome sizes, given that the number of paralogs present in the genome correlates strongly with overall size (Gevers *et al.*, 2004) (see Chapters 5 and 6). It has been suggested in the past that entire genome duplications have contributed to genome size increases in bacteria (e.g., Wallace and Morowitz, 1973; Herdman, 1985), but the recent analysis by Gevers *et al.* (2004) points to the input of small-scale duplication events only.

Genome Reductions in Obligate Parasites and Endosymbionts

One of the clearest patterns in bacterial genome size is that species engaged in obligate associations with hosts—that is, parasites and endosymbionts, in particular intracellular ones—have smaller genomes than free-living species (Mira *et al.*, 2001; Stepkowski and Legocki, 2001; Moran, 2002; Islas *et al.*, 2004; Klasson and Andersson, 2004) (Fig. 10.13). Wallace and Morowitz (1973) suggested that *Mycoplasma* spp. (which have the smallest bacterial genomes) represent the original bacteria, and that all other lineages had experienced increases in genome size (via large duplications). However, it has long been recognized that the small genomes of *Mycoplasma* spp. and other parasites are derived features, meaning that there have been secondary reductions in genome size in parasitic and endosymbiotic lineages (e.g., Herdman, 1985). Indeed, it is clear that such reductions have occurred independently many times in bacteria, for example in *Wolbachia* and *Rickettsia* within the α-proteobacteria (Stepkowski and Legocki, 2001; Wu *et al.*, 2004); in obligate intracellular symbionts of insects such as *Baumannia* (in leafhoppers), *Blochmannia* (in ants), *Buchnera* (in aphids), and *Wigglesworthia* (in tsetse flies) within the γ-proteobacteria (Moran and Mira, 2001; Wernegreen *et al.*, 2002; Moran *et al.*, 2003a; van Ham *et al.*, 2003); in the Chlamydiae (Read *et al.*, 2000); in the Gram-positive intracellular plant pathogen *Phytoplasma asteris* (Oshima *et al.*, 2004); and in parasitic spirochetes such as *Borrelia* (Fraser *et al.*, 1997). Extreme genome reduction can also be found in the hyperthermophilic nanoarchaeon *Nanoarchaeum equitans*, which has the smallest prokaryote genome so far sequenced and appears to be a parasite of the crenarchaeon *Igniococcus*, with which it lives in obligate co-culture (Waters *et al.*, 2003; Fraser, 2004b). In each of these cases, genome size reduction has been carried out primarily via the loss of a large number of genes.[6]

Reductions in genome size among parasites and symbionts have often been attributed to selection for more rapid replication, but several observations make it clear that this is not the explanation: (1) there is no correlation between genome size and doubling time in bacteria, (2) individual deletions on the order of single genes have not been shown to affect cell division rate (let alone those of only 3 bp, the median size observed in reduced genomes), (3) some of the reduced genomes in pathogens and endosymbionts contain a high proportion of noncoding pseudogenes, (4) the small genomes of host-dependent species do not appear any more tightly packed with regard to intergenic spacers than those of free-living species, and (5) species such as *Buchnera* have highly reduced genomes but maintain a relatively large amount of DNA in each cell by being highly polyploid (Mira *et al.*, 2001; Moran and Mira, 2001; Moran, 2002, 2003).

Part of the explanation for this extensive gene loss relates to the fact that a shift from a free-living to a host-dependent lifestyle renders many genes obsolete, meaning that they are then no longer maintained by selection, will be deactivated by mutation, and then subsequently lost early on by large deletions, followed by

---

[6]In an interesting exception that proves the rule, the small (0.2–0.5 × 0.5–2.5 μm) δ-proteobacterial species *Bdellovibrio bacteriovorus* which, as its name implies, is an obligate parasite of other Gram-negative bacteria, maintains a surprisingly large genome of 3,782,950 bp containing 3584 genes (Rendulic *et al.*, 2004). This is explained by the predator's remarkably complex life cycle, which includes eight different stages involving a variety of specialized mechanisms of motility, prey recognition, invasion, and digestion (Rendulic *et al.*, 2004; Sebaihia *et al.*, 2004). Intriguingly, despite its genome size and direct access to the genetic material of the bacteria on which it feeds, there is no evidence of recent horizontal transfer to *B. bacteriovorus* from its prey (Rendulic *et al.*, 2004).

more gradual mutational erosion (Mira *et al.*, 2001; Moran and Mira, 2001; Moran, 2002, 2003; Wernegreen, 2002). In fact, the significant bias toward deletions over insertions in bacterial genomes indicates that DNA generally will not be retained unless selection favors its presence (Andersson and Andersson, 2001; Mira *et al.*, 2001; Moran, 2002). *Mycobacterium leprae* and *Rickettsia prowazekii* appear to represent snapshots of this phenomenon in progress, because large fractions of their genomes consist of pseudogenes that have yet to be eliminated (Andersson and Andersson, 1999a,b; Moran, 2002). Genome reduction by this route is a long-term and still ongoing process (Andersson and Andersson, 1999a,b; van Ham *et al.*, 2003), and evidence from *Buchnera*, *Rickettsia*, and other species suggests that pseudogenes may remain largely intact in the genome for very long periods of time (Mira *et al.*, 2001).

Interestingly, it clearly is not the same set of genes that is lost in each genome reduction event, because the collection of universal orthologs shared among intracellular bacteria is limited (Moran, 2002; Wernegreen, 2002). Nevertheless, there are some general trends. Notably, genes commonly lost include those underlying the synthesis of amino acids and other products that can be acquired from the host (Moran, 2002; Klasson and Andersson, 2004). The major exception to this is found in *Buchnera*, which maintains genes that allow it to provision its aphid host with essential amino acids that are not commonly found in the plant phloem sap upon which the insects feed (although genes coding for amino acids that the host can produce itself have been lost) (Moran *et al.*, 2003b). In fact, some of the genes involved in the synthesis of tryptophan and leucine have been recruited to plasmids in *Buchnera*, presumably as a means to up-regulate their expression for the host's benefit (Moran and Baumann, 2000). *Buchnera* has lost important DNA repair genes (e.g., *recA*), but *Nanoarchaeum equitans*, which grows at 90°C, retains a large repertoire of genes involved in DNA repair despite having the smallest known genome of any cellular organism (Fraser, 2004b).

Other genes commonly lost include regulatory elements and genes involved in nucleotide and vitamin biosynthesis (Moran, 2002; Klasson and Andersson, 2004). Again there is an exception to this: *Wigglesworthia glossinidia*, the primary endosymbiont of the blood-feeding tsetse fly, retains genes that allow it to provision its host with B-complex vitamins not available from vertebrate blood (Akman *et al.*, 2002). *Wigglesworthia* also retains genes coding for the production of flagella, which may have been adapted to facilitate the transmission of bacteria from adult flies to larvae (Akman *et al.*, 2002). In general, the genes lost tend to differ between intracellular pathogens versus endosymbionts, with the latter maintaining biosynthesis-related sequences necessary for provisioning the host with nutrients, and the former possessing transport protein genes needed for the theft of metabolic products from the host and cell-surface proteins required for avoidance of the host's immune system (Andersson, 2000; Tamas *et al.*, 2001). Overall, endosymbiosis may involve a greater loss of genes, perhaps reflecting the more intimate association with the host than parasitism (Fig. 10.13). For example, Group C and D *Wolbachia* (which are mutualists of filarial nematodes) have genomes about 30% smaller than strains in Group A (which, along with Group B, are parasites of arthropods) (Sun *et al.*, 2001).

The simple "use it or lose it" scenario does not provide the entire story, however, because many presumably beneficial genes (e.g., for replication, transcription, translation, recombination, and repair) also are often lost (Moran, 2002). This can happen because movement into a host represents a bottleneck in which effective population size is significantly reduced; this will be particularly acute in maternally inherited intracellular bacteria, which are passed on in limited numbers from one host generation to the next. In this case, genetic drift may come to outweigh selection in such a way that even useful (but not essential) genes can be lost by mutation (Andersson and Kurland, 1998; Moran and Mira, 2001; Moran, 2002; Wernegreen, 2002). There is also the possibility that bacteria living inside hosts are exposed to lower levels of horizontal gene transfer and therefore take up less DNA than free-living species (Lawrence *et al.*, 2001; Mira *et al.*, 2001; Ortutay *et al.*, 2003; Silva *et al.*, 2003). To be sure, the phylogenetic congruence between intracellular bacteria and their hosts indicates that little or no gene transfer occurs between strains occupying different hosts (e.g., Wernegreen, 2002). Overall, it is apparent that genome reduction via gene loss involves several different mechanisms, including the deactivation and deletion of genes that become superfluous in a host environment, a weakening of selection maintaining useful but nonessential genes, and a reduced level of gene replacement by horizontal transfer (Mira *et al.*, 2001).

Other genomic changes are typical of intracellular bacteria. For example, many pathogens and symbionts have very low $G+C$ contents, which helps to explain the positive correlation with genome size (Moran, 2002; Wernegreen, 2002) (Fig. 10.8). This may reflect a mutational bias resulting from the loss of genes encoding DNA repair enzymes that would normally correct the error-related tendency toward $A+T$ enrichment (e.g., by the accidental deamination of cytosine to uracil, which is then replaced by thymine) (Akman *et al.*, 2002; Moran, 2002; Wernegreen, 2002). Observable mutation rates in general are greatly accelerated in intracellular bacteria, reflecting not only the reduced repair mechanisms, but possibly also a relaxation of selection in the stable environment of a host cell or (more likely) the effects of increased genetic drift in small populations that allow more mutations to become fixed (Moran and Baumann, 2000; Moran, 2002; Wernegreen, 2002).

BACTERIA, ORGANELLES, OR SOMETHING IN BETWEEN?

It is important to recognize that the evolution of the eukaryotic hosts, and not just that of the prokaryotes they harbor, has been greatly influenced by the occurrence of intracellular bacteria. As noted previously, parasitic *Wolbachia* have evolved mechanisms to alter host reproduction, which could provide a mechanism of reproductive isolation and thereby influence host speciation (e.g., Werren, 1998). The diversification of various insects has clearly been affected by intracellular symbionts as well, given that more than 10% of all insect species are thought to be dependent on obligate bacterial symbionts for successful growth and reproduction (Wernegreen, 2002). Certainly, the invasion of niches based on nutritionally deficient food sources such as plant sap and vertebrate blood would not have been possible without the contribution of intracellular bacteria.

The associations between *Wigglesworthia* and *Buchnera* and their hosts began long ago (50–100 and 150–200 million years ago, respectively) and have since evolved into relationships of strong mutual dependency. The closeness of the interactions is reflected in the fact that these bacteria live in specialized host cells called "bacteriocytes" (or "mycetocytes"), often concentrated within a distinct organ called the "bacteriome." *Wigglesworthia* has lost genes involved in the initiation of DNA replication, which suggests that the endosymbiont may be dependent on the tsetse fly host for its very reproduction (Akman *et al.*, 2002), and *Buchnera* seems to be in the process of losing the capacity for cell wall biosynthesis (Tamas *et al.*, 2001). Irreversible changes like these make the intracellular bacteria even more dependent on their hosts (Tamas *et al.*, 2001; Wernegreen *et al.*, 2002).

Taken together, this long-running, highly intimate, and irreversible mutual dependency between bacteria and host has raised the question as to whether *Buchnera, Wigglesworthia,* and their ilk should actually be considered organelles rather than separate bacteria (Andersson, 2000; Douglas and Raven, 2002). The determining feature in this case would be whether genes had been transferred from the bacteria to the host nucleus, which remains an open question at present (Douglas and Raven, 2002). The only comparison available thus far along these lines is between *Wolbachia* and *Drosophila,* in which transfer to the host appears not to have happened (Wu *et al.*, 2004), but this is a case of intracellular parasitism, not mutualism. It is acknowledged that the localization of endosymbionts like *Buchnera* inside specialized bacteriocytes that are not part of the germline, meaning that transfers to the nuclei of these cells would not be passed on, renders such gene transfers less likely (Douglas and Raven, 2002). However, alternative routes to gene transfer may exist (Douglas and Raven, 2002), and it is possible that examples will be found if the hosts' genomes are sequenced. Until then, the most suitable compromise may be to consider *Buchnera* and others as "proto-organelles," with genomes "at the interface between bacteria and organelles" (Douglas and Raven, 2002).

THE MINIMAL GENOME CONCEPT

The discovery that the human genome contains a mere ~30,000 genes indicated that even the most complex life-forms can function with far fewer coding sequences than previously anticipated (see Chapter 9). This finding reinforced a question that had been contemplated for several years prior, namely the minimum number of genes necessary to sustain the simplest forms of life (Mushegian, 1999; Koonin, 2000; Reich, 2000; Peterson and Fraser, 2001). This concept of a "minimal genome" is being actively investigated from a variety of perspectives using prokaryote genome data.

Breaking it Down

The most obvious way of determining the minimal gene set necessary for cellular life is to determine how many genes can be removed from existing genomes without rendering the cell inviable. As discussed previously, nature has undertaken this process to a significant degree in certain bacterial lineages; researchers are now attempting to carry it to its minimalist conclusion.

Natural Examples

Ten years before the first complete bacterial genome was sequenced, Herdman (1985) considered whether the genome of the sugar-fermenting bacterium *Zymomonas mobilis* "perhaps defines the minimum genetic information content required for a free-living existence," based on its small genome size and specialized lifestyle. This line of inquiry has continued to the present day, although the focus is now on parasites and endosymbionts, which have the most thoroughly reduced genomes of any cellular life-forms. In this regard, it is clear that a host-dependent organism may function with as few as 480 protein-coding genes, as in *Mycoplasma genitalium* (Fraser et al., 1995), and possibly only 400 genes, as in some strains of *Buchnera* (Gil et al., 2002). Although any naturally occurring genome will not represent the true minimum gene set possible in theory owing to the presence of prophage-encoded genes, gene duplicates, and other redundant or otherwise nonessential elements, these do set an "upper limit" on what the smallest number of required genes must be. Based on a practical understanding of cell biology, it can be assumed that the minimum gene set will include at least 100 genes (Koonin, 2000), which means that even with no further analysis the minimal genome can be defined in rough outline to a range of about 100 to 450 genes.

Computational Approaches

To obtain a more refined estimate of the minimal gene set, Mushegian and Koonin (1996) made comparisons of the gene content in the newly sequenced genomes of *Haemophilus influenzae* and *Mycoplasma genitalium*. Under the assumption that any genes common to these two distantly related parasites are likely to be essential to a bacterial cell, they calculated a minimum required complement of 256 genes.

Such computational methods have been expanded in the time since these two genomes were sequenced, and in general have been consistent in suggesting a minimal set of about 250–300 essential genes (Mushegian and Koonin, 1996; Mushegian, 1999; Koonin, 2000).

Experimental Gene Inactivation

A more direct means of assessing whether a gene is essential is to inactive it and observe whether the cell remains viable. When performed on a genomewide scale, this approach can provide insight into the total subset of genes necessary for survival. Itaya (1995) made use of a method of "transformational mutagenesis" (i.e., disrupting existing genes by inserting a foreign sequence with the use of restriction enzymes) to knock out genes at 79 random loci in *Bacillus subtilis*, and found that only six of these were lethal. Cell viability could even be maintained after the simultaneous inactivation of 33 loci. Based on these results, Itaya (1995) calculated that 318 to 562 kb of the *B. subtilis* genome was essential, which, given an average bacterial gene length of ~ 1 kb, translates to about 300–500 genes.

In a more detailed experiment, Hutchison *et al.* (1999) used "global transposon-mediated mutagenesis" (i.e., knocking out genes by transposon insertions) to disrupt one gene at a time in *Mycoplasma genitalium* and found that only 265 to 350 of the 480 genes present (including about 100 of unknown function) were essential for cell viability. A similar transposon-mediated mutagenesis study in *Haemophilus influenzae* by Akerley *et al.* (2002) indicated that only 259 genes were essential for survival, of which 54% lacked clearly defined functional roles. Working with *Staphylococcus aureus*, Ji *et al.* (2001) used antisense RNA to inhibit individual gene expression, and identified slightly more than 150 critical genes (30% of which were of unknown function), although it must be noted that this latter method is limited to genes for which sufficient expression of inhibitory antisense DNA can be achieved (Kobayashi *et al.*, 2003).

Most recently, Kobayashi *et al.* (2003) used the insertion of a nonreplicating plasmid by a single crossover recombination to deactivate individual genes in *B. subtilis*. Of the roughly 4100 genes in the genome, only 192 were shown to be essential, with another 79 predicted to be indispensable, for a total of 271. Of these, about $1/2$ are involved in information processing, $1/5$ play a role in the synthesis of the cell envelope and the determination of cell shape and division, and $1/10$ relate to cell energetics (Kobayashi *et al.*, 2003).

In sum, the various analytical approaches that have been brought to bear on the question of the minimal genome seem to converge on an estimate of about 250–300 genes. However, although this represents a useful working range, it must be borne in mind that each of these methods has limitations that may lead to either over- or underestimation of the size of the minimal gene set. For example, comparative computational approaches might underestimate the minimal gene set because this only includes genes that remain similar enough to be identified as shared and ignores the fact that some common essential functions may be fulfilled by different genes in unrelated species (Kobayashi et al., 2003). Transposon-mediated mutagenesis may miss essential genes that can tolerate TE insertion, or may overestimate by counting as essential genes whose inactivation only slows but does not arrest growth (Kobayashi et al., 2003). More generally, all of the experimental methods based on knocking out individual genes are limited because this approach indicates only whether a given gene is needed in combination with functional copies of all other genes. It does not reveal what the actual minimum workable *combination* of genes would be, which is a question that cannot be addressed experimentally at present because the number of possible gene combinations is prohibitively large.

Perhaps more important are the conceptual limitations of minimal genome research. It is clear that the minimal gene sets determined in the various species studied are not composed of the same genes (Koonin, 2000), and that one must also address the question of "minimal in which environment?" (e.g., Peterson and Fraser, 2001). In the case of experimental studies, the environment is an idealized one enriched with all necessary nutrients and lacking environmental stresses (Koonin, 2000; Peterson and Fraser, 2001). In this sense, a paring down of the genome is not so much a simplification of genetics as it is a transfer of the support system required for survival from within the organism itself to its external environment. As Peterson and Fraser (2001) noted, the minimal genome is a theoretical construct only and is not something that could ever be found in nature. Nonetheless, this may still be useful as the minimal gene set determined under these unrealistically favorable conditions would presumably be needed in any natural environment (Peterson and Fraser, 2001).

BUILDING IT UP

In November of 2002, Craig Venter announced his intention to assemble a minimal bacterial genome from scratch and then insert it into an existing cell from which the DNA had been removed. If the cell functions, then future work could involve adding genes useful for hydrogen fuel production or environmental remediation (Check, 2002; Marshall, 2002). However, the main reason for conducting the project, Venter conceded, is to determine the minimal gene set that can support cellular life (Marshall, 2002). Many researchers remain skeptical as to the project's chances of succeeding, but there is already evidence that artificial genome synthesis is possible, at least for the relatively small genomes of viruses. In fact, two functional viral genomes have already been constructed from scratch: that of the poliovirus (Cello et al., 2002) and the phage φX174 (Smith et al., 2003). This latter project took only 14 days to complete using a technique known as "polymerase cycle assembly" (PCA; an adaptation of the classic PCR method) and seems an especially appropriate choice, given that phi X's was the first genome ever fully deconstructed by complete sequencing.

GENOMIC INSIGHTS INTO PROKARYOTIC ABUNDANCE AND DIVERSITY

The naming of prokaryote species is controlled by the International Code of Nomenclature of Bacteria (Lapage *et al.,* 1992), which includes rules regarding valid publication, legitimacy, and priority of naming. According to the DSMZ database (www.dsmz.de), as of the spring of 2004 there were 6544 validly published bacterial species names in 1227 genera, although ~20% of these are synonyms. Despite their much lower overall diversity, nearly 10 times more vertebrates have been described than prokaryotes. To appreciate the scale of this discrepancy, one need only consider that 500–1000 prokaryotic species reside within the intestinal tract of each human being alone, represented by a staggering number of cells (6×10^{13} individual prokaryotes *per colon* versus 6×10^9 humans on the planet) and collectively encoding far more genes (2–4 million versus 30,000) in this "microbiome" than are found in the human genome (Xu *et al.,* 2003; Oren, 2004).

One of the main reasons that so much prokaryote diversity (probably representing a majority of life's total) remains unrecorded is that taxonomically relevant features are very difficult to discern among tiny single-celled organisms, even in the small minority (<1%) of species that can be cultured and studied in detail in the lab (Rosselló-Mora and Amann, 2001; Hugenholtz, 2002; Torsvik *et al.,* 2002). Genomic data are not limited in the same way and are now providing the first glimpses of the overwhelming complexity of this formerly invisible microbial world.

WHAT IS A PROKARYOTIC SPECIES?

The first crucial step in the assessment of prokaryotic species diversity is to determine what constitutes a prokaryotic species. Unfortunately, species definitions are contentious subjects and represent one of the most enduring and frustrating problems in evolutionary biology—even more so in prokaryotes than eukaryotes. Most eukaryotic species definitions involve irreversible reproductive (i.e., genetic) and phylogenetic isolation among cohesive groups. Given a potentially very high level of horizontal gene transfer and a predominantly asexual mode of reproduction in the Bacteria and Archaea, these criteria pose substantial difficulties for prokaryote taxonomists (e.g., Lawrence, 2002). Cohan (2002) made the interesting observation that bacterial systematists broke away from their colleagues working with eukaryotes on the definition of species at about the same time that this broader cohesion criterion took hold during the development of the Modern Synthesis (see Chapter 11). A detailed discussion of prokaryotic species concepts falls well outside the scope of this chapter,[7] but the one most widely in use is a pragmatic "phylo-phenetic species concept" in which a prokaryotic species is defined as "a monophyletic and genomically coherent cluster of individual organisms that show a high degree of overall similarity with respect to many independent characteristics, and is diagnosable by a discriminative phenotypic property" (Rosselló-Mora and Amann, 2001). This inclusion of both genomic and phenotypic data is known as the "polyphasic approach," and is generally seen as necessary for the naming of new species[8] (Rosselló-Mora and Amann, 2001; Stackebrandt et al., 2002).

GENETIC DELINEATION OF SPECIES

Molecular techniques have been used to characterize prokaryote species since the 1960s, and now represent the dominant approach (Rosselló-Mora and Amann, 2001; Stackebrandt et al., 2002). At first, only basic genomic data such as G + C content (as estimated by biochemical methods) were available, but these could only be used to show that two bacteria were not from the same species; they could not be used to establish conspecifics. The first major molecular breakthrough came in the 1970s with the advent of DNA–DNA hybridization methods, which gave a crude measure of the overall similarity of two genomes (see Rosselló-Mora and Amann, 2001). Johnson (1973) established that strains of the same species consistently shared at least 70% of their genome content, whereas representatives of any two randomly chosen species tended to show less than 70% similarity. Thus a 70% DNA–DNA hybridization similarity became the standard for clustering individuals into the same species. However, it has been argued that this cutoff is arbitrary and too inclusive; as Staley (1997) pointed out, a 70% cutoff would place humans, the other great apes, and lemurs all in the same species. Such a problem arises in the genus *Neisseria*, for example, in which *N. gonorrhoeae*, *N. meningitis*, *N. lactamica*, and *N. polysaccharea* would all be considered the same species by this measure (Dykhuizen, 1998). On the other hand, a flexible definition may be appropriate for bacteria, given that some strains of the same species may differ as much as 25% in genome size (Stackebrandt, 2003) (Table 10.1).

---

[7]For discussions of prokaryotic species concepts and delineation criteria, the reader is referred to Palys et al. (1997), Staley (1997), Ward (1998), Cohan (2001, 2002), Lan and Reeves (2001), Rosselló-Mora and Amann (2001), and Oren (2004).

[8]Putative prokaryote "species" that have been well characterized genomically but for which a formal description is not available are given the taxonomic designation "Candidatus" before the proposed species name (Murray and Stackebrandt, 1995).

In the 1980s, sequencing of rRNA genes became common, and 16S rRNA in particular began to contribute substantially to prokaryote taxonomy.[9] In this case, it was found that the cutoff for within-species variation was 97% sequence similarity (Stackebrandt and Goebel, 1994). 16S rRNA sequences are now provided in prokaryotic classifications as a matter of course (Rosselló-Mora and Amann, 2001). Together, the 97% rRNA and 70% DNA–DNA hybridization similarity cutoffs have allowed for the discovery and classification of many new prokaryote species that have resisted culturing, although it should be noted that these two measures do not always agree (Stackebrandt, 2003; Oren, 2004). It is likely that other types of genomic data (e.g., sequencing of multiple housekeeping genes and DNA profiling) will be added to the repertoire of prokaryote taxonomy methods in the near future (Stackebrandt *et al.,* 2002; Oren, 2004), and methods involving microarray and other hybridization approaches have already been implemented to this end (e.g., Cho and Tiedje, 2001). Additional molecular methods, including DNA fingerprinting, multilocus sequence typing (MLST), and numerous others, are already commonly used to assess strain diversity within species as well (Rosselló-Mora and Amann, 2001; Cohan, 2002; Stackebrandt, 2003).

GENOMIC PERSPECTIVES ON PROKARYOTE DIVERSITY

The question "how many prokaryotes are there?" can be answered in two ways: either in terms of abundance (the number of individuals) or diversity (the number of species). Recent estimates using cell counts in various environments suggest that there are about 1.2×10^{29} individual prokaryotes in aquatic habitats (ocean, lakes, rivers, etc.), 2.6×10^{29} in the soil, 2.5–25×10^{29} in terrestrial subsurface areas, and 3.6×10^{30} in oceanic subsurface regions—for a global grand total of about 5×10^{30} individual prokaryotic cells (Whitman *et al.,* 1998). To put this in perspective, the estimated number of stars in the known universe is "only" 7×10^{22}.

Genomic data have already vastly expanded the realm of known prokaryote groups (Fig. 10.14) and are the only hope for addressing the issue of how many species are represented in the enormous global population of prokaryotic cells. This is a question that has most recently been addressed by the large-scale analysis of genomic data in what is now known as "ecogenomics," "environmental genomics," or "metagenomics" (Béjà, 2004). Beginning in 1990, several studies were conducted using a method of purifying and amplifying bacterial rRNA genes from environmental samples to assess the number of species present. Specifically, the mixture of 16S rDNA PCR fragments so amplified is cloned and a predescribed

---

[9]A dataset of over 100,000 rRNA gene sequences is now available, and can be accessed from the Ribosomal Database Project II (RDP-II) (rdp.cme.msu.edu/index.jsp) (Cole *et al.,* 2003).

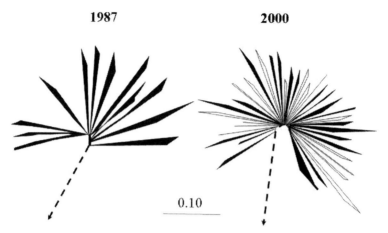

FIGURE 10.14 Schematic representation of a major increase in recognized prokaryote diversity between 1987 (left) and 2000 (right). According to N.R. Pace (personal communication), the diversity of known bacteria has nearly doubled again since the first such comparison was made in the late 1990s. The solid branches in the 2000 tree indicate groups that have cultivated representatives, and open branches reflect groups known only on the basis of environmental rRNA gene sequences. Branch width roughly equates to recognized species diversity (on the basis of available rRNA gene sequences). The dashed branch in both trees leads to the Archaea and Eukaryota. The scale bar at the bottom indicates genetic distance and the branches in the diagrams are drawn to this scale. Based on a figure first presented by Pace (1997), reproduced by permission (© American Association for the Advancement of Science); see also Hugenholtz et al. (1998) and DeLong and Pace (2001). The 1987 diversity diagram was based on data originally compiled by Woese (1987).

number of clones picked and sequenced. The resulting sequences are then analyzed using phylogenetic methods to group them according to relatedness and similarity, and thereby to allow an estimation of the number of species that have contributed sequences to the sample (Hugenholtz, 2002). By noting the lack of correspondence between these newly obtained rRNA sequences and those of known bacteria in existing sequence databases, it was initially estimated that only about 1–2% of species had been described. Larger environmental rRNA surveys performed over the ensuing decade implied that current descriptions actually cover only 0.1% of bacterial species (see Oren, 2004). Based on the number described thus far, this would suggest that there are more than 6 million species of prokaryotes.

DNA–DNA hybridization has also been used to estimate prokaryote diversity in particular environments. For example, Torsvik et al. (2002) reported that 30–100 cm^3 samples of soil contain 3000–11,000 different genomes, corresponding to 10,000 species by the 70% cutoff criterion. Dykhuizen (1998) estimated the number of bacterial species in 30 grams of forest soil at more than half a million, and calculated that globally there are at least 1 billion species. Although such environmental DNA–DNA hybridization methods probably significantly overestimate the number of real species (Stackebrandt, 2003), it is clear that both of the prominent molecular classification methods imply an enormous amount of prokaryotic diversity.

Genomic studies have also revealed a surprisingly high level of diversity among the Archaea (Stein and Simon, 1996; Pace, 1997; Oren, 2004). For example, environmental rRNA sampling in extreme habitats such as the hot springs of Yellowstone National Park in Wyoming showed archaeal communities to be far more diverse than anticipated (Barns *et al.*, 1994, 1996). Two new major taxonomic groups, the Korarchaeota and Nanoarchaeota, have also been identified by such methods. Even more surprisingly, archaeons were found to be present in large numbers in nonextreme environments, including the open ocean, coastal marine waters, a polar sea, continental shelf sediments, salt marshes, freshwater and saline lakes, soils, and in association with eukaryotes (DeLong and Pace, 2001; Takai *et al.*, 2001). It has now been estimated that Archaea constitute 20% of prokaryote cells in the ocean, whereas a decade ago they were not known to exist there at all (DeLong and Pace, 2001).

Using "a meaningful difference in the sequence of the 16S RNA gene" as the defining indicator of bacterial "species," coupled with the standard ecological tool known as "log-normal species abundance curves," Curtis *et al.* (2002) were able to provide one of the most promising estimates of global prokaryote diversity available thus far (see also Ward, 2002; Nee, 2003). Specifically, they estimated that there are 160 species of prokaryotes in 1 milliliter of ocean water, 6400 to 38,000 species per gram of soil, and at most 70 species per milliliter of sewage. Unfortunately, it is not immediately obvious how these estimates of diversity in small samples translate to global diversity—as Nee (2003) noted, 70 species per milliliter of sewage could mean anything from 70 to several billion species per sewage plant, depending on how diversity changes through space. Moreover, these analyses were admittedly based on limited data and unproven assumptions about species distributions (Curtis *et al.*, 2002). Nevertheless, Curtis *et al.* (2002) provided a rough estimate of about 2 million prokaryote species in the oceans and 4 million in soil, which is compatible with previous suggestions that to date less than 0.1% of species have been given names. The finding by Curtis *et al.* (2002) and others that there are about twice as many prokaryote species (and individuals) in soil than in oceans is especially intriguing, but has so far defied explanation. As Nee (2003) noted, "understanding this difference is as important as understanding the latitudinal gradients [in biodiversity] observed for macroscopic organisms."

SHOTGUN FIRST, ASK QUESTIONS LATER

In the past, limitations of technology made it possible to assess only small fragments of DNA, especially 16S rRNA genes, from species sampled from microbial communities. Although this provided important insights into the diversity of unculturable and therefore previously unknown species, it does not grant any more detailed information about their physiology or biochemistry (Béjà et al., 2000). Fortunately, recent developments in genomic library construction, high-throughput sequencing, and genome assembly methods have begun to allow much more substantial DNA sequences to be examined from large environmental samples (DeLong, 2002).

Bacterial artificial chromosomes (BACs) have been used to create genomic libraries of large DNA fragments from environmental samples of prokaryotes, from both soil and marine environments (Béjà et al., 2000; Rondon et al., 2000; DeLong, 2002; Béjà, 2004; Daniel, 2004). In both cases, 16S rRNA genes were examined and a small subset of the larger clones was sequenced. Subsequent analyses of BAC clones from unculturable marine prokaryotes revealed a surprising diversity of types among photosynthetic bacteria (Béjà et al., 2002; Béjà, 2004).

By far the most ambitious example of environmental genomics is the approach used recently by Venter et al. (2004) to shotgun sequence entire samples collected from the Sargasso Sea in the North Atlantic Ocean off the coast of Bermuda. In this case, ~2 million cloned fragments 2 to 6 kb in size were sequenced randomly, covering a total of just over 1 billion nonredundant base pairs of DNA sequence. In this sample, 1.2 million new genes were identified (about three times as many as had been characterized in the past decade) (Fig. 10.5), and 1800 genomic "species" were delineated, including 148 previously unknown bacterial phylotypes (Venter et al., 2004). Amazingly, the Sargasso Sea was chosen specifically to simplify the analysis because it is nutrient-impoverished—and yet, 1 liter of water from this site contains as much unique genetic material as the entire human genome (Whitfield, 2003).

There are still some challenges to be overcome before this approach can be applied more broadly. First, such large-scale sequencing is only possible at a few dedicated institutes, and is still well beyond the reach of the typical individual laboratory (Béjà, 2004). Second, reassembling shredded genomic fragments from so many species (which is essentially akin to assembling 1800 jigsaw puzzles from a single box containing 2 million pieces) (Whitfield, 2003) is very difficult even with the most powerful computers available (Béjà, 2004). Third, this method generates a substantial amount of redundancy—about 50% of the clones represented DNA from just the two most abundant species (Falkowski and de Vargas, 2004). Indeed, Venter et al. (2004) were only able to reassemble (mostly) complete genomes for two species in their sample. Finally, estimates of biodiversity based on such a method are complicated by the fact that rare species are overlooked; it would take an order of magnitude more sequences to be sure that 95% of species present were accounted for (Falkowski and de Vargas, 2004).

These problems aside, the random environmental shotgun sequencing method holds extraordinary promise for the large-scale characterization of diversity, which, only a few years ago, would have been totally hidden to science. As Venter noted, "Based on the data we already have, we're predicting that this will become the number one way for characterizing the environment" (quoted in Whitfield, 2003).

APPLICATIONS OF PROKARYOTE GENOMICS

Comparative genomics projects are often instigated and funded with specific potential applications in mind. In other cases, possible applied uses become apparent only after the genome sequence is available. In combination, this has resulted in a varied collection of potential benefits to be derived from prokaryote genomics, ranging from the obvious to the highly imaginative.

MEDICINE

A large fraction of species sequenced to date are pathogens of humans and domestic animals (see Fraser, 2004a), which clearly reveals the hope that genomic data will contribute substantial medical benefits. While genomics-based approaches have yet to pay any large dividends in this area, and although these benefits will not come automatically but instead will require additional computational, comparative, and experimental work, it is evident that genome research has the potential to influence the future of medicine in several important ways (Black and Hare, 2000; Read *et al.*, 2001; Sassetti and Rubin, 2002; Hughes, 2003; Mills, 2003).

Understanding the Mechanisms and Evolution of Virulence

Comparative genomic data have allowed significant advances to be made in the study of bacterial virulence, as noted in several previous sections (see also Weinstock, 2000; Weinstock *et al.*, 2000; Fitzgerald and Musser, 2001; Schoolnik, 2002a; Hughes, 2003). Comparisons between related virulent and nonpathogenic species (e.g., in the genus *Vibrio*) or strains (e.g., in *E. coli*) can provide important information about the mechanisms and historical pathways by which benign bacteria have become pathogenic. This is a significant problem, because the number of pathogenic forms is on the rise (Tang and Moxon, 2001). Microarrays developed from whole genome sequencing projects can be used to determine which genes are expressed in a pathogenic bacterium as compared to any nonpathogenic close relative and in different environments or during different phases of infection (Weinstock *et al.*, 2000; Schoolnik, 2002b; Hughes, 2003; Miesel *et al.*, 2003). The discovery and characterization of individual virulence genes and pathogenicity islands that can be acquired via transformation or transmitted by plasmids, prophages, and transposons reveals how these are passed from one bacterial genome to another. The existence in sequence databases of genes from pathogens can also allow genomic data to be used for diagnostic purposes (Weinstock *et al.*, 2000; Fitzgerald and Musser, 2001). Most recently, high-throughput genotyping methods have been used in clinical microbiology to determine the genetic and phylogenetic relatedness of microbes in clinical samples, and to quickly identify the presence of pathogenic forms (e.g., Terletski et al., 2003; van Belkum, 2003; Wong et al., 2003). All of this information is of use in preventing or combating the outbreak of disease, and it is clear that such work will continue to form a significant component of medical microbiology in the future.

Antibiotic Discovery and Resistance

Antibiotics play a crucial role in modern medicine, and it has been estimated that their use has raised the average life span by 10 years in the United States (Hughes, 2003). However, the increasing commonality of antibiotic-resistant strains has threatened to create major problems for human health. It is therefore of great practical significance that data derived from complete genome sequencing can contribute to the effective development and application of antibiotics, which they can do in at least three ways. First, genomic data provide critical insights into the various means by which resistance to antibiotics emerges and is spread among bacterial species (Hughes, 2003). Second, with the genome sequence of natural antibiotic-producing species like *Streptomyces* in hand, it will be possible to characterize, isolate, and potentially modify the genes responsible for producing these useful pharmaceutical compounds. Third, genomic data allow the determination of genes essential for the growth and division of pathogenic bacteria, which can then be targeted in the development of new antibiotics (Tang and Moxon, 2001; Waterman, 2001; Fraser, 2002; Miesel et al., 2003; Mills, 2003). A proof-of-principle example involving the peptide deformylase (*def*) gene has already showed the potential utility of this approach (Mills, 2003). Targeted genetic strategies would also allow only virulent strains to be affected, without harming beneficial types (Hughes, 2003). Microarrays can also be of use in this context, by revealing the expression patterns of genes following different antibiotic treatments (Tang and Moxon, 2001; Miesel et al., 2003; Mills, 2003).

Vaccine Development

Vaccines, which enhance the immune system's own antimicrobial capacities rather than attacking bacteria chemically, have proven highly effective in nearly or completely eradicating formerly devastating diseases such as those caused by the smallpox and polio viruses. They have also been effective against some pathogenic bacteria, but unfortunately vaccines are only available for a minority of disease-causing strains. Not surprisingly, there has already been much discussion of the prospects for using genomic data in the design of antibacterial vaccines (Fraser, 2002, 2004a; Klade, 2002; Grandi, 2003; Sampson et al., 2003; Saunders and Butcher, 2003; Zagursky et al., 2003). In the first application of this approach, Pizza et al. (2000) used the entire genome sequence of serogroup B *Neisseria meningitidis* (a major cause of meningitis) to identify 570 putative genes coding for cell surface-expressed or secreted proteins (see also Fraser *et al.*, 2000; Grandi, 2003; Fraser, 2004a). These were cloned into *E. coli* vectors, and 61% were expressed successfully in *E. coli* and were used to immunize mice. Immune sera from the mice were screened for bactericidal activity and cell surface binding, and based on this, seven representative proteins were evaluated for variability in sequence among different serogroups of *N. meningitidis*. Two of these proteins were found to be highly conserved among serotypes and were subjected to further study as possible antigens to be used in a vaccine. A similar approach has since been used to investigate possible vaccines for *Chlamydia pneumoniae* (Grandi, 2003).

Biological Control of Prokaryotic Pathogens

Biological agents, including both parasites and predators, are frequently used to control undesirable animals because this is often considered preferable to the use of pesticides or other toxic chemicals. Genomic studies are now raising the possibility of using biological controls to combat pathogenic bacteria as well. For example, although prophages are involved in the transmission of pathogenicity genes in many cases, bacteriophages have also long been used to attack virulent bacteria (e.g., Merril *et al.*, 1996). This method fell out of favor with the advent of chemical antibiotics, but may become popular once again now that genome sequences for both phages and bacteria are available. It has also been suggested that the genetic study of how phages lyse bacterial cells could lead to the development of new antibiotics that replicate this effect (Senior, 2001). The recent completion of the genome sequence of *Bdellovibrio bacteriovorus*, which is predatory on other bacteria, has raised hopes that insights can be gleaned regarding the mechanisms that make it such an effective "living antibiotic" (Rendulic *et al.*, 2004; Sebaihia *et al.*, 2004).

Biological Control of Eukaryotic Vectors and Parasites

Many infectious diseases are caused by protozoan and invertebrate animal pathogens. Genome sequencing is playing a role in the fight against such diseases as well, and has included projects involving both the pathogens themselves and their vectors (e.g., the malaria parasite *Plasmodium falciparum* and its mosquito vector *Anopheles gambiae*) (Gardner *et al.*, 2002; Holt *et al.*, 2002) (see Chapter 9). Interestingly, it has been suggested that information from prokaryote genomics may also be useful in combating eukaryotic pathogens. For example, a complete genome sequence is now available for *Wigglesworthia glossinidia brevipalpis*, the required endosymbiont of the tsetse fly (Akman *et al.*, 2002). This is significant because this insect is the vector for the pathogenic protist *Trypanosoma brucei*, which causes African sleeping sickness, suggesting that treatments attacking *Wigglesworthia* would adversely affect the vector and therefore help to lower the spread of infection. Similarly, it has been suggested that genome-based treatments against symbiotic *Wolbachia* required by filarial nematodes could cut down on the incidence of river blindness (Wren, 2002). Many other such applications may become possible in the future, given that obligate endosymbionts appear to be quite common among invertebrates (Wernegreen, 2002).

INDUSTRY AND THE ENVIRONMENT

As with medicine, much hope has been expressed regarding the potential bene-fits of genomics to industry and environmental protection, in terms of both assuaging the negative impacts of some bacteria and further developing the posi-tive effects of others. For example, the complete genome sequence has recently been published for the anaerobic sulfate-reducing δ-proteobacterium *Desulfovibrio vulgaris,* which produces toxic and highly reactive sulfide as an end-product of its metabolism (Heidelberg *et al.,* 2004). Species such as these are responsible for the microbially influenced corrosion (MIC, or "biocorrosion") of ferrous metals, which can, for example, cause significant problems for pipelines and machinery in the petroleum industry. As such, it is hoped that genomic analyses will provide insights into the genes involved in biocorrosion and how to combat this effect.

On the positive side, *D. vulgaris* may be of some use in the bioremediation of toxic metals (Heidelberg *et al.,* 2004). In this sense, it joins a large number of other species whose genome sequences have been published or are currently under way that are used in the bioremediation of oil spills and other organic pol-lution (Tiedje, 2002; Pieper *et al.,* 2004). Other species with potential utility in this capacity whose genomes have now been sequenced include the extraordinar-ily radiation-tolerant *Deinococcus radiodurans* (White *et al.,* 1999), and *Shewanella oneidensis,* which can reduce metal contaminants like uranium and chromium dis-solved in groundwater to insoluble oxides (Heidelberg *et al.,* 2002). Several other metal-reducing species (e.g., *Geobacter* spp. and *Ralstonia metallidurans*) have already been sequenced as well (Tiedje, 2002). It has also been pointed out that bacterial genomics tools can be used in environmental impact assessments and monitoring programs, which would be aimed at preventing environmental damage from occurring in the first place (Purohit *et al.,* 2003).

Genomic and metagenomic data may permit useful modifications of species that are already used in industry to produce enzymes, to clean up oil spills, and to extract valuable metals with less pollution (e.g., Tiedje, 2002; Daniel, 2004). It is also possible that genomes that do not yet exist may assist in the protection of the environment. The construction from scratch of the φX174 genome described previ-ously (Smith *et al.,* 2003) was only the beginning of a larger project undertaken by the Institute for Biological Energy Alternatives (IBEA) (www.bioenergyalts.org) with $3 million in funding from the U.S. Department of Energy. One of the stated goals of this program is to create a minimal bacterial genome that can be customized with genes that allow it to digest CO_2 from industrial emissions or to produce large quantities of hydrogen for use in pollution-free fuel cells (Check, 2002). Some of the novel genes characterized from the environmental sampling of the Sargasso Sea by the IBEA group may prove useful in this regard (Whitfield, 2003). Finally, it appears that genomic data may contribute to the development of a biological power cell that harnesses the electrochemical output of bacteria in ocean sediments. The genome sequence of *Geobacter sulfurreducens*, a relative of some of the "electrode bugs" involved, has already been completed (Methé *et al.*, 2003), and it has been suggested that "genetically engineered superbugs, capable of ultra-efficient electrode electron transfer, could end up in a wide range of environmentally fuelled and environmentally friendly fuel cells" (DeLong and Chandler, 2002).

AGRICULTURE AND FOOD PRODUCTION

Most of the discussion of the relevance of genomics to agriculture and food production relates to the sequencing of crop plants such as wheat and rice and the controversy surrounding genetically modified organisms. However, there are several ways in which prokaryote genomics can be important in this context as well. For example, sequencing projects of nitrogen-fixing bacteria (e.g., *Bradyrhizobium japonicum*, *Mesorhizobium loti*, *Sinorhizobium meliloti*) may provide important insights into how these species contribute to the fitness of the leguminous plants with which they live in symbiotic association. The genomic study of pathogenic bacteria that cause disease in livestock (e.g., *Brucella suis*, *Mycobacterium bovis*) and crop plants (e.g., *Agrobacterium tumefaciens*) is likewise clearly of interest, as are species that contaminate and spoil foods or cause food poisoning (e.g., *E. coli* O157:H7, *Salmonella enterica*, *Bacillus cereus*, *Campylobacter jejuni*) (de Vos, 2001).

Other less obvious applications include the possible improvement of bacteria used in the fermentation of milk products (e.g., *Lactobacillus plantarum*, *Lactococcus lactis*, *Streptococcus thermophilus*) and the protection of these useful species from attacks by phages (Pridmore *et al.*, 2000). Some species that can be used as beneficial gut bacterium supplements (probiotics), such as *Lactobacillus*, *Bacteroides*, and *Bifidobacterium*, have also been studied from a genomics perspective (Pridmore *et al.*, 2000; de Vos, 2001; Xu *et al.*, 2003). Finally, it is conceivable that genomic insights could allow the development of treatments aimed at the obligate endosymbiotic bacteria of plant-destroying pest insects, as has been considered for animal pathogens and disease vectors.

EVOLUTIONARY BIOLOGY

Genomic data have provided crucial insights into the evolutionary histories of prokaryotes at all scales, be it involving conspecific strains, members of closely related species, higher taxonomic assemblages, or the deepest branches in the phylogeny of cellular life (e.g., Eisen, 2000b; Brown, 2001; Bansal and Meyer, 2002). It has been suggested that large-scale environmental genome sequencing, such as that undertaken in the Sargasso Sea by Venter *et al.* (2004), may allow the reconstruction of the ancient metabolic pathways that emerged billions of years ago in a long-extinct early microbial world (Falkowski and de Vargas, 2004).

Genome sequencing in prokaryotes is also assisting in the study of eukaryote evolution at several scales. For example, genomic analyses have been used in an attempt to identify the closest extant relatives of the formerly free-living bacteria that became organelles early in the evolution of eukaryotic cells (e.g., Andersson *et al.*, 1998; Gray, 1998; Wu *et al.*, 2004). The study of modern intracellular symbionts like *Buchnera* and *Wigglesworthia*, which are at the interface between bacteria and organelles, provides additional insight into this process that would be impossible to obtain without large-scale sequence information (Douglas and Raven, 2002). Prokaryote genomics also has the potential to enlighten the study of recent human evolution. Specifically, genomic analyses of *Helicobacter pylori* have shown genetic variation among strains of this ulcer-causing Gram-negative gut bacterium to correlate with both ancient and recent patterns of human migration (Ghose *et al.*, 2002; Disotell, 2003; Falush *et al.*, 2003; Robinson, 2003; Spratt, 2003). In a detailed analysis, Falush *et al.* (2003) sequenced a set of eight genes from 370 strains of *H. pylori* from 27 geographic, ethnic, and linguistic groups of people from around the world, and used these to construct a phylogeny of the *H. pylori* strains. Several major groupings of *H. pylori* strains were identified, with their modern distributions reflecting such historical events as the colonization of the Americas by Native Americans that began at least 12,000 years ago, the migration of Neolithic farmers into Europe from North Africa and the Near East between 5000 and 8000 years ago, and the forced movement of people from West Africa to the United States during the time of the American slave trade within the past few hundred years (Disotell, 2003; Falush *et al.*, 2003; Robinson, 2003; Spratt, 2003).

CONCLUDING REMARKS AND FUTURE PROSPECTS

Even though prokaryote comparative genomics is still essentially in the beginning stages, it is clear from the review presented here that the ongoing genomic revolution has fundamentally altered the science of microbiology. It is not difficult to identify the major impact that genomic information has already had on studies of prokaryote taxonomy, phylogeny, and diversity. However, with the genomes of about 170 prokaryote species published and 370 new projects under way at the time of this writing (not even counting multiple strains), and with the rate of complete genome sequencing continuing to accelerate, the "state of the art" in prokaryote comparative genomics is a fast-moving target that cannot be captured in any short summary. Likewise, it is very difficult to make confident predictions about the future course of prokaryote genomics, especially because the advance of the field has consistently surpassed expectations.

Some areas of research are sure to make up a significant fraction of future work in prokaryote genomics, especially the investigation of applications for medicine, agriculture, and industry. Large-scale genome sequencing of both select species and large environmental samples are likely to continue to enable the discovery of vast numbers of new genes, and the achievement of a better understanding of prokaryote ecology, evolution, physiology, and diversity. In fact, the prospects for future genomic work are so open-ended at this stage that it may be more appropriate to ask not what can or probably will be done, but rather what might *not* be done based on either practical constraints or policy choices.

WHICH PROKARYOTES CAN (OR CANNOT) BE SEQUENCED?

In light of the increasing ease and cost effectiveness of large-scale sequencing, it is evident that hundreds or even thousands of prokaryote genomes could be sequenced over the next few decades. This is important, because the extraordinary diversity of prokaryotes makes it clear that few generalizations can be extended from the study of small numbers of species. As Nierman *et al.* (2000) put it, "The idea of a 'model organism' in the microbial world may not be a valid concept given the vast [genomic] differences that we have observed, even between related species." That said, financial cost is not the only factor that will determine the breadth of the resulting dataset, and already there seem to be more significant practical constraints than simple issues of funding.

Most of the prokaryotic genome sequencing projects carried out to date have begun with the isolation of genomic DNA via the usual molecular methods from cells cultured in the lab. However, the vast majority of prokaryote species cannot be (or at least have not yet been) successfully grown in culture and so are not amenable to techniques of this sort. The question therefore remains as to whether there are species of prokaryotes, possibly vast numbers of them, that simply cannot be isolated and sequenced. Three different approaches have been taken to sequencing members of the more recalcitrant majority, which inspires some confidence that the practical limitation of unculturability can be overcome in many cases.

The first approach is to develop unique methods of isolating bacteria to be sequenced. For example, individual species or cells of bacteria can be isolated from soil samples using micromanipulators, optical tweezers, centrifugation, filtration, flow cytometric cell sorting, and sample dilution (Hugenholtz, 2002). Once separated from other "weedier" species that dominate in growth media (but not necessarily in the natural environment), some of these may prove capable of growing in culture. In some cases, simulating the natural environment, rather than using standard growth media, may allow otherwise unculturable species to be grown in the lab (Kaeberlein *et al.*, 2002). Bacteria that live in obligate association with eukaryotes may not be culturable under any circumstances, and it is therefore necessary to isolate them from their hosts. In many cases, pathogenic strains can be obtained in sufficient numbers from infected animals and plants, for example. Even intracellular symbionts can be obtained from hosts if the appropriate methods are employed; by way of example, Shigenobu *et al.* (2000) dissected the specialized bacteriocytes from host aphids in order to obtain samples of *Buchnera* for sequencing.

The second approach involves cloning large fragments of DNA from environmental samples, potentially covering entire genomes, into BAC libraries (Béjà, 2004). Once stored in vectors, specific genes or entire fragments can be sequenced as desired. The third approach is similar but involves the direct random shotgun sequencing of entire genomes derived from large environmental samples (Venter *et al.*, 2004). In these cases, it is not necessary to culture the species at all, although it is obviously much more difficult to reassemble and identify whole genomes from a pool of sequences representing hundreds of species. Presumably, advances in computer technology and assembly algorithms will help to overcome this challenge in the relatively near future.

POLICY ISSUES: FROM SEQUENCE COMPLETENESS TO BIOTERRORISM

Policy issues, as well as practical ones, will play a role in influencing the future of prokaryote genome sequencing. For example, in 1998, the U.S. Department of Energy changed its policy for the Joint Genome Institute to provide only high-coverage draft sequences for organisms of interest, rather than completing the sequences fully. This policy has been criticized as generating less information per genome, an error rate that may be 100 times higher in the available sequences, and a much greater risk of contamination from foreign sources of DNA (Fraser *et al.*, 2002). These problems have led some researchers to insist that complete (and not only draft) sequences should be preferred as a general policy (Fraser *et al.*, 2002), even though this may mean that fewer species are covered overall.

From the standpoint of subsequent analytical insights, there are several reasons that completely sequenced, accurate, and annotated prokaryotic genomes should be maintained as the goal of microbial genomics (Fraser *et al.*, 2002). First, functional genomics relies on the interpretation of subtle sequence differences in genes coding for proteins and would suffer greatly from inaccurate sequencing. Second, sequences that are not closed fail to reveal important features of large-scale organization, such as the presence of plasmids and multiple chromosomes and their circular versus linear arrangement. Furthermore, incomplete genomes do not provide any reliable information about the absence of a given gene from a genome and can obscure analyses regarding the order of those genes that are found. For example, the detection of X-alignments, which reflect symmetric inversions around the replication origin and terminus (Eisen *et al.*, 2000) (Fig. 10.6), would be impeded by inaccurate or incomplete sequence information. Finally, a policy of complete sequencing and closure is important for the future, in terms of both promoting the continued development of tools to facilitate these tasks and the utility of the data for types of analyses that may not even have been conceived yet.

In addition to deciding which species are to be sequenced and to what level of completeness, there has been debate in the last few years over who should have access to the data generated in such projects. This applies especially to the issue of biological weapons, which have been in use for centuries—including both infectious agents themselves (bacteria or viruses) and toxins produced by bacteria and eukaryotes—and which have been targeted at people as well as essential crops and domestic animals (e.g., Boissinot and Bergeron, 2003; Fraser, 2004a). Many have voiced concern over the possibility that genomic data could be used to render such agents more dangerous, and as a result it has been suggested that biologists should consider allowing only limited access to genomic information for potential bioweapons agents such as *Bacillus anthracis* and *Brucella suis* (see, e.g., Aldous, 2001).

Constraints on the availability of genomic data have been resisted by many in the genomic community because such a policy would run counter to the established protocol of open and unlimited access for all researchers. More importantly, some microbiologists note that most of the steps involved in creating and using bioweapons (e.g., growth, storage, dispersal) would not be furthered with genomic information. Rather, sequence data are much more likely to be useful in the development of defenses against bioweapon attack (e.g., rapid detection and identification methods, effective antibiotics and vaccines), such that limiting access to these data would primarily hinder biodefense, not bioterrorism (e.g., Fraser and Dando, 2001; Read and Parkhill, 2002; Slezak *et al.*, 2003; Fraser, 2004a).

In some cases, it is the researchers themselves, against the general opinion of their colleagues, who have called for possible limitations on information sharing. Stirring up more controversy following the lack of free access to the human genome sequence produced by Celera Genomics (see Chapter 9), Craig Venter has suggested that perhaps the methods involved in creating a minimal bacterial genome from scratch should not be made completely public, lest this approach be used to create biological weapons (Check, 2002). However, many others apparently feel that such information is unlikely to pose a threat, because the modification of existing pathogens is a far more likely route by which new bioweapons would be engineered (e.g., Powledge, 2002).

Overall, it seems that the general opinion among genome biologists and microbiologists is that complete, fully closed, and annotated genome sequences are desirable, and that access to this information should not be restricted. Indeed, accurate and freely available genomic data are considered one of the best resources for defending against the malicious use of microorganisms.

PROKARYOTE GENOMICS: THE END OF THE BEGINNING

Over the span of less than one decade, complete genome sequencing has gone from an ambitious hypothetical concept—with only a few viral genomes completely sequenced, detailed mapping projects taking years even for average-sized bacterial genomes, and a novel (and now standard) method of random shotgun sequencing rejected as unworkable—to one of the most extensive sources of biological data ever assembled. In under 10 years, the cost of sequencing has been reduced by more than an order of magnitude, and the number of prokaryote sequencing projects has gone from zero to several hundred, without doubt making this one of the most explosive revolutions in the history of scientific inquiry.

Many discussions of prokaryote genomics place an emphasis on the practical benefits that the field may provide, such as potential applications in medicine and industry. But as the opening stage of comparative genomics draws to a close, it is important to note what the field has already accomplished. Comparative prokaryote genomics has already provided a great deal of knowledge regarding bacterial and archaeal genome structure, content, organization, and size, and has highlighted exceptions to long-held assumptions in each of these areas. It has resulted in the discovery of hundreds of thousands of new genes and has shed light on the vast array of metabolic and ecological means by which prokaryotes make their livings. It has provided important insights into the evolution of prokaryotes at all taxonomic levels, from the rise of pathogens and obligate symbionts within lineages to what may be the deepest phylogenetic divergence of them all. It has given some clues regarding the minimal genome required for life and the probable widespread occurrence of horizontal gene transfer during the earliest stages of cellular evolution. And, perhaps most importantly, genomic analyses are providing the first glimpses of the incredible diversity of the previously unseen prokaryotic majority among life on Earth.

The growth of comparative genomics shows no signs of slowing—indeed, all indications suggest it will continue to accelerate for the foreseeable future. It is therefore likely that the next decade will provide even more exciting and unexpected findings than have the last 10 years. Given the impressive litany of technical, analytical, and exploratory achievements already generated, this next stage promises to be an era of extraordinary biological discovery.

REFERENCES

Achaz G, Coissac E, Netter P, Rocha EPC. 2003. Associations between inverted repeats and the structural evolution of bacterial genomes. *Genetics* 164: 1279–1289.

Ajdić D, McShan WM, McLaughlin RE, *et al.* 2002. Genome sequence of *Streptococcus mutans* UA159, a cariogenic dental pathogen. *Proc Natl Acad Sci USA* 99: 14434–14439.

Akerley BJ, Rubin EJ, Novick VL, *et al.* 2002. A genome-scale analysis for identification of genes required for growth or survival of *Haemophilus influenzae*. *Proc Natl Acad Sci USA* 99: 966–971.

Akman L, Yamashita A, Watanabe H, *et al.* 2002. Genome sequence of the endocellular obligate symbiont of tsetse flies, *Wigglesworthia glossinidia*. *Nat Genet* 32: 402–407.

Aldhous P. 2001. Biologists urged to address risk of data aiding bioweapon design. *Nature* 414: 237–238.

Andersson JO. 2000. Is *Buchnera* a bacterium or an organelle? *Curr Biol* 10: R866–R868.

Andersson JO, Andersson SGE. 1999a. Genome degradation is an ongoing process in *Rickettsia*. *Mol Biol Evol* 16: 1178–1191.

Andersson JO, Andersson SGE. 1999b. Insights into the evolutionary process of genome degradation. *Curr Opin Genet Dev* 9: 667–671.

Andersson JO, Andersson SGE. 2001. Pseudogenes, junk DNA, and the dynamics of *Rickettsia* genomes. *Mol Biol Evol* 18: 829–839.

Andersson JO, Doolittle WF, Nesbø CL. 2001. Are there bugs in our genome? *Science* 292: 1848–1850.

Andersson SGE, Kurland CG. 1998. Reductive evolution of resident genomes. *Trends Microbiol* 6: 263–268.

Andersson SGE, Karlberg O, Canbäck B, Kurland CG. 2003. On the origin of mitochondria: a genomics perspective. *Philos Trans R Soc Lond B* 358: 165–179.

Andersson SGE, Zomorodipour A, Andersson JO, *et al.* 1998. The genome sequence of *Rickettsia prowazekii* and the origin of mitochondria. *Nature* 396: 133–143.

Aravind L, Tatusov RL, Wolf YI, *et al.* 1998. Evidence for massive gene exchange between archaeal and bacterial hyperthermophiles. *Trends Genet* 14: 442–444.

Aravind L, Tatusov RL, Wolf YI, *et al.* 1999. Reply. *Trends Genet* 15: 299–300.

Avery OT, MacLeod CM, McCarty M. 1944. Studies on the chemical nature of the substance inducing transformation of pneumococcal types. I. Induction of transformation by a desoxyribonucleic acid fraction isolated from Pneumococcus type III. *J Exp Med* 79: 137–158.

Bachellier S, Clément J-M, Hofnung M. 1999. Short palindromic repetitive DNA elements in enterobacteria: a survey. *Res Microbiol* 150: 627–639.

Bailey KA, Reeve JN. 1999. DNA repeats and archaeal nucleosome positioning. *Res Microbiol* 150: 701–709.

Baillie L, Read TD. 2001. *Bacillus anthracis*, a bug with attitude! *Curr Opin Microbiol* 4: 78–81.

Banks DJ, Beres SB, Musser JM. 2002. The fundamental contribution of phages to GAS evolution, genome diversification and strain emergence. *Trends Microbiol* 10: 515–521.

Bansal AK, Meyer TE. 2002. Evolutionary analysis by whole-genome comparisons. *J Bacteriol* 184: 2260–2272.

Barns SM, Delwiche CF, Palmer JD, Pace NR. 1996. Perspectives on archaeal diversity, thermophily and monophyly from environmental rRNA sequences. *Proc Natl Acad Sci USA* 93: 9188–9193.

Barns SM, Fundyga RE, Jeffries MW, Pace NR. 1994. Remarkable archaeal diversity detected in a Yellowstone National Park hot spring environment. *Proc Natl Acad Sci USA* 91: 1609–1613.

Baumann C, Judex M, Huber H, Wirth R. 1998. Estimation of genome sizes of hyperthermophiles. *Extremophiles* 2: 101–108.

Béjà O. 2004. To BAC or not to BAC: marine ecogenomics. *Curr Opin Biotechnol* 15: 187–190.

Béjà O, Suzuki MT, Heidelberg JF, *et al.* 2002. Unsuspected diversity among marine aerobic anoxygenic phototrophs. *Nature* 415: 630–633.

Béjà O, Suzuki MT, Koonin EV, *et al.* 2000. Construction and analysis of bacterial artificial chromosome libraries from a marine microbial assemblage. *Environ Microbiol* 2: 516–529.

Belay N, Mukhopadhyay B, Conway de Macario E, *et al.* 1990. Methanogenic bacteria in human vaginal samples. *J Clin Microbiol* 28: 1666–1668.

Bendich AJ. 2001. The form of chromosomal DNA molecules in bacterial cells. *Biochimie* 83: 177–186.

Bendich AJ, Drlica K. 2000. Prokaryotic and eukaryotic chromosomes: what's the difference? *BioEssays* 22: 481–486.

Bensasson D, Zhang D-X, Hartl DL, Hewitt GM. 2001. Mitochondrial pseudogenes: evolution's misplaced witnesses. *Trends Ecol Evol* 16: 314–321.

Bentley SD, Chater KF, Cerdeno-Tarraga AM, *et al.* 2002. Complete genome sequence of the model actinomycete *Streptomyces coelicolor* A3(2). *Nature* 417: 141–147.

Beres SB, Sylva GL, Barbian KD, *et al.* 2002. Genome sequence of a serotype M3 strain of group A *Streptococcus*: phage-encoded toxins, the high-virulence phenotype, and clone emergence. *Proc Natl Acad Sci USA* 99: 10078–10083.

Berg OG, Kurland CG. 2000. Why mitochondrial genes are most often found in nuclei. *Mol Biol Evol* 17: 951–961.

Bernal A, Ear U, Kyrpides N. 2001. Genomes OnLine Database (GOLD): a monitor of genome projects world-wide. *Nucleic Acids Res* 29: 126–127.

Bernander R. 2000. Chromosome replication, nucleoid segregation and cell division in Archaea. *Trends Microbiol* 8: 278–283.

Black T, Hare R. 2000. Will genomics revolutionize antimicrobial drug discovery? *Curr Opin Microbiol* 3: 522–527.

Blanchard JL, Lynch M. 2000. Organellar genes: why do they end up in the nucleus? *Trends Genet* 16: 315–320.

Blattner FR, Plunkett G, Bloch CA, *et al.* 1997. The complete genome sequence of *Escherichia coli* K-12. *Science* 277: 1453–1474.

Boissinot M, Bergeron MG. 2003. Génomique et bioterrorisme. *Médicine/Sciences* 19: 967–971.

Boucher Y, Douady CJ, Papke RT, *et al.* 2003. Lateral gene transfer and the origins of prokaryotic groups. *Annu Rev Genet* 37: 283–328.

Boyd EF, Brüssow H. 2002. Common themes among bacteriophage-encoded virulence factors and diversity among the bacteriophages involved. *Trends Microbiol* 10: 521–529.

Boyd EF, Davis BM, Hochhut B. 2001. Bacteriophage–bacteriophage interactions in the evolution of pathogenic bacteria. *Trends Microbiol* 9: 137–144.

Bresler V, Fishelson L. 2003. Polyploidy and polyteny in the gigantic eubacterium *Epulopiscium fishelsoni*. *Mar Biol* 143: 17–21.

Brochier C, Bapteste E, Moreira D, Philippe H. 2002. Eubacterial phylogeny based on translational apparatus proteins. *Trends Genet* 18: 1–5.

Brosch R, Gordon SV, Eiglmeier K, *et al.* 2000. Comparative genomics of the leprosy and tubercle bacilli. *Res Microbiol* 151: 135–142.

Broudy TB, Pancholi V, Fischetti VA. 2001. Induction of lysogenic bacteriophage and phage-associated toxin from Group A Streptococci during coculture with human pharyngeal cells. *Infect Immun* 69: 1440–1443.

Brown JR. 2001. Genomic and phylogenetic perspectives on the evolution of prokaryotes. *Syst Biol* 50: 497–512.

Brown JR. 2003. Ancient horizontal gene transfer. *Nat Rev Genet* 4: 121–132.

Brüssow H, Hendrix RW. 2002. Phage genomics: small is beautiful. *Cell* 108: 13–16.

Budd A, Blandin S, Levashina EA, Gibson TJ. 2004. Bacterial α_2-macroglobulins: colonization factors acquired by horizontal gene transfer from the metazoan genome? *Genome Biol* 5: R38.1–R38.13.

Bult CJ, White O, Olsen GJ, *et al.* 1996. Complete genome sequence of the methanogenic archaeon, *Methanococcus jannaschii. Science* 273: 1058–1073.

Cairns J. 1963. The chromosomes of *Escherichia coli. Cold Spring Harb Symp Quant Biol* 28: 43–46.

Canchaya C, Proux C, Fournous G, *et al.* 2003. Prophage genomics. *Microbiol Mol Biol Rev* 67: 238–276.

Case RB, Chang Y-P, Smith SB, *et al.* 2004. The bacterial condensin MukBEF compacts DNA into a repetitive, stable structure. *Science* 305: 222–227.

Casjens S. 1998. The diverse and dynamic structure of bacterial genomes. *Annu Rev Genet* 32: 339–377.

Casjens S. 2003. Prophages and bacterial genomics: what have we learned so far? *Mol Microbiol* 49: 277–300.

Casjens S, Palmer N, van Vugt R, *et al.* 2000. A bacterial genome in flux: the twelve linear and nine circular extrachromosomal DNAs in an infectious isolate of the Lyme disease spirochete *Borrelia burgdorferi. Mol Microbiol* 35: 490–516.

Cavalier-Smith T. 1998. A revised six-kingdom system of life. *Biol Rev* 73: 203–266.

Cavalier-Smith T. 2002. The neomuran origin of archaebacteria, the negibacterial root of the universal tree and bacterial megaclassification. *Int J Syst Evol Microbiol* 52: 7–76.

Cello J, Paul AV, Wimmer E. 2002. Chemical synthesis of poliovirus cDNA: generation of infectious virus in the absence of a natural template. *Science* 297: 1016–1018.

Chalmers R, Blot M. 1999. Insertion sequences and transposons. In: Charlebois RL ed. *Organization of the Prokaryotic Genome.* Washington, DC: American Society for Microbiology Press, 151–169.

Charlebois RL. 1999. Archaea: whose sister lineage? In: Charlebois RL ed. *Organization of the Prokaryotic Genome.* Washington, DC: American Society for Microbiology Press, 63–76.

Check E. 2002. Venter aims for maximum impact with minimal genome. *Nature* 420: 350.

Chen CW. 1996. Complications and implications of linear bacterial chromosomes. *Trends Genet* 12: 192–196.

Chen CY, Wu KM, Chang YC, *et al.* 2003. Comparative genome analysis of *Vibrio vulnificus*, a marine pathogen. *Genome Res* 13: 2577–2587.

Cho J-C, Tiedje JM. 2001. Bacterial species determination from DNA–DNA hybridization by using genome fragments and DNA microarrays. *Appl Environ Microbiol* 67: 3677–3682.

Clark MA, Baumann L, Thao ML, *et al.* 2001. Degenerative minimalism in the genome of a psyllid endosymbiont. *J Bacteriol* 183: 1853–1861.

Claverys J-P, Prudhomme M, Mortier-Barrière I, Martin B. 2000. Adaptation to the environment: *Streptococcus pneumoniae*, a paradigm for recombination-mediated genetic plasticity? *Mol Microbiol* 35: 251–259.

Cohan FM. 2001. Bacterial species and speciation. *Syst Biol* 50: 513–524.

Cohan FM. 2002. What are bacterial species? *Annu Rev Microbiol* 56: 457–487.

Cole JR, Chai B, Marsh TL, *et al.* 2003. The Ribosomal Database Project (RDP-II): previewing a new autoaligner that allows regular updates and the new prokaryotic taxonomy. *Nucleic Acids Res* 31: 442–443.

Cole S, Saint-Girons I. 1999. Bacterial genomes—all shapes and sizes. In: Charlebois RL ed. *Organization of the Prokaryotic Genome.* Washington, DC: American Society for Microbiology Press, 35–62.

Cole ST, Brosch R, Parkhill J, *et al.* 1998. Deciphering the biology of *Mycobacterium tuberculosis* from the complete genome sequence. *Nature* 393: 537–544.

Cole ST, Eiglmeier K, Parkhill J, *et al.* 2001. Massive gene decay in the leprosy bacillus. *Nature* 409: 1007–1011.

Curtis TP, Sloan WT, Scannell JW. 2002. Estimating prokaryotic diversity and its limits. *Proc Natl Acad Sci USA* 99: 10494–10499.

Daniel R. 2004. The soil metagenome—a rich resource for the discovery of novel natural products. *Curr Opin Biotechnol* 15: 199–204.

Daubin V, Perrière G. 2003. G+C₃ structuring along the genome: a common feature in prokaryotes. *Mol Biol Evol* 20: 471–483.

de Vos WM. 2001. Advances in genomics for microbial food fermentations and safety. *Curr Opin Biotechnol* 12: 493–498.

Dehal P, Satou Y, Campbell RK, *et al.* 2002. The draft genome of *Ciona intestinalis*: insights into chordate and vertebrate origins. *Science* 298: 2157–2167.

DeLong EF. 1992. Archaea in coastal marine environments. *Proc Natl Acad Sci USA* 89: 5685–5689.

DeLong EF. 2002. Microbial population genomics and ecology. *Curr Opin Microbiol* 5: 520–524.

DeLong EF, Chandler P. 2002. Power from the deep. *Nat Biotechnol* 20: 788–789.

DeLong EF, Pace NR. 2001. Environmental diversity of Bacteria and Archaea. *Syst Biol* 50: 470–478.

Deppenmeier U, Johann A, Hartsch T, *et al.* 2002. The genome of *Methanosarcina mazei*: evidence for lateral gene transfer between Bacteria and Archaea. *J Mol Microbiol Biotechnol* 4: 453–461.

Disotell TR. 2003. Discovering human history from stomach bacteria. *Genome Biol* 4: 213.211–213.214.

Domenech P, Barry CE, Cole ST. 2001. *Mycobacterium tuberculosis* in the post-genomic age. *Curr Opin Microbiol* 4: 28–34.

Doolittle RF. 2002. Microbial genomes multiply. *Nature* 416: 697–700.

Doolittle WF. 1999. Phylogenetic classification and the universal tree. *Science* 284: 2124–2128.

Douglas AE, Raven JA. 2002. Genomes at the interface between bacteria and organelles. *Philos Trans R Soc Lond B* 358: 5–18.

Dutta C, Pan A. 2002. Horizontal gene transfer and bacterial diversity. *J Biosci* 27 (Suppl. 1): 27–33.

Dykhuizen DE. 1998. Santa Rosalia revisited: why are there so many species of bacteria? *Antonie van Leeuwenhoek* 73: 25–33.

Dziejman M, Balon E, Boyd D, *et al.* 2002. Comparative genomic analysis of *Vibrio cholerae*: genes that correlate with cholera endemic and pandemic disease. *Proc Natl Acad Sci USA* 99: 1556–1561.

Eggington JM, Haruta N, Wood EA, Cox MM. 2004. The single-stranded DNA-binding protein of *Deinococcus radiodurans*. *BMC Microbiol* 4: 2.1–2.12.

Eisen JA. 2000a. Horizontal gene transfer among microbial genomes: new insights from complete genome analysis. *Curr Opin Genet Dev* 10: 606–611.

Eisen JA. 2000b. Assessing evolutionary relationships among microbes from whole-genome analysis. *Curr Opin Microbiol* 3: 475–480.

Eisen JA. 2001. Gastrogenomics. *Nature* 409: 463–466.

Eisen JA, Heidelberg JF, White O, Salzberg SL. 2000. Evidence for symmetric chromosomal inversions around the replication origin in bacteria. *Genome Biol* 1: research0011.0011–0011.0019.

Enserink M. 2001. Taking anthrax's genetic fingerprints. *Science* 294: 1810–1812.

Euzéby JP. 1997. List of bacterial names with standing in nomenclature: a folder available on the Internet. *Int J Syst Bacteriol* 47: 590–592.

Faguy DM. 2003. Lateral gene transfer (LGT) between Archaea and *Escherichia coli* is a contributor to the emergence of novel infectious disease. *BMC Infect Dis* 3: 13.11–13.14.

Falkowski PG, de Vargas C. 2004. Shotgun sequencing in the sea: a blast from the past? *Science* 304: 58–60.

Falush D, Wirth T, Linz B, *et al.* 2003. Traces of human migrations in *Helicobacter pylori* populations. *Science* 299: 1582–1585.

Ferretti JJ, McShan WM, Ajdic D, *et al.* 2001. Complete genome sequence of an M1 strain of *Streptococcus pyogenes*. *Proc Natl Acad Sci USA* 98: 4658–4663.

Fitzgerald JR, Musser JM. 2001. Evolutionary genomics of pathogenic bacteria. *Trends Microbiol* 9: 547–553.

Fleischmann RD, Adams MD, White O, *et al.* 1995. Whole-genome random sequencing and assembly of *Haemophilus influenzae* Rd. *Science* 269: 496–512.

Fleischmann RD, Alland D, Eisen JA, *et al.* 2002. Whole-genome comparison of *Mycobacterium tuberculosis* clinical and laboratory strains. *J Bacteriol* 184: 5479–5490.

Forterre P, Brochier C, Philippe H. 2002. Evolution of the Archaea. *Theor Pop Biol* 61: 409–422.

Frangeul L, Nelson KE, Buchrieser C, et al. 1999. Cloning and assembly strategies in microbial genome projects. *Microbiology* 145: 2625–2634.

Fraser CM. 2002. Microbial genome sequencing: prospects for development of novel vaccines and anti-microbial compounds. *Sci World J* 2: 1–2.

Fraser CM. 2004a. A genomics-based approach to biodefence preparedness. *Nat Rev Genet* 5: 23–33.

Fraser CM. 2004b. All things great and small. *Trends Microbiol* 12: 7–8.

Fraser CM, Dando MR. 2001. Genomics and future biological weapons: the need for preventive action by the biomedical community. *Nat Genet* 29: 253–256.

Fraser CM, Casjens S, Huang WM, et al. 1997. Genomic sequence of a Lyme disease spirochaete, *Borrelia burgdorferi*. *Nature* 390: 580–586.

Fraser CM, Eisen JA, Nelson KE, et al. 2002. The value of complete microbial genome sequencing (you get what you pay for). *J Bacteriol* 184: 6403–6405.

Fraser CM, Eisen JA, Salzberg SL. 2000. Microbial genome sequencing. *Nature* 406: 799–803.

Fraser CM, Gocayne JD, White O, et al. 1995. The minimal gene complement of *Mycoplasma genitalium*. *Science* 270: 397–403.

Gaasterland T. 1999. Archaeal genomics. *Curr Opin Microbiol* 2: 542–547.

Galagan JE, Nusbaum C, Roy A, et al. 2002. The genome of *M. acetivorans* reveals extensive metabolic and physiological diversity. *Genome Res* 12: 532–542.

Gamieldien J, Ptitsyn A, Hide W. 2002. Eukaryotic genes in *Mycobacterium tuberculosis* could have a role in pathogenesis and immunomodulation. *Trends Genet* 18: 5–8.

Garcia-Vallvé S, Romeu A, Palau J. 2000. Horizontal gene transfer in bacterial and archaeal complete genomes. *Genome Res* 10: 1719–1725.

Gardner MJ, Hall N, Fung E, et al. 2002. Genome sequence of the human malaria parasite *Plasmodium falciparum*. *Nature* 419: 498–511.

Garnier T, Eiglmeier K, Camus JC, et al. 2003. The complete genome sequence of *Mycobacterium bovis*. *Proc Natl Acad Sci USA* 100: 7877–7882.

Gevers D, Vandepoele K, Simillion C, Van de Peer Y. 2004. Gene duplication and biased functional retention of paralogs in bacterial genomes. *Trends Microbiol* 12: 148–154.

Ghose C, Perez-Perez GI, Dominguez-Bello M-G, et al. 2002. East Asian genotypes of *Helicobacter pylori* strains in Amerindians provide evidence of its ancient human carriage. *Proc Natl Acad Sci USA* 99: 15107–15111.

Gil R, Sabater-Muñoz B, Latorre A, et al. 2002. Extreme genome reduction in *Buchnera* spp.: toward the minimal genome needed for symbiotic life. *Proc Natl Acad Sci USA* 99: 4454–4458.

Glaser P, Rusniok C, Buchrieser C, et al. 2002. Genome sequence of *Streptococcus agalactiae*, a pathogen causing invasive neonatal disease. *Mol Microbiol* 45: 1499–1513.

Gogarten JP, Doolittle WF, Lawrence JG. 2002. Prokaryotic evolution in light of gene transfer. *Mol Biol Evol* 19: 2226–2238.

Gogarten JP, Taiz L. 1992. Evolution of proton pumping ATPases: rooting the Tree of Life. *Photosynth Res* 33: 137–146.

Gould SJ. 1996. *Full House*. New York: Harmony Books.

Graham DE, Kyrpides N, Anderson IJ, et al. 2001. Genome of *Methanocaldococcus* (*Methanococcus*) *jannaschii*. *Methods Enzymol* 330: 40–123.

Graham MR, Smoot LM, Lei B, Musser JM. 2001. Toward a genome-scale understanding of group A *Streptococcus* pathogenesis. *Curr Opin Microbiol* 4: 65–70.

Grandi G. 2003. Rational antibacterial vaccine design through genomic technologies. *Int J Parasitol* 33: 615–620.

Gray MW. 1998. *Rickettsia*, typhus and the mitochondrial connection. *Nature* 396: 109–110.

Gregory TR. 2004. Insertion–deletion biases and the evolution of genome size. *Gene* 324: 15–34.

Gupta RS. 1998a. Protein phylogenies and signature sequences: a reappraisal of evolutionary relationships among Archaebacteria, Eubacteria, and Eukaryotes. *Microbiol Mol Biol Rev* 62: 1435–1491.

Gupta RS. 1998b. What are the archaebacteria: life's third domain or monoderm prokaryotes related to Gram–positive bacteria? A new proposal for the classification of prokaryotic organisms. Mol Microbiol 29: 695–707.

Hallin PF, Ussery D. 2004. CBS Genome Atlas Database: a dynamic storage for bioinformatic results and sequence data. Bioinformatics (in press).

Hayashi T, Makino K, Ohnishi M, et al. 2001. Complete genome sequence of enterohemorrhagic Escherichia coli O157:H7 and genomic comparison with a laboratory strain K-12. DNA Res 8: 11–22.

Heidelberg JF, Eisen JA, Nelson WC, et al. 2000. DNA sequence of both chromosomes of the cholera pathogen Vibrio cholerae. Nature 406: 477–483.

Heidelberg JF, Paulsen IT, Nelson KE, et al. 2002. Genome sequence of the dissimilatory metal ion-reducing bacterium Shewanella oneidensis. Nat Biotechnol 20: 1118–1123.

Heidelberg JF, Seshadri R, Haveman SA, et al. 2004. The genome sequence of the anaerobic, sulfate-reducing bacterium Desulfovibrio vulgaris Hildenborough. Nat Biotechnol 22: 554–559.

Heinemann JA, Sprague GF. 1989. Bacterial conjugative plasmids mobilize DNA transfer between bacteria and yeast. Nature 340: 205–209.

Herdman M. 1985. The evolution of bacterial genomes. In: Cavalier-Smith T ed. The Evolution of Genome Size. Chichester, UK: John Wiley & Sons, 37–68.

Hill CW. 1999. Large genomic sequence repetitions in bacteria: lessons from rRNA operons and Rhs elements. Res Microbiol 150: 665–674.

Hollingshead SK, Briles DE. 2001. Streptococcus pneumoniae: new tools for an old pathogen. Curr Opin Microbiol 4: 71–77.

Holste D, Weiss O, Grosse I, Herzel H. 2000. Are noncoding sequences of Rickettsia prowazekii remnants of "neutralized" genes? J Mol Evol 51: 353–362.

Holt RA, Subramanian GM, Halpern A, et al. 2002. The genome sequence of the malaria mosquito Anopheles gambiae. Science 298: 129–149.

Hood DW, Deadman ME, Allen T, et al. 1996. Use of the complete genome sequence information of Haemophilus influenzae strain Rd to investigate lipopolysaccharide biosynthesis. Mol Microbiol 22: 951–965.

Hooper SD, Berg OG. 2003. Duplication is more common among laterally transferred genes than among indigenous genes. Genome Biol 4: R48.41–R48.49.

Hoskins J, Alborn WE, Arnold J, et al. 2001. Genome of the bacterium Streptococcus pneumoniae strain R6. J Bacteriol 183: 5709–5717.

Howe CJ, Barbrook AC, Koumandou VL, et al. 2003. Evolution of the chloroplast genome. Philos Trans R Soc Lond B 358: 99–107.

Huber H, Hohn MJ, Rachel R, et al. 2002. A new phylum of Archaea represented by a nanosized hyperthermophilic symbiont. Nature 417: 63–67.

Hugenholtz P. 2002. Exploring prokaryotic diversity in the genomic era. Genome Biol 3: reviews0003.0001–0003.0008.

Hugenholtz P, Goebel BM, Pace NR. 1998. Impact of culture-independent studies on the emerging phylogenetic view of bacterial diversity. J Bacteriol 180: 4765–4774.

Hughes D. 2000. Evaluating genome dynamics: the constraints on rearrangements within bacterial genomes. Genome Biol 1: reviews0006.0001–0006.0008.

Hughes D. 2003. Exploiting genomics, genetics and chemistry to combat antibiotic resistance. Nat Rev Genet 4: 432–441.

Hutchison CA, Peterson SN, Gill SR, et al. 1999. Global transposon mutagenesis and a minimal Mycoplasma genome. Science 286: 2165–2169.

Ikeda H, Ishikawa J, Hanamoto A, et al. 2003. Complete genome sequence and comparative analysis of the industrial microorganism Streptomyces avermitilis. Nat Biotechnol 21: 526–531.

International Human Genome Sequencing Consortium. 2001. Initial sequencing and analysis of the human genome. Nature 409: 860–921.

Islas S, Becerra A, Luisi PL, Lazcano A. 2004. Comparative genomics and the gene complement of a minimal cell. *Orig Life Evol Biosph* 34: 243–256.

Itaya M. 1995. An estimation of minimal genome size required for life. *FEBS Lett* 362: 257–260.

Ivanova N, Sorokin A, Anderson I, *et al.* 2003. Genome sequence of *Bacillus cereus* and comparative analysis with *Bacillus anthracis. Nature* 423: 87–91.

Iwabe N, Kuma K, Hasegawa M, *et al.* 1989. Evolutionary relationship of archaebacteria, eubacteria, and eukaryotes inferred from phylogenetic trees of duplicated genes. *Proc Natl Acad Sci USA* 86: 9355–9359.

Jackson PJ, Walthers EA, Kalif AS, *et al.* 1997. Characterisation of the variable-number tandem repeats in vrrA from different *Bacillus anthracis* isolates. *Appl Environ Microbiol* 63: 1400–1405.

Jain R, Rivera MC, Lake JA. 1999. Horizontal gene transfer among genomes: the complexity hypothesis. *Proc Natl Acad Sci USA* 96: 3801–3806.

Jain R, Rivera MC, Moore JE, Lake JA. 2002. Horizontal gene transfer in microbial genome evolution. *Theor Pop Biol* 61: 489–495.

Jensen LJ, Friis C, and Ussery DW. 1999. Three views of microbial genomes. *Res Microbiol* 150: 773–777.

Ji Y, Zhang B, Van Horn SF, *et al.* 2001. Identification of critical staphylococcal genes using conditional phenotypes generated by antisense RNA. *Science* 293: 2266–2269.

Jin Q, Yuan Z, Xu J, *et al.* 2002. Genome sequence of *Shigella flexneri* 2a: insights into pathogenicity through comparison with genomes of *Escherichia coli* K12 and O157. *Nucleic Acids Res* 30: 4432–4441.

Johnson JL. 1973. Use of nucleic-acid homologies in the taxonomy of anaerobic bacteria. *Int J Syst Bacteriol* 23: 308–315.

Jordan IK, Makarova KS, Spouge JL, *et al.* 2001. Lineage-specific gene expansions in bacterial and archaeal genomes. *Genome Res* 11: 555–565.

Kaeberlein T, Lewis K, Epstein SS. 2002. Isolating "uncultivable" microorganisms in pure culture in a simulated natural environment. *Science* 296: 1127–1129.

Kaneko T, Nakamura Y, Sato S, *et al.* 2002. Complete genomic sequence of nitrogen-fixing symbiotic bacterium *Bradyrhizobium japonicum* USDA110. *DNA Res* 9: 189–197.

Keim PK, Kalif AS, Schupp J, *et al.* 1997. Molecular evolution and diversity in *Bacillus anthracis* as detected by amplified fragment length polymorphism markers. *J Bacteriol* 179: 818–824.

Kidwell MG. 2002. Transposable elements and the evolution of genome size in eukaryotes. *Genetica* 115: 49–63.

Kim JI, Cox MM. 2002. The RecA proteins of *Deinococcus radiodurans* and *Escherichia coli* promote DNA strand exchange via inverse pathways. *Proc Natl Acad Sci USA* 99: 7917–7921.

Kjems J, Larsen N, Dalgaard JZ, *et al.* 1992. Phylogenetic relationships amongst the hyperthermophilic Archaea determined from partial 23S rRNA gene sequences. *Syst Appl Microbiol* 15: 203–208.

Klade CS. 2002. Proteomics approaches towards antigen discovery and vaccine development. *Curr Opin Mol Ther* 4: 216–223.

Klasson L, Andersson SGE. 2004. Evolution of minimal-gene-sets in host-dependent bacteria. *Trends Microbiol* 12: 37–43.

Klotz MG, Loewen PC. 2003. The molecular evolution of catalytic hydroperoxidases: evidence for multiple lateral transfer of genes between Prokaryotes and from Bacteria to Eukaryota. *Mol Biol Evol* 20: 1098–1112.

Kobayashi K, Ehrlich SD, Albertini A, *et al.* 2003. Essential *Bacillus subtilis* genes. *Proc Natl Acad Sci USA* 100: 4678–4683.

Kobryn K, Chaconas G. 2001. The circle is broken: telomere resolution in linear replicons. *Curr Opin Microbiol* 4: 558–564.

Kolisnychenko V, Plunket G, Herring CD, *et al.* 2002. Engineering a reduced *Escherichia coli* genome. *Genome Res* 12: 640–647.

Komaki K, Ishikawa H. 1999. Intracellular bacterial symbionts of aphids possess many genomic copies per bacterium. *J Mol Evol* 48: 717–722.

Kondo N, Nikoh N, Ijichi N, *et al.* 2002. Genome fragment of *Wolbachia* endosymbiont transferred to X chromosome of host insect. *Proc Natl Acad Sci USA* 99: 14280–14285.

Koonin EV. 2000. How many genes can make a cell: the minimal-gene-set concept. *Annu Rev Genomics Hum Genet* 1: 99–116.

Koonin EV, Makarova KS, Aravind L. 2001. Horozontal gene transfer in prokaryotes: quantification and classification. *Annu Rev Microbiol* 55: 709–742.

Korbel JO, Snel B, Huynen MA, Bork P. 2002. SHOT : A web server for the construction of genome phylogenies. *Trends Genet* 18: 158–162.

Koski LB, Morton RA, Golding GB. 2001. Codon bias and base composition are poor indicators of horizontally transferred genes. *Mol Biol Evol* 18: 404–412.

Kulik EM, Sandmeier H, Hinni K, Meyer J. 2001. Identification of archaeal rDNA from subgingival dental plaque by PCR amplification and sequence analysis. *FEMS Microbiol Lett* 196: 129–133.

Kunin V, Ouzounis CA. 2003. The balance of driving forces during genome evolution in prokaryotes. *Genome Res* 13: 1589–1594.

Kunst F, Ogasawara N, Moszer I, *et al.* 1997. The complete genome sequence of the Gram-positive bacterium *Bacillus subtilis*. *Nature* 390: 249–256.

Kurland CG, Canbäck B, Berg OG. 2003. Horizontal gene transfer: a critical view. *Proc Natl Acad Sci USA* 100: 9658–9662.

Kyrpides N. 1999. Genomes OnLine Database (GOLD): a monitor of complete and ongoing genome projects world wide. *Bioinformatics* 15: 773–774.

Kyrpides NC, Olsen GJ. 1999. Archaeal and bacterial hyperthermophiles: horizontal gene exchange or shared ancestry? *Trends Genet* 15: 298–299.

Kyrpides NC, Olsen GJ, Klenk HP, *et al.* 1996. *Methanococcus jannaschii* genome: revisited. *Microb Comp Genomics* 1: 329–338.

Lakshminarayan MI, Aravind L, Coon SL, *et al.* 2004. Evolution of cell–cell signaling in animals: did late horizontal gene transfer from bacteria have a role? *Trends Genet* 20: 292–299.

Lan R, Reeves PR. 2001. When does a clone deserve a name? A perspective on bacterial species based on population genetics. *Trends Microbiol* 9: 419–424.

Lapage SP, Sneath PHA, Lessel EF eds. 1992. *International Code of Nomenclature of Bacteria (1992 Revision)*. Washington, DC: American Society for Microbiology Press.

Lawrence JG. 1999. Gene transfer, speciation, and the evolution of bacterial genomes. *Curr Opin Microbiol* 2: 519–523.

Lawrence JG. 2002. Gene transfer in bacteria: speciation without species? *Theor Pop Biol* 61: 449–460.

Lawrence JG, Hendrickson H. 2003. Lateral gene transfer: when will adolescence end? *Mol Microbiol* 50: 739–749.

Lawrence JG, Ochman H. 1998. Molecular archaeology of the *Escherichia coli* genome. *Proc Natl Acad Sci USA* 95: 9413–9417.

Lawrence JG, Ochman H. 2002. Reconciling the many faces of lateral gene transfer. *Trends Microbiol* 10: 1–4.

Lawrence JG, Hendrix RW, Casjens S. 2001. Where are the pseudogenes in bacterial genomes? *Trends Microbiol* 9: 535–540.

Lepp PW, Brinig MM, Ouverney CC, *et al.* 2004. Methanogenic Archaea and human periodontal disease. *Proc Natl Acad Sci USA* 101: 6176–6181.

Levin-Zaidman S, Englander J, Shimoni E, *et al.* 2003. Ringlike structure of the *Deinococcus radiodurans* genome: a key to radioresistance? *Science* 299: 254–256.

Liò P. 2002. Investigating the relationship between genome structure, composition, and ecology in prokaryotes. *Mol Biol Evol* 19: 789–800.

Loferer-Krössbacher M, Witzel K-P, Psenner R. 1999. DNA content of aquatic bacteria measured by densitometric image analysis. *Arch Hydrobiol Spec Iss Adv Limnol* 54: 185–198.

Logsdon JM, Faguy DM. 1999. *Thermotoga* heats up lateral gene transfer. *Curr Biol* 9: R747–R751.

Lucier TS, Brubaker RR. 1992. Determination of genome size, macrorestriction pattern polymorphism, and nonpigmentation-specific deletion in *Yersinia pestis* by pulsed-field gel electrophoresis. *J Bacteriol* 174: 2078–2086.

Lynch M, Conery JS. 2003. The origins of genome complexity. *Science* 302: 1401–1404.

Makarova KS, Koonin EV. 2003. Comparative genomics of archaea: how much have we learned in six years, and what's next? *Genome Biol* 4: 115.111–115.117.

Makarova KS, Aravind L, Galperin MY, *et al.* 1999. Comparative genomics of the Archaea (Euryarchaeota): evolution of conserved protein families, the stable core, and the variable shell. *Genome Res* 9: 608–628.

Makarova KS, Aravind L, Wolf YI, *et al.* 2001. Genome of the extremely radiation-resistant bacterium *Deinococcus radiodurans* viewed from the perspective of comparative genomics. *Microbiol Mol Biol Rev* 65: 44–79.

Makino K, Oshima K, Kurokawa K, *et al.* 2003. Genome sequence of *Vibrio parahaemolyticus*: a pathogenic mechanism distinct from that of *V. cholerae*. *Lancet* 361: 743–749.

Margulis L, Sagan D. 2002. *Acquiring Genomes*. New York: Basic Books.

Marshall E. 2002. Venter gets down to life's basics. *Science* 298: 1701.

Martin AP, Costello EK, Meyer AF, *et al.* 2004. The rate and pattern of cladogenesis in microbes. *Evolution* 58: 946–955.

Maynard Smith J, Szathmáry E. 1995. *The Major Transitions in Evolution*. Oxford: Oxford University Press.

Mayr E. 1998. Two empires or three? *Proc Natl Acad Sci USA* 95: 9720–9723.

Merril CR, Biswas B, Carlton R, *et al.* 1996. Long-circulating bacteriophage as antibacterial agents. *Proc Natl Acad Sci USA* 93: 3188–3192.

Messer W. 2002. The bacterial replication initiator DnaA. DnaA and oriC, the bacterial mode to initiate DNA replication. *FEMS Microbiol Rev* 26: 355–374.

Methé BA, Nelson KE, Eisen JA, *et al.* 2003. Genome of *Geobacter sulfurreducens*: metal reduction in subsurface environments. *Science* 302: 1967–1969.

Miesel L, Greene J, Black TA. 2003. Genetic strategies for antibacterial drug discovery. *Nat Rev Genet* 4: 442–456.

Miller TL, Wolin MJ. 1982. Enumeration of *Methanobrevibacter smithii* in human feces. *Arch Microbiol* 131: 14–18.

Mills SD. 2003. The role of genomics in antimicrobial discovery. *J Antimicrob Chemother* 51: 749–752.

Mira A, Klasson L, Andersson SGE. 2002. Microbial genome evolution: sources of variability. *Curr Opin Microbiol* 5: 506–512.

Mira A, Ochman H, Moran NA. 2001. Deletional bias and the evolution of bacterial genomes. *Trends Genet* 17: 589–596.

Mock M, Fouet A. 2001. Anthrax. *Annu Rev Microbiol* 55: 647–671.

Moran NA. 2002. Microbial minimalism: genome reduction in bacterial pathogens. *Cell* 108: 583–586.

Moran NA. 2003. Tracing the evolution of gene loss in obligate bacterial symbionts. *Curr Opin Microbiol* 6: 512–518.

Moran NA, Baumann P. 2000. Bacterial endosymbionts in animals. *Curr Opin Microbiol* 3: 270–275.

Moran NA, Mira A. 2001. The process of genome shrinkage in the obligate symbiont *Buchnera aphidicola*. *Genome Biol* 2: 0054.0051–0054.0012.

Moran NA, Dale C, Dunbar H, *et al.* 2003a. Intracellular symbionts of sharpshooters (Insecta: Hemiptera: Cicadellinae) form a distinct clade with a small genome. *Environ Microbiol* 5: 116–126.

Moran NA, Plague GR, Sandström JP, Wilcox JL. 2003b. A genomic perspective on nutrient provisioning by bacterial symbionts of insects. *Proc Natl Acad Sci USA* 100: 14543–14548.

Mourier T, Hansen AJ, Willerslev E, Arctander P. 2001. The Human Genome Project reveals a continuous transfer of large mitochondrial fragments to the nucleus. *Mol Biol Evol* 18: 1833–1837.

Murray RG, Stackebrandt E. 1995. Taxonomic note: implementation of the provisional status Candidatus for incompletely described procaryotes. *Int J Syst Bacteriol* 45: 186–187.

Mushegian A. 1999. The minimal genome concept. *Curr Opin Genet Dev* 9: 709–714.

Mushegian AR, Koonin EV. 1996. A minimal gene set for cellular life derived by comparison of complete bacterial genomes. *Proc Natl Acad Sci USA* 93: 10268–10273.

Nakagawa I, Kurokawa K, Yamashita A, et al. 2003. Genome sequence of an M3 strain of *Streptococcus pyogenes* reveals a large-scale genomic rearrangement in invasive strains and new insights into phage evolution. *Genome Res* 13: 1042–1055.

Nakamura Y, Nishio Y, Ikeo K, Gojobori T. 2003. The genome stability in *Corynebacterium* species due to lack of the recombinational repair system. *Gene* 317: 149–155.

Nee S. 2003. Unveiling prokaryotic diversity. *Trends Ecol Evol* 18: 62–63.

Nelson KE, Clayton RA, Gill SR, et al. 1999. Evidence for lateral gene transfer between Archaea and bacteria from genome sequence of *Thermotoga maritima*. *Nature* 399: 323–329.

Nelson KE, Paulsen IT, Heidelberg JF, Fraser CM. 2000. Status of genome projects for nonpathogenic bacteria and archaea. *Nat Biotechnol* 18: 1049–1054.

Nierman W, Eisen JA, Fraser CM. 2000. Microbial genome sequencing 2000: new insights into physiology, evolution and expression analysis. *Res Microbiol* 151: 79–84.

Ochman H, Lawrence JG, Groisman EA. 2000. Lateral gene transfer and the nature of bacterial innovation. *Nature* 405: 299–304.

Ogata H, Audic S, Renesto-Audiffren P, et al. 2001. Mechanisms of evolution in *Rickettsia conorii* and *R. prowazekii*. *Science* 293: 2093–2098.

Okinaka R, Cloud K, Hampton O, et al. 1999a. Sequence, assembly and analysis of pX01 and pX02. *J Appl Microbiol* 87: 261–262.

Okinaka RT, Cloud K, Hampton O, et al. 1999b. Sequence and organization of pXO1, the large *Bacillus anthracis* plasmid harboring the anthrax toxin genes. *J Bacteriol* 181: 6509–6515.

Oren A. 2004. Prokaryote diversity and taxonomy: current status and future challenges. *Philos Trans R Soc Lond B* 359: 623–638.

Ortutay C, Gáspári Z, Tóth G, et al. 2003. Speciation in *Chlamydia*: genomewide phylogenetic analyses identified a reliable set of acquired genes. *J Mol Evol* 57: 672–680.

Oshima K, Kakizawa S, Nishigawa H, et al. 2004. Reductive evolution suggested from the complete genome sequence of a plant-pathogenic phytoplasma. *Nat Genet* 36: 27–29.

Pace NR. 1997. A molecular view of microbial diversity and the biosphere. *Science* 276: 734–740.

Palys T, Berger E, Mitrica I, et al. 2000. Protein-coding genes as molecular markers for ecologically distinct populations: the case of two *Bacillus* species. *Int J Syst Evol Microbiol* 50: 1021–1028.

Palys T, Nakamura LK, Cohan FM. 1997. Discovery and classification of ecological diversity in the bacterial world: the role of DNA sequence data. *Int J Syst Bacteriol* 47: 1145–1156.

Perna NT, Plunkett G, Burland V, et al. 2001. Genome sequence of enterohaemorrhagic *Escherichia coli* O157:H7. *Nature* 409: 529–533.

Pertsemlidis A, Fondon JW. 2001. Having a BLAST with bioinformatics (and avoiding BLASTphemy). *Genome Biol* 2: reviews2002.2001–2002.2010.

Peterson JD, Umayam LA, Dickinson TM, et al. 2001. The comprehensive microbial resource. *Nucleic Acids Res* 29: 123–125.

Peterson SN, Fraser CM. 2001. The complexity of simplicity. *Genome Biol* 2: comment2002.2001–2002.2008.

Philippe H, Douady CJ. 2003. Horizontal gene transfer and phylogenetics. *Curr Opin Microbiol* 6: 498–505.

Pieper DH, Martins dos Santos VAP, Golyshin PN. 2004. Genomic and mechanistic insights into the biodegradation of organic pollutants. *Curr Opin Biotechnol* 15: 215–224.

Pizza M, Scarlato V, Masignani V, et al. 2000. Identification of vaccine candidates against serogroup B meningococcus by whole-genome sequencing. Science 287: 1816–1820.

Planet PJ, Kachlany SC, Fine DH, et al. 2003. The widespread colonization island of Actinobacillus actinomycetemcomitans. Nat Genet 34: 193–198.

Postow L, Crisona NJ, Peter BJ, et al. 2001. Topological challenges to DNA replication: conformations at the fork. Proc Natl Acad Sci USA 98: 8219–8226.

Powledge TM. 2002. Minimal controversy. The Scientist, November 22.

Pridmore RD, Crouzillat D, Walker C, et al. 2000. Genomics, molecular genetics and the food industry. J Biotechnol 78: 251–258.

Pupo GM, Lan R, Reeves PR. 2000. Multiple independent origins of Shigella clones of Escherichia coli and convergent evolution of many of their characteristics. Proc Natl Acad Sci USA 97: 10567–10572.

Purohit HJ, Raje DV, Kapley A, et al. 2003. Genomics tools in environmental impact assessment. Environ Sci Technol 37: 357A–363A.

Ragan MA. 2001. On surrogate methods for detecting lateral gene transfer. FEMS Microbiol Lett 201: 187–191.

Ramisse V, Patra G, Vaissaire J, Mock M. 1999. The Ba813 chromosomal DNA sequence effectively traces the whole Bacillus anthracis community. J Appl Microbiol 87: 224–228.

Rasko DA, Ravel J, Okstad OA, et al. 2004. The genome sequence of Bacillus cereus ATCC 10987 reveals metabolic adaptations and a large plasmid related to Bacillus anthracis pXO1. Nucleic Acids Res 32: 977–988.

Read TD, Parkhill J. 2002. Restricting genome data won't stop bioterrorism. Nature 417: 379.

Read TD, Brunham RC, Shen C, et al. 2000. Genome sequences of Chlamydia trachomatis MoPn and Chlamydia pneumoniae AR39. Nucleic Acids Res 28: 1397–1406.

Read TD, Gill SR, Tettelin H, Dougherty BA. 2001. Finding drug targets in microbial genomes. Drug Discov Today 6: 887–892.

Read TD, Peterson SN, Tourasse N, et al. 2003. The genome sequence of Bacillus anthracis Ames and comparison to closely related bacteria. Nature 423: 81–86.

Read TD, Salzberg SL, Pop M, et al. 2002. Comparative genome sequencing for discovery of novel polymorphisms in Bacillus anthracis. Science 296: 2028–2033.

Redenbach M, Scheel J, Schmidt U. 2000. Chromosome topology and genome size of selected actinomycetes species. Antonie van Leeuwenhoek 78: 227–235.

Reich KA. 2000. The search for essential genes. Res Microbiol 151: 319–320.

Reid SD, Herbellin CJ, Bumbaugh AC, et al. 2000. Parallel evolution of virulence in pathogenic Escherichia coli. Nature 406: 64–67.

Rendulic S, Jagtap P, Rosinus A, et al. 2004. A predator unmasked: life cycle of Bdellovibrio bacteriovorus from a genomic perspective. Science 303: 689–692.

Rivera MC, Lake JA. 2004. The ring of life provides evidence for a genome fusion origin of eukaryotes. Nature 431: 152–155.

Robinson R. 2003. Gut up and go. The Scientist, March 7.

Rocha EPC. 2003. DNA repeats lead to the accelerated loss of gene order in bacteria. Trends Genet 19: 600–603.

Rocha EPC, Danchin A, Viari A. 1999. Functional and evolutionary roles of long repeats in prokaryotes. Res Microbiol 150: 725–733.

Roelofs J, Van Haastert PJ. 2001. Genes lost during evolution. Nature 411: 1013–1014.

Rondon MR, August PR, Betterman AD, et al. 2000. Cloning the soil metagenome: a strategy for accessing the genetic and functional diversity of uncultured microorganisms. Appl Environ Microbiol 66: 2541–2547.

Roselló-Mora R, Amann R. 2001. The species concept for prokaryotes. FEMS Microbiol Rev 25: 39–67.

Rowe-Magnus DA, Guérout A-M, Mazel D. 1999. Super-integrons. Res Microbiol 150: 641–651.

Rowe-Magnus DA, Mazel D. 1999. Resistance gene capture. *Curr Opin Microbiol* 2: 483–488.

Saccone C, Pesole G. 2003. *Handbook of Comparative Genomics*. Hoboken, NJ: John Wiley & Sons.

Salzberg SL, White O, Peterson J, Eisen JA. 2001. Microbial genes in the human genome: lateral transfer or gene loss? *Science* 292: 1903–1906.

Sampson SL, Rengarajan J, Rubin E. 2003. Bacterial genomics and vaccine design. *Expert Rev Vaccines* 2: 437–445.

Sanger F, Air GM, Barrell BG, *et al.* 1977. Nucleotide sequence of bacteriophage phi X174 DNA. *Nature* 265: 687–695.

Sassetti C, Rubin EJ. 2002. Genomic analyses of microbial virulence. *Curr Opin Microbiol* 5: 27–32.

Saunders NJ, Butcher S. 2003. The use of complete genome sequences in vaccine design. *Methods Mol Med* 87: 301–312.

Savva N, McAllen CJ, Giddins GE. 2003. Current approaches to whole genome phylogenetic analysis. *Brief Bioinform* 4: 63–74.

Schoolnik GK. 2002a. Functional and comparative genomics of pathogenic bacteria. *Curr Opin Microbiol* 5: 20–26.

Schoolnik GK. 2002b. Microarray analysis of bacterial pathogenicity. *Adv Microb Physiol* 46: 1–45.

Sebaihia M, Thomson N, Crossman L, Parkhill J. 2004. Bacterial minimalism. *Nat Rev Microbiol* 2: 274–275.

Selosse M-A, Albert B, Godelle B. 2001. Reducing the genome size of organelles favours gene transfer to the nucleus. *Trends Ecol Evol* 16: 135–141.

Senior K. 2001. Bacteriophages: a rich store of new antibiotics? *Drug Discov Today* 6: 865–866.

Sensen CW. 1999. Sequencing microbial genomes. In: Charlebois RL ed. *Organization of the Prokaryotic Genome*. Washington, DC: American Society for Microbiology Press, 1–9.

Sherratt DJ. 2003. Bacterial chromosome dynamics. *Science* 301: 780–785.

Shigenobu S, Watanabe H, Hattori M, *et al.* 2000. Genome sequence of the endocellular bacterial symbiont of aphids *Buchnera* sp. APS. *Nature* 407: 81–86.

Shuter BJ, Thomas JE, Taylor WD, Zimmerman AM. 1983. Phenotypic correlates of genomic DNA content in unicellular eukaryotes and other cells. *Am Nat* 122: 26–44.

Silva FJ, Latorre A, Moya A. 2003. Why are the genomes of endosymbiotic bacteria so stable? *Trends Genet* 19: 176–180.

Slezak T, Kuczmarski T, Ott L, *et al.* 2003. Comparative genomics tools applied to bioterrorism defence. *Brief Bioinform* 4: 133–149.

Smith DR, Doucette-Stamm LA, Deloughery C, *et al.* 1997. Complete genome sequence of *Methanobacterium thermoautotrophicum* ΔH: functional analysis and comparative genomics. *J Bacteriol* 179: 7135–7155.

Smith HO, Hutchison CA, Pfannkoch C, Venter JC. 2003. Generating a synthetic genome by whole genome assembly: φX174 bacteriophage from synthetic oligonucleotides. *Proc Natl Acad Sci USA* 100: 15440–15445.

Smith HO, Tomb JF, Dougherty BA, *et al.* 1995. Frequency and distribution of DNA uptake signal sequences in the *Haemophilus influenzae* Rd genome. *Science* 269: 468–470.

Smith M, Brown NL, Air GM, *et al.* 1977. DNA sequence at the C termini of the overlapping genes A and B in bacteriophage phiX174. *Nature* 265: 702–705.

Smoot JC, Barbian KD, Van Gompel JJ, *et al.* 2002. Genome sequence and comparative microarray analysis of serotype M18 group A *Streptococcus* strains associated with acute rheumatic fever outbreaks. *Proc Natl Acad Sci USA* 99: 4668–4673.

Snel B, Bork P, Huynen MA. 2002. Genomes in flux: the evolution of archaeal and proteobacterial gene content. *Genome Res* 12: 17–25.

Spratt BG. 2003. Stomachs out of Africa. *Science* 299: 1528–1529.

Stackebrandt E. 2003. The richness of prokaryotic diversity: there must be a species somewhere. *Food Technol Biotechnol* 41: 17–22.

Stackebrandt E, Goebel BM. 1994. Taxonomic note: a place for DNA:DNA reassociation and 16S rRNA sequence analysis in the present species definition in bacteriology. *Int J Syst Bacteriol* 44: 846–849.

Stackebrandt E, Frederiksen W, Garrity GM, et al. 2002. Report of the ad hoc committee for the re-evaluation of the species definition in bacteriology. *Int J Syst Evol Microbiol* 52: 1043–1047.

Staley JT. 1997. Biodiversity: are microbial species threatened? *Curr Opin Biotechnol* 8: 340–345.

Stanhope MJ, Lupas A, Italia MJ, et al. 2001. Phylogenetic analyses do not support horizontal gene transfers from bacteria to vertebrates. *Nature* 411: 940–944.

Stein JL, Simon MI. 1996. Archaeal ubiquity. *Proc Natl Acad Sci USA* 93: 6228–6230.

Stepkowski T, Legocki AB. 2001. Reduction of bacterial genome size and expansion resulting from obligate intracellular lifestyle and adaptation to soil habitat. *Acta Biochim Pol* 48: 367–381.

Sun LV, Foster JM, Tzertzinis G, et al. 2001. Determination of *Wolbachia* genome size by pulsed-field gel electrophoresis. *J Bacteriol* 183: 2219–2225.

Takai K, Moser DP, DeFlaun M, et al. 2001. Archaeal diversity in waters from deep South African gold mines. *Appl Environ Microbiol* 67: 5750–5760.

Takami H, Horikoshi K. 2000. Analysis of the genome of an alkaliphilic *Bacillus* strain from an industrial point of view. *Extremophiles* 4: 99–108.

Tamas I, Klasson L, Canbäck B, et al. 2002. 50 million years of genomic stasis in endosymbiotic bacteria. *Science* 296: 2376–2379.

Tamas I, Klasson LM, Sandström JP, Andersson SGE. 2001. Mutualists and parasites: how to paint yourself into a (metabolic) corner. *FEBS Lett* 498: 135–139.

Tanaka N, Hirahata M, Miyazaki S, Sugawara H. 2003. The status quo of microbial genomic data available in the public domain: archaea and bacteria. *World Fed Cult Collect Lett* 36: 13–20.

Tang CM, Moxon ER. 2001. The impact of microbial genomics on antimicrobial drug development. *Annu Rev Genomics Hum Genet* 2: 259–269.

Tatusov RL, Fedorova ND, Jackson JD, et al. 2003. The COG database: an updated version includes eukaryotes. *BMC Bioinformatics* 4: 41.41–41.14.

Tatusov RL, Koonin EV, Lipman DJ. 1997. A genomic perspective on protein families. *Science* 278: 631–637.

Tatusov RL, Natale DA, Garkavtsev IV, et al. 2001. The COG database: new developments in phylogenetic classification of proteins from complete genomes. *Nucleic Acids Res* 29: 22–28.

Terletski V, Schwarz S, Carnwath J, Niemann H. 2003. Subtracted restriction fingerprinting—a tool for bacterial genome typing. *Biotechniques* 34: 304–310.

Tettelin H, Masignani V, Cieslewicz MJ, et al. 2002. Complete genome sequence and comparative genomic analysis of an emerging human pathogen, serotype V *Streptococcus agalactiae*. *Proc Natl Acad Sci USA* 99: 12391–12396.

Tettelin H, Nelson KE, Paulsen IT, et al. 2001. Complete genome sequence of a virulent isolate of *Streptococcus pneumoniae*. *Science* 293: 498–506.

Thompson CJ, Fink D, Nguyen LD. 2002. Principles of microbial alchemy: insights from the *Streptomyces coelicolor* genome sequence. *Genome Biol* 3: reviews1020.1021–1020.1024.

Tiedje JM. 2002. *Shewanella*—the environmentally versatile genome. *Nat Biotechnol* 20: 1093–1094.

Torsvik V, Øvreås L, Thingstad TF. 2002. Prokaryotic diversity—magnitude, dynamics, and controlling factors. *Science* 296: 1064–1066.

Trevors JT. 1996. Genome size in bacteria. *Antonie van Leeuwenhoek* 69: 293–303.

van Belkum A. 2003. High-throughput epidemiologic typing in clinical microbiology. *Clin Microbiol Infect* 9: 86–100.

van Ham RC, Kamerbeek J, Palacios C, et al. 2003. Reductive genome evolution in *Buchnera aphidicola*. *Proc Natl Acad Sci USA* 100: 581–586.

Vas A, Leatherwood J. 2000. Where does DNA replication start in archaea? *Genome Biol* 1: reviews1020.1021–1020.1024.

Vellai T, Vida G. 1999. The origin of eukaryotes: the difference between prokaryotic and eukaryotic cells. *Proc R Soc Lond B* 266: 1571–1577.

Vellai T, Takács K, Vida G. 1998. A new aspect to the origin and evolution of eukaryotes. *J Mol Evol* 46: 499–507.

Venkatesan MM, Goldberg MB, Rose DJ, *et al*. 2001. Complete DNA sequence and analysis of the large virulence plasmid of *Shigella flexneri*. *Infect Immun* 69: 3271–3285.

Venter JC, Remington K, Heidelberg JF, *et al*. 2004. Environmental genome shotgun sequencing of the Sargasso Sea. *Science* 304: 66–74.

Vinogradov AE. 1998. Genome size and GC-percent in vertebrates as determined by flow cytometry: the triangular relationship. *Cytometry* 31: 100–109.

Vinogradov AE. 2004. Testing genome complexity. *Science* 304: 389–390.

Volff JN, Altenbuchner J. 1998. Genetic instability of the *Streptomyces* chromosome. *Mol Microbiol* 27: 239–246.

Wagner PL, Waldor MK. 2002. Bacteriophage control of bacterial virulence. *Infect Immun* 70: 3985–3993.

Wallace DC, Morowitz HJ. 1973. Genome size and evolution. *Chromosoma* 40: 121–126.

Wang L, Qu W, Reeves PR. 2001. Sequence analysis of four *Shigella boydii* O-antigen loci: implication for *Escherichia coli* and *Shigella* relationships. *Infect Immun* 69: 6923–6930.

Ward BB. 2002. How many species of prokaryotes are there? *Proc Natl Acad Sci USA* 99: 10234–10236.

Ward DM. 1998. A natural species concept for prokaryotes. *Curr Opin Microbiol* 1: 271–277.

Waterman SR. 2001. Bacterial genomics as a potential tool for discovering new antimicrobial agents. *Am J Pharmacogenomics* 1: 263–269.

Waters E, Hohn MJ, Ahel I, *et al*. 2003. The genome of *Nanoarchaeum equitans*: insights into early archaeal evolution and derived parasitism. *Proc Natl Acad Sci USA* 100: 12984–12988.

Wayne LG, Brenner DJ, Colwell RR, *et al*. 1987. Report of the ad hoc committee on reconciliation of approaches to bacterial systematics. *Int J Syst Bacteriol* 37: 463–464.

Wei J, Goldberg MB, Burland V, *et al*. 2003. Complete genome sequence and comparative genomics of *Shigella flexneri* serotype 2a strain 2457T. *Infect Immun* 71: 2775–2786.

Weinstock GM. 2000. Genomics and bacterial pathogenesis. *Emerg Infect Dis* 6: 496–504.

Weinstock GM, Smajs D, Hardham J, Norris SJ. 2000. From microbial genome sequence to applications. *Res Microbiol* 151: 151–158.

Welch RA, Burland V, Plunkett G, *et al*. 2002. Extensive mosaic structure revealed by the complete genome sequence of uropathogenic *Escherichia coli*. *Proc Natl Acad Sci USA* 99: 17020–17024.

Wernegreen JJ. 2002. Genome evolution in bacterial endosymbionts of insects. *Nat Rev Genet* 3: 850–861.

Wernegreen JJ, Lazarus AB, Degnan PH. 2002. Small genome of *Candidatus* Blochmannia, the bacterial endosymbiont of *Camponotus*, implies irreversible specialization to an intracellular lifestyle. *Microbiology* 148: 2551–2556.

Werren JH. 1998. *Wolbachia* and speciation. In: Berlocher SH ed. *Endless Forms: Species and Speciation*. Oxford: Oxford University Press, 245–260.

Westberg J, Persson A, Holmberg A, *et al*. 2004. The genome sequence of *Mycoplasma mycoides* subsp. *mycoides* SC type strain PG1T, the causative agent of contagious bovine pleuropneumonia (CBPP). *Genome Res* 14: 221–227.

White MF, Bell SD. 2002. Holding it together: chromatin in the Archaea. *Trends Genet* 18: 621–626.

White O, Eisen JA, Heidelberg JF, *et al*. 1999. Genome sequence of the radioresistant bacterium *Deinococcus radiodurans* R1. *Science* 286: 1571–1577.

Whitfield J. 2003. Genome pioneer sets sights on Sargasso Sea. *Nature Science Update*, April 30.

Whitman WB, Coleman DC, Wiebe WJ. 1998. Prokaryotes: the unseen majority. *Proc Natl Acad Sci USA* 95: 6578–6583.

Woese CR. 1987. Bacterial evolution. *Microbiol Rev* 51: 221–271.

Woese CR. 1998. The universal ancestor. *Proc Natl Acad Sci USA* 95: 6854–6859.

Woese CR. 2000. Interpreting the universal phylogenetic tree. *Proc Natl Acad Sci USA* 97: 8392–8396.

Woese CR. 2002. On the evolution of cells. *Proc Natl Acad Sci USA* 99: 8742–8747.

Woese CR, Fox GE. 1977. Phylogenetic structure of the prokaryotic domain: the primary kingdoms. *Proc Natl Acad Sci USA* 74: 5088–5090.

Woese CR, Gupta R. 1981. Are archaebacteria merely derived 'prokaryotes'? *Nature* 289: 95–96.

Wolf YI, Rogozin IB, Grishin NV, Koonin EV. 2002. Genome trees and the Tree of Life. *Trends Genet* 18: 472–479.

Wong DA, Yip PC, Tse DL, *et al.* 2003. Routine use of a simple low-cost genotypic assay for the identification of mycobacteria in a high throughput laboratory. *Diagn Microbiol Infect Dis* 47: 421–426.

Wren BW. 2002. Deciphering tsetse's secret partner. *Nat Genet* 32: 335–336.

Wu M, Sun LV, Vamathevan J, *et al.* 2004. Phylogenomics of the reproductive parasite *Wolbachia pipientis* wMel: a streamlined genome overrun by mobile genetic elements. *PLoS Biol* 2: 0327–0341.

Xu J, Bjursell MK, Himrod J, *et al.* 2003. A genomic view of the human–*Bacteroides thetaiotaomicron* symbiosis. *Science* 299: 2074–2076.

Yeramian E, Buc H. 1999. Tandem repeats in complete bacterial genome sequences: sequence and structural analyses for comparative studies. *Res Microbiol* 150: 745–754.

Young DB. 1998. Blueprint for the white plague. *Nature* 393: 515–516.

Zagursky RJ, Olmsted SB, Russell DP, Wooters JL. 2003. Bioinformatics: how it is being used to identify bacterial vaccine candidates. *Expert Rev Vaccines* 2: 417–436.

PART VI

The Genome in Evolution

CHAPTER 11

Macroevolution and the Genome

T. RYAN GREGORY

The topics of genomics and evolutionary biology are deeply interconnected. For most of the issues discussed in the previous chapters (genome size, parasitic elements, gene duplications, polyploidy), this is obvious. In none of these cases can the genomic features in question be understood without reference to both their own evolutionary histories and to their implications for the organisms and/or species possessing them. Even the detailed study of (mostly) complete genome sequences is best carried out in an evolutionarily comparative context, and in turn provides important new insights into evolutionary relationships. However, so far in this volume there has not been any explicit discussion of the framework of evolutionary theory in which all of these analyses must take place.

There are many ways of studying the mechanisms and outcomes of evolution, ranging from genetics and genomics at the lowest scales through to paleontology at the highest. Unfortunately, the division into specialties according to scale has often led to protracted disagreement among evolutionary theorists from different disciplines regarding the nature of the evolutionary process. Although the resulting debate has undoubtedly led to a refinement of the various theoretical approaches employed, it has also prevented the development of a complete and unified theory of evolution. Without such a theory, all evolutionary phenomena, including those involving features of the genome, will remain at best only partially understood.

This chapter provides a discussion of some important conceptual and empirical links between the lowest levels of evolutionary analysis (the genome and its components) and the highest (large-scale patterns in deep evolutionary time). This issue is approached from two very different perspectives, with the chapter bisected in the following way: The first part introduces some basic concepts of higher-level evolutionary theory developed mostly by paleontologists, and then uses them to understand various aspects of genome evolution. The second part describes the role of "nonstandard" genome-level processes (that is, ones other than small-scale variations in protein-coding genes) in helping to shape the major evolutionary transitions of particular interest to human evolutionary history. The goal throughout this chapter is to provide an expansion, not a refutation, of existing evolutionary theory, and to build some much-needed bridges across traditionally disparate disciplines. In this, the following discussion marks only a beginning.

PART ONE—MACROEVOLUTIONARY THEORY AND GENOME EVOLUTION

A BRIEF HISTORY OF EVOLUTIONARY THEORY

FROM DARWIN TO NEO-DARWINISM

Charles Darwin did not invent the concept of evolution ("descent with modification" or "transmutation," in the terminology of his time). In fact, the notion of evolutionary change long predates Darwin's (1859) contributions in *On the Origin of Species*, which were essentially twofold: (1) providing extensive evidence, from a variety of sources, for the fact that species are related by descent, and (2) developing his theory of natural selection to explain this fact. Although quite successful in establishing the fact of evolution (the subsequent Creationist movement in parts of North America notwithstanding), Darwin's explanatory mechanism of natural selection received only a lukewarm reception in contemporary scientific circles.

By the beginning of the 20th century, Darwinian natural selection had fallen largely out of favor, having been overshadowed by several other proposed mechanisms including *mutationism*, whereby species form suddenly by single mutations, with no intermediates; *saltationism*, in which major chromosomal rearrangements generate new species suddenly; *neo-Lamarckism*, which supposed that traits are improved directly through use and lost through disuse; and *orthogenesis*, under which inherent propelling forces drive evolutionary changes, sometimes even to the point of being maladaptive. Mutationism, in particular, gained favor after the rediscovery of Mendel's laws of inheritance by Hugo de Vries and others, which showed heredity to be "particulate"—with individual traits passed on intact, even if hidden for a generation—rather than "blending," as Darwin had believed. Particulate inheritance was taken by de Vries and others to imply that discontinuous variation in traits would be much more important than the continuous variability expected under gradual Darwinian selection.

The problem faced by proponents of Darwinism was to reconcile the concept of discrete hereditary units with the graded variation required by natural selection. This issue was settled in the 1930s and 1940s with the advent of population genetics, which provided mathematical models to describe the behavior of genic variants ("alleles") within populations, and showed that a particulate mechanism of inheritance did not prohibit the action of natural selection. This new theoretical framework is generally known as "neo-Darwinism" or the "Modern Synthesis," because it sought to synthesize (i.e., combine) Mendelian genetics and Darwinian natural selection.

The first stage in the development of population genetics was to determine how alleles segregate within populations under "equilibrium" conditions. This issue was addressed by H.G. Hardy and Wilhelm Weinberg, resulting in what is now known as the "Hardy-Weinberg equilibrium," a null hypothesis about the behavior of alleles in populations that are not subject to *natural selection, genetic drift* (random changes in allele frequencies, for example by the accidental loss of a subset of the population, passage through a population bottleneck, or the founding of a new population by an unrepresentative sample of the parental population), *gene flow* (an influx of alleles from other populations by migration), or *mutation* (the generation of new alleles). When populations are not in Hardy-Weinberg equilibrium, one can begin to investigate which of these processes is (or are) responsible. More complex population genetics models were developed for dealing with this issue, most notably by Ronald Fisher, Sewall Wright, and J.B.S. Haldane. Others, like Theodosius Dobzhansky and G.L. Stebbins, established that natural populations contain sufficient genetic variation for these new models to work.

According to Provine (1988), the Modern Synthesis was really more of a "constriction" than an actual "synthesis," in which a major goal was the elimination of the non-Darwinian alternatives listed previously, and the associated restoration of selection to prominence in evolutionary theory. In at least one important sense, the term "synthesis" is clearly a misnomer, given that there remained a highly acrimonious divide between Fisher, who favored models based on large populations with a dominant role for selection, and Wright, whose "adaptive landscape" model dealt primarily with small populations and emphasized genetic drift. Despite these divisions, neo-Darwinians did succeed in narrowing the range of explanatory approaches to those involving mutation, selection, drift, and gene flow.

Genomes, Fossils, and Theoretical Inertia

As far as genetics is concerned, evolutionary theory has always been far ahead of its time. Darwin's theory of natural selection was developed in the absence of concrete knowledge of hereditary mechanisms, and the mathematical framework of neo-Darwinism was assembled before the structure of DNA had been established (and even before DNA was identified as the molecule of inheritance). As a consequence, numerous surprises, puzzles, and conflicts have emerged from new discoveries in genetics and genomics. Consider, for example, the recent findings of deep genetic homology undergirding "analogous" features of unrelated organisms, the role of clustered master control genes in regulating development (see Chapter 5), the remarkably low gene numbers in humans (see Chapter 9), the collapse of the "one gene–one protein" model, the extraordinary abundance of transposable elements in the genomes of humans and other species (see Chapter 3), and the increasing evidence for a role of large-scale genome duplications in evolution (see Chapter 6). Also recognized for decades (and still the subject of healthy debate) are the importance of smaller-scale gene duplications (see Chapter 5), the role of recurrent hybridization and polyploidy (see Chapters 7 and 8), the preponderance of neutral evolution at the molecular level, and the initially quite alarming disconnect between genome size and organismal complexity (see Chapters 1 and 2). Advances in genetics and genomics have also provided revolutionary insights into the relationships among organisms, from the smallest scales (e.g., human–chimpanzee genetic similarity) to the largest (e.g., deep divergences between "prokaryote" groups) (see Chapter 10). None of these was (or indeed, could have been) predicted or expected by the accepted formulation of evolutionary theory that preceded it. The historical record in evolutionary biology is that theories are developed under assumptions about the existence—or perhaps more commonly, the *absence*—of certain genetic mechanisms, and must later be revised as new knowledge comes to light regarding genomic structure, organization, and function. This mode of progress is not necessarily problematic, except when theoretical inertia forestalls the acceptance of the new information and its implications.

Genomics is not the only field to have faced theoretical inertia. For decades, prominent paleontologists have argued that their observations of the fossil record fail to fit the expectations of strict Darwinian gradualism. Darwin's view of speciation, sometimes labeled as "phyletic gradualism," was based on the slow, gradual (but not necessarily constant) evolution of one species or large segments thereof into another through a series of imperceptible changes, often without any splitting of lineages (i.e., by "anagenesis"). By contrast, the theory of "punctuated equilibria" ("punk eek" to afficionados, "evolution by jerks" to some critics) proposes that most species experience pronounced morphological stasis for most of their tenure, with change occurring only in geologically rapid bursts associated with speciation events (Eldredge and Gould, 1972; Gould and Eldredge, 1977, 1993; Gould, 1992, 2002) (Fig. 11.1). Moreover, speciation in this second case involves the branching off of new species ("cladogenesis") via small, peripherally isolated populations rather than the gradual transformation of the parental stock itself.

Based on differences such as these, many of those who study evolutionary patterns in deep time have developed alternative theoretical approaches to account

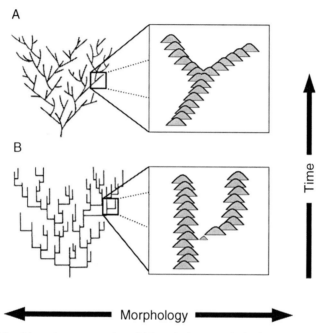

FIGURE 11.1   Schematic representation of (A) gradualism, in which change at the species level is gradual (though not necessarily constant in rate), possibly occurring without splitting (i.e., by "anagenesis"), and (B) punctuated equilibria, whereby long periods of stasis are punctuated by periods of rapid change associated with speciation events. Under punctuated equilibria, speciation is always by branching ("cladogenesis"), with parent and daughter species coexisting for a period of time. A small subsection is highlighted for each tree, showing the changes in the distributions of morphological parameters within populations, with each vertical bell curve marking a snapshot in time. When splitting occurs under the gradualist model, it involves the division of one large population into two, which then diverge to eventually become two new species. Under punctuated equilibria, a small peripheral isolate diverges from the parental stock and a new daughter species is formed quickly, which then coexists with the original species. Punctuated equilibria makes it possible to define births (by speciation), life spans (stasis), deaths (extinction), and reproductive capacities (the production of daughter species) at the species level. This raises the possibility that species themselves may participate as evolutionary entities, thereby providing a mechanism of evolution that operates at the highest scales in addition to those occurring within populations.

for the large-scale features of evolution. This, too, has generally proceeded with a minimal consideration of genomic information, and as such there is a need for increased communication between these two fields. In fact, despite their residence at opposite ends of the spectrum in evolutionary science, there is great potential for integration between genomics and paleontology because ultimately both are concerned with variation among species and higher taxa.

IS A THEORY OF MACROEVOLUTION NECESSARY?

MICROEVOLUTION, MACROEVOLUTION, AND EXTRAPOLATIONISM

The extent to which processes observable within populations and tractable in mathematical models can be extrapolated to explain patterns of diversification occurring in deep time remains one of the most contentious issues in modern evolutionary biology. This is a debate with a lengthy pedigree, extending back more than 75 years, and therefore long predating any of the issues of genome evolution discussed in this book. Nevertheless, genomes reside at an important nexus in this debate by containing the genes central to population-level discussions, but also having their own complex large-scale evolutionary histories.

Writing as an orthogeneticist in 1927, prior to the Modern Synthesis when Darwinian natural selection was largely eclipsed as a mechanism of evolutionary change, Iurii Filipchenko made the following argument:

> Modern genetics doubtless represents the veil of the evolution of Jordanian and Linnaean biotypes (microevolution), contrasted with the evolution of higher systematic groups (macroevolution), which has long been of central interest. This serves to underline the above-cited consideration of the absence of any intrinsic connection between genetics and the doctrine of evolution, which deals particularly with macroevolution. [Translation as in Hendry and Kinnison, 2001].

In modern parlance, *microevolution* represents the small-scale changes in allele frequencies that occur within populations (as studied by population geneticists and often observable over the span of a human lifetime), whereas *macroevolution* involves the generation of broad patterns above the species level over the course of Earth history (as studied in the fossil record by paleontologists, and with regard to extant taxa by systematists). In contrast to Filipchenko, Dobzhansky (1937, p. 12) noted that because macroevolution could not be observed directly, "we are compelled at the present level of knowledge reluctantly to put a sign of equality between the mechanisms of macro- and micro-evolution." However, although Dobzhansky was tentative in his assertion of micro–macro equivalence, the doctrine of "extrapolationism" was embraced as fact by many other architects and early adherents of the Modern Synthesis. Thus as Mayr (1963, p. 586) later explained, "the proponents of the synthetic theory maintain that *all* evolution is due to the accumulation of *small genetic changes,* guided by natural selection, and that trans-specific evolution is *nothing but* an extrapolation and magnification of the events that take place within populations and species" (emphasis added). There was an obvious reason for this strict adherence to extrapolationism at the time, namely the belief that if micro- and macroevolution "proved to be basically different, the innumerable studies of micro-evolution would become relatively unimportant and would have minor value in the study of evolution as a whole" (Simpson, 1944, p. 97). As such, only proponents of non-Darwinian mechanisms, most notably the much-maligned saltationist Richard Goldschmidt (1940, p. 8), argued at the time that "the facts of microevolution do not suffice for an understanding of macroevolution."

Obviously, the "present level of knowledge" is not the same today as it was in Dobzhansky's time. A great deal of new information has since been gleaned—and continues to accrue—regarding the mechanisms of heredity and the major patterns of evolutionary diversification. Considering Mayr's statement, it is now clear that not all relevant genetic changes are small (cf., genome duplications), nor is all change guided by natural selection (cf., neutral molecular evolution), nor do all relevant processes operate within populations and species (cf., hybridization). In one of the more notorious exchanges on the subject, Gould (1980) went so far as to declare this simple version of the neo-Darwinian synthesis as "effectively dead, despite its persistence as textbook orthodoxy."[1] To be more specific, this applies not to the Modern Synthesis at large, but to strict extrapolationism. Using a far less aggressive tone, another prominent macroevolutionist put it as follows: "That advances in molecular biology contribute to the need for a formal expansion

---

[1] Of course, far from simply mourning their loss, microevolutionists responded to this charge with some vigor (e.g., Stebbins and Ayala, 1981; Ayala, 1982; Charlesworth *et al.,* 1982; Hecht and Hoffman, 1986), perhaps overlooking the fact that only the strict extrapolationist definition given by Mayr (1963), and not the synthesis in its entirety, was proclaimed deceased (see Gould, 2002). Although some may argue that Mayr's (1963) definition was already outdated by this time, and that Gould's (1980) criticism was therefore misplaced, it bears noting that such a definition had been in common use throughout the period in question and well beyond (e.g., Mayr, 1980; Ruse, 1982; Hecht and Hoffman, 1986). As for Gould's (1980) claim of "textbook orthodoxy," one may consider Freeman and Herron's (1998) recent textbook, which considers the Modern Synthesis to be composed of two main postulates: "[1] Gradual evolution results from small genetic changes that are acted upon by natural selection. [2] The origin of species and higher taxa, or macroevolution, can be explained in terms of natural selection acting on individuals, or microevolution." Futuyma's (1998) more advanced text provides a much more detailed description of the Modern Synthesis but the fundamental extrapolationist point remains.

of evolutionary theory is an exigency we can hardly hold against the early archi-tects of the synthesis" (Eldredge, 1985, p. 86). Indeed, it is interesting to imagine the view that Fisher, Dobzhansky, Haldane, Wright, or even Darwin might have taken had they been privy to modern genomic insights.

Simpson's (1944) account of the threat to the relevance of microevolution is also in need of revision. It is simply not the case that a mechanistic disconnect between micro- and macroevolution would render microevolutionary study obsolete. Far from it, because any genomic changes, regardless of the magnitude of their effects, must still pass through the filters of selection and drift to reach a sufficiently high frequency if they are to be of evolutionary significance. So, even if understanding this filtration process does not, by itself, provide a complete understanding of macroevolution, it would still be a crucial component of an expanded evolutionary theory. Consider, for example, the topic of major developmental regulation genes, which involves at least four different questions, all mutually compatible, studied by four different disciplines: (1) *Evolutionary developmental biology* ("evo-devo")—How do such genes act to produce observed phenotypes? (2) *Comparative genomics*—What is the structure of these genes, and what role did processes like gene (or genome) duplication play in their evolution (see Chapters 5 and 6)? (3) *Population genetics*—How would such genes have been filtered by selection, drift, and gene flow to reach their current state of fixation (e.g., Johnson and Porter, 2001)? (4) *Paleontology*—What is the relevance of these genes for understanding the emer-gence of new body plans and thus new macroevolutionary trajectories (e.g., Carroll, 2000; Erwin, 2000; Jablonski, 2000; Shubin and Marshall, 2000)?

Though the protagonists have often been divided along these professional lines, the micro–macro debate is not between paleontologists and population geneticists per se. Rather, it is between strict extrapolationists who argue that all evolution can be understood by studying population-level processes and those who argue that there are additional factors to consider. Members of this latter camp may come from all quarters of evolutionary biology, from genome biologists to paleontologists, although the latter have been by far the most vocal proponents of an expanded outlook. For strict extrapolationists, there may be little value in pursuing this debate. But for those open to a more pluralistic approach who seek a resolution to the issue, there is much value in understanding the arguments presented in favor of a distinct macroevolutionary theory that coexists with, but is not subsumed by, established microevolutionary principles.

CRITIQUES OF STRICT EXTRAPOLATIONISM

At least five major criticisms have been levied against the extrapolationist approach. In each case, the relevant point has been that small-scale genetic changes within populations are not sufficient to explain all observations at macroevolutionary scales, regardless of their necessity as part of evolutionary theory. In the words of one of extrapolationism's harshest critics, "nothing about microevolutionary population genetics, or any other aspect of microevolutionary theory, is wrong or inadequate at its level ... but it is not everything" (Gould, 1982).

(1) *There is more to evolution than "a change in allele frequencies."* Simply put, microevolutionary theory deals with the ways in which genetic variants ("alleles") succeed or disappear in populations. Although the mathematical models of population genetics were developed with small-scale variations of genes in mind, the question of what "alleles" actually are was necessarily left open because the physical basis of genes was not known. In principle, it is possible to define any genetic variant, including entire genomes if necessary, as "alleles." Even genetic variants that do not segregate in a Mendelian fashion (another assumption of the earliest population genetics models), such as segregation distorters or transposable elements, can be accommodated in an expanded microevolutionary theory. Such flexibility in the definition of "alleles" has allowed an application of basic population genetics principles to a wide array of genetic phenomena. However, one common criticism is that such models necessarily involve fundamental simplifications in order to remain mathematically manageable. To many paleontologists, this fact may raise considerable doubts about the ability of such models to cope with the enormously contingent process(es) of macroevolution. Developmental and genome biologists may similarly question whether models based on a small number of parameters accurately reflect the level of complexity now being revealed in their fields.

More worrisome than this oversimplification, from the point of view of macroevolutionists, is the tendency of many microevolutionists to define "evolution" itself as no more than "a change in allele frequencies." In strict application, such a definition would make population genetics the exclusive means by which to understand "evolution," so not surprisingly this does not sit well with those working in other evolutionary disciplines. For example, whereas the origin of a new allele may be considered a change in "frequency" (from zero to something else), this says nothing about the processes by which these new alleles come into being—an issue that genome biologists also consider to be a crucial component of "evolution." Indeed, as discussed in the second part of this chapter, this mechanistic consideration is very important for understanding macroevolutionary patterns, even if microevolutionary analyses can treat all manner of alleles in essentially the same way.

More generally, this definition of evolution as "a change in allele frequencies" does not end the micro–macro debate simply because this is decidedly *not* what most people mean by the term "evolution." As noted by Filipchenko (1927) at the very beginning of this debate, to the majority of people (friends and foes alike) "evolution" means *macro*evolution—large-scale patterns in diversification above the species level, and the passage through major transitions that allowed said diversification to happen as it did (see Part Two of this chapter). No one, not even the most ardent Creationist, denies—but neither do systematists reconstruct nor philosophers contemplate—the occurrence of allele frequency changes within populations. As Ernst Mayr himself has noted: "Geneticists said that evolution is a change in gene frequencies among populations. But this is nonsense. Changes in gene frequencies are the *result* of evolution, not the *mechanism* [much less the *definition*, by obvious implication]" (quoted in Shermer and Sulloway, 2000, italics in original; see also Mayr, 1982). It is a simple truism that if there is more to evolution than changes in allele frequencies, then there must be more to evolutionary theory than population genetics.

(2) *Microevolutionary theory is too focused on the organism level.* It is important to note that all evolutionists (including macroevolutionists) recognize complex adaptations at the organism level (e.g., the eye) to be the products of natural selection operating among individual organisms (or perhaps genes, depending on one's viewpoint). As far as such adaptive evolution is concerned, the study of population-level processes is quite appropriate. However, and as is increasingly acknowledged, not all features of organisms are complex adaptations, meaning that macroevolution and organismal adaptation cannot be equated. The reason is that an important contribution to organismal features is made by preexisting design limitations, correlations of growth, cooptions, and other such factors (e.g., Gould and Lewontin, 1979; Gould, 2002). In order to understand these aspects of organismal evolution, it is necessary to consider factors operating at levels both above and below the individual organism, such as phylogenetic and developmental constraints. Genomic considerations are also relevant here, as shown by the underlying genetic "homology" of morphologically "analogous" traits, which suggests that there are genetic constraints against the evolution of certain features (the eye being a prime example) in groups lacking the genetic prerequisites. According to Gould (1983a, 2002), the Modern Synthesis began as explicitly pluralistic but later "hardened" into a much more rigid adaptationist formulation, which set the stage for the strict extrapolationism exemplified by Mayr's summary quoted earlier. The classic critique by Gould and Lewontin (1979) was directed at the adherents of this hardened theory who appeared to see all features of organisms as adaptations, and therefore took all of evolution to be the result of adaptive microevolution (see Pigliucci and Kaplan, 2000, for a recent update on this issue).

When considering macroevolutionary patterns, the need for an expanded mechanistic view is even clearer. Mass extinctions, for example, may profoundly alter the course of evolutionary diversification and the resulting distribution of species yet have little to do with the usual process of selection among organisms within populations. Climate change, meteorite impacts, continental drift, and other large-scale external processes must also be taken into consideration if the patterns observed in the fossil record are to be understood, and these obviously lie outside the domain of microevolutionary study. As will be argued in the second part of this chapter, an identification of the particular genomic processes involved in shaping macroevolutionary patterns is likewise necessary, and these are also peripheral to microevolutionary theory in some important ways.

(3) *Laboratory observations of microevolution do not address the issue of macroevolution.* An important part of Darwin's argument for the efficacy of natural selection involved drawing parallels with the process of artificial selection imposed by breeders of domestic animals and plants. Today, artificial selection is employed directly in evolutionary biology, with experimental evolution programs using bacterial, fungal, plant, or insect models allowing hypotheses about evolutionary change (albeit simple ones) to be tested explicitly. As a result of such work, there is no longer any question that microevolutionary forces can substantially alter the genetic makeup, or even certain phenotypic aspects, of populations.

However, as relentlessly pointed out by Creationists, although microevolution may operate very effectively within populations, it has never been shown to relate directly to the origin of new and substantially divergent taxa, by either natural or artificial selection. Of course, the relevance of this observation to Creationists hinges on their erroneous belief that challenging a proposed explanation refutes the underlying fact to be explained—note that apples did not suddenly begin floating when Newton's theory of gravity was proved false by Einstein. Though for very different reasons, this point has also been made by some macroevolutionary theorists, as when Goldschmidt (1940, p. 183) argued in the earliest days of the debate that "microevolution ... is a process which leads to diversification strictly within species." Whatever their utility in dealing with population-level questions, laboratory demonstrations of natural selection in action cannot lend credence to the extrapolationist approach because they do not involve the generation of new species.

It must also be borne in mind that studies of experimental evolution are carried out under highly simplified conditions that may not accurately reflect the situation in nature. Notably, despite these simplifications, the results of such experiments may still be inconsistent from one study to the next (e.g., Harshman and Hoffman, 2000). More seriously, it has been argued that both the selective pressures applied and the mutations examined may be far too severe to be of the type in operation under real-life evolutionary conditions (e.g., Lewontin, 2000). (On the other hand, the second part of this chapter will discuss some genomic events of major effect that have been largely overlooked or even dismissed in standard neo-Darwinian theory.)

Interestingly, some of the harshest criticism of undue extrapolationism based on small-scale population-level observations has come from the architects of the Modern Synthesis themselves. As G.G. Simpson (1944, p. xvii) famously said of those geneticists who attempt an overzealous extrapolation of such observations to long-term evolutionary processes, "They may reveal what happens to a hundred rats in the course of ten years under fixed and simple conditions, but not what happened to a billion rats in the course of ten million years under the fluctuating conditions of earth history. Obviously, the latter problem is much more important." Ernst Mayr has recently made a similar point in stating that "geneticists studying guinea pigs in cages cannot see the evolutionary process" (quoted in Shermer and Sulloway, 2000). Again, the validity of these criticisms is dependent on how one defines the "evolutionary process." If by this is meant the basic ways in which genetic variants are filtered within populations (microevolution), then guinea pigs in cages may provide many important insights. If, however, such experimental work is assumed to be a direct way of studying the "evolutionary process" that generated guinea pigs and their kin in the first place (macroevolution), then Mayr's critique is justified.

(4) *Field observations of microevolution do not address the issue of macroevolution.* Even if one accepts the above critique that experimental evolutionary studies lack sufficient realism, there remains a voluminous literature describing the action of natural selection in the wild. Some of the most famous and impressive examples include a measurable change in leg length among *Anolis* lizards transplanted to different habitats (Losos *et al.*, 1997), and the detailed and ongoing study of Darwin's finches on the Galápagos Islands by Peter and Rosemary Grant. In the latter case, observations over more than 20 years have clearly demonstrated the capacity of natural selection to modulate features such as beak proportions and body size in response to changes in environmental conditions (see Weiner, 1994). In this sense, it is an extremely useful study as an illustration of the principle of natural selection as a force of change within populations.

The source of conflict comes from the assertion by the authors of this sort of work that "macroevolution may just be microevolution writ large—and, consequently, insight into the former may result from the study of the latter" (Losos *et al.*, 1997). There are, however, significant reasons to doubt the validity of such extrapolationist conclusions based on these kinds of data. Field studies, like those performed in the lab, show only changes within populations and are therefore subject to the same criticisms given previously with regard to their supposed implications for macroevolution. More particularly—and not a little ironically—it has been pointed out that the examples of natural microevolution observed in these classic studies "are *vastly too rapid* to represent the general modes of change that build life's history through geological ages" (Gould, 2000, p. 343, italics in original). This is what Gould (2000) calls "the paradox of the visibly irrelevant"— that evolutionary change so easily documented during a short period of human observation cannot be the kind of change responsible for the evolutionary patterns seen in the fossil record. It may be more reasonable to consider such changes as intrapopulational noise than the early stages of speciation; this is especially true where the phenotypic changes reverse direction as conditions fluctuate, as is apparent in the case of Darwin's finches.

(5) *Microevolutionary explanations are too expansive.* In an additional piece of irony, it may be argued that one of the problems with neo-Darwinian extrapolationism is that it is *too broadly applicable* to evolutionary observations at various scales. Consider, for example, the assertion that microevolutionary theory is fully compatible with both punctuated and gradualistic macroevolutionary patterns (e.g., Stebbins and Ayala, 1981; Charlesworth *et al.*, 1982; Newman *et al.*, 1985; Turner, 1986), coupled with the claim that "reported instances of little or no adaptive change within populations are not in conflict with neo-Darwinian theory because stasis or maladaptation can be explained by selection, micromutation, gene flow, and genetic drift" (Hendry and Kinnison, 2001). Of course, microevolutionary models may indeed have the capacity to account for all possible observations at both population- and species-level scales, even distinctly opposite ones (punctuated versus gradual speciation patterns, adaptation versus maladaptation, continuous change versus stasis), as implied by these statements. Although it would be rather extreme to equate the flexibility of neo-Darwinism with unfalsifiability, it does bear noting that complete *post hoc* compatibility with all possible observations greatly limits the ability of the theory to distinguish between even diametrically opposed alternatives. A theory that accounts for everything, so it is said, explains nothing. Even worse is the predictive power of such an all-encompassing model. Indeed, although observations of punctuationism, stasis, and maladaptation may have been rendered compatible with neo-Darwinian theory after the fact, they were not predicted by it.[2] Finally, it should be obvious that *compatibility* with such observations does not automatically imply that the standard organism-level processes on which microevolutionary theory is based are the *cause* of all or any of them.

Again, none of the various genome-level observations listed earlier in this chapter were expected under standard neo-Darwinian theory, even if compatibility was (or is being) established after their discovery. Along these lines, it is important to consider the notion that "any general theory of genomic architecture evolution must account for the peculiar molecular attributes of various genetic elements, in addition to being compatible with the principles of population genetics" (Lynch and Conery, 2003). Insofar as the "principles of population genetics" simply refer to filtering by selection and drift, then this premise seems very reasonable (and has been accepted throughout this chapter). However, in another sense, the phrasing

---

[2]There is an important exception to this. The possibility of punctuated patterns of change was discussed explicitly by Mayr (1954) in his treatment of allopatric speciation in island birds of the New Guinea region, in which he argued that "rapidly evolving peripherally isolated populations may be the place of origin of many evolutionary novelties. Their isolation and comparatively small size may explain phenomena of rapid evolution and lack of documentation in the fossil record, hitherto puzzling to the palaeontologist." See also Mayr (1992).

of this statement seems backward with regard to the source of the necessary compatibility. Specifically, it must be remembered that population genetics is a *mathematical model,* meant to describe and predict the behavior of genetic variants within populations under various conditions. It is *not* a set of physical laws, akin to thermodynamics, that can never be violated by any valid scientific theory. Thus, whenever an observation is found to contradict the expectations of the model, it is the model that must be adjusted or expanded. In terms of the genomic processes in question, compatibility with population genetics was achieved (in some cases, only after a battle against theoretical inertia) by broadening the definition of "alleles" and greatly expanding the conceptual framework to accommodate many different kinds of "mutations." The same is true of the population- and species-level observations noted earlier, which also required a revision of the model (e.g., to emphasize the importance of small peripheral populations). In this sense, the study of subjects outside the realm of microevolutionary theory, often using very different conceptual outlooks, remains critical for the improvement of evolutionary theory, including microevolutionary models.

REDUCTIONISM IN THE POST-GENOMIC ERA

In the most general terms, the debate between extrapolationists and macroevolutionists reflects the larger issue of reductionism as a philosophical approach to science, and of its utility in fields dealing with complex, historically based subjects like evolutionary biology. Obviously, extrapolationists see no problem with an application of reductionistic principles to the study of evolution (e.g., Ayala, 1985), whereas their opponents disagree that macroevolutionary questions are fully accessible to microevolutionary analysis.

Reductionism is a key issue in genome biology as well. To be sure, the rapid progress of genomics—from the structure of DNA to the sequence of the human genome in less than 50 years—can be taken as a (but not *the*) triumph of the reductionistic mode. However, at the dawn of the post-genomic era, it is becoming increasingly clear that strict reductionism will not, by itself, reveal the deepest secrets of genome function and evolution. It should be patently obvious, despite considerable media hype to the contrary, that having the genome sequence of *Homo sapiens* in hand hardly reveals the meaning of life, the nature of humanity, or, for that matter, even the basic workings of developmental genetics. In fact, even with complete annotation whereby the position and protein product(s!) of every gene are identified, still very little will be known about how the recipe of the genome translates into the cake of a human being. Sequencing the human genome did not provide conclusive answers to old biological mysteries—instead, it showed that the right questions had not even begun to be asked.

The question of whether phenomena can and should always be studied at the lowest possible level of explanation applies to all scientific disciplines, not just evolutionary biology and genomics. For example, one may just as readily ask whether, say, chemistry can or should be "understood" solely in terms of physics. To quote Nobel Prize–winning physicist Steven Weinberg (2001),

> Almost any physicist would say that chemistry is explained by quantum mechanics and the simple properties of electrons and atomic nuclei. But chemical phenomena will never be entirely explained in this way, and so chemistry persists as a separate discipline. Chemists do not call themselves physicists; they have different journals and different skills from physicists. It's difficult to deal with complicated molecules by the methods of quantum mechanics, but still we know that physics explains why chemicals are the way they are.

In like fashion, developmental biologists, physiologists, and ecologists are unlikely to accept the notion that "all biology is molecular biology" (George Wald, quoted in Mayr, 1997, p. 120). If this were true, then why stop at molecular biology? Is not all molecular biology really chemistry, which in turn is physics? Pointing out this *reductio ad absurdum* is certainly not meant to imply that more complex fields are not compatible with, or perhaps even reducible *in principle,* to more general fields (on some level they must be, or science fails). The issue is only whether such complete reductionism affords a full understanding of processes at higher scales. Chemical reactions are necessarily compatible with the principles of quantum physics, but the existence of chemistry as a healthy, independent discipline suggests that they are not best studied or explained in these lower-level terms. Similar arguments can be applied to the study of macroevolution, whether or not it can be reduced in principle to microevolution, and to the study of the genome, even if analyzing individual sequences remains a fruitful enterprise.

THE STRUCTURE OF MACROEVOLUTIONARY THEORY

In general, the alternative interpretations of the macroevolutionary process offered by paleontologists and other theorists seek to expand (but by no means exclusively shift) the operation of natural selection to levels other than organisms within populations. In particular, this "multilevel" or "hierarchical" selection theory has traditionally been applied to groups and larger assemblages, especially species. These are complex (and in some cases controversial) issues, and detailed treatments are available elsewhere (e.g., Vrba, 1989; Grantham, 1995; Lieberman and Vrba, 1995; Eldredge, 1998; Gould, 1998, 2002; Sober and Wilson, 1998; Erwin, 2000; Gregory, 2004). For the purposes of the present discussion, a brief review will suffice to set the basic context for applying these ideas to levels *below* the organism level, most notably the genome.

GROUP SELECTION

The concept of group selection has a long history in evolutionary biology—Darwin himself invoked it to explain human sociality in 1871—but has always generated much controversy (see Sober and Wilson, 1998, for review). The classical problem of naïve group selectionism is that groups of altruistic organisms would be taken over by "selfish" individuals who share the benefits but not the costs of altruism, and therefore enjoy a distinct fitness advantage over legitimate altruists. In other words, though a sucker may be born every minute, exploiters are born even faster.

To the extent that groups are totally independent units cut off from all other such entities, this problem is very difficult to escape. But, as Sober and Wilson (1998) pointed out, groups often interact as part of a "metapopulation" with a shared gene pool. This fact makes it possible for altruists to increase in frequency in the metapopulation even if they are outcompeted within their respective groups. The reason is that groups with a high proportion of altruists may do better as a whole, and therefore contribute more to the metapopulation, than groups composed of selfish individuals. Under experimental conditions, group selection of this sort can be quite effective, although the frequency of its occurrence in nature is still the subject of debate (e.g., Goodnight and Stevens, 1997; Sober and Wilson, 1998).

SPECIES SELECTION: CONCEPTS AND CHALLENGES

Selection at higher levels, namely among species, has been less controversial in theory but much more difficult to establish in practice than has group selection. There are three major reasons for this. The first was outlined as follows by R.A. Fisher in the second edition of his founding document of population genetics, *The Genetical Theory of Natural Selection* (originally published in 1930):

> There would, however, be some warrant on historical grounds for saying that the term Natural Selection should include not only the selective survival of individuals of the same species, but of mutually competing species of the same genus or family. The relative unimportance of this as an evolutionary factor would seem to follow decisively from the small *number* of closely related species which in fact do come into competition, as compared to the number of individuals in the same species; and from the vastly greater *duration* of the species compared to the individual. (Fisher, 1958, p. 50).

That is to say, species selection could be an important evolutionary force, were it not so easily overshadowed by the much more powerful action of selection on the more numerous and faster-reproducing organisms of which species are composed (see Gould, 1998, 2002, for additional discussion).

The second major difficulty involves the concept of species itself. Are species snapshots of an ever-changing continuum (as expected under strict anagenesis), or are they real entities with the defining properties of "individuals"? More precisely, do they possess the necessary features to qualify them as Darwinian individuals on which selection can act—i.e., spatio-temporal boundedness and heritable variation? In plain terms, do species have births, deaths, cohesive life spans, and offspring? Under strict anagenetic models, in which a species gradually evolves as a whole into a new species, the answer is "no." But under models of punctuated equilibria, in which species are "born" rapidly (in geological terms), remain largely static for the duration of their tenure, produce and coexist with daughter species by cladogenesis, and die by extinction, the answer is clearly "yes" (Fig. 11.1).

In any case, it is not enough to have offspring that share many of one's traits; this heritable variation must come with fitness consequences. Thus the third problem for species selection, and one that again raises the issue of reductionism, relates to the necessity of identifying species-level traits that can affect the evolutionary success of "individuals" at the species level. The duplication of developmental control genes provides a useful illustration of the issues involved. On the microevolutionary level, the resulting "allele" (a clustered set of regulatory genes) must pass through the filter of selection at the population level when it first occurs, and so is not very different from other mutations in this sense. However, on a macroevolutionary scale, the increased flexibility of the developmental program engendered by such a duplication may grant a heightened ability for speciation and/or avoidance of extinction—that is, it may improve the "reproductive rate" and/or "survivability" of the lineage possessing it. If so, then this would be an important consideration in explaining why there are so many species that exhibit this genomic configuration relative to those that do not. Of course, whereas such an example illustrates the concept of higher-level fitness effects, it does not necessarily indicate species selection per se. In fact, the identification of features that are involved in legitimate species-level selection has been a point of contention even among paleontologists who advocate a hierarchical theory of selection.

AGGREGATE VERSUS EMERGENT CHARACTERS

One of the defining features of the biological hierarchy is that individuals are composed of parts and in turn assemble into collectivities. In the archetypal example, organisms are composed of cells and combine to form populations and species. Although this nesting makes hierarchical selection possible, it also poses one of the greatest challenges for the identification of selection processes operating above the default level of the organism. Species may indeed qualify as Darwinian individuals, and a given species may "reproduce" (speciate) more or "live" (avoid extinction) longer than another, but to what extent is this really selection among species, and not just among the organisms that make up the species in question?

When dealing with a conglomerate of component entities like populations or species, it is necessary to differentiate between "aggregate" and "emergent" characters. Aggregate characters are those that are just the "sum of the parts" and are ultimately reducible to the properties of the component organisms. Thus a feature such as generation time may influence the rate of speciation, but in fact is only a product of the individual reproductive timings of the constituent organisms. Emergent characters, by contrast, are not reducible to the properties of organisms and appear only at the higher level of organization. The ratio of males to females, for example, may have fitness consequences for higher-level groupings, and is obviously a property that emerges only at the group level because individual organisms cannot have sex ratios.

SELECTION VERSUS SORTING

This question of aggregate versus emergent characters raises another important consideration, namely the distinction between *patterns* and *processes* observed at the species level. In particular, it has been pointed out that whereas there may often be a clear pattern of differential survival and reproduction among species, this need not be caused by the process of selection. It is therefore necessary to distinguish between *sorting*—the observed pattern—and *selection*—one possible cause of this pattern (Vrba and Gould, 1986; Lieberman and Vrba, 1995). Even at the organism level, it is clear that sorting can be caused by processes other than selection (drift, for example). At the species level, this distinction is important because it means that patterns of sorting need not be caused by selection at the species level, but rather could result from selection (or other processes) at the standard organismal level. The question, therefore, is: At what point can selection, and not just sorting, be said to operate at the species level? This, too, is an issue on which not all macroevolutionists agree.

BOTTOM-UP PROCESSES: THE EFFECT HYPOTHESIS VERSUS EMERGENT FITNESS

A bottom-up process is one in which the combined properties of lower-level entities (e.g., organisms), either by emergence or simply in the aggregate, influence the patterns of sorting at higher levels (e.g., groups or species). When heritable emergent characters generate differential survival and reproduction among groups or species, there is no controversy: this definitely qualifies as higher-level selection. Likewise, when the input of aggregate traits can be completely reduced to effects at the organism level, the conclusion is just as obvious: this is definitely *not* higher-level selection. The debate arises over the hazy gray area in between these extremes, where aggregate characters exert an influence on species level sorting.

According to Vrba's (1989) "effect hypothesis," true species selection occurs only when emergent characters are involved. Aggregate characters may still have important "effects" at the higher level (hence the name), and the process may still best be studied at this higher level (in Weinberg's chemistry-is-not-just-physics sense), but under the effect hypothesis this would not represent a causal process of selection at the species level. As a more flexible alternative, Lloyd and Gould (1993) have developed an "emergent fitness" approach that "requires only that a trait have a specified relation to fitness in order to support the claim that a selection process is occurring at that level." In other words, if the fitness of the species is not reducible to the sum fitnesses of its component organisms, then this would still qualify as species selection, even if only aggregate characters are involved (Gould, 2002).

One of the criticisms of the effect hypothesis is that its requirement of emergent characters for species selection is too restrictive and unnecessarily instigates a search for "adaptations" at the species level. Under the emergent fitness approach, no such higher-level adaptations are necessary, because the features may still be evolving owing to selection at the lower level, but then are also filtered by selection at the higher level. In either case, it is acknowledged that "the acid test of a higher level selection process [i.e., among groups or species] is whether it can in principle oppose selection at the next lower level [i.e., among organisms, which are the default level]" (Vrba, 1989). As a prime example, an observed female-biased sex ratio, which might be favored at the group level but which should be undone by selection at the organism level, is usually taken as good evidence of higher-level selection in action (Wilson and Colwell, 1981).

It has also been pointed out that stasis at one level may often reflect a dynamic tension between opposing pressures from above and below (Gould, 1998, 2002). However, antagonism need not always define the interplay among levels of selection, and it is possible that in a high frequency of cases, the evolutionary processes operating simultaneously on several levels may do so synergistically or simply orthogonally (e.g., Vrba, 1989; Gould, 1998, 2002). Of course, multilevel selection tends only to be noticed when antagonism dominates and is otherwise dismissed in the name of parsimony. But this preference for explaining a selection process automatically and exclusively in terms of the default level is based on a philosophical choice, not a logical necessity.

TOP-DOWN PROCESSES: CONTEXT-DEPENDENT SORTING

Bottom-up processes are an important part of macroevolutionary theory, but as Eldredge (1985) noted, "sorting of lower-level individuals, in general, has a far less profound effect on upper-level individuals than the converse." Such top-down processes are best appreciated within the hierarchical framework in terms of Vrba's (1989) model of "context-dependent sorting." Organisms do not exist in a vacuum, but instead hold membership in groups and species, and as such their individual fitnesses can be influenced by the properties of these higher-level assemblages. In an apt analogy (and the reason context-dependent sorting is also sometimes called "Mustapha Mond sorting," after *Brave New World*), Vrba (1989) described the top-down control of fitness among the citizenry of different political states: "To the extent that a national ruler or law dictates that members of the population with certain characteristics may have more children than others, or must die at different ages, sorting among humans depends on whether they live in that nation or in another more liberal one."

Macroevolutionary Theory: A Summary

Higher-level selection processes have a strong conceptual basis, but have been difficult to demonstrate in practice, especially at the species level. Part of the debate revolves around the distinction between the specific process of selection and the general pattern of sorting. Some macroevolutionary theorists consider species selection to have operated only when emergent characters can be identified at the species level. Under this view, in situations in which a relevant species-level character is merely aggregate (i.e., no more than the sum of the parts), this sorting counts only as an "effect." Others ascribe species selection to any cases where there are emergent species-level fitness consequences, be they based on emergent or aggregate characters. In addition to these bottom-up processes, there is the top-down effect of "context-dependent sorting," in which the fitness of a Darwinian individual is affected by the properties of the higher-level assemblage of which it is a part (Fig. 11.2). With these concepts in mind, it is now possible to consider whether hierarchical selection processes are also at play in genome evolution.

A MACROEVOLUTIONARY LOOK AT THE GENOME

Did the Genome Originate by Group Selection?

It has often been useful in evolutionary theory to consider selection as operating at the level of genes and not just on organisms, although even in the most extreme gene-centered approach, this is considered a different way of looking at the same process rather than a new theory of selection (e.g., Dawkins, 1989). Of course,

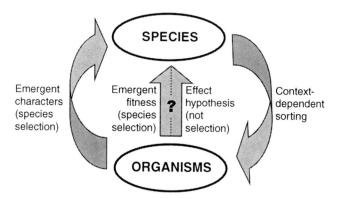

FIGURE 11.2 Summary of interactions between the organism and species levels involving sorting. When the combined properties of organisms create emergent characters (more than the sum of the parts) at the species level, species-level selection can occur. In this case, there is no ambiguity as to whether the selection process can be reduced to the organism level. When organisms combine to form aggregate characters (equal to the sum of the parts), the situation is less clear. Some authors do not consider this to allow selection at the species level, but do recognize that this can have effects on the sorting of species (the "effect hypothesis"). Others argue that so long as the fitness consequences are exerted at the species level, and cannot be reduced to the individual fitnesses of the component organisms, then this will also allow species selection to operate (the "emergent fitness" approach). There are also top-down effects, in which the properties of species influence the fitnesses of individual organisms by providing the context in which they live ("context-dependent sorting").

genes predate the emergence of "organisms" in the standard sense, so a gene-level perspective is almost certainly necessary when considering the origins of organized genomes—that is, of linking formerly independent genes into the first chromosome.

The theoretical problem to be overcome is that "chromosomes will have a competitive disadvantage within the cell, since it takes longer to replicate them than unlinked genes" (Maynard Smith and Szathmáry, 1995, p. 114). To account for the origin of genomes despite this difficulty, Maynard Smith and Szathmáry (1995, p. 114) outlined the following scenario from a gene's-eye view:

> Let A and B be two complementary genes [i.e., both are needed by the protocell], and AB the chromosome. Is an A gene better off on its own, or as part of a chromosome? On its own, it replicates faster within a cell. However, an A gene that is part of a chromosome is certain to find itself, in the next generation, in a cell that also contains a B gene: that is, it is certain to be in a fit cell, whereas an isolated A gene may find itself in an unfit cell, with no B gene. A second advantage of linkage is the synchronization of replication, and hence the elimination of within-cell competition between different genes.

It is also important to recognize that the protocells themselves are reproducing in this scenario, and that the behavior of the genes they contain can impact this reproduction. In keeping with the gene-based interpretation, Maynard Smith (2002) suggests that "selection will favor an AB 'chromosome' over independently replicating A and B genes, even if the independent genes replicate more rapidly than the chromosome within a cell, provided that a cell containing both A and B genes grows substantially faster than one lacking one or the other gene." Again, this is based on the necessity of keeping two complementary genes together, and it is the genes that are ultimately seen to suffer when they are separated.

There is another (and not mutually exclusive) way to consider this issue. Specifically, there is a metapopulation of genes divided into groups inhabiting individual protocells. Some of these genes are autonomous and fast-replicating and others are linked into cooperative protochromosomes and replicate more slowly. Independent genes maintain a within-protocell advantage but may negatively impact cell division by their failure to cooperate. Linked genes have lower individual fitness within a given protocell, but the cells containing a high proportion of cooperative genes would reproduce more successfully than cells containing mostly unlinked genes. Although it has been applied to a different level of organization, this is precisely the model of group selection proposed by Sober and Wilson (1998). Independent genes may increase in frequency within each protocell, but cells containing cooperative genes will contribute more to the population of cells and therefore to the metapopulation of genes. Linkage of genes into genomes could be favored not only because this prevents the separation of complementary genes, but also because genic cooperation has positive fitness consequences at the higher level of the cell.

Genomic Parasites Require a Hierarchical Interpretation

There is at least one point on which prominent theorists from both genome biology and paleontology have already reached consensus: that the existence of "selfish DNA" necessitates a multilevel theory of natural selection. Thus Doolittle (1989) noted that "although genes are made of DNA, much DNA is not genes, and it is not clear that we can so easily understand all of the structures and evolutionary behaviors of DNA without some further theoretical expansion," and in particular that "much of the data of modern molecular biology might better be understood as revealing the operation of selection at several levels in a real biological hierarchy." Gould (1992) made an even stronger point when he noted that "punctuated equilibrium is but one pathway to the elaboration of hierarchy, and probably not the best or most persuasive; that role will probably fall to our new understanding of the genome and the need for gene-level selection embodied in such ideas as 'selfish DNA.'"

The logic behind these assertions is straightforward: "Selfish DNA" persists, even if it serves no function or even imposes a cost, because it is favored by intragenomic selection, thereby adding one level upon which selection acts. However, as shown by the lesson of species selection—which is also sound in logic but has been very difficult to establish in practice—logical consistency does not, in itself, prove actual importance in evolution. As Doolittle (1987) pointed out, "many logically possible evolutionary processes do not actually often occur, and it was important to the selfish DNA argument (not in terms of the logic but in terms of the biology) to show that some real DNAs actually are selfish." Transposable elements and B chromosomes provided clear examples of DNA segments that could be selected intragenomically, but having established the reality of selfish DNA, the question became one of relative frequencies. Thus Gould (1983b, p. 176) asked: "How much repetitive DNA is self-centered DNA? If the answer is 'way less than one percent' because conventional selection on [organisms] almost always overwhelms selection among genes, then self-centered DNA is one more good and plausible idea scorned by nature. If the answer is 'lots of it,' then we need a fully articulated hierarchical theory of evolution." In this age of complete genome sequencing, it has become abundantly clear that the answer is, in fact, "lots and lots and lots of it" (see Chapter 3).

Even in its simplest formulation, the selfish DNA theory automatically requires a recognition of selection operating on at least two levels and therefore validates the underlying principles of hierarchy theory (Gould, 1983b, 1992; Vrba and Eldredge, 1984; Doolittle, 1987, 1989; Gregory, 2004) (Fig. 11.3). As discussed in Chapter 4, this was first recognized 60 years ago by Östergren (1945) in reference to B chromosomes. However, as noted in Chapters 3 and 4, strict selfishness is only one way in which transposable elements and B chromosomes can come to interact with their host genomes. Both almost certainly begin life as parasites, but it may be far more accurate to consider purely selfish parasitism as only one extreme of a continuum that also ranges through neutrality all the way to mutualism (Kidwell and Lisch, 2001). To be sure, and as with selective processes operating above the level of the organism, antagonism is the most easily observable expression of hierarchy in action with regard to subgenomic elements. But synergy among levels is also possible, and cooperation can be just as hierarchical as competition.

It has become increasingly apparent that transposable element evolution extends far beyond the simple concept of selfish propagation, and in so doing actually provides an even stronger case for an interpretation of genome evolution based on hierarchical interactions. Some of the cellular, organismal, and macroevolutionary consequences of these processes will be addressed in a later section, but for the time being it is worthwhile to consider some of the most important interactions that occur from the sequence to chromosomal levels.

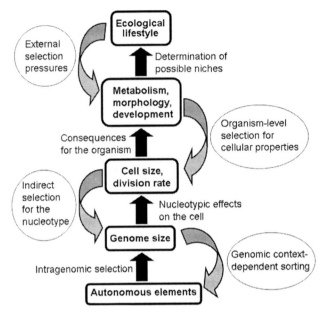

FIGURE 11.3 A simplified schematic of the hierarchical relationships between subgenomic elements, genomes, cells, and organisms. Beginning with bottom-up influences (black arrows): At the lowest level, intragenomic selection can lead to a proliferation or limitation of autonomous DNA elements (e.g., transposable elements and B chromosomes). This in turn can affect genome size, which itself exerts an influence on cell size and cell division rate. At the organism level, these cytological effects may be expressed as impacts on metabolic rate, body size, or developmental rate, depending on the biology of the group in question. These organismal properties likewise act to determine the available niches and ecological lifestyles available. There are also top-down interactions (curved arrows): External ecological pressures influence the evolution of metabolic, morphological, and developmental features at the organism level, which can involve selection for changes in cellular parameters. These may be met by changes in genome size, which in turn involve undoing, limiting, or even promoting the spread of autonomous subgenomic elements. Note that this is a highly simplified description of the hierarchical processes in operation, because it does not consider interactions between different transposable elements or between these and the host genomes, influences on genome structure and organization, impacts on coding genes, and so on.

Interactions with Other Transposable Elements

Transposable elements (TEs) do not operate in a void, but live and reproduce in a genomic ecosystem along with large numbers of other transposable elements (both "conspecifics" and unrelated sequences). Like the members of any ecosystem, cohabitating transposable elements may interact in a variety of ways. Thus transposable elements may, in some cases, be required to compete with one another for "resources" (e.g., preferred insertion sites, materials for replication). At other times, they may contribute to each other's spread, as with the generation of processed pseudogenes by long interspersed nuclear elements (LINEs) (Brosius, 1999; Esnault et al., 2000). At the other extreme, some elements may be seen to parasitize their fellow parasites, a prime example of which is the hitchhiking of short interspersed nuclear elements (SINEs) on LINEs (Zeyl and Bell, 1996; Prak and Kazazian, 2000) (see Chapter 3).

Interactions with the Host Genome

As with all symbionts and their hosts, there is a dynamic coevolutionary interaction between TEs and the genomes in which they reside. Again, under the simplest version of the selfish DNA theory, this interaction is strictly one of parasitism, although there may of course be selection for reduced virulence (e.g., Sapienza and Doolittle, 1981; Doolittle et al., 1984; Charlesworth and Langley, 1986). However, there is growing evidence that SINEs are at least partly functional in host genomes (e.g., Makalowski, 1995; Schmid and Rubin, 1995), and that transposable elements of various types have been coopted to serve vital regulatory functions (Britten, 1996a,b, 1997; Brosius, 1999; Kidwell and Lisch, 2001) (see Chapter 3). In terms of gene structure, transposable elements are capable of broad-scale mutator activity during both insertion and excision events (Kidwell and Lisch, 1997, 2002) (see Chapter 3). Most importantly from a macroevolutionary perspective, this can include the generation of regulatory mutations of major developmental effect (Finnegan, 1989; McDonald, 1990, 1995). Various genomic mechanisms are known for the suppression of selfish DNA activity, but in many cases this is not necessary, because some TEs preferentially insert into noncoding heterochromatic regions (e.g., Dimitri and Junakovic, 1999; Hutchison et al., 1999) (see Chapter 3). To reiterate, even if they begin as strictly selfish, transposable elements may experience a variety of evolutionary fates, ranging from active suppression by the host to exaptation and integration into the regulatory mechanisms of the genome itself. Which of the available outcomes befalls any given TE will be dependent on its impacts at each of several levels of organization and the net result of the hierarchical interactions it engenders.

Interactions with the Chromosomal Level

At the level above genes and other DNA sequences, transposable elements interact in important ways with the chromosomes. In this case, the influence proceeds in both directions. For example, TEs may incite chromosomal rearrangements by virtue of their transpositional activity (Gray, 2000; Kidwell and Lisch, 2000, 2001), whereas in other cases chromosome rearrangements may lead to the loss of the transposable elements. An illustration of this latter phenomenon is provided by the case of barley, in which the repetitive nature of long terminal repeat (LTR) retrotransposons promotes homologous and/or illegitimate recombination and in the process can cause most of the element to be deleted (Kalendar *et al.*, 2000; Bennetzen, 2002) (see Chapter 2). Formerly parasitic elements may have other important impacts on chromosome structure itself, as will be discussed in more detail in Part Two of this chapter.

GENOME SIZE, EMERGENT FITNESS, AND AN UPSIDE-DOWN ACID TEST

If transposable elements exert an influence on gene structure and expression, then these effects will be played out at the levels of the cellular and organismal phenotypes. Most obviously, transposable elements may have "accidental" deleterious effects by causing mutations in coding regions that lead to various genetic diseases (Labuda *et al.*, 1995; Kidwell and Lisch, 2001). In some instances, mutations caused by transposable elements may remove the top-down controls on cell-level selection, the tragic consequences of which are known as cancer (e.g., Weinstein *et al.*, 1988; Labuda *et al.*, 1995).

As important as the insights discussed earlier may be, they tend to speak only to the basic premise that selection can operate on more than the organism level and not to the validity of the more specific features of modern hierarchy theory. Moreover, they remain reductionistic by focusing only on the elements themselves, and not on their contributions to genomes at large. As discussed in detail in Chapters 1 and 2, genome size itself can have important consequences for organismal fitness through its effects on cell size and division rate. In the language of hierarchy theory, genome size represents an *aggregate character*, being the sum of all the sequences contained within the chromosome set, but one which exerts *emergent fitness* effects on cells and organisms (Gregory, 2004) (Fig. 11.3).

An emergent fitness approach to genome size carries with it some important implications. For example, this means that the origin of the DNA sequences contributing to genome size are unimportant with regard to the fitness consequences felt at the higher level. Selection of various forms, or indeed simply chance, may be the reason behind any given sequence's presence within the genome, but regardless of how it came to be there, it will contribute to the emergent fitness effects of genome size. Intragenomic selection is of special importance because such a large portion of eukaryotic genomes is comprised of transposable elements, but pseudogenes, introns, coding genes, and other sequences will all participate in this process.

Under the simplest "mutation pressure" theories, genome size increases by either intragenomic selection or chance, and only becomes relevant once it begins to exert adverse effects on organismal fitness (see Chapter 1). At the opposite extreme, strict "optimal DNA" theories propose that genome size is modulated adaptively in response to cellular and organismal needs (see Chapter 1). In either case, genome size exerts emergent fitness consequences, but ultimately it is organism-level selection alone that shapes it.

Conflict between organism- and species-level selection, with the latter emerging victorious, is, as noted by Vrba (1989), the acid test of a higher-level selection process. Again, this is because the organism is the default level of explanation, in part because turnover (and therefore the expected strength of selection) is much higher among organisms than among species. In terms of lower-level selection processes, the acid test would have to be reversed, with genome-level processes overwhelming selection at the default organism level. To the extent that genome size is influenced by the quantity of transposable elements, themselves amplified by intragenomic selection, this would involve finding adverse emergent effects of DNA content that persist despite what should logically have been powerful organism-level selective constraints.

Indeed, some very interesting examples of such a bottom-up victory were identified in Chapter 1. Most notable among these are the combined effects of increased genome size and drastic body size miniaturization in salamanders in the tribe Bolitoglossini of the family Plethodontidae. This has resulted in the possession of a very small cranial capacity in which to contain large and slowly differentiating neurons—the net effect being a remarkable reduction in brain complexity (Roth *et al.*, 1994). This was obviously of substantial relevance to fitness, because it has compromised the visual processing centers needed for an active predation strategy. However, instead of shrinking genome sizes, selection at the organism level led to adaptations to accommodate this effect, including the adoption of a lie-in-wait predation strategy and the evolution of a highly specialized projectile tongue (Roth and Schmidt, 1993). There may also have been a negative effect in terms of the physical problem of circulating enormous erythrocytes in tiny blood vessels in these animals. But again, instead of reducing their genome size in response to this organism-level pressure, it seems that the outcome has been to adjust by evolving a significant proportion of enucleated red bloods cells (R.L. Mueller, in preparation) (see Chapter 1).

Such emergent genome size effects on the trajectory of morphological and behavioral evolution in salamanders clearly pass the (upside-down) acid test. In fact, it is becoming increasingly apparent that such bottom-up effects extend well beyond the level of the organism. For example, although they may be more tolerant to drought, frost, and high CO_2 (Castro-Jimenez *et al.*, 1989; Wakamiya *et al.*, 1993, 1996; Jasienski and Bazzaz, 1995; MacGillivray and Grime, 1995), it seems that large-genomed plants are typically excluded from harsh environments, be they hot or cold (Knight and Ackerly, 2002; Knight *et al.*, 2005). Indeed, moving into a harsh environment frequently entails a shift from perennial to annual lifestyle, with a concomitant reduction in genome size (e.g., Watanabe *et al.*, 1999) (see Chapter 2). The implication here is that genome size also exerts an indirect influence at the level of ecological communities (Fig. 11.3). Even the very highest level in the hierarchy, the global distribution of species and larger clades, seems to be affected by variation in genome size, given that large genome size is associated with heightened extinction risk in plants (Vinogradov, 2003).

CONTEXT-DEPENDENT SORTING OF GENES AND NONGENES

DNA sequences are part of larger genomic communities contained within cells, which in turn exist within the bodies of organisms. This assemblage into genomes can have important consequences for higher levels in the hierarchy, as noted earlier. The reverse is also true, meaning that features at the cell, organism, or higher levels can exert top-down influences on these genetic elements. In the language of hierarchy theory, this is an example of context-dependent sorting, with the context being the genome, cell, organism, population, and species in which a given element finds itself.

Beginning at the highest levels, it is well known that species-level traits such as sexual versus asexual reproduction may directly affect the ability of genomic parasites to spread and be maintained within genomes (e.g., Zeyl and Bell, 1996; Burt and Trivers, 1998; Wright and Schoen, 1999; Arkhipova and Meselson, 2000; Schön and Martens, 2000). Population-level happenings, such as the invasion of new habitats, also appear to have downward effects on transposable element activity (e.g., Biémont *et al.,* 2001; Vieira *et al.,* 2002).

Selective constraints on genome size can also influence the fitness of TEs and other elements within genomes. For example, the high metabolic demands of powered flight in birds, which require small cells and therefore impose strong constraints on genome size, provide a prime example (Gregory, 2002a). Similar constraints can be seen with regard to metamorphosis in amphibians and insects, and annual lifestyle in plants (Gregory, 2002b) (see Chapters 1 and 2). Conversely, the loss of flight in certain bird lineages, the enucleation of erythrocytes in mammals, and the evolution of neoteny in some salamanders, all appear to have been followed by secondary increases in genome size (see Chapter 1). Such changes at the organismal level alter the nature of the constraints on subgenomic elements, generating what might be considered the genomic version of an "ecological release." Or, to return to Vrba's (1989) political analogy, transposable elements living within the genomes of flighted birds suffer under a strict regime, whereas those in neotenic amphibians enjoy extraordinary freedom.

Genes, too, must operate within a larger genomic mileu, as has long been recognized even among those with a strongly reductionistic view of genetics. The interactions among protein-coding genes, as well as their regulation by other sequences, have become a primary focus of modern developmental genetics, thereby emphasizing the importance of context at the genic level. Even more holistically, it has been recognized that chromatin structure and distribution within the nucleus can affect gene transcription (discussed in Gregory, 2001a). Genome-level context dependence also applies to transposable elements. Most simply, because TE insertion will be selected against if it disrupts gene function, a large genome with much noncoding DNA will provide more "safe" insertion sites and be more conducive to further expansion. Genome-level regulatory mechanisms may also place limits on the capacity of transposable elements to propagate (see Chapter 3).

As Vrba (1989) has pointed out, context-dependent sorting "at a given level most often occurs together with selection at that level," which would certainly apply to the case of transposable elements being selected intragenomically. However, this is a process of context-dependent *sorting*, which means that it can also operate in the absence of selection at the lower levels and therefore applies to nonselective processes within the genome. For example, the context set by selection at the organism level via the emergent fitness consequences of genome size will also impact the accumulation of pseudogenes ("junk DNA," *sensu stricto*), which spread by chance duplication rather than intragenomic selection.

GENOMES IN THE EVOLUTIONARY HIERARCHY

Many of the biological hierarchies described in the past include genes and some even include cells, but genomes or chromosomes are almost always lacking. As Vrba and Eldredge (1984) explained, "We have difficulty in seeing chromosomes and cells per se engaging in evolutionarily important birth and death processes that need to be considered separately from the dynamics at genome and phenotype levels" (with genomes generally defined in such cases as collections of genes). However, such a perspective ignores the ongoing evolution of B chromosomes (see Chapter 4) as well as the immense number of single-celled organisms still in existence. Cancer also provides a persistent reminder that cell-level selection is suppressed with some difficulty within multicellular organisms. Finally, because genomes as a whole can have important fitness consequences for levels above (cells and organisms, if not species and clades), it is important to include them in the hierarchy of interacting evolutionary entities (Gregory, 2004).

Recognizing the need to include genomes in the biological hierarchy is a crucial first step, but this will not be easy to implement in practice. As noted in Chapter 1, both the etymology and the exact meaning of the term "genome" remain somewhat ambiguous. First, "genome" may be taken to mean either all the DNA within a given chromosome set or the total collectivity of the genes, both of which may be historically accurate definitions even though the existence of so much noncoding DNA makes them incompatible. In addition, the "genome" may refer at various times to the mass of nuclear DNA that exists in at least one copy per cell, the specific DNA sequence of an individual organism, or the collective DNA complement of a species (e.g., "the human genome"). Furthermore, under some of these definitions, the "genome" is a highly ephemeral entity, being scrambled by recombination in every generation among sexually reproducing organisms. Finally, it must be noted that whereas clades consist of species, which are composed of populations, which are made up of organisms, which in turn are comprised of cells, this connection breaks down at lower levels because cells are not made of genomes. Even the occasional inclusion of genes in biological hierarchies adds to the confusion, because genomes are made of far more than genes. Such ambiguities create a conceptual quagmire when attempting to place genomes within biological and evolutionary hierarchies (Gregory, 2004). Nevertheless, this is an important issue that will warrant careful consideration in the future.

PART TWO—"NONSTANDARD" GENOMIC PROCESSES AND MAJOR EVOLUTIONARY TRANSITIONS

Until now, the focus of this chapter has been on a top-down application of the concepts and terminology of macroevolutionary theory to processes operating at the genome level. This second part explores an opposite (but by no means opposing) issue—namely, whether some of the genomic processes discussed earlier in this chapter and treated in detail in the various preceding chapters can influence macroevolutionary patterns from the bottom up. Two genome-level phenomena, both considered "nonstandard" in that they do not involve small-scale variation in existing coding genes, will be emphasized in particular: the evolution of "selfish" genetic elements (including transposable elements and spliceosomal introns) and gene/genome duplications.

Because any genome-level change—that is, an "allele" very broadly defined—must pass through the filters of selection and drift at the population level, the identification of new kinds of "mutation" is unlikely to greatly affect the view of how *microevolution* works. In terms of microevolutionary theory, it may be of little consequence whether macroevolution results exclusively from the cumulative effects of intrapopulational changes in the frequencies of small-scale gene variants, or involves an essential input of nonstandard (indeed often nongenic) "mutational" mechanisms at key stages. To *macroevolutionists*, by contrast, it is of critical importance to determine the types of genomic events necessary for the evolution of novel features. In fact, if nonstandard processes play a crucial role during major evolutionary transitions, then a complete understanding of

macroevolution will simply be impossible without acknowledging their oper
and elucidating their nature, even if the most this does to *microevolutic*
necessitate a tweaking of definitions.

The remainder of this chapter is devoted to examining the roles played by n
standard genomic processes in shaping some major evolutionary transitions, w
particular reference to human evolutionary history. The choice to focus c
humans should in no way be construed as implying that evolution is directed o
progressive in any global sense, or that there is any central macroevolutionary
tendency toward humans. That said, the specific pathway with humans at the
current terminus is justifiably of special interest to those whose ancestry it
represents, which, as far as is known, includes all readers of this volume.
Fortunately, all but the most recent of these transitions also apply to many other
animals and can therefore be used to illustrate more general points about the
issue at hand. Because of space constraints (itself a testament to the number of
available examples), each item can be discussed only briefly; more details about
any given case can be found in the references cited.

THE ORIGIN OF GENOMES AND CELLS

In some ways, the origins of DNA genomes and of cellular life are part of a single
overarching question, even though they may not have occurred simultaneously.
Here, the concept of independently active ("selfish") genes is particularly important to a proper understanding of the issue, because what must ultimately be
explained is how cohesion prevailed over competition (which may have been by
gene-level or "group" selection, or both, as discussed previously). However, this
role of autonomous elements may not be restricted to egoistic genes: for example,
Jurka (1998) has proposed that transposable elements may trace their history all
the way back to the RNA world and could have played a role as "genome builders."
Put more directly, "as we move back in time towards 'primitive genomes,' the
distinction between the host genome and genomic 'selfish' elements becomes
obsolete" (Miller *et al.*, 1999). As an interesting alternative, but one that also suggests a prominent role for rogue genetic elements, Forterre (2002) has proposed
that viruses were the first to use DNA containing uracil (U-DNA) in an effort to
elude host defenses and thereby ended up playing a major role in the emergence
of DNA genomes from RNA ones.

THE EVOLUTION OF SEX

The evolution of meiosis has long been a puzzle in evolutionary biology, at least
when considered from the traditional organism level where it seems paradoxical

that sex, which involves the passage of only 50% of one's genes to the next generation, could have been favored over asexual reproduction. Many possible explanations have been offered for this outcome, usually involving a search for an organism-level benefit to counteract the inherent genetic cost. Well-known examples of this line of reasoning include claims for reduced competition or improved capacity to tolerate environmental change among the resulting offspring. Perhaps the most widely accepted view is that sex provides a means of countering the rapid evolution of pathogens. Because viruses are either descended from transposable elements (e.g., Temin, 1980) or perhaps coevolved with cells as independent and partially selfish genetic units (Hendrix *et al.*, 2000), the evolution of sex for this reason would clearly have been influenced by the action of nonstandard subgenomic processes.

Of more direct relevance to the present discussion is the notion that meiosis may have evolved as a means of ensuring the "fair" replication of potentially selfish genes and to suppress their uncooperative tendencies (Hurst, 1995). In other words, sex may have evolved as a way to dampen or eliminate the effects of within-genome selection, of which there are nevertheless still ghosts in the form of segregation distorters and various other intragenomic "cheaters" (Pomiankowski, 1999). In this sense, sex is considered an extension of the process that allowed genomes to form in the first place—by the coordination of previously selfish elements. From an opposite perspective, it has been proposed that the genetic machinery involved in meiosis may have originated from the domestication of former transposable elements, and that TEs could have initiated the process of sex in order to facilitate their own spread among hosts (Hickey, 1982; Bell, 1993). This latter possibility would represent a bottom-up process that has had a tremendous impact on subsequent higher-level sorting. Under either the top-down or bottom-up scenarios, the subgenomic processes discussed here would be essential to an understanding of one of the most important watersheds in evolutionary history.

THE ORIGIN OF EUKARYOTES

LINEAR CHROMOSOMES

The single circular chromosome with only one replicon origin found in most prokaryotes imposes strong constraints on the size of the genome that can be replicated and therefore places a limit on the number of genes that can be contained within it (Maynard Smith and Szathsmáry, 1995) (see Chapter 10). Indeed, limitations on genetic complexity imposed by genome size constraints may explain why bacteria appear to deviate so little from their earliest phenotypes (Vellai and Vida, 1998, 1999). A shift in large-scale genomic structure, from a single circular chromosome to several linear ones, would therefore have been necessary for the transition from prokaryotes to eukaryotes and the subsequent increase in genetic and morphological complexity among members of the latter group.

Telomeres—the tightly compacted caps of DNA located at the tips of eukaryotic chromosomes—are vital to the maintenance of linear chromosome integrity (e.g., Bell, 1993). Malfunctioning or shortened telomeres have been implicated in cancers (e.g., Artandi et al., 2000), DNA damage response-related senescence (d'Adda di Fagagna et al., 2003), reduced immune and other stress responses (e.g., Rudolph et al., 1999), and severe developmental anomalies in plants (Riha et al., 2001), so not surprisingly their structure and length is tightly regulated in healthy cells. It is now becoming clear that transposable elements contribute significantly to chromosome structure in this regard. In the intriguing case of Drosophila, for example, telomeres are maintained directly by the serial insertion of transposable elements rather than by the usual activity of telomerase (Levis et al., 1993; Pardue et al., 1997; Casacuberta and Pardue, 2002). More broadly, it seems that telomerase itself may have evolved from transposable elements (Eickbush, 1997; Pardue et al., 1997) (see Chapter 3).

Centromeres, too, may be the byproducts of transposable element activity, and it is notable in this regard that newly formed centromeres may act as segregation distorters—that is, they apparently have not yet abandoned their selfish leanings (Kidwell and Lisch, 2001, 2002) (see Chapter 3). Recall also that some autonomous B chromosomes may be little more than independent centromeres (see Chapter 4). Both telomeres and centromeres play important roles in meiosis, and therefore in generating genetic complexity by meiotic recombination (e.g., Ishikawa and Naito, 1999; Carroll, 2002). Finally, it has also recently been proposed that transposable elements play a role in maintaining chromosome structure by participating in the process of double-strand break repair (Labrador and Corces, 1997; Eickbush, 2002; Morrish et al., 2002).

INCREASED GENETIC COMPLEXITY

Knoll (1995) reported that the Proterozoic fossil record reveals the progressive appearance of lineages of which extant members display larger and more numerous introns. As Carroll (2002) argued, "this suggests that introns became more and more prevalent during the evolution of complex unicellular eukaryotes, and that this is associated with more and more complex proteins." Moreover, increased exon shuffling during this time by unequal crossing over (which is facilitated by TEs), probably contributed to an increase in genomic complexity among these early eukaryotes, as well as among their metazoan descendents (Patthy, 1999; Carroll, 2002) (see later section).

It has been pointed out that the evolution of eukaryotic cells, with their higher gene contents and more complex physiologies, could not have evolved without the development of more sophisticated mechanisms of gene regulation. This transition from prokaryotes to eukaryotes was probably partly dependent on the evolution of structured chromatin and the nuclear membrane (Bird, 1995; Bird and Tweedie, 1995). It has been suggested that transposable elements played a direct role in the evolution of heterochromatin, and therefore contributed in a crucial way to the origin of eukaryotes (McDonald, 1998). Finally, it bears noting that detailed genomic analyses have suggested that eukaryotic genomes are actually chimeras, containing gene sequences derived from more than one ancestral lineage (Katz, 1999). To the extent that such genomic mixtures can be generated by horizontal transfer, this implies another potentially important role for transposons and other independent genetic elements in shaping eukaryotic genome diversity.

THE ORIGIN OF MULTICELLULARITY AND THE EMERGENCE OF COMPLEX METAZOA

TRANSPOSABLE ELEMENTS AND GENE REGULATION

As with the prokaryote–eukaryote transition, the evolution of multicellular organisms with specialized tissues required the increased ability of genomes to regulate and coordinate gene expression. Gene silencing, in particular, is of direct relevance to the production of a complex multicellular organism whose differentiated cells contain the same genetic information. It is now becoming widely accepted that the more prominent gene silencing mechanisms in animals evolved as defenses against RNA viruses and transposable elements (Matzke *et al.*, 1999; Plasterk, 2002). In addition, the cooption of formerly selfish genetic elements for roles in gene regulation is probably also an important part of this increased capacity for tissue specialization (e.g., Britten, 1996a,b, 1997; Kidwell and Lisch, 2000, 2001) (see Chapter 3).

SPLICEOSOMAL INTRONS AND EXON SHUFFLING

Multicellularity requires the generation of proteins that serve roles in the extracellular matrix, in cell adhesion, and as cellular receptors. The majority of such proteins are modular, having been produced by exon shuffling. As Patthy (1999) pointed out, "most modular proteins produced by exon-shuffling are associated with, and are absolutely essential for, multicellularity of metazoa." Exon shuffling of this type is mediated by recombination among middle repetitive DNAs located within spliceosomal introns, which in turn probably evolved from Group II mobile introns sometime shortly before the Cambrian explosion (Patthy, 1999). Without the emergence of these subgenomic elements, the origin of complex metazoa may never have been possible at all (Lundin, 1999; Patthy, 1999).

Subsequent increases in complexity among animals may have been dependent on similar processes. Notably, a comparison of the gene numbers for various species whose genomes have been sequenced demonstrates that gene number does not correlate well with overall complexity (see Chapter 9). As such, much of the diversification of animal complexity must relate to alternative means of generating protein variation. One such mechanism is to splice the exons of the same gene sequence together in different ways in order to generate different protein-products. As Graveley (2001) recently noted, "it is becoming increasingly clear that alternative splicing has an extremely important role in expanding protein diversity." Inasmuch as alternative splicing is dependent on the existence of spliceosomal introns, it is fair to say that much of the diversification of metazoans has relied on the nonstandard genome-level processes emphasized here.[3]

GENE DUPLICATION AND DEVELOPMENTAL COMPLEXITY

The importance of both small- and large-scale gene duplications was discussed in detail in Chapters 5 and 6, and so will not be reviewed again here. With regard to the topic of this chapter, it is highly relevant to note that genes encoding proteins for cell–cell communications, which are necessary for maintaining the integrity of multicellular organisms, arose by gene duplication (Suga *et al.*, 1999). Several gene families are also known to have expanded by duplication as part of the origin and/or early diversification of the metazoa (Lundin, 1999), indicating that both of these major transitions involved a primary influence of gene duplication. From the present perspective, the most important among them are those that play key roles in animal development, most notably homeobox genes, the best known of which are the *Hox* genes of the Bilataria. Again, these exist in one or more clusters (depending on the group), and exert a crucial influence on body patterning during development. The *Hox* clusters are now believed to be quite ancient, having evolved

---

[3]As an interesting tie-in with Part One of this chapter, it is worth noting that lineages with alternative splicing mechanisms may be more evolutionarily flexible, and thus more likely to diversify and/or avoid extinction, than lineages lacking such mechanisms. As Doolittle (1987) pointed out, one could "easily argue that introns are, because of their effects on species reproduction and survival, more prevalent among species than they otherwise would have been, and thus are to that extent maintained by species selection."

even before the divergence of the Bilataria and Cnidaria (Ferrier and Holland, 2001). As Carroll (2000) pointed out, "The origin of multicellularity and complex body plans among animals was a unique phenomenon, dependent on the evolution of *Hox* genes near the end of the Precambrian (late Neoproterozoic). Once evolved, their subsequent duplication and divergent change in adaptively distinct lineages established the basis for the radiation of the many metazoan phyla." Indeed, it has been proposed that, along with exon shuffling, gene duplication may have helped to fuel the Cambrian explosion (Lundin, 1999; Patthy, 1999; but see Suga *et al.*, 1999).

As an interesting footnote to this discussion, it has recently been found that ~8–10% of the examples of alternative splicing in humans, *Drosophila*, and *C. elegans* have been associated with exon duplications (Kondrashov and Koonin, 2001; Letunic *et al.*, 2002). It has also been shown in some cases that transposable elements can be involved in alternative splicing (Hughes, 2001; Lev-Maor *et al.*, 2003; Makalowski, 2003), perhaps even by serving as alternatively spliced exons, as in the case of the enormously abundant *Alu* elements in the human genome (Kreahling and Graveley, 2004). This provides an intriguing case in which all of the genome-level processes emphasized here may converge to exert similar effects on evolutionary diversification.

THE EVOLUTION OF IMMUNITY

Complex multicellular organisms are subject to infection by pathogens (including selfish genetic elements like viruses) and therefore require defensive immune systems. Based as it is on the interactions of dozens of multiplied immunoglobulin genes, it is clear that the adaptive immune system of animals evolved as a result of gene (or chromosome, or even genome) duplication (Kasahara *et al.*, 1997; Zhang, 2003) (see Chapter 5). Less well appreciated, but certainly no less important, is the role that transposable elements have played in the emergence of this system.

In vertebrates, the flexibility of the immune system is maintained by the shuffling of the variable (V), joining (J), and in some cases diversity (D) segments of the immunoglobulin genes in lymphocytes by a process known as "V(D)J recombination" (see Gellert, 2002, for a detailed review). The first phase of this recombination process is mediated by the protein-products of the recombination-activating genes *RAG1* and *RAG2*. Intriguingly, it now appears that these critical genes share many key properties with transposable elements (Agrawal *et al.*, 1998; Hiom *et al.*, 1998; Gellert, 2002), strongly suggesting that "a fundamental component of the vertebrate immune system probably evolved from a transposon, whose capacity for DNA rearrangement was exploited to produce rapid somatic variability in specific host cells" (Kidwell and Lisch, 2001). As an interesting twist on this, it seems that the evolution of viviparity (as found most notably among placental mammals) required a *reverse* input from transposable elements. In particular, the development of a "foreign" body inside the mother requires a mechanism to suppress the parental immune system, a capability that seems to have been borrowed from the action of endogenous retroviruses (Venables *et al.*, 1995; Villarreal, 1997; Kidwell and Lisch, 2001). As yet another twist, it has been suggested that the possession of an adaptive immune system may limit the number of genes that can be contained within a genome and thereby hinder future evolutionary diversification (George, 2002). As Smit (1999) noted with regard to the role of TEs in immune system evolution, "One can not ask for a better example of a contribution of transposons and of the quirky nature of evolution."

A working genome-level immune system is also necessary for defense against mutagenic TE and virus activity. Two important mechanisms of genomic immunity are DNA methylation and RNA silencing. DNA methylation plays an important role in the epigenetic modification of gene expression, as seen in X inactivation and genomic imprinting (Jaenisch and Bird, 2003), but may also act as a defense against selfish invaders by selectively methylating TE sequences (Yoder *et al.*, 1997). Currently it is not clear whether both functions are ancient and evolved in parallel with one another, whether DNA methylation first evolved for the modification of gene expression and was later coopted as a genomic defense mechanism, or whether the genomic defense came first and was coopted for gene regulation, but all are interesting possibilities. If the latter turns out to be the case, then this would add another prime example of genomic exaptation with important consequences for gene regulation.

THE ORIGIN OF VERTEBRATES

GEN(OM)E DUPLICATION

There is now little doubt that the duplication of *Hox* clusters and other genes played a crucial role in the emergence of the vertebrate body plan (Holland *et al.*, 1994; Graham, 2000). Some question does remain as to whether this was brought about by small- or large-scale gene, or even complete genome, duplication, but the implications for this major evolutionary transition are clear in either case. As discussed in detail in Chapter 6, there is considerable evidence that a genome duplication event occurred early in vertebrate evolution (and another in the teleost fish lineage). Although polyploidy has been far from ubiquitous in more recently derived vertebrate lineages (see Chapter 8), it does seem that the phenomenon played an important role at the base of the entire group. Though treated only very briefly here, the profound importance of this fact should not be underestimated.

Silencing and/or Splicing

Based on a comparison of available genome sequences, it is apparent that vertebrates possess more coding genes than invertebrates. This implies the need for more sophisticated regulatory systems, which, again, probably involve former transposable elements. More generally, an increased number of genes requires a mechanism for tissue-specific silencing, and it has been postulated that, as with the evolution of heterochromatin in the origin of eukaryotes, the emergence of methylation as a silencing mechanism was important in the transition from invertebrates to vertebrates (Bird, 1995; Bird and Tweedie, 1995; Bird et al., 1995). And once again, the evolution of this regulatory mechanism has been envisioned as the cooption of a mechanism of genomic defense against transposable element activity (McDonald, 1998; Matzke et al., 1999). However, it should be noted that there are more recent suggestions that methylation remains primarily a mechanism of genomic defense against TEs (Yoder et al., 1997), although on the other hand some authors have argued that methylation as a gene regulatory mechanism extends to invertebrates as well (Regev et al., 1998).

In any case, it is now well known from the human genome sequence that the gene number discrepancy between vertebrates and invertebrates is much less profound than had initially been assumed (International Human Genome Sequencing Consortium, 2001; Venter et al., 2001), so gene silencing is probably not the primary consideration when dealing with this transition. What is needed in this case is a mechanism to account for the high number of protein transcripts that must be derived from a relatively small number of protein-coding genes. As noted previously, it is becoming increasingly apparent that much of this proteomic diversity is generated by alternative splicing involving noncoding DNA in introns (Kondrashov and Koonin, 2003).

HUMAN UNIQUENESS

Human intelligence is without equal (at least on this planet) and differs markedly from that of even the next most closely related primates. It is a well-known fact that humans and chimpanzees are very similar genetically,[4] indicating that differences

---

[4] The exact level of genetic similarity between humans and chimpanzees depends considerably on what is being measured. Wildman et al. (2003) recently estimated a 98.4% similarity based on an analysis of synonymous substitutions in coding genes, and 99.4% in terms of nonsynonymous changes. If insertions and deletions are included in an analysis of noncoding DNA, the similarity may be closer to 95% (Britten, 2002; Britten et al., 2003). On a broader genomic scale, chimpanzees have both more DNA per genome (3.75 pg vs. 3.50 pg) and more chromosomes (2n = 48 vs. 2n = 46) than humans (Gregory, 2001b). It has been argued on genetic grounds that humans and chimps should be classified in the same genus (e.g., Wildman et al., 2003), although this remains controversial (e.g., Marks, 2002).

in gene expression and perhaps a few genes of major effect account for the pronounced morphological, cognitive, and linguistic differences between the species (e.g., Enard *et al.*, 2002a). In some cases, as with the accelerated rates of molecular evolution in at least one speech-related gene (*FOXP2*), this may be the result of "standard" genic processes (e.g., Enard *et al.*, 2002b; Zhang *et al.*, 2002). On the other hand, there is already some indication that the nonstandard genome-level processes emphasized here have played a crucial role in the emergence of human characteristics. For example, BC200, a brain-specific RNA that is part of a ribonucleoprotein complex preferentially located in the dendrites of all anthropoid primates, appears to have been derived about 35–55 million years ago from an *Alu* transposable element (Skryabin *et al.*, 1998; Smit, 1999).

DIVERSITY IN GENE EXPRESSION

As discussed earlier in this chapter and in Chapter 3, there is a growing list of transposable elements that have taken on key functions in gene regulation. In this regard, it is of obvious relevance that upward of 25% of human regulatory genes appear to be derived from former transposable elements (Jordan *et al.*, 2003). It has also been estimated that at least 35% of all human genes undergo alternative splicing (Mironov *et al.*, 1999; Brett *et al.*, 2000), which, again, is mediated by spliceosomal introns derived from autonomous mobile elements. This mechanism is capable of producing substantial proteomic diversity from a relatively small number of genes and is therefore of general importance in the evolution of complex animals such as humans. It should be noted, however, that the proportion of these alternatively spliced configurations that are functional remains a subject of debate (e.g., Sorek *et al.*, 2004). There are nevertheless specific cases in which the process has evidently been important to features of relevance to human uniqueness. For example, alternative splicing has been shown to play a key role in axon guidance, synaptogenesis, receptor specificity, neurotransmission, ion channel function, and learning and memory processes in the nervous system (Grabowski and Black, 2001; Graveley, 2001).

THE ROLE OF *ALU* ELEMENTS

In a general sense, *Alu* elements (which are specific to primates) may also have played a role in primate diversification through their ability to generate exonic variation by alternative splicing (Lev-Maor *et al.*, 2003; Makalowski, 2003) and by serving as recombinational hotspots involved in segmental duplications (Makalowski, 1995; Bailey *et al.*, 2003). In some cases, *Alu* elements cause deletion mutations that are specific to humans, as in one exon deletion found in the tropoelastin gene (Szabó *et al.*, 1999). However, because another exon had already been deleted from this gene prior to the divergence of the great apes, this particular case probably does not relate to human uniqueness (Hayakawa *et al.*, 2001). There is, however, at least one very intriguing case that may indeed be of relevance in this context.

In vertebrates, the sialic acids represent a family of acidic sugars found on the cell surface. The most common forms in mammals are N-acetylneuraminic acid (Neu5Ac) and its hydroxylated derivative, N-glyconeuraminic acid (Neu5Gc); the conversion between these two has impacts on a wide array of biological features via effects on cellular sialic acid receptors (Chou *et al.*, 1998). Although Neu5Gc is common in most tissues in mammals, its production is typically down-regulated in the brain. In humans Neu5Gc is entirely absent from all tissues because of an inactivation of the CMP-N-acetylneuraminic acid hydroxylase enzyme gene that occurred after the divergence from the other great apes but before the split between modern humans and Neanderthals—that is, shortly before the expansion of the brain in the human lineage (Chou *et al.*, 1998, 2002). As Gagneux and Varki (2001) pointed out, "it cannot be ruled out that Neu5Gc loss alters glycoprotein function in human brains and potentially affects brain development." Neu5Gc is also exploited by several viral and bacterial pathogens to gain entry into host cells, so it is likely that its loss in early hominids allowed an expansion of habitat range and probably also the domestication of other vertebrates carrying such diseases (Hayakawa *et al.*, 2001). Importantly, the inactivation of the CMP-N-acetylneuraminic acid hydroxylase enzyme gene in humans was accomplished by a 92-bp exonic deletion caused by the insertion of a human-specific *Alu* element (Fig. 11.4).

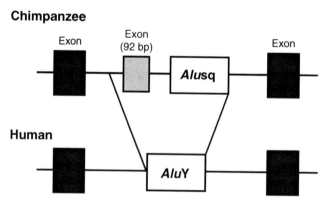

FIGURE 11.4 Differences in the CMP-Neu5Ac hydroxylase gene between chimpanzees (top) and humans (bottom). In humans, a 92-bp exon has been deleted by the insertion of an *AluY* element, rendering the gene nonfunctional. This TE insertion may have had wide-ranging effects on human biology. From Hayakawa *et al.* (2001), reproduced by permission (© National Academy of Sciences USA).

The possibility that many of the genetic differences between humans and other primates have resulted from transposable element activity—through the action of regulatory regions descended from TEs (Makalowski, 1995; Jordan *et al.*, 2003) and/or via TE-mediated exon splicing (Lev-Maor *et al.*, 2003; Makalowski, 2003) and deletion mutations (Chou *et al.*, 1998, 2002; Szabó *et al.*, 1999; Hayakawa *et al.*, 2001)—represents an important and exciting avenue for future investigation.

Gene Duplication and Back Again

Gene duplications have occurred frequently in the primate lineage, with many duplication-related differences apparent between humans and other apes (Gagneux and Varki, 2001; Hacia, 2001). In fact, the human lineage in particular may have experienced a major increase in the rate of gene duplications (Eichler, 2001; Gu *et al.*, 2002; Courseaux *et al.*, 2003), with recent estimates indicating that 700–1800 duplications have occurred in the human genome since the split with chimpanzees (Zhang, 2003). Duplications and deletions of genes, among other mechanisms, may generate the variation in gene expression observed between chimps and humans, especially in the brain (Enard *et al.*, 2002a). The evolution of trichromatic color vision, which is found in no mammals other than primates, was also crucially dependent on the duplication of opsin genes (Dulai *et al.*, 1999). Moreover, it appears that recombination mediated by *Alu* elements may have contributed to the occurrence of this important duplication event (Dulai *et al.*, 1999). This shift in visual capability seems to have been directly associated with a loss of the vomeronasal pheromone sensitivity found in most other mammals (owing to the pseudogenization of pheromone receptor genes), which in turn probably incited profound changes in sexual, aggressive, and other behaviors among certain primate lineages, including the one leading to humans (Gilad *et al.*, 2003, 2004; Zhang and Webb, 2003).

In a final example, it is notable that ape hair keratin is encoded in clustered gene families, meaning that they were produced by duplication. Interestingly, it appears that in humans a type I hair keratin pseudogene has been inactivated by a single point mutation that has not occurred in other apes, among whom the orthologous gene copy remains functional (Winter *et al.*, 2001). In this case, a mutation that *undid* the effects of an earlier gene duplication may have played a role in generating one of the most obvious human traits, the loss of body hair.

Nonstandard Genomic Processes: A Summary

The list of examples provided in the previous sections is not exhaustive, and it is very likely that many others will be discovered in the near future. Similar discussions may also be possible with regard to major events in the evolution of plants and other groups. The point of this discussion is not to provide complete answers to such questions as: How did eukaryotes evolve from prokaryotes?, How did complex immune systems evolve?, or How did the vertebrate body plan emerge?. Rather, it is to emphasize that information about mutational mechanisms— including, but certainly not only, "nonstandard" ones—is a crucial component of a complete answer to such questions. It hardly bears mentioning that in no case does "by a change in allele frequencies" provide a satisfactory solution, whether technically true in a microevolutionary sense or not. Moreover, many of these key transitions have involved the input of genetic elements that have their own evolutionary histories within the genome. Knowledge of these intriguing (and in most cases, wholly unexpected) genomic inputs may have little bearing on the current understanding of microevolutionary processes, but it has substantial implications for the issue of how specific macroevolutionary patterns come to be. It also greatly bolsters the view that macroevolutionary patterns arise by the operation of selection and other processes at several levels of organization, both above and, in this case, below the level of the individual organism.

CONCLUDING REMARKS AND FUTURE PROSPECTS

FROM REDUCTIONISM TO INTEGRATIONISM

As with life itself, evolutionary biology has passed through numerous major transitions that opened up new and previously inaccessible trajectories. The first came nearly 150 years ago with Darwin's (1859) publication of his theory of natural selection. The (re)discovery of Mendelian genetics around the beginning of the 20th century marks another, as does the subsequent development of population genetics a few decades later. The molecular revolution of 50 years ago represents one of the most important events, enabling a host of analyses that would otherwise have been impossible, if not inconceivable, and paving the way for the genome-level studies described throughout this book.

The current transition into the post-genomic era has the potential to be the most monumental watershed in the history of biology. In the span of only a few years, comparisons of complete genome sequences have already provided major new insights into the organization and regulation of genes, the role of duplications, and the evolution of nongenic elements. It is very likely that this new ability to assess patterns and processes on a genome-wide scale will change the outlook on how genomes, species, and biotas evolve even more than any of the previous revolutions in evolutionary science. The task of evolutionary biologists from all disciplines, ranging from geneticists to paleontologists, will be to embrace this transition as an opportunity for a level of integration that has never before been imaginable. The reductionistic approach to genomics has been extraordinarily successful thus far, but perhaps its greatest contribution has yet to come in the form of stimulating a shift to true integrationism.

A POST-GENOMIC EVOLUTIONARY SYNTHESIS

Understanding the evolution of any given biological feature (not to mention of life at large) involves answering numerous broad questions, including: What selective pressures and other forces account for its origin, and which account for its current maintenance? When did it evolve, by what historical route, and according to what underlying developmental and genetic principles? How has its evolution been influenced by historical contingency and preexisting constraints, and what have been the subsequent evolutionary and ecological consequences of its emergence? How can its evolution be understood in terms of processes operating at the subgenomic, genomic, cellular, organismal, populational, species, and global levels? No single set of analytical and conceptual tools, nor any approach based on only one level of analysis, can address more than a minority of these component issues. The net result is that evolutionary theory must be broad enough to recognize and deal with these various topics—if it is not so currently, then it must be formally expanded.

The conceptual transition being provoked by the genomic revolution is not unlike the one experienced a century ago with the rediscovery of Mendel's laws of inheritance. The vast new knowledge emerging from the study of genomes and their components, and the attendant appreciation of their roles in influencing both ontogenetic and phylogenetic patterns, indicate the need for a new "Post-Genomic Synthesis" in evolutionary biology. Unlike the "Modern Synthesis" of 60 to 70 years ago, this post-genomic version will be based primarily on expansion instead of constriction.

As a first step, it will be necessary to recognize that micro- and macro-evolutionary theory are not competing alternatives; rather, the two deal with very different sets of questions. They are, however, linked by the common evolutionary principles of mutation, selection, chance, contingency, and constraint, even though the specific expressions of these principles differ greatly between the micro- and macroevolutionary scales. A second step in the development of a post-genomic synthesis will be a broadening of how these principles are defined and understood. This chapter has provided but one example of this, namely an expansion of the concept of "mutations" to include the roles of nonstandard genome-level processes. Contingency, constraint, and each of the other principles could be treated from a similar genome-level perspective, and should likewise be dealt with from the point of view of other levels of organization. A third critical step in the

Post-Genomic Synthesis will be to implement the aforementioned philosophical shift from reductionism to integrationism. Importantly, this will entail a policy of perpetual synthesis, in which information gleaned from a variety of disciplines is combined as a matter of course in order to provide multifaceted but unified answers to complex biological questions.

GENOMES AND THE FUTURE OF BIOLOGY

Evolutionary theory is but one aspect of biological science that stands to be significantly influenced by the insights emerging from the study of genomes. As many of the preceding chapters have stressed, an understanding of the large-scale features of genomes, as well as a detailed inventory of the genes and other elements contained therein, has the potential to make a substantial impact on medicine, agriculture, pharmacology, and a host of other applied biosciences. More generally, genomic data promise to shed considerable light on the diversity of life on Earth and how that diversity came to exist.

That said, it is important to note that genomes are only one of many important levels of biological organization, and that genomic data are only truly useful when viewed in the context of morphological, cytological, developmental, physiological, and ecological information. By itself, the possession of even a complete genome sequence does not provide any great insight into the workings of living systems, let alone reveal the place or essence of humanity. However, when properly integrated with a wide range of other biological knowledge, the detailed study of the evolution of the genome will greatly assist in the age-old pursuit of answers to such questions.

REFERENCES

Agrawal A, Eastman QM, Schatz DG. 1998. Implications of transposition mediated by V(D)J-recombination proteins RAG1 and RAG2 for origins of antigen-specific immunity. *Nature* 394: 744–751.

Arkhipova I, Meselson M. 2000. Transposable elements in sexual and ancient asexual taxa. *Proc Natl Acad Sci USA* 97: 14473–14477.

Artandi SE, Chang S, Lee S-L, et al. 2000. Telomere dysfunction promotes non-reciprocal translocations and epithelial cancers in mice. *Nature* 406: 641–645.

Ayala FJ. 1982. Microevolution and macroevolution. In: Bendall DS ed. *Evolution from Molecules to Men.* Cambridge, UK: Cambridge University Press, 387–402.

Ayala FJ. 1985. Reduction in biology: a recent challenge. In: Depew DJ, Weber BH eds. *Evolution at a Crossroads.* Cambridge, MA: MIT Press, 65–79.

Bailey JA, Liu G, Eichler EE. 2003. An *Alu* transposition model for the origin and expansion of human segmental duplications. *Am J Hum Genet* 73: 823–834.

Bell G. 1993. The sexual nature of the eukaryote genome. *J Hered* 84: 351–359.

Bennetzen JL. 2002. Mechanisms and rates of genome expansion and contraction in flowering plants. *Genetica* 115: 29–36.

Biémont C, Vieira C, Borie N. 2001. Éléments transposables et évolution du génome d'une espèce invasive: le cas de *Drosophila simulans*. *Genet Sel Evol* 33 (Suppl. 1): S107–S120.

Bird AP. 1995. Gene number, noise reduction and biological complexity. *Trends Genet* 11: 94–100.

Bird A, Tweedie S. 1995. Transcriptional noise and the evolution of gene number. *Philos Trans R Soc Lond B* 349: 249–253.

Bird A, Tate P, Nan X, et al. 1995. Studies of DNA methylation in animals. *J Cell Sci* (Suppl.) 19: 37–39.

Brett D, Hanke J, Lehmann G, et al. 2000. EST comparison indicates 38% of human mRNAs contain possible alternative splice forms. *FEBS Lett* 474: 83–86.

Britten RJ. 1996a. DNA sequence insertion and evolutionary variation in gene regulation. *Proc Natl Acad Sci USA* 93: 9374–9377.

Britten RJ. 1996b. Cases of ancient mobile element DNA insertions that now affect gene regulation. *Mol Phylogenet Evol* 5: 13–17.

Britten RJ. 1997. Mobile elements inserted in the distant past have taken on important functions. *Gene* 205: 117–182.

Britten RJ. 2002. Divergence between samples of chimpanzee and human DNA sequences is 5%, counting indels. *Proc Natl Acad Sci USA* 99: 13633–13635.

Britten RJ, Rowen L, Williams J, Cameron RA. 2003. Majority of divergence between closely related DNA samples is due to indels. *Proc Natl Acad Sci USA* 100: 4661–4665.

Brosius J. 1999. RNAs from all categories generate retrosequences that may be exapted as novel genes or regulatory sequences. *Gene* 238: 115–134.

Burt A, Trivers R. 1998. Selfish DNA and breeding system in flowering plants. *Proc R Soc Lond B* 265: 141–146.

Carroll RL. 2000. Towards a new evolutionary synthesis. *Trends Ecol Evol* 15: 27–32.

Carroll RL. 2002. Evolution of the capacity to evolve. *J Evol Biol* 15: 911–921.

Casacuberta E, Pardue M-L. 2002. Coevolution of the telomeric retrotransposons across *Drosophila* species. *Genetics* 161: 1113–1124.

Castro-Jimenez Y, Newton RJ, Price HJ, Halliwell RS. 1989. Drought stress responses of *Microseris* species differing in nuclear DNA content. *Am J Bot* 76: 789–795.

Charlesworth B, Langley CH. 1986. The evolution of self-regulated transposition of transposable elements. *Genetics* 112: 359–383.

Charlesworth B, Lande R, Slatkin M. 1982. A neo-Darwinian commentary on macroevolution. *Evolution* 36: 474–498.

Chou H-H, Hayakawa T, Diaz S, et al. 2002. Inactivation of CMP-N-acetylneuraminic acid hydroxylase occurred prior to brain expansion during human evolution. *Proc Natl Acad Sci USA* 99: 11736–11741.

Chou H-H, Takematsu H, Diaz S, et al. 1998. A mutation in human CMP-sialic acid hydroxylase occurred after the *Homo-Pan* divergence. *Proc Natl Acad Sci USA* 95: 11751–11756.

Courseaux A, Richard F, Grosgeorge J, et al. 2003. Segmental duplications in euchromatic regions of human chromosome 5: a source of evolutionary instability and transcriptional innovation. *Genome Res* 13: 369–381.

d'Adda di Fagagna F, Reaper PM, Clay-Farrace L, et al. 2003. A DNA damage checkpoint response in telomere-initiated senescence. *Nature* 426: 194–198.

Darwin C. 1859. *On the Origin of Species by Means of Natural Selection, or the Preservation of Favoured Races in the Struggle for Life*. London: John Murray.

Darwin C. 1871. *The Descent of Man, and Selection in Relation to Sex*. London: John Murray.

Dawkins R. 1989. *The Selfish Gene*. 2d ed. Oxford: Oxford University Press.

Dimitri P, Junakovic N. 1999. Revisiting the selfish DNA hypothesis: new evidence on accumulation of transposable elements in heterochromatin. *Trends Genet* 15: 123–124.

Dobzhansky T. 1937. *Genetics and the Origin of Species*. New York: Columbia University Press.

Doolittle WF. 1987. What introns have to tell us: hierarchy in genome evolution. *Cold Spring Harb Symp Quant Biol* 52: 907–913.

Doolittle WF. 1989. Hierarchical approaches to genome evolution. *Can J Philos* 14 (Suppl.): 101–133.

Doolittle WF, Kirkwood TBL, Dempster MAH. 1984. Selfish DNAs with self-restraint. *Nature* 307: 501–502.

Dulai KS, von Dornum M, Mollon JD, Hunt DM. 1999. The evolution of trichromatic color vision by opsin gene duplication in new world and old world primates. *Genome Res* 9: 629–638.

Eichler EE. 2001. Recent duplication, domain accretion and the dynamic mutation of the human genome. *Trends Genet* 17: 661–669.

Eickbush TH. 1997. Telomerase and retrotransposons: which came first? *Science* 277: 911–912.

Eickbush TH. 2002. Repair by retrotransposition. *Nat Genet* 31: 126–127.

Eldredge N. 1985. *Unfinished Synthesis*. Oxford: Oxford University Press.

Eldredge N. 1998. *The Pattern of Evolution*. New York: W.H. Freeman & Co.

Eldredge N, Gould SJ. 1972. Punctuated equilibria: an alternative to phyletic gradualism. In: Schopf TJM ed. *Models in Paleobiology*. San Francisco, CA: Freeman, Cooper, & Co., 82–115.

Enard W, Khaitovich P, Klose J, et al. 2002a. Intra- and interspecific variation in primate gene expression patterns. *Science* 296: 340–343.

Enard W, Przeworski M, Fisher SE, et al. 2002b. Molecular evolution of *FOXP2*, a gene involved in speech and language. *Nature* 418: 869–872.

Erwin DH. 2000. Macroevolution is more than repeated rounds of microevolution. *Evol Dev* 2: 78–84.

Esnault C, Maestre J, Heidmann T. 2000. Human LINE retrotransposons generate processed pseudogenes. *Nat Genet* 24: 363–367.

Ferrier DEK, Holland PWH. 2001. Ancient origin of the Hox gene cluster. *Nat Rev Genet* 2: 33–38.

Filipchenko IA. 1927. *Variabilität und Variation*. Berlin: Gebrüder Borntraeger.

Finnegan DJ. 1989. Eukaryotic transposable elements and genome evolution. *Trends Genet* 5: 103–107.

Fisher RA. 1958. *The Genetical Theory of Natural Selection*. 2d ed. New York: Dover.

Forterre P. 2002. The origin of DNA genomes and DNA replication proteins. *Curr Opin Microbiol* 5: 525–532.

Freeman S, Herron JC. 1998. *Evolutionary Analysis*. Upper Saddle River, NJ: Prentice Hall.

Futuyma DJ. 1998. *Evolutionary Biology*. 3d ed. Sunderland, MA: Sinauer Associates Inc.

Gagneux P, Varki A. 2001. Genetic differences between humans and great apes. *Mol Phylogenet Evol* 18: 2–13.

Gellert M. 2002. V(D)J recombination. In: Craig NL, Craigie R, Gellert M, Lambowitz AM eds. *Mobile DNA II*. Washington, DC: American Society for Microbiology Press, 705–729.

George AJ. 2002. Is the number of genes we possess limited by the presence of an adaptive immune system? *Trends Immunol* 23: 351–355.

Gilad Y, Bustamente CD, Lancet D, Pääbo S. 2003. Natural selection on the olfactory receptor gene family in humans and chimpanzees. *Am J Hum Genet* 73: 489–501.

Gilad Y, Wiebe V, Przeworski M, Pääbo S. 2004. Loss of olfactory receptor genes coincides with the acquisition of full trichromatic vision in primates. *PLoS Biol* 2: 0120–0125.

Goldschmidt RB. 1940. *The Material Basis of Evolution*. New Haven, CT: Yale University Press.

Goodnight CJ, Stevens L. 1997. Experimental studies of group selection: what do they tell us about group selection in nature? *Am Nat* 150 (Suppl.): S59–S79.

Gould SJ. 1980. Is a new and general theory of evolution emerging? *Paleobiology* 6: 119–130.

Gould SJ. 1982. The meaning of punctuated equilibrium and its role in validating a hierarchical approach to macroevolution. In: Milkman R ed. *Perspectives on Evolution*. Sunderland, MA: Sinauer, 83–104.

Gould SJ. 1983a. The hardening of the modern synthesis. In: Grene M ed. *Dimensions of Darwinism*. Cambridge, UK: Cambridge University Press, 71–93.

Gould SJ. 1983b. *Hen's Teeth and Horse's Toes*. New York: W.W. Norton & Co.

Gould SJ. 1992. Punctuated equilibrium in fact and theory. In: Somit A, Peterson SA eds. *The Dynamics of Evolution*. Ithaca, NY: Cornell University Press, 54–84.

Gould SJ. 1998. Gulliver's further travels: the necessity and difficulty of a hierarchical theory of selection. *Philos Trans R Soc Lond B* 353: 307–314.

Gould SJ. 2000. *The Lying Stones of Marrakech*. New York: Harmony Books.

Gould SJ. 2002. *The Structure of Evolutionary Theory*. Cambridge, MA: Harvard University Press.

Gould SJ, Eldredge N. 1977. Punctuated equilibria: the tempo and mode of evolution reconsidered. *Paleobiology* 3: 115–151.

Gould SJ, Eldredge N. 1993. Punctuated equilibrium comes of age. *Nature* 366: 223–227.

Gould SJ, Lewontin RC. 1979. The spandrels of San Marco and the Panglossian paradigm: a critique of the adaptationist program. *Proc R Soc Lond B* 205: 581–598.

Grabowski P, Black DL. 2001. Alternative RNA splicing in the nervous system. *Prog Neurobiol* 65: 289–308.

Graham A. 2000. The evolution of the vertebrates—genes and development. *Curr Opin Genet Dev* 10: 624–628.

Grantham TA. 1995. Hierarchical approaches to macroevolution: recent work on species selection and the "effect hypothesis." *Annu Rev Ecol Syst* 26: 301–321.

Graveley BR. 2001. Alternative splicing: increasing diversity in the proteomic world. *Trends Genet* 17: 100–107.

Gray YHM. 2000. It takes two transposons to tango: transposable-element-mediated chromosomal rearrangements. *Trends Genet* 16: 461–468.

Gregory TR. 2001a. Coincidence, coevolution, or causation? DNA content, cell size, and the C-value enigma. *Biol Rev* 76: 65–101.

Gregory TR. 2001b. *Animal Genome Size Database*. www.genomesize.com.

Gregory TR. 2002a. A bird's-eye view of the C-value enigma: genome size, cell size, and metabolic rate in the class Aves. *Evolution* 56: 121–130.

Gregory TR. 2002b. Genome size and developmental complexity. *Genetica* 115: 131–146.

Gregory TR. 2004. Macroevolution, hierarchy theory, and the C-value enigma. *Paleobiology* 30: 179–202.

Gu X, Wang Y, Gu J. 2002. Age distribution of human gene families shows significant roles of both large- and small-scale duplications in vertebrate evolution. *Nat Genet* 31: 205–209.

Hacia JG. 2001. Genome of the apes. *Trends Genet* 17: 637–645.

Harshman LG, Hoffman AA. 2000. Laboratory selection experiments using *Drosophila*: what do they really tell us? *Trends Ecol Evol* 15: 32–36.

Hayakawa T, Satta Y, Gagneux P, et al. 2001. *Alu*—mediated inactivation of the human CMP-N-acetylneuraminic acid hydroxylase gene. *Proc Natl Acad Sci USA* 98: 11399–11404.

Hecht MK, Hoffman A. 1986. Why not neo-Darwinism? A critique of paleobiological challenges. *Oxford Surv Evol Biol* 3: 1–47.

Hendrix RW, Lawrence JG, Hatfull GF, Casjens S. 2000. The origins and ongoing evolution of viruses. *Trends Microbiol* 8: 504–508.

Hendry AP, Kinnison MT. 2001. An introduction to microevolution: rate, pattern, process. *Genetica* 112–113: 1–8.

Hickey DA. 1982. Selfish DNA: a sexually-transmitted nuclear parasite. *Genetics* 101: 519–531.

Hiom K, Mele M, Gellert M. 1998. DNA transposition by the RAG1 and RAG2 proteins: a possible source of oncogenic translocations. *Cell* 94: 463–470.

Holland PWH, Garcia-Fernàndez J, Williams NA, Sidow A. 1994. Gene duplications and the origins of vertebrate development. *Development* 120 (Suppl.): 125–133.

Hughes DC. 2001. Alternative splicing of the human VEGFGR-3/FLT4 gene as a consequence of an integrated human endogenous retrovirus. *J Mol Evol* 53: 77–79.

Hurst LD. 1995. Selfish genetic elements and their role in evolution: the evolution of sex and some of what it entails. *Philos Trans R Soc Lond B* 349: 321–332.

Hutchison CA, Peterson SN, Gill SR, *et al.* 1999. Global transposon mutagenesis and a minimal *Mycoplasma* genome. *Science* 286: 2165–2169.

International Human Genome Sequencing Consortium. 2001. Initial sequencing and analysis of the human genome. *Nature* 409: 860–921.

Ishikawa F, Naito T. 1999. Why do we have linear chromosomes? A matter of Adam and Eve. *Mutat Res* 434: 99–107.

Jablonski D. 2000. Micro- and macroevolution: scale and hierarchy in evolutionary biology and paleobiology. *Paleobiology* 26 (Suppl.): 15–52.

Jaenisch R, Bird A. 2003. Epigenetic regulation of gene expression: how the genome integrates intrinsic and environmental signals. *Nat Genet* 33 (Suppl.): 245–254.

Jasienski M, Bazzaz FA. 1995. Genome size and high CO_2. *Nature* 376: 559–560.

Johnson NA, Porter AH. 2001. Toward a new synthesis: population genetics and evolutionary developmental biology. *Genetica* 112: 45–58.

Jordan IK, Rogozin IB, Glazko GV, Koonin EV. 2003. Origin of a substantial fraction of human regulatory sequences from transposable elements. *Trends Genet* 19: 68–72.

Jurka J. 1998. Repeats in genomic DNA: mining and meaning. *Curr Opin Struct Biol* 8: 333–337.

Kalendar R, Tanskanen J, Immonen S, *et al.* 2000. Genome evolution of wild barley (*Hordeum spontaneum*) by *BARE*–1 retrotransposon dynamics in response to sharp microclimatic divergence. *Proc Natl Acad Sci USA* 97: 6603–6607.

Kasahara M, Nakaya J, Satta Y, Takahata N. 1997. Chromosomal duplication and the emergence of the adaptive immune system. *Trends Genet* 13: 90–92.

Katz LA. 1999. Changing perspectives on the origin of eukaryotes. *Trends Ecol Evol* 13: 493–497.

Kidwell MG, Lisch D. 1997. Transposable elements as sources of variation in animals and plants. *Proc Natl Acad Sci USA* 94: 7704–7711.

Kidwell MG, Lisch DR. 2000. Transposable elements and host genome evolution. *Trends Ecol Evol* 15: 95–99.

Kidwell MG, Lisch DR. 2001. Transposable elements, parasitic DNA, and genome evolution. *Evolution* 55: 1–24.

Kidwell MG, Lisch DR. 2002. Transposable elements as sources of genomic variation. In: Craig NL, Craigie R, Gellert M, Lambowitz AM eds. *Mobile DNA II* Washington, DC: American Society for Microbiology Press, 59–90.

Knight CA, Ackerly DD. 2002. Variation in nuclear DNA content across environmental gradients: a quantile regression analysis. *Ecol Lett* 5: 66–76.

Knight CA, Molinari N, Petrov DA. 2005. The large genome constraint hypothesis: evolution, ecology, and phenotype. *Ann Bot* (in press).

Knoll AH. 1995. Proterozoic and Early Cambrian protists: evidence for accelerating evolutionary tempo. In: Fitch WM, Ayala FI eds. *Tempo and Mode in Evolution*. Washington, DC: National Academy Press, 63–83.

Kondrashov FA, Koonin EV. 2001. Origin of alternative splicing by tandem exon duplication. *Hum Mol Genet* 10: 2661–2669.

Kondrashov FA, Koonin EV. 2003. Evolution of alternative splicing: deletions, insertions and origin of functional parts of proteins from intron sequences. *Trends Genet* 19: 115–119.

Kreahling J, Graveley BR. 2004. The origins and implications of Aluternative splicing. *Trends Genet* 20: 1–4.

Labrador M, Corces VG. 1997. Transposable element–host interactions: regulation of insertion and excision. *Annu Rev Genet* 31: 381–404.

Labuda D, Zietkiewicz E, Mitchell GA. 1995. Alu elements as a source of genomic variation: deleterious effects and evolutionary novelties. In: Maraia RJ ed. *The Impact of Short Interspersed Elements (SINEs) on the Host Genome*. New York: Springer, 1–24.

Letunic I, Copley RR, Bork P. 2002. Common exon duplication in animals and its role in alternative splicing. *Hum Mol Genet* 11: 1561–1567.

Lev-Maor G, Sorek R, Shomron N, Ast G. 2003. The birth of an alternatively spliced exon: 3' splice–site selection in *Alu* exons. *Science* 300: 1288–1291.

Levis RW, Ganesan R, Houtchens K, *et al.* 1993. Transposons in place of telomeric repeats at a *Drosophila* telomere. *Cell* 75: 1083–1093.

Lewontin R. 2000. *The Triple Helix*. Cambridge, MA: Harvard University Press.

Lieberman BS, Vrba ES. 1995. Hierarchy theory, selection, and sorting. *BioScience* 45: 394–399.

Lloyd EA, Gould SJ. 1993. Species selection on variability. *Proc Natl Acad Sci USA* 90: 595–599.

Losos JB, Warheit KI, Schoener TW. 1997. Adaptive differentiation following experimental island colonization in *Anolis* lizards. *Nature* 387: 70–73.

Lundin L-G. 1999. Gene duplications in early metazoan evolution. *Semin Cell Dev Biol* 10: 523–530.

Lynch M, Conery JS. 2003. The origins of genome complexity. *Science* 302: 1401–1404.

MacGillivray CW, Grime JP. 1995. Genome size predicts frost resistance in British herbaceous plants: implications for rates of vegetation response to global warming. *Funct Ecol* 9: 320–325.

Makalowski W. 1995. SINEs as a genomic scrap yard: an essay on genomic evolution. In: Maraia RJ ed. *The Impact of Short Interspersed Elements (SINEs) on the Host Genome*. New York: Springer, 81–104.

Makalowski W. 2003. Not junk after all. *Science* 300: 1246–1247.

Marks J. 2002. *What it Means to be 98% Chimpanzee*. Berkeley, CA: University of California Press.

Matzke MA, Mette MF, Aufsatz W, *et al.* 1999. Host defenses to parasitic sequences and the evolution of epigenetic control mechanisms. *Genetica* 107: 271–287.

Maynard Smith J. 2002. The major transitions in evolution. In: Pagel M ed. *Encyclopedia of Evolution, Vol. 1*. Oxford: Oxford University Press, E17–E22.

Maynard Smith J, Szathmáry E. 1995. *The Major Transitions in Evolution*. Oxford, UK: Oxford University Press.

Mayr E. 1954. Change of genetic environment and evolution. In: Huxley JS, Hardy AC, Ford EB eds. *Evolution as a Process*. London: George Allen & Unwin, 157–180.

Mayr E. 1963. *Animal Species and Evolution*. Cambridge, MA: Harvard University Press.

Mayr E. 1980. Some thoughts on the history of the evolutionary synthesis. In: Mayr E, Provine WB eds. *The Evolutionary Synthesis*. Cambridge, MA: Harvard University Press.

Mayr E. 1982. Speciation and macroevolution. *Evolution* 36: 1119–1132.

Mayr E. 1992. Speciational evolution or punctuated equilibria. In: Somit A, Peterson SA eds. *The Dynamics of Evolution*. Ithaca, NY: Cornell University Press, 21–53.

Mayr E. 1997. *This is Biology*. Cambridge, MA: Harvard University Press.

McDonald JF. 1990. Macroevolution and retroviral elements. *BioScience* 40: 183–191.

McDonald JF. 1995. Transposable elements: possible catalysts of organismic evolution. *Trends Ecol Evol* 10: 123–126.

McDonald JF. 1998. Transposable elements, gene silencing and macroevolution. *Trends Ecol Evol* 13: 94–95.

Miller WJ, McDonald JF, Nouaud D, Anxolabéhère D. 1999. Molecular domestication—more than a sporadic episode in evolution. *Genetica* 107: 197–207.

Mironov AA, Fickett JW, Gelfland MS. 1999. Frequent alternative splicing of human genes. *Genome Res* 9: 1288–1293.

Morrish TA, Gilbert N, Myers JS, *et al.* 2002. DNA repair mediated by endonuclease-independent LINE-1 retrotransposition. *Nat Genet* 31: 159–165.

Newman CM, Cohen JE, Kipnis C. 1985. Neo-Darwinian evolution implies punctuated equilibria. *Nature* 315: 400–401.

Östergren G. 1945. Parasitic nature of extra fragment chromosomes. *Bot Notiser* 2: 157–163.

Pardue ML, Danilevskaya ON, Traverse KL, Lowenhaupt K. 1997. Evolutionary links between telomeres and transposable elements. *Genetica* 100: 73–84.

Patthy L. 1999. Genome evolution and the evolution of exon shuffling—a review. *Gene* 238: 103–114.

Pigliucci M, Kaplan J. 2000. The fall and rise of Dr Pangloss: adaptationism and the *Spandrels* paper 20 years later. *Trends Ecol Evol* 15: 66–70.

Plasterk RH. 2002. RNA silencing: the genome's immune system. *Science* 296: 1263–1265.

Pomiankowski A. 1999. Intragenomic conflict. In Keller L ed. *Levels of Selection in Evolution*. Princeton, NJ: Princeton University Press, 121–152.

Prak ETL, Kazazian HH. 2000. Mobile elements and the human genome. *Nat Rev Genet* 1: 134–144.

Provine WB. 1988. Progress in evolution and meaning in life. In: Nitecki MH ed. *Evolutionary Progress*. Chicago: University of Chicago Press, 49–74.

Regev A, Lamb MJ, Jablonka E. 1998. The role of DNA methylation in invertebrates: developmental regulation or genome defense? *Mol Biol Evol* 15: 880–891.

Riha K, McKnight TD, Griffing LR, Shippen DE. 2001. Living with genome instability: plant responses to telomere dysfunction. *Science* 291: 1797–1800.

Roth G, Schmidt A. 1993. The nervous system of plethodontid salamanders: insight into the interplay between genome, organism, behavior, and ecology. *Herpetologica* 49: 185–194.

Roth G, Blanke J, Wake DB. 1994. Cell size predicts morphological complexity in the brains of frogs and salamanders. *Proc Natl Acad Sci USA* 91: 4796–4800.

Rudolph KL, Chang S, Lee H-W, *et al.* 1999. Longevity, stress response, and cancer in aging telomerase-deficient mice. *Cell* 96: 701–712.

Ruse M. 1982. *Darwinism Defended*. Reading, MA: Addison-Wesley.

Sapienza C, Doolittle WF. 1981. Genes are things you have whether you want them or not. *Cold Spring Harb Symp Quant Biol* 45: 177–182.

Schmid CW, Rubin CM. 1995. Alu: what's the use? In: Maraia RJ ed. *The Impact of Short Interspersed Elements (SINEs) on the Host Genome*. New York: Springer, 105–123.

Schön I, Martens K. 2000. Transposable elements and asexual reproduction. *Trends Ecol Evol* 15: 287–288.

Shermer M, Sulloway FJ. 2000. The grand old man of evolution: an interview with evolutionary biologist Ernst Mayr. *Skeptic* 8: 76–82.

Shubin NH, Marshall CR. 2000. Fossils, genes, and the origin of novelty. *Paleobiology* 26.

Simpson GG. 1944. *Tempo and Mode in Evolution*. New York: Columbia University Press.

Skryabin BV, Kremerskothen J, Vassilacopoulou D, *et al.* 1998. The BC200 RNA gene and its neural expression are conserved in Anthropoidea (Primates). *J Mol Evol* 47: 677–685.

Smit AFA. 1999. Interspersed repeats and other momentos of transposable elements in mammalian genomes. *Curr Opin Genet Dev* 9: 657–663.

Sober E, Wilson DS. 1998. *Unto Others*. Cambridge, MA: Harvard University Press.

Sorek R, Shamir R, Ast G. 2004. How prevalent is functional alternative splicing in the human genome? *Trends Genet* 20: 68–71.

Stebbins GL, Ayala FJ. 1981. Is a new evolutionary synthesis necessary? *Science* 213: 967–971.

Suga H, Koyanagi M, Hoshiyama D, *et al.* 1999. Extensive gene duplication in the early evolution of animals before the parazoan–eumetazoan split demonstrated by G proteins and protein tyrosine kinases from sponge and hydra. *J Mol Evol* 48: 646–653.

Szabó Z, Levi–Minzi SA, Christiano AM, *et al.* 1999. Sequential loss of two neighboring exons of the tropoelastin gene during primate evolution. *J Mol Evol* 49: 664–671.

Temin HM. 1980. Origin of retroviruses from cellular moveable genetic elements. *Cell* 21: 599–600.

Turner JRG. 1986. The genetics of adaptive radiation: a neo-Darwinian theory of punctuational evolution. In: Raup DM, Jablonski D eds. *Patterns and Processes in the History of Life*. Berlin: Springer-Verlag, 183–207.

Vellai T, Vida G. 1999. The origin of eukaryotes: the difference between prokaryotic and eukaryotic cells. *Proc R Soc Lond B* 266: 1571–1577.

Vellai T, Takács K, Vida G. 1998. A new aspect to the origin and evolution of eukaryotes. *J Mol Evol* 46: 499–507.

Venables PJ, Brookes SM, Griffiths D, *et al.* 1995. Abundance of an endogenous retroviral envelope protein in placental trophoblasts suggests a biological function. *Virology* 211: 589–592.

Venter JC, Adams MD, Myers EW, *et al.* 2001. The sequence of the human genome. *Science* 291: 1304–1351.

Vieira C, Nardon C, Arpin C, *et al.* 2002. Evolution of genome size in *Drosophila*. Is the invader's genome being invaded by transposable elements? *Mol Biol Evol* 19: 1154–1161.

Villarreal LP. 1997. On viruses, sex, and motherhood. *J Virol* 71: 859–865.

Vinogradov AE. 2003. Selfish DNA is maladaptive: evidence from the plant Red List. *Trends Genet* 19: 609–614.

Vrba ES. 1989. Levels of selection and sorting with special reference to the species level. *Oxford Surv Evol Biol* 6: 111–168.

Vrba ES, Eldredge N. 1984. Individuals, hierarchies and processes: towards a more complete evolutionary theory. *Paleobiology* 10: 146–171.

Vrba ES, Gould SJ. 1986. The hierarchical expansion of sorting and selection: sorting and selection cannot be equated. *Paleobiology* 12: 217–228.

Wakamiya I, Newton RJ, Johnston JS, Price HJ. 1993. Genome size and environmental factors in *Pinus*. *Am J Bot* 80: 1235–1241.

Wakamiya I, Price HJ, Messina MG, Newton RJ. 1996. Pine genome size diversity and water relations. *Physiol Plant* 96: 13–20.

Watanabe K, Yahara T, Denda T, Kosuge K. 1999. Chromosomal evolution in the genus *Brachyscome* (Asteraceae, Astereae): statistical tests regarding correlation between changes in karyotype and habit using phylogenetic information. *J Plant Res* 112: 145–161.

Weinberg S. 2001. Can science explain everything? Anything? *New York Review of Books*, May 31.

Weiner J. 1994. *The Beak of the Finch*. New York: Vintage Books.

Weinstein IB, Hsiao WL, Hsieh L-L, *et al.* 1988. A possible role of retrotransposons in carcinogenesis. In: Lambert ME, McDonald JF, Weinstein IB eds. *Eukaryotic Transposable Elements as Mutagenic Agents*. New York: Cold Spring Harbor Laboratory Press, 289–297.

Wildman DE, Uddin M, Liu G, *et al.* 2003. Implications of natural selection in shaping 99.4% nonsynonymous DNA identity between humans and chimpanzees: enlarging genus *Homo*. *Proc Natl Acad Sci USA* 100: 7181–7188.

Wilson DS, Colwell RK. 1981. Evolution of sex ratio in structured demes. *Evolution* 35: 882–897.

Winter H, Langbein L, Krawczak M, *et al.* 2001. Human type I hair keratin pseudogene φhHaA has functional orthologs in the chimpanzee and gorilla: evidence for recent inactivation of the human gene after the *Pan-Homo* divergence. *Hum Genet* 108: 37–42.

Wright SI, Schoen DJ. 1999. Transposon dynamics and the breeding system. *Genetica* 107: 139–148.

Yoder JA, Walsh CP, Bestor TH. 1997. Cytosine methylation and the ecology of intragenomic parasites. *Trends Genet* 13: 335–340.

Zeyl C, Bell G. 1996. Symbiotic DNA in eukaryotic genomes. *Trends Ecol Evol* 11: 10–15.

Zhang J. 2003. Evolution by gene duplication: an update. *Trends Ecol Evol* 18: 292–298.

Zhang J, Webb DM. 2003. Evolutionary deterioration of the vomeronasal pheromone transduction pathway in catarrhine primates. *Proc Natl Acad Sci USA* 100: 8337–8341.

Zhang J, Webb DM, Podlaha O. 2002. Accelerated protein evolution and origins of human-specific features: FOXP2 as an example. *Genetics* 162: 1825–1835.

INDEX

Page numbers followed by f indicate figures; page numbers followed by t indicate tables

A

Accessory chromosomes, *see* B chromosomes
Aedes
  genome size in 18, 25
  transposable elements in 208
Aggregate characters 695-697, 698, 704
Agnatha, *see* Jawless fishes
Allopolyploidy, *see* Polyploidy
Allozymes, *see* Isozymes and allozymes
Alternative splicing and exon shuffling 207,
  551-552, 551f, 712-714, 716-718
Alu elements 172f, 173t, 188, 192, 205, 207,
  213, 548, 714, 717-718, 718f
Ambystoma 44f, 430f, 469-471, 493
Amphibians, *see also* Frogs, Salamanders
  B chromosomes in 229t, 231, 235-238t, 251
  cell size in 42f, 44f, 45, 57-58, 64-65,
    463-464
  genome size in 8f, 11t, 15, 40, 41, 45,
    46-47f, 58-59, 59f, 60-61, 62, 63f,
    64-65, 71, 73, 461, 472
  polyploidy in 443, 461-472, 491, 493, 502
Amphiuma 7, 42f, 48f, 58, 59f
Aneuploidy
  adverse effects of 298, 299f, 334, 431f, 437,
    450, 523–524f
  and B chromosomes 230, 247, 250-252,
    263, 273
  and genome size 24, 31, 135, 136, 146
  as mechanism of duplication 298, 333-334
Angiosperms
  B chromosomes in 228-229t, *see also*
    B chromosomes
  frequency of polyploidy in 379-380, 440,
    see also Polyploidy
genome size in 90, 91f, 92, 95–96, 97t,
  98–99, 100f, 101t, 102f, 103, 104f, 117,
  118f, 120–121, 120f, 124-127, 125f,
  128f, 130, 136-137, 150
Animal Genome Size Database 10-12, 11t,
  64f, 631
Annelids
  genome size in 8f, 11t, 22, 54, 60, 490
  polyploidy in 442, 443, 490-493, 492f, 497
Anopheles
  gene duplication in 302, 563
  genome size in 18, 71, 211t, 539t, 561
  number of genes in 211t, 539t
  sequencing of genome 71, 539t, 562-563,
    569f, 653
  transposable elements in 195-196, 211t, 561
Arabidopsis
  gene and ancient genome duplication in 110,
    297f, 301, 320, 338, 339f, 340, 341f,
    342f, 343, 346, 349, 354-356, 359-360,
    382-383, 382f, 404, 410, 414, 501, 560
  genome size in 101t, 106, 113, 114, 126,
    147, 148-149, 149f, 211t, 539t, 541,
    559, 560
  number of genes in 211t, 539t, 543f,
    559-560
  sequencing of genome 106, 147, 537, 539t,
    540f, 559-560, 569f
  transposable elements in 107, 109, 113,
    173t, 178, 190t, 195, 202, 203, 204,
    208, 211t, 547-549
Arachnids
  B chromosomes in 229t, 256
  genome size in 8f, 11t, 21, 40
  polyploidy in 496

Breinigsville, PA USA
02 February 2011
254604BV00006B/69/P